Equipment Commonly Used in the Organic Chemistry Laboratory

Filter flask

Büchner funnel

Hirsch funnel

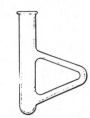

Thiele tube

Separatory funnel (Squibb type)

Separatory funnel with ground-glass joints

Bent adapter

Vacuum adapter with ground-glass joints

West condenser

Hempel fractionating column

Round-bottomed flask

Claisen connecting tube

Still head

Thermometer adapter with neoprene fitting

FOURTH EDITION

modern experimental organic chemistry

Royston M. Roberts • John C. Gilbert *University of Texas, Austin*

Lynn B. Rodewald • Alan S. Wingrove *Towson State University*

Saunders Golden Sunburst Series

Saunders College Publishing *Philadelphia • New York • Chicago • San Francisco • Montreal • Toronto • London • Sydney • Tokyo • Mexico City • Rio de Janeiro • Madrid*

Address orders to:
383 Madison Avenue
New York, NY 10017

Address editorial correspondence to:
West Washington Square
Philadelphia, PA 19105

Text Typeface: 10/12 Times Roman
Compositor: York Graphic Services, Inc.
Acquisitions Editor: John Vondeling
Project Editor: Joanne Fraser
Copyeditor: Janis Moore
Art Director: Carol C. Bleistine
Art/Design Assistant: Virginia A. Bollard
Text Design: Adrianne Onderdonk Dudden
Cover Design: Lawrence R. Didona
Text Artwork: ANCO/Boston
Production Manager: Tim Frelick
Assistant Production Manager: Maureen Iannuzzi

Library of Congress Cataloging in Publication Data

Main entry under title:
Modern experimental organic chemistry.

Includes index.

1. Chemistry, Organic—Laboratory manuals.
I. Roberts, Royston M.
QD261.M63   1985   547'.007'8   84-14151
ISBN 0-03-063018-5

MODERN EXPERIMENTAL ORGANIC CHEMISTRY   ISBN 0-03-063018-5

5678   016   98765432

CBS COLLEGE PUBLISHING
Saunders College Publishing
Holt, Rinehart and Winston
The Dryden Press

preface

In the fourth edition of this book, we continue to have the same goal as in the previous editions: to provide students of organic chemistry an essentially self-contained laboratory **textbook** that requires little or no supplementation, in the form of reference to lecture textbooks, for a complete understanding of the experiments. The experiments have been selected to illustrate many of the highlights of the subjects covered in contemporary organic lecture courses. Each of the experiments is preceded by a thorough discussion of the theoretical as well as the practical aspects of the experiment.

Every experiment and every section of the third edition of the book have been subjected to critical examination by the authors, with consideration of our experience with our own students and of the information provided to us by many of the instructors in the more than 200 other universities and colleges in which the previous editions have been used. As a result, **this fourth edition may be briefly characterized as follows:** (1) all experiments that have proved satisfactory and valuable to thousands of students using the previous editions have been retained; (2) experiments that have been found not to have wide use have been removed; (3) all discussions and experiments in the text have been updated; (4) new discussions and experiments have been added; (5) **safety in the laboratory** has continued to receive strong emphasis, both in the form of expanded cautions of the ''Do It Safely'' type and by exclusion of dangerous chemicals from the experiments; and (6) the book has been reorganized extensively to make it more convenient for both instructors and students to use. Highlights of the major changes are given in the following paragraphs.

Introduction: Laboratory Safety and Laboratory Notebook The first chapter contains an introduction to experimental organic chemistry and an expanded discussion of laboratory safety measures. A detailed description of how to prepare a laboratory notebook is also given, including new examples demonstrating calculations of limiting reagents and theoretical and actual percent yields.

Techniques and Apparatus The second chapter is completely new. It describes all the common experimental techniques that will be encountered in later chapters, including the material on drying agents and heating and stirring techniques that was contained in appendices in the third edition. We believe that the integrated treatment of the fundamental techniques early in the book will be advantageous to the students. The other sections of the Appendix in the third edition have similarly been incorporated into the body of the fourth edition at appropriate places.

Solids and Liquids   Chapters 3 and 4 deal with the *properties* of organic compounds in the two most commonly encountered physical states and the important *techniques* of melting-point and boiling-point determinations and purification by recrystallization and distillation. The material of Chapter 1 of the third edition is thus separated and expanded into two chapters in the new edition.

Natural Products and Other Compounds of Biological Importance   This edition contains an increased number of experiments on natural products and synthetic compounds of biological importance. For example, the isolation of citral now appears in Chapter 4 as a demonstration of the use of steam distillation, and the isolation of trimyristin from nutmeg is presented in Chapter 5 as an example of the technique of extraction. The chapters on amino acids and proteins and on carbohydrates, which are valuable for students in the life sciences, have been retained, including a section on classification tests for carbohydrates that was omitted in early printings of the third edition. A new addition is a multistep synthesis of the important local anesthetic lidocaine in Chapter 23, which also contains the popular synthesis of sulfanilamide and a synthesis of a 4-aminoquinoline, which is related to antimalarial drugs.

Stereoisomers and Stereochemistry   A revised chapter on stereochemistry now appears much earlier in the book, in Chapter 7, in line with the earlier treatment in most contemporary lecture texts. In addition to the experiments retained from the third edition, new experiments have been added to demonstrate the difference in properties (TLC retention times) of simple diastereoisomers such as *cis*- and *trans*-1,2-cyclohexanediol and the difference in the properties of enantiomers in chiral environments, demonstrated dramatically by the odors of (+)- and (−)-carvone, natural products that are isolated by vacuum distillation of the essential oils in which they occur.

Spectroscopy   The coverage of spectroscopy has been greatly increased in this edition, the single chapter of the previous editions having now been expanded into four chapters. Chapter 8 deals with Infrared Spectroscopy and Chapter 9 with Nuclear Magnetic Resonance, including a new section on carbon-13 magnetic resonance (**CMR**) as well as proton magnetic resonance (**PMR**). These chapters contain a collection of IR and PMR spectra of compounds containing most of the common functional groups. Many of these compounds are those encountered in the experiments of this text. The expanded discussion of IR and PMR spectra in Chapters 8 and 9 includes more solved examples of spectra interpretation, and Chapter 10 presents a discussion of how IR and PMR information may be used in combination as powerful tools for organic structure determination. These chapters should be especially useful to students expected to utilize spectroscopy in conjunction with classical qualitative organic analysis. Chapter 11 describes Ultraviolet/Visible Spectroscopy and Mass Spectroscopy, the latter subject being a new addition in this edition.

Qualitative Organic Analysis   In accordance with the encouragement we have received from many colleagues, Chapter 27 has been considerably expanded, with the inclusion of fuller discussion of the systematic approach to compound identification. *All* the qualitative functional group tests, some of which were distributed throughout various chapters in the previous editions, have now been placed in Chapter 27. The integration of spectroscopic analysis with classical ("wet") qualitative analysis methods has been given more attention. The tables of compounds that are possible "unknowns" have been greatly expanded, and a number of new tables have been added, so now it should seldom be necessary for students to require supplementary texts for the identification of their "unknowns."

The Literature of Organic Chemistry   This chapter has been thoroughly updated to include the latest editions of the books and journals described, including a new example demonstrating how to use the newest volumes of *Beilstein* to locate organic compounds. The latest information on accessing *Chemical Abstracts* by means of computers is given.

Other Deletions and Additions in This Edition   Several other deletions and additions in this edition may be noted. Some of these changes were the result of a comprehensive user survey made by our publisher. It was very helpful to learn which experiments had been found to be most useful and successful and which either caused some difficulties or perhaps were just not seen as appropriate. The chapter on carbenes and arynes has been deleted, as have the chapters on nonbenzenoid aromatic compounds and photochemical dimerization of 4-methylbenzophenone. The preparations of many of the derivatives of carboxylic acids have been eliminated, but the preparation of an aromatic carboxylic acid by oxidation of an arene has been retained in the chapter on Oxidation Reactions (Chapter 19). The esterification of benzoic acid now appears in Chapter 22, where the ester is used in reaction with a Grignard reagent. The multistep synthesis of 1-bromo-3-chloro-5-iodobenzene has been eliminated and replaced by the synthesis of lidocaine. An additional experiment involving the Diels-Alder reaction has been added: The reaction of maleic anhydride with a mixture of *cis*- and *trans*-1,3-pentadiene illustrates the selectivity of the Diels-Alder reaction with geometric isomers. The electrophilic addition of unsymmetrical reagents to unsymmetrical alkenes is a major topic of discussion in organic lecture courses, yet there have been, up to now, no simple laboratory experiments to demonstrate the operation of Markovnikov's Rule. We have corrected this deficiency by presenting the addition of hydrogen bromide to 1-hexene using a phase-transfer catalyst. Another new experiment in Chapter 21 involves a conjugate addition to an α, β-unsaturated ketone, followed by an acid-catalyzed cyclic aldol condensation; this reaction procedure incorporates a simple and clever water trap of the Dean-Stark type. It also illustrates the scale of the chemicals used in some of the experiments; 86 millimoles of 2-methylpropanal and 67 millimoles of 3-buten-2-one are used in this experiment. As an example of the thought taken to remove objectionable, if not dangerous, chemicals from the experiments, the time-honored aldol condensation that produces benzalacetophenone has been modified by the use of other compounds that give analogous products that are not skin irritants.

Pre-lab and Post-lab Exercises   Many of the exercises that were included at the end of each experiment in the previous editions have been retained, some of them have been modified, and many new ones have been added. As a new feature of this edition, we have divided the exercises into two groups: ''Pre-lab Exercises'' and ''Post-lab Exercises.'' The latter group, referred to in the text simply as ''Exercises,'' will serve the same purpose as in previous editions, to reinforce the concepts illustrated by the experiments performed. Some of the more difficult Exercises have been designated with an asterisk (*). The ''Pre-lab Exercises,'' contained in a paperback ancillary book that accompanies the text, are designed to serve a new pedagogical purpose. Throughout many years of supervising the organic laboratory, we authors and, we suspect, many other instructors, have frequently been frustrated by students who come to lab with no acquaintance with or appreciation of the experiments to be performed, in spite of a thorough exposition of the theory and the practical considerations presented in the laboratory text. This is not only unfortunate for the unprepared student, who will learn little from the experiment, but may actually be dangerous for all in the laboratory if this student is not aware of the ways to ''Do It Safely.'' In our considerations of how we might alleviate this situation, we were helped by a suggestion from John Vondeling, Associate Publisher of Saunders College Publishing, the result being the creation of these new ''Pre-lab Exercises,'' which are designed to test the student's knowledge of the experiment in advance of the actual performance. At the option of the instructor, the completion in writing of these ''Pre-lab Exercises'' may be required in advance of an assigned experiment, and because the sheets on which they are printed are perforated, they may be removed from the text and handed in for grading. In general, these exercises are easy to grade and pertain to the knowledge of the experimental operations and the reasons for them, rather than to the results of the experiment, which are more the domain of the post-lab "Exercises."

We hope that in this fourth edition of the textbook students will find the experiments challenging and illuminating, and we hope that instructors will benefit from the fact that discussions are thorough and allow students to be independent in their laboratory work.

Instructor's Manual   An Instructor's Manual is available at no charge from the publisher to all instructors who choose to use this textbook. It is similar to those provided with the earlier editions of the book, containing lists of recommended desk equipment, shelf reagents, and solvents, lists of required chemicals by chapter and alphabetically, notes on the Experimental Procedures, including time requirements, and **answers to all exercises, both pre- and post-lab.** The answers are set apart from the rest of the material in the Manual so that they may be reproduced readily for posting by instructors, if so desired. It also contains interpretation of the CMR spectral data that have been given throughout the textbook.

We are pleased to acknowledge again our gratitude to many persons who have helped in the preparation of this edition. The following colleagues reviewed the third edition and offered valuable suggestions for its revision: Richard E. Bozak, California State University, Hayward; Thomas J. Cogdell, University of Texas, Arlington; J. D. Heinrich, Southwestern College, Chula Vista, California; Michael M. King, George Washington University; Terry Kruger, Ball State University, Muncie, Indiana; and Robert J. Olsen, Wabash College, Crawfordsville, Indiana. In addition, we also profited from the answers from many other colleagues who responded to the comprehensive user survey referred to earlier. Paul M. Puckett, Texas A and M University, made a careful critique of the present edition in manuscript form that was very useful to us. Brent Blackburn, Ahmed El-Khawaga, Aubrey Skinner, and Michael Vuper of the University of Texas at Austin, and Linda Sweeting of Towson State University, provided us with valuable critical and technical assistance. We are indebted to hundreds of undergraduate students and dozens of graduate students at the University of Texas and Towson State University for serving as "guinea pigs" and teaching assistants during the years in which many of the experiments were developed and tested.

Finally, we are pleased to thank John Vondeling, our publisher, and Joanne Fraser, our Project Editor, for their invaluable and capable assistance in bringing this project to completion.

Royston M. Roberts • John C. Gilbert
University of Texas at Austin, Austin, Texas 78712

Lynn B. Rodewald • Alan S. Wingrove
Towson State University, Baltimore, Maryland 21204

contents overview

contents

IR, NMR, AND UV SPECTRA

COMPOUND	IR (Fig. No.)	PMR (Fig. No.)	CMR (Fig. No.)	UV (Fig. No.)
p-Acetamidobenzenesulfonyl chloride	23.7			
Acetanilide	23.5	9.24	23.6	
Acetophenone	8.23a	20.10a	20.10b	
Adduct, maleic anhydride and dienes	15.18	15.19a	15.19b	
Adipic acid (*See* Hexanedioic acid)				
Aniline	8.32a	23.4a	23.4b	
trans-Anisalacetophenone	21.12	21.13		21.14
trans-Anisalacetophenone dibromide	21.15	21.16		21.17
Anisaldehyde	21.10	21.11a	21.11b	
Anisole	8.29b			
Benzaldehyde	8.22a	19.4a	19.4b	
Benzamide	8.26a			
Benzanilide	8.26b, 10.3a	10.3b		
Benzoic acid	8.24a	19.5a	19.5b	
Benzoic anhydride	8.28			
Benzonitrile	8.31			
p-Benzoquinone	15.6		15.7	
Benzoyl chloride	8.27			
Benzyl alcohol (phenylmethanol)	8.13	19.6a	19.6b	
Benzyl chloride	21.1		21.2	
Bromobenzene	22.2		22.3	
1-Bromohexane	13.26	13.27a	13.27b	
2-Bromohexane	13.28	13.29a	13.29b	
Butanal (*n*-butyraldehyde)		9.19		
1-Butanol	18.1	9.9	18.2	
2-Butanone (methyl ethyl ketone)		9.20	9.25	
Butanoic acid (butyric acid)		9.21		
3-Buten-2-one	21.34	21.35a	21.35b	
Butyl acetate (*See* Butyl ethanoate)				
t-Butylbenzene	12.13	12.14a	12.14b	
p-tert-Butylphenol	10.6a	10.6b		
sec-Butyl-*p*-xylene	10.8a	10.8b		
Butyl ethanoate (butyl acetate)	8.25b			
n-Butyraldehyde (*See* Butanal)				

IR, NMR, AND UV SPECTRA

COMPOUND	IR	PMR	CMR	UV
	(Fig. No.)	(Fig. No.)	(Fig. No.)	(Fig. No.)
Butyric acid (*See* Butanoic acid)				
Carvone	7.15ab, 27.3a	7.16a, 27.3b	7.16b	
α-Chloroacetyl chloride	23.18		23.19	
1-Chlorobutane	12.5	12.6a	12.6b	
Chlorocyclohexane	12.8		12.9	
α-Chloro-2,6-dimethylacetanilide	23.20	23.21		
2-Chloro-2-methylbutane	8.18	13.2a	13.2b	
1-Chloro-2-methylpropane		9.7		
Cinnamaldehyde	20.5	20.6a	20.6b	
Citral	10.12a	10.12b		11.3
p-Cresol	8.21			
Cumene (*See* Isopropylbenzene)				
Cyclobutanecarboxylic acid	21.28		21.29	
Cyclobutane-1,1-dicarboxylic acid	8.7			
Cyclohexane	8.15		12.7	
Cyclohexane-*cis*-1,2-dicarboxylic acid	20.3		20.4	
cis-1,2-Cyclohexanediol	7.2		7.1	
trans-1,2-Cyclohexanediol	7.3		7.4	
Cyclohexanol	8.19	9.15	13.21	
Cyclohexanone	8.23a	16.6a	16.6b	
Cyclohexene	13.22	13.23a	13.23b	
4-Cyclohexene-*cis*-1,2-dicarboxylic acid	20.1		20.2	
4-Cyclohexene-*cis*-1,2-dicarboxylic anhydride	15.10	15.11		
1,3-Cyclopentadiene	15.2	15.3a	15.3b	
1,3-Dibromopropane	21.24	21.25a	21.25b	
1,x-Dichlorobutanes, mixture		12.4		
Diethyl ether	8.29a			
Diethyl ethoxymethylenemalonate		23.12a	23.12b	
2,6-Dimethylaniline	23.16	23.17a	23.17b	
2,3-Dimethylbutane	8.14a			
Dimethyl cyclobutane-1,1-dicarboxylate	21.26	21.27		
4,4-Dimethyl-2-cyclohexene-1-one	21.36	21.37		21.39
Dimethyl fumarate	7.8	7.9a	7.9b	

IR, NMR, AND UV SPECTRA

COMPOUND	IR (Fig. No.)	PMR (Fig. No.)	CMR (Fig. No.)	UV (Fig. No.)
Dimethyl maleate	7.6		7.7	
Dimethyl maleate and dimethyl fumarate, mixture		7.5		
Dimethyl malonate	21.22	21.23a	21.23b	
2,6-Dimethylnitrobenzene	23.14	23.15a	23.15b	
trans,trans-Diphenyl-1,3-butadiene	21.8	21.9		
Ethyl acetate (*See* Ethyl ethanoate)				
Ethylbenzene	12.10	12.11a	12.11b	
Ethyl ethanoate (ethyl acetate)		9.22		
Ethyl phenyl ether (phenetole)		9.16		
2-Furaldehyde	16.3	16.4a	16.4b	16.5
3-(2-Furyl)-1-(3-nitrophenyl) propenone	21.20	21.21		
Hexanedioic acid (adipic acid)	8.24b		19.3	
1-Hexene	13.24	13.25a	13.25b	
Hydrocinnamic acid (*See* 3-Phenyl-propanoic acid)				
Isobutylchloride (*See* 1-Chloro-2-methylpropane)				
Isobutyl formate	10.7a	10.7b		
Isobutyraldehyde (*See* 2-Methylpropanal)				
Isophorone (*See* 3,5,5-Trimethylcyclo-hex-2-enone)				
Isopropylbenzene (cumene)	8.19	9.14	12.12	
Isopropyl-*p*-xylene	17.8b	9.8 17.9b		
Lidocaine	23.22	23.23		
Maleic anhydride	15.4		15.5	
2-Methoxyethyl acetate	27.2a	27.2b		
Methyl benzoate	8.25a	22.1	9.26	
Methylbutane	8.14b			
2-Methyl-2-butanol (*t*-pentyl alcohol)	18.3	18.4a	18.4b	
3-Methyl-2-butanone			9.30	
2-Methyl-1-butene	8.16	9.12	13.5	
2-Methyl-2-butene	13.3	13.4a	13.4b	

IR, NMR, AND UV SPECTRA

COMPOUND	IR	PMR	CMR	UV
	(Fig. No.)	(Fig. No.)	(Fig. No.)	(Fig. No.)
Methylbutenes, reaction mixtures from dehydrohalogenation of 2-chloro-2-methylbutane		13.6 13.7		
2-Methyl-3-butyn-2-ol	14.1	9.13	14.2	
3-Methyl-4-cyclohexene-*cis*-1,2-di-carboxylic anhydride	15.16	15.17		
3-Methyl-2-cyclohexenone	8.3			
Methylcyclopentane		9.11		
Methyl ethyl ketone (*See* 2-Butanone)				
2-Methyl-3-heptanol	22.5	22.6a	22.6b	
3-Methyl-3-hydroxy-2-butanone	14.3	14.4a	14.4b	
4-Methyl-2-pentanol	13.9	13.10a	13.10b	
2-Methyl-1-pentene	13.19	13.20		
2-Methyl-2-pentene	13.12	13.13a	13.13b	
4-Methyl-1-pentene	8.34	13.11a	13.11b	
cis-4-Methyl-2-pentene	13.17	13.18a	13.18b	
trans-4-Methyl-2-pentene	13.15	13.16a	13.16b	
4-Methyl-3-penten-2-one				11.2
2-Methylpropanal (isobutyraldehyde)	8.22b	19.2a	19.2b	
Methyl propanoate (methyl propionate)	10.2a		10.2b	
2-Methyl-1-propanol (isobutyl alcohol)	10.1a	10.1b	19.1	
2-Methyl-2-propanol (*t*-butyl alcohol)	10.5a		10.5b	
Methyl propionate (*See* Methyl propanoate)				
Methyl vinyl ketone (*See* 3-Buten-2-one)				
Mineral oil	8.6a			
N-Cinnamylidene-*m*-nitroaniline	20.7			
N-Cinnamyl-*m*-nitroaniline	20.8	20.9		
N,*N*-Dimethylacetamide	8.26c			
N,*N*-Dimethylaniline	8.32c			
N-Ethylaniline		9.23		
N-Methylaniline	8.32b			

IR, NMR, AND UV SPECTRA

COMPOUND	IR (Fig. No.)	PMR (Fig. No.)	CMR (Fig. No.)	UV (Fig. No.)
m-Nitroacetophenone	21.18	21.19a	21.19b	
Nitrobenzene	8.30		23.3	
4-Nitrobromobenzene	17.11	9.17	17.12	
1-Nitropropane		9.4		
1,3-Pentadiene, commercial mixture of isomers	15.12	15.13		
cis-1,3-Pentadiene	15.16	15.17a	15.17b	
trans-1,3-Pentadiene	15.14	15.15a	15.15b	
Pentane		9.10		
t-Pentyl alcohol (*See* 2-Methyl-2-butanol)				
Perfluorokerosene	8.6b			
Phenetole (*See* Ethyl phenyl ether)				
Phenol		9.18		
Phenylacetylene (*See* Phenylethyne)				
1-Phenylethanol	20.11	20.12		
Phenylethyne (phenyl acetylene)	8.17			
Phenylmethanol (*See* Benzyl alcohol)				
3-Phenylpropanoic acid (hydrocinnamic acid)	10.4a	10.4b		
Polystyrene	8.12			
n-Propyl-*p*-xylene	17.8a	17.9a		
cis-Stilbene	21.3	21.5a	21.5b	21.7
trans-Stilbene	21.4	21.6		21.7
Styrene	24.2		24.3	24.4
Sulfanilamide	23.8	23.9a	23.9b	
3-Sulfolene (2,5-dihydrothiophene-1,1-dioxide)	15.8	15.9a	15.9b	
3,3,5,5-Tetramethylcyclohexanone	22.9	22.10a	22.10b	
3,5,5-Trimethylcyclohex-2-enone	22.7	22.8a	22.8b	
Triphenylmethanol		22.4		
Triptycene	8.5			
p-Xylene	17.6	17.7a	17.7b	

experimental procedures

chapter **1**

introduction, laboratory safety, and laboratory notebook

1.1 Introduction

The laboratory part of an introductory course in organic chemistry complements the lecture part of the course. In the laboratory, you will learn firsthand that the compounds and reactions discussed in lecture are more than abstract notations. Many of the theoretical concepts discussed in lecture can be investigated experimentally at the introductory level. You will be able to collect and interpret data that will add reality to the theories of organic chemistry.

One important purpose of the organic chemistry laboratory is to introduce students to the experimental techniques of the practicing organic chemist. You will learn to handle organic chemicals and to manipulate apparatus, both of which are highly important to the education of a chemist. The learning of the scientific approach to laboratory is equally important, as is the learning of good experimental technique. In this section, we outline the approach to developing good experimental technique.

Never start any experiment until you understand its overall purpose and the reasons for carrying out each operation. This requires that you *study* and *not* just read the complete experiment *before coming into the laboratory.* Your performance in laboratory will be better if you prepare in advance, and you will benefit much more from the experiments, in knowledge as well as in grades! To encourage this advance preparation for the laboratory work, a set of **Pre-lab exercises** is provided on perforated pages in a paperback booklet that accompanies this text. Your instructor may require you to turn in answered exercises before you carry out each of the assigned experiments. Even if not required to complete and hand in these exercises, you will find that considering them carefully will be valuable preparation for the experimental work.

Good technique involves neatness in working in the laboratory and in keeping complete and neat records of the work that you do. Experimental organic chemistry is *potentially* dangerous, since many of the chemicals are highly flammable and/or toxic. Careless handling of chemicals and assembly of the apparatus pose potential danger both to you and to those around you. For this reason, throughout this text, we emphasize the importance of safe handling of chemicals and the proper assembly of apparatus.

The steps we provide in all of the Experimental Procedures in this text should be followed closely. There is a good reason why each operation must be carried out exactly as described, even though the reason may not be immediately obvious. The Experimental Procedures given in the early chapters of the book are rather specific. Those given in the later chapters are a little less specific since the experimental techniques should be more familiar to you by then, and we want to avoid your feeling that you are always following a "cookbook." As you gain familiarity and confidence in working in the laboratory, you may want to formulate alternative procedures for carrying out a reaction or purifying a product. For safety reasons, *always check alternative plans with your instructor before carrying out the work*. In advanced organic laboratory courses and in graduate school, you are expected to use your chemical intuition more freely. For example, the literature of organic chemistry often provides exceedingly brief experimental directions and requires that the chemist supply many of the experimental details and techniques to be used.

After performing various reactions and different techniques in the laboratory, you should appreciate the problems facing the manufacturers of chemicals and drugs. Nearly all organic reactions occur with less than 100% conversion of the starting compounds into the desired product. Organic reactions occur with side reactions that reduce the amount of product obtained. Much of the loss occurs during purification of the desired compound.

Organic experiments tend to be less precise than quantitative inorganic analysis. An Experimental Procedure in this text may call for diluting a solution with 10 mL of water. Unless specifically directed otherwise, *precisely* 10 mL of water does not have to be used, but only *approximately* that amount, say 8–12 mL, could be added without adverse effect.

Stars (★) are inserted periodically in many of the Experimental Procedures in this text to indicate places where the experiment can be interrupted without affecting its final outcome. This should help you make the best use of the assigned laboratory time. For example, an experiment may be started toward the end of the laboratory period, carried through to a starred point (★), and then discontinued until the next period. Chemicals and equipment should be safely stored until the next period. However, stars (★) have not been placed at every possible place, because in some cases it is obvious that the experiment can be halted without any problem.

Laboratory safety and the preparation of the **laboratory notebook** are exceedingly important aspects of any experimental laboratory science. Laboratory safety is important to you, to other students in the laboratory, to your instructor, and even to the facilities in which you work. The laboratory notebook is the place where records of all your work are kept, and it is important to develop good habits in record keeping. After leaving college, you may be working for a chemical company or in a research environment. Your employer or supervisor will demand safe laboratory work and accurate record keeping. In the next sections of this chapter, laboratory safety and the laboratory notebook are discussed.

1.2 Laboratory Safety

Concerns about the possible health-related dangers of chemicals have increased in recent years. Any chemistry laboratory is potentially dangerous because most laboratories contain flammable liquids, fragile glassware, and poisonous chemicals. Many also are equipped with high-pressure gas cylinders, and while some of these gases—for example, nitrogen and helium—are not dangerous themselves, the fact that they are under high pressure poses potential danger. Proper precautions must always be taken and safe experimental procedures must be followed; if this is done, the laboratory is no more dangerous than a kitchen or bathroom. One must always carefully consider what equipment to use and make sure that the equipment is safe. Common sense must prevail at all times. The beginning student cannot be expected to possess extensive knowledge of what constitutes safe laboratory practice. This must be developed and learned in the same way that the theoretical aspects of organic chemistry must be mastered.

"Do It Safely" sections are included in this text and serve to call attention to special safety features that are needed for most experiments. These sections should be read carefully *before* setting up and performing an experiment. They indicate potential safety problems about the equipment and/or chemicals to be used and represent some of the thinking that an experienced organic chemist takes for granted when working in the laboratory. We hope you read and accept these safety discussions with a positive attitude, for they should help you develop safe laboratory technique with the ultimate goal of instilling "common sense" in your work. We present some important *general* safety considerations in the paragraphs that follow. Read them carefully, and practice them faithfully until they become "second nature."

DO IT SAFELY: GENERAL DISCUSSION

1 When checking into laboratory for the first time, examine all of the glassware carefully. Look for small cracks, chips, or other imperfections in the glass that might weaken it. It is especially important to check round-bottomed flasks and condensers. Organic experiments often require heating a round-bottomed flask, and, if it contains any cracks or imperfections, it may crack during heating, thus causing its contents, usually potentially dangerous and/or flammable chemicals, to be spilled. Cracks in the joints or at the ring seals of condensers (a ring seal is the point at which the inner tube of the condenser and the water jacket are joined) may allow water to drain into the flask below, which may contain reagents that react violently with water. Replace cracked glassware immediately. Take time to store glassware carefully in your locker or drawer. Carelessly stored glassware may become cracked as a result of opening and closing drawers. **Develop the habit of examining glassware before each use.**

2 Nearly all organic substances are **flammable.** Avoid using flames whenever possible! Sometimes Bunsen burners or open flames must be used, as for example in the distillation of very high-boiling liquids. The "Do It Safely" sections indicate when open flames must be used and suggest special precau-

tions for their use. When flames must be used, observe the following precautions:

a Use a burner to heat a flammable liquid only when it is contained in a flask that is protected by a condenser. **Never heat a flammable liquid in an open container with a burner.** Use a steam bath, hot plate, or similar electrical heating device for heating liquids in open containers. Table 3.1 provides information about the flammability of many commonly used organic solvents. Do not assume that a solvent is nonflammable just because it is not listed there. *Never* use a burner in the laboratory without the permission of your instructor. Further information about heating methods and Bunsen burners is provided in Section 2.5.

b Do not pour flammable liquids (either hot or cold) from one container to another when there are open flames within several feet.

c Do not pour any organic solvents into any drains or sinks, as they may be carried to places further down the bench or into other labs where there are open flames.

3 Most chemicals are *at least* mildly toxic. Some are *very* toxic, corrosive, or caustic. **Handle all chemicals with caution.** It is good practice to wear some form of hand protection, such as rubber or plastic gloves, whenever handling chemicals. Avoid breathing the vapors of all compounds. While we have endeavored in experiments presented in this text to eliminate the use of all chemicals that are known to possess appreciable toxicity, it is best to assume that *all chemicals are toxic*.

4 **Never taste anything** unless specifically told to do so.

5 **Always wear safety glasses or goggles in the laboratory.** Even if you are only washing glassware and are not doing any work yourself, other students working nearby may have an accident that involves you! *Do not wear contact lenses in the laboratory*, even if wearing safety glasses or goggles over them, because chemicals may, in the event of an accident, get behind them. If this should happen, you are unlikely to be able to get the contacts off before the chemicals have done damage to your eyes. Students who wear correction glasses should ascertain that they contain shatter-proof lenses. Unlike many chemical reactions, *eye damage is irreversible*. The few hours of discomfort that wearing safety glasses may impart per week is a small price to pay compared to impaired vision or blindness for the rest of your life. The constant wearing of eye protection cannot be overemphasized to all persons who work in a chemistry laboratory.

6 **Avoid inhaling the vapors of all organic compounds.** If possible, use a fume hood when pouring and handling dangerous substances and for carrying out reactions that release noxious gases. If this is not feasible, attach appropriate gas traps to the apparatus, as directed in the Experimental Procedures.

7 Use appropriate equipment and glassware to handle or to transfer chemicals from one container to another. **Avoid allowing chemicals to come in contact with skin.** If chemicals are spilled on your skin, they can usually be removed by thoroughly washing with soap and water. This should be done promptly following spillage. Avoid using organic solvents such as acetone or ethanol to remove chemicals from your skin. These solvents may actually hasten absorption of the chemical into your skin.

8 **Become familiar with the layout of the laboratory room.** Know the location of fire extinguishers, fire blankets, safety showers, and eye-wash fountains. Make sure your instructor explains the location, operation, and purpose of each of these safety devices. Read the ''First Aid in Case of Accident'' section on the inside front cover of this book.

9 **Learn the location of the nearest exits from your laboratory to the outside of the building.** In case of fire in the building, use the stairs to exit; never use the elevator. Remain calm and walk, do not run.

10 **Never work in the laboratory alone.** Your instructor should advise you about working in lab at times other than the regular laboratory period, if this is permitted. Someone else should always be in the room when you are working. The importance of this is indicated by the fact that most chemical companies require at least two people in a lab at all times; employees who do not obey this regulation are subject to dismissal.

(a) Disposal of Chemicals and Glassware

The safe disposal of chemicals is of great importance to the preservation of our environment. Your laboratory should have waste containers for used chemicals. Plastic containers should not be used for organic chemicals, since many of them will react with or dissolve plastic. Liquid and solid waste chemicals are best kept in separate *glass* containers. The use of these special containers prevents the buildup of solids and flammable liquids in drain traps and pipes. Care must be taken not to mix together different organic materials, owing to the possible chemical reactions that may take place between them. To avoid this problem, all waste containers should be marked with labels describing the contents. Aqueous solutions usually can be discarded in the sink along with ample running cold water. Do not mix organic waste and aqueous waste, since acidic, basic, or other types of material possibly present in the solutions may initiate chemical reactions in the container.

Solid inorganic waste should be discarded in a manner directed by your instructor. Most of it should *not* be thrown in waste cans as this would allow the housekeeping staff to be exposed unknowingly to possibly dangerous chemicals.

Many laboratories have a special place to discard broken glassware, again to protect the housekeeping staff. Consult your instructor to determine if this is the case in your laboratory. Broken thermometers present a special problem since the mercury they contain is extremely toxic, even in small quantities. They should *never* be thrown in a waste can, and any spilled mercury should be cleaned up immediately. Ask your instructor about cleaning up mercury spills.

REFERENCES

1. Sax, N. I. *Dangerous Properties of Industrial Materials,* Reinhold Publishing Corporation, New York, 1957.

2. *Merck Index of Chemicals and Drugs,* 10th ed., Merck and Company, Rahway, N.J., 1983.

3. *Handbook of Laboratory Safety*, N. V. Steere, ed., The Chemical Rubber Company, Cleveland, Ohio.

4. *Manual of Hazardous Chemical Reactions*, 4th ed., National Fire Protection Association, Boston, 1971.

5. *Suspected Carcinogens*, 2d ed., National Institute for Occupational Safety and Health, Bethesda, Md., 1976.

6. *Safety in the Chemical Laboratory*, 4 vols., Journal of Chemical Education, Easton, Pa.

1.3 The Laboratory Notebook

One of the most important characteristics of a successful scientist is the habit of keeping good records of work that has been done. A scientist, including the organic chemist, must be *observant in all stages* of experimental work. Did a precipitate form? Was there a color change when a reagent was added to a solution? Is a reaction exothermic? These observations may appear insignificant at the time but later may prove vital to the correct interpretation of an experimental result. Your future employer will *require* you to keep accurate records, so get into the habit of doing so now! For example, many chemical patents have been awarded on the basis of accurate record keeping, especially if one company can demonstrate that they had done the work before another one.[1]

Some important features of a good laboratory notebook are provided here:

1 Record all observations and data in a notebook *at the time they are obtained*. Scraps of paper should *never* be used for keeping records, since they are often lost or mixed up so that the experimenter does not know which data belong to which experiment. There is no reason to record data on odd pieces of paper and transcribe the information into a notebook later. Neatness is desirable but less important than a complete, accurate notebook. Recopying experimental data takes time, cannot possibly improve it, and may even worsen it because of errors made in recopying. *The information recorded at the time of performance constitutes the primary record of the work*. Do not trust your memory or think that you can remember several hours later what you have done.

2 The criterion for what should be included in the notebook is that the record should be so thorough and well organized that anyone who reads your notebook will understand what has been done and can repeat it in precisely the same way the original work was done.

3 A *bound notebook* is the best permanent laboratory record. The binding makes it less likely that pages will tear out and be lost. Specially designed laboratory notebooks are available that contain pairs of identically numbered pages so that carbon paper can be inserted between the pages, thus permitting a copy of the information to be recorded. The copy can be torn out and stored in another place, thus avoiding the possible loss of valuable data in the event of a fire or other disaster. Many professional scientists use this type of notebook.

4 Record all entries *in ink*. Nothing should ever be deleted from a notebook. If a mistake is made, it should be crossed out and the correct information added. There are many instances known where information which was *considered* incorrect was later determined to have been correct. Many significant scientific discoveries and inventions have resulted from having all observations and data in a notebook!

---

[1] An interesting article describing the importance of maintaining good laboratory records is "Keeping a Laboratory Notebook" by Anne Eisenberg in *Journal of Chemical Education*, **1982,** *59,* 1045.

5 Use the first page of the notebook as a title page, and leave several additional pages for a Table of Contents. The pages should be numbered; most laboratory notebooks have already been numbered by the manufacturer.

6 The overriding requirement of a good laboratory notebook is completeness of observation and recording. The results should be summarized, and conclusions should be drawn for each experiment. An explanation should be provided if the results are not those expected.

7 When performing an experiment using specific directions, it is usually unnecessary to copy the complete experimental procedure into the notebook. However, every such experiment should include a specific *reference* so that another chemist could perform the experiment later if desired. The professional chemist seldom copies an entire experimental procedure in a notebook if it is given in detail in the literature; to do so is a waste of valuable time. However, any *variations* from the given procedure should be noted, along with the reasons for doing so. If this is not done, it will be impossible to reconstruct later on what was done and why it was done.

8 The laboratory notebook should therefore be a *complete* log of all laboratory operations. *Dates, times, and other pertinent information should be entered regularly.* Many patent cases have been won on the basis of the dates and times entered in a laboratory notebook.

Your instructor will probably provide specific directions for the preparation and use of your laboratory notebook.

(a) Types of Organic Experiments

There are two broad classes of organic chemistry experiments contained in this text: **investigative experiments** and **preparative experiments.** Investigative experiments usually involve making observations and learning essential experimental techniques that are common to most organic chemistry laboratory work. Examples include distillation, recrystallization, extraction, purification, spectroscopy, and qualitative organic analysis. On the other hand, preparative experiments involve the conversion of one compound into another.

These types of experiments usually require a slightly different notebook format, which we discuss in the next several sections. These are only suggestions, and your instructor may provide specific instructions regarding notebook format.

Successful laboratory work requires *preparing for the experiment in advance*. To learn anything from an experiment, you must *understand* what you are doing while you do it. Read the theory and experimental procedure before coming to lab. Failure to do so may mean that you will not complete the work in the allotted amount of time and, more important, that you will probably not appreciate the meaning of the experiment. Your instructor may require a certain amount of advance preparation, including Pre-lab exercises, and completion of certain parts of the laboratory notebook before you start work.

(1) INVESTIGATIVE-TYPE EXPERIMENTS AND THE LABORATORY NOTEBOOK

A possible format for investigative-type experiments is discussed here. Each experiment should start on a fresh page, which should contain a title and reference at the top. Dates should be entered throughout the notebook. Important information to be included in the notebook follows:

1 *Introduction*. Give a brief introduction to the experiment, and state clearly the purpose(s) of it. This should take up no more than half of the page.

2 *Experiments and Results*. Provide a brief one- or two-line statement for each part of an experiment. Leave sufficient room to enter results as they are obtained. In general, do *not* recopy the entire experimental procedure from the text, but provide a line to identify each part of the experiment. This section is best completed before coming to lab to ensure that you have fully in mind what you have to do. This *advance* preparation will help ensure that you do all parts of the experiment and also will help you to gain more understanding from your work.

3 *Conclusions*. Upon completion of the experiment, state briefly the conclusions reached, which are based upon the results obtained. If identifying an unknown, summarize the results of your findings here.

4 *Answers to Exercises*. If your instructor has assigned some of the exercises from the text and desires them to be part of your laboratory report or notebook, a logical place to answer them is at the end of the experiment. Many of the exercises in this text bear directly on the experiments and thus merit serious and thoughtful consideration.

(2) PREPARATIVE-TYPE EXPERIMENTS AND THE LABORATORY NOTEBOOK

Preparative-type experiments can be identified easily since they involve converting a compound into a new one using a sequence of one or more reactions. A suggested general notebook format for preparative experiments is provided; however, your instructor may provide other specific instructions for notebook format. Start each new experiment on a fresh notebook page and include the title and reference for the experiment, along with the dates of performance. Some of the information that may be required in your laboratory notebook is listed below.

1. *Introduction*. Prepare a brief statement describing the experiment.
2. *Main Reaction(s) and Mechanism(s)*. Write equations for the main reaction(s) for the conversion of the starting compound(s) to the product(s). The equations must be *balanced* in order to calculate the theoretical yield of product to be expected. The calculation of theoretical yield is discussed in part 4 below. When possible, provide the mechanisms of the reactions that are being performed.
3. *Table of Reactants and Products*. A convenient method for summarizing the amount of reactants to be used and of product to be formed involves setting up a Table of Reactants and Products. This table should contain *only* the reactants used initially; many other reagents may be used in the purification process, but they should **not** be included in the table. This type of table might contain the following entries:
   a. The name and/or structure of each starting material and the product(s).
   b. The molecular weight of each compound.
   c. The weight in grams of each starting material used.
   d. The moles of each starting material used, which can be calculated from (2) and (3) above. A brief review of the method of calculating moles is given in Section 1.3c.
   e. The theoretical mole ratio, expressed in whole numbers, for the reactants and products. This ratio comes from the *balanced* equation for the reaction.
   f. Physical properties of the reactants and products. This entry contains information such as boiling point, melting point, density, and color.
4. *Yield Data*. The maximum expected amount of product, called the **theoretical yield,** can be calculated from the Table of Reactants and Products. From the number of moles of

each reactant used and the mole ratio of reactants in the balanced equation, the reactant that is the **limiting reagent** can be determined. The reaction stops when the limiting reagent is consumed; that is, one reagent will limit the maximum amount of product formed, no matter how much of the other reactants remains. The theoretical yield of product, in moles, can be calculated from the number of moles of product formed, as determined from the number of moles of limiting reagent and the balanced equation. The theoretical yield of product in grams can be calculated by multiplying the theoretical yield of product (in moles) by the molecular weight of the product.

Another method of expressing the efficiency of a reaction is **percent yield.** After completing the reaction and obtaining the product in pure form, the **actual yield** of product, in grams, is determined. The percent yield can be calculated from the theoretical yield and the actual yield, using equation 1.

$$\text{Percent yield} = \frac{\text{actual yield (g)}}{\text{theoretical yield (g)}} \times 100 \qquad \textbf{(1)}$$

Unless the starting materials and product have been weighed on an analytical balance, percent yields are rounded off to whole numbers since the data do not warrant any greater accuracy. One advantage of expressing the yield as a percent is that it does not specify the weights involved. It gives an indication of how "efficient" the reaction is. Percent yields greater than 80% are considered excellent for organic reactions.

In this text, we often give the *expected yield* in preparative experiments. This value is never used in any calculations but is provided to give an indication of the percent yield you might anticipate for the preparation. Yields greater than this indicate above-average technique, and lower yields suggest that the technique needs improvement.

5. *Observed Properties of the Product.* Enter the physical properties of the product that you obtained from the experiment, for example, melting point or boiling point, color, odor, crystalline form, and related data. The *reported* melting point or boiling point of the product, which can be looked up in various reference books such as the *Handbook of Chemistry and Physics,* should be recorded. Compare the reported melting point or boiling point with the one you obtained.

6. *Side Reactions.*[2] List all the possible side reactions that may occur in the reaction. Consult additional sources, such as lecture notes or lecture textbook, to determine these. It is important to deduce what side reactions might occur, since the side products must be removed in the purification process.

7. *Other Methods of Preparation.*[2] Alternate methods for preparing a desired compound are often available and may be better in some way than the method being used. Although we have selected experiments that we consider to be most appropriate for an undergraduate laboratory course, it is important when working as a professional chemist to consider alternatives. Thus, it is good practice to look up and consider other methods of preparation.

8. *Method of Purification.*[2] A very important phase of preparative organic chemistry is the purification of the product. The most difficult aspect of preparing a compound is frequently the isolation of the product in pure form, free from side products and unchanged starting materials. Only after considering the properties of the possible side products and the starting materials can a purification scheme be devised to permit the removal of these impurities. A flow diagram is useful to show the various steps taken in the purification process to remove unchanged starting materials and side products.

[2]The inclusion of these parts in your lab notebook is *optional,* although we indicate advantages of doing so. Your instructor will indicate whether or not you should include them.

9. *Answers to Exercises.*[2] Your instructor may assign some of the exercises from this text, the answers to which should appear here.

A detailed example of a preparative write-up is given in Figure 1.1 for the chlorination of cyclohexane (Section 12.1A). Even though you may not do this experiment, follow this example through to understand how the nine items described above are applied. Various entries in Figure 1.1 are labeled with **bold** numbers, for example, (1), and are discussed further here.

(1) Begin on a new notebook page. Include the title of the experiment, the reference to the lab text where the experiment is found, and the date.

(2) This is a brief description of the preparation.

(3) List the main reaction(s). In this example, there is only one reaction, but many preparations involve more than one step, and equations for all main reactions should be given. Be sure to *balance* the equation(s), as the information so provided will be used in the Table of Reactants and Products. The mechanism for this reaction has been intentionally omitted.

(4) This illustrates a useful format for setting up the Table of Reactants and Products.

(5) Give the name of each reactant and of the product; the structures could also be given here if desired.

(6) Calculate the molecular weight, M.W., of each reactant and product.

(7) Enter the weight, in grams, of each starting reactant.

(8) Calculate the moles of each reactant using the data entered in (6) and (7).

(9) Enter the theoretical, whole-number ratio of moles for the reactants and products. These numbers come from the balanced equation for the main reaction.

(10) This entry contains pertinent physical properties of reactants and products. This information must be obtained from a reference book.

(11) The *limiting reagent* can be determined as follows. From (8), we see that we started with 0.2 mole of sulfuryl chloride and 0.4 mole of cyclohexane. From the theoretical, whole-number ratio of moles in (9) (recall that this ratio comes from the balanced equation for the main reaction), we see that one mole of sulfuryl chloride requires one mole of cyclohexane. From the actual amounts used, we can deduce that 0.2 mole of sulfuryl chloride would require 0.2 mole of cyclohexane, but we started with 0.4 mole of cyclohexane. Since we started with *twice* as much cyclohexane as we actually needed, sulfuryl chloride is the limiting reagent. We can reach the same conclusion from another direction. Suppose we started by asking how much sulfuryl chloride is required for 0.4 mole of cyclohexane, and the main reaction tells us that these are required in a 1:1 ratio. We would need 0.4 mole of sulfuryl chloride, but we started with only 0.2 mole of it, so we conclude again that sulfuryl chloride is the limiting reagent.

(12) Here we determine the theoretical yield of chlorocyclohexane, first in moles and then in grams. Since the limiting reagent is sulfuryl chloride and the balanced main reaction tells us that one mole of sulfuryl chloride gives one mole of chlorocyclohexane, we deduce that the maximum possible amount of chlorocyclohexane that can be produced is 0.2 mole, since we started with 0.2 mole of sulfuryl

Figure 1.1 Sample notebook format for preparative experiments.   Your Name
                                                                          Date

CHLORINATION OF CYCLOHEXANE

(1) Reference: *Modern Experimental Organic Chemistry*, 4th edition, by Roberts, Gilbert, Rodewald and Wingrove. Section 12.1A page 324.

(2) 1. INTRODUCTION

Chlorocyclohexane is to prepared from cyclohexane and sulfuryl chloride in the presence of an initiator.

(3) 2. MAIN REACTION(S) AND MECHANISM(S)

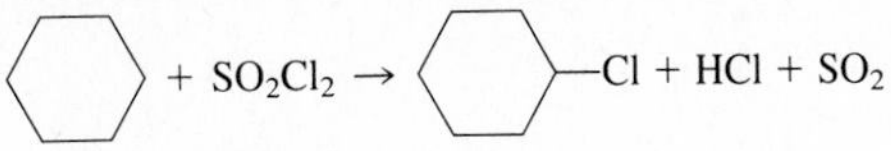

$$\bigcirc + SO_2Cl_2 \rightarrow \bigcirc\!-Cl + HCl + SO_2$$

(Mechanism intentionally omitted)

(4) 3. TABLE OF REACTANTS AND PRODUCTS

(5) Compound	(6) M.W.	(7) Weight Used (g)	(8) Moles Used	(9) Ratio of Moles: Theory	(10) Other Data
Cyclohexane	84	33.6	0.4	1	density = 0.779; bp 81°C
Sulfuryl chloride	135	27.0	0.2	1	density = 1.667; bp 69°C
Chlorocyclohexane	118.5			1	density = 1.016; colorless; bp 142.5°C
Azobisisobutyronitrile		0.1	(initiator)		

(11) Limiting reagent: *Sulfuryl chloride*

(12) 4. YIELD DATA

See discussion in text. Moles sulfuryl chloride = moles chlorocyclohexane = 0.2.
Theoretical yield of chlorocyclohexane = moles $C_6H_{11}Cl$ × M.W. $C_6H_{11}Cl$
= 0.2 moles × 118.5 g/mole
= 23.7 g

Actual yield = 15.0 g

Percent yield = $\dfrac{15.0\ g}{23.7\ g} \times 100 = 63\%$

(13) 5. OBSERVED PROPERTIES OF PRODUCT

bp 138–140°C; colorless liquid; insoluble in water

(14) 6. SIDE REACTIONS

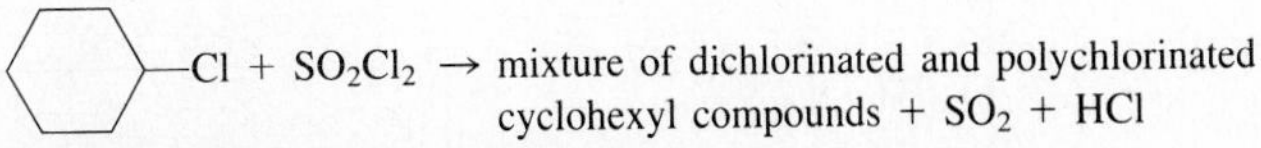

$$\bigcirc\!-Cl + SO_2Cl_2 \rightarrow \text{mixture of dichlorinated and polychlorinated cyclohexyl compounds} + SO_2 + HCl$$

(15) 7. OTHER METHODS OF PREPARATION

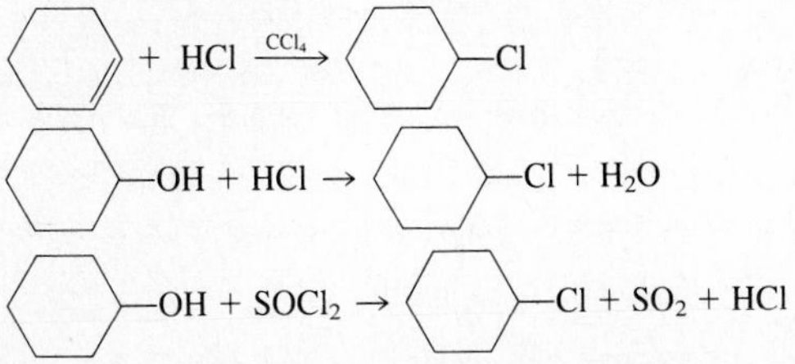

$$\bigcirc + HCl \xrightarrow{CCl_4} \bigcirc\!-Cl$$

$$\bigcirc\!-OH + HCl \rightarrow \bigcirc\!-Cl + H_2O$$

$$\bigcirc\!-OH + SOCl_2 \rightarrow \bigcirc\!-Cl + SO_2 + HCl$$

(16)   8. METHOD OF PURIFICATION

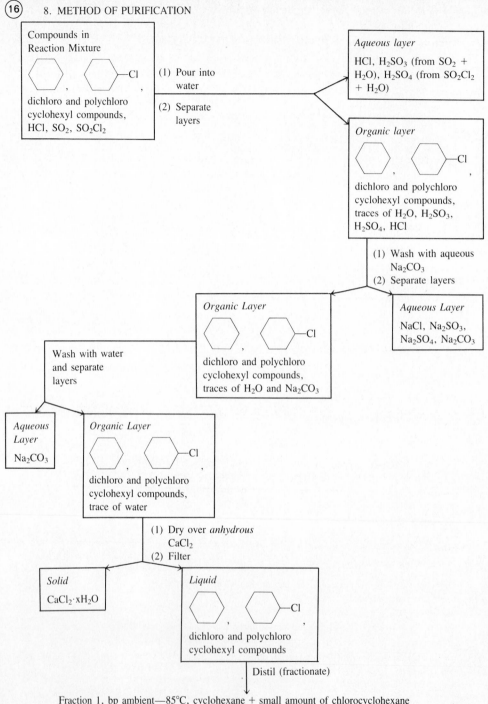

Compounds in
Reaction Mixture

dichloro and polychloro
cyclohexyl compounds,
HCl, SO_2, SO_2Cl_2

(1) Pour into
    water

(2) Separate
    layers

Aqueous layer

HCl, H_2SO_3 (from SO_2 +
H_2O), H_2SO_4 (from SO_2Cl_2
+ H_2O)

Organic layer

—Cl

dichloro and polychloro
cyclohexyl compounds,
traces of H_2O, H_2SO_3,
H_2SO_4, HCl

(1) Wash with aqueous
    Na_2CO_3
(2) Separate layers

Aqueous Layer

NaCl, Na_2SO_3,
Na_2SO_4, Na_2CO_3

Organic Layer

—Cl

dichloro and polychloro
cyclohexyl compounds,
traces of H_2O and Na_2CO_3

Wash with water
and separate
layers

*Aqueous
Layer*

Na_2CO_3

Organic Layer

—Cl

dichloro and polychloro
cyclohexyl compounds,
trace of water

(1) Dry over *anhydrous*
    $CaCl_2$
(2) Filter

Solid

$CaCl_2 \cdot xH_2O$

Liquid

—Cl

dichloro and polychloro
cyclohexyl compounds

Distil (fractionate)

Fraction 1, bp ambient—85°C, cyclohexane + small amount of chlorocyclohexane

Fraction 2, bp 85—145°C, chlorocyclohexane + possibly small amount of cyclohexane

Residue, bp > 145°C, dichloro and polychloro cyclohexyl products

(17)   9. ANSWERS TO EXERCISES (if any assigned)

chloride. The theoretical yield of chlorocyclohexane (in grams) is calculated by multiplying the maximum number of moles that can be formed, 0.2 mole, by its molecular weight.

If we obtained 15.0 g of pure chlorocyclohexane in an actual experiment, the percent yield is calculated by dividing the actual yield by the theoretical yield, both expressed in grams, and multiplying the result by 100. The percent yield could also be calculated by dividing the actual yield, in moles, by the theoretical yield, in moles, and multiplying the result by 100. This involves one additional calculation since the weight of chlorocyclohexane obtained must first be converted into moles.

(13) These are the properties actually observed when doing the experiment. The reported boiling point of chlorocyclohexane could also be entered here for comparison purposes.

(14) Several of the possible side products are given here.

(15) Other possible methods for preparing chlorocyclohexane are looked up in a textbook and written here.

(16) This entry starts with the possible components of the reaction mixture that might be present *after* the main reaction has been carried out. The flow diagram illustrates where and how each of the inorganic and organic side products and the unchanged starting materials are removed by the various experimental steps of the purification process. At the end of the flow chart, chlorocyclohexane remains as a pure product.

(17) Answer any Exercises that may have been assigned from the lab text.

(b) Examples of Table of Reactants and Products; Theoretical and Percent Yields

Several examples showing how to set up a Table of Reactants and Products and how to calculate the theoretical and percent yields are given here.

Example 1

Problem Suppose we start with 25.0 g of $CH_3CH_2CH_2OH$ and 10.0 g of HBr, which react according to the equation

$$CH_3CH_2CH_2OH + HBr \rightarrow CH_3CH_2CH_2Br + H_2O$$
$$\text{1-Propanol} \qquad\qquad \text{1-Bromopropane}$$

From this reaction, we obtain 11.0 g of $CH_3CH_2CH_2Br$. Set up the table and calculate the theoretical and percent yields.

Solution Start by noting that the equation given above is already balanced and that each compound has the molar coefficient of "1." From this information, the Table of Reactants and Products given in Table 1.1 can be constructed.

TABLE 1.1 **Table of Reactants and Products for the Reaction**
$CH_3CH_2CH_2OH + HBr \rightarrow CH_3CH_2CH_2Br + H_2O$

Compound	M.W.	Weight Used (g)	Moles Used	Ratio of Moles: Theory
$CH_3CH_2CH_2OH$	60	25.0	0.42	1
HBr	80	10.0	0.10	1
$CH_3CH_2CH_2Br$	123	*	*	1

*Intentionally left blank, as this is the product. Theoretical yield of product is determined by the procedure given in the discussion.

Now determine the limiting reagent. We started with 0.12 mole of HBr and 0.42 mole of $CH_3CH_2CH_2OH$. The 0.12 mole of HBr would require 0.12 mole of $CH_3CH_2CH_2OH$, but we started with 0.42 mole of it. Therefore, HBr is the limiting reagent.

To determine the theoretical yield of $CH_3CH_2CH_2Br$, start by noting that the balanced equation indicates that 1 mole of HBr produces 1 mole of $CH_3CH_2CH_2Br$. We started with just 0.12 mole of HBr, so we expect the maximum number of moles of $CH_3CH_2CH_2Br$ to be 0.12. To calculate the theoretical yield of $CH_3CH_2CH_2Br$ in grams, multiply 0.12 by the molecular weight of $CH_3CH_2CH_2Br$: 0.12 mole $\times$ 123 g/mole = 14.7 g. To determine the percent yield of $CH_3CH_2CH_2Br$, divide the actual yield (11 g) by the theoretical yield and multiply by 100, to give

$$\text{Percent yield} = \frac{11.0 \text{ g}}{14.7 \text{ g}} \times 100 = 75\%$$

Example 2

Problem In the following reaction, sulfuric acid serves only as a catalyst, so although it must be present for the reaction to occur it is not consumed. This example shows how a catalyst is treated in yield calculations.

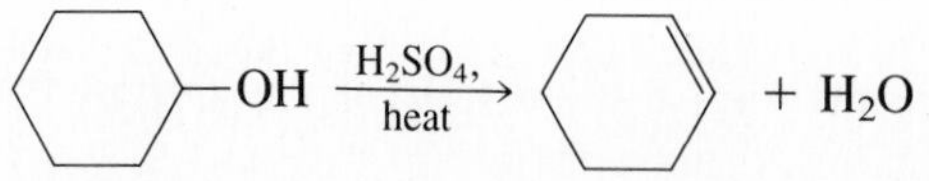

   1   **2**
  Cyclohexanol   Cyclohexene

Suppose we start with 10.0 g of cyclohexanol (**1**, molecular formula $C_6H_{12}O$) and 10.0 g of sulfuric acid and that we isolate 5.0 g of cyclohexene (**2**, molecular formula C_6H_{10}). Set up the Table of Reactants and Products and compute the theoretical and percent yields.

Solution The equation is already balanced, with each reactant and product having a molar coefficient of "1." We can now set up the Table of Reactants and Products given in Table 1.2.

TABLE 1.2 Table of Reactants and Products for the Reaction

$$\text{(cyclohexanol)}\text{—OH} \xrightarrow{\text{H}_2\text{SO}_4} \text{(cyclohexene)} + \text{H}_2\text{O}$$

1   2

Compound	M.W.	Weight Used (g)	Moles Used	Ratio of Moles: Theory
$C_6H_{12}O$ (**1**)	102	10.0	0.098	1
C_6H_{10} (**2**)	82	*	*	1

*Intentionally left blank, as this is the product. Theoretical yield of product is determined by the procedure given in the discussion.

To determine the limiting reagent, recall that sulfuric acid is a catalyst (it is not consumed in the reaction), so it does not enter into calculation of the limiting reagent. Hence, the limiting reagent is **1**. From the balanced equation, note that 1 mole of **1** gives 1 mole of **2**. Since we started with 0.098 mole of **1**, the theoretical yield of **2** is 0.098 mole.

Calculate the theoretical yield in grams by multiplying the number of moles of **2** by its molecular weight: 0.098 mole × 82 g/mole = 8.0 g. The percent yield is calculated by dividing the actual yield (5.0 g) by the theoretical yield (8.0 g) and multiplying the result by 100:

$$\text{Percent yield} = \frac{5.0 \text{ g}}{8.0 \text{ g}} \times 100 = 63\%$$

Example 3

Problem Suppose we start with 20.0 g of $CH_3CH_2CH_2OH$ and 20.0 g of PBr_3, which react according to the unbalanced equation

$$CH_3CH_2CH_2OH + PBr_3 \rightarrow CH_3CH_2CH_2Br + P(OH)_3$$

1-Propanol   1-Bromopropane

We obtain 5.0 g of pure $CH_3CH_2CH_2Br$ from this experiment. Set up the Table of Reactants and Products and determine the theoretical and percent yields.

Solution Start by balancing the equation to give

$$3 \text{ CH}_3\text{CH}_2\text{CH}_2\text{OH} + 1 \text{ PBr}_3 \rightarrow 3 \text{ CH}_3\text{CH}_2\text{CH}_2\text{Br} + 1 \text{ P(OH)}_3$$

We now construct the Table of Reactants and Products given in Table 1.3. This table indicates that we started with 0.074 mole of PBr_3 and 0.33 mole of $CH_3CH_2CH_2OH$, and the balanced equation reveals that they react in the ratio of 1 mole PBr_3 to 3 moles $CH_3CH_2CH_2OH$. Thus, 0.074 mole of PBr_3 would require three times as many moles of $CH_3CH_2CH_2OH$ or 3 × 0.074 mole or 0.222 mole. Since we actually started with 0.33 mole of $CH_3CH_2CH_2OH$, we have more of it than needed, and PBr_3 is the limiting reagent.

TABLE 1.3 **Table of Reactants and Products for the Reaction**
3 CH$_3$CH$_2$CH$_2$OH + PBr$_3$ → 3 CH$_3$CH$_2$CH$_2$Br + P(OH)$_3$

Compound	M.W.	Weight Used (g)	Moles Used	Ratio of Moles: Theory
CH$_3$CH$_2$CH$_2$OH	60	20.0	0.33	3
PBr$_3$	271	20.0	0.074	1
CH$_3$CH$_2$CH$_2$Br	123	*	*	3

*Intentionally left blank, as this is the product. Theoretical yield of product is determined by the procedure given in the discussion.

Now calculate the theoretical yield of CH$_3$CH$_2$CH$_2$Br in moles and grams. From the equation, we see that 1 mole of PBr$_3$ yields 3 moles of CH$_3$CH$_2$CH$_2$Br, but we started with 0.074 mole of PBr$_3$ and not 1 mole. Thus, the theoretical yield of CH$_3$CH$_2$CH$_2$Br is 3 × 0.074 mole, or 0.222 mole. The theoretical yield of CH$_3$CH$_2$CH$_2$Br in grams is determined by multiplying the theoretical yield in moles by its molecular weight: 0.222 mole × 123 g/mole = 27.3 g. To determine the percent yield, divide the actual yield (5.0 g) by the theoretical yield (27.3 g) and multiply the result by 100 to get

$$\text{Percent yield} = \frac{5.0 \text{ g}}{27.3 \text{ g}} \times 100 = 18\%$$

(c) Review of Chemical Calculations: Moles

The calculation of yield data makes extensive use of the mole concept, which is reviewed briefly here for solids, liquids, and aqueous solutions.

1. Moles from Weights Given a weight of a pure solid or liquid, the number of moles can be determined by the formula

$$\text{Moles} = \frac{\text{grams}}{\text{molecular weight}}$$

Given any two of the three—moles, grams, or molecular weight—the other quantity can be calculated. This type of calculation was used earlier to prepare a Table of Reactants and Products.

2. Moles from Volumes: Pure Liquids Given a volume of a pure liquid and its density, both the weight of the liquid and the number of moles it represents can be calculated. Density is defined as mass per unit volume and is commonly reported in handbooks as g/mL. Given a volume of solution in mL and its density in g/mL, the weight can be computed by multiplying the density by the volume. Volume must be expressed in mL. For example, the weight of 25 mL of a liquid whose density is 1.2 g/mL is 25 mL × 1.2 g/mL, or 30 g. The number of moles can be computed as in part 1 above once the weight is known.

3. Moles from Volume and Molarity: Aqueous Solutions The molarity, M, of a solution is defined as the number of moles of solute per liter of solution. If the molarity and the volume are known, the number of moles in the sample can be computed by multiplying the molarity by the volume, expressed in liters, L. For example, the number of moles of HCl in 50 mL of a 0.5 M solution of HCl is 0.5 mole/liter $\times$ 0.05 L, or 0.025 mole. The weight of HCl can be determined from the number of moles, as described above in 1.

4. Moles from Weight and Weight Percent: Aqueous Solutions Given the weight of a solution and the percentage by weight of a certain compound in that solution, the number of moles can be calculated. This is illustrated by the following example.

Given 30 g of a sulfuric acid solution that contains 80% by weight of H_2SO_4, calculate the weight and moles of sulfuric acid present in the solution. Of the 30 g of solution, 80% of it (by weight) is sulfuric acid, so it contains 30 g $\times$ 0.80, or 24 g of sulfuric acid. The number of moles of H_2SO_4 present is 24 g divided by 98 g/mole (its molecular weight), or 0.245 mole.

(d) Writing Laboratory Reports

The following comments provide a general approach for correctly writing *any* type of *formal* laboratory report.

1. Reports should be written *in ink* or typed, using one side of the paper only.
2. References to ''standard'' procedures should be included as footnotes, either placed on the page where the footnote number appears or collected together at the end of the report as ''References.''
3. Reports should be written in the *impersonal form;* the words ''I'' and ''we'' should *NOT* appear in a formal scientific report. The following examples show a few incorrect phrases commonly found in laboratory reports and some suggestions for rewriting them to avoid the use of the personal form.
   a. INCORRECT: I added 12 g of water to . . .
      CORRECT: Twelve grams of water were added to . . .
   b. INCORRECT: I shook the separatory funnel . . .
      CORRECT: The separatory funnel was shaken . . .
   c. INCORRECT: I boiled the mixture . . .
      CORRECT: The mixture was boiled . . .
   d. INCORRECT: You told me that . . .
      CORRECT: The instructor indicated that . . .
   e. INCORRECT: We set up a distillation . . .
      CORRECT: A distillation was set up . . .
   f. INCORRECT: I determined that . . .
      CORRECT: It was determined that . . .

In general, avoid using personal pronouns, and rewrite the sentences so that the meaning is retained. With a little thought and practice, the student will soon become familiar with the proper form used in writing reports. Examination of scientific literature reveals that this writing style is used exclusively in articles and papers.

techniques and apparatus

This chapter presents and discusses briefly some of the many techniques commonly used in the organic chemistry laboratory, and the apparatus associated with these techniques is shown and described. The theory of some of the techniques is discussed in later chapters.

In many of the experiments presented in this text, reference is made to the appropriate technique and/or apparatus as described in this chapter. Thus, you will have cause to refer to this chapter often as you perform the assigned experiments.

2.1 Glassware: Precautions and Cleaning

Many experiments involve the use of glassware, and the following safety precautions regarding the proper and safe use of glassware should be read carefully.

The cardinal rule in handling and using laboratory glassware is *never apply undue pressure or strain to any piece of glassware*. This rule applies to insertion of thermometers or glass tubes into rubber or cork stoppers or rubber tubing. **If you have to force it, don't do it!** Either make the hole larger or use a smaller piece of glass. Another convenient method of inserting glass into corks or tubing is to lubricate the glass with a little water or water containing soap or glycerol. Always grasp the glass piece very close to the rubber or cork part when trying to insert glass into it. It is wise to wrap a towel around the glass and cork before trying to insert one into the other. This usually prevents a serious cut in the event the glass happens to break.

These admonitions become less necessary if ground-jointed glassware, called *standard-taper glassware*, is used in your laboratory course. Other considerations then come into play, for you must avoid applying undue pressure or strain to this type of glassware as well. Be careful to ensure that strain does not develop because of carelessly positioned glassware components. Strained glassware may break when heated or even upon standing.

It is best to clean glassware *immediately* after use. It is good practice to wipe off the lubricant from ground-glass joints with a towel or tissue moistened with a solvent such as acetone or hexane before washing the equipment. If this is not done, the lubricant will stick to the brush used to wash the glassware and thus will be carried onto all surfaces the brush touches. Most chemical residues can be removed by washing the glassware with soap and water, and special laboratory soap or detergent is available in most laboratories for cleaning equipment. Common organic solvents, such as petroleum ether, pentane, toluene, and acetone, can also be used to clean glassware. However, acetone should not be used to clean equipment that contains residual amounts of bromine, since a powerful lachrymator, bromoacetone, may form. A *lachrymator* is a chemical that adversely affects the eyes and causes crying. It also can affect one's lungs. The villains in industrial or photochemical smog include chemicals that are lachrymators.

Stubborn residues may require the application of more powerful cleaning solutions. Chromic acid (made from concentrated sulfuric acid and chromic anhydride or potassium dichromate) is sometimes effective, but it must be used *with great care*. When using chromic acid, wear rubber gloves and pour it *carefully* into the equipment. When through, pour the chromic acid solution into a specially designated bottle and *not* into the sink. Another powerful cleaning solution can be made by putting some ethanol into the equipment, adding a few pellets of solid potassium hydroxide, and warming the solution gently while swirling the solution around inside the glassware. *Before using any cleaning solutions other than soap and water, consult your instructor for permission and directions concerning their safe handling.*

Some organic reactions result in the formation of a layer of carbon in the glassware, and this can be removed by carefully scraping the glassware with a bent spatula in the presence of soap and water. If this fails to remove the carbon, try acetone. If acetone fails to do the job, ethanol containing potassium hydroxide (see preceding paragraph) can be used.

If you have used manganese-containing compounds in a reaction, brown stains of manganese dioxide are sometimes left in the glassware. These can generally be removed by rinsing the apparatus with a 30% (4 M) aqueous solution of sodium bisulfite, $NaHSO_3$. If this fails to work, wash the equipment with water and then add a small amount of 6 M HCl. This must be done in a good fume hood since chlorine gas is evolved; try this only after obtaining permission from your instructor.

2.2 Standard-taper Glassware

Standard-taper glassware has the advantages of convenience and safety, and possible contamination of the reaction being performed is also avoided. Before the advent of standard-taper glassware, chemists had to bore corks or rubber stoppers to fit each type of equipment being connected. This was time-consuming, and many chemicals reacted with the cork.

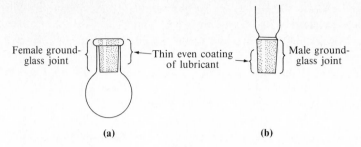

Figure 2.1 Standard-taper joints: (a) female, (b) male.

Moreover, corks had to be bored for each different reaction to avoid cross-contamination, since corks tend to absorb reagents.

There are many manufacturers of standard-taper glassware, but regardless of the manufacturer, a given standard-taper joint size will fit another of the same size. Hence, the meaning of "standard" in the name of the glassware becomes evident. The term "taper" implies the joints are tapered, which they are. A standard-taper joint is depicted in Figure 2.1. Standard-taper joints come in a number of sizes. They are designated by the symbol $\overline{T}\!\!\!S$ followed by two sets of numbers, separated by a slash (for example, 14/20, 19/22, 24/40). The first number is the diameter of the joint in mm at its widest point, and the second number is the length of the joint in mm. A standard-taper joint designated as $\overline{T}\!\!\!S$ 19/22 therefore has a widest diameter of 19 mm and a length of 22 mm. This is also shown in Figure 2.1.

When using ground-jointed glassware, the joints must be properly lubricated so that they do not freeze together and become difficult, if not impossible, to separate. Lubrication is accomplished by spreading a thin layer of joint grease (lubricant) around the outside of the male joint, mating the two joints, and then rotating them together to cover the surfaces of the joints with a thin coating of the lubricant. Using the proper amount of lubricant is important. If too much is used, the reaction contents in the apparatus may become contaminated, and if too little is used, the joints will freeze and cannot be separated. Care must be exercised in setting up equipment using standard-taper glassware to ensure that no strain is put on the joints. The cleaning of standard-taper glassware is discussed in Section 2.1.

Apparatus that contains standard-taper glassware is illustrated throughout this text. However, if this type of glassware is not available, non-standard-taper glassware fitted with bored corks may be substituted.

2.3 Melting-Point Methods and Apparatus

The theory and use of melting points is discussed in Section 3.2. Many different methods can be used to determine the melting point of a solid organic compound. One major limitation of this technique is availability of sample, and therefore the **micro melting-point** method, which uses a very small amount of sample, is discussed. These methods are not too exact but are convenient and easy to use. The basic concept of determining a melting point involves heating a solid and determining the temperature at which it melts. Many different types of heating devices can be used, but most equipment utilizes a capillary tube for the sample. The preparation of the capillary tube and several different types of melting-point apparatus are discussed here.

(a) Capillary Tubes and Sample Preparation

Commercially available capillary tubes have one end already sealed and are open at the other end to permit introduction of the sample. The sample is put into the tube as follows. (See also Figure 2.2.) Place a small amount of the solid whose melting point you wish to determine on a clean watch glass and tap the open end of the capillary tube into the solid on the glass so that a small amount is forced about 2–3 mm into the tube. To get the solid to the closed end of the tube, take a piece of 6–8 mm glass tubing about 1 m (3 ft) long, place this tube vertically on a hard surface (bench top or floor), and drop the capillary tube (sealed end down) through the large tube several times. Surprising as it might seem, the capillary tube does not break, and the solid ends up packed at the sealed end of the capillary tube!

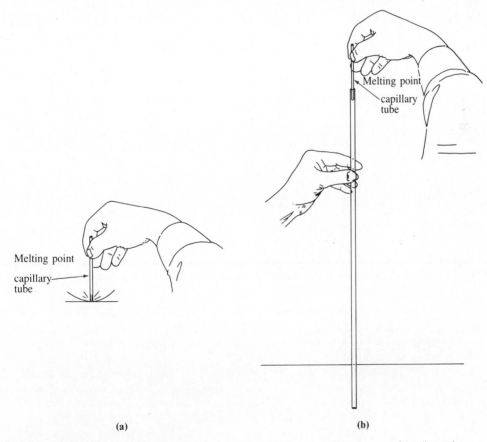

Melting point
capillary tube

Melting point
capillary tube

(a)   (b)

Figure 2.2 (a) Filling a capillary melting point tube. (b) Packing sample at bottom of capillary tube.

(b) Melting-Point Determination

The determination of the melting point involves taking the capillary tube containing the sample and heating it in an appropriate apparatus until the solid melts. Best results are obtained by heating the sample at the rate of about two degrees per minute. Many organic compounds undergo a change in crystalline structure just before melting, usually as a consequence of the release of solvent of crystallization. The solid takes on a softer, perhaps "wet," appearance, which may also be accompanied by a shrinkage of the sample in the capillary tube. These changes in the sample should *not* be interpreted as the beginning of the melting process. Wait for the first tiny drop of liquid to appear. Melting invariably occurs over a temperature range, and the **melting-point range** is defined as the temperature at which the first tiny drop of liquid appears up to and including the temperature at which the solid has melted completely.

(c) Thiele Tube Apparatus

A simple type of melting-point apparatus is the **Thiele tube,** which is shown in Figure 2.3. This tube is shaped so that heat applied to a liquid in the sidearm by a burner is distributed to all parts of the vessel by convection currents in the heating liquid so that stirring is not required. Temperature control is accomplished by adjusting the flame produced by the burner; this may seem difficult at first but can be mastered with practice.

Proper use of the Thiele tube is required to obtain reliable melting points. Secure the capillary tube to the thermometer at the position indicated in Figure 2.3; use either a rubber

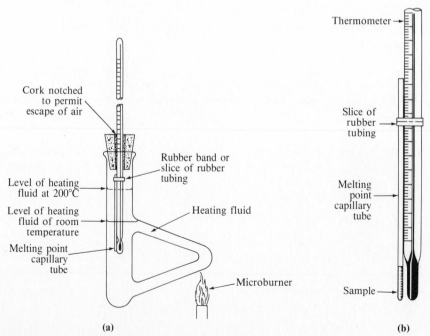

Figure 2.3 (a) Thiele melting-point apparatus. (b) Arrangement of sample and thermometer for determination of melting point.

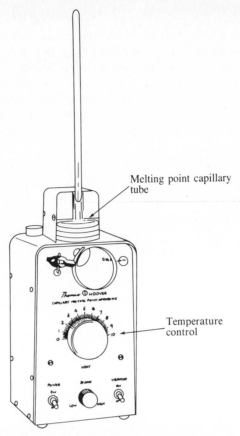

Melting point capillary
tube

Temperature
control

Figure 2.4 Thomas-Hoover melting-point apparatus. (Courtesy of
Arthur H. Thomas Company.)

band or a small segment of rubber tubing for this purpose. Be sure that the band used to hold
the capillary tube on the thermometer is as close to the top of the tube as possible. Support
the thermometer, with the capillary tube containing the sample already attached, in the
apparatus with a cork, as shown in Figure 2.3, or by carefully clamping the thermometer so
that it is immersed in the oil. The thermometer and capillary tube must *not* contact the glass
of the Thiele tube in any way. Make sure that the height of the heating fluid is approximately
at the level indicated in Figure 2.3. The oil will expand on heating. For this reason, the
rubber band should be in the position indicated. Otherwise, the hot oil will come in contact
with the rubber and melt it, and the sample tube will fall into the oil.

Heat the Thiele tube at the rate of one to two degrees per minute in order to determine the
melting point. The maximum temperature to which the apparatus can be heated is dictated by
the nature of the heating fluid, a topic that is discussed in Section 2.5d.

(d) Electric Melting-Point Apparatus

One type of electric melting-point apparatus is the Thomas-Hoover melting-point unit,
shown in Figure 2.4. This unit makes use of an electrically stirred and heated oil bath. Inside

the casing is a container of either mineral oil or (better) silicone oil, into which is immersed an electrical resistance heater. By varying the voltage across the heating element using the large knob in the front of the apparatus, the oil may be heated at a controlled, slow rate. A motor inside drives a stirrer in the container, and the rate of stirring is controlled by a knob at the bottom of the unit. Some models are equipped with a movable magnifying lens system that allows the user to view the thermometer better while viewing the sample in the capillary tube. The capillary tube containing the sample is placed in the location indicated on the figure. This particular unit allows for the determination of the melting points of up to five samples at one heating.

Another electric unit, the Mel-Temp apparatus shown in Figure 2.5, utilizes a heated metal block rather than a liquid for transferring the heat to the capillary tube. A thermometer is inserted into a hole bored into the block, and the thermometer gives the temperature of the

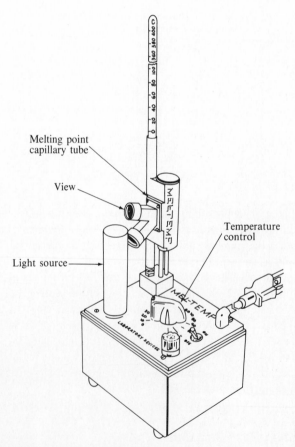

Figure 2.5 Mel-Temp melting-point apparatus. (Courtesy of Laboratory Devices.)

block and the capillary tube. Heating is accomplished by controlling the voltage applied to the heating element contained within the block.

Further description and discussion of these and other types of melting-point devices are available.[1]

2.4 Boiling-Point Apparatus

Just as micro melting-point techniques are used to determine the melting points of small quantities of solids, techniques have also been developed for determining the boiling points of small amounts of liquids. A simple **micro boiling-point** apparatus may be constructed in the following manner. First, prepare a capillary ebullition tube by taking a standard melting-point capillary tube, which is already sealed at one end, and make a seal in it about 1 cm from the open end, using a hot flame (Figure 2.6a). This is most easily done using a glass-blowing torch, and such tubes may be prepared for your use by your instructor. Alterna-

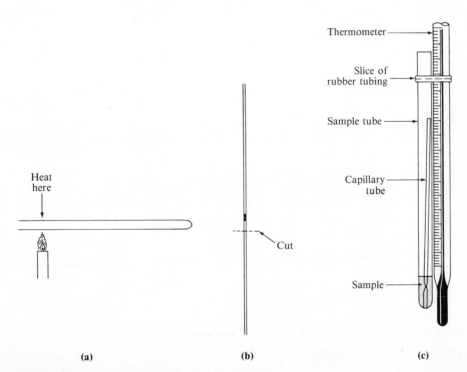

Figure 2.6 Micro boiling-point apparatus: (a) using a single capillary tube; (b) joining two capillary tubes and cutting one end off; (c) assembly of micro boiling-point apparatus, showing correct placement of ebullition tube, sample tube, and thermometer.

[1]Skau, E. L.; Arthur, J. C., Jr., in *Technique of Chemistry,* A. Weissberger and B. W. Rossiter, editors, Wiley-Interscience, New York, 1971, Vol. 1, Part 5, Chapter 3, pp. 105ff.

tively, two melting-point capillary tubes can be joined by heating the closed ends in a hot flame. Then make a clean cut about 1 cm below the point where the tubes have been joined (Figure 2.6b).

Second, seal a piece of 4–6 mm glass tubing at one end and cut it to a length about 1 cm shorter than the prepared capillary ebullition tube. These tubes may also be prepared for your use by your instructor.

Attach the 4–6 mm tube to a thermometer with a rubber ring near the top of the tube. The bottom of the tube should be even with the mercury bulb of the thermometer. Place the capillary ebullition tube in the larger glass tube, and with a pipet add the liquid whose boiling point you wish to determine until the level of the liquid is about 2 mm above the seal of the capillary tube (Figure 2.6c).

Immerse the thermometer and the attached tubes in a heating bath. A Thiele tube (Figure 2.3) is convenient for this purpose, but any other suitable heating bath can be used. *Be sure that the rubber ring is well above the level of the oil in the heating bath or Thiele tube.* Heat the oil bath rather quickly until a *rapid* and *continuous* stream of bubbles comes out of the capillary ebullition tube. Before this occurs, you may see some bubbles evolve in an erratic fashion. This is due to expansion of air trapped in the capillary tube. You should see a marked change from the slow evolution of air bubbles to the rapid evolution of bubbles resulting from the liquid boiling as the boiling point of the liquid is reached. *However, this is NOT the boiling point!* Remove the heating source and allow the bath to cool slowly. As the rate of bubbling decreases and the liquid starts to rise into the capillary tube, note the temperature of the thermometer. *This is the boiling point of the liquid.* If the liquid rises sufficiently slowly into the capillary tube, note the temperatures at which the liquid starts to rise and at which the capillary tube is full. This will be the **boiling-point range** of the liquid.

Remove the capillary ebullition tube and expel the liquid from the small end by gentle shaking. Replace it in the sample tube and repeat the heating and cooling process. The first trial in a boiling-point determination is often done quickly to determine the approximate boiling point. Greater accuracy is obtained when the determination is repeated more slowly. Observed boiling points may be reproduced to within 1 or 2°C.

The physical basis of this technique is interesting. Before the liquid is heated, the capillary tube is filled with air. As the bath is heated, the air in the capillary tube is driven out and replaced with the vapor of the liquid. When the apparatus is heated until vigorous boiling of the liquid is observed, the actual boiling point of the liquid has been *exceeded*. The air in the capillary tube has also been replaced completely with the vapor of the liquid. On cooling, the vapor pressure of the liquid becomes equal to the external pressure, thus allowing the liquid to rise into the capillary tube. The temperature at which this occurs is, by definition, the boiling temperature of the liquid; see Section 4.1.

2.5 Heating Methods

Heating is an important laboratory technique, and it serves a variety of functions. It increases the rate of chemical reactions and is used in the distillation of liquids and in the dissolution of solids during the course of recrystallization and purification. Some of the common heating techniques along with the advantages and disadvantages of the various methods are discussed in the following paragraphs.

(a) Burners

Most chemistry laboratories are supplied with natural gas for use with various types of burners. A burner provides the convenience of a rapid and reasonably inexpensive source of heat. Since nearly all organic substances are flammable, care and good judgment should always be exercised when considering the use of a burner for heating, especially when low-boiling, volatile solvents are to be heated. Table 3.1 provides a list of common solvents, their boiling points, and their flammability. *Never use a burner to heat flammable materials in open containers, such as beakers or Erlenmeyer flasks.* Doing so demonstrates a lack of common sense, equivalent to using a match to look into one's automobile gas tank! Burners *can* be conveniently and safely used to heat *aqueous* solutions containing no volatile and flammable solutes. They are also appropriate for heating higher-boiling flammable liquids that are contained in round-bottomed flasks fitted with a reflux condenser (Section 2.19) or equipped for distillation (Sections 2.7 and 2.8). In these instances, it is important to lubricate the joints of the apparatus with a hydrocarbon or silicone grease to avoid the leakage of vapors through the joints. If an alternate mode of heating is available, choose it in preference to a burner. The experimenter must be aware of the type of work others are doing in the laboratory. The person using a burner may be performing a completely safe experiment, but someone else nearby may be working with a very volatile, flammable material. Many of the volatile organic compounds, especially solvents, expel vapors into the room that may be ignited by a nearby open flame.

Two common types of laboratory burners are pictured in Figure 2.7. The classic **Bunsen burner,** named after its inventor, is shown in Figure 2.7a. The needle valve at the bottom of the burner serves as a fine adjustment of the gas flow. Turning the barrel of the burner controls the amount of air input into the flame. Movement of the barrel in one direction opens the air ports and lets in more air, and movement in the other direction closes them and reduces the amount of air. By proper adjustment of the gas and air flow, a hot flame can be

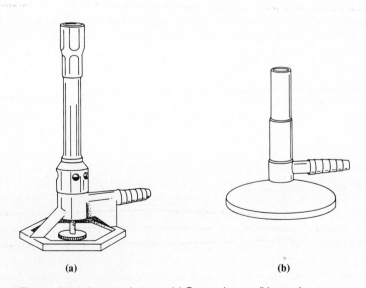

(a)   (b)

Figure 2.7 Laboratory burners: (a) Bunsen burner; (b) microburner.

obtained in which there is a rather sharply pointed blue cone. The hottest portion of the flame is at the *top* of the cone.

Figure 2.7b shows what is commonly called a **microburner,** which produces a smaller flame that allows for easier control of the rate of heating. This type of burner is particularly useful for heating purposes in small equipment. Air flow is adjusted at the baffle at the bottom of the burner. The gas flow is adjusted at the gas valve on the laboratory bench.

Heating a flask with a burner often produces "hot spots" because most of the heat is supplied to a small area on the bottom of the flask. The heat must then be dispersed throughout the liquid in the flask by convection or through turbulence associated with boiling of the liquid. Hot spots can lead to severe bumping in the heated flask. This situation can be minimized by placing a wire gauze between the flame and the flask, which serves to diffuse the heat reaching the flask and provides more uniform heating. The gauze is normally supported with an iron ring.

Burners can provide a convenient method for producing a hot-water bath. Temperatures from ambient to about 90°C may be attained by immersing a flask into a water bath held at the desired temperature by either intermittent or slow heating with a burner. The temperature of the water bath can be monitored by inserting a thermometer in the water. A hot plate (Section 2.5e) can also be used for heating water baths. Temperatures between 90 and 100°C can be attained by steam heating (Section 2.5c), particularly if electrical heating is not available.

(b) Electrical Heating

Heating with electricity is very convenient in the laboratory, and it is especially safe since it avoids the use of open flames. Electrical heating can be used in several ways, as described here. Two essential parts of this type of heating are a resistance element in which electrical energy is transformed into thermal energy, and a method for controlling the voltage reaching the resistance element, which indirectly controls the rate of heating. Voltage control is accomplished by using a **vari**able **AC** transformer, which is commonly called a **Variac** (Figure 2.8b). The scale on the voltage control knob is *not* necessarily the actual voltage passing out of the transformer, even though the scale on most is from 0 to 120 or 130.

(c) Heating Mantles

A widely used device for electrical heating of *round-bottomed flasks* is a *woven-glass **heating mantle*** such as the Glas-Col shown in Figure 2.8a. These mantles have an electrical resistance coil imbedded in the woven-glass fabric. Most heating mantles are designed with a hemispherical cavity *so that a different mantle is required for each different size of flask.* Some mantles are conically shaped to accommodate different styles of flasks. Owing to the nature of their construction, heating mantles must not come in contact with water, and care must be exercised to avoid spilling any type of liquid on them. A special cord is used to connect the heating mantle to the Variac, and *heating mantles must NEVER be plugged directly into the wall outlet.* The cost of a transformer and several common sizes of heating mantles is substantial, but they are probably the most convenient and safest form of heating in the organic laboratory.

Heating mantles usually have a pair of small wires coming out of them in addition to the plug that is used to supply electricity. These small wires are connected to a thermocouple inside the mantle and may be connected to equipment that allows for measurement of the temperature *inside* the mantle. However, most organic chemists do not take advantage of this feature.

Heating mantles have some inconveniences. They initially heat up rather slowly, they are hotter than the contents of the flask, and they have a rather high heat capacity. It is difficult to obtain a given temperature with heating mantles, unless one utilizes the thermocouple mentioned previously. It is also hard to maintain a constant temperature with a heating mantle. If one needs to discontinue heating suddenly (for example, if a reaction begins to get out of control), it is not sufficient either to lower the voltage or to turn off the electricity. To discontinue heating, *the heating mantle must be removed immediately from below the flask* to allow the flask to cool, either on its own or by means of a cooling bath. After the mantle is removed, the electricity should be turned off at the transformer. One should always allow for the possibility of having to remove the heating mantle *quickly;* therefore, apparatus should *not* be assembled so that the mantle rests directly on the laboratory bench. Instead, the mantle should be supported above the bench either by an iron ring or a laboratory jack (Figure 2.8b) and the apparatus assembled accordingly.

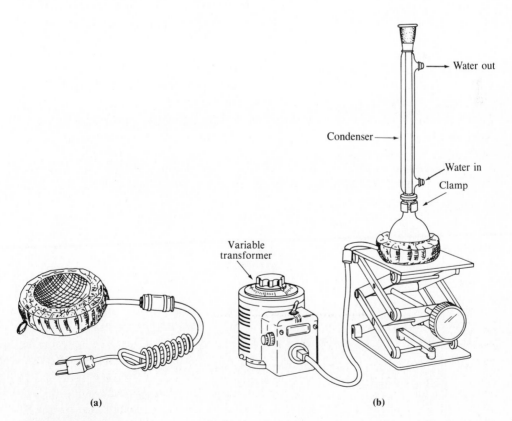

Figure 2.8 (a) Woven-glass heating mantle. (b) Use of a heating mantle and variable transformer to heat a reaction mixture under reflux.

The amount of heat supplied by a heating mantle is moderated by the boiling liquid contained in the flask because the hot vapors of the liquid carry heat away from the mantle. If the flask becomes dry or nearly so, the mantle may become sufficiently hot to melt the resistance wire inside, thus causing the mantle to "burn out." Most mantles are marked with a maximum voltage to be supplied, and this should *not* be exceeded. If nearly all of the contents of a flask have boiled out, as in a distillation, for example, then the electricity should be turned off at the transformer and heating discontinued.

Heating mantles are constructed of nonferrous material and can be used in conjunction with magnetic stirring (Section 2.6b), thus permitting simultaneous heating and stirring of the contents of the flask.

(d) Oil Baths

Electrically heated liquid baths are commonly employed in the laboratory. The liquid used in these baths is usually mineral oil or silicone oil, so they generally are called **oil baths.** Other liquids can also be used. The liquids are heated either by placing the bath container on a hot plate or by inserting a coil of resistance wire in the bath. In the latter case, the resistance wire (the heating element) is attached to a transformer by means of an electrical cord and plug (Figure 2.9). In most instances, the heating element coil is wound using about 3 meters of 26-gauge Nichrome wire. Heating baths offer several advantages. First, the temperature of the bath can be determined with a thermometer inserted in the liquid, and a desired bath temperature may be obtained and accurately maintained by careful adjustment of the transformer. Second, heat is transferred smoothly and uniformly to the full surface of the flask to the depth of its immersion in the bath, so there are no hot spots.

Some inconveniences are encountered with heating baths. First, there is some limitation in the maximum temperature that can be attained safely, this being a function of the type of heating liquid being used; the upper temperature limit is usually between 200 and 250°C.[2]

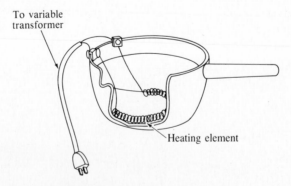

To variable transformer

Heating element

Figure 2.9 An electrically heated oil bath.

[2]Temperatures higher than this can be obtained using special types of heating fluid. One example is "Wood's metal," an alloy that is solid at room temperature (mp 60°C), and can be heated in excess of 400°C with safety.

Second, because the volume of the heating liquid is normally fairly large, it takes considerable time to reach the desired bath temperature. Finally, the heating bath has a high heat capacity, so if the desired temperature is exceeded, some time is required to cool down to it. This may be counteracted by reducing the depth of immersion of the flask in the bath, since the rate of heat transferral is proportional to the surface area of contact between the flask and the heating fluid.

A minor nuisance associated with oil heating baths is the cleaning of apparatus coated with mineral oil or silicone oil, both of which are water-insoluble. Generally they can be removed by either hydrocarbon or chlorinated solvents. Alternatively, polyethylene glycols can be used as the heating liquids. They have the advantage of being water-soluble, so that glassware is easily cleaned. For example, Carbowax 600, a polyethylene glycol, is a liquid at room temperature and may be used up to about 160°C with safety.

There are some other factors of importance concerning the liquids used in heating baths. Silicone oils such as Dow Silicone 550 are more expensive but are generally preferable to mineral oils because they can be heated well above the 200°C temperature range without danger of reaching the flash point (the temperature at which a liquid can burst into flame) and without thickening through decomposition. Mineral oil should *not* be heated above about 200°C because it will begin to smoke, and there is the potential danger of flash ignition of the vapors. This danger is most prevalent with darkened and used mineral oil. Residual amounts of water must not be present in mineral and silicone oils. If drops of water are present in these oils and they are heated to about 100°C, the water will boil and produce bumping in the oil bath, resulting in the *spattering of hot oil*. These baths should be examined regularly, and if water drops are clearly present in the oil, the heating fluid should be changed and the container cleaned and dried before refilling.

(e) Hot Plates

Hot plates (Figure 2.10) are frequently convenient when a *flat-bottomed* container such as a beaker or an Erlenmeyer flask must be heated. The flat upper surface of the hot plate is heated by electrical resistance coils to a temperature that is controlled by a built-in device on the front of the unit. A hot plate should generally be limited to heating liquids such as water (or its solutions) and nonflammable organic solvents such as chloroform and carbon tetra-

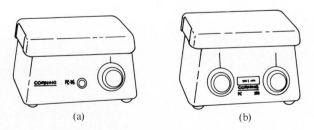

(a)   (b)

Figure 2.10 (a) Hot plate. (b) Stirring hot plate.

chloride. Organic solvents should be heated in the hood to avoid filling the room with vapors from these solvents. *Under no circumstances should a hot plate be used to boil and/or evaporate highly flammable organic solvents.* The vapors of these solvents may ignite as they billow onto the hot surface of the hot plate or come into contact with the electrical resistance coils. Many hot plates use as a control a relay that turns the electricity on and off as needed to maintain the desired temperature, and these relays are sometimes not explosion-proof. If a flammable solvent must be evaporated, use either a steam bath *in the hood* or set up a distillation apparatus for this purpose. As an alternative to working in a hood, an inverted funnel attached to a vacuum source (Figure 2.11c) can be used to remove volatile solvents from the room.

Hot plates are available in combination with built-in magnetic stirrers, and these are called **stirring hot plates** (Figure 2.10b). This unit is convenient for heating and stirring a solution or a reaction mixture simultaneously. However, *round-bottomed flasks* cannot be heated effectively with a hot plate or a stirring hot plate, since there is contact between the flask and the hot surface only at one small point. If a round-bottomed flask is to be heated with a hot plate or a stirring hot plate, then a heating bath must be used to permit uniform transmission of heat to the flask.

(f) Steam Heating

Steam provides a useful source of heat when temperatures up to 100°C are desired. The steam outlet is connected to either a **steam bath** or a **steam cone** (Figure 2.11), both of

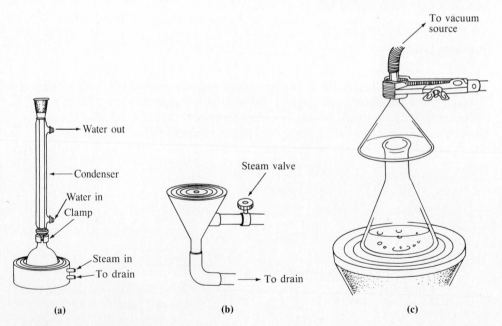

Figure 2.11 (a) Steam bath being used to heat a reaction mixture. (b) Steam cone. (c) Heating an Erlenmeyer flask on a steam cone, with inverted funnel attached to vacuum source to remove solvent vapors and keep them out of the laboratory.

which have an outlet at the bottom to drain the condensed water. When the steam valve is first turned on, several minutes are usually required for the condensed water to drain out of the steam lines. Once the steam is issuing smoothly, the steam valve should be adjusted to provide a *slow, steady* flow of steam. No benefit is gained from having a *fast* flow of steam, other than to give everyone in the lab a sauna bath! Regardless of how fast the steam is flowing, the temperature will never exceed 100°C. Another disadvantage of a fast flow of steam is that water will condense on the equipment and may even find its way into the flask or reaction mixture being heated.

The tops of steam baths and cones are typically fitted with a series of overlapping concentric rings that may be removed in succession until the opening accommodates the size of the flask being heated. The proper-sized opening is normally dictated by the requirements of the operation being performed. For example, if a rapid rate of heat transfer is desired, the rings are removed until perhaps up to one-half of the surface of the flask is immersed in the steam. Another technique for increasing the rate of heat transfer involves wrapping cloth towels around the flask and having the towels in contact with the steam bath. If a slower rate of heating is desired, the opening should be smaller so that less of the flask is in contact with the steam.

Figure 2.11a shows a round-bottomed flask being heated on a steam bath. The same principles apply for heating beakers or Erlenmeyer flasks, except that the top opening should be small enough so that the container sits directly on top of the steam bath with only its lower surface exposed to steam. When heating a flask containing a volatile solvent, as in recrystallization (Section 3.1), an inverted funnel that is attached to a vacuum source can be placed over the top of the container (Figure 2.11c) to remove vapors and keep them from entering the room.

If it is necessary to maintain *anhydrous* conditions within the flask being heated, the flask or apparatus to which it is attached should be protected with a *drying tube* (Section 2.23).

2.6 Stirring Methods

Stirring is often required in organic experiments and is most effectively done by means of *mechanical or magnetic stirring,* as described below. The cost factor prevents many undergraduate laboratories from being equipped with such stirrers, so various alternative procedures are described that provide satisfactory stirring. Note that a boiling reaction mixture often requires no additional stirring (except in some cases of heterogeneous mixtures) because the turbulence and motion of boiling are often sufficient to maintain reasonable mixing.

(a) Swirling

If mechanical or magnetic stirrers are not available, the best simple alternative for mixing the contents of a reaction flask is *swirling*. This is accomplished by loosening the clamp(s) that support the flask and attached apparatus and swirling the contents by manually rocking the flask with a circular motion. This may be done periodically during the course of the reaction. If the entire apparatus is supported by clamps attached to a *single* ring stand, the clamp(s) attached to the flask do not have to be loosened. Instead, *if one has made sure all the clamps are tight,* the ring stand may be picked up and the contents of the flask swirled by gently moving the entire assembly.

(b) Magnetic Stirring

Magnetic stirring equipment consists of a **magnetic stirrer,** which contains a large bar magnet rotated by a variable-speed electric motor, and a **magnetic stirring bar,** which is placed in the flask whose contents are to be stirred. The stirring bar is usually coated with a chemically inert substance such as Teflon or glass. A flat-bottomed container (beaker or Erlenmeyer flask) may be placed directly on top of the stirrer (Figure 2.12a), and a round-bottomed flask may be clamped directly above the center of the stirrer (Figure 2.12b). The magnetic stirring bar rotates in phase with the motor-driven magnet in the stirrer by interaction of their respective magnetic fields, thus effectively stirring the contents of the flask.

    The following should be considered when using magnetic stirring. First, the flask containing the magnetic stirring bar should be *centered* on the magnetic stirrer. If this is not done, the stirring bar will wobble around aimlessly. Second, the revolution rate of the stirrer's magnet should be adjusted so that the stirring bar rotates at a reasonable speed without wobbling inside the flask. Third, the stirring bar should be placed *gently* in the flask, not dropped in directly, as this will often cause the flask to crack. The stirring bar is best

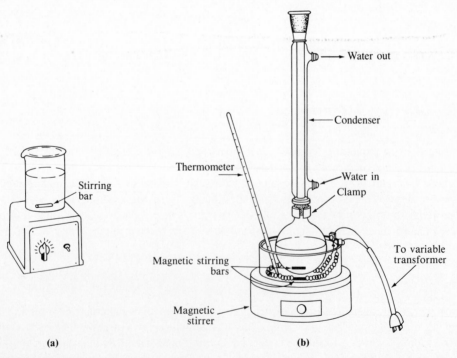

(a)　　　　　　　　　　　　　　　　　　　　　　　　**(b)**

Figure 2.12 (a) Magnetic stirring of the contents of a beaker. (b) Use of an oil bath with magnetic stirring of a reaction mixture.

placed in the flask before adding any reagents to it by tilting the flask and letting the stirring bar slide gently down the side of the flask. Fourth, magnetic stirring is not effective for *viscous* liquids or reaction mixtures. Other stirring methods, such as mechanical stirring [see (c) below], must then be employed.

Magnetic stirring can also be used in conjunction with heating. Figure 2.9b shows an example in which the contents of the flask are being heated with an oil bath and stirred magnetically at the same time. The contents of both the flask *and* the heating bath can be stirred using a single magnetic stirrer; either a large stirring bar or paper clip is used in the bath. Stirring the heating bath is desirable to maintain a homogeneous temperature in the heating fluid. A heating mantle can be used in place of the heating bath, but this has the disadvantage of making it difficult to determine the temperature of the heating source.

(c) Mechanical Stirring

A variable-speed electric motor may be used to drive a stirring shaft that extends directly into the flask whose contents are to be stirred, where a paddle at the end of the shaft provides agitation of the contents of the flask (Figure 2.13). This is the preferred and most efficient method for stirring *heterogeneous* or very *viscous* reaction mixtures. The stirrer shaft is usually constructed of glass and is fitted with a paddle stirrer made of Teflon or glass.

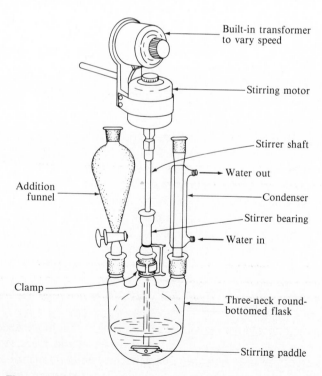

Figure 2.13 Flask equipped for mechanical stirring with a Trubore stirrer.

Normally either type of paddle can be used, except that a glass paddle must be used to stir a reaction mixture containing active metals such as sodium or potassium. The paddle is easily removed from the shaft to facilitate cleaning, and different-sized paddles can be used according to the size of the flask. The glass shaft and the inner bore of the standard-taper bearing are ground to fit each other precisely and constitute what is called a *Trubore stirrer*. A cup at the top of the bearing is used to hold a few drops of silicone or mineral oil, which lubricates the shaft and provides an effective seal.

The stirrer shaft is connected to the motor by means of a short length of heavy-walled rubber tubing. The motor and shaft *must* be carefully aligned to avoid wear on the glass surfaces of the shaft and the bearing and to minimize vibration of the apparatus. The bearing is held in the flask by means of either twisted copper wire and a rubber band or a clamp to keep it from working loose while the motor is running. The rate of stirring is controlled by varying the speed of the motor with either a built-in or separate variable transformer.

Mechanical stirring has some limitations, depending on the type of stirring motor being used. The motors should have "high torque," which means that they have enough power to turn the shaft and stir even viscous reaction mixtures. Some mechanical stirring motors are not explosion-proof and must not be used when the flask contains flammable, volatile organic compounds or solvents. The best mechanical stirring motors are explosion-proof and have high torque.

As illustrated in Figure 2.13, various operations can be performed while also using mechanical stirring. For example, the flask, called a *three-neck standard-taper round-bottomed flask,* is equipped with an addition funnel and a condenser. The apparatus could thus be used in cases where dropwise addition of some reagent to a heated and stirred reaction mixture is needed.

2.7 Simple Distillation

The theory of simple distillation is discussed in Section 4.3, and the use of this technique is described here. Figure 2.14 shows the apparatus needed for carrying out a simple distillation. The thermometer measures the temperature of the vapors that are ultimately collected as liquid in the receiver at the end of the condenser. The location of the thermometer is *very* important: It should be positioned so that the *top* of the *mercury bulb* is approximately parallel with the *bottom of the side-arm outlet.* If a *pure* liquid is being distilled, the temperature read on the thermometer, the **head temperature,** will be identical to the temperature of the liquid boiling in the distilling vessel, the **pot temperature,** provided that the liquid is not superheated. The head temperature thus corresponds to the boiling point of the liquid, and it will remain constant throughout the distillation. Distillation of a mixture of substances, however, will result in differences being observed between the head and pot temperatures (Section 4.4).

Proper assembly of the glassware is important to avoid possible breakage and spillage. Joints must fit snugly to avoid expelling flammable vapors into the room. Apparatus that is too rigidly clamped may be stressed to the point that it may break during the distillation. The comments here are general. *Read these comments completely before starting to set up a simple distillation apparatus.*

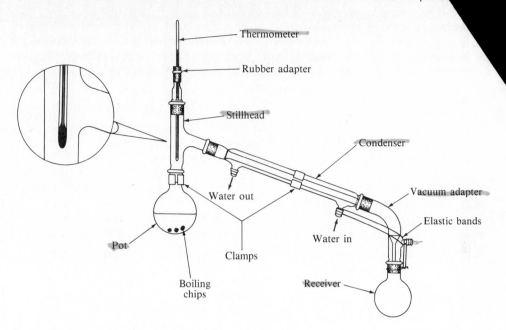

Figure 2.14 Typical apparatus for simple distillation, either at atmospheric pressure or under vacuum. Inset shows correct placement of thermometer in stillhead.

1. Set up the equipment so that the distillation flask is elevated 15 cm (6 in) or so above the bench to allow placement of a heating mantle or a heating bath around the flask.
2. Place several boiling stones in the distillation flask (stillpot) along with the liquid being distilled *before* attaching the rest of the glassware. *Start assembling the equipment by clamping the stillpot in position,* and attach the rest of the glassware by next putting the stillhead in place, then the condenser, and finally the vacuum adapter at the end of the condenser. Place a thin film of lubricant (Section 2.2) on each of the standard-taper glass joints before mating them.
3. Note the location of the clamps in Figure 2.14. A common tendency of beginning students is to ''overclamp'' apparatus. Too many clamps may result in an unsafe situation because of the increased likelihood of stressing and possibly breaking the glassware. Avoid applying undue pressure on the glassware; tighten the clamps only enough to hold each piece in place. Align the jaws of the clamp *parallel* to the piece of glass being clamped. This enables the clamp to be tightened without twisting the glass and either breaking it or pulling another joint loose. *Do not tighten the clamp until the piece of glassware is correctly positioned and the clamp is properly aligned.*
4. The clamp on the condenser does not have to be *rigidly* tight. Tighten it only enough to prevent the joint between the condenser and the stillhead from slipping apart.
5. The vacuum adapter may be held in place by a strong rubber band. Rubber has a tendency to deteriorate in the presence of organic vapors, so use only rubber bands that are uncracked and have sufficient elasticity and strength. The receiving flask can be fastened to the vacuum adapter by twisting a short piece of stiff wire about the neck of the flask and fastening the flask to the vacuum adapter by a rubber band. However, *do not depend on rubber bands to support relatively heavy weights.* Additional support should be provided for flasks of 250-mL capacity or larger, as well as for 100-mL flasks that are expected to

han half full during a distillation. Instead of clamping the receiver, use an
g a piece of wire gauze to support the receiving flask from underneath.
of the "water in" and "water out" nipples on the condenser. The tube
ing water is *always* attached to the lower point, which ensures that the
with water at all times (as long as you remember to turn it on!).
w through the condenser to a modest flow rate. No benefit is gained
the increased pressure in the apparatus may cause a piece of rubber
result in your fellow students or instructor receiving an unexpected
ood practice to *wire* the hoses to the condenser and to the water faucet to
possibility that they may fall off and cause a flood.

2.8 Fractional Distillation

The theory of fractional distillation is given in Section 4.4. An apparatus used for fractional
distillation at atmospheric pressure or below (vacuum) is shown in Figure 2.15. A fractional

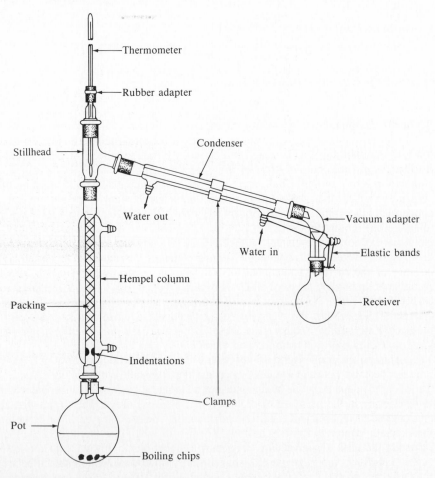

Figure 2.15 Typical apparatus for fractional distillation, either at atmospheric pressure or
under vacuum.

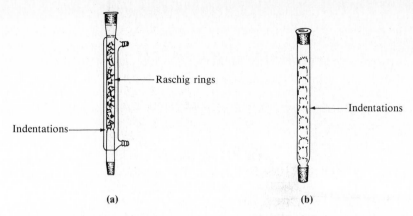

Figure 2.16 Fractional distillation columns. (a) Hempel column filled with Raschig rings (or other type of packing material); (b) Vigreux column.

distillation column (Figure 2.16) looks similar to a condenser, the major difference being that the former has a larger outside water jacket and some indentations at the male end of the column to hold the column packing in place. Unpacked distillation columns of the type shown here can be used as condensers, but condensers cannot be used conveniently for packed distillation columns.

Before assembling the apparatus, clean and dry the inner tube of the distillation column but *not* the water jacket, since traces of water in it will not affect the distillation. Pack the fractional distillation column by adding the desired column packing, a small quantity at a time, through the top of the column while holding it vertical. The column packings are inert material such as glass, ceramic, or metal pieces in a variety of shapes (helices, "saddles," woven mesh, etc.). Some column packings are sufficiently large that they will stop at the indentations. If the packing falls through the column, put a small piece of glasswool or wire screen (gauze) into the column just above the indentations by pushing it down the column with a piece of glass tubing. The column packing should extend to the top of the water jacket, but should not be packed too tightly.

Assemble the apparatus using the general guidelines for simple distillation apparatus (Section 2.7). Start by putting the liquid to be distilled, along with several boiling chips, in the distillation flask, then clamp the flask in place. Do not fill the flask more than half full. Attach the distillation column and make sure that it is as nearly *vertical* as possible. Lubricate and tighten all the joints after the stillhead and condenser are put in place. Clamp only the distillation flask and condenser. *Do not run water through the jacket of the distillation column.*

2.9 Vacuum Distillation

The boiling point of a liquid is the temperature at which the total vapor pressure of the liquid is equal to the external pressure. It is most convenient to distill liquids under conditions such that the external pressure is atmospheric pressure (760 torr). However, a high-molecular-weight compound being distilled may decompose, oxidize, or undergo molecular rearrange-

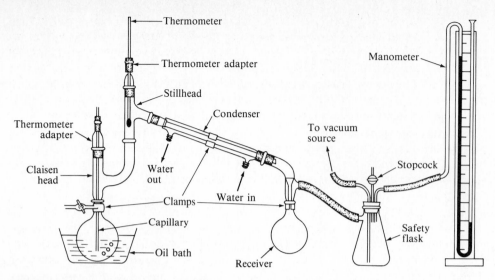

Figure 2.17 Apparatus for vacuum distillation.

ment at temperatures below its atmospheric boiling point. These problems may be alleviated by carrying out the distillation at pressures *less* than atmospheric, which lowers the boiling temperature. This technique is called **vacuum distillation,** or distillation under reduced pressure.

Reduced pressures may be obtained by connecting a water aspirator ("water pump") or a mechanical vacuum pump to the vacuum distillation apparatus (Figure 2.17). The source of the reduced pressure is dictated by availability and the magnitude of the reduction of pressure needed. An aspirator will commonly reduce the pressure to about 25 torr (mm of Hg) and an oil pump to below 1 torr. The pressure obtained in the lower range by a vacuum pump is dependent on the condition of the pump, its oil, and on the tightness of the connections to the distillation apparatus. Some laboratories are equipped with "house vacuum" lines that are connected to a large central vacuum pump. House vacuum systems seldom reduce the pressure to less than 50 torr. The vacuum produced by a water aspirator is limited by the vapor pressure of the water and hence by its temperature. In cold climates, pressures as low as 8–10 torr may be obtained from a water aspirator, whereas in warmer climates the minimum pressure may be no less than 25–30 torr.

Two useful *approximations* of the effect of lowered pressure on boiling points are as follows:

1 Reduction from atmospheric pressure to 25 torr lowers the boiling point of a compound boiling at 250–300°C at atmospheric pressure by about 100–125°C.

2 Below 25 torr, the boiling point is lowered by about 10°C each time the pressure is cut in half.

More accurate estimates of the effect of pressure upon boiling point may be made by use of charts and nomographs.[3] Calculations may also be made using the Clausius-Clapeyron equation found in physical chemistry textbooks.

[3]References: Jordan, T. E., *Vapor Pressure of Organic Compounds,* Interscience Publishers, New York, 1954, and Lippincott, S. B.; Lyman, M. M., "Vapor Pressure–Temperature Nomographs," *Industrial and Engineering Chemistry,* **1946,** *38,* 320.

A typical apparatus for vacuum distillation is shown in Figure 2.17. The distilling flask should not be more than half full and is heated by an oil bath or some other suitable heating device such as a heating mantle. The flask should be immersed to a depth above the level of the liquid in the flask to help prevent bumping. The flask should *never* be heated *directly* with a flame because this may produce hot spots and intensifies the possibility of bumping; however, it may be heated directly *if* wire gauze is placed under the flask to diffuse the heat and avoid hot spots. (Various heating techniques are discussed in Section 2.5.) If a water aspirator is used, the vacuum adapter is connected to a safety flask, which serves as a trap to prevent the backup of water into the apparatus. This may occur if there is a sudden decrease in the water pressure in the aspirator. The safety flask, when properly equipped with a stopcock release valve, provides a means of connecting the apparatus and the manometer to the source of the vacuum and of releasing the vacuum when desired.

The manometer measures the pressure at which the distillation is being conducted. This measured pressure value is an important part of a reported boiling point at reduced pressure. For example, benzaldehyde boils at 180°C at atmospheric pressure and at 87°C at 35 torr. The correct way to report these two boiling points is: bp 180°C (760 torr) and 87°C (35 torr). Hence, the pressure at which a liquid boils should be carefully determined. The reliability of pressure measurements is dependent on the rate of distillation, as discussed below.

Two significant problems are encountered in a vacuum distillation. First, the volume of the vapor formed by the volatilization of a given amount of liquid is pressure-dependent. For example, the volume of vapor formed from vaporization of a drop of liquid will be about 20 times as great at 38 torr as it would be at 760 torr. As a consequence, serious bumping problems may occur in the distilling flask during vacuum distillation because large bubbles of vapor may escape from the liquid. This causes vigorous and sometimes violent splashing and splattering. The insertion of a Claisen connecting tube between the distillation flask and the stillhead (Figure 2.17) will partially solve this problem by preventing the liquid in the distillation flask from splashing into the side arm leading to the condenser. Boiling chips are generally ineffective in preventing bumping in vacuum distillations. One method for producing regular and even ebullition of the vapors is to extend a *very thin,* flexible capillary tube into the bottom of the distillation flask. This tube introduces a fine stream of air bubbles at the bottom of the flask, and the air bubbles provide nuclei for the regular production of bubbles of vapor. The volume of air so introduced is small compared with the evacuating capacity of a water or oil pump, so this "leak" has no significant effect on the pressure of the system. The capillary is drawn from a piece of 6-mm glass tubing and should be fine enough to allow only a slow stream of fine bubbles when air is blown through the tube with its capillary end inserted into a test tube containing acetone. Alternatively, a thin wooden applicator stick may be used in place of a capillary, or magnetic stirring (Section 2.11b) can be used in the distillation flask to promote smooth ebullition.

The second problem associated with lower vapor densities at reduced pressure is that the measured pressure is greatly affected by the rate of distillation. Since a drop of condensate forms from a much larger volume of vapor at low pressure than at atmospheric pressure, the velocity of vapor entering the condenser is tremendously increased by the reduced pressure. The back pressure caused by high velocities of vapor provides a higher pressure in the distillation column than that read on the manometer, which is beyond the condenser and receiver and is therefore insensitive to the back pressure. The difference between the actual and apparent (measured) pressure is directly dependent on the rate of distillation and on the application of heat to the distillation flask and can be minimized by:

1 Maintaining a slow but steady rate of distillation.

2 Avoiding superheating of the vapor by maintaining the oil bath at a temperature no more than 15–25°C higher than the head temperature.

The following paragraphs provide a general procedure for carrying out a vacuum distillation using the apparatus shown in Figure 2.17.

1. *Never use glassware with cracks, or thin-walled vessels, especially those with flat bottoms, such as Erlenmeyer flasks, in vacuum distillations.* Pressures of many hundred pounds may be exerted on the exterior surfaces of systems under reduced pressure, even with systems of only moderate size under water pump evacuation. Weak or cracked glassware may *implode*, and the air rushing in will shatter the glassware in a manner little different from that of an explosion. Additional dangers are burns from the hot oil of the bath and the possibility of starting a serious fire. *Examine the glassware carefully and always wear safety glasses when doing a vacuum distillation.*

2. Lubricate and seal all glass joints carefully during assembly of the apparatus. This will help avoid air leaks and provide lower pressure. The rubber fittings holding the thermometer and capillary in place must be tight.[4] The Neoprene fittings normally used with the thermometer adapters may be replaced, if necessary, with short pieces of heavy-walled tubing. Do not use rubber stoppers elsewhere in the apparatus, since rubber in direct contact with the hot vapors during distillation may cause contamination. The three-holed rubber stopper in the safety flask should fit snugly and provide tight connections to the pieces of glass tubing. The safety trap must be made of heavy-walled glass. Wrap the trap with electrical tape to protect the experimenter and other workers nearby. *Heavy-walled vacuum tubing must be used for all vacuum connections.* Check the completely assembled apparatus to make sure that all joints and connections are tight.

3. Place the liquid to be distilled in the flask. The tip of the capillary must extend nearly to the bottom of the flask. *Do not heat the flask until the system is fully evacuated.* Turn on the vacuum. The release valve on the safety flask should be *open*. *Slowly* close the release valve, but be prepared to reopen it if necessary. If the liquid contains small quantities of low-boiling solvents, as it very likely will if the liquid has been obtained from a solution by evaporation of solvent, foaming and bumping are likely to occur in the distillation flask. If this occurs, reopen the release valve until the foaming abates. This may have to be done several times until the solvent has been removed completely. When the surface of the liquid in the flask is relatively quiet, fully evacuate the system to the desired pressure. The release valve may have to be opened slightly until the desired pressure is obtained. Check the manometer for constancy of pressure. Note and record the pressure under which the distillation is being carried out. The pressure should be monitored even when fractions are not being taken as it may change during the course of the distillation. Then begin heating the flask, and maintain a bath temperature only as high as necessary to provide distillate at the rate of 3–4 drops per minute.

4. If it is necessary to use multiple receivers to collect fractions of different boiling ranges, the distillation must be interrupted to change flasks. Remove the heating source with caution and allow the stillpot to cool somewhat. *Slowly* open the vacuum release valve to readmit air to the system. When atmospheric pressure is attained, change receivers, close the release valve, reevacuate the system to the same pressure as previously, reapply heat and continue distilling. The operation may result in a different pressure in the fully evacuated system. Periodically monitor and record the head temperature and the pressure, particularly just before and after changing receivers.

[4]Thermometers and capillary tubes that contain standard-taper joints are available. These are preferable to using rubber connections.

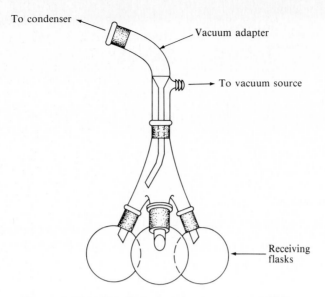

To condenser

Vacuum adapter

To vacuum source

Receiving
flasks

Figure 2.18 Multiple-flasked receiver for vacuum distillation.

5. After the distillation is complete, discontinue heating, allow the pot to cool somewhat, slowly release the vacuum, and turn off the source of the vacuum.

One of the most inconvenient aspects of the above procedure is the disruption of the distillation in order to change receivers. More elaborate apparatus eliminates this problem. For example, the ''cow'' receiver (Figure 2.18) has three or four round-bottomed flasks attached to it. These flasks are successively employed as receivers by rotating them into the receiving position.

2.10 Steam Distillation

The theory of steam distillation is discussed in Section 4.7. Steam distillation provides a method for removing volatile liquid or solid organic compounds that are insoluble or nearly insoluble in water under comparatively mild conditions from compounds that are nonvolatile. This technique is not applicable to substances that react with water, decompose on prolonged contact with steam or hot water, or have a vapor pressure of 5 torr or less at 100°C. Steam distillation may be accomplished in one of two ways. The first and usually most efficient method involves placing the organic compounds to be distilled in a round-bottomed flask equipped with a Claisen head, a stillhead, and a water-cooled condenser (Figure 2.19). The Claisen head helps prevent splattering of the mixture into the condenser during distillation. Steam can be produced externally in a generator (Figure 2.20) and then introduced into the bottom of the distillation flask via the tube shown in Figure 2.19. Alternatively and preferably, it can be obtained from a laboratory steam line. If this is done, a trap (Figure 2.21) is usually placed between the steam line and the distillation flask to permit removal of

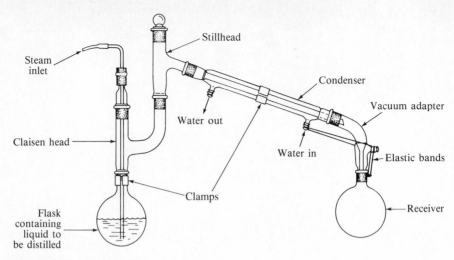

Figure 2.19 Apparatus for steam distillation. The steam tube is replaced by a stopper if steam is generated by direct heating.

any water and/or impurities present in the steam. Water may condense in the distillation flask and fill it to undesirable levels. This problem can be circumvented by gently heating the flask with a Bunsen burner or a heating mantle.

If only a small amount of steam is necessary to separate a mixture completely, another method of steam distillation may be employed. This consists of direct addition of water to the

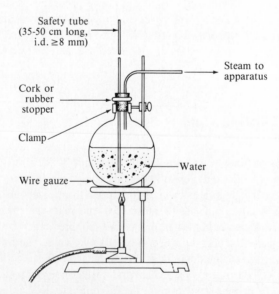

Figure 2.20 Steam generator. The round-bottomed flask is initially half-filled with water, and boiling chips are added before heating. The safety tube serves to relieve internal pressure if steam is generated at too rapid a rate.

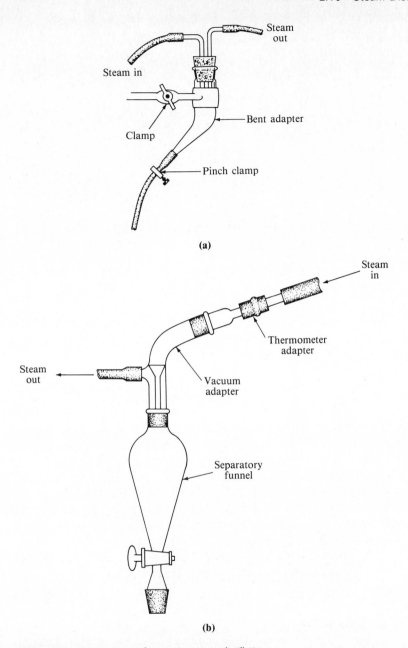

Figure 2.21 Water traps for use in steam distillations.

distillation flask along with the organic compounds to be distilled. The flask is heated with a Bunsen burner or heating mantle. This approach is generally not applicable for distillations that require large amounts of steam, since the water in the flask would have to be replenished frequently by using an addition funnel, or an inconveniently large flask would have to be used.

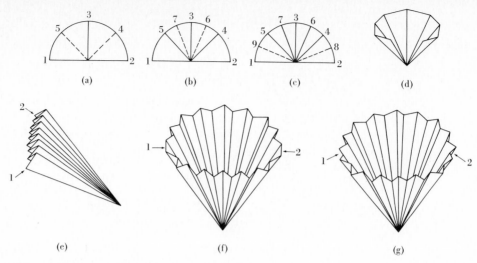

Figure 2.22 Folding filter paper to produce fluted filter paper.

2.11 Fluted Filter Paper

"Fluted" filter paper is used in many filtering operations. The filter paper is fluted to increase the surface area and thus the rate and ease of filtration. Although fluted filters are available commercially, one of several possible methods of preparing them is shown in Figure 2.22. Fold the paper in half, and then into quarters. Fold edge 2 onto 3 to form edge 4, and then 1 onto 3 to form 5 (Figure 2.22a). Now fold 2 onto 5 to form 6, and 1 onto 4 to form 7 (Figure 2.22b). Continue by folding 2 onto 4 to form 8, and 1 onto 5 to form 9 (Figure 2.22c). The paper now appears as shown in Figure 2.22d. All folds thus far have been in the same direction. Do not crease the folds tightly at the center because this might weaken the paper and cause it to tear during filtration. Now make folds in the *opposite* direction between 1 and 9, 9 and 5, 5 and 7, and so on, to produce the fanlike appearance shown in Figure 2.22e. Open the paper (Figure 2.22f) and fold each of the sections 1 and 2 in half with reverse folds to form paper that is ready to use (Figure 2.22g).

2.12 Gravity Filtration

Gravity filtration is useful to remove solid materials, which are often impurities, drying agents (Section 2.21), or decolorizing carbon (Section 2.14) from liquids. Gravity filtration involves first wetting the filter paper with solvent and then pouring a solution containing a solid onto filter paper or fluted filter paper (Section 2.16) that has been placed on a filter funnel. The solution passes through the paper, and the solid remains on the filter paper. Gravity filtration at room temperature is shown in Figure 2.23.

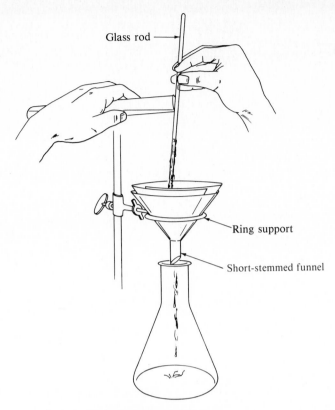

Glass rod

Ring support

Short-stemmed funnel

Figure 2.23 Gravity filtration.

On occasion, it may be necessary to remove a finely divided or colloidal solid that passes through the filter paper and is thus not removed, as for example in the removal of decolorizing carbon. This problem may be overcome by adding 1 to 2 grams of **filter-aid** to the solution before filtration. Filter-aid is a finely divided, inert material that will adsorb colloidal substances and prevent them from passing through the filter paper or clogging its pores; of course the filter-aid must not be too finely divided or it too would tend to clog the paper. It is usually a finely divided silica, frequently diatomaceous earth, and is sold under different trade names such as Super Cel and Celite.

2.13 Hot Gravity Filtration

Hot gravity filtration is used to filter solid impurities or decolorizing carbon (Section 2.14) from solutions that contain organic compounds that separate and solidify at room temperature. It is most often used in the recrystallization of solid compounds when the use of gravity filtration at room temperature would cause the desired compound to separate in the funnel along with the solid impurities. Hot gravity filtration is especially useful for removing decolorizing carbon from *hot* mixtures, and even then filter-aid (Section 2.12) may be required for a successful filtration.

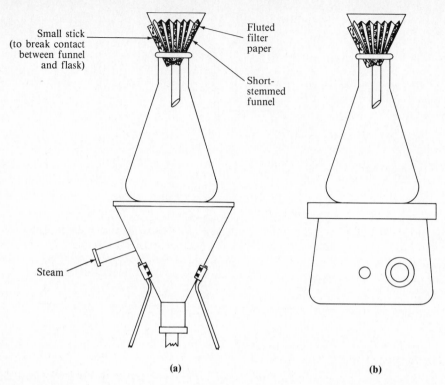

Figure 2.24 Hot gravity filtration: (a) using steam cone for heating; (b) using hot plate for heating.

The apparatus for performing a hot gravity filtration is shown in Figure 2.24. A little of the *pure* solvent is placed in an Erlenmeyer flask on top of which is a short-stemmed or stemless filter funnel containing a piece of fluted filter paper. The flask is heated until the solvent boils and rises up the sides of the flask and around the funnel; this has the effect of heating the filter funnel. A small stick or piece of thick paper is inserted between the flask and the funnel to allow the hot solvent to rise up around the outside of the funnel. The flask can be heated with either a steam bath or hot plate. A steam bath is preferred for solvents that boil below 90°C and is required if the solvents are flammable, whereas a hot plate can be used for nonflammable solvents or those boiling above 90°C.

When the flask and funnel are hot, the hot solution is poured onto the filter paper. Swirling may be necessary to ensure that the solid is transferred with the liquid. After the hot solution has been completely poured onto the filter paper, several milliliters of hot solvent are added to the flask that contained the original recrystallization solute and solvent. The additional hot solvent is poured onto the filter paper to ensure that as much material as possible has been transferred to the flask containing the filtered solution.

2.14 Decolorizing Carbon

A colored solution is sometimes observed during the purification of an organic solid or liquid. For purposes of this discussion, it is assumed that the solid is already dissolved in a suitable solvent. The color can result from the compound itself being colored or from the presence of colored impurities in the liquid or solution. If in doubt about the cause of the color, an attempt to remove it is done by adding **decolorizing carbon** to the liquid or solution and then filtering off the carbon. Decolorizing carbon is commercially available under a variety of trade names, such as Norite, Nuchar, vegetable charcoal, and so forth. All decolorizing carbon is essentially the same and possesses the common property of being *very* finely divided carbon with an extremely large surface area. Its addition to an organic liquid or organic solid dissolved in a solvent serves to adsorb the colored impurities on its surface. After the decolorizing carbon is removed, the liquid or solution should be free of colored impurities. However, the liquid or solution may still be colored after treatment with decolorizing carbon if the compound itself is colored. This process will result in the loss of some of the desired compound by adsorption on the carbon.

Decolorization is done by adding decolorizing carbon to the liquid or solution contained in an Erlenmeyer flask and swirling the flask gently. Only a small amount of decolorizing carbon is added. There are no hard and fast rules concerning the amount of carbon to add, but a good first approximation is to add about 0.5 to 1 g per 100 mL of solution. When decolorizing an organic liquid, gravity filtration can be used, but hot gravity filtration must be used when an organic solid is dissolved in a hot solvent. If the compound is known to be colorless and if the decolorization process does not remove the color, the process should be repeated.

One major experimental problem likely to be encountered is the incomplete removal of the decolorizing carbon, which is best accomplished by the use of fluted filter paper. (Vacuum filtration should not be used.) Care must be exercised when removing the carbon by filtration, and the filter paper must be properly folded and contain no tears or holes. Pouring the solution or liquid containing the carbon so that any of it runs down the side of the funnel should be avoided. The solution or liquid should be refiltered if it contains any small black specks of carbon or any other solid material. Alternatively, filter-aid (Section 2.12) can be added to the solution before filtration to ensure complete removal of the carbon.

2.15 **Decanting Solutions**

A viable alternative to gravity filtration for removing a solution or organic liquid from a *small* amount of solid is **decantation.** Decanting a liquid is especially useful in removing drying agents (Section 2.19) or other compact, dense solids, but it cannot be used to remove decolorizing carbon. To decant a liquid from a solid, the solid should first be allowed to settle to the bottom of the container. The container is then *carefully* tilted and the liquid is *slowly* poured into a clean container. If done carefully, the solid will remain in the original container. Decantation is preferable to gravity filtration when working with a very volatile organic liquid, since filtration is likely to result in a great deal of evaporation and loss of material. For decantation from an Erlenmeyer flask, a *loosely packed* ball of glasswool can be put in the neck of the flask to help keep the solid in the flask.

One major disadvantage of decantation is that some liquid will remain in the flask and be lost. However, the loss will still be less than using gravity filtration if the liquid is very volatile.

2.16 Vacuum Filtration

Vacuum filtration is used to remove crystalline solids from solvents. This technique makes use of a Büchner funnel and a vacuum filter flask attached to an aspirator or house vacuum line through a trap (Figure 2.25). The trap prevents water from the aspirator from backing up into the filter flask in case of loss of water pressure. The trap also provides a place to collect any solvent that may overflow the filter flask, thus preventing loss of the solvent and protecting the vacuum system. *Both the filter flask and the filter trap should be cleaned before doing the filtration,* since there may be need to save the filtrate. The trap should be a *heavy-walled* Erlenmeyer flask or bottle, or a second vacuum filter flask. If a filter flask is used, it should be equipped with a two-holed stopper and its side arm attached to the water aspirator or house vacuum. Before filtration, the proper size filter paper, which should lie *flat* on the funnel plate, is selected. The paper should cover all the small holes in the funnel and should *not* extend up the sides of the funnel. A vacuum is applied to the system, and the filter paper is "wet" with a small amount of pure solvent in order to seal it to the funnel. This keeps crystals from going through the holes in the filter and serves to hold the paper tightly on the funnel.

To filter the solution containing the crystals, suction (vacuum) is applied to the apparatus, and the solution containing the crystals is transferred to the funnel. For the transfer process, the flask containing the crystals is swirled and the solution containing the crystals is poured slowly onto the funnel. A stirring rod or spatula may be used to aid the transfer. The last of the crystals may be transferred by washing them from the flask with some of the filtrate, which is called the **mother liquor.** When all the solution has passed through the filter, the suction is released by opening the screw clamp or stopcock on the trap. The crystals are washed to remove adhering mother liquor, which contains impurities, by adding to the funnel *cold, pure solvent* just sufficient to cover the crystals. Suction is reapplied to remove the

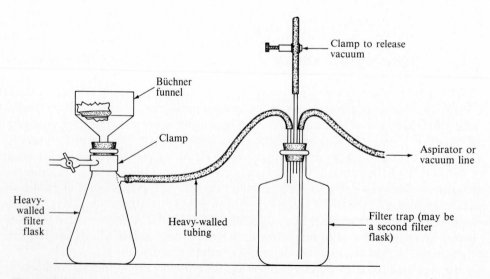

Figure 2.25 Apparatus for vacuum filtration of solids.

wash solvent, and the crystals are pressed as dry as possible with a clean cork or a spatula while the funnel is under suction. Contamination of the crystals can be minimized by placing a clean piece of filter paper on top of the crystals *before* pressing them dry.

Most of the solvent may be evaporated from the crystals by allowing the suction to pull air through the crystals on the funnel for a few minutes. A clean spatula is used to transfer the crystals from the filter funnel to a clean watch glass. The filter paper must not be torn, as this would contaminate the crystals. The crystals may be dried completely by allowing them to air-dry for a few hours or by leaving them loosely spread on a filter paper in a locker until the next period. The drying process may be accelerated by placing the crystals in an oven. If this is done, the oven temperature must be at least 20°C below the melting point of the crystals. Heat drying may not be used for crystals that sublime readily at atmospheric pressure (Section 2.17). A vacuum desiccator can also be used to hasten the drying process. If one is available, your instructor will advise you concerning its use. Vacuum desiccators are frequently used in the research lab for drying crystals. Specially designed desiccators permit heating samples under vacuum, but as with drying ovens, the temperature must be kept about 20°C below the melting point of the solid.

Vacuum filtration can also be used to remove undesired solids from a solution but may be ineffective if the solid is finely divided or colloidal. Filter-aid (Section 2.12) can be used to ensure complete removal of the solid as follows. A pad of filter-aid can be formed on the filter funnel by first making a slurry of 0.5 to 1 gram of filter-aid in a few milliliters of the solvent being used. The slurry is poured onto a filter paper contained in a Büchner funnel attached to a *clean* filter flask; vacuum is then slowly applied to remove the solvent from the slurry to leave a thin, even pad of filter-aid. The solution containing the solid is then filtered as described above. Note, however, that this technique is useful *only* when it is desired to keep the *solution* that contains the substance of interest; it can be used to remove solid impurities but not to remove a desired solid from solution.

2.17 Sublimation

An alternative to crystallization for purifying some solids is **sublimation.**[5] This method uses to advantage the differences in vapor pressures of solids in a way analogous to simple distillation. The impure sample is vaporized directly from the solid state by heating it at a temperature below its melting point, and the vapor is condensed (crystallized) directly back to the solid state on a cold surface. Sublimation occurs *without* the intermediacy of the liquid state.

Two types of sublimation apparatus, shown in Figure 2.26, have these common characteristics: (1) a chamber that may be evacuated using a water aspirator, house vacuum, or vacuum pump and (2) a fingerlike projection in the center of the vacuum chamber that is cooled to provide the surface upon which the sublimed crystals may form. In each apparatus, the impure sample is placed at the bottom of the chamber. The "cold finger" in Figure 2.26a is cooled by circulating water and that in Figure 2.26b by a medium such as ice water or Dry

---

[5]A reference to this technique is: Tipson, R. S., in *Technique of Organic Chemistry*, A. Weissberger, editor, Interscience Publishers, New York, 1965, Vol. IV, Chapter 8.

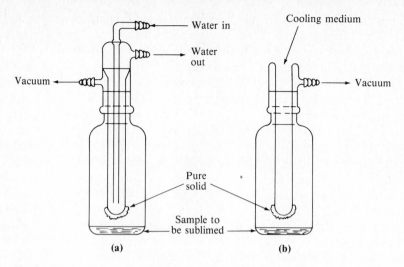

Figure 2.26 Sublimation apparatus: (a) using water as the coolant; (b) using low-temperature coolants.

Ice in acetone. Figure 2.27 shows a simple sublimation apparatus that may be assembled inexpensively from two test tubes (one with a side arm), rubber stoppers, and glass tubing.

Two criteria must be met for effective purification by sublimation: (1) the impurities must have vapor pressures substantially lower than that of the compound that is to be obtained in pure form, and (2) the compound to be purified must have a relatively high vapor pressure. Should these criteria not be satisfied, either recrystallization (Section 3.1) or chromatographic techniques (Chapter 6) must be used.

The technique of purification by sublimation involves placing the impure solid in the

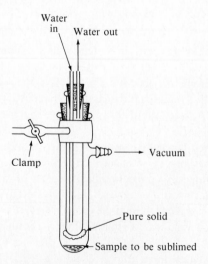

Figure 2.27 Test tube sublimator.

sublimation chamber and heating it to a temperature higher than that of the surface (cold finger) on which the pure material will condense but lower than the melting point of the solid. The solid will be vaporized and transferred via the vapor phase to the colder surface. The crystals that form on the cold surface will usually be very pure, since the impurities will not be incorporated into the condensate provided that they do not sublime themselves. Should the vapor pressure of the impurities be similar to that of the compound being purified, sublimation will *not* be an effective method of purification.

Very few organic solids exhibit vapor pressures high enough for sublimation at atmospheric pressure. Reduced pressure (vacuum) is generally needed to increase the rate of vaporizing the solid and thus the rate of sublimation. This is analogous to using vacuum distillation for high-boiling liquids.

Sublimation is generally restricted to relatively nonpolar substances having fairly symmetrical structures. In these cases, crystal forces are lower and vapor pressures tend to be higher. The ease with which a molecule may escape from the solid to the vapor phase is determined by the strength of the intermolecular attractive forces between the molecules of the solid. The most important attractive force is of an electrostatic nature. Symmetrical structures will have relatively symmetrical distributions of electron density and will have smaller dipole moments than less symmetrical molecules, so a smaller dipole moment implies a higher vapor pressure. Van der Waals attractive forces are also important but less so than electrostatic attractions. In general, van der Waals forces increase in magnitude with increasing molecular weight. Thus, large molecules, even if symmetrical, are less adaptable to purification by sublimation.

2.18 Separatory Funnels and Their Use

Separatory funnels are available in many different shapes, ranging from almost spherical to elongated pear shape (Figure 2.28). The more elongated the funnel, the longer the time required for the two liquid phases to separate from one another after the funnel is shaken. When liquids have similar densities, the more spherical separatory funnels are preferred

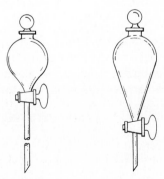

Figure 2.28 Separatory funnels: (a) conical, (b) pear-shaped.

since their use requires less time for the separation of the layers. Separatory funnels are equipped with either a glass or Teflon stopcock; in general, the latter are preferred since they do not require lubrication and thus minimize contamination of the solutions being separated.

Separatory funnels can be used for extractions (Chapter 5), for separating mixtures that already contain two immiscible liquid phases, and for "washing" organic layers in the process of purifying organic compounds. Only the *general* techniques for using separatory funnels are discussed here:

1. **Filling separatory funnels.** The stopcock should be closed before any liquids are added to the funnel. It is good technique to place a clean beaker under the funnel before adding any liquids to the funnel in the event that the stopcock leaks or is not completely closed.
2. A separatory funnel should never be more than three-quarters full, especially when doing an extraction.
3. The upper opening of the funnel is stoppered either with a ground-glass stopper, for which most separatory funnels are fitted, or with a rubber stopper.
4. **Holding and using separatory funnels.** If the contents of the funnel are to be shaken, it is held in a rather specific manner. If the user is right-handed, the stopper should be placed against the base of the index finger of the left hand and the funnel grasped with the first two fingers and the thumb and turned so that the first two fingers of the right hand can be curled around the handle of the stopcock. This permits the stopper and the stopcock to be held tightly in place during the shaking process (Figure 2.29). A left-handed user would find it easier to use the opposite hand for each position.
5. **Shaking separatory funnels.** The purpose of shaking a separatory funnel and its contents is to permit the immiscible liquids to be mixed as intimately as possible. The shaking process greatly increases the surface area of contact between the liquids so that the equilibrium distribution of the solute between the solvents will be attained as quickly as possible. The funnel must be "vented" every few seconds to avoid the potential buildup of pressure within the funnel. Venting is accomplished by *inverting the funnel with the stopcock pointing upward* and carefully opening it to release any pressure. It is good sense to point the stopcock-end of the funnel away from one's self and one's neighbors, in case of accidental venting of liquid along with the vapor. Venting is particularly important when using volatile, low-boiling solvents or when an acid is brought into contact with either sodium carbonate or sodium bicarbonate, as CO_2 is released. If venting is not done, the stopper may be blown out and the contents of the separatory funnel lost. There is also potential danger of the funnel blowing up or its contents being splattered on the user and other workers in the lab. If the funnel is held as described, the stopcock may be opened by twisting the fingers curled around it without readjusting the grip on the funnel. At the end of the shaking period (1 to 2 min are usually sufficient if the shaking is vigorous), the funnel is vented a final time. It is then supported on an iron ring (Figure 2.30), and the layers are allowed to separate. When an iron ring is used to support a separatory funnel, the ring should be covered with a length of rubber or Tygon tubing to prevent breakage. This may be accomplished by slicing the tubing on the side and slipping the tubing over the ring. Wire, preferably copper, may be used to fix the tubing permanently in place. The lower layer in the separatory funnel is carefully drawn into a flask through the stopcock. It is sometimes difficult to see the interface between two clear layers, but with practice one can become proficient in doing so.
6. **Layer identification.** The correct identification of layers requires care and thought. Layers will usually separate so that the solvent of greater density is on the bottom. A knowledge of the densities of the liquids being separated is thus important for this identification; the densities of several water-insoluble solvents may be found in Table 3.1. However,

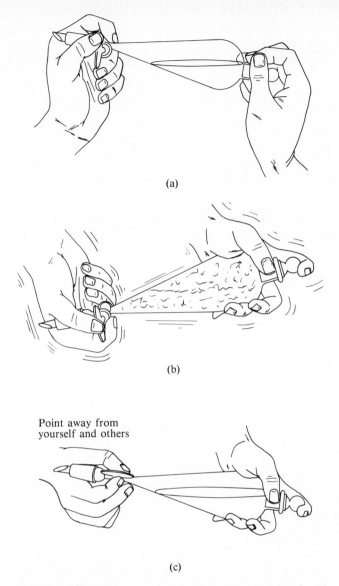

Figure 2.29 (a) Efficient method of holding a separatory funnel. (b) Shaking separatory funnel. (c) Venting separatory funnel.

this generalization is not foolproof because the nature and concentration of solute may be such as to invert the relative densities of the two liquids. It is extremely important not to confuse the identity of the two layers in the funnel so that the wrong layer is discarded. *Both layers should be saved until there is no doubt about the identity of each*. In most instances, one of the layers is aqueous and the other is organic. A good, foolproof method of ascertaining which layer is which is the following: Withdraw a few drops of the upper layer with a dropper or pipet and add these drops to about 0.5 mL of water in a test tube.

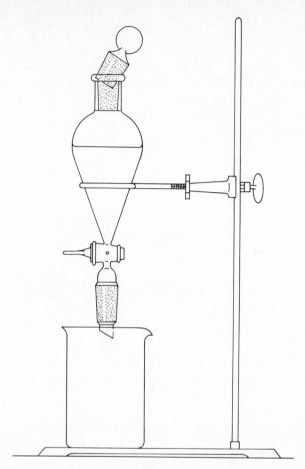

Figure 2.30 Separatory funnel positioned on iron ring with beaker underneath to collect funnel contents in case of funnel malfunction.

If the upper layer is aqueous, these drops will be miscible with and will dissolve in the water in the test tube. If the upper layer is organic, the droplets will remain undissolved.

7. **Emulsions.** Occasionally the two immiscible liquids will not separate cleanly into two layers after shaking, and the result is an **emulsion** that appears as a colloidal mixture of the two layers. Encountering an emulsion during an experiment can be a very frustrating experience because there are no infallible procedures for "breaking" emulsions. An emulsion left unattended for an extended period of time sometimes separates. However, it is usually more expedient to attempt one or more of the following remedies:

a. Add a few milliliters of brine (*saturated* aqueous sodium chloride solution) to the funnel and gently reshake the contents. This increases the ionic strength of the water layer, which helps force the organic material into the organic layer. This process can be repeated, but if it does not work the second time, try the following.

b. Filter the heterogeneous mixture by gravity filtration and return the filtrate to the separatory funnel. Sometimes emulsions are caused by small amounts of gummy organic materials. Their removal by filtering may allow the layers to separate.

c. Add a *small* quantity of water-soluble detergent to the mixture and reshake the mixture. This method is not as desirable as the first two, particularly if the desired compounds are in the water layer, because the detergent adds an impurity that must be removed later.

d. If all these procedures fail, it may be necessary to select a different extraction solvent.

Small quantities of insoluble material sometimes collect at the phase interface and make it difficult to observe the true boundary between layers. This difficulty may be overcome by removing any solids along with the *undesired* liquid layer. A small amount of the desired layer is inevitably lost by this procedure. An alternative procedure is to filter the mixture by gravity before separating layers.

2.19 Heating Under Reflux

The term **heating under reflux** means that a given solvent or reaction mixture is heated at its boiling point in a flask equipped with a **reflux condenser** without the boiling temperature being specified. A typical apparatus for this technique is shown in Figure 2.31a. The top of the condenser is not stoppered, since *a closed system is never heated*. The apparatus should be set up 15 cm (6 in) or more from the benchtop to allow for fast removal of the heating source from the flask if necessary. The solvent or reactants are placed in a boiling flask, along with several boiling chips, and the flask is equipped with a heating device and reflux condenser. The condenser must have water *slowly* running through it. The materials being heated vaporize and reliquefy in the condenser. The condensate returns to the boiling flask, where it is revaporized. Heating the contents of a flask under reflux means that no solvent is removed or lost. This technique is especially useful for reactions that occur more rapidly at elevated temperature but which require considerable time for completion.

After the reflux apparatus is set up and the materials are added to the boiling flask, heat is slowly applied to the flask until boiling commences. The vapors should rise 3 to 6 cm (1 to 2 in) above the bottom of the condenser. If they rise to the top of the condenser or if they are "seen" being emitted into the room, either too much heat is being supplied to the flask or the flow of water through the condenser is insufficient. On occasion, the condenser may not be of sufficient size to permit adequate cooling, and either a larger-capacity condenser must be used or two condensers should be hooked together. If the latter option is selected, clamps are required on each condenser.

Upon completion of the reflux period, heat is removed, and the flask is allowed to come to room temperature. This ensures that all of the material in the condenser has returned to the flask.

Many chemical reactions are carried out by adding all the reagents, including solvent and catalyst (if any), to a reaction flask and heating the resulting mixture under reflux, as described above. However, sometimes it is necessary to add one or more reagents to the reaction mixture gradually over a period of time. For example, this might be necessary if a reaction is highly exothermic, and the rate of reaction may be controlled by adding one of the reagents a little at a time. Other reactions may require that one of the reagents be present in high dilution in order to minimize the formation of side products. Figure 2.31b shows a typical apparatus that allows a liquid reagent or solution to be added to the reaction flask by means of an addition funnel; the temperature in the flask can be maintained at room temperature or below with an ice-water bath, or the flask can be heated to reflux during the addition.

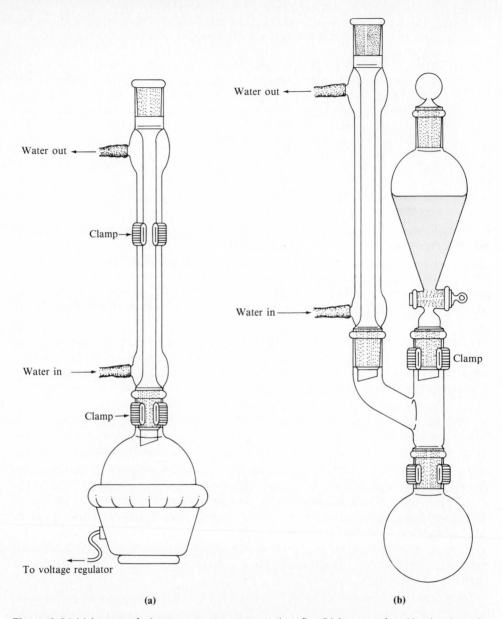

Water out ←

Clamp →

Water in →

Clamp →

To voltage regulator

(a)

Water out ←

Water in →

Clamp

Clamp

(b)

Figure 2.31 (a) Apparatus for heating a reaction mixture under reflux. (b) Apparatus for adding liquid reagent or solution by means of addition funnel to a reaction flask equipped with condenser.

The standard-taper separatory funnel available in many glassware kits can also serve as an addition funnel, as shown in Figure 2.31b. Its use for this purpose requires some provision for air to be admitted above the surface of the liquid to equalize the pressure inside; this can be accomplished by not placing the stopper on top of the funnel or by inserting a small piece of filter paper between the stopper and the funnel.

Although not discussed here, techniques and apparatus are also available for adding solids or gases to a reaction flask during the time that the reaction is in progress.

2.20 Gas Traps

Some organic reactions release noxious gases whose presence in the laboratory is undesirable. Other reactions involve materials that react with the moisture in the air. These problems can be solved with moderate success by utilizing a **gas trap** (Figure 2.32), which can be attached to the top of a reflux condenser or directly to a reaction flask. Granular, anhydrous calcium chloride in the vacuum adapter removes moisture from the air as it is "pulled" through the adapter by *gently* applying vacuum from either a water aspirator or house vacuum system. Any gas formed by the reaction is removed by the dry air passing through the gas trap.

2.21 Drying Agents, Desiccants, and Drying Liquids

Drying a reagent, solvent, or product will be encountered at some stage of nearly every reaction performed in the organic chemistry laboratory. The techniques of drying solids and liquids are described here and in the next section.

(a) Drying Agents and Desiccants

In purifying organic liquids that have been in contact with water, any traces of moisture that are present must be removed. Most organic liquids are distilled at the end of the purification process, and any moisture present either may react with the compound at or below the

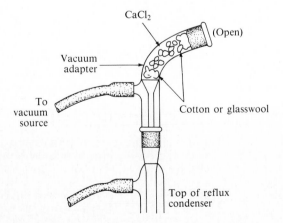

Figure 2.32 Gas trap for removal of gases from a reaction mixture.

temperature of the distillation or may co- or steam-distil with the liquid and thereby contaminate the distillate. Drying agents, called **desiccants,** are used to remove these small traces of moisture before distillation.

The drying of some organic solvents makes profitable use of the codistillation of water with the solvent in a process called **azeotropic distillation** (Section 4.5). For example, benzene or toluene can be dried reasonably well in this manner. Water can also be removed from 95% ethanol by adding benzene and distilling off the azeotropic mixture of benzene:water to give pure ethanol, which is called *denatured* ethanol owing to the presence of small amounts of toxic benzene. Solids can sometimes be dried by dissolving them in a suitable solvent, removing any water by azeotropic distillation, and then recovering the solid by removal of the solvent.

TABLE 2.1   Table of Common Drying Agents, Their Properties, and Uses

Drying Agent	Acid-Base Properties	Product(s) with Water	Comments[1]
$CaCl_2$	Neutral	$CaCl_2 \cdot H_2O$ $CaCl_2 \cdot 2\,H_2O$ $CaCl_2 \cdot 6\,H_2O$	High capacity and fast action; reasonable efficiency; good preliminary drying agent; readily separated from dried solution because $CaCl_2$ is available as large granules; cannot be used to dry either alcohols and amines (because of compound formation) or phenols, esters, and acids [because drying agent contains some $Ca(OH)_2$]; hexahydrate decomposes (loses water) above 30°C.
Na_2SO_4	Neutral	$Na_2SO_4 \cdot 7\,H_2O$ $Na_2SO_4 \cdot 10\,H_2O$	Inexpensive, high capacity; relatively slow action and low efficiency; good general preliminary drying agent; physical form is that of a powder, so filtration required for removal of drying agent from dried solution; decahydrate decomposes above 33°C.
$CaSO_4$	Neutral	$CaSO_4 \cdot \frac{1}{2}\,H_2O$	Low capacity but somewhat higher efficiency than Na_2SO_4 and $MgSO_4$; preliminary drying of solution with drying agent of higher capacity strongly recommended; hemihydrate can be dehydrated by heating at 235°C for 2–3 hr.
K_2CO_3	Basic	$K_2CO_3 \cdot 1\frac{1}{2}\,H_2O$ $K_2CO_3 \cdot 2\,H_2O$	Fair efficiency and capacity; good for esters, nitriles, and ketones; cannot be used with acidic organic compounds.
$MgSO_4$	Weakly acidic	$MgSO_4 \cdot H_2O$ $MgSO_4 \cdot 7\,H_2O$	About equivalent to Na_2SO_4 as a general drying agent; requires filtration for removal of drying agent from dried solution; heptahydrate decomposes above 48°C.

H_2SO_4	Acidic	$H_3O^+HSO_4^-$	Good for alkyl halides and aliphatic hydrocarbons; cannot be used with even such weak bases as alkenes and ethers; high efficiency.
P_2O_5	Acidic	H_2PO_3 $H_4P_2O_7$ H_3PO_4	See comments under H_2SO_4; also good for ethers, aryl halides, and aromatic hydrocarbons; generally high efficiency; preliminary drying of solution recommended; dried solution can be distilled from drying agent.
CaH_2	Basic	$H_2 + Ca(OH)_2$	High efficiency with both polar and nonpolar solvents, although inexplicably it fails with acetonitrile; somewhat slow action; good for basic, neutral, or *weakly* acidic compounds; cannot be used for base-sensitive substances; preliminary drying of solution is recommended; dried solution can be distilled from drying agent.
Na or K	Basic	$H_2 + NaOH$ or KOH	Good efficiency but slow action; cannot be used on compounds sensitive to alkali metals or to base; care must be exercised in destroying excess drying agent; preliminary drying *required*; dried solution can be distilled from drying agent.
BaO or CaO	Basic	$Ba(OH)_2$ or $Ca(OH)_2$	Slow action but high efficiency; good for alcohols and amines; cannot be used with compounds sensitive to base; dried solution can be distilled from drying agent.
KOH or NaOH	Basic	Solution	Rapid and efficient but use limited almost exclusively to drying of amines; has potential for other nonacidic solvents such as *p*-dioxane.
Molecular Sieve #3A or #4A[2]	Neutral	Water strongly adsorbed	Rapid and generally highly efficient; preliminary drying recommended; dried solution can be distilled from drying agent if desired; Molecular Sieve is the trade name for aluminosilicates whose crystal structure contains a network of pores of uniform diameter; the pore sizes of sieves #3A and 4A are such that only water and other small molecules such as ammonia can pass into the sieve; water is strongly adsorbed as water of hydration; hydrated sieves can be reactivated by heating at 300–320°C under vacuum or at atmospheric pressure.

[1]Capacity, as used in this table, refers to the amount of water that can be removed by a given weight of drying agent; efficiency refers to the amount of water, if any, in equilibrium with the hydrated desiccant.
[2]The numbers refer to the nominal pore size, in Ångstrom units, of the sieve.

Two general requirements for a drying agent are as follows: (1) neither it nor its hydrolysis product will react chemically with the organic liquid being dried, and (2) it can be *completely* and *easily* removed from the dry liquid. A drying agent must be efficient in its action so that most or all of the water will be removed by the desiccant in a *reasonably* short period of time.

Some commonly used drying agents and their properties are listed in Table 2.1. Of the agents listed, *anhydrous* calcium chloride, sodium sulfate, and magnesium sulfate generally suffice for the needs of the introductory organic chemistry laboratory course.

The desiccants in Table 2.1 function in one of two ways: (1) the drying agent interacts *reversibly* with water by the process of adsorption or absorption (equation 1), or (2) it reacts irreversibly, with water serving as an acid or a base.

In reversible hydration, a certain amount of water will remain in equilibrium with the hydrated drying agent. The lesser the amount of water left at equilibrium, the greater the *efficiency* of the desiccant. For example, sodium sulfate is more efficient in its drying power for *p*-dioxane than is magnesium sulfate, but it is less efficient than calcium chloride.[6]

$$\text{Drying agent (solid)} + x\ H_2O\ \text{(liquid)} \rightleftarrows \text{Drying agent} \cdot x\ H_2O\ \text{(solid)} \quad \textbf{(1)}$$
$$\quad\text{Anhydrous} \hspace{7cm} \text{Hydrate}$$

A drying agent that forms a hydrate (equation 1) must be *completely* removed by gravity filtration (Section 2.12) or by decantation (Section 2.15) *before* the dried liquid is distilled. Most hydrates decompose with loss of water at temperatures above 30–40°C.

Drying agents that remove water by irreversible chemical reaction have very high efficiencies but are generally more expensive than other types of drying agents. They are sometimes more difficult to handle. For example, phosphorus pentoxide, P_2O_5, removes water by reacting with it to form phosphoric acid (equation 2, which is unbalanced).

$$P_2O_5\ \text{(solid)} + H_2O\ \text{(liquid)} \rightarrow H_3PO_4 \quad \textbf{(2)}$$

Desiccants such as calcium hydride (CaH_2), sodium metal, or phosphorus pentoxide (P_2O_5) react vigorously with water. A *preliminary* drying with a less reactive and efficient desiccant that has a high water-removing capacity is normally done before using these desiccants. When CaH_2 or Na metal is used as a drying agent, hydrogen gas is evolved, and appropriate precautions must be taken to vent the hydrogen and prevent buildup of this highly flammable gas.

Do *not* use an unnecessarily large quantity of drying agent when drying a liquid, since the desiccant may adsorb or absorb the organic liquid along with the water. Mechanical losses on filtration or decantation of the dried solution may also become significant. The amount of drying agent required depends upon the quantity of water present, the capacity of the drying agent, and, indirectly, the amount of liquid to be dried.

(b) Experimental Procedure for Drying Liquids

Use an Erlenmeyer flask for drying a liquid and select a flask size that will be no more than half full with liquid. In general, a portion of drying agent that covers the bottom of the flask

[6]Burfield, D. R.; Lee, K.-H.; Smithers, R. H., *Journal of Organic Chemistry*, **1983**, *48*, 2420.

should be sufficient. Start by adding a small amount of drying agent and swirling the flask gently. If the *liquid* still appears cloudy, add some more drying agent. Repeat this process until the liquid appears clear. Remove the drying agent by gravity filtration (Section 2.12) if the liquid is moderately high-boiling and nonvolatile or by decantation (Section 2.15) if it is low-boiling and volatile.

Swirling the flask of liquid and desiccant enhances the rate of drying with most desiccants because it hastens the establishment of equilibrium in the heterogeneous mixture. Agitation also appears to hasten the rate of drying even in cases where reaction with water is irreversible. This is thought to be due to a breakdown in the size of the desiccant particles, which results in an increase in the net surface area of the drying agent.

2.22   Drying Solids

Solid organic compounds must be dried because the presence of water or organic solvents in them may affect their melting points, quantitative elemental analysis, and spectra. Difficulties also arise if the solid is wet and if it is to be used in a subsequent reaction whose success depends on the absence of small amounts of water or other liquids. A solid that has been recrystallized from a volatile organic solvent can usually be dried to an extent satisfactory for most purposes by allowing it to air-dry at room temperature. The drying process can be accelerated by spreading the solid on a piece of filter paper on a clean watch glass and allowing it to stand overnight or longer. The filter paper absorbs traces of water or any excess solvent present. The rate of drying can be increased by collecting the solid on a Büchner funnel held on a filter flask, pressing it as dry as possible with a clean cork, and pulling air through the solid by means of a water aspirator or house vacuum. See Section 2.16 for more details of this technique.

If the sample is **hygroscopic,** meaning that the solid absorbs water from moisture contained in the air, or if it has been recrystallized from water or a high-boiling solvent, the substance must usually be dried in an oven operating at a temperature below the melting or decomposition point of the sample. The oven-drying process can be done at atmospheric pressure or under vacuum. Air-sensitive solids must be dried either in an inert atmosphere, such as in nitrogen or helium, or under vacuum. Samples to be submitted for quantitative elemental analysis are normally dried to constant weight by heating under vacuum.

Dry organic solids are conveniently stored in desiccators containing desiccants such as silica gel, phosphorus pentoxide, or calcium chloride. Tightly stoppered bottles containing one of these desiccants make good storage containers. A desiccator can even be used to dry an organic solid that is wet with water if it contains one of the aforementioned drying agents. Moreover, a solid contaminated with a hydrocarbon solvent can be dried by placing it in a desiccator containing a block of paraffin, which absorbs the hydrocarbon.

2.23   Drying Tubes

The organic chemist is often faced with the problem of protecting a reaction mixture from moisture. There are advanced techniques for doing so, such as using an inert, dry atmosphere

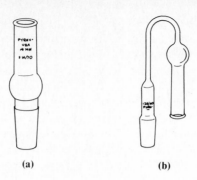

(a)   (b)

Figure 2.33 Drying tubes: (a) straight, (b) bent, or U-shaped.

of nitrogen or argon gas. A simple and moderately effective procedure utilizes a drying tube containing a drying agent.

Two types of drying tubes are shown in Figure 2.33; the straight tube is more common in the introductory organic laboratory. A drying tube is prepared for use by placing a very loose plug of glasswool at the bottom of the tube and filling the tube with a *granular* drying agent. Anhydrous calcium chloride is commonly used. It is usually not necessary to "cap" the straight tube with glasswool, whereas with a bent drying tube, a loose plug of glasswool must also be placed in the other end of the tube to prevent the drying agent from falling out. Figure 2.33 shows both types of filled tubes.

2.24 Drying of Apparatus

Before a reaction that must be carried out under anhydrous conditions is started, as much moisture as possible must be removed from the apparatus. This can be done by setting up the glassware without any reagents present and gently warming the apparatus with a Bunsen burner or microburner, a process called **flaming.** The apparatus must have at least one opening to the atmosphere before warming. Start warming at the point most remote from the opening to the atmosphere and work towards the opening. The moisture in the apparatus will be driven through the opening. If a condenser is part of the apparatus, make sure no water is in or running through it. When the apparatus is warm, insert a filled drying tube (Section 2.23) in the opening and allow the system to cool to room temperature. As cooling occurs, air from the atmosphere is drawn into the apparatus through the drying agent, which removes the moisture. This technique accomplishes two things: first, moisture is driven from the apparatus by the heat, and second, the air drawn back into the equipment is free of moisture.

solids
recrystallization and melting points

The organic chemist most often works with substances in two physical states, liquid and solid. The purpose of this and the next two chapters is to present an introduction to the theory and practice of the most important methods used to separate and purify organic liquids and solids.

It is entirely possible that a compound believed to be pure may actually contain a small amount of another substance. In fact, some of the work reported before the 1950s has been subsequently shown to be incorrect, owing to the presence of impurities in what were believed to be pure compounds. The use of recently developed chromatographic and spectral techniques has aided the chemist in ascertaining that a solid or liquid is pure.

A pure substance contains only one type of molecule, and it is essential to ensure that a compound is as pure as possible before conducting studies on it. It is a rare occasion when an organic chemist obtains a desired compound in pure form, whether that substance is the product of a chemical reaction or is isolated from some natural source. For example, a chemical transformation designed to produce a particular target compound almost invariably yields a reaction mixture that contains a number of contaminants. These may include products of side reactions that proceed concurrently with the main reaction, varying amounts of unchanged starting materials, inorganic materials, and solvents. Chemists devote a great deal of effort to the separation of the desired product from such impurities. Even chemicals purchased commercially are often not more than 95% pure, owing to the expense to the supplier of the final purification process or to decomposition prior to use. The ultimate goal is to obtain a substance that cannot be separated further by any known experimental techniques. This chapter discusses the purification of solids by recrystallization and the physical property of melting points.

3.1 Recrystallization

One of the most valuable and useful techniques to be mastered by the organic chemist is **recrystallization.** This technique is the one most often used to effect the purification of *solids*. Other methods that can be used for this purpose are sublimation (Sec. 2.17), extraction (Chapter 5) and chromatography (Chapter 6). Even when one of these alternative methods has been used, a careful worker will frequently perform a final recrystallization of the material to achieve added confidence that it is pure.

The process of recrystallization involves, first, the complete disruption of the crystal structure by dissolving the solid in a suitable solvent and, second, regrowth of the crystals so that, ideally, the impurities are left in solution. This technique is called **solution recrystallization** and is discussed below. An alternative to recrystallization involves melting the solid in the absence of any solvent and then allowing the crystals to regrow so that the impurities are left in the melt. This method is seldom used because the crystals usually form in the presence of a rather viscous oil that contains the impurities and from which it is difficult to separate the desired solid.

Solution crystallization takes advantage of the fact that nearly all solids are *more* soluble in a *hot* than in a *cold* solvent. Thus, if the crystals are dissolved in a volume of hot solvent that is insufficient to dissolve them when cold, and if that hot solution is then allowed to cool, crystals should form from the cooling solution. The extent to which the crystals form depends on the difference in solubility between the temperature extremes. The high temperature extreme is limited by the boiling point of the solvent being used; the low temperature limit is usually dictated by experimental convenience. An ice-water bath is often utilized for cooling the solution. However, the freezing point of the solvent can be the determining factor.

If the impurities present in the original solid mixture have dissolved and *remain dissolved* after the solution is cooled, filtration of the crystals that have formed on cooling should *ideally* provide pure material. On the other hand, the impurities may remain undissolved in the hot solution and can then be filtered from the solution *before* it is allowed to cool, so that the crystals that subsequently form on cooling should be purer than the original solid mixture. Solution recrystallization is seldom so simple, but this scenario should provide an understanding of the general principles of the technique.

Application of the technique of solution recrystallization involves some or all of the following experimental steps:

1 selection of an appropriate solvent
2 dissolution of the solid to be purified in the solvent near or at its boiling point
3 decolorization to remove colored impurities and filtration of the hot solution to remove insoluble impurities and decolorizing carbon
4 formation of crystals from the solution as it cools
5 filtration of the purified crystals from the cooled solution and removal of adhering solvent
6 drying the crystals

Each step of the recrystallization process is discussed in detail below, and the experimental procedures are presented at the end of the discussion.

After an impure solid has been recrystallized, the crystals that are obtained may or may not be pure. One of the methods that can be used to determine their purity is to determine the melting point of the recrystallized solid. This technique is discussed in Section 3.2.

1. Selecting a Solvent A solvent must satisfy certain criteria in order to be used for recrystallization: (a) The substance being purified must be *insoluble* or *nearly insoluble* in cold solvent. This solubility is often tested at room temperature. (b) The compound being purified should ideally be soluble in the hot solvent but insoluble or nearly so when cooled. The impurities must remain at least moderately soluble in the cold solvent, otherwise both the desired compound and the impurities would recrystallize. This solubility property of the solvent with respect to the solubility of the solute and impurities is called the *temperature coefficient* of the solvent, and it must be favorable for a successful recrystallization. Another possibility is that the impurities are insoluble in the hot solution, from which they may be filtered. (c) The boiling point of the solvent should be low enough so that it can be removed easily from the crystals in the final drying step. (d) In general, the boiling point of the solvent should be lower than the melting point of the solid being purified. (e) The solvent should not react chemically with the substance being purified.

If a compound has been prepared previously, the chemical literature describing its synthesis will usually give information concerning a suitable recrystallization solvent. If the compound has *not* been studied previously, it will be necessary to resort to trial-and-error techniques in order to determine a suitable solvent for recrystallization, and it is desirable to use *small* amounts of material for this purpose. Some solubility generalizations can be utilized in selecting a solvent. Normally, polar compounds are soluble in polar solvents and insoluble in nonpolar solvents. Conversely, nonpolar compounds are more soluble in nonpolar solvents. These solubility properties are frequently summarized by the adage "like dissolves like." A highly polar compound is unlikely to be soluble in a hot nonpolar solvent, but it may be very soluble in a cold, very polar solvent, so that a solvent of intermediate polarity *may* be satisfactory.

Common recrystallization solvents have a wide range of polarity, and the *dielectric constants* (ϵ) listed in Table 3.1 provide a measure of this property. Those solvents with dielectric constants in the range of 2–3 should be regarded as nonpolar and those with constants above 10 as polar. Solvents in the 3–10 range are of intermediate polarity. The structures of most of the solvents in Table 3.1 are provided above the table. The name of one of these solvents can be confusing and deserves special mention. *Petroleum ether* is a mixture of aliphatic hydrocarbons obtained from petroleum refining and is *not* an *ether* at all. Such a mixture may vary in composition and boiling range, depending on the distillation "cut" taken. In order to define the liquid being used, the boiling range is usually given, for example, petroleum ether (bp 60–80°C). To confuse the issue further, the name *ligroin* is occasionally used in place of petroleum ether.

Sometimes the organic chemist is unable to find a single pure solvent that is suitable for a particular recrystallization, so that it is necessary to use **mixed solvents.** This technique involves using a mixture of two (or more) solvents and is done as follows. One solvent is selected in which the compound is insoluble when cold and another in which the compound is soluble when cold. The compound being purified is placed in the solvent in which it is insoluble and then heated near the boiling point of the solvent. Then the second solvent is added in small portions until the solid dissolves. If crystals form on cooling, a suitable

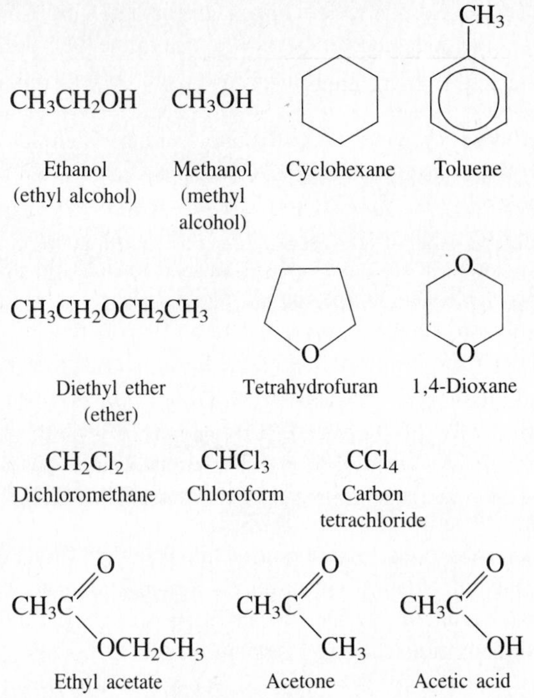

$$CH_3CH_2OH \quad CH_3OH$$

| Ethanol | Methanol | Cyclohexane | Toluene |
| (ethyl alcohol) | (methyl alcohol) | | |

$$CH_3CH_2OCH_2CH_3$$

| Diethyl ether (ether) | Tetrahydrofuran | 1,4-Dioxane |

$$CH_2Cl_2 \qquad CHCl_3 \qquad CCl_4$$

| Dichloromethane | Chloroform | Carbon tetrachloride |

| Ethyl acetate | Acetone | Acetic acid |

TABLE 3.1 Common Recrystallization Solvents[a]

Solvent[b]	Boiling Point (°C)	Freezing Point (°C)[c]	Water Soluble	Dielectric Constant (ϵ)	Flammable	Specific Gravity[d]
Water*	100	0	—	78.54	No	1.000
95% Ethanol*	78		Yes	24.6	Yes	
Methanol	65		Yes	32.63	Yes	
Petroleum ether*	Variable		No	1.9	Yes	About 0.7
Cyclohexane	81	6	No	2.02	Yes	0.779
Toluene	111		No	2.38	Yes	0.867
Diethyl ether	35		Slightly	4.34	Yes	0.714
Tetrahydrofuran	65		Yes	7.58	Yes	
1,4-Dioxane	101	11	Yes	2.21	Yes	
Dichloromethane[e]	41		No	9.08	No	1.335
Chloroform*[e]	61		No	4.81	No	1.492
Carbon tetrachloride[e]	77		No	2.23	No	1.594
Ethyl acetate*	77		Yes	6.02	Yes	
Acetone	56		Yes	20.7	Yes	
Acetic acid	118	17	Yes	6.15	Yes	

[a]Benzene has been omitted from this list, owing to its toxicity. Cyclohexane can often be successfully substituted for it.

[b]The solvents marked with asterisks should normally be used first in a trial-and-error search for the best recrystallization solvent.

[c]Freezing points not listed are below 0°C.

[d]Only the specific gravities of water-insoluble solvents are included.

[e]As a general rule, avoid use of chlorocarbon solvents such as dichloromethane, chloroform, and carbon tetrachloride, *if another equally good solvent can be found*. If not, take care to avoid excessive inhalation of their vapors.

solvent mixture may have been found, but trial-and-error with other solvent mixtures may be required before pure crystals are obtained.

A major requirement for using mixed solvents is that the solvents themselves be completely soluble (miscible) in one another. For example, information given in Table 3.1 indicates that carbon tetrachloride is water-insoluble, so that a mixture of water and CCl_4 could not be used. Some frequently used mixed-solvent pairs are ethanol-water, toluene–petroleum ether, acetic acid–water, diethyl ether–alcohol, and diethyl ether–petroleum ether. Mixtures of toluene-ethanol are infrequently used and then only with absolute ethanol, since the presence of water causes solvent separation, particularly on cooling. Mixed solvents are also used in paper and thin-layer chromatography (Chapters 6 and 26).

2. Dissolving the Solid The following operations are best carried out within a ventilation hood to avoid inhalation of the solvent vapors and the possibility of fire. If this is not possible, an inverted funnel should be clamped over the flask, and the stem of the funnel should be connected to a vacuum source such as a water aspirator or the house vacuum system.

The solid to be purified should be weighed and placed in an appropriately sized *Erlenmeyer* flask. *A beaker should never be used for recrystallization*. A few crystals of the impure solid should be saved, for they may be needed later to induce crystallization. A few milliliters of the desired solvent are added to the flask. With *constant stirring or swirling*, the mixture of the solid and solvent is then heated to boiling, preferably using a steam bath, hot water bath, or hot plate. *Open flames should never be used with flammable solvents* (solvent flammability is given in Table 3.1). More solvent is added to the boiling mixture in *small* portions until just enough solvent is present to dissolve the solid. Enough time is allowed for the solution to boil again after adding each portion of the solvent, for some solids dissolve slowly, especially if they are not finely divided. After all of the solid has just dissolved, it is generally desirable to add 2 to 5% additional solvent to prevent premature crystallization during the hot filtration, if this step is necessary. A large excess of solvent should be avoided, as this would minimize the amount of pure crystals that will form on cooling. Solids remain soluble to some extent even in cool solution, and the recovery will be reduced by an amount that depends on both this solubility and the volume of solvent present. If, near the end of the dissolution process, it appears that additional solvent is not dissolving any more of the solid, *particularly when only a relatively small quantity of solid remains,* it is probable that enough solvent has been added. The remaining solid is likely to consist of insoluble impurities that will be removed in the hot filtration step. The recovery of only a small amount of crystals can most often be attributed to having added a volume of solvent far in excess of that needed.

The use of mixed solvents involves the same general approach as that described above. As an alternative to the procedure outlined under ''Selecting a Solvent,'' the solid to be purified can first be dissolved in a minimum volume of the hot solvent in which it is soluble. The second solvent is then added to the *boiling* solution until it turns *cloudy*. If this solvent has a lower boiling point than the first, the solution must be cooled *below* the boiling point of the second solvent before it is added. The second solvent decreases the dissolving ability of the solvent mixture, and when the solubility limit is reached, the solute begins to come out of solution and produces the cloudy appearance. Finally, more of the first solvent is added *dropwise* until the solution becomes clear again. It is advantageous to carry out the hot filtration step, if needed, before adding the second solvent to prevent crystallization during filtration. The alternative procedure just described has the potential disadvantage of requiring unduly large volumes of solvent.

3. Decolorization and Hot Filtration After the solid mixture is dissolved in hot solvent, the solution may be colored, owing to the presence of colored impurities. If the hot solution is colorless, the decolorization process may be omitted. If it is colored, the colored impurities may often be removed by adding a small amount of decolorizing carbon (activated vegetable charcoal) to the hot, *but not boiling,* solution. Carbon added to the boiling solution is likely to cause the liquid to boil over the top of the flask. The amount of decolorizing carbon required is usually small, and typically that which can be held on the tip of a small spatula will suffice. Colored impurities and the compound being purified are adsorbed on the surface of the carbon particles, but the colored substance, being present in minor amounts, is completely removed, whereas the uncolored one is not. The decolorizing carbon is subsequently removed by hot filtration. If too much carbon is added, the amount of solid recovered will be reduced because the compound being purified is also adsorbed on the surface of the charcoal. After the charcoal is added, the solution should be heated to boiling for a few minutes while being continuously stirred or swirled to prevent bumping of the boiling mixture.

To remove insoluble impurities, including dust and decolorizing carbon (if used), the hot solution is filtered by gravity filtration (Sections 2.12 and 2.13). If no insoluble impurities are present and the solution is clear, this step may be omitted. Gravity filtration is preferred to vacuum filtration because the latter will both cool and concentrate the solution, owing to evaporation of the solvent, and this may result in premature crystallization. A short-stemmed or stemless glass funnel and a fluted filter paper should be used for gravity filtration into a second Erlenmeyer flask. The top of the paper should not extend above the top of the funnel because of the possibility of the liquid flowing over the top of the funnel. The use of *fluted* filter paper (Section 2.11) allows for more rapid filtration.

The hot solution is poured onto the middle to upper portion of the filter paper in order to maximize the efficiency of the filtration. In this way the solution will come in contact with a larger area of the filter paper and thus will be filtered more rapidly. However, no solution should be allowed to pass between the side of the paper and the funnel.

Decolorizing carbon is very finely divided and is not always completely removed by filtration, as evidenced by the presence of a dark tint in the filtrate. If the carbon is not completely removed, a small amount of "filter-aid" may be added to the filtrate. The solution is heated again just to boiling and refiltered using a fresh fluted filter paper. The carbon is trapped in the filter-aid and thus removed.

4. Crystallization The hot filtrate is allowed to cool slowly by standing at room temperature, which should result in the formation of crystals. Rapid cooling by immersion of the flask in water is undesirable because the crystals formed will tend to be very small, and their resulting large surface area may facilitate adsorption of impurities from the solution. Generally the solution should not be agitated while cooling, since this will also lead to formation of small crystals. On the other hand, the formation of crystals larger than about 2 mm should also be avoided, since this may cause *occlusion* (trapping) of solution *within* the crystals. Such crystals are difficult to dry and when dried may contain deposits of impurities. If large crystals appear to be forming, the flask may be agitated to lower the average crystal size. With experience, it should become easier to judge the proper crystal size.

If crystallization does not occur after the flask has been allowed to cool for a while, either too much solvent was used during dissolution or the solution is supersaturated. A supersaturated solution can usually be made to form crystals by *seeding*. A tiny crystal of the original solid is added to the cooled solution, and crystals may form quite rapidly. Another method that may be used to induce crystal formation is to scratch the inside surface of the flask *at or just above the surface of the solution* with a glass stirring rod. If this is done, *do not apply undue pressure to the stirring rod* in order to avoid breaking the rod or the flask.

Occasionally the solute will separate from the solution as an "oil" rather than as crystals, which may be due to insufficient solvent having been added during the dissolution. This type of separation is not as selective as crystallization, so these oils generally contain significant amounts of impurities and their formation is undesirable. There are two somewhat different types of oiling problems. (1) Oils may persist on full cooling with no evidence of crystallization. A remedy that frequently works in such cases is to scratch the oil against the side of the flask with a glass stirring rod in the presence of the **mother liquor** in order to induce crystallization. If this fails, several *small* seed crystals of the original impure solid may be added to the oil and the mixture allowed to stand for a period of time, perhaps until the next class period. If these alternatives fail to work, it may be necessary to separate the oil from the mother liquor and crystallize it from a different solvent. (2) When recrystallizing certain compounds, such as acetanilide from water, cooling of the hot solution may result in the separation of an oil that freezes into a compact mass at a lower temperature. After the oil appears, relatively pure crystals may start to form in the residual solution. The oil is not a pure liquid solute but is a liquid solution of solute, solvent and perhaps other impurities whose freezing point is below the temperature at which the oil formed. Therefore, the crystals, oil, and/or solid mass from the oil should *not* be collected, because they are likely to be impure. Problems of this type may usually be remedied by reheating the mixture to boiling, adding a few milliliters of additional pure solvent, and allowing the solution to cool again. This should either solve the problem or at least reduce the amount of oil formed.

A second crop of crystals can be obtained as discussed under "Filtration" below.

5. Vacuum Filtration and Removal of Solvent from Crystals The cool mixture of crystals and solution is filtered by vacuum filtration, and the solution is collected in a *clean, dry filter flask*. The crystals should be washed with a small amount of pure, cold solvent, and then as much solvent as possible should be removed from the crystals. The experimental details for vacuum filtration and solvent removal are described in Section 2.16.

After filtering the crystals, a second "crop" of crystals can usually be obtained by further cooling of the filtrate, often referred to as the mother liquors, in an ice-water bath. The mother liquors can also be concentrated by boiling away some of the solvent and cooling the residual solution. The crystals obtained as a second or even third crop may not be as pure as the ones obtained from the initial crystallization, and the various crops of crystals should *not* be combined until their purity has been assessed. As discussed in the next section, purity can be determined by the use of melting points.

6. Drying the Crystals   After vacuum filtration and solvent removal, the crystals are removed from the filter funnel and dried to remove residual amounts of solvent. The technique for drying solids is presented in Section 2.22.

Pre-lab exercises for Section 3.1, Recrystallization, are found on page PL. 1.

EXPERIMENTAL PROCEDURE

DO IT SAFELY

1 Do not use a burner in these procedures unless you have been instructed to do so. Most of the solvents used for recrystallization are flammable.

2 If you are using a hot plate, do not use it at the highest setting. A moderate setting will prevent overheating and the consequent bumping and splashing of hot solvents and materials from the flask during heating. Hot plates should *not* be used for heating volatile or flammable solvents; the use of a steam bath or a hot water bath is preferable for these types of solvents.

3 Avoid excessive inhalation of solvent vapors. If a ventilation hood is not available for your use, clamp an inverted funnel just above the Erlenmeyer flask in which you will be heating solvents. Attach this funnel by means of rubber tubing to a water aspirator or to a house vacuum line. This will serve to lower the concentration of vapors in your work area.

4 When pouring or transferring solutions, either wear rubber gloves or be careful to avoid getting these solutions on your skin. Organic compounds are much more rapidly absorbed through the skin when they are in solution, particularly in water-soluble solvents such as acetone and ethanol. For this reason you should never rinse organic materials off your skin with solvents such as acetone, but instead should wash your hands thoroughly with hot water and soap.

5 If you use decolorizing carbon, **never add it to a boiling solution** because it may cause the solution to boil out of the flask. The carbon should be added when the temperature of the solvent is **well below** the boiling temperature. If it is added when the boiling temperature has been reached, but vigorous ebullition has not yet begun, the addition of the carbon may cause violent boiling so that the mixture overflows the flask.

A. Selection of Solvent for Recrystallization

Heat a beaker of water to a gentle boil on a hot plate and use the hot water to serve as a heating medium; replenish the water as necessary. Alternatively, a steam bath may be used (Section 2.5f).

Several options are available for selecting the solvent for a recrystallization. This may be done on known compounds or on an unknown compound. Your instructor will tell you which option to use.

1. Known Compounds The procedures given below involve several common organic crystalline compounds: resorcinol, naphthalene, benzoic acid, and acetanilide. Look up the structures of these compounds in an organic textbook.

Place about 20 mg (a small spatula-tip full) of finely crushed resorcinol in a small test tube and add about 0.5 mL of water to the tube. Stir with a glass rod and determine whether resorcinol is soluble in water at room temperature. Record your observations in your notebook, using the following definitions: *soluble*—20 mg of solute will dissolve in 0.5 mL of solvent; *slightly soluble*—some but not all of the 20 mg of solute will dissolve in 0.5 mL of solvent; *insoluble*—none of the solute appears to dissolve. If the resorcinol is not completely soluble at room temperature, place the test tube in the hot-water bath, and with stirring or swirling of the tube observe whether resorcinol is soluble in hot water. Record your observations on the basis of the definitions above.

Repeat the solubility test for resorcinol using 95% ethanol and then petroleum ether (bp 60–80°C), and record your observations in your notebook. If additional practice is desired, determine in like fashion the solubility properties of naphthalene, benzoic acid, and acetanilide in water, in ethanol, and in petroleum ether.

If any of these solutes is soluble in the hot solvent but only slightly soluble or insoluble in the cold solvent, allow the hot solution to cool slowly to room temperature and compare the quantity, size, color, and form of the resulting crystals with the original solid material. Note which solvent you would consider best suited for recrystallization of each of the solutes.

2. Unknown Compounds If you are given an unknown solid to recrystallize, first determine its solubility in various common solvents. This should be done on a *small* scale, and once the most suitable solvent has been selected, then the process can be "scaled" up, as described in part B below. Table 3.1 lists various solvents that are generally useful for recrystallization of organic solids, and some descriptive information is provided for each solvent. It should not be necessary to test *all* the solvents, but you should consider trying those solvents in the table that are denoted with an asterisk. Your instructor may indicate which solvents to try.

A systematic approach to testing the solubility of an unknown compound is encouraged. Clean and dry enough small test tubes so that one can be used for each solvent that will be tested. Place about 20 mg (a small spatula-tip full) of finely crushed unknown in each test tube and add about 0.5 mL of each solvent to different tubes containing the solid. Stir each with a glass rod and determine whether the solid is soluble in each solvent at room temperature. Use the definitions of *soluble, slightly soluble,* or *insoluble* given in part A, above. If the unknown is not soluble in a given solvent, place the test tube in the hot-water bath. Stir or swirl the tube, and observe whether the unknown is soluble in hot solvent. If the solid is soluble in the hot solvent but only slightly soluble or insoluble at room temperature, allow the hot solution to cool slowly to room temperature. If crystals form in the cool solution, compare their quantity, size, color, and form with the original solid material and with the other samples. It may be helpful to construct a table containing the solubility data, from which you should be able to decide the solvent that appears best suited for recrystallization. Generally it is a good idea to test the solubility in a variety of solvents, even though you may find crystals formed in the first solvent you try; another solvent might prove better if it provides either more or better crystals.

If these solubility tests produce no clear solvent choice, mixed solvents should be considered; review the previous discussion in this section for the philosophy of using a mixture of two solvents. Before trying pairs of solvents in combination, you should take about 0.5 mL of each *pure* solvent being considered and mix them to ensure that they are soluble in one another. If they are not miscible, another pair of solvents should be considered.

B. Recrystallization of Impure Solids

Carefully read the discussion provided earlier in this section to become familiar with the techniques for carrying out the following steps in the recrystallization procedure. This information should help you avoid or surmount most of the problems that you may encounter in effecting the purification of impure solids by recrystallization. Only abbreviated directions are provided in the experiments that follow.

Obtain one or more samples of impure solids for recrystallization. Among those that might be assigned are (1) benzoic acid, (2) acetanilide, (3) naphthalene, or (4) an unknown compound. If you are assigned an unknown, determine the appropriate solvent(s) to use as in part A and then proceed by following the general instructions given in the discussion part of this section.

For each compound that is recrystallized, determine the melting point, weight, and percent recovery of the pure crystals that are obtained. Melting points and the procedure for determining them are discussed in Section 3.2. Percent recovery can be calculated from the weight of the impure solid with which you started and the weight of the pure substance that you obtain. Equation 1 illustrates this calculation:

$$\text{Percent recovery} = \frac{\text{weight of pure crystals recovered}}{\text{weight of original sample}} \times 100 \qquad \textbf{(1)}$$

For example, suppose you started with 1.0 g of an unknown solid and obtained 0.45 g of pure crystals. The percent recovery would be $(0.45 \text{ g}/1.0 \text{ g}) \times 100 = 45\%$.

1. Benzoic Acid   Place 1 g of impure benzoic acid in a clean 50-mL Erlenmeyer flask. Measure 25 mL of water in a graduated cylinder and add a 10-mL portion of it to the benzoic acid. Heat to gentle boiling with a microburner or hot plate. Add water in 1-mL portions until no more solid appears to dissolve in the boiling solution. Record the total volume of water used. No more than 20 mL should be required.

Since pure benzoic acid is colorless, a colored solution should be treated with decolorizing carbon. (*Caution:* Do not add decolorizing carbon to a *boiling* solution!) Cool the solution slightly, add approximately 0.1 g of carbon, and reheat to boiling for a few minutes. To aid in the filtration of the finely divided carbon, allow the solution to cool slightly, add a small amount of filter-aid, and reheat.

Perform a hot filtration according to the directions provided in Section 2.13. Rinse the empty flask with 1 or 2 mL of *hot* water and filter this wash solution into the main solution. If the filtered solution remains colored, repeat the treatment with decolorizing carbon. Cover the flask with a watch glass or inverted beaker,★ and allow the filtrate to stand undisturbed until it has cooled to room temperature and no more crystals form. To complete the crystallization place the flask in ice water for at least 15 min.

Collect the white crystals by suction filtration and wash the filter cake with two small portions of *cold* water. With a clean spatula or cork press the crystals as dry as possible on the funnel. Spread the crystals onto a piece of filter paper or, better, a watch glass and allow them to air-dry completely. Determine the weight and melting point of the purified product. Calculate the percent recovery.

2. Acetanilide Place 5 g of impure acetanilide in a 250-mL Erlenmeyer flask. Measure 100 mL of water into a graduated cylinder and add a 50-mL portion to the crude acetanilide. Boil the mixture gently with the aid of a burner or hot plate.

Note the formation of an oil layer consisting of a solution of water in acetanilide that forms a separate phase. This second liquid phase forms at temperatures only above 83°C in mixtures whose compositions lie between 5.2% and 87% acetanilide. Acetanilide, however, is sufficiently soluble in water at temperatures slightly above 100°C, the boiling point of the solution, to form solutions of greater than 5.2% acetanilide. Thus, a boiling solution prepared with the *minimum* quantity of water to effect solution will yield an oil on cooling to 83°C, and crystals below 83°C. Reread the section on crystallization of compounds that form oils (Section 3.1, part 4).

Continue adding water in small portions (3–5 mL) to the boiling solution until the oil has completely dissolved. Note that any solid present at this point must consist of insoluble impurities. Once the acetanilide has just dissolved, add an additional 5 mL of water to prevent formation of oil during the crystallization step. If oil forms at this time, reheat the solution and add a little more water. Record the total volume of water used.

Allow the solution to cool below boiling and add about 0.1 g of decolorizing carbon; with stirring, gently boil the solution for a few minutes. Cool the solution, then add a small amount of filter-aid, stirring thoroughly. Reheat to boiling, and perform a hot filtration according to the directions given in Section 2.13. Cover the flask with a watch glass or, better, an inverted beaker,[★] and allow the filtrate to stand undisturbed while cooling to room temperature; when crystallization apparently is complete, cool the flask in ice water for at least 15 min to complete the crystallization.

Collect the crystals by vacuum filtration and wash the filter cake with two small portions of *cold* water. With a spatula or cork press the crystals as dry as possible on the funnel. Spread the crystals onto a piece of filter paper or, better, a watch glass and allow them to air-dry completely. Determine the weight and melting point of the purified acetanilide. Calculate the percent recovery.

3. Naphthalene Naphthalene may be conveniently recrystallized from either methanol, 95% ethanol, or 2-propanol. Because these solvents are either toxic or flammable, or both, proper precautions should be taken. Operations through the hot filtration step should be carried out in a ventilation hood. If this is not available, an inverted funnel connected by tubing to an aspirator or vacuum line and positioned *no more than an inch or two* above the mouth of the flask in which the solvent is being heated will afford reasonable protection.

Place 5 g of impure naphthalene in a 250-mL Erlenmeyer flask and dissolve it in the minimum required amount of boiling alcohol. (*Caution:* Use a steam bath for heating; do not use a burner.) Add 2 or 3 mL of additional solvent. A colored solution should be treated with decolorizing carbon. Perform a hot filtration, and cover the flask containing the hot filtrate with either a watch glass or an inverted beaker,[★] and allow the filtrate to stand undisturbed while cooling to room temperature. Collect the crystals by vacuum filtration, wash them with two small portions of *cold* solvent, and press dry. Transfer the crystals to a piece of filter paper or a watch glass and allow them to air-dry. Determine the weight and melting point of the purified naphthalene. Calculate the percent recovery.

4. Unknown Compound   Accurately weigh 1–2 g of the unknown compound and place it in a 125-mL Erlenmeyer flask. Measure about 10 mL of the solvent you have selected for the recrystallization and place it in the flask. Bring the mixture to a gentle boil with a steam or hot-water bath, add a 5-mL portion of the solvent, and allow the solution to boil again. Continue adding 5-mL portions of solvent, one portion at a time, until the solid has completely dissolved. Allow the solution to boil after adding each portion of solvent. Record the total volume of solvent that was added.

If the solution containing the solute is colored, cool the solution slightly, add about 0.1 g of decolorizing carbon, and reheat to boiling for a few minutes. To aid in the filtration of the finely divided carbon, allow the solution to cool slightly, add about 0.2 g of filter-aid, and reheat. Filter the mixture by gravity as directed below.

If the solution is colorless but contains small particles of insoluble material, omit the treatment with decolorizing carbon but filter the mixture by gravity as directed below.

Solutions that have been treated with decolorizing carbon or which are colorless but contain small particles of insoluble material are filtered using *hot gravity filtration*. The procedure for this process is given in Section 2.13. Rinse the empty flask with 1–2 mL of *hot* solvent, pour it through the filter, and allow this solvent to collect in the same flask that contains the original filtrate. If the filtered solution remains colored or still contains insoluble particles, repeat the treatment with decolorizing carbon and refilter by hot gravity filtration.

Cover the flask containing the filtrate with a clean watch glass or clean inverted beaker, and allow the filtrate to stand undisturbed until it has cooled to room temperature and no more crystals form.★ Complete the crystallization by placing the flask in an ice-water bath for at least 15 min.

Collect the crystals by vacuum filtration using the technique described in Section 2.16. Wash the crystals that have been collected on the filter with two 5-mL portions of *cold, pure* solvent, and allow them to dry as thoroughly as possible on the filter funnel. Dry the crystals as described in Section 2.22.

Determine the weight and melting point of the purified product and calculate the percent recovery.

EXERCISES

1. Define or describe each of the following terms as applied to recrystallization:
   a. solution recrystallization
   b. temperature coefficient of a solvent
   c. the relationship between dielectric constant and polarity of a solvent

d. petroleum ether	j. fluted filter paper	o. occlusion
e. mixed solvents	k. mother liquor	p. seeding
f. solvent selection	l. filtrate	q. washing crystals
g. decolorization	m. solute	r. drying crystals
h. hot gravity filtration	n. solvent	s. percent recovery
i. vacuum filtration		

2. Briefly explain how a colored solution may be decolorized.

3. Briefly explain how insoluble particles can be removed from a hot solution.

4. List five criteria that should be used in selecting a solvent for a recrystallization.

5. When is hot gravity filtration used and when is vacuum filtration used in recrystallization? What is the purpose of each type of filtration?

6. When might filter-aid be used in recrystallization, and why is it used?

7. In hot gravity filtration, what might happen if the filter funnel is not preheated before the solution is poured through it? *Crystallize at funnel*

8. Briefly describe how a mixture of sand and benzoic acid, which is soluble in hot water, might be separated to provide pure benzoic acid.

9. Describe each of the following pieces of equipment and its use:
   a. filter flask
   b. filter trap
   c. Büchner funnel

10. The following solvent selection data were collected for two different impure solids:

Solid A

Solvent	Solubility at Room Temp	Solubility When Heated	Crystals Formed When Cooled
Methanol	Insoluble	Insoluble	—
Chloroform	Insoluble	Soluble	Very few
Cyclohexane	Insoluble	Soluble	Many
Toluene	Insoluble	Soluble	Very few

Solid B

Solvent	Solubility at Room Temp	Solubility When Heated	Crystals Formed When Cooled
Water	Soluble	—	—
Ethanol	Soluble	—	—
Carbon tetrachloride	Insoluble	Insoluble	—
Petroleum ether	Insoluble	Insoluble	—
Toluene	Insoluble	Insoluble	—

Based on these results, what solvents or mixture of solvents might you consider using to recrystallize solids *A* and *B*?

11. List each of the steps in the systematic procedure for recrystallization. Indicate briefly the functional purpose of each of these steps in accomplishing the purification of the originally impure solid.

12. The goal of the recrystallization procedure is to obtain *purified* material with a *maximized recovery*. For each of the items listed, explain why this goal would be adversely affected.

a. In the solution step, an unnecessarily large volume of solvent is used.

b. The crystals obtained by vacuum filtration are not washed with fresh cold solvent before drying. This step is omitted.

c. The crystals referred to in b are washed with fresh *hot* solvent.

d. A large excess of decolorizing carbon is used.

e. Crystals are obtained by breaking up the solidified mass of an oil that originally separated from the hot solution.

f. Crystallization is accelerated by immediately placing the flask of hot solution in ice water.

13. A second crop of crystals may be obtained by concentrating the vacuum filtrate and cooling. Why is this crop of crystals probably less pure than the first crop?

14. Explain why the rate of dissolution of a crystalline substance may depend on the *size* of its crystals.

15. The solubility of benzoic acid at 0°C is 0.02 g per 100 mL of water and that of acetanilide is 0.53 g per 100 mL of water. If you performed either of these recrystallizations, calculate, with reference to the total volume of water you used in preparing the hot solution, the amount of material in your experiment that was unrecoverable by virtue of its solubility at 0°C.

16. Assuming that either solvent is otherwise acceptable in a given instance, what advantages does ethanol have over 1-octanol as a crystallization solvent? hexane over pentane? water over methanol?

17. Look up the solubility of benzoic acid in hot water. According to the published solubility, what is the minimum amount of water in which 1 g of benzoic acid can be dissolved?

18. Why is it important to

a. break the vacuum before turning off the water pump when employing the vacuum filtration equipment shown in Figure 2.25?

b. avoid the inhalation of vapors of organic solvents?

c. know the position and procedure of operation of the nearest fire extinguisher when employing diethyl ether as a crystallization solvent?

d. use a *fluted* filter paper for hot filtration?

REFERENCE

Tipson, R. S., in *Technique of Organic Chemistry,* 2nd ed., A. Weissberger, editor, Interscience Publishers, New York, 1956, Vol. III, Part I, Chapter 3.

3.2 Physical Constants: Melting Points

(a) Physical Constants

Physical constants of organic compounds are numerical values associated with measurable properties of these substances. These properties are invariant and are useful in the identification and characterization of substances encountered in the laboratory as long as accurate measurements are made under specified conditions such as temperature and pressure. Some of the more commonly measured physical properties are melting point (mp), boiling point (bp), index of refraction (n), density (d), and specific rotation [α]. Melting and boiling points are the most commonly encountered properties. Melting points are discussed in part (b) below, and boiling points are discussed in Section 4.1. Index of refraction and density are mentioned in Chapter 27. Specific rotation is applicable only to molecules that are *optically active*, and this topic is discussed in Chapters 7 and 25.

More than one compound *may* exhibit the same constant for one or two of the common physical properties, but it is extremely unlikely that two different compounds will display the same constants for several measurable properties. A list of physical constants is therefore regarded as a highly useful means of characterizing a substance. Several extensive compilations of the physical constants are available (see Chapter 28). One of the most convenient is the Chemical Rubber Company's *Handbook of Chemistry and Physics,* in which a very large number of inorganic and organic compounds with their known physical constants and properties are tabulated. The Chemical Rubber Company also publishes the *Handbook of Tables for Organic Compound Identification,* which is especially useful for organic compounds. Neither of these books is comprehensive; rather, they contain listings only of the more common organic and inorganic substances. There are so many known compounds whose physical properties have been reported in the literature that multi-volume sets of books are required to list them (see Chapter 28).

It must be emphasized that physical constants are useful only in the identification of *previously known* compounds, because it is not possible to predict accurately such properties as melting point and boiling point of *newly* synthesized or isolated compounds. Moreover, the observed melting or boiling point of a substance may provide information about the purity of the sample under consideration. Other properties such as color, odor, and crystal form are also useful and should be recorded in your laboratory notebook.

(b) Melting Point of a Pure Substance

The melting point of a substance is defined as the temperature at which the liquid and solid phases exist in equilibrium with one another without change of temperature. If heat is added to a mixture of the solid and liquid phases of a pure substance at the melting point, ideally no rise in temperature will occur until all the solid has melted and been converted to liquid. If heat is removed from such a mixture, the temperature will not drop until all the liquid solidifies. Thus, the melting point and freezing point of a pure substance are identical.

The relationship between phase composition, total supplied heat, and temperature for a pure compound is shown in Figure 3.1. It is *assumed* that heat is being supplied to the compound at a *slow* and *constant rate*, and the elapsed time of heating is a cumulative measure of the supplied heat. At temperatures below the melting point, the compound exists in the solid phase, and the addition of heat causes the temperature of the solid to rise. As the melting point is reached, the first small amount of liquid appears, and equilibrium is established between the solid and liquid phases. As more heat is added, the temperature does not change, since the additional heat causes the solid to be converted to liquid, with both phases remaining in equilibrium. When the last of the solid melts, the heat subsequently supplied causes the temperature to rise linearly at a rate that depends on the rate of heating.

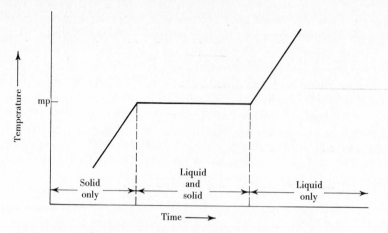

Figure 3.1 Phase changes with time and temperature.

The interconversion of the liquid and solid phases may be described in terms of their respective vapor pressures, which are directly related to the rates at which the molecules pass from one phase to the other. In terms of melting points, only the transfer of molecules between the solid and liquid phases is considered. The term **vapor pressure** may appear to be a misnomer in this case, but the same "escaping tendency" that produces the measurable equilibrium vapor pressure between liquid and gas phase is responsible for the direct transfer of molecules between the solid and liquid phases. Figure 3.2 represents, by means of reversible arrows, the equilibrium that exists between the solid, liquid, and gas phases. At the melting point, the vapor pressures of the solid and liquid phases are equal, and there is no *net* transfer of molecules from one phase to the other unless there is a change in the heat supplied to the system.

The equilibria, which are illustrated in Figure 3.2, are present any time there is some of each phase present, but in reality the most important equilibria depend on the substance being studied. For melting points, the most important equilibrium is that between the solid and liquid, since the vapor pressure of the liquid is *very* low. Boiling points (Section 4.1) involve equilibrium between liquid and gas because there is no solid present. For a limited number of substances, equilibrium exists between the solid and gas phases, a phenomenon that is observed in **sublimation** (Section 2.17), where the liquid phase is not present. Examples of sublimation, which involves the *direct* conversion of the solid phase into the gaseous phase

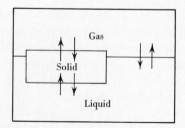

Figure 3.2 Phase equilibria between the solid, liquid, and gas phases.

without the intervention of the liquid phase, include the evaporation of solid carbon dioxide (Dry Ice) and the conversion of iodine crystals into gaseous iodine vapor.

The above discussion regarding the theory of melting points assumes that the experimental apparatus for melting-point determinations allows one to make and measure *extremely small changes in the heat supplied to a solid sample.* In reality, this is difficult and very time-consuming to accomplish. However, melting points of accuracy sufficient for most purposes can be determined in a short amount of time, as outlined in the Experimental Procedure at the end of this section.

The melting point of a *pure* compound is characteristic of the substance being studied. In practice, most melting-point apparatus is designed for ease of use and does not permit the determination of the *exact* temperature at which a compound melts, that is, the melting *point*. Instead, a **melting range** is actually measured and is commonly reported as a "melting point" for most compounds. If a solid substance is *pure*, it should melt over a very *narrow* range, which should be about 1°C if it is determined carefully. The melting range reported for many pure compounds may be greater than 1°C because a given compound was not quite pure or the melting point was not determined carefully.

The melting-point apparatus available in most organic chemistry laboratories is not sufficiently sensitive to allow reporting of a single melting-point temperature. Instead, the melting point should be expressed as the temperature range at which the solid *started* to melt and at which it was completely converted to liquid. Melting may actually be initiated when the solid starts to "soften" prior to melting. Since softening is difficult to observe, melting is defined as the temperature at which the first *liquid* is observed. Unless an exceedingly expensive and elaborate melting-point apparatus is used, it is improper and incorrect to report a single temperature for a melting point (for example, 118°C); a melting-point range should be recorded (for example, 118–119°C or 118–120°C).

The observation of a "sharp" melting-point range carries with it the important implication that the compound is *pure*. For example, the previous section describes a procedure for purifying a solid mixture by recrystallization with the hope that a pure solid compound will be obtained. One method for ascertaining if the recrystallized solid is pure involves the determination of its melting point. If the observed melting point is sharp (of a narrow range), it may tentatively be concluded that the recrystallization was successful and that the substance is pure. Part (c) below describes a situation in which a sharp melting-point range is observed, yet the solid is impure.

(c) Effect of Impurities on Melting Points; Mixtures

The melting point of a pure compound, which exhibits a sharp melting-point range, was discussed in part (b) above. Many solid substances are impure, however, and the effect of impurities on melting point are now considered.

Consider a mixture of a pure solid, *A*, in equilibrium with liquid *A* at its melting temperature. If a small amount of a second pure material, *B*, is dissolved in liquid *A*, solid *A* will begin to melt. This occurs because the addition of *B* lowers the vapor pressure of liquid *A* in accordance with Raoult's law (equation 2, Section 4.3), but the vapor pressure of solid *A* remains the same. Thus the rate at which molecules of solid *A* pass into the liquid phase is greater than the rate at which liquid *A* solidifies. If the temperature is kept constant by supplying the heat required for this melting process, all of solid A will melt, but if no heat is added, the temperature will drop because heat is consumed by melting the solid. Since the vapor pressure of a solid decreases more rapidly than the vapor pressure of its solution as the temperature drops, the vapor pressures will become equal at the lower temperature and equilibrium will be reestablished at that temperature. This lower temperature represents the melting (freezing) point of the mixture of A and B, which can be considered to be "impure A." As more of the impurity B is added, the melting point of the mixture will continue to decrease up to a point and then will start to increase again. *In general, the addition of an impurity to a pure solid results in a decrease in its melting point.*

Binary mixtures of most organic substances exhibit this kind of behavior, and the addition of a second substance progressively lowers the melting point of the first, as represented by the melting point–composition diagram in Figure 3.3. In this diagram *a* represents the melting point of pure A and *b* the melting point of pure B. Point F represents the temperature (*f*) at which crystals of pure A are in equilibrium with a molten solution consisting of 80 mole % A and 20 mole % B. If a mixture of this composition is prepared and melted and if the external heat is lowered to temperature *f*, pure crystals of A will form in the melt. As additional heat is removed, A will continue to crystallize, and the percentage of A in the melt will be reduced. The equilibrium temperature will correspondingly decrease because of the reduced vapor pressure of A in the molten solution. When the equilibrium temperature reaches *e*, at which temperature the melt has the composition indicated at E, then and *only then* does B, the impurity, also begin to crystallize. A and B will now crystallize in the

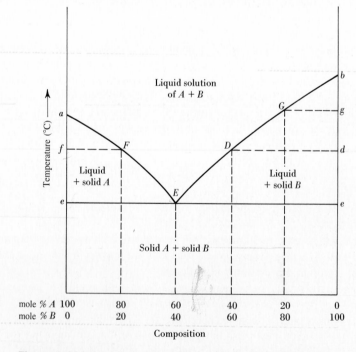

Figure 3.3 Melting point–composition diagram for two hypothetical solids, A and B.

constant ratio of 60 mole % A and 40 mole % B, and the melt composition will change no further. During this stage the melt of the specified composition is in equilibrium with both solid A and solid B. The temperature remains constant until the entire sample has crystallized, and if no heat is supplied the temperature falls as the solid cools. Point E is called the **eutectic point,** and it defines the **eutectic mixture,** which is the composition at which A and B can cocrystallize. The temperature at which the eutectic mixture freezes or melts is designated by e.

The reverse process, the melting of a solid mixture containing 80 mole % A and 20 mole % B, is of more interest to the organic chemist. Heat is applied to the sample, and the temperature rises. When it reaches e, A and B will melt together at a constant ratio corresponding to the eutectic composition because the eutectic melt is in equilibrium with both solid A and solid B. Eventually, B will melt entirely because it is the "impurity" present in minor amount and will leave only solid A in equilibrium with the melt of eutectic composition. As more heat is applied, the remaining solid A will continue to melt, and the percentage of A in the molten solution above the eutectic composition will increase. Since the vapor pressure of A in the solution is thus increased, the temperature at which solid A is in equilibrium with the molten solution will rise. The relationship between the equilibrium temperature and the composition of the molten solution is represented by curve EF. All of solid A has melted when the temperature reaches f. Thus, this impure sample A exhibits a depressed melting "point" that occurs over the relatively broad temperature range e–f.

Thus far we have considered the case in which substance A contains B as an impurity. If the composition of the solid mixture had been to the right of point E in Figure 3.3, we would conversely speak of substance B containing A as the impurity. The rising temperature during the melting process would follow curve ED or EG, and the melting range would be e–d or e–g.

In practice, it is very difficult to observe the initial melting point, particularly with the capillary tube melting-point technique presented in the Experimental Procedure. This is because only a small amount of liquid is produced in the early stages of melting near the eutectic temperature if only a small amount of impurity is present. However, the temperature at which the last crystals melt (points d and g) can be determined quite accurately, so that the mixture containing the smaller amount of impurity will have both a higher final melting point and a narrower *observed* melting-point range.

A sample whose composition is exactly that of the eutectic mixture will exhibit a *sharp* melting point at the eutectic temperature. Thus, a eutectic mixture can be mistaken for a pure compound, since both have a sharp melting point. It is easy to determine whether or not a substance is a pure compound or a eutectic mixture of two solids. Assuming that one knows the identity of at least one of the components of a mixture, a small amount of this known component is added to the original sample, and the melting point is redetermined. If the original sample was pure, the melting points will be identical, but if the original sample was a mixture, the melting "point" of the original sample will increase.

In general, the melting-point range of a pure compound or a eutectic mixture is *sharp* and that of a non-eutectic mixture is *broad*. The experimental observation of narrow melting ranges almost always indicates the substance is pure, since the likelihood of having a eutectic mixture is not great; for the same reasons, it is most likely that a mixture will have a broad melting-point range.

As we have discussed, the introduction of an impurity into a pure compound lowers the melting point and produces a wide melting-point range. This may be used to advantage in the identification of a pure substance by a technique known as a **mixture melting-point** determination, often ungrammatically referred to as a ''mixed melting point'' determination. This technique is perhaps best illustrated by the following example. Assume that a compound melts at 133°C, and you suspect it is either urea, H_2NCONH_2, or *trans*-cinnamic acid, $C_6H_5CH{=}CHCO_2H$, both of which melt at 133°C. If the compound is mixed intimately with urea and the melting point of this mixture is found to be lower than that of the pure compound and pure urea, then urea is acting as an impurity, and the compound *cannot* be urea. If the mixture melting-point is identical to that of the pure compound and of urea, the compound is *tentatively* identified as urea. The possibility of having a eutectic mixture can be excluded by mixing urea in at a different proportion and redetermining the melting point. This procedure is occasionally useful in identifying compounds only when samples of the likely possibilities are available, and this severely limits its applicability.

Melting point–composition diagrams, such as the one shown in Figure 3.3, are different for various binary mixtures and must be determined experimentally for each pair of compounds. This technique involves preparing many samples of different mole percent composition and determining the melting point ranges for each sample. The data are plotted, and the best curve is drawn to yield the diagram.

A specific example of the melting point–composition diagram for the binary mixture of naphthalene and camphor is shown in Figure 3.4, from which the following information can be gleaned. The melting points of pure camphor and pure naphthalene are 179°C and 80.1°C, respectively. The eutectic mixture is composed of 58 mole % camphor and 42 mole % naphthalene and melts at 32.3°C. A mixture of any composition of the two will be liquid if heated above 179°C and solid at temperatures less than 32.3°C. Now suppose we have a sample containing a mixture of these two compounds, determine its melting-point range to be 50–55°C, and desire to know the composition of this mixture. The composition cannot be determined from this melting-point range alone because there are two different mole percent compositions, labeled *A* and *B* on Figure 3.4, that would melt at 50–55°C. This question can be answered, however, by taking the mixture, adding a small amount of pure naphthalene or camphor to it, mixing the sample intimately, and determining the melting-point range of the resulting sample. If we added some camphor and found that the melting-point range increased, the composition of the original mixture would correspond to *A* and would be 63% camphor: 37% naphthalene. This conclusion can be reached by examining Figure 3.4 and observing that increasing the mole percent of camphor would result in a higher melting-point range. On the other hand, if the melting-point range decreased on adding some camphor, the composition of the original sample would correspond to *B* and would be 38% camphor: 62% naphthalene, because Figure 3.4 shows that increasing the mole percent of camphor starting at *B* would result in a lower melting-point range. Similar reasoning would apply if a little naphthalene were added to the original mixture; the observation of a higher melting-point range would indicate the original mixture had composition *B*, and a lower range would indicate composition *A*. Note that the melting points of pure naphthalene and pure camphor and of the eutectic mixture are *not* expressed as a melting-point range, since they were determined starting with ultra-pure compounds and with precision generally not possible with the melting-point apparatus likely to be available in most organic chemistry laboratories.

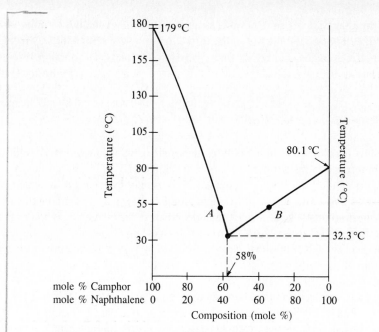

Figure 3.4 Melting point–composition diagram for a binary mixture of camphor and naphthalene.

(d) Other Kinds of Melting Behavior of Mixtures

Although the melting-point behavior just described is typical of many binary mixtures, there are exceptions to this general pattern. A second component does not always lower the melting temperature of an organic compound. Figure 3.5 illustrates some unusual melting point–composition diagrams. In Figure 3.5a the eutectic composition contains so little of compo-

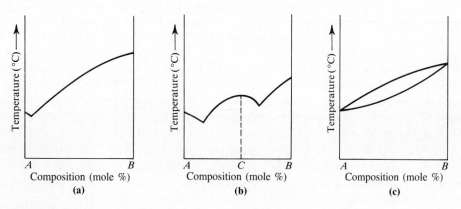

Figure 3.5 Unusual melting point–composition diagrams.

nent *B* that even a fairly small amount of *B* as an impurity may actually raise the observed melting point of *A*. Figure 3.5b illustrates the formation of a new compound (*C*) from *A* and *B*, and Figure 3.5c is representative of one type of solid solution formation. For further information, consult the reference at the end of this section or a physical chemistry textbook.

(e) Micro Melting-Point Methods

The determination of an exact melting point of an organic compound requires enough sample so that equilibrium can be established between the liquid and solid phase and so that the temperature at the equilibrium state can be measured, usually by means of repeated cooling and heating cycles that show a plateau in a temperature-time plot. The amount of material required for such procedures is often more than the chemist has available, so micro methods have been developed that are not so exact but that are convenient and require negligible amounts of sample. There is less emphasis today on extremely precise melting-point determinations as a result of the fairly recent development of spectroscopic methods (Chapters 8–11) for the characterization of organic compounds.

The most commonly used capillary-tube melting-point procedure is described below. The *microscopic hot stage* and simpler variations of it are also popular, an example of which is the Kofler hot-stage technique described on page 155 of the reference at the end of this section. Although the microscopic hot-stage apparatus is somewhat more expensive, it permits the satisfactory use of still smaller quantities of crystals. Chemists faced with performing numerous melting-point determinations are likely to use an ''automatic'' microprocessor-controlled apparatus that displays the melting points on a digital readout.

The melting point is actually determined as a melting *range* in all micro melting-point methods. When properly carried out with a *pure* solid, a range of no more than 0.5–1.0°C should be observed, but very few commercially available compounds exhibit such a narrow melting range.

The observed melting-point range depends on several factors: sample size, state of subdivision of the sample, heating rate, and purity and identity of the sample. The first three factors can cause the observed melting-point range to differ from the actual melting point because of the time lag in the transfer of heat from the heating fluid to the sample and because of heat conduction within the sample. If heating is too fast, the thermometer reading will lag behind the actual temperature of the heating fluid and yield measurements that are *low*.

Some compounds decompose on melting, and when this occurs discoloration of the sample is usually evident. The decomposition products constitute impurities in the sample, and the melting point is actually lowered as a result of decomposition. When this behavior is observed, the melting point should be reported so as to indicate that decomposition occurred, for example, 195°C *d* or 195°C (dec).

The accuracy of any type of temperature measurement depends ultimately on the thermometer. In some temperature ranges a given thermometer may provide accurate readings, and in others the reading may be a degree or two off, either high or low. The following Experimental Procedure describes thermometer calibration.

Pre-lab exercises for Section 3.2, Melting Points, are found on page PL. 3.

EXPERIMENTAL PROCEDURE

DO IT SAFELY

1 If you use a burner in this experiment, make sure that no flammable solvents are nearby. Keep the rubber tubing leading to the burner away from the flame. Turn off the burner when it is not actually in use.

2 Some kinds of melting-point apparatus, such as the Thiele tube, use mineral or silicone oils as the heat transfer medium. These oils may **not** be heated safely if they are contaminated with even a few drops of water. Heating these oils above 100°C, the boiling point of water, may produce splattering of hot oil as a result of steam formation. Fire can also result if splattered oil comes in contact with open flames. Examine your Thiele tube for evidence of water droplets in the oil. If any are observed, either change the oil or exchange tubes. Give the contaminated tube to your instructor.

3 Mineral oil is a mixture of high-boiling hydrocarbons and should not be heated above about 200°C because of the possibility of spontaneous ignition, particularly when a burner is used for heating. Silicone oils may be heated to about 300°C without danger.

4 Be careful to avoid contact of chemicals with your skin. Clean up any spilled chemicals immediately with a brush or paper towel.

5 If you use a Thiele tube, be very careful when handling it when you are through. These tubes cool slowly, and care should be exercised when removing them from their support in order to avoid being burned.

A. Calibration of Thermometer

Your instructor should advise you if you are required to calibrate your thermometer. Calibration of a thermometer involves the measurement of temperature at a series of known points within the range of the thermometer and the comparison of these actual readings with expected temperatures. The difference between the actual and the expected readings provides a correction that must be applied to the actual thermometer reading in order to correct for thermometer error. These corrections can be plotted in your notebook as deviations from zero versus the temperature over the range of the thermometer. Thus you can tell at a glance, for example, that at about 130°C your thermometer gives readings that are 2°C too low, or that at 190°C the readings are about 1.5°C too high. These corrections are valid only for the thermometer used in the calibration; if the thermometer is broken, a new one must be calibrated. These corrections should then be applied to all temperature measurements taken during the course.

To calibrate a thermometer, *carefully* determine the melting points of a series of standard substances. A list of suitable standards is provided in Table 3.2. Your instructor may suggest other or additional standards to be used. The temperatures given in Table 3.2 correspond to the upper limit of the melting range for pure samples of these standards.

TABLE 3.2 **Standards for Thermometer Calibration**

Compound	Melting Point (°C)
Ice water	0
3-Phenylpropanoic acid	48.6
Acetamide	82.3
Acetanilide	114
Benzamide	133
Salicylic acid	159
4-Chloroacetanilide	179
3,5-Dinitrobenzoic acid	205

B. Determination of Capillary-Tube Melting Points

1. General Procedure The determination of a capillary-tube melting point involves the following steps, the techniques of which are presented in Section 2.3. This section should be read carefully before undertaking experimental work.

> Sample preparation: Section 2.3a
> Melting-point determination: Section 2.3b
> Thiele tube apparatus and use: Section 2.3c
> Electric melting-point apparatus and use: Section 2.3d

It is recommended that you also reread the discussion in Section 3.2 concerning melting points and their determination.

2. Mixture Melting Points In the preparation of a sample for mixture melting-point determination, the two components *must* be thoroughly and intimately mixed. This is best accomplished by grinding them together with a small mortar and pestle. If these are not available, a small, clean watch glass and glass stirring rod may be used. Be careful, however, not to apply too much pressure to the glass rod, because it is more fragile than a pestle and may break. Do not do this on filter paper because particles of paper may contaminate the sample.

3. Experiments You may be assigned one or more of the following melting-point experiments, which should be performed using the General Procedure given above. Your instructor should advise you about the type of apparatus to use.

A great deal of time can be spent waiting for an unknown compound to melt when using slow heating. It is usually convenient to prepare two capillary tubes containing the compound being studied and to determine the approximate melting point by rapidly heating one of the tubes. The heating fluid is then allowed to cool to 10–15°C below this approximate melting point, the second sample is inserted, and gentle heating is resumed to obtain an accurate melting point. In the event that both Thiele tubes and electric melting-point apparatus are available, it is convenient to determine the approximate melting point using a Thiele tube and the accurate melting-point range with the electric equipment, which may be available in limited quantity.

At the completion of this experiment and at the direction of your instructor, return any unconsumed samples or dispose of them in an appropriately labeled container for solid organic waste. Do not discard these solids in a waste can or wash them down the drain. Properly dispose of used capillary tubes in a designated container; do not leave them in the area of the melting-point apparatus and do not throw them on the floor or in sinks.

(a) Known Compounds Select one or two compounds from a list of supplied compounds of *known* melting point. Determine the melting-point ranges for each of these substances, using the capillary melting-point procedure. Repeat as necessary until you obtain accurate results and are confident with the procedure. Since the melting points of these compounds are known, it is not necessary to first determine their approximate melting points.

(b) Mixture Melting Points Using one of the compounds whose melting range was determined in (a), introduce 5–10% of a second substance as an impurity. Thoroughly mix the two components of the mixture and determine its melting range in order to learn the effect of impurities on the melting range of a previously pure compound.

(c) Unknown Compound Accurately determine the melting range of an unknown pure compound supplied by your instructor. It is advisable to determine the approximate melting point by rapid heating before obtaining an accurate melting range.

(d) Unknown Compound Obtained from Recrystallization In the event that you were given an impure mixture to purify by recrystallization (Section 3.1), accurately determine the melting range of the recrystallized product. It is advisable to determine the approximate melting range by rapid heating before obtaining an accurate melting range.

If a broad melting range was observed, it may be caused by one of the following problems: The compound may not be free of solvent and should be dried for a longer time. On the other hand, the recrystallization may not have been successful, and the solid should be recrystallized again using the same solvent, or the original mixture should be recrystallized using a different solvent.

EXERCISES

1. Describe errors in procedure that may cause an observed capillary melting-point of a pure compound
   a. to be *lower* than the correct melting point
   b. to be *higher* than the correct melting point
   c. to be *broad* in range (over several degrees)

2. Briefly define the following terms:
   a. vapor pressure as applied to melting
   b. melting-point range
   c. mixture melting point
   d. eutectic point
   e. eutectic mixture
   f. melting point

3. Filter paper is usually a poor material on which to powder a solid sample before introducing it into a capillary melting point tube because small particles of paper end up in the tube along with the sample. Why is this undesirable, and how might the presence of paper in the sample make the melting-point determination difficult?

4. Criticize the following statements by indicating whether each is true or false, and if false, by explaining why.
   a. An impurity always lowers the melting point of an organic compound.
   b. A sharp melting point for a crystalline organic substance always indicates a pure single compound.
   c. If the addition of a sample of compound A to compound X does not lower the melting point of X, X must be identical to A.
   d. If the addition of a sample of compound A lowers the melting point of compound X, X and A cannot be identical.

5. The melting points of pure benzoic acid and pure 2-naphthol are 122.5°C and 123°C, respectively. Given a pure sample that is known to be either pure benzoic acid or 2-naphthol, describe a procedure you might use to determine the identity of the sample.

6. A student used the Thiele micro melting-point technique to determine the melting point of an unknown and reported it to be 182°C. Is this value believable? Explain why or why not.

7. The melting point–composition diagram for two substances, Q and R, is provided in Figure 3.6, which should be used to answer the following questions.
   a. What are the melting points of pure Q and R?
   b. What are the melting point and the composition of the eutectic mixture?
   c. Would a mixture of 20 mole % Q and 80 mole % R melt if heated to 120°C? to 160°C? to 75°C?

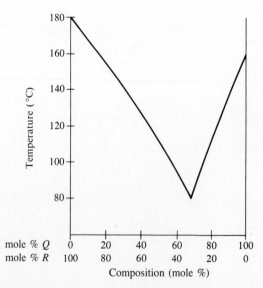

Figure 3.6 Melting point–composition diagram for Exercise 7.

    d. A mixture of Q and R was observed to melt at 105–110°C. What can be said about the composition of this mixture? Explain briefly.

8. For the following melting points, indicate what might be concluded regarding the purity of the sample:

    a. 120–122°C

    b. 46–60°C

    c. 147°C (dec)

    d. 162.5–163.5°C

REFERENCE

Skau, E. L.; Arthur, J. C., Jr., in *Technique of Chemistry*, A. Weissberger and B. W. Rossiter, editors, Wiley-Interscience, New York, 1971, Vol. 1, Part 5, Chapter 3, pp. 105ff.

liquids
distillation and boiling points

The purification of solids by recrystallization and the use of melting points as a criterion of their purity were discussed in Chapter 3. The techniques used for the purification of liquids, namely, *simple, fractional, steam, and vacuum distillation*, are topics of this chapter. Boiling points are also discussed as a physical property that can be used as one means of determining the purity of liquids.

4.1 Boiling Points of Pure Liquids

(a) Discussion

The molecules of a liquid are in constant motion. When some of these molecules are located at the surface, they are able to escape into space above the liquid. Consider a liquid contained in a closed, *evacuated* system. The number of the molecules in the gas phase can increase until the rate at which they reenter the liquid becomes equal to the rate of their escape; the rate of reentry is proportional to the number of molecules in the gas phase. At this point, no further *net* change is observed in the system, which is said to be in a state of dynamic equilibrium.

The molecules in the gas phase are in rapid motion and are continually colliding with the walls of the vessel, which results in the exertion of pressure against the walls. The magnitude of this vapor pressure at a given temperature is called the **equilibrium vapor pressure** of the particular liquid substance at that temperature. Vapor pressure is temperature-dependent, as shown in Figure 4.1, and can be understood in terms of the escaping tendency of the mole-

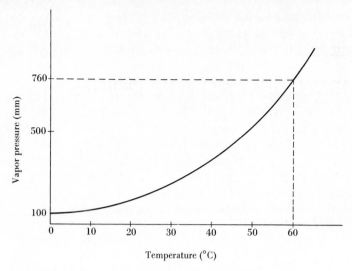

Figure 4.1 Graph showing the dependence of vapor pressure on temperature for a liquid.

cules from the liquid. As the temperature of the system increases, the average kinetic energy of the liquid molecules increases, thus facilitating their escape into the gas phase. The rate of reentry of gaseous molecules into the liquid state also increases, and equilibrium is soon established at the higher temperature. However, there are more molecules in the gas phase now than there were at the lower temperature, so the vapor pressure is greater.

Now suppose that a liquid sample at a given temperature is placed in an *open* container so that the molecules of the vapor above the liquid may escape into the atmosphere. The vapor above such a sample is composed of molecules of air as well as of the sample. The *total* (external) pressure above the liquid is defined by *Dalton's law of partial pressures* (equation 1) and is equal to the sum of the partial pressures of the sample and of the air.

$$P_{\text{total}} = P_{\text{sample}} + P_{\text{air}} \tag{1}$$

The partial pressure of the sample is equal to its *equilibrium vapor pressure* at the given temperature. If the temperature of the liquid is raised, the equilibrium vapor pressure of the sample increases, and the number of gas molecules from the liquid will immediately increase in the space above the liquid and will have the net effect of displacing some of the air. At the higher temperature the partial pressure of the sample will be a larger *percentage* of the *total* pressure. This trend will continue as the temperature of the liquid is further increased until the equilibrium vapor pressure becomes equal to the external pressure, at which point all the air will have been displaced from the vessel. This process is termed **evaporation,** and further evaporation will have the effect of gaseous molecules of the sample displacing others.

Consideration of these facts leads to the conclusion that the equilibrium vapor pressure of the sample has an upper limit that is dictated by the external pressure. At this temperature the rate of evaporation increases dramatically, and bubbles can be seen to form in the liquid; this process is commonly known as **boiling.** The temperature at which this occurs is called the **boiling point** of the liquid, and it is defined as *the temperature at which the vapor pressure* *of a liquid just equals the external pressure*. Since the boiling point is dependent upon the external pressure, the pressure should be specified when boiling points are reported, for example, ''bp 152°C (752 torr).'' The bp of water is 100°C *only* when the external pressure is 760 torr, or 1 atm. The **normal boiling point** of a liquid is measured at 760 torr (1 atm) and is shown in Figure 4.1 by the dashed lines to be 60°C. This figure also allows one to determine what the boiling point of the liquid will be at various external pressures. For example, at 500 torr the boiling point will be 50°C.

The dependence of vapor pressure upon external pressure, which is reflected by the boiling point of a liquid, can be used to advantage in the following way. Suppose that a certain liquid has such a high standard boiling point—for example, 250°C—that it decomposes appreciably when heated to its boiling point. Reduction in the external pressure reduces its boiling point to a lower temperature at which it no longer decomposes. This technique, called *vacuum distillation,* is discussed in Section 2.9.

The effect of external pressure upon boiling point has several familiar consequences. For example, it is well known that the ''3-minute egg'' takes longer to cook at high altitudes, which can be attributed to the fact that the external (atmospheric) pressure decreases as elevations above sea-level are reached, and this lowers the boiling point of water. Hence, the maximum cooking temperature is lower than 100°C so that the required cooking time is increased. On the other hand, foods cook faster in a pressure cooker because the internal pressure is greater than 1 atm, which causes water to boil at a *higher* temperature. The vapor inside the cooker (steam) is in equilibrium with the boiling water and at a temperature higher than it would be at atmospheric pressure.

The discussions so far have considered the boiling points of pure liquids and the effect of pressure upon them. However, it is well known that the addition of certain *nonvolatile* compounds to water increases its boiling point. Antifreeze contains ethylene glycol, $HOCH_2CH_2OH$, which is essentially nonvolatile compared with water. It increases the boiling point and decreases the freezing point of water and thus prevents the water in automobile radiators from boiling in the summer or freezing in the winter. Boiling is prevented in two ways. First, the water in an automobile cooling system is contained in a *closed* system, which causes the pressure inside to increase as the temperature rises and prevents the water from evaporating; as discussed above, this causes the boiling point of the water to increase. Second, the added ethylene glycol causes the boiling point of water to *increase*. This effect can be understood by recalling that boiling is caused by the ability of molecules to escape from the liquid. When ethylene glycol is added to water, the surface of the *liquid* is now composed of both water molecules and ethylene glycol molecules. That is, there are *fewer* water molecules at the surface of the liquid containing the additive than there are in pure water. In order for the water to boil, its vapor pressure must equal the external pressure so that more heat must be supplied to the liquid mixture to cause enough water molecules to *escape*. Both the increased pressure that results from the closed system in a radiator and the addition of antifreeze causes water to boil at a higher temperature. This allows the automobile motor to operate more efficiently at higher temperatures.

The effects of volatile and nonvolatile impurities upon boiling points are discussed in Section 4.3.

Boiling points are useful for identification of pure liquids and some low-melting solids. In general, a *pure* liquid will boil at a constant temperature, provided the pressure remains constant. On the other hand, most mixtures of liquids boil over a fairly wide temperature range at constant pressure; this effect is discussed in Section 4.4.

(b) Determination of Micro Boiling Points

Pre-lab exercises for Section 4.1, Boiling Points of Pure Liquids, can be found on page PL. 5.

EXPERIMENTAL PROCEDURE

DO IT SAFELY

1 Burners are used in this experiment. Exercise care to avoid using them near volatile organic liquids. Read Section 2.5a.

2 Check the heating fluid in the Thiele tube apparatus for the presence of moisture; see Section 2.5d.

3 Spilled liquids should carefully be absorbed onto a paper towel, which should then be discarded as directed by your instructor. Avoid contact of organic liquids with your skin, and if this happens, wash the affected area thoroughly with soap and water.

Determine the boiling point of the liquid(s) assigned by your instructor. Follow the technique presented in Section 2.4 for using the micro boiling-point apparatus. A Bunsen burner or a microburner is used for heating; Section 2.5a should be consulted regarding the proper use of burners. Before using the Thiele tube apparatus, read Section 2.5d concerning the heating fluid it contains.

In the event you are assigned an unknown liquid whose boiling point is to be determined, it is suggested that you *first* determine an approximate boiling-point range by heating the Thiele tube *fairly rapidly*. Repeat the determination by heating the tube until the temperature is 20–30°C *below* the approximate boiling point, and then heat the sample *slowly* to obtain an accurate value. It may be desirable to repeat this procedure in order to obtain a reliable boiling point.

EXERCISES

1. Refer to Figure 4.1 and answer the following:
   a. What external pressure would be required in order for the liquid to boil at 30°C?
   b. At what temperature would the liquid boil when the external pressure is 100 torr?

2. Describe the relationship between escaping tendency of liquid molecules and vapor pressure.

3. Define the following terms:
   a. boiling point
   b. normal boiling point
   c. Dalton's law of partial pressure
   d. equilibrium vapor pressure

4. Why should no droplets of water be present in the mineral oil of a heating bath (Thiele tube)?

5. Why is the micro boiling-point technique not applicable for boiling points in excess of 200°C if mineral oil is the liquid in the Thiele tube?

4.2 **Purification of Liquids by Distillation: A Survey**

The most commonly used method of purifying *liquids* is distillation, a process that consists of vaporizing the liquid by heating and condensing the vapor into a separate vessel to yield a **distillate.** After the distillation is complete, any liquid remaining in the original distillation flask is called the **pot residue.** The common types of distillation are *simple distillation, fractional distillation, vacuum distillation,* and *steam distillation.* An overview of each is provided here, followed by a more detailed discussion in later sections of this chapter.

(a) Simple Distillation

This technique can be used to remove a pure liquid from nonvolatile impurities that will be left behind in the pot residue at the completion of the distillation. For example, a homogeneous mixture of potassium iodide, KI, and acetone, $(CH_3)_2C{=}O$, can be subjected to simple distillation. The acetone distils, leaving the potassium iodide in the distillation flask.

Simple distillation can sometimes be used to separate a mixture of liquids, provided the *difference* between the boiling points of each pure substance is greater than 20–30°C. For example, a mixture of diethyl ether, bp 35°C, and toluene, bp 111°C, could be separated by simple distillation, with the ether distilling first. This technique is often used to remove volatile organic solvents that may have been used for purification and extraction as part of an experimental procedure.

Simple distillation is discussed further in Section 4.3.

(b) Fractional Distillation

If the *difference* in the boiling points of the pure liquids that may be present in a mixture is less than 20–30°C, fractional distillation is normally used to obtain each liquid in pure form. For example, a mixture of methyl acetate, bp 57°C, and ethyl acetate, bp 77°C, is difficult to separate by simple distillation but easy to do by fractional distillation. Mixtures of more than two liquids can also be separated by fractional distillation, although this may be experimentally more difficult.

Fractional distillation is discussed in more detail in Section 4.3.

(c) Vacuum Distillation

It is most convenient to distil liquids under conditions such that the external pressure is the atmospheric pressure. However, as noted earlier, some substances decompose, oxidize, or undergo other undesirable reactions at their normal boiling point. When one or more of these situations arise, it is necessary to carry out the distillation at pressures less than atmospheric so that the boiling temperature is reduced. This is the technique of vacuum distillation, which is discussed in Section 2.9. Both simple and fractional distillation can be carried out under reduced pressure.

(d) Steam Distillation

The separation and purification of *volatile* organic compounds that are immiscible or nearly immiscible with water can often be accomplished by steam distillation. This technique involves the codistillation of a mixture of water and organic liquids; some solids can also be separated and purified by this means. Of the various distillation methods, steam distillation is utilized less frequently, owing to the rather stringent limitations on the types of substances for which it can be used. The theory of this topic is presented in Section 4.5.

4.3 Simple Distillation

As indicated earlier, simple distillation allows separation of a pure liquid from substances that are nonvolatile or nearly so. It can also be utilized to separate a mixture of volatile liquids whose boiling points differ by more than 20–30°C.

The effect of the addition of ethylene glycol on the boiling point of water was discussed in Section 4.1. Now consider another situation in which a nonvolatile impurity, such as sugar, is added to a pure liquid, such as water, which dissolves the sugar and produces a homogeneous solution. The impurity reduces the vapor pressure of the liquid because the nonvolatile component lowers the concentration of the volatile constituent. The consequence of this is shown graphically in Figure 4.2. In this figure, curve 1 corresponds to the dependence of the temperature upon the vapor pressure of a *pure* liquid and intersects the 760-torr line at 60°C. Curve 2 is for a *solution* of the same liquid and a nonvolatile impurity. Note that the vapor pressure at any temperature is reduced by a constant amount by the presence of the impurity (Raoult's law, discussed later in this section). The temperature at which this curve intersects the 760-torr line is higher because of the lower vapor pressure, and the temperature of the boiling solution is higher, 65°C.

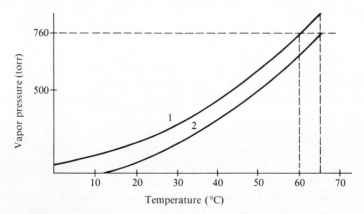

Figure 4.2 Diagram showing dependence of vapor pressure on temperature. Curve 1 represents a pure liquid and curve 2 a pure liquid to which a nonvolatile impurity has been added.

It is recommended that you examine Figure 2.14 while reading the following discussion, which integrates the theory and technique of simple distillation. The experimental procedures and apparatus for simple distillation are provided in Section 2.7.

A thermometer determines the boiling temperature of the distillate, which is the material collected in the receiver. In the case of a pure liquid, the temperature shown on the thermometer—the **head temperature**—will be identical to the temperature of the liquid boiling in the distilling vessel—the **pot temperature**—if the liquid is not superheated. The head temperature thus corresponds to the boiling point of the liquid and will remain constant throughout the distillation process. A simple distillation is an alternative to the micro boiling-point method (Section 4.1b) for determining the boiling point of a pure liquid. Unlike the micro boiling-point method, which requires a very small amount of liquid, simple distillation requires a much larger sample of liquid. The boiling point is also observed whenever the final step in the synthesis of a liquid compound requires a simple distillation.

When a nonvolatile impurity is present in a liquid being distilled, the head temperature will be the same as for the pure liquid, since the material condensing on the thermometer bulb is uncontaminated by the impurity. However, as shown in Figure 4.2, the pot temperature will be *elevated*, owing to the decreased vapor pressure of the solution. The pot temperature will progressively rise during the course of the distillation because the concentration of the impurity in the pot will increase steadily as the volatile component is distilled away, further lowering the vapor pressure of the liquid. However, the head temperature will remain constant, as in the case of the distillation of a pure liquid.

The quantitative relationship between vapor pressure and composition of homogeneous liquid mixtures is known as *Raoult's law* and may be expressed as shown in equation 2.

$$P_R = P_R^\circ N_R \tag{2}$$

P_R represents the partial pressure of component R, which, at a given temperature, is equal to the vapor pressure of pure R at that temperature (P_R°) times the mole fraction of R (N_R) in the mixture. The mole fraction of R is defined as the fraction of *all* molecules present in the liquid mixture that are molecules of R. It is obtained by dividing the number of moles of R in a mixture by the sum of the number of moles of all components (equation 3).

$$N_R = \frac{n_R}{n_R + n_S + n_T + \cdots} \tag{3}$$

Raoult's law is applicable only for *ideal solutions,* which are defined as those in which the interactions between *like* molecules are the same as those between *unlike* molecules. Fortunately, many organic solutions approximate the behavior of ideal solutions.

Note that the partial vapor pressure of R above an ideal solution depends *only* on its *mole fraction* in solution and is in no way dependent upon the vapor pressures of the other volatile components in the solution. If all components other than R are nonvolatile, the total vapor pressure of the mixture will be equal to the partial pressure of R, since the vapor pressure of nonvolatile compounds may be taken as zero. Thus the distillate from such a mixture will always be pure R. However, if two or more of the components are volatile, the *total* vapor pressure will equal the sum of the partial vapor pressures of each volatile component. This is known as *Dalton's law* and is expressed by equation 4, where R, S, and T refer to the vapor pressures of the volatile components only.

$$P_{\text{total}} = P_R + P_S + P_T + \cdots \tag{4}$$

The process of distilling such a liquid mixture is significantly different from that of simple distillation, because here the distillate may contain each of the volatile components. Separation in this case will usually require fractional distillation, which is discussed in the next section.

Experiments involving simple distillation are presented at the end of Section 4.4.

4.4 Fractional Distillation of Ideal Solutions

(a) Theory and Discussion

For simplicity we shall consider here only binary, ideal solutions, which contain two volatile components, R and S. Solutions containing more than two volatile components are often encountered, but their behavior on distillation may be understood by extending the following principles for binary systems. Some examples of liquid mixtures that deviate significantly from ideality are discussed in Section 4.5.

Because vapor pressure is a measure of the ease with which molecules escape the surface of a liquid, the number of molecules of component R in a given volume of the vapor above a liquid mixture of R and S is proportional to the partial vapor pressure of component R. The same is true for component S, as may be seen from equation 5, where N'_R/N'_S is the ratio of the mole fractions of R and S in the *vapor* phase.

$$\frac{N'_R}{N'_S} = \frac{P_R}{P_S} = \frac{P^{\circ}_R N_R}{P^{\circ}_S N_S} \tag{5}$$

The mole fraction of each component may be calculated from the equations $N'_R = P_R/(P_R + P_S)$ and $N'_S = P_S/(P_R + P_S)$. The partial vapor pressures, P_R and P_S, are determined by the composition of the liquid solution (Raoult's law; equation 2). Since the solution will boil when the sum of the partial vapor pressures of R and S is equal to the external pressure, the boiling temperature of the solution is determined by its composition.

The relationship between temperature and the composition of the liquid and vapor phases of ideal binary solutions may be illustrated by the diagram in Figure 4.3 for mixtures of benzene (bp 80°C) and toluene (bp 111°C). The lower curve is called the **liquid line** and gives the boiling points of all mixtures of these two compounds. The upper curve, which is called the **vapor line,** is calculated using Raoult's law and gives the composition of the vapor phase in equilibrium with the boiling liquid phase at the same temperature. For example, a mixture whose composition is 58 mole % benzene and 42 mole % toluene will boil at 90°C, as shown by point A in Figure 4.3. The composition of the vapor in equilibrium with the solution when it *first* starts to boil can be determined by drawing a horizontal line from the *liquid line* to the *vapor line;* in this case, the vapor has the composition 78 mole % benzene

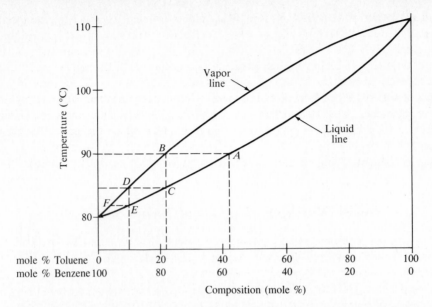

Figure 4.3 Boiling point–composition diagram for a binary mixture of benzene and toluene.

and 22 mole % toluene, as shown by point *B* in Figure 4.3. At any given temperature, the *vapor phase is much richer in the more volatile component than the boiling liquid with which it is in equilibrium.* In general, the boiling of a binary mixture results in the first vapor produced initially being richer in the more volatile component, which is the component having the lower boiling point. This behavior provides the basis of fractional distillation, which is discussed in the following two paragraphs.

When the liquid mixture containing 58 mole % benzene and 42 mole % toluene is heated to its boiling point (90°C), the first vapor formed contains 78 mole % benzene and 22 mole % toluene. If this first vapor is condensed, the liquid would have this composition and would be much richer in benzene than the original liquid mixture from which it was distilled. After this first vapor is removed from the original mixture, the liquid remaining in the distilling flask will contain a smaller mole % of benzene and a greater mole % of toluene because more benzene was removed by vaporization than toluene. The boiling point of the liquid remaining in the distilling flask will rise. If distillation is continued, the boiling point of the mixture will continue to rise until it approaches or reaches the boiling point of pure toluene. As the amount of heat supplied to the distilling flask is increased, the composition of the distillate will change and will consist of pure toluene at the end of the distillation.

Now let us return our attention to the first few drops of distillate which could be obtained by condensing the first vapor that formed from the original mixture. If this condensed liquid, which has the same composition as the vapor that was initially formed from the original mixture, is collected separately and then redistilled, its boiling point will be the temperature at point *C*, 85°C. Since the compositions of the vapor and of the liquid that results from condensing it are equal, the boiling point of the condensate is determined by drawing a vertical line from the vapor line at point *B* to the liquid line at point *C* and noting the temperature that corresponds to *C* (85°C). When this liquid is redistilled, the first distillate would have the composition D, 90 mole % benzene and 10 mole % toluene; this composition is obtained by drawing a horizontal line from point C on the liquid line to the vapor line. In theory, this process could be repeated again and again to give a *very small amount* of pure benzene. Similarly, collecting the *last* small fraction of each distillation and redistilling it in the same stepwise manner would yield a very small amount of pure toluene. If larger amounts of the initial and final distillates were collected, reasonable amounts of materials could be separated, but a large number of individual simple distillations would be required. This process would be extremely tedious and time-consuming. Fortunately, the repeated distillation can be accomplished almost automatically in a single step by using a **fractional distillation column,** the theory and use of which are described next.

(b) Fractional Distillation Columns and Their Operation

The experimental procedure and apparatus for fractional distillation are provided in Section 2.8, and it is recommended that you refer to Figures 2.15 and 2.16 as you read the following.

There are many types of fractional distillation columns, but all can be discussed in terms of a few fundamental characteristics. The column provides a vertical path through which the vapor must pass from the distilling flask to the condenser before being collected in the receiver (Figure 2.15). This path is significantly longer than in a simple distillation apparatus. As the vapor from the distilling flask passes up through the column, some of it condenses *in the column* and falls back into the distilling flask. *If the lower part of the distilling column is maintained at a higher temperature than the upper part of the column,* the condensate will be partially revaporized as it drains down the column. The uncondensed vapor, together with that produced by revaporization of the condensate in the column, rises higher and higher in the column and undergoes a series of repeated condensations and revaporizations. This amounts to a number of simple distillations having been performed *within* the column, and the vapor phase produced in each step becomes richer in the more volatile component, whereas the condensate, which drains down the column, becomes richer in the less volatile component.

Each step along the route $A-B-C-D-E-F$ of Figure 4.3 represents a single ideal distillation. One type of fractional distillation column, a "bubble-plate" column, was designed to bring about one such step for each "plate" it contained, and the efficiency of any fractional distillation column is often described in terms of its equivalency to such a column in "theoretical plates." Another index of the design efficiency of a fractional distillation column is the *HETP*, which stands for *height equivalent to a theoretical plate* and is the vertical length of a column that is necessary to obtain a separation efficiency of one theoretical plate. For example, a column 60 cm long with an efficiency of 30 plates has an HETP value of 2 cm. Such a column would usually be better for research purposes than a 60-plate column that is 300 cm long (HETP = 5 cm) because of the smaller liquid capacity and "holdup" of the shorter column. Holdup refers to the condensate that remains in a column during and after distillation. When small amounts of material are to be distilled, a column must be chosen that has an efficiency (HETP) adequate for the desired separation and also a moderate to low holdup.

As stated earlier, equilibrium between liquid and vapor phases must be established in a fractional distillation column in order that the more volatile component is selectively carried to the top of the column and into the condenser, where the vapor condenses into the liquid state. After all of the more volatile component is distilled, the less volatile one remains in the column and distilling flask; the heat supplied to the distilling flask is then increased in order to distil the second component. The most important requirements for performing a successful fractional distillation are (1) intimate and extensive contact between the liquid and the vapor phases in the column, (2) maintenance of the proper temperature gradient along the column, (3) sufficient length of the column, and (4) sufficient difference in the boiling points of the components of the liquid mixture. Each of these factors is discussed below.

1. The desired contact between the liquid and vapor phases can be obtained by filling the column with some inert material that has a large surface area. Examples of appropriate packing material include glass, ceramic, or metal pieces in a variety of shapes, including helices, "saddles," and woven mesh. Figure 2.16a shows a Hempel column packed with Raschig rings, which are pieces of glass tubing approximately 6 mm long. This type of column will have from 2 to 4 theoretical plates per 30 cm of length, depending on whether the distillation is carried out sufficiently slowly to maintain equilibrium conditions. An alternate type of fractional distillation column, which is useful for small-scale distillations of small amounts of liquid where low holdup is necessary, is the Vigreux column, shown in Figure 2.16b. Although a Vigreux column about the same length as a 30-cm Hempel column will have only about 1 to 2 theoretical plates and consequently will be less efficient, it has the advantage of a holdup of less than 1 mL. The holdup, and consequent loss of distillate, of the Raschig ring–filled Hempel column is 2–3 mL.

   A particularly effective type of fractional distillation apparatus is the *spinning-band column*, which contains a helical band of stainless steel or Teflon rotating at a high speed within the column. This band forces the condensed liquid *down* the column while at the same time leaving a very thin film of liquid on the inside surface of the column. This provides for *rapid* and very effective equilibration between the liquid and vapor. A well-insulated spinning-band column, 60 cm in length, may have a holdup of about 0.2 mL and a rating of 125 theoretical plates, and it can separate the components of a mixture having only a 2°C difference in boiling points.

   Fractional distillation columns that are many stories tall exist at places such as the National Bureau of Standards and permit the separation of liquids whose boiling points differ by as little as 0.1°C. The temperature gradient obtained in these columns is microprocessor-controlled.

2. **Temperature gradient** refers to the difference in temperature between the top and bottom of the column. Particularly important for an effective fractional distillation is the maintenance of the proper temperature gradient within the column. Ideally, the temperature at the bottom of the column should be approximately equal to the boiling temperature of the solution in the pot, and it should decrease continually in the column until it reaches the boiling point of the more volatile component at the head of the column. The significance of the temperature gradient may be visualized by reference to Figure 4.3, where the boiling temperature of the distillate decreases with each succeeding step, for example, *A* (90°C) to *C* (80°C) to *E* (82°C). The necessary temperature gradient from pot to stillhead will, in most distillations, be established automatically by the condensing vapors *if* the rate of distillation is properly adjusted. Frequently, this gradient can be maintained only by insulating the column with a material such as glasswool or, most effectively, with a silver-coated vacuum jacket around the outside of the column. Insulation helps reduce heat losses from the column to the atmosphere. For longer columns, particularly when inadequate insulation is used, it may be necessary to supply additional heat by means of electrical resistance wire wrapped around the column. Even when the column is insu-

lated, an insufficient amount of vapor may be produced to heat the column if the rate of heating the pot is too slow, so that little or no condensate will then reach the head. The rate of heating the pot must then be increased, but it must be kept below the point where the column is flooded. A flooded column may be identified by a "column" of *liquid* that is observed within the distillation column.

Factors directly affecting the temperature gradient in the column are the rate of heating of the pot and the rate at which vapor is removed at the stillhead. If the heating is too vigorous or the vapor is removed too rapidly, the whole column will heat up almost uniformly, and there will be no fractionation and thus no separation of the components. On the other hand, if the pot is heated too vigorously and if the vapor is removed too slowly at the top, the column will flood with returning condensate. Proper operation of a fractional distillation column requires careful control of the heat supplied to the distilling flask and of the rate at which the distillate is removed at the stillhead. This rate should be *no more than* 1 drop every 2 to 3 sec.

The ratio of the amount of vapor that condenses and returns down the column to the amount of vapor taken off as distillate during a given time period is defined as the **reflux ratio.** In general, higher reflux ratios provide for a more efficient fractional distillation. Elaborate distillation apparatuses are equipped with a total reflux–partial takeoff stillhead and provide higher reflux ratios; however, this type of equipment is not likely to be available in the introductory organic chemistry laboratory.

3. Correct column length is difficult to determine in advance of performing a fractional distillation. The trial-and-error technique must sometimes be used, and if a particular column does not efficiently separate a certain mixture, a longer column or a different type of column or column packing should be utilized.

4. The difference in boiling points between the two pure components of a mixture should be no less than 20–30°C in order for a fractional distillation to be successful when a Hempel column packed with Raschig rings or a similar type of packing is used. As mentioned previously, modifications in column length and type may result in the successful separation of mixtures having smaller boiling-point differences.

In summary, the most important variables that can be controlled experimentally in a fractional distillation are correct selection of the column and column packing, adequate insulation of the column, and careful control of the rate of heating so as to provide the most favorable temperature gradient within the column.

(c) Experiments Involving Simple and Fractional Distillation

The following experiments have been designed to demonstrate the techniques of simple and fractional distillation and the relative efficiencies of the different types of apparatus that are used to separate mixtures of volatile components. Experimental procedures are given for the purification of ethyl acetate by simple distillation and the fractional separation of a mixture containing ethyl acetate and *n*-butyl acetate, but other pure compounds and mixtures may be assigned by your instructor.

Note the following differences in procedure for the simple distillation of a liquid having a single volatile component and for the fractional distillation of a mixture having two volatile components. In the former, simple apparatus is used, while more complex fractional distillation equipment is used for the latter. A faster rate of distillation is used in the simple distillation of a liquid containing one volatile component, whereas fractional distillation of a binary mixture must be performed more slowly and carefully. This slower rate is necessary to maintain equilibrium conditions and the proper temperature gradient in the distillation column. In simple distillation the entire distillate is collected in a single receiving flask, whereas in fractional distillation, receiving flasks are changed as the composition of the distillate changes, as indicated by the head temperature.

In the simple distillation of a single volatile component, the head temperature should rise to a temperature that corresponds to the normal boiling point of the pure liquid and should remain constant until the distillation is complete and no more distillate is obtained. In the fractional distillation of a binary liquid mixture, ideally the head temperature should rise to the normal boiling point of the more volatile component and remain there until that component is completely removed. The head temperature should then drop some, indicating that the more volatile component has been removed. At this point, insufficient heat is being supplied to the distillation flask for the higher-boiling component to distil. Consequently, the heat should be increased until the latter starts to distil, and the now higher head temperature should remain constant until all of the second component has been distilled. In other words, two different temperature *plateaus* should be observed. This idealized process will not be observed *unless* the amount of heat supplied to the distilling flask is *very carefully* adjusted, and this requires constant attention during the course of the distillation.

The fractional distillation of a binary mixture, such as one containing ethyl acetate and *n*-butyl acetate, poses an interesting question about the effectiveness of the separation. Ideally, two distillation fractions should be obtained, one composed of pure ethyl acetate and the other of pure *n*-butyl acetate. A convenient technique that can be used to determine the actual compositions of the fractions is gas chromatography (GC, Chapter 6). This simple technique permits one to determine *what* components are present in each distillation fraction and, if more than one is present, the percent of each present. Gas chromatography provides an excellent adjunct to the fractional distillation experiment, but its use may depend on equipment availability and other considerations.

Pre-lab exercises for Sections 4.2 to 4.4, Simple and Fractional Distillation, are found on page PL. 7.

EXPERIMENTAL PROCEDURES

DO IT SAFELY

1 Although the use of burners in the organic laboratory is to be discouraged except in rare instances, the conditions in your laboratory may necessitate it in this experiment. Ethyl acetate and *n*-butyl acetate are both flammable, and great care should be exercised if they are to be heated with a burner.

2 Examine your glassware for cracks and other weaknesses before assembling the distillation apparatus. Look particularly for "star cracks" in round-bottomed flasks, because these may cause a flask to break while being heated.

3 Proper assembly of glassware is important in order to avoid possible breakage and spillage, as well as the release of distillate vapors into the room. Have your instructor examine your apparatus after it is assembled and before proceeding with the distillation.

4 The apparatus used in these experiments *must* be open to the atmosphere at the receiving end of the condenser. **Never heat a closed system.** If heated, a closed system will build up pressure that may cause the apparatus to explode.

5 Be certain that the water hoses are securely fastened to your condensers so that they will not ''pop'' off and cause a flood. If heating mantles or oil baths are to be used for heating in this experiment, water hoses that come loose may cause water to spray onto electrical connections, into the heating mantle, or into *hot* oil, each of which is potentially dangerous to you and those who work around you.

6 Avoid excessive inhalation of organic vapors at all times.

7 The organic substances that remain in the distillation flask or are collected as distillates should be disposed of properly. They should be placed in suitably marked waste containers and *should not be discarded in the sink.* The placement of these volatile substances in the drain is a potential fire hazard and abuses the environment.

The experiments described below utilize ethyl acetate and *n*-butyl acetate. Other pure substances or liquid mixtures may be assigned; if so, different boiling temperatures will be observed, and distillation cuts must be taken at different temperatures.

A. Simple Distillation of a Pure Liquid

Before starting, read Section 2.1 concerning the precautions and correct handling of glassware. Place 40 mL of ethyl acetate in a 100-mL round-bottomed flask and add two or three boiling chips to ensure smooth boiling. Assemble the simple distillation apparatus as directed in Section 2.7 and shown in Figure 2.14. To collect the distillate, use as a receiving flask either an Erlenmeyer flask or a graduated cylinder of sufficient volume. Have your instructor check your apparatus before you start heating the distilling flask. Use the method of heating specified by your instructor; details of various heating methods are discussed in Section 2.5, which should be consulted *before* you begin heating.

Begin heating the distillation flask, and as soon as the liquid begins to boil *and the condensing vapors have reached the thermometer bulb,* regulate the heat supply so that distillation continues steadily at a rate of *2 to 4 drops per second.* As soon as the distillation rate is adjusted and the thermometer has reached a constant temperature, note and record it. Continue the distillation and periodically record the head temperature. Discontinue heating when only 2–3 mL of ethyl acetate remains *in the distillation flask.* Determine and record the volume of distilled ethyl acetate that you obtained.

Unless directed otherwise, return the distilled and undistilled ethyl acetate to a bottle marked ''Recovered Ethyl Acetate.''

B. Fractional Distillation of a Binary Mixture

This experiment provides for the separation of a mixture of ethyl acetate and *n*-butyl acetate by fractional distillation, but similar results may be obtained with other mixtures such as carbon tetrachloride–toluene or methanol-water. Your instructor may wish to give you a multicomponent mixture of unknown composition for distillation. Regardless of your knowledge of the boiling points of the various components of the mixture you are to separate, successful completion of this experiment depends on your understanding that the distillate should be collected in *separate* receiving flasks, these flasks being changed when the thermometer readings indicate different *plateaus*. For example, a two-component mixture will require *three* receiving flasks, two for the main fractions and one for an intermediate fraction that is obtained as the temperature rises from the first plateau to the second. A three-component mixture will require *five* receiving flasks, three for the three main fractions and two for the two intermediate fractions that are obtained between the three plateaus. If the composition of the mixture given to you is known, look up the respective boiling points in a suitable handbook.

The distillate can be collected by one of the three possible procedures described (I, II or III); your instructor will inform you of the procedure to be used. Procedures II and III are advantageous if the distillation is to be repeated using different equipment, as outlined under Part C below, for they require data to be collected and plotted on a graph. If you are to follow either II or III, *before coming to laboratory* prepare a table in your notebook to allow for recording approximately 25 successive measurements of temperature as a function of total accumulated volume of distillate.

1. FRACTIONAL DISTILLATION WITH PACKED COLUMN

Place 30 mL of ethyl acetate and 30 mL of *n*-butyl acetate in a 100-mL round-bottomed flask, and add two or three boiling chips to ensure smooth boiling. Equip this flask for fractional distillation as shown in Figure 2.15; it is recommended that you read the discussion given in Section 2.8. Pack a Hempel or similar distillation column, using the type of packing specified by your instructor. The technique for packing columns is discussed in Section 2.8. When packing the column, *be careful not to break off the glass support indentations at the base of the column*. Do not pack the column too tightly, for vapors cannot pass through a tightly packed column; heating a fractional distillation apparatus equipped with a column that is too tightly packed is analogous to heating a closed system. The position of the thermometer in the stillhead is particularly important; the *top* of the mercury thermometer bulb should be on a level with the *bottom* of the sidearm of the distillation head. Have your instructor check your assembled apparatus before supplying heat to the distillation flask.

If you have been told to collect the distillate by Procedure I, clean and dry three 50-mL containers for use as receiving flasks, which may be bottles or Erlenmeyer flasks, and label them *A*, *B*, and *C*. If you are to use Procedure II, clean, dry, and label three 50-mL graduated cylinders. Procedure III requires one clean, dry 100-mL graduated cylinder.

Begin heating the distillation flask using the heating method specified by your instructor. See Section 2.5 for a discussion about various heating methods. As soon as the liquid mixture begins to boil and the vapors have reached the thermometer, regulate the heat so that distillation continues steadily at a rate *no faster* than *1 drop of distillate every 1 to 2 sec*. Collect the distillate using the procedure assigned by your instructor.

2. COLLECTION OF DISTILLATE

Note: The temperature "cuts" given below apply only for a mixture of ethyl acetate and *n*-butyl acetate and are the normal boiling points at 760 torr of these two substances. If different mixtures are being distilled, the cuts should be taken at the boiling points of each pure component of the mixture.

Procedure I

Place receiver A so that the tip of the vacuum adapter extends inside the neck of the container, as this will minimize evaporation of the distillate. As the mixture is heated, the temperature will rise to 81°C and remain there for a period of time, but after a while either the head temperature will start to rise above 81°C or will drop slightly. Receiver A should be left in place until this increase or decrease is observed. As soon as the temperature deviates from 81°C by more than ±5°C, change to receiver B, and increase the amount of heat supplied to the distillation flask. The temperature will start to increase again, and more liquid will distil; leave receiver B in place until the temperature reaches 123°C and then change to receiver C. (Note that these temperature cuts are for a mixture of ethyl acetate and n-butyl acetate only.) Continue the distillation until 1–2 mL of liquid remains in the pot, and then discontinue heating. Measure the volumes of the distillate collected in each receiver by means of a graduated cylinder, and record them. Allow the liquid in the column to drain into the distillation flask, then measure and record the volume of this liquid.

If instructed to do so, submit or save 1-mL samples of each fraction A, B, and C for gas chromatographic analysis, which is discussed in Section 6.6. Columns containing either silicone gum rubber or SF-96 as the stationary phase give good separation of ethyl acetate and n-butyl acetate.

Unless directed otherwise, place the organic liquids that were collected in suitably labeled organic waste containers.

Procedure II

Place graduated cylinder A so that the tip of the vacuum adapter extends inside the neck of the cylinder. This will minimize evaporation of the distillate. As soon as the distillation rate is adjusted to a rate of approximately 1 drop every 1 to 2 sec, note and record the head temperature and the *total accumulated* volume of distillate in the receiving cylinder. Continue the distillation, recording the temperature and total volume of each additional increment of approximately 2 mL of distillate. As the mixture is heated, the temperature will rise to 81°C and remain there for a period of time, but after a while the head temperature will start to rise above 81°C or will drop slightly. Graduated cylinder A should be left in place until this increase or decrease is observed. As soon as the temperature deviates from 81°C by more than ±5°C, change to graduated cylinder B, and increase the amount of heat supplied to the distillation flask. The temperature will start to increase again, and more liquid will distil; leave graduated cylinder B in place until the temperature reaches 123°C, and then change to graduated cylinder C. (Note that these temperature cuts are for a mixture of ethyl acetate and n-butyl acetate only.) Continue the distillation until 1–2 mL of liquid remains in the pot, and then discontinue heating. Record the volumes of the distillate collected in each graduated cylinder. Allow the liquid in the column to drain into the distillation flask, then measure and record the volume of this liquid.

If instructed to do so, save or submit 1-mL samples of each fraction A, B, and C for gas chromatographic analysis, which is discussed in Section 6.6. Columns containing either silicone gum rubber or SF-96 as the stationary phase give good separation of ethyl acetate and n-butyl acetate.

Unless directed otherwise, place the organic liquids that were collected in suitably labeled organic waste containers.

Treatment of Data   Plot the distillation data on a *completely labeled* graph, with the head temperature, in °C, being the vertical coordinate and the volume of distillate, in mL, being the horizontal coordinate. Answer the following questions concerning the distillation results:

1 On the basis of the graph, indicate what can be learned about the efficiency of the separation of the mixture.

2 If gas chromatographic analysis of the three fractions is available, calculate the percentage of ethyl acetate and *n*-butyl acetate in each fraction. What can be said regarding the purity of the three fractions?

3 Based on the volume and composition of the distillate collected in each fraction, does fractional distillation appear to provide a good method for separating ethyl acetate and *n*-butyl acetate?

Procedure III

Place a 100-mL graduated cylinder so that the tip of the vacuum adapter extends inside the neck of the cylinder. This will minimize evaporation of the distillate. *All* of the distillate will be collected in the *same* cylinder. As soon as the distillation rate is adjusted to a rate of approximately 1 drop every 1 to 2 sec, note and record the head temperature and the *total accumulated* volume of distillate in the receiving cylinder. Continue the distillation, recording the temperature and total volume of each additional increment of approximately 2 mL of distillate. Note that you must be prepared to collect samples as described in the next paragraph, if your instructor has assigned this procedure. As the mixture is heated, the temperature will rise to 81°C and remain there for a period of time, but after a while, the head temperature will start to rise above 81°C or will drop slightly. As soon as the temperature deviates from 81°C by more than ±5°C, increase the amount of heat supplied to the distillation flask. The temperature will start to increase again and more liquid will distil. (Note that these temperatures are for a mixture of ethyl acetate and *n*-butyl acetate only.) Continue the distillation until 1–2 mL of liquid remains in the pot and then discontinue heating. Allow the liquid in the column to drain into the distillation flask, then measure and record the volume of this liquid.

If instructed to do so, collect 0.5-mL samples of distillate in clean, dry, labeled sample vials at the following total accumulated volumes: 10 mL, 20 mL, 30 mL, 40 mL, 50 mL, and at the point where the distillation is nearly complete, as evidenced by the presence of only several mL of liquid in the distillation flask. These samples may be obtained by temporarily replacing the graduated cylinder with the sample vial at the desired points and collecting 0.5 mL of liquid. Save or submit these samples for gas chromatographic analysis, which is discussed in Section 6.6. Columns containing either silicone gum rubber or SF-96 as the stationary phase give good separation of ethyl acetate and *n*-butyl acetate.

Unless directed otherwise, place the organic liquids that were collected in suitably labeled organic waste containers.

Treatment of Data   Plot the distillation data on a *completely labeled* graph, with the head temperature, in °C, being the vertical coordinate and the volume of distillate, in mL, being the horizontal coordinate. Answer the following questions concerning the distillation results:

1 On the basis of the graph, indicate what can be learned about the efficiency of the separation of the mixture.

2 If gas chromatographic analysis of the six samples is available, calculate the percentage of ethyl acetate and *n*-butyl acetate in each. What can be said regarding the purity of the samples that were collected at various distillation volumes?

3 Based on the composition of the distillate collected in each vial, does fractional distillation appear to provide a good method for separating ethyl acetate and *n*-butyl acetate?

C. Distillation of a Binary Liquid Mixture Using Alternate Equipment

It is interesting to compare the results of distilling a binary mixture, such as ethyl acetate and *n*-butyl acetate, using different types of equipment. For example, this mixture can be distilled by simple distillation or by fractional distillation in which the distillation column is *not* packed. If assigned to perform either or both of these distillations, proceed as follows.

1. SIMPLE DISTILLATION

Using the procedure provided in Part A (Simple Distillation) above, distil a mixture consisting of 30 mL of ethyl acetate and 30 mL of *n*-butyl acetate. You may be asked to collect the distillate by one of the collection procedures given in Part B above and to save or submit samples of distillate for gas chromatographic analysis.

2. FRACTIONAL DISTILLATION WITH UNPACKED COLUMN

Using the procedure provided in Part B above, distil a mixture consisting of 30 mL of ethyl acetate and 30 mL of *n*-butyl acetate. You may be asked to collect the distillate by one of the procedures given and to save or submit samples of distillate for gas chromatographic analysis.

3. TREATMENT OF DATA

If you followed procedure II or III for collecting the distillate, it is convenient to plot the data from two or more distillations on the same piece of graph paper so that the curves obtained for each distillation involving different equipment will overlay one another. Compare the results of each type of distillation equipment, and draw conclusions suggested by the data concerning the relative efficiencies of each type of distillation. The results from gas chromatographic analyses, if performed, should be incorporated into your conclusions.

EXERCISES

1. Define the following terms:

a. simple distillation
b. fractional distillation
c. head temperature
d. pot temperature
e. Raoult's law
f. ideal solution

g. mole fraction
h. height equivalent to a theoretical plate (HETP)
i. temperature gradient
j. Dalton's law
k. reflux ratio

2. Sketch and completely label the apparatus required for (a) simple distillation and (b) fractional distillation.

3. Explain why a packed fractional distillation column is more efficient at separating two closely boiling liquids than an unpacked column.

4. If heat is supplied to the distillation flask too rapidly, the ability to separate two liquids by fractional distillation may be drastically reduced. In terms of the theory of distillation presented in the discussion, explain why this is so.

5. Explain why the column of a fractional distillation apparatus should be aligned as nearly vertical as possible.

6. Explain the role of the boiling chips that are normally added to a liquid that is to be heated to boiling.

7. The top of the mercury bulb of the thermometer placed at the head of a distillation apparatus should be adjacent to the exit opening to the condenser. Explain the effect on the observed temperature reading if the bulb is placed (a) below the opening to the condenser and (b) above the opening.

8. a. A mixture of 80 mole % n-propylcyclohexane and 20 mole % n-propylbenzene is distilled through a *simple distillation apparatus* (assume that no fractionation occurs during the distillation). The boiling temperature is found to be 157.3°C as the first *small* amount of distillate is collected. The standard vapor pressures of n-propyl-cyclohexane and n-propylbenzene are known to be 769 torr and 725 torr, respectively, at 157.3°C. Calculate the percentage of each of the two components in the first few drops of distillate.

   b. A mixture of 80 mole % benzene and 20 mole % toluene is distilled under exactly the same conditions as in part a. Using Figure 4.3, determine the distillation temperature and the percentage composition of the first few drops of distillate.

   c. The normal boiling points of n-propylcyclohexane and n-propylbenzene are 156.9°C and 159°C, respectively. Compare the distillation results in parts a and b. Which of the two mixtures would require the more efficient fractional distillation column for separation of the components? Why?

9. Examine the boiling point–composition diagram for mixtures of toluene and benzene given in Figure 4.3.

   a. Assume you are given a mixture of these two liquids of composition 80 mole % toluene and 20 mole % benzene and that it is necessary to effect a fractional distillation that will afford *at least some* benzene of greater than 99% purity. What would be the *minimum* number of theoretical plates required in the fractional distillation column chosen to accomplish this separation?

   b. Assume that you are given a 20-cm Vigreux column having an HETP of 10 cm in order to distil a mixture of 58 mole % benzene and 42 mole % toluene. What would be the composition of the first small amount of distillate that you obtained?

10. At 100°C, the vapor pressures (in torr) for water, methanol and ethanol are 760, 2625, and 1694, respectively. Which compound has the highest normal boiling point and which the lowest?

11. At 50°C, the vapor pressures (in torr) for methanol and ethanol are 406 and 222, respectively. Given a mixture at 50°C that contains 0.1 mole of methanol and 0.2 mole of ethanol, compute the partial pressures of each liquid and the total pressure.

12. Figure 4.4 shows a temperature-composition diagram for a mixture of carbon tetrachloride and toluene. Answer the following questions, assuming that these substances form an ideal solution:

a. What are the boiling points of pure carbon tetrachloride and toluene?

b. Answer the following questions for a mixture that contains 50 mole % carbon tetrachloride and 50 mole % toluene:

(1) At what temperature will it boil?

(2) What is the composition, in mole %, of the first small amount of vapor that forms?

(3) If the vapor in (2) is condensed and the resulting liquid is heated to its boiling point, what is the composition, in mole %, of the new vapor?

c. A mixture of unknown composition is heated until it just starts to boil, and the composition of the resulting vapor is found to be 50 mole % toluene and 50 mole % carbon tetrachloride. What is the percent composition of the original mixture, and at what temperature does the mixture start to boil?

d. Calculate the amount of carbon tetrachloride and toluene in the liquid and vapor phases in a boiling mixture of the two at 85°C and at 100°C. Express your answer both in mole percent and in mole fraction.

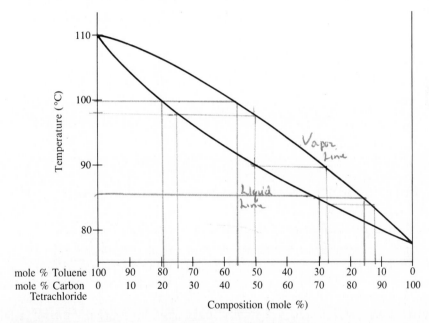

Figure 4.4 Temperature-composition diagram for Exercise 12.

13. The data given in the table below were obtained from separate distillations on 40 mL of a mixture containing equal volumes of ethyl acetate and *n*-butyl acetate. The following types of distillation apparatus were used: simple distillation, fractional distillation with a packed column, and fractional distillation with an unpacked column.

a. Graph these data on the same piece of graph paper and plot cumulative volume (horizontally) vs. head temperature (vertically).

b. With reference to the graph, which distillation was the most effective in separating the mixture and which the least effective?

c. Suggest the type of apparatus that was used for distillations *A*, *B*, and *C*.

Distillation Data

Total Accumulated Volume (mL)	Head Temperatures (°C) for Distillation		
	A	B	C
2	80	80	80
4	82	82	82
6	83	83	83
8	83	85	84
10	83	86	85
12	83	88	86
14	83	89	87
16	83	92	89
18	85	95	91
20	95	98	95
22	110	104	103
24	122	112	110
26	123	117	115
28	123	120	119
30	123	122	119
32	123	123	122
34	123	123	123
36	123	123	123
38	123	123	123
40	123	123	123

4.5 Fractional Distillation of Nonideal Solutions

Most homogeneous organic liquid mixtures behave as ideal solutions, but examples are known in which the behavior is nonideal. In these solutions, the unlike molecules are not indifferent to the presence of one another. The result is the observation of deviations from Raoult's law in either of two directions. Some solutions display greater vapor pressures than expected and are said to exhibit *positive deviations;* others display lower vapor pressures than expected and are said to exhibit *negative deviations*. The discussions that follow are limited to binary mixtures, although some three-component systems exhibit similar deviations.

In the case of positive deviations, the forces of attraction between the molecules of the two components are weaker than those between the identical molecules of each component, with the result that, in a certain composition range, the combined vapor pressure of the two components is greater than the vapor pressure of the pure, more volatile component. Mixtures in this composition range, between *X* and *Y* in Figure 4.5, have boiling temperatures *lower* than the boiling temperature of either pure component. The *minimum-boiling* mixture,

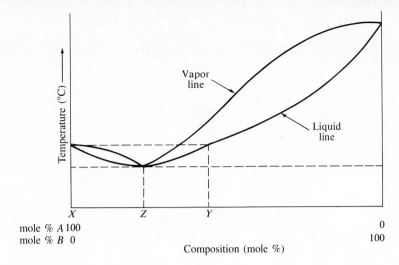

Figure 4.5 Temperature-composition diagram for a minimum-boiling azeotrope.

composition Z in Figure 4.5, must be considered as if it were a third component. It has a constant boiling point because the vapor in equilibrium with the liquid has the same composition as the liquid itself. This mixture is called a **minimum-boiling azeotropic mixture,** or **azeotrope.** Fractional distillation of such mixtures will not yield both of the components in pure form but only the azeotrope and the component present in excess of the azeotropic composition. For example, pure ethanol cannot be obtained by fractional distillation of aqueous solutions containing less than 95.57% ethanol, the azeotropic composition, even though the boiling point of this azeotrope is only 0.15°C lower than the boiling point of pure ethanol. Since optimum fractional distillations of aqueous solutions containing less than 95.57% ethanol yield this azeotropic mixture, "95% ethyl alcohol" is the common composition of the commercial solvent ethanol. Pure or "absolute" ethanol can be obtained by chemically removing the water or by a distillation procedure that involves the use of a ternary mixture of ethanol-water-benzene. While binary azeotropic mixtures are central to our discussion, the ternary mixture of water (7.4%), ethanol (18.5%) and benzene (74.1%) is the minimum-boiling azeotropic mixture that permits the removal, by distillation, of water from 95% ethanol to which benzene has been added. The ethanol so obtained is one form of *denatured ethanol* and is unsafe for human consumption, owing to the presence of traces of toxic benzene. On the other hand, pharmaceutical-grade absolute ethanol is obtained by chemically removing the water.

In the case of a negative deviation from Raoult's law, the forces of attraction between the different molecules of the two components are stronger than those between identical molecules of each component, with the result that, in a certain composition range, the combined vapor pressure of the two components is less than the vapor pressure of the pure, less volatile component (see Figure 4.6). Mixtures in this composition range, between X and Y in Figure 4.6, boil at temperatures *higher* than the boiling temperature of *either* pure component. There is one particular composition, Z in Figure 4.6, that corresponds to a **maximum-boiling azeotrope.** Fractional distillation of mixtures of any composition other than that of the azeotrope will result in the distillation from the mixture of whichever of the two compo-

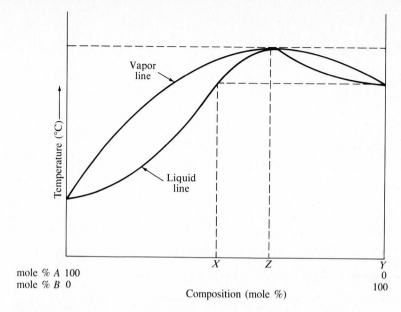

Figure 4.6 Temperature-composition diagram for a maximum-boiling azeotrope.

nents is present in excess of the azeotropic composition, Z; the composition of the pot residue will therefore approach that of Z. As an illustration, formic acid and water form a maximum-boiling azeotrope having the composition 77.5% formic acid–22.5% water. This azeotrope of pure formic acid (bp 100.7°C) and water (bp 100.0°C) boils at 107.3°C. The distillation of any mixture having other than the azeotropic composition will result in either water or formic acid—whichever is in excess—being distilled away, and the azeotropic mixture will remain as the pot residue.

Azeotropes, like pure liquids, exhibit constant boiling points. As a result, they may be confused with pure compounds *if* boiling points alone are used as the basis for determining purity. Data on other minimum-boiling or maximum-boiling azeotropes may be found in the references at the end of the chapter, and extensive tables are given in the handbooks listed there.

EXERCISES

1. Define the following terms:
   a. azeotrope
   b. minimum-boiling azeotrope
   c. maximum-boiling azeotrope
   d. positive deviation of vapor pressure
   e. negative deviation of vapor pressure

2. A certain liquid boils at a constant temperature, for example, 127°C. Based on this boiling point alone, is it a pure liquid? Explain.

3. Suppose that equimolar amounts of water (bp 100°C) and ethanol (bp 78.32°C) are mixed and subjected to fractional distillation.

a. The first liquid to distil boils at 78.17°C. Explain this result. What does this distillate contain?

b. After all the material boiling at 78.17°C has distilled, the next fraction to be collected boils at 100°C. No liquid boiling at 78.32°C is collected. Explain this result.

4.6 Vacuum Distillation; Distillation Under Reduced Pressure

The effect of external pressure upon the boiling point of a pure liquid was discussed in Section 4.1. Distillations, either simple or fractional, are easiest to perform at atmospheric pressure (defined as 760 torr), and under these conditions the observed boiling temperature is the *normal boiling point*. As indicated in Section 4.1, the attempted distillation of some very high-boiling liquids is often difficult or impossible, owing to possible decomposition or undesired side reactions. If this situation arises, it will be necessary to perform a simple or fractional distillation at reduced pressure, a technique called *vacuum distillation,* which is described in Section 2.9.

4.7 Steam Distillation

(a) Theory and Discussion

The separation and purification of *volatile* organic compounds that are immiscible with (insoluble in) water, or nearly so, is often accomplished by *steam distillation,* a technique that involves the codistillation of a mixture of water and organic substances. The virtues and limitations of this technique can be illustrated by consideration of the principles that underlie steam distillation.

The partial pressure (P_i) of each component (i) of a mixture of immiscible, volatile substances at a given temperature is equal to the vapor pressure (P_i°) of the pure compound at the same temperature (equation 6) and does not depend on the mole fraction of the compound in the mixture; that is, each component of the mixture vaporizes independently of the others.

$$P_i = P_i^\circ \qquad\qquad \text{(6)}$$

This behavior is in sharp contrast to that exhibited by solutions of miscible liquids, for which the partial pressure of each constituent of the mixture depends on its mole fraction in the solution (Raoult's law, equation 2). Now, the total pressure, P_T, of a mixture of gases, according to Dalton's law (equation 4), is equal to the sum of the partial pressures of the constituent gases, so that the total vapor pressure of a mixture of immiscible, volatile compounds is given by equation 7:

$$P_T = P_a^\circ + P_b^\circ + \cdots P_i^\circ \qquad\qquad \text{(7)}$$

This expression indicates that the total vapor pressure of the mixture at any temperature is always greater than the vapor pressure of even the most volatile component at that temperature, owing to the contributions of the vapor pressures of the other constituents in the mixture. The boiling temperature of a mixture of immiscible compounds must then be *lower* than that of the lowest-boiling component.

Demonstration of the principles just outlined is available from the following discussion about the steam distillation of an immiscible mixture of water (bp 100°C) and bromobenzene (bp 156°C). Figure 4.7 shows a plot of the vapor pressure versus temperature for each pure substance and for a mixture of these compounds. This graph shows that the mixture should boil at about 95°C, the temperature at which the total vapor pressure equals the atmospheric pressure. As theory predicts, this temperature is below the boiling point of water, which is the lowest-boiling component in this example. The ability to distil a compound at the relatively low temperature of 100°C or less by means of a steam distillation is often of great use, particularly in the purification of substances that are heat-sensitive and would therefore decompose at higher temperatures. It is useful also in the separation of compounds from reaction mixtures that contain large amounts of nonvolatile residues, such as the notorious tars so often formed in many organic reactions.

The composition of the condensate from a steam distillation depends upon the molecular weights of the compounds being distilled and upon their respective vapor pressures at the temperature at which the mixture steam distils. Consider a mixture of two immiscible components, A and B. If the vapors of A and B behave as ideal gases, the ideal gas law can be applied, and equations 8a and 8b are obtained.

$$P_A^\circ V_A = (g_A/M_A)(RT) \qquad \textbf{(8a)}$$

$$P_B^\circ V_B = (g_B/M_B)(RT) \qquad \textbf{(8b)}$$

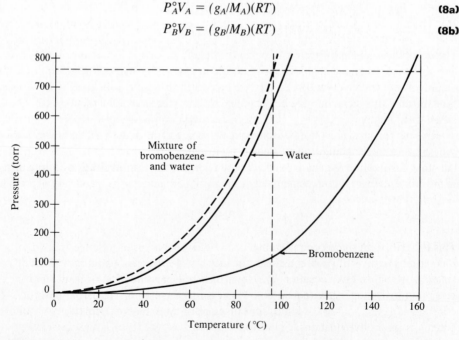

Figure 4.7 Vapor pressure–temperature graph showing pure bromobenzene, pure water and a mixture of bromobenzene and water.

In these two expressions, $P°$ is the vapor pressure of the pure liquid, V is the volume in which the gas is contained, g is the weight in grams of the component in the gas phase, M is its molecular weight, R is the gas constant, and T is the absolute temperature in kelvins (°K). Dividing equation 8a by equation 8b gives equation 9.

$$\frac{P_A^\circ V_A}{P_B^\circ V_B} = \frac{g_A M_B (RT)}{g_B M_A (RT)}$$ (9)

Because the RT factors in the numerator and denominator are identical and because the volume in which the gases are contained is the same for both ($V_A = V_B$), these terms cancel in equation 9 to yield equation 10.

$$\frac{\text{grams of } A}{\text{grams of } B} = \frac{(P_A^\circ)(\text{molecular weight of } A)}{(P_B^\circ)(\text{molecular weight of } B)}$$ (10)

For example, consider a mixture of bromobenzene and water, whose molecular weights are 157 g/mole and 18 g/mole, respectively, and whose vapor pressures, as determined from Figure 4.7, are 120 torr and 640 torr, respectively, at 95°C. The composition of the distillate at 95°C can be calculated from equation 10 as follows:

$$\frac{g_{\text{bromobenzene}}}{g_{\text{water}}} = \frac{(120)(157)}{(640)(18)} = \frac{1.64}{1}$$

On the basis of weight, this calculation indicates that more bromobenzene than water is contained in the steam distillate, even though the vapor pressure of the bromobenzene is much lower at the temperature of the distillation. Because organic compounds generally have molecular weights much higher than that of water, it is possible to steam-distil compounds having vapor pressures of only about 5 torr at 100°C with a fair efficiency on a weight-to-weight basis. Even solids can be purified by steam distillation.

In summary, steam distillation provides a method for separating and purifying moderately volatile liquid and solid organic compounds that are insoluble or nearly insoluble in water from nonvolatile compounds. Relatively mild conditions are used in steam distillation, but even so, it cannot be used for substances that decompose on prolonged contact with steam or hot water, that react with water, or that have vapor pressures of 5 torr or less at 100°C. The technique for carrying out a steam distillation is presented in Section 2.10, and an experiment that uses this method is given in part (b) below.

Steam distillation involves passing steam through a mixture of one or more volatile compounds that are immiscible with water and collecting the resulting distillate, which contains water and the desired compound(s). Some procedures involve the isolation of a desired product using a form of steam distillation in which the steam is *not* obtained from an external source, such as a steam line, but is generated internally. Several preparative experiments appearing later in this text utilize this form of steam distillation: (a) 1-bromobutane (Section 18.2), (b) aniline (Section 23.2b), and (c) cyclohexene (Section 13.2). In each of these syntheses a mixture containing water, the desired compound, and nonvolatile impurities is subjected to distillation, yielding a distillate that consists of water and the desired compound in moderately pure form.

(b) Natural Products; Isolation of Citral from Lemon Grass Oil by Steam Distillation

(1) GENERAL DISCUSSION OF NATURAL PRODUCTS

Naturally occurring organic compounds, which are found in and produced by living organisms, have been a source of fascination for centuries. The interest in these substances exists for a variety of reasons, ranging from the practical applications of such compounds in daily life to the scientific challenges they present. "Natural products" have been used to alleviate pain, to cure diseases, to make colorful dyes, to flavor foods, and even to cause death. From the standpoint of science it was the natural products, because of their ready availability, that provided chemists with one of their first experimental challenges during the period when chemistry was emerging from alchemy into a more exact science. Today the isolation, identification, and synthesis of natural products still present some of the greatest challenges to modern organic chemists.

As the scientific field now labeled "natural products chemistry" evolved, these substances were placed in one of four general categories, which are defined on the basis of common structural features found in most natural products. These four categories are the carbohydrates (sugars), the acetogenins, the terpenes and steroids, and the alkaloids. Carbohydrates are the topic of Chapter 25, so they are not discussed here.

The *acetogenins* are a group of compounds that share the common distinction that their *biosynthesis* involves the head-to-tail polymerization of the two-carbon acetate unit to generate a *linear* polyacetyl chain, **1.** "Biosynthesis" refers to the synthesis of a chemical compound by a living organism. As a class, the acetogenins are characterized by the absence of extensive branching in their carbon skeletons. Further transformations of the basic structure **1** can produce substances as diverse as stearic acid **(2)**, a fatty acid, and rhamnetin **(3)**, a yellow pigment of the flavone type.

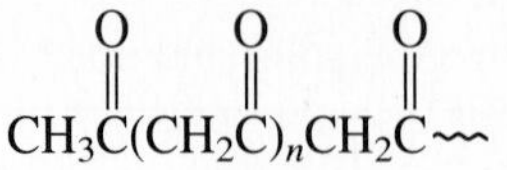

$$CH_3\overset{O}{\overset{\|}{C}}(CH_2\overset{O}{\overset{\|}{C}})_n CH_2\overset{O}{\overset{\|}{C}}\sim$$

1

A polyacetyl chain

$$CH_3CH_2(CH_2)_{14}CH_2CO_2H$$

2

Stearic acid

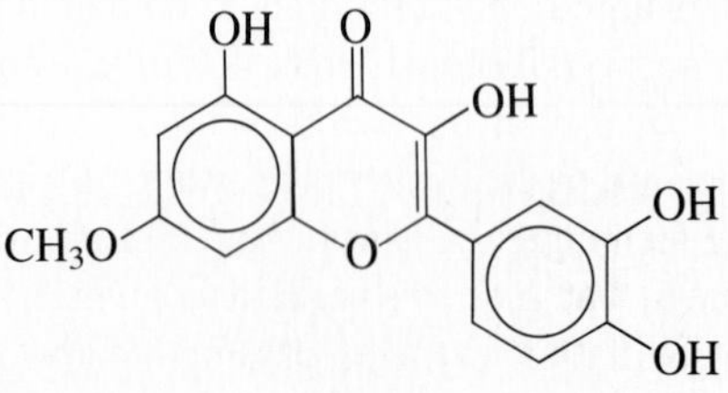

3

Rhamnetin

The substances that fall into the category designated *terpenes* and *steroids* can be viewed as having carbon skeletons constructed by the 1,4-polymerization of the five-carbon, branched-chain isopentyl unit, **4.** The active form of this unit in biosynthesis is isopentenyl pyrophosphate **(5)**, which is readily produced in nature from three two-carbon acetate units, with the loss of one carbon atom as carbon dioxide. In contrast to the generally linear structure of acetogenins, the terpenes and steroids possess carbon skeletons that are branched at regular intervals, as illustrated by the structure of vitamin A (**6**). Substances in the terpene class contain two or more of the five-carbon *isoprene* units. For example, vitamin A contains a total of 20 carbon atoms and therefore *four* isoprene units. Camphor (**7**) comes from two isoprene units to give the 10 carbon atoms it contains, and cholesterol (**8**) is produced biologically from six isoprene units by a complex sequence of reactions that results in the loss of three carbon atoms.

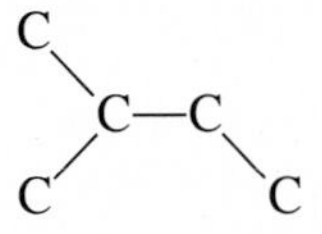

4

Isopentyl unit

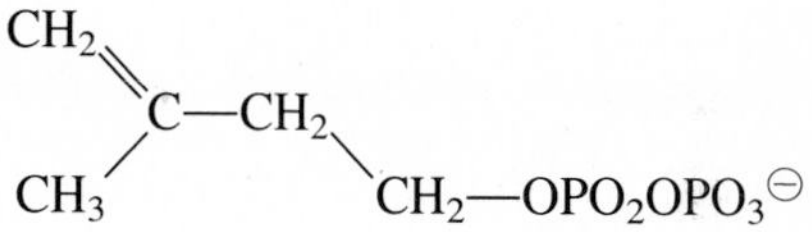

5

Isopentenyl pyrophosphate

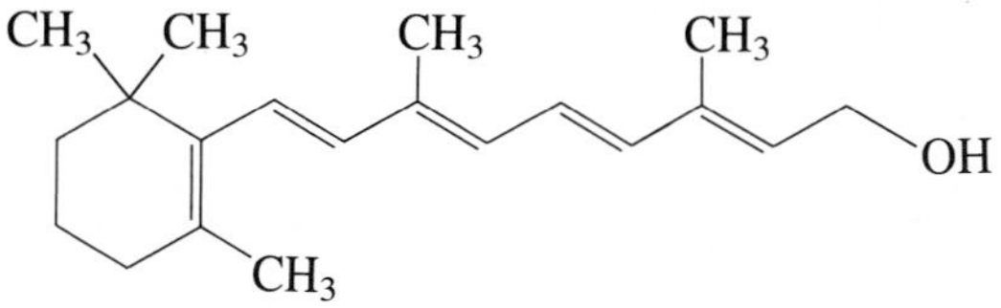

6

Vitamin A

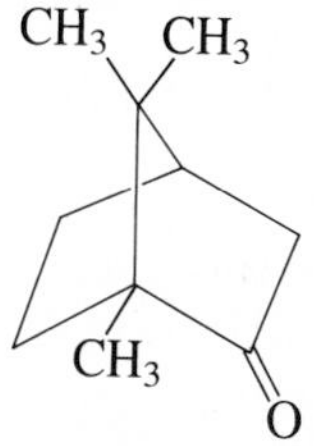

7

Camphor

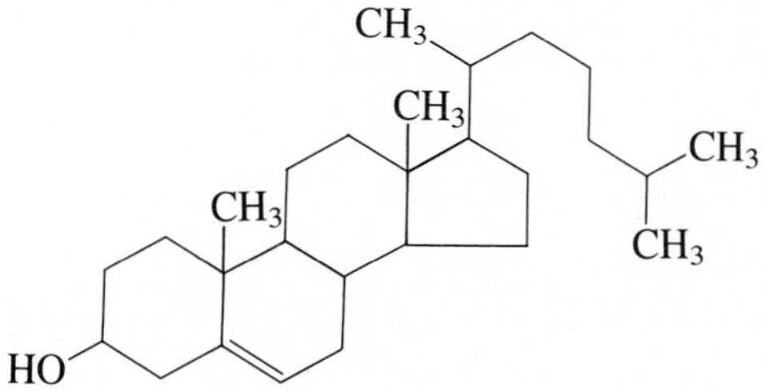

8

Cholesterol

The final category of natural products considered here, the *alkaloids,* probably contains the greatest number of compounds and is the most diverse in terms of structural type. This is a consequence of the fact that all natural products that are *basic* are classified as alkaloids. The basicity of these substances results from the presence of one or more nitrogen atoms in them, the nitrogen arising from the incorporation of one or more α-amino acid units **(9)** into alkaloids during their biosynthesis. Nicotine **(10)** and strychnine **(11)** are examples of alkaloids.

$$R—CHCO_2H$$

$$: NH_2$$

9

An α-amino acid

10

Nicotine

11

Strychnine

Historically, most natural products have been extracted from plants rather than from animals. The natural products from plants generally have simpler structures that are easier to elucidate. Microorganisms are now assuming an increased importance as sources of natural products.

The isolation of a natural product in pure form normally presents a considerable challenge to the experimenter, mainly because even the simplest plants and microorganisms contain mixtures of many compounds. The general approach used to isolate and purify a natural product may be summarized as follows. The plant or microorganism is ground or homogenized into fine particles, and the resulting material is then usually extracted with a solvent or mixture of solvents in which the desired natural product is expected to dissolve. An example is the extraction of trimyristin from nutmeg using diethyl ether (Section 5.3). Volatile natural products in the extract can be detected and possibly even isolated by gas chromatography. More often, however, the natural product is relatively nonvolatile, and removal of the solvent from the extraction yields an oil or gum that requires further separation in order to obtain its various components in pure form.

A more common procedure for natural products isolation involves treatment of the residual oil or gum with an acid or base in order to separate basic and acidic components, respectively, from neutral substances. Slightly volatile compounds might be separated from nonvolatile ones by subjecting the residue to steam distillation. The principles of acid and base extraction, along with typical experimental procedures, are discussed in Section 5.4.

Various chromatographic techniques are also used to isolate and purify natural products. These include thin- and thick-layer, liquid-liquid, and gas-liquid chromatography. Chapter 6 introduces many of these techniques and includes several experiments involving chromatographic methods.

The next stage facing the natural products chemist is determination of the structure of the isolated pure product. Traditional techniques, such as qualitative tests for various functional groups (Chapter 27), and chemical degradations to known, simpler compounds, were and are of great importance. More recently, spectroscopic techniques such as mass spectrometry and ultraviolet, infrared, and nuclear magnetic resonance spectroscopy have greatly facilitated the determination of structure. The theory and application of various spectroscopic methods are discussed in Chapters 8–11.

In many instances, the final goal of chemists working in this field is the total *laboratory* synthesis of a natural product. The synthesis of some natural products represents mainly an intellectual challenge and/or an opportunity to demonstrate the utility of new synthetic techniques. In cases where the natural product has medicinal uses, the development of an efficient synthetic route may be of great importance, especially when the supply of the material from the natural source is severely limited or its isolation is difficult. Chapter 23 provides examples of multistep syntheses of two medically important compounds, sulfanilamide and lidocaine.

(2) ISOLATION OF CITRAL FROM LEMON GRASS OIL

Terpenes are responsible for the characteristic flavors, odors, and colors of many substances encountered in nature. For example, citral (**12**) is a terpene that possesses a pleasant lemonlike odor and taste. Although it evokes pleasant odor and taste responses in humans, it apparently is less attractive to other forms of life. Certain insects such as ants are known to secrete citral in order to ward off potential predators, that is, it serves as a defense pheromone. Citral is of commercial importance as a constituent of perfumes in which a lemonlike essence is desired; it is also used as an intermediate for the synthesis of vitamin A (**6**).

$$CH_3 \diagup C = CH \diagdown \diagup C = CH - C \diagdown^{O}_{H}$$
$$CH_3 \qquad CH_3$$

12
Citral

Owing to the commercial importance of citral, an extensive search for its presence in natural products has been carried out. One source is the oil from the skins of lemons and oranges, although it is only a minor component of this oil. However, citral is the major constituent of the oil that results from the pressing of lemon grass, and, in fact, 75–85% of this oil is the desired natural product.

Citral contains two carbon-carbon double bonds, one of which is conjugated with the carbon-oxygen double bond of the aldehyde group. The aldehyde group of citral is readily oxidized to a carboxylic acid, a reaction typical of aldehydes, and owing to the conjugation, the molecule is also subject to facile polymerization. Consequently, citral is a chemically labile substance under conditions that induce its polymerization and/or oxidation. Heat, light, and reagents such as acids, bases, and oxidizing agents promote these types of reactions. The isolation of pure citral therefore can present a significant challenge to the experimenter. This task is greatly simplified, however, by the fact that citral is relatively volatile (bp 229°C at 760 torr) and has a low solubility in water. These properties make it a suitable candidate for steam distillation, a technique that allows distillation of citral from crude lemon grass oil at a temperature less than 100°C, which is far below the normal boiling point of citral (229°C). Neutral conditions are maintained in steam distillation, as is the partial exclusion of oxygen, so that the possibility of polymerization and/or oxidation of citral is minimal. This emphasizes the value of steam distillation, namely, that it is often the method of choice when reactive, volatile substances are to be separated from nonvolatile and/or water-soluble contaminants.

The citral that will be isolated in this experiment is actually a mixture of two geometric isomers, geranial (**12a**) and neral (**12b**). It is extremely difficult to separate these two isomers, and this will not be attempted in this experiment.

12a
Geranial

12b
Neral

The steam distillation can be performed by two different methods. One involves the use of an external steam source, either a laboratory steam line or a steam generator, passing the steam through a mixture of lemon grass oil and water; this method has some advantages but is experimentally more difficult. The other method involves heating a mixture of lemon grass oil and water and collecting the distillate. Although simpler from the experimental standpoint, there are some limitations to the application of this procedure. For example, in the steam distillation of only slightly volatile substances, a large initial volume of water will be required, or water must be added as the distillation proceeds, perhaps by means of an addition funnel. Procedures using both alternatives are provided below.

After steam distillation of the lemon grass oil, the distillate contains a mixture of citral and water, from which the citral is isolated by extraction with diethyl ether. The ether extracts are dried by a drying agent, anhydrous calcium chloride, and the ether is removed by evaporation under vacuum. Mild conditions are used for the extraction and evaporation in order to minimize the polymerization and/or oxidation of citral.

Structure Determination for Citral   Various spectroscopic and chemical analyses can be performed on the citral that is isolated from lemon grass oil by steam distillation. These analyses are mentioned briefly here, and references are given to later parts of the text where they are discussed in more detail.

The structure of citral can be confirmed using infrared (IR), proton magnetic resonance (PMR), and/or ultraviolet (UV) spectroscopy; its IR and PMR spectra are given in Exercise 3 at the end of Chapter 10, and its UV spectrum is presented in Figure 11.3. Citral can also be analyzed using gas chromatography (Chapter 6), which allows the assessment of the nature and purity of the product that has been isolated.

Various chemical methods can be used to characterize citral. For example, qualitative tests for the carbon-carbon double bonds and for the aldehyde group can be performed, and solid derivatives can be prepared by utilizing reactions of the aldehyde group (Chapter 27).

Pre-lab exercises for Section 4.7, Steam Distillation of Citral, are found on page PL. 9.

EXPERIMENTAL PROCEDURE

DO IT SAFELY

1 Steam distillation involves the use of glassware that becomes very hot. Exercise care when handling hot glassware.

2 If a house line is used as a source of external steam, take care to blow out the condensed water in the line before attaching the hose to the water trap. Do this by holding the end of the hose in the sink or drain and opening the steam valve until no more water is ejected.

3 When passing steam from a house line through the lemon grass, control the rate of steam flow carefully so that pressure does not build up in the apparatus. This minimizes the potential for glass joints to "pop" apart and release the steam dangerously or for the explosion of the apparatus.

4 If the flow of steam from a house line is interrupted, a partial vacuum in the line and water trap will be created, and the mixture of water and lemon grass oil will be sucked back into the trap. If this occurs, it will be necessary to disassemble the apparatus and pour the mixture in the trap back into the distillation flask before resuming the steam distillation.

When the steam distillation is completed, the drain from the water trap must be opened and the steam inlet tube removed from the distilling flask. Otherwise, the hot liquid in the flask will "back up" into the water trap.

5 Be certain that the steam distillate is cooled below 30°C before extracting it with diethyl ether; otherwise excessive pressure may develop in the separatory funnel and may blow out the stopper.

6 Diethyl ether is **extremely flammable.** Take care that no flames are in your vicinity during its use and its removal from the citral.

Option 1: Distillation Using an External Steam Source   Determine the weight of 10 mL of lemon grass oil, and then place it along with 100 mL of water in a 250-mL round-bottomed flask. Equip this flask for steam distillation as described in Section 2.10 and as shown in Figure 2.19. Steam may be obtained from a steam line, if this is available in your laboratory, or may be generated using the apparatus shown in Figure 2.20. Regardless of the steam source, a water trap, shown in Figure 2.21, must be placed *between* the steam source and the steam distillation apparatus. Steam-distil the mixture as rapidly as possible, and continue the distillation until droplets of oil no longer appear in the distillate; approximately 250-mL of distillate should be collected. Drain condensed water from the water trap whenever necessary; do not allow the water trap to fill up.★ Proceed to "Isolation of Citral," below.

Option 2: Steam Distillation Using Internal Source of Steam   Determine the weight of 10 mL of lemon grass oil, and then place it along with 250 mL of water in a 500-mL round-bottomed flask. Equip this flask for fractional distillation (Section 2.8), but *do not pack the distillation column*. Heat the flask to boiling with a burner or heating mantle, as directed by your instructor. Adjust the heat source so that the distillation proceeds as rapidly as possible, but avoid applying excess heat to the distillation flask, especially when the volume of water in the flask has been reduced below about 100 mL. Continue the distillation until oil droplets no longer appear in the distillate, which should occur after about 150–200 mL of distillate has been collected. Allow the distillate to cool to room temperature or below, using an ice-water bath if necessary.★ Proceed to "Isolation of Citral," below.

Isolation of Citral Transfer 50–75 mL of the distillate to a separatory funnel containing 50 mL of technical-grade diethyl ether, shake the funnel, and allow the layers to separate in the funnel. The proper use of a separatory funnel is described in Section 2.18. The funnel should be vented *frequently* to avoid the buildup of pressure inside. Carefully draw off the bottom layer, which is the aqueous layer, and leave the upper layer, which is the ether layer, in the funnel. Add another portion of the distillate to the ether layer in the funnel, shake the funnel, and separate the layers as before. Repeat these steps until all the distillate has been shaken, in small portions, with the ether, so that the citral has been removed from the aqueous phase. After the extraction is complete, all of the aqueous layers may be discarded.

Transfer the organic layer that remains in the separatory funnel to a 125-mL Erlenmeyer flask, and add about 1 g of anhydrous calcium chloride. Allow the contents of the flask to remain in contact with the drying agent until the organic layer is dry, as evidenced by observing that it is completely clear. If the experiment is stopped at this point, the flask containing the ether should be loosely stoppered and stored in a ventilation hood; *never leave flasks containing ether in your locker drawer*.★ Decant (Section 2.15) the dried organic solution into a 250-mL round-bottomed flask, and evaporate the ether solvent under vacuum, using a water aspirator or the house vacuum. When vacuum is applied, the ether is likely to start foaming; if this occurs, reduce the vacuum until foaming stops. It will be advantageous to place the flask in a pan of water *at room temperature* during the evaporation of the ether. After the ether is completely removed, the residue that remains is citral, bp 229°C (760 torr). Determine its weight, and calculate the percentage recovery of citral, based on the weight of the original sample of lemon grass oil.

If requested to do so, save or submit a sample of citral for gas chromatographic or spectral (IR, PMR, and/or UV) analysis; the latter are discussed in Chapters 10 and 11.

Chemical Characterization of Citral You may be asked to perform one or more of the following tests on the citral you have isolated.

1 Test for carbon-carbon double bond. Perform the tests for unsaturation described in Section 27.5b.

2 Test for an aldehyde. Perform the chromic acid test outlined in Section 27.5a, p. 701.

3 Preparation of solid derivatives. Prepare the 2,4-dinitrophenylhydrazone (see Section 27.5a, p. 704) and the semicarbazone (see Section 27.5a, p. 704) of citral. The melting points of these derivatives are given below.

Compound	Melting Point of 2,4-Dinitrophenylhydrazone	Melting Point of Semicarbazone
12a	134–135°C	164–165°C
12b	171–172°C	125–126°C

Recrystallization of either derivative of citral usually removes any derivative resulting from the minor amount of **12b** contained in citral. After recrystallization, the derivatives that you obtain should have the melting points given for **12a.**

REFERENCES

1. *Technique of Organic Chemistry,* A. Weissberger, editor, Interscience Publishers, New York, 1965, Vol. IV.

2. *Azeotropic Data,* American Chemical Society, Washington, D.C., 1952.

3. Jordan, T. E., *Vapor Pressure of Organic Compounds,* Interscience Publishers, New York, 1954.

extraction

The most common methods of isolating and purifying organic compounds, recrystallization for solids and distillation for liquids, are described in Chapters 3 and 4. Two other widely applicable and relatively simple methods of separating a desired compound from its impurities, or of isolating each of the individual components of a mixture, are **extraction** and **chromatography,** both of which are based on the principle of *phase distribution*. These methods are the subjects of this chapter and Chapter 6, respectively.

The technique of extraction involves the *selective removal* of one or more of the components of a gaseous, liquid, or solid mixture by contact of the mixture with a second phase. A substance may establish an equilibrium distribution between two *insoluble* phases with which it is in contact, in a ratio dependent upon its relative stability in each of those phases.

The process on which such a distribution depends may be one of two varieties: (1) *partitioning,* based on differing relative solubilities of the components in immiscible solvents (selective dissolution) and (2) *adsorption,* based on the selective adherence of the components of a liquid or gaseous mixture to the surface of a solid phase. The various techniques of chromatography involve both of these processes, whereas the techniques of extraction involve only the former.

5.1 Theory of Extraction

A widely used method of separating organic compounds from mixtures in which they are found or produced is that of *liquid-liquid* extraction. In fact, virtually every organic synthesis or isolation of a natural product requires one or more extractions of this type at some stage for the purification of a product.

In its simplest form, extraction involves the distribution of a solute between two immiscible solvents. The distribution is expressed quantitatively in terms of the *distribution* (or *partition*) *coefficient, K* (equation 1). This expression indicates that a solute, A, in contact with a mixture of two immiscible liquids, S and S', will be distributed (partitioned) between the liquids so that at equilibrium the ratio of concentrations of A in each phase will be constant, at constant temperature. For simplicity we shall define S as that solvent in which the solute is more soluble. Therefore in the remaining discussion the value of K will by definition always be greater than 1.0.

$$K = \frac{\text{concentration of } A \text{ in } S}{\text{concentration of } A \text{ in } S'} \tag{1}$$

Ideally, the distribution coefficient of A is equal to the ratio of the individual solubilities of A in pure S and in pure S'. In practice, however, this correspondence is generally only approximate, since no two liquids are completely immiscible. The extent to which they dissolve in each other alters their solvent characteristics and thus slightly affects the value of K.

It is evident that for A to dissolve completely in one or the other of two immiscible liquids, the value of K must be infinity or zero. Neither of these limiting values is actually attained. However, so long as K is larger than 1.0, and the volume of solvent S is equal to or larger than the volume of solvent S' (see equation 3), the solute will be found in *greater amounts* in solvent S. The amount of solute that will remain in the other solvent, S', will depend on the value of K.

Equation 1 may be rewritten as shown in equations 2 and 3, given below:

$$K = \frac{\text{grams of } A \text{ in } S/\text{mL of } S}{\text{grams of } A \text{ in } S'/\text{mL of } S'} \tag{2}$$

$$K = \frac{\text{grams of } A \text{ in } S}{\text{grams of } A \text{ in } S'} \times \frac{\text{mL of } S'}{\text{mL of } S} \tag{3}$$

Note that when the volumes of S and S' are equal,[1] the ratio of the grams of A in S and in S' will equal the value of K. If the volume of S is doubled and the volume of S' is kept the same, the ratio of the grams of A in S to the grams of A in S' will be *increased* by a factor of two. This follows because K is a *constant*. Therefore, if A is to be recovered by extraction into solvent S, the amount of A recovered will be increased by using larger quantities of solvent S.

A further consequence of the distribution law (equation 1) is of practical importance in performing an extraction. If a given total volume of solvent S is to be used to separate a solute from its solution in S', it can be shown to be more efficient to effect several successive extractions with portions of that volume than one extraction with the full volume of solvent.

---

[1]To be strictly correct, the volumes of *solution* should be used in this expression. However, when the solutions are reasonably dilute, volumes of *solvent* may be used without appreciable error.

Thus, for example, *more* butanoic acid will be removed from water solution by two successive extractions with 50-mL portions of diethyl ether than will be removed in a single extraction with 100 mL of diethyl ether. Three successive extractions with 33-mL portions would be still more efficient (see Exercise 3 at the end of this chapter). There is, however, a point beyond which the further effort of additional extractions no longer yields a commensurate return. The larger the distribution coefficient, the fewer the number of repetitive extractions that are necessary to separate the solute effectively. This is an important consideration because it is desirable to keep the total volume of extracting solvent to a minimum, not only for reasons of expense but also because of the time involved in eventual removal of the solvent by distillation.

Consider now a solution of two compounds in solvent S'. For effective separation of these two compounds by extraction with solvent S, the distribution coefficient of one should be significantly greater than 1.0, whereas the distribution coefficient of the other should be significantly smaller than 1.0. If these conditions are met, one compound will be mainly distributed in solvent S and the other in solvent S', at equilibrium. Physical separation of the two liquid layers would then result in at least a partial separation of the two compounds.

When the coefficients are of similar magnitude, separation by the extraction technique may be quite ineffective, since the relative concentrations of the compounds in each of the two liquid phases may be little changed from those of the original mixture. In this case other methods of separation should be used, such as adsorption chromatography (Section 6.1), fractional distillation (Section 4.4), or fractional recrystallization (Section 3.1). The specialized techniques of *fractional extraction* and *countercurrent distribution* can also be used to separate compounds with similar but not identical distribution coefficients. For the interested student, details of these techniques are supplied in the references at the end of the chapter.

When choosing an extracting solvent for the isolation of a component from a solution, some general principles should be kept in mind: (1) The solvent must of course be immiscible with the solvent of the solution. (2) The solvent chosen must have the most favorable distribution coefficient for the component that is to be separated and must have unfavorable coefficients for the impurities or other components. Since distribution coefficients for most compounds in various solvents are not available, extraction solvents used in experiments in this book will be suggested to you. However, as you gain experience and understanding about the solubility principles of organic compounds, you should be able to suggest appropriate solvents. (3) A solvent must be chosen that does not react chemically with the components of the mixture, just as in recrystallization. (4) Following the extraction, the solvent should be readily separable from the solute. Usually the solvent is removed by distillation, so that use of relatively volatile solvents is advantageous.

5.2 Continuous Extraction

Another type of experimental problem often encountered is that involving separation of one component that is only slightly soluble in the extracting solvent from a mixture whose other components are essentially insoluble. Large quantities of solvent would have to be used in order to effect the separation in only one or two extractions, and the handling of such quantities may be extremely unwieldy. Alternatively, it would be tedious to do a very large number of extractions with smaller quantities of solvent. The method of *continuous extraction,* in which a relatively small volume of extracting solvent is used, is a possible solution to this problem. The solution of extracting solvent and solute is continuously separated into a boiling flask from the mixture being extracted. The solution is subjected to continuous distillation, and the condensed distillate is returned as fresh extracting solvent to the extraction vessel and reused. In the process the extracted material builds up in an increasingly concentrated solution in the boiling flask. This is because a dilute solution of it is continuously draining into the flask, while at the same time the amount of solvent is kept constant.

A continuous liquid-liquid extraction is performed by percolating the extracting solvent, as condensate from the condenser, through the solution containing the desired compound. The direction of solvent flow through the solution depends on the relative densities of the two liquids, and specialized equipment is available for either situation. For example, apparatus such as shown in Figures 5.1 and 5.2 are designed for use with solvents less and more dense, respectively, than the solution containing the solute being extracted.

For separation of the components of a solid mixture by continuous *solid-liquid* extraction, a Soxhlet extraction apparatus (Figure 5.3) is convenient. The solid is placed in a

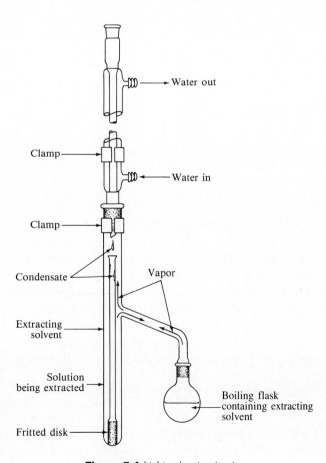

Figure 5.1 Light-solvent extractor.

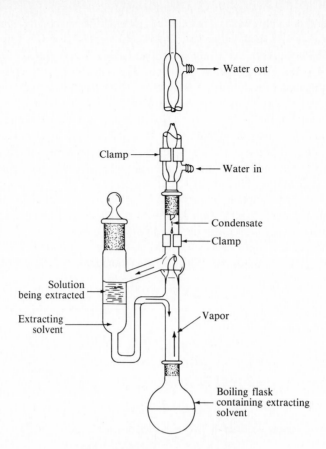

Figure 5.2 Heavy-solvent extractor.

porous thimble in the chamber, as shown, and the extracting solvent is added to the boiling flask below. The solvent is heated to reflux, and the distillate, as it drops from the condenser, collects in the chamber. By coming in contact with the solid in the thimble, the liquid effects the extraction. After the chamber fills to the level of the upper reach of the siphon arm, the solution empties from this chamber into the boiling flask by a siphoning action. This process may be continued automatically and without attendance for as long as is necessary for effective removal of the desired component, which will then be contained in the solvent in the boiling flask.

5.3 Extraction of Natural Products

The isolation of citral from lemon grass oil by steam distillation is described in Section 4.7, and mention is made of the fact that extraction by a solvent is often used as an alternative way

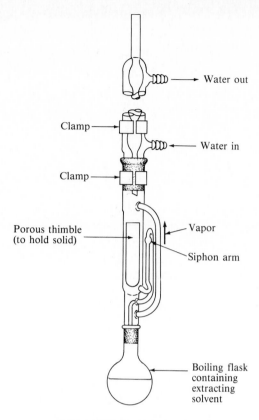

Water out

Clamp

Water in

Clamp

Porous thimble
(to hold solid)

Vapor

Siphon arm

Boiling flask
containing
extracting
solvent

Figure 5.3 Soxhlet extractor.

of isolating a desired substance from its natural source. Some very common examples of this procedure are the extractions of caffeine and the accompanying flavor components from ground coffee beans and from tea leaves in the everyday preparation of coffee and tea. In these cases, hot water is used as the extracting solvent. For the extraction of water-insoluble materials, various organic solvents are used, such as diethyl ether, ethanol, acetone, and dichloromethane.

As mentioned in Section 4.7, extracts obtained by solvent extraction are usually a complex mixture of compounds that must be separated by additional procedures. Further extractions may be carried out with acids or bases to separate acidic and basic natural products from each other and from neutral compounds (consult Section 5.4), and chromatographic techniques (Section 6.1) may be required. Thus, the isolation of natural products is usually a complex and tedious process. There are, however, some exceptions to this general rule, and the isolation of trimyristin from nutmeg is one example.

Trimyristin, or glyceryl trimyristate, **1,** is the principal lipid found in nutmegs, the fruit buds of the nutmeg tree. Nutmeg has been a valued spice ever since the Portuguese sea captains brought it back from the ''Spice Islands'' of Indonesia in the 16th century.

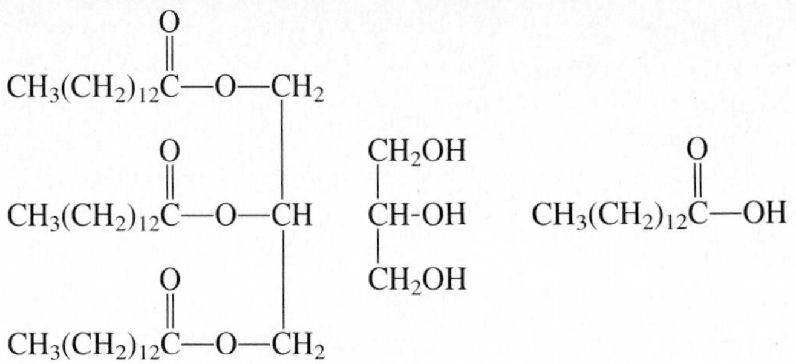

$$CH_3(CH_2)_{12}\overset{\displaystyle O}{\overset{\|}{C}}—O—CH_2$$

$$CH_3(CH_2)_{12}\overset{\displaystyle O}{\overset{\|}{C}}—O—CH$$

$$CH_3(CH_2)_{12}\overset{\displaystyle O}{\overset{\|}{C}}—O—CH_2$$

CH$_2$OH

CH-OH

CH$_2$OH

$$CH_3(CH_2)_{12}\overset{\displaystyle O}{\overset{\|}{C}}—OH$$

1
Trimyristin

2
Glycerol

3
Myristic acid

What is unusual about the presence of trimyristin in nutmeg is that it occurs in high concentration as the triester of glycerol (2) and tetradecanoic (myristic) acid (3), without the presence of other closely related esters. This unique fact allows the isolation of myristin in an almost-pure state by a simple extraction with diethyl ether, a procedure that is illustrative but, because of its simplicity, not quite representative of the isolation of natural products by extraction.

Pre-lab Exercises for Section 5.3, Extraction of Natural Products, can be found on page PL. 11.

EXPERIMENTAL PROCEDURE

DO IT SAFELY

1 Diethyl ether and acetone are both volatile and flammable solvents. No flames should be allowed in the laboratory during this experiment.

2 When heating the diethyl ether under reflux to extract the trimyristin from the nutmeg, which should be done using a steam or hot-water bath, take care that the boiling is so gentle that the ring of condensate in the reflux condenser is not above the lower third of the condenser. Very little heat will be required to maintain this level of refluxing. This will ensure that no vapors of ether escape through the top of the condenser.

Place about 40 g of ground nutmeg and several boiling stones in a 250-mL round-bottomed flask and add 100 mL of diethyl ether. Attach a reflux condenser and start a stream of cold water through it. Using a steam or hot-water bath, gently heat the mixture to reflux, observing the precautions described in the Do It Safely section. Continue the refluxing for about 1 h, then allow the reaction mixture to cool to room temperature.★ Separate the ether solution from the residual nutmeg by filtration through a fluted filter paper; rinse any nutmeg remaining in the flask onto this filter with the aid of an additional 10 to 15 mL of diethyl ether.★

Remove the diethyl ether from the trimyristin by simple distillation, using a steam or hot-water bath for heating. Dissolve the residue in 50 mL of acetone by heating over the steam or hot-water bath, then pour the hot solution into a 125-mL Erlenmeyer flask to cool.★ Allow the crystallization to proceed at room temperature for about 0.5 h and then cool the mixture in an ice-water bath for 0.5 h. Collect the crystals of trimyristin by vacuum filtration, allow them to dry in air, and weigh them to determine the percent recovery, based on the original weight of nutmeg. Determine the melting point of the trimyristin, whose reported melting point is 58°C.

EXERCISES

1. A certain plant material known to contain mainly trimyristin and tripalmitin (look up the structure of this compound), mp 66°C, in approximately equal amounts was extracted with diethyl ether. After removal of the diethyl ether from the extract, an oil remained which was difficult to crystallize. Explain. [*Hint:* Consult Section 3.2c.]

2. Why was diethyl ether rather than acetone chosen as the extraction solvent?

3. Why was acetone rather than diethyl ether chosen as the recrystallization solvent?

4. Would water be a good choice as an extracting solvent for trimyristin? Consider the anticipated solubility properties of trimyristin.

5. Judging from your examination of trimyristin, is it responsible for the odor and taste of nutmeg?

6. Why is gravity rather than vacuum filtration used to separate the ether extract from the residual trimyristin?

7. Why is a fluted filter paper rather than a plain filter paper used?

5.4 Acid and Base Extractions

The solubility characteristics of organic acids and bases in water are quite dramatically affected by the adjustment of pH. For example, except for the lower molecular weight members, most organic acids, such as carboxylic acids and phenols, are either insoluble or only slightly soluble in water. However, these same acids are found to be soluble in dilute aqueous sodium hydroxide solution in which the pH is decidedly *above* 7. The reason for this behavior is illustrated in equation 4.

$$
\underset{\text{(Insoluble in } H_2O)}{R-\overset{\overset{\displaystyle O}{\|}}{C}-O-H} + NaOH \rightarrow \underset{\text{(Soluble in } H_2O)}{R-\overset{\overset{\displaystyle O}{\|}}{C}-O^{\ominus}Na^{\oplus}} + H_2O \qquad \text{(4)}
$$

The organic acid undergoes *deprotonation* in an acid-base reaction with the sodium hydroxide to produce its corresponding conjugate base, which, with the sodium ion, constitutes a *salt*. Because of its ionic character, the salt of the organic acid is soluble in water even though the acid itself is not. If this alkaline solution is then neutralized or made slightly acidic by the subsequent addition of a mineral acid such as hydrochloric acid, the conjugate base will be reprotonated to form the original organic acid, which, because of its insolubility in water, will precipitate from solution or form a second layer if a liquid (equation 5). The organic acid may then be recovered from this heterogeneous mixture, usually by filtration.

$$R-\overset{\overset{\displaystyle O}{\|}}{C}-O^{\ominus}Na^{\oplus} + HCl \rightarrow R-\overset{\overset{\displaystyle O}{\|}}{C}-O-H + Na^{\oplus}Cl^{\ominus} \tag{5}$$

(Soluble in H_2O)   (Insoluble in H_2O)

Similarly, an organic base that is insoluble in water (pH 7) will generally be found to be quite soluble in dilute hydrochloric acid solution in which the pH is decidedly *below* 7. In this case the increase in solubility of the organic base in the aqueous medium relies on the *protonation* of that base by the HCl to produce the ionic and therefore water-soluble conjugate acid, a substituted ammonium ion, which, with the chloride ion, constitutes a salt (equation 6). If this acidic solution is then neutralized or made slightly basic by the addition of aqueous sodium hydroxide, the conjugate acid will be *deprotonated* to produce the original organic base, which is insoluble in water and precipitates from the solution or forms a second layer if a liquid, leading to its recovery (equation 7).

$$R-\overset{..}{N}H_2 + HCl \rightarrow R-\overset{\overset{\displaystyle H}{|}}{\underset{\oplus}{N}}H_2\ Cl^{\ominus} \tag{6}$$

(Insoluble in H_2O)   (Soluble in H_2O)

$$R-\overset{\overset{\displaystyle H}{|}}{\underset{\oplus}{N}}H_2\ Cl^{\ominus} + NaOH \rightarrow R-\overset{..}{N}H_2 + H_2O + Na^{\oplus}Cl^{\ominus} \tag{7}$$

(Soluble in H_2O)   (Insoluble in H_2O)

Thus an organic acid may be selectively removed from a mixture containing other neutral or basic materials by dissolving that mixture in an organic solvent such as diethyl ether or dichloromethane and then extracting the solution with dilute sodium hydroxide solution, which will *selectively* remove the organic acid into the aqueous phase through formation of its conjugate base. Or an organic base may be selectively removed from a similar mixture by extraction with dilute hydrochloric acid solution.

In the following Experimental Procedure, the separation of three organic compounds is described. Benzoic acid **(1)**, *p*-nitroaniline **(2, a base)**, and azobenzene **(3, a neutral com-**

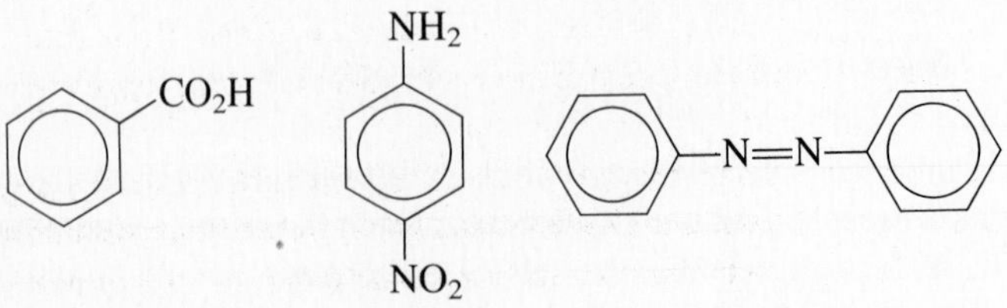

        1   **2**   **3**
    Benzoic acid   *p*-Nitroaniline   Azobenzene

pound), each of which is only very slightly soluble in cold water, have been chosen to demonstrate the principles of separations by extractions based on acid-base reactions.

One should note that similar techniques are used throughout the book in many of the preparative-type experiments.

Pre-lab Exercises for Section 5.4, Acid and Base Extractions, are found on page PL. 13.

EXPERIMENTAL PROCEDURE

DO IT SAFELY

1 Take care not to let any of the chemicals in this experiment come in contact with your skin. If they do, wash the affected areas with soap and water. If you wear rubber gloves, be careful in handling wet glassware because it will be slippery in contact with the rubber.

2 Remember to vent the separatory funnel frequently during its use to avoid buildup of pressure.

In preparation for this experiment, study the detailed instructions for using a separatory funnel in Section 2.18.

Prepare or obtain from your instructor a mixture composed of 2 g of benzoic acid, 1 g of p-nitroaniline, and 1 g of azobenzene. Dissolve this mixture in about 75 mL of dichloromethane. If *small* quantities of solid do not dissolve, filter the mixture into a separatory funnel. Extract the solution with two 50-mL portions of 6 M hydrochloric acid. Combine the two aqueous layers (be sure you know which layer is aqueous—consult Section 2.18) from the two extractions in a flask labeled "acidic extract." Return the organic layer, which has now been extracted twice with hydrochloric acid solutions, to the separatory funnel and extract it twice more, using 50-mL portions of 3 M sodium hydroxide solution each time. Combine these two aqueous layers in a second flask labeled "basic extract."

Dry the organic layer by adding 4–5 g of *anhydrous* sodium sulfate to the liquid that has been placed in a third flask labeled "organic solution."★

While the "organic solution" is drying over sodium sulfate, cool each of the aqueous extracts in an ice-water bath. Neutralize the "acidic extract" with 3 M sodium hydroxide and add a little excess base to make the solution distinctly basic to litmus paper. Neutralize the "basic extract" with 6 M hydrochloric acid and add a little excess acid to make the solution distinctly acidic to litmus paper. Upon neutralization, a precipitate should appear in each flask. (What are these precipitates?)

Collect the precipitates separately by vacuum filtration. Wash each of the precipitated solids *on the Büchner funnel* with *cold* distilled water. Preweigh ("tare") three small labeled vials and place the solid materials obtained thus far in two of them. (What should the labels read?) Either allow these samples to air-dry until the next laboratory period or place them in an oven at a temperature of no more than 90°C for 1.5 h. If the samples are to be allowed to air-dry until the next laboratory period, place them in your drawer and cover the tops of the vials loosely with filter paper or absorbent cotton.

Separate the "organic solution" from the sodium sulfate by gravity filtration and remove the solvent by simple distillation, using a steam or hot-water bath for heating. Discontinue the distillation when only a small amount of material remains in the stillpot, even if no crystals are evident in the residue. Allow the stillpot to cool and then attach it to a source of vacuum for a few minutes in order to remove the last traces of the dichloromethane solvent. Transfer the solid residue to the third tared vial and allow it to air-dry; do not use an oven to dry this sample. (Why not?)

After the samples have dried, reweigh each of the vials to obtain the weights of the separated materials. Determine the melting points of each of them. For final purification, your instructor may wish you to recrystallize one or all of them.

The reported melting points of the pure original components of the mixture are as follows:

benzoic acid, 121°C
p-nitroaniline, 147°C
azobenzene, 68°C

Optional Experiment

Your instructor may assign an unknown mixture to be separated using the preceding procedure.

EXERCISES

1. On the basis of what you learned in this experiment, plus the information that phenol is soluble in sodium hydroxide solution but not in sodium hydrogen carbonate solution, whereas benzoic acid is soluble in both solutions, show how you could separate a mixture containing phenol, benzoic acid, *p*-nitroaniline, and azobenzene.

*2. (a) The pK_a of benzoic acid is 4.19. Show mathematically that this acid is 50% ionized at pH 4.19. (b) Use the result of part a to explain why precipitation of benzoic acid is incomplete if the pH of an aqueous solution of benzoate ion is lowered only to pH 7 by addition of acid.

3. Given 400 mL of an aqueous solution containing 12 g of compound **A,** from which it is desired to separate **A,** how many grams of **A** could be removed in a single extraction with 200 mL of diethyl ether? (Assume the distribution coefficient for **A** in diethyl ether/water is equal to 3.0.)

   How many *total* grams of **A** can be removed with *three successive extractions* of 67 mL each?

   [*Note:* In solving problems of this type, one should recognize that equation 3 applies to the situation pertaining *after* equilibrium has been reached; for example, a practical form of the equation is

   $$K = \frac{x}{a - x} \times \frac{\text{mL of } S'}{\text{mL of } S}$$

   where a = grams of **A** originally present in water (S') and x = grams of **A** present in diethyl ether (S) after extraction.]

4. Why is anhydrous sodium sulfate added to the ''organic solution''?

5. Why are the ''acidic extract'' and the ''basic extract'' cooled before neutralization?

6. Why was the water used for washing the precipitates on the funnels specified to be *cold*?

REFERENCES

1. Craig, L. C.; Craig, D., in *Technique of Organic Chemistry,* 2nd ed., A. Weissberger, editor, Interscience Publishers, New York, 1956, Vol. III, Part 1, Chapter 2.

2. Frank, F.; Roberts, T.; Snell, J.; Yates, C.; Collins, J., *Journal of Chemical Education,* **1971,** *48,* 255.

chromatography

Chromatography, like extraction, is based on the principles of phase distribution, as is mentioned in the introduction to Chapter 5. Reduced to its fundamentals, chromatography involves the *selective removal* of the components of one phase from that phase as it is flowing through a secondary *stationary* phase. The partitioning of components into the stationary phase is an equilibrium process, so that molecules contained in that phase continuously reenter the moving phase. Separation of two or more components in the moving phase will result when the equilibrium constants for the partitioning of these components between the two phases differ. Expressed simply, the more tenaciously one component is held by the stationary phase, the higher the percentage of molecules of that component that will be held *immobile*. A second component, less strongly held, will provide a higher percentage of molecules in the *mobile* phase than will the first component. Therefore, on the average, the molecules of the component that is held less strongly will move over the stationary phase, in the direction of flow, at a higher rate than the other, resulting in a migration of the components into separate regions, or bands, of the stationary phase, as shown in Figure 6.1.

The major types of chromatography are characterized by the difference in the nature of the two phases, mobile and stationary, that interact in the separation process. **Column** chromatography, **ion-exchange** chromatography, **thin-layer** chromatography (**TLC),** and **high-pressure liquid** chromatography (**HPLC**) are all characterized by *liquid-solid* phase interactions. **Gas** chromatography (**GC,** also called gas-liquid partition chromatography, GLPC) involves a mobile *gas* phase and a stationary *liquid* phase. **Paper** chromatography is essentially a *liquid-liquid* multiple extraction process. These different types will be described in more detail in the following sections. Experimental procedures to illustrate some of them

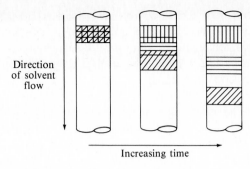

Direction
of solvent
flow

Increasing time

Figure 6.1 Separation of components by column chromatography.

are presented in this chapter, and references are made to applications of some of them in other sections of the book.

6.1 Column Chromatography

In column chromatography the stationary phase is a solid, which separates the components of a liquid passing through it by selective *adsorption* on its surface. The types of interactions that cause adsorption are the same as those that cause attractions between any molecules, that is, electrostatic attraction, complexation, hydrogen bonding, and van der Waals forces.

A column such as that shown in Figure 6.2 is used to separate a mixture by column

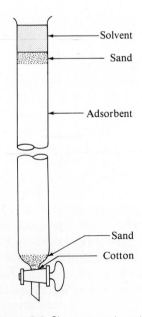

Solvent

Sand

Adsorbent

Sand

Cotton

Figure 6.2 Chromatography column.

chromatography. The column is packed with a finely divided solid such as alumina or silica gel, which serves as the stationary phase, and a sample of the mixture to be separated is applied at the top. If the mixture is solid, it must be dissolved in a solvent and then applied. The sample will initially be adsorbed at the top of the column, but when an eluting solvent, the mobile phase, is allowed to flow through the column, it will carry with it the components of the mixture. Owing to the selective adsorption power of the solid phase, the components ideally will move down the column at different rates. A more weakly adsorbed compound will be eluted more rapidly than a more strongly adsorbed compound, because the former will have a higher percentage of molecules in the mobile phase. The progressive separation of the components will appear as in Figure 6.1.

The separated components may be recovered in two ways: (1) The solid packing may be extruded and the portion of the solid containing the desired band cut out and extracted with an appropriate solvent. (2) Solvent can be passed through the column until the bands are eluted from the bottom of the column and collected in different containers, since they will be eluted at different times. The second method is the more commonly used; the first is rather difficult because the solid must be extruded from the column in one piece.

With colored materials the band may be directly observed as they pass down the column. The word chromatography (Gr. *chromatos,* a color) was originally coined to describe this technique from such observations. With colorless materials the changes cannot be observed directly. Many materials fluoresce when irradiated with ultraviolet light, however, and this provides a method of observing the bands in these cases. Usually the progress of a column chromatographic experiment is followed by collecting a series of fractions of eluent of equal volume, for example, 25 mL. The solvent is then evaporated from each to see if any solute is present. If the volume of each fraction is kept relatively small—for example, less than 10% of the volume of the column—the different bands will usually be obtained in different flasks, although each component may be distributed among several flasks. Another convenient method of following the separation is to analyze the eluent at intervals by thin-layer chromatography (Section 6.2). The composition of the material eluted from the column may also be ascertained by determination of its physical constants (Chapters 3 and 4) or its spectra (Chapters 8–11) and comparison of these with those of known materials.

A few of the solid adsorbents commonly used include alumina, silica gel, Florisil, charcoal, magnesia, calcium carbonate, starch, and sugar. The organic chemist usually finds alumina, silica gel, and Florisil (activated magnesium silicate) of the greatest utility.

Alumina (Al_2O_3) is a highly active, strongly adsorbing, polar compound coming in three forms: neutral, and base and acid washed. Basic and acidic alumina offer good separating power for acids and bases, respectively. For compounds sensitive to chemical reaction under acidic or basic conditions, neutral alumina should be used. Being highly polar itself, alumina adsorbs polar compounds quite tenaciously, so that they may be difficult to elute from the column. The activity (adsorptivity) of alumina may be reduced by the addition of small amounts of water; the weight percentage of water present determines the activity grade of the alumina. Silica gel and Florisil are also polar, but less so than alumina.

For the greatest effectiveness, the solid adsorbent should be of uniform particle size and of *high specific area,* a property that promotes more rapid equilibrium of the solute between the two phases. This is important for producing narrow bands. Good grades of alumina and silica gel have very high specific areas, on the order of several hundred m^2/g.

The strength of adsorption depends on the adsorbate as well as on the adsorbent. It has been found that the strength of adsorption for compounds having the following types of polar functional groups increases on any given adsorbent in the following order:

$$Cl-, \ Br-, \ I- \ < \ \underset{/}{\overset{\backslash}{C}}=\underset{\backslash}{\overset{/}{C}} \ < \ -OCH_3 \ < \ -CO_2R \ < \ \underset{/}{\overset{\backslash}{C}}=O \ < \ -CHO \ <$$

$$-SH \ < \ -NH_2 \ < \ -OH \ < \ -CO_2H$$

The nature of the liquid phases (solvents) to be used is an important consideration when designing a chromatographic experiment. The solvent may also be adsorbed on the solid, thereby competing with the solute for the adsorptive sites on the surface. If the solvent is more polar and more strongly adsorbed than the components of the mixture, these components will remain almost entirely in the mobile liquid phase, and little separation will occur during the experiment. For effective separation, the eluting solvent must be significantly less polar than the components of the mixture. Furthermore, the components must be soluble in the solvent; if they are not, they will remain permanently adsorbed on the stationary phase of the column. The eluting powers of various solvents, that is, their ability to move a given substance down a column, are generally found to occur in the order shown.

Petroleum ether (bp 30–60°C)
Hexane
Carbon tetrachloride
Toluene
Dichloromethane   Increasing
Chloroform   eluting
Diethyl ether   power
Ethyl acetate
Acetone
1-Propanol
Ethanol
Methanol
Water

The amount of sample that can be applied to a column depends on the difficulty of the separation and the quantity of adsorbent being used. Typically, a ratio of 1 g of sample per 25 g of adsorbent allows satisfactory separations to be achieved, but if the components of the mixture have very similar distribution coefficients, a lower ratio of sample to adsorbent, such as 1:100, may be required. When a mixture of solids in solution is introduced at the top of a column, a minimum amount of solvent should be used in order to minimize the width of the initial band, thus enhancing the potential for a successful separation. Using too much solvent could, for example, distribute the sample mixture over the upper 25% of the column prior to initiation of the elution operation. When applying a liquid sample, regardless of whether it is a mixture of liquids only or a solution of solid materials, one must take care that the liquid covers the entire surface of the adsorbent at the top of the column to avoid having the sample adsorbed only where it touches the adsorbent. As an aid to maintaining even distribution of the sample at the top of the column, a layer of sand is often placed above the adsorbent at the top of the column, where it prevents disruption of the packing material as solvent is added.

The method of packing the column is very important, because a poorly packed column will have minimal resolution. The adsorbent is usually "packed wet" by partially filling the column with an inert liquid and then adding the adsorbent, allowing it to settle through the liquid while tapping the column so as to provide even distribution of the adsorbent and elimination of entrapped air bubbles. Throughout the subsequent use of the column, the liquid level should never be allowed to reach the top of the column so that any part of the adsorbent becomes dry, as this would lead to channeling during the elution process, which would give rise to ragged bands during the development of the chromatogram.

In a *simple elution* experiment the sample of the mixture to be separated is placed on the column, and a *single* solvent is used throughout the separation. The optimum solvent choice will be that which produces the greatest band separation. Since the best solvent will most likely be found by trial and error, it is sometimes convenient to use the techniques of thin-layer chromatography (Section 6.2) in the selection of a solvent for column chromatography. A series of thin-layer chromatographic experiments using various solvents may be performed in a relatively short time. The best solvent or mixture of solvents found in this way will usually be appropriate for the column chromatography.

A procedure known as *stepwise* or *fractional elution* is most commonly used. In this method a *series* of increasingly more polar solvents is used to develop the chromatogram. Starting with a nonpolar solvent such as hexane, one band may move down and off the column while the others remain very near the top. Ideally, then, the solvent would be changed to one of slightly greater polarity in the hope that one more band will be eluted while the others remain behind. If too large a jump in polarity is attempted, all the remaining bands may come off at once. Therefore small systematic increases in solvent polarity should be effected at each step. This is best accomplished not by changing solvents entirely but by using *mixed* solvents.[1] For example, an appropriate quantity of hexane could be passed through the column, followed by a quantity of solvent having the composition 95% hexane–5% chloroform. Solvent mixtures containing 10, 15, 20, 40, and 80% chloroform could then be used in succession, followed by pure chloroform. A similar series utilizing chloroform with diethyl ether or acetone could then be employed. This technique has proved to be quite efficient with regard both to column resolution and to time invested and is very commonly employed in research laboratories. A rule of thumb to use in determining the amount of solvent to pass through before changing to one of higher polarity is to use a volume equal to approximately three times the packed volume of the column. There are exceptions, but this is a reasonable suggestion in the absence of more specific knowledge.

In the following procedure, specific details for the preparation of a typical chromatographic column are given, followed by an example of its use in separating the geometric isomers of azobenzene, colored compounds whose separation can be followed visually. Commercial azobenzene is mainly in the form of the *anti*-isomer, **1,** which is orange, but it contains a smaller amount of the *syn*-isomer, **2,** which is yellow. The different colors of the two isomers allow one easily to follow the progress of a chromatographic separation on a column.

[1]Refer to Section 17.3, in which 2- and 4-nitrobromobenzene are separated by column chromatography.

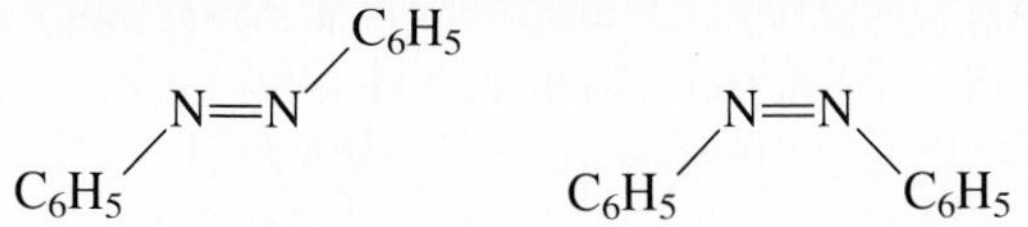

$$
\begin{array}{cc}
\underset{C_6H_5}{\overset{C_6H_5}{\diagdown}}N\!=\!N & \underset{C_6H_5}{\overset{N=N}{\diagup}}\diagdown_{C_6H_5}
\end{array}
$$

$$
\begin{array}{cc}
\textbf{1} & \textbf{2} \\
\textit{anti}\text{-Azobenzene} & \textit{syn}\text{-Azobenzene}
\end{array}
$$

Pre-lab Exercises for Section 6.1, Column Chromatography, are found on page PL. 15.

EXPERIMENTAL PROCEDURE

DO IT SAFELY

Petroleum ether, bp 30–60°C, is highly volatile and flammable. Have no flames in the vicinity during the preparation and development of the chromatographic column.

A. Preparation of Column

Clamp in a vertical position a 50-mL buret with its stopcock closed but ungreased. Fill the buret to approximately the 40-mL mark with petroleum ether (bp 30–60°C) and insert a small plug of glasswool into the bottom of the buret by means of a long piece of glass tubing. Introduce into the buret enough clean sand to form a 1-cm layer on top of the glasswool plug. *Slowly* add 15 g of alumina to the buret while continually tapping the column with a "tapper" made from a pencil and a one-holed rubber stopper. The agitation of the column while the alumina is being sifted in assures an evenly packed column. With a little additional petroleum ether, wash down the inner walls of the buret to loosen any adhered alumina. Introduce a 1-cm layer of sand at the top of the column. Open the stopcock and allow the solvent to drain just to the top of the upper layer of sand. The column should now be ready to receive the sample.

B. Separation and Purification of *anti*-Azobenzene

Obtain 1–2 mL of a half-saturated solution of commercial azobenzene in petroleum ether (bp 30–60°C). Also obtain a small amount of the solid commercial azobenzene and accurately determine its melting range. To a column prepared as described in part A using 15 g of alumina, apply 1.0 mL of the azobenzene solution. The solution may be transferred to the column with a pipet, allowing the solution to run evenly down the inside surface of the buret. Open the stopcock and allow the liquid to drain to the top of the sand. In a like manner introduce 1–2 mL of fresh petroleum ether to wash down the inner surface of the buret. Again allow the liquid to drain to the top of the sand. Fill the buret with fresh petroleum ether. Open the stopcock; as the solvent runs through the column, a broad orange band of the *anti*-isomer of azobenzene should be observed to move slowly down the column. A narrow yellow band of the *syn*-isomer should remain very near the top of the column. Allow petroleum ether to run through the column until the orange band is eluted and collected at the bottom of the column in a small Erlenmeyer flask. Remove most of the solvent by simple distillation and then transfer the final 1–2 mL of the concentrated eluent to a small round-bottomed flask. Attach the flask to an aspirator or house vacuum line in order to remove the last small amount of solvent under evacuation. When the crystals of the purified *anti*-azobenzene are completely dry, accurately determine their melting point. The reported melting point of *anti*-azobenzene is 68°C.

If desired, after the *anti*-isomer has been eluted from the column, the buret may be refilled with methanol and the *syn*-isomer eluted from the column. Typically, however, an insufficient amount of the *syn*-isomer is obtained for adequate characterization.

EXERCISES

1. Why is it preferable to use an ungreased rather than a greased stopcock on a buret used for column chromatography?

2. Why should care be exercised in the preparation of the column to prevent air bubbles from being trapped in the adsorbent?

3. Why is a layer of sand placed above the cotton plug prior to the addition of the column packing material?

4. Why does *syn*-azobenzene move down the column faster when methanol is used?

5. Why does the proper technique for column chromatography involve changing from a less polar to a more polar solvent rather than the reverse procedure?

6. Normally when increasing the polarity of the eluent, the increase is made gradually. However, in the suggested procedure for the isolation of *syn*-azobenzene, pure methanol is added directly to the column, rather than a mixture of petroleum ether–methanol. Why is this variation from the normal technique permissible in this case?

*7. Explain why *anti*-azobenzene is less strongly adsorbed onto alumina than the *syn*-isomer.

6.2 Thin-Layer Chromatography

Thin-layer chromatography involves the same principles as column chromatography, and it also is a form of solid-liquid adsorption chromatography. In this case, however, the solid adsorbent is spread as a thin layer (approximately 250 μ) on a plate of glass or rigid plastic. A drop of the solution to be separated is placed near one edge of the plate, and the plate is placed in a container, called a developing chamber, with enough of the eluting solvent to come to a level just below the "spot." The solvent migrates *up* the plate, carrying with it the components of the mixture at different rates. The result may then be a series of spots on the plate, falling on a line perpendicular to the solvent level in the container (see, for example, Figure 6.3b). The retention factor (R_f) of a component can then be measured as indicated in Figure 6.3.

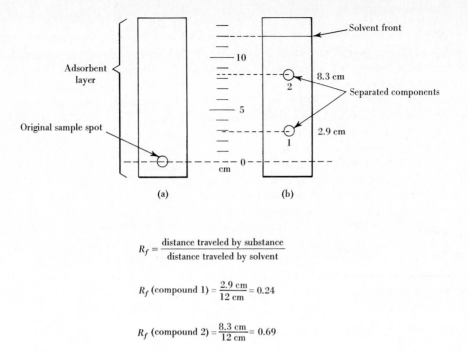

$$R_f = \frac{\text{distance traveled by substance}}{\text{distance traveled by solvent}}$$

$$R_f \text{ (compound 1)} = \frac{2.9 \text{ cm}}{12 \text{ cm}} = 0.24$$

$$R_f \text{ (compound 2)} = \frac{8.3 \text{ cm}}{12 \text{ cm}} = 0.69$$

Figure 6.3 Thin-layer chromatogram: (a) original plate; (b) developed chromatogram.

This chromatographic technique is very easy and rapid to perform. It lends itself well to the routine analysis of mixture composition and may also be used to advantage in determining the best eluting solvent for subsequent column chromatography. However, it should be borne in mind that volatile compounds (bp $<\sim 100°$C) cannot be analyzed by TLC.

The same solid adsorbents used for column chromatography may be employed for TLC; silica gel and alumina are the most widely used. The adsorbent is usually mixed with a small amount of "binder"—for example, plaster of Paris, calcium sulfate, or starch—to ensure proper adherence of the adsorbent to the plate. The plates may be prepared before use, or commercially available prelayered plastic sheets may be used. In either case, the plates or sheets should be dried for an hour or more in an oven at 110°C before use. This procedure removes moisture that is adsorbed on the adsorbent and produces a surface that is more effective in adsorbing and separating the components of the mixture being analyzed.

The relative eluting abilities of the various solvents that may be used are the same as those given in Section 6.1. It should be remembered that the eluting power required of the solvent is directly related to the strength of adsorption of the components of the mixture on the adsorbent.

A distinct advantage of TLC is the very small quantity of sample required. A lower limit of detection of 10^{-9} g is possible in some cases. The spot of sample must be applied to the TLC plate with care; it should not be large, for the same reason that the amount of sample placed at the top of a column should not be too large. However, larger samples such as 0.5 mg may be used on larger TLC plates, which have thicker coats of adsorbent. In these cases isolation of the components may be achieved by scraping the separated spots from the plate and eluting (extracting) them with an appropriate solvent. It is usually necessary to repeat this process a number of times in order to obtain even several milligrams of material. Such a procedure would offer a method of identification of the various components, however, because enough material could be collected to allow spectra to be obtained.

Often it is useful for the purpose of identification to run TLC chromatograms of knowns and unknowns side by side. Other useful applications involve running multiple aliquots of samples collected from a chromatographic column or samples taken from a reaction mixture in order to follow reaction progress as a function of time.

Detection of spots on the chromatogram is easy for colored materials, and a number of procedures are available for locating spots of colorless materials. For example, irradiation of the plate with ultraviolet light will permit location of the spots of compounds that fluoresce. Alternatively, the solid adsorbent may be impregnated with an otherwise inert, fluorescent substance. Spots of materials that absorb ultraviolet light but do not fluoresce will show up as black spots against the fluorescing background when the plate is irradiated with ultraviolet light. Other detecting agents are more often used. These agents may be sprayed onto the chromatograms, causing the spots to become readily apparent. Examples of detecting agents used in this way are sulfuric acid, which causes many organic compounds to char, and potassium permanganate solution. Iodine is another popular detecting agent. In this case the plate is placed in a vessel whose atmosphere is saturated with iodine vapor. Iodine is adsorbed by many organic compounds, and their spots on the chromatogram become colored (usually brown). Since these spots usually fade, it is a good idea to circle the spots with a pencil while they are still visible in order to have a permanent record of the chromatogram.

Under a given set of conditions (adsorbent, solvent, layer thickness, and homogeneity) the rate of movement of a compound with respect to the rate of movement of the solvent front, R_f, is a property of that compound. The value is determined by measuring the distance traveled by a substance from a starting line (Figure 6.3). The property has the same significance as retention time in a GC experiment.

Two experiments are presented here to demonstrate the TLC technique. The first is the separation of *syn-* and *anti-*azobenzene isomers. Commercial azobenzene exists mainly in the *anti* form, as mentioned in Section 6.1. The *anti* isomer, which is the more stable form, can be isomerized to the higher-energy and less-stable *syn-*isomer by irradiating it with ultraviolet light in the form of sunlight. The different colors of the two isomers allow a visual identification of them, and by spotting a sample of the original material side by side with a sample of irradiated material it is possible to assess the effectiveness of the irradiation in producing the isomerization.

In the second experiment the pigments present in green leaves are separated. Leaves from a variety of sources may be used, such as grass and various types of trees and shrubs, but waxy leaves and ivy leaves are not suitable. The interested student may wish to perform this experiment with more than one type of leaf and compare the results.

Another application of TLC is presented in Section 7.2. The diastereomers of 1,2-cyclohexanediol are separated. It may be worth mentioning that although these *diastereomers* have very similar physical properties and are difficult to separate by fractional crystallization, they can be separated successfully by chromatography. On the other hand, it might be valuable to demonstrate that the carvone *enantiomers*, which are purified by fractional distillation of two naturally occurring essential oils (Section 7.5), cannot be separated by chromatography.

There are various other opportunities to use TLC in analyzing the products of experiments in this book, such as those in Sections 7.2, 17.3, and 27.2c, Part B.

Pre-lab Exercises for Section 6.2, Thin-Layer Chromatography, can be found on page PL. 17.

EXPERIMENTAL PROCEDURE

A. Separation of *syn*- and *anti*-Azobenzenes by TLC

Obtain from your instructor a 10-cm strip of silica gel chromatogram sheet (without fluorescent indicator) and about 0.5 mL of a 10% toluene solution of commercial azobenzene. Place a spot of this solution on the TLC plate about 1 cm from one edge and about 2 cm from the bottom, using a capillary tube to apply the spot. The spot should be 1 or 2 mm in diameter. Allow the spot to dry and then expose the plate to sunlight for one or two hours. Alternatively, the plate may be placed beneath a sunlamp for about 20 min. After this time, apply another spot of the *original* solution on the plate at the same distance from the bottom as the first and leave about 1 cm between the spots. Again, allow the plate to dry.

A wide-mouth bottle with a tightly fitting screw-top cap may be used as a developing chamber. Alternatively, a beaker covered with a watch glass may be used. Prepare a 9:1 (by volume) mixture of hexane:chloroform and place a 1-cm layer of this mixture in the bottom of the developing chamber. Fold a piece of filter paper, as shown in Figure 6.4a, and place it into the developing chamber, as shown in Figure 6.4b. Saturate the chamber with the vapors of the solvent by shaking. This inhibits the evaporation of solvent from the plate during the development of the chromatogram. The piece of filter paper helps maintain this saturated state.

Place the chromatogram plate in the chamber, being careful not to splash the solvent onto the plate. The spots *must be above* the solvent level. Allow solvent to climb to within about 1.5 cm of the top of the plate and then remove the plate and allow it to air-dry. Note the

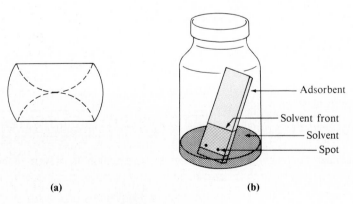

(a)   (b)

Figure 6.4 TLC chamber.

number of spots arising from each of the two original spots and compare the intensities of the two spots nearest the starting point, which correspond to *syn*-azobenzene.

B. Separation of Green Leaf Pigments by TLC

DO IT SAFELY

Have no flames in the vicinity of petroleum ether when you use it to extract the green pigments, as it is extremely flammable.

Place in a mortar several fresh spinach leaves and a few milliliters of a 2:1 mixture of petroleum ether (bp 30–60°C) and ethanol, and grind the leaves well with a pestle. By means of a pipet, transfer the liquid extract to a small separatory funnel and *swirl* with an equal volume of water; shaking may cause formation of an emulsion. Separate and discard the lower aqueous phase. Repeat the water washing twice, discarding the aqueous phase each time. The water washing serves to remove the ethanol as well as other water-soluble materials that have been extracted from the leaves. Transfer the petroleum ether layer to a small Erlenmeyer flask and add 2 g of *anhydrous* sodium sulfate. After a few minutes decant the solution from the sodium sulfate, and if the solution is not deeply and darkly colored, concentrate it by evaporating part of the petroleum ether, using a gentle stream of air.

Obtain from your instructor a 10-cm × 2-cm strip of silica gel chromatogram sheet. Place a spot of the pigment solution on the sheet about 1.5 cm from one end, using a capillary tube to apply the spot. Avoid allowing the spot to diffuse to a diameter of no more than 2 mm during application of the sample. Allow the spot to dry, and develop the chromatogram according to the general directions given in the second paragraph of part A, but use chloroform as the developing solvent.

It is sometimes possible to observe as many as eight colored spots. In order of decreasing R_f values, these spots have been identified as the carotenes (two spots, orange), chlorophyll *a* (blue-green), the xanthophylls (four spots, yellow), and chlorophyll *b* (green).

Calculate the R_f values of any spots observed on your developed plate. Also, as an aid in maintenance of a permanent record of the plate, draw to scale a picture of the developed plate in your notebook.

EXERCISES

1. Which of the two isomers of azobenzene would you expect to be the more thermodynamically stable? Why?

2. From the results of the TLC experiment with the azobenzenes, describe the role of sunlight.

3. In a TLC experiment why must the *spot* not be immersed in the solvent in the developing chamber?

4. Explain why the solvent must not be allowed to evaporate from the plate during the development.

5. Why is anhydrous sodium sulfate added to the petroleum ether extract in part B?

6. Why is a spot of the original mixture put on the plate in part A?

*7. Explain why the R_f of *anti*-azobenzene is greater than the R_f of *syn*-azobenzene.

8. Two components, A and B, were separated by TLC. When the solvent front had moved 6.5 cm above the level of the original sample spot, the spot of A was 5.0 cm and that of B was 3.6 cm above the original spot. Calculate the R_f for A and for B.

6.3 Paper Chromatography

Paper chromatography bears a resemblance to thin-layer chromatography (Section 6.2) but is slightly different in principle. Small spots of the mixture to be separated are placed near the bottom of a strip of paper; the end of the paper strip (but not the spot) is placed in solvent, and as the solvent ascends the paper, the components of the mixture are separated into individual spots (Figure 6.5). Paper chromatography involves distribution of the sample between a polar liquid phase, normally water, which is strongly adsorbed on the cellulose fibers of the paper, and an eluting solvent. It is thus an example of *liquid-liquid* partition chromatography, essentially a multiple extraction process.

Compounds may be identified by comparison of their R_f values (defined in Section 6.2). R_f values are far more reproducible with paper chromatography than with TLC and thus are more valuable in this application.

The great utility of paper chromatography (and TLC as well) is in the rapid analysis of reaction mixtures rather than in separation of compounds on a useful preparative scale. The minute quantities of sample required to accomplish an analysis and the ease and rapidity with which qualitative analysis of reaction mixtures can be performed make this chromatographic technique a valuable tool for the organic chemist and especially the biochemist.

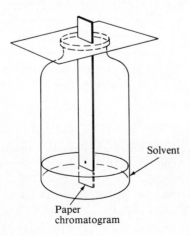

Solvent

Paper
chromatogram

Figure 6.5 A simple paper chromatography chamber.

The application of paper chromatography to the separation and identification of α-amino acids is described in Section 26.2.

6.4 Ion-Exchange Chromatography

Ion-exchange chromatography is a special type of liquid-solid chromatography in which strong ionic attractions are operative rather than the relatively weaker adsorption forces on which ordinary column and thin-layer chromatography depend. Columns may be filled with solid acidic material such as sulfonated polystyrene beads (Amberlite IR-120, for example), which will donate protons to any bases in the liquid phase, so that separation of bases of different strengths can be effected. Basic column material is also available in the form of polystyrene beads containing free amino groups (Dowex 3, for example), which can be used to separate mixtures of organic acids.

Ion-exchange chromatography is considerably less wide in its applicability than other forms of chromatography. It is useful in water-softening and in the production of de-ionized water.

6.5 High-Pressure Liquid Chromatography

In classical applications of column chromatography, *low flow rates* of the mobile liquid phase over the stationary solid phase are necessary for reasonable efficiency in the separation. This requirement is directly related to the *slow rates of diffusion* that are characteristic in liquid phases. If flow rates are too high, the desired chromatographic equilibration of the components of the sample between the two phases is not attained, and the components are not resolved. Thus the conditions required for efficient separations generally result in rather time-consuming experiments.

One approach to the solution of this problem is to decrease the *distance* through which molecules must diffuse. This may be accomplished by using much smaller solid particles for column packings. Not only does this dramatically increase the total surface area of the stationary phase, but the particles pack more tightly, resulting in a significant reduction of the interstitial volumes of liquid between the particles. Under these conditions equilibration between phases is established in a much shorter time, theoretically allowing much higher flow rates of the mobile phase through the column. However, the much tighter packing of the solid particles leads to a restriction of flow, so that in order to attain flow rates equivalent to those in ordinary (gravity) column chromatography (10–15 mL/min), the eluting solvent must be introduced into the column under pressure. This requirement has been met by the development of pulse-free pumping systems that provide inlet pressures up to 10,000 psi in small (2 to 3 mm) diameter columns. This has made possible an improved form of column chromatography called **high-pressure** (or ''high-performance'') **liquid chromatography, HPLC.** Although the liquid phase enters the column under high pressure, it emerges from the column at atmospheric pressure and can be either evaporated in separate fractions to provide the desired components (preparative HPLC) or simply passed through a detector of

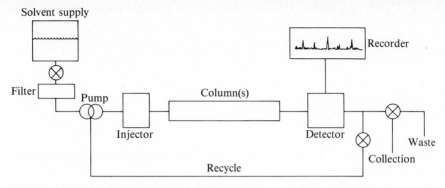

Figure 6.6 Block diagram of HPLC system.

some type, as described below (analytical HPLC). Preparative columns with diameters of 3 to 4 cm are available for separating mixtures of 15 to 20 g of material. If a single pass through the column does not produce a satisfactory separation, the sample can be recycled through the pump, column, and detector by means of a closed loop in some types of HPLC apparatus (Figure 6.6).

Packing materials of uniformly sized particles as small as 10 microns are available for use in HPLC columns; such materials composed of alumina or silica gel are discussed in Section 6.4. Another useful type of column material consists of glass beads 30–50 μm in diameter coated with a layer (1–2 μ) of porous material. These coated beads are referred to as *pellicular beads*. They may be used directly as a solid stationary phase, or the porous material may in turn be coated with a very thin layer of liquid to serve as the stationary phase. If because of the nature of this liquid and the mobile liquid phase there is the possibility of the liquid coating being washed away during the operation of the column, the coating can be chemically bonded to the glass beads.

The components of the mixture in the effluent from the column are detected in a flow-through cell in which some property of the components is monitored continuously. The properties most commonly used are ultraviolet absorption, fluorescence, and refractive index. The detector signal is fed to a chart recorder to provide a chromatogram similar to one produced by a gas chromatograph (Section 6.6).

The improvement in operating time achieved by the new HPLC systems is significant. Separations that would require many hours in gravity columns may be made in a matter of minutes in high-pressure equipment. There are other advantages over other forms of chromatography as well. Efficiency is much higher than in TLC, and whereas gas chromatography is not practical for high-molecular-weight (above 200 M.W.), nonvolatile compounds, HPLC can be applied to high-molecular-weight compounds very successfully. Another valuable aspect is the fact that HPLC is normally performed at room temperature, so that there is no danger of the decompositions or molecular rearrangements that may be induced in thermally unstable compounds by gas chromatography.

HPLC is a development that is becoming extremely valuable, particularly in the field of natural products. It has been credited with making important contributions, for example, to the synthesis of vitamin B_{12}, a complex high-molecular-weight compound, and to the isolation of the sex attractant of the American cockroach.

Although HPLC equipment is generally much more expensive than that for gas chromatography, less expensive instruments that operate at lower pressures (about 100 psi), but that have some of the advantages of HPLC, have been developed. Further improvements in this form of chromatography can be expected.

6.6 Gas Chromatography

Gas chromatography (GC), which is also called gas-liquid partition chromatography, or vapor-phase chromatography (abbreviated, respectively, GLPC and VPC), is widely used as a means of separating volatile compounds that may differ in boiling point by less than one-half degree. It is used both analytically and preparatively. In the latter form it has largely supplanted fractional distillation, even the most efficient type such as the use of spinning-band columns, for separations and purifications on a small scale.

Theory of Gas Chromatography In gas chromatography, the mixture to be separated is *vaporized* and carried along a column by a flowing *inert* gas such as nitrogen or helium, which is called the carrier gas. The gaseous mixture is the *mobile* phase. The column is packed with a solid, finely divided substance, on the surface of which is coated a liquid of relatively low volatility. This liquid serves as the *stationary* phase. Because of selective phase distribution of the components of the mixture between the mobile and stationary phases, these components may move through the column at different rates and thus be separated. The physical process involved in the separation of the components of a mixture in the GC column is the *partitioning* of the components between the gas and liquid phases.

A large variety of gas chromatographs are commercially available. However, the basic features of these instruments are quite similar and are represented by the schematic shown in Figure 6.7. Parts 1–5 are needed to supply dry carrier gas at a controlled flow rate. The column (7) is connected to the gas supply and is contained within an oven (8); the tempera-

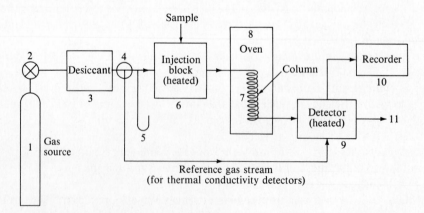

Figure 6.7 Schematic diagram of apparatus for gas chromatography. (1) Carrier gas supply. (2) Pressure-reducing valve. (3) Desiccant. (4) Fine-control valve. (5) Flowmeter. (6) Heated injection port. (7) Column. (8) Oven. (9) Detector. (10) Electronic recorder. (11) Exit port.

ture within the oven is controlled by a thermostat and heating elements. The sample to be separated is introduced by means of a syringe into the flow system at the injection port (6), which is individually heated to facilitate vaporization of the sample. The vaporized sample is then swept into the column by the carrier gas. As the sample passes through the column, its components separate into individual "bands" in the carrier gas, which then pass through the detector (9). The detector produces an electronic signal whose voltage is proportional to the amount of material different from the carrier gas itself present in the gas stream. The recorder (10) plots this voltage as a function of time to give the gas chromatogram (see Figure 6.9, for example). The vapors then pass from the detector into either the atmosphere or a collecting system at the exit port (11).

The elapsed time necessary for a given component to pass from the injection port to the detector is known as its *retention time*. Four important experimental factors affect the retention time of a compound: (1) the *nature* of the stationary liquid phase, (2) the *length* of the column, (3) the *temperature* at which the column is maintained, and (4) the *rate of flow* of the inert gas. *The retention time is independent of the presence (or absence) of other components in the mixture*. For any given column and set of conditions (temperature and flow rate), the retention time is a property of the compound in question and may be used to identify it.

Stationary and Mobile Phases   The choice of a stationary liquid phase best suited for a GC experiment depends for the most part on the types of compounds to be separated. Although a very large number of liquid phases are available, only a few (Table 6.1) are

TABLE 6.1   GC Stationary Phases

Liquid Phase	Type	Property	Maximum Temperature Limit, °C	Used for Separating
Squalane	Hydrocarbon grease	Nonpolar	100	Hydrocarbons, general application
Apiezon-L	Hydrocarbon grease	Nonpolar	300	Hydrocarbons, general application
Carbowax 20M	Hydrocarbon wax	Polar	250	Alcohols, C_6–C_{18} aldehydes, sulfur compounds
DC-550	Silicone oil	Intermediate polarity	275	C_1–C_5 aldehydes, C_6–up hydrocarbons, halogen compounds
QF-1	Silicone (fluoro)	Intermediate polarity	250	Polyalcohols, alkaloids, halogen compounds, pesticides, steroids
SE-30	Silicone gum rubber	Nonpolar	375	C_5–C_{10} hydrocarbons, pesticides, steroids
Diethyleneglycol succinate (DEGS)	Polyester	Polar	190	Esters, fatty acids
Butanediol succinate	Polyester	Intermediate polarity	225	Esters, fatty acids

widely used. The best choice will be that liquid giving rise to the largest differences in the partition coefficients of the components to be separated. Obviously, then, the solubility principles of organic compounds need to be considered.

The solubility of a gas in a liquid depends to some extent on its vapor pressure. In general, lower-boiling, more highly volatile components will move through the column faster and exit sooner than those of lower volatility because, to a rough approximation, the higher the vapor pressure of a gas, the lower its solubility in a liquid. Factors other than volatility (and thus, indirectly, molecular weight) are important, however, in determining the rate at which a compound will move through the column. The solubility of a substance in a liquid is also influenced by polar interactions between the molecules of solute and solvent, for example, hydrogen bonding and other electrostatic interactions.

An age-old rule of thumb in chemistry is ''like dissolves like.'' This would suggest that if it is desired to separate nonpolar types of compounds, a nonpolar liquid substrate should be used. On the other hand, if the components to be separated are of the polar variety, a nonpolar substrate would not be a very good choice, because none of the components would be very soluble in it, and all would tend to pass unhindered through the column with little if any separation. Thus, it is best to use polar stationary liquid phases for polar samples.

The most common means of supporting the liquid phase within the column is as a thin film coating on an inert solid support. The support should be of small, evenly meshed granules, so that a large surface area is available. This provides a correspondingly large film area in contact with the vapor phase, which is necessary for efficient separation. Some common types of solid supports are given in Table 6.2. The liquid is coated on the solid support by dissolving the liquid in a suitable low-boiling solvent and mixing this solution with the solid. The low-boiling solvent is then evaporated, leaving the solid granules evenly coated, and the column is filled with these coated granules.

TABLE 6.2   Solid Supports

Chromosorb P	Pink diatomaceous earth (surface area: 4–6 m^2/g)
Chromosorb W	White diatomaceous earth (surface area: 1–3.5 m^2/g)
Crushed Firebrick Chromosorb T	40/60 mesh Teflon 6

Capillary Columns   An alternative method of supporting the liquid phase is used in capillary columns of the Golay type. These columns are very long (300 m is not unusual) and of very small diameter (0.1–0.2 mm). In these columns the liquid is coated directly on the inner walls of the tubing. These types of columns are highly efficient and relatively expensive.

Practical Considerations   In general, column efficiency increases with increasing path length and decreases with increasing diameter. As suggested in the introductory discussion, the separation between bands increases with increasing path length. Therefore, with a longer column the likelihood of separating two components will be increased, and their retention time difference will be larger. The diameter of the column will affect the band *width*. A

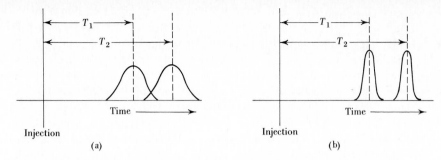

Figure 6.8 Effect of band width on resolution: (a) wide bands; (b) narrow bands.

smaller-diameter column will give rise to narrower bands and greater efficiency. It is evident that with a small band separation (measured from the band centers), wide bands are more likely to *overlap* (Figure 6.8a) than narrow bands (Figure 6.8b). Therefore a smaller-diameter column will effect a more efficient separation; that is, the *resolution* will be greater.

The temperature at which the column is maintained is another important factor in a GC experiment. Retention times may be shortened by using higher temperatures, because the solubility of gases in liquids decreases with increasing temperature. The partition coefficients are thus affected, and the bands move through the column at a faster rate. This is sometimes desirable because of the convenience of a faster experiment, but it will also usually mean that the band separation will decrease, resulting in lower resolution.

There is a maximum temperature limit for each liquid phase. This point is determined by the stability and volatility of the liquid being used. At higher temperatures the liquid phase may volatilize to some extent and be carried off the column by the carrier gas, which is of course undesirable. With some liquids there is a minimum temperature, governed by the melting point of the substance or its viscosity. If the liquid should be partially solidified or quite viscous, it will be very inefficient at dissolving the components of the gaseous mixture.

Just as with higher temperatures, higher flow rates of the carrier gas will also cause retention times to decrease. In spite of the decreased resolution obtained at higher temperatures and flow rates, these conditions are sometimes necessary for substances otherwise having very long retention times.

Analytical and Preparative Gas Chromatography There are two basic types of experiments to which gas chromatography may be applied: (1) qualitative and/or quantitative analysis of the sample and (2) preparative experiments for the purpose of separating and purifying the components of the sample (preparative GC).

Under a given set of conditions (column, temperature, and flow rate, for example), the retention time of a given compound is a property of that compound and may be used to identify it. Often, identification may be made by injecting and obtaining the retention time, under the same conditions, of material *known* to be the compound in question. If the retention time of the known compound matches that of a peak of the chromatogram, this is *necessary* evidence that the two compounds may be identical; it is not *sufficient* evidence, however, because more than one compound may exhibit the same retention time. Since one usually has some indication of what compounds are present in the mixture, the identification is reasonably certain. Furthermore, if it is assumed that the voltage output of the detector and therefore the pen response of the recorder is proportional to the mole fraction of the material being detected in the vapor, then the relative *areas* under two peaks of the chromatogram will equal the percentage ratio of these two components in the mixture. Although the relative detector response may not necessarily correspond exactly to the relative mole fractions of two components in a mixture, the deviation is usually fairly small. Thus peak-area measurements will provide a reasonably accurate *quantitative* analysis of a mixture.

An example of the analytical use of GC is shown in Figure 6.9. These sets of peaks represent gas chromatographic analyses of the distillation fractions of a mixture of ethyl acetate and *n*-butyl acetate similar to those obtained in the distillation experiment of Chapter 4. The notations *A*, *B*, and *C* refer to the three fractions taken in that experiment. Comparison of retention times with those of pure ethyl acetate and pure *n*-butyl acetate

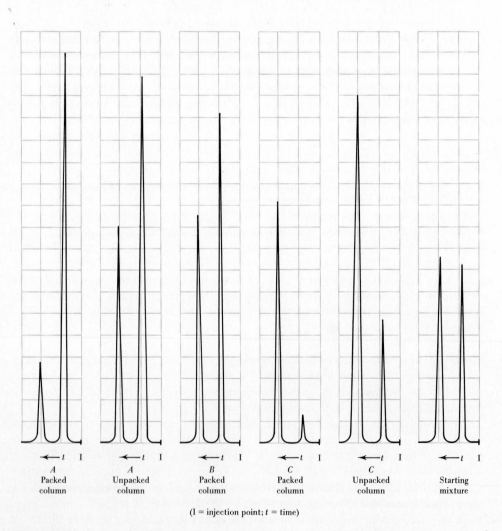

←*t* I	←*t* I	←*t* I	←*t* I	←*t* I	←*t* I
A	*A*	*B*	*C*	*C*	
Packed column	Unpacked column	Packed column	Packed column	Unpacked column	Starting mixture

(I = injection point; *t* = time)

Figure 6.9 Gas chromatographic analysis of the distillation fractions from the distillation experiment of Chapter 4.

indicates that the first peak, with shorter retention time, in each case is ethyl acetate, and the other is *n*-butyl acetate. Comparison of the relative areas for the peaks of fractions *A* and *C* using the packed column with the results for fractions *A* and *C* for the unpacked column demonstrates the greater efficiency of the packed fractional distillation column. Clearly, the introduction of packing in the Hempel column has increased the number of theoretical plates in the column.

There are a number of methods of obtaining the peak areas necessary for quantitative evaluation of the chromatogram. (1) The detector output may be *electronically integrated* to provide a digital readout of the area. However, this instrumentation is relatively expensive and is not likely to be encountered outside of research laboratories. (2) The areas may be measured by means of a *planimeter,* a tracing device that provides a number that is proportional to the area of a region whose boundary has been traced. (3) When peaks are *symmetrical,* such as those shown in Figure 6.9, their areas may be geometrically approximated by considering them to be *triangles.* Thus the area of a symmetrical peak is approximately equal to its height times its width at half-height, as is illustrated in Figure 6.10. (4) Since chart paper is of reasonably uniform thickness and density, peaks may be carefully cut out with scissors and their areas assumed to be proportional to their weight, as measured on an analytical balance. Since one usually wishes to save the original chromatogram as a permanent record, it is best to cut the peaks from a photocopy of the chromatogram.

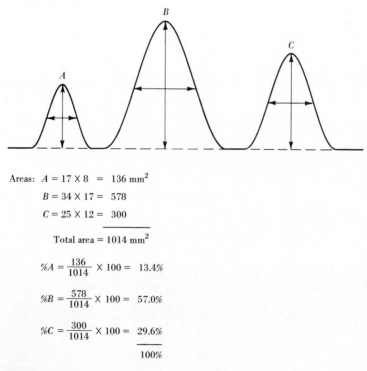

Areas: $A = 17 \times 8 = 136 \text{ mm}^2$

$B = 34 \times 17 = 578$

$C = 25 \times 12 = 300$

Total area $= 1014 \text{ mm}^2$

$\%A = \dfrac{136}{1014} \times 100 = 13.4\%$

$\%B = \dfrac{578}{1014} \times 100 = 57.0\%$

$\%C = \dfrac{300}{1014} \times 100 = 29.6\%$

100%

Figure 6.10 Determination of percentage composition of a mixture by gas chromatography.

Once the peak areas have been obtained, the percentage of each component in the mixture may be calculated as the area of the peak corresponding to that component, expressed as a percentage of the sum of the areas of all peaks in the chromatogram. Sample calculations of this type are shown in Figure 6.10.

When particularly accurate quantitative analyses are required, one must remain aware that a given detector is not necessarily equally sensitive or responsive to each of the components of a mixture. Thus, even though two components may be present in equal amounts in a sample mixture, their respective peaks in the chromatogram may have slightly different areas. As an example, the chromatogram of a 50:50 (wt:wt) mixture of toluene and benzene does not show equal-sized peaks, but instead shows a peak for toluene that is about 2% smaller than the peak for benzene. (This value is for a thermal conductivity detector; the difference is 1% when a flame ionization detector is used.) To obtain an accurate analysis, the peak area for toluene must be multiplied by 1.02 before calculating the percentages of the components of the mixture.

The relative response factors of many common organic compounds with different detectors are given in various reference sources, such as reference 8 at the end of this chapter. The factors for a few of the materials that students may encounter in this course are provided in Table 6.3. These factors are determined by preparing a standard solution containing known *weights* of the substances whose factors are desired and a known weight of a standard substance, which is benzene for those compounds given in Table 6.3. The density of this solution is determined so that one can calculate the exact weight of each component, including the standard, which will be injected in, for example, a 1-microliter (μL) sample. The areas of the various peaks are then measured, and each is divided by the exact weight of that component in the injected sample. The inverse of each of the resulting numbers is then normalized to the standard by dividing each by the similarly attained number for the standard.

In order to analyze quantitatively a mixture of substances whose weight factors are known, the peak area for each component of the mixture is *multiplied* by the weight factor for that particular compound, and the resulting *corrected* areas are utilized in calculating the

TABLE 6.3 Weight (W_f) and Molar (M_f) Correction Factors for Some Representative Substances[a]

Substance	Thermal Conductivity		Flame Ionization	
	W_f	M_f	W_f	M_f
Benzene	1.00	1.00	1.00	1.00
Toluene	1.02	0.86	1.01	0.86
Ethylbenzene	1.05	0.77	1.02	0.75
Isopropylbenzene	1.09	0.71	1.03	0.67
Ethyl acetate	1.01	0.895	1.69	1.50
n-Butyl acetate	1.10	0.74	1.48	0.995
Heptane	0.90	0.70	1.10	0.86
o-Xylene	1.08	0.79	1.02	0.75
m-Xylene	1.04	0.765	1.02	0.75
p-Xylene	1.04	0.765	1.02	0.75
Ethanol	0.82	1.39	1.77	3.00
Water	0.71	3.08	—	—

[a]Taken from reference 8 at the end of this chapter.

percentage composition of the mixture according to the procedure outlined in Figure 6.10. Note that the use of these factors provides the composition on a *weight percentage* basis. To obtain mole factors, M_f, which would provide the composition on a mole percentage basis, the weight factors are divided by the molecular weights of each component of the standard solution and the resulting numbers normalized to the standard.

A sample calculation utilizing these correction factors is provided here in the analysis of a mixture of ethanol, heptane, benzene, and ethyl acetate on a gas chromatograph equipped with a thermal conductivity detector. The last column shows the percentage composition calculated directly from the measured peak areas, without correction for detector response. The dramatic differences in the calculated composition, with and without this correction, as noted in the last two columns, should underscore the necessity of this correction when accurate quantitative results are desired.

Compound	Area (A) (mm²)	M_f	A × M_f	Mole % $\left(\dfrac{A \times M_f}{194} \times 100\right)$	Uncorrected % $\left(\dfrac{A}{207.1} \times 100\right)$
Ethanol	44.0	1.39	61.16	31.5%	21.2%
Heptane	78.0	0.70	54.6	28.1%	37.7%
Benzene	23.2	1.00	23.2	11.9%	11.2%
Ethyl acetate	61.9	0.895	55.4	28.5%	29.9%
	Total = 207.1		Total = 194	100%	100%

In *preparative* GC, the purified components of the mixture are obtained by collecting and condensing the vapors as they come from the exit port. The recorder indicates when the material is passing through the detector; by making allowance for the time required for the material to pass from the detector to the exit port, one can estimate when to collect the condensed vapor. As different peaks are observed, different collection vessels may be used.

Many instruments are designed specifically for analytical determinations. Relatively small columns are used, and thus only very small samples may be injected (approximately 0.1–5 μL). With such sample sizes, it is evident that a very large number of repetitive injections would be needed before any sufficient quantity of the components could be collected in a preparative experiment. Instruments designed specifically for preparative experiments are also available; these utilize much larger columns (some of the larger ones are 30 cm in diameter) in order to handle larger samples. Larger quantities of purified materials can thus be obtained with fewer repetitions of the injection-separation-collection cycle.

Pre-lab Exercises for Section 6.6, Gas Chromatography, are found on page PL. 19.

EXPERIMENTAL PROCEDURE

DO IT SAFELY

The solvents and other liquids that may be used in sample mixtures for gas chromatography analysis are flammable and volatile. The usual precautions against having flames in the vicinity of these liquids should be exercised.

NOTES ON INJECTION OF SAMPLES INTO A GAS CHROMATOGRAPH

Liquid samples are injected into the heated port of a gas chromatograph by means of a syringe with a capacity of 1 to 10 microliters for analytical work and 50 microliters or more for preparative work. The sample is either injected neat or, in the case of less-volatile materials, dissolved in a volatile liquid such as diethyl ether or pentane. The sample should not contain nonvolatile substances that may eventually clog the injection port or contaminate the stationary phase of the column.

Syringes are precision-made and expensive; they should be used with care. Fill the syringe with slightly more than the desired volume of sample, neat or in solution, then point the syringe needle-up, push out the excess liquid, and wipe the tip of the needle with a tissue. Insert the needle into the rubber injection septum as far as it will go, push the plunger all the way in with one quick movement, and remove the syringe from the septum while holding the plunger in place. If the sample is injected slowly the sample will be spread along the column and the peaks will be broadened.

It is important to clean the syringe immediately after use. Rinse it with a volatile solvent such as acetone and dry it by drawing a stream of air through it.

Since there is considerable variation in the operating procedures of different commercial gas chromatographs, students should consult their instructors for specific directions in the use of an instrument.

A. Identification of an "Unknown" Compound by Gas Chromatography

Obtain a known mixture of compounds and prepare a gas chromatogram as directed by your instructor. A suitable mixture would be one containing some or all of the following compounds: ethyl acetate, toluene, n-butyl acetate, ethylbenzene, and isopropylbenzene.

Obtain an "unknown" compound and prepare a gas chromatogram under the same instrumental conditions used for the known mixture.

Identify the "unknown" compound as follows: (1) Compare the retention time of the "unknown" compound with the retention times of the known components of the mixture. (2) Prepare a "sample mixture" by adding about one volume of your "unknown" compound to two volumes of the known mixture, and prepare a gas chromatogram of this "sample mixture." (3) Confirm the identity of the "unknown" compound by observing which of the peaks of the known mixture has been amplified in the chromatogram of the "sample mixture."

B. Quantitative Analysis of a Mixture by Gas Chromatography

Obtain a mixture of three or more of the compounds listed in part A and prepare a gas chromatogram as directed by your instructor. Using one of the quantitative methods described in the preceding discussion section, determine the relative amounts, in percentages, of the components of the mixture. Consult with your instructor in order to know which, if any, correction factors to use.

C. Other Analyses

Some of the mixtures encountered in other experiments in this text that are appropriate for demonstrating gas chromatographic analysis, which may be assigned by your instructor, are as follows:

1 The fractions collected from the fractional distillation of mixtures of ethyl acetate and *n*-butyl acetate in Section 4.4.

2 The mixture of chloroheptanes produced in the experiment of Section 12.1.

3 The mixture of alkenes produced by the elimination experiment of Section 13.2.

4 The isomeric propyl-*p*-xylenes produced in the Friedel-Crafts alkylation experiment of Section 17.2.

5 The impure 2-methylpropanal prepared in the oxidation experiment of Section 19.2, which contains unchanged starting material.

EXERCISES

1. Define: (a) stationary phase, (b) mobile phase, (c) carrier gas, (d) retention time, (e) solid support, (f) thermal conductivity.

2. Refer to the GC traces given in Figure 6.9. These are analyses of the various fractions taken in the fractional distillation of the mixture of ethyl acetate and *n*-butyl acetate. Utilizing the GC correction factors (thermal conductivity detector) provided in Table 6.3, determine both the weight percent and the mole percent compositions of the fractions *A*, *B*, and *C* obtained using a packed distillation column.

3. Benzene (20 g, 0.25 mole) is subjected to Friedel-Crafts alkylation with 1-chloropropane (19.6 g, 0.25 mole) and $AlCl_3$. The product (24.0 g) is subjected to analysis on a gas chromatograph equipped with a thermal conductivity detector. The chromatogram shows two product peaks identified as *n*-propylbenzene (area $= 60$ mm^2; $W_f = 1.06$) and isopropylbenzene (area $= 108$ mm^2; $W_f = 1.09$). Calculate the percent yield of each of the two isomeric products obtained in this reaction. [Since each of the products has the same molecular weight (120), the use of weight factors gives both weight and mole percent composition.]

REFERENCES

1. *Chromatographic Science, A Series of Monographs,* Vol. 23, "Liquid Chromatography Detectors," T. M. Vickrey, editor, Marcel Dekker, Inc., New York, 1983. See previous volumes for other topics in chromatography.

2. Stock, R.; Rice, C. B. F., *Chromatographic Methods,* 3rd edition, Chapman and Hall, Halsted Press Division, Wiley, New York, 1974.

3. *Chromatography,* 3rd edition, E. Heftmann, editor, Von Nostrand Reinhold, New York, 1975.

4. Scott, R. P. W., "Contemporary Liquid Chromatography," in *Techniques of Chemistry,* A. Weissberger, editor, Wiley-Interscience, New York, 1976, Vol. 11.

5. *Techniques in Liquid Chromatography,* C. F. Simpson, editor, John Wiley and Sons, New York, 1982.

6. Kirchner, J. G., "Thin-Layer Chromatography," in *Techniques of Chemistry,* E. S. Perry and A. Weissberger, editors, Wiley-Interscience, New York, 1978, Vol. 14.

7. Schupp, O. E., III, "Gas Chromatography," in *Technique of Organic Chemistry,* E. S. Perry and A. Weissberger, editors, Interscience, New York, 1968, Vol. 13.

8. McNair, H. M.; Bonelli, E. J., *Basic Gas Chromatography,* 3rd ed., Varian Aerograph, Walnut Creek, Calif., 1967.

stereoisomers

7.1 Introduction

Stereoisomers are defined as those isomers having identical molecular constitutions but differing in the orientation of their atoms in space. One way in which the atoms of covalent molecules may assume different positions in space is by the rotation of parts of the molecule about single bonds, for example, the C_2—C_3 bond of butane. The differences in energy separating the **conformational isomers** (conformers) produced by such rotations are usually too small to allow for their isolation. Compounds capable of assuming two or more distinct conformations, such as cyclohexane, are typically found as equilibrium mixtures of the conformational isomers.

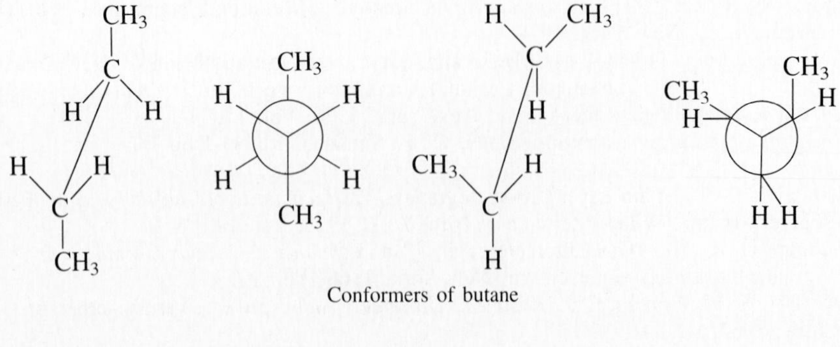

Conformers of butane

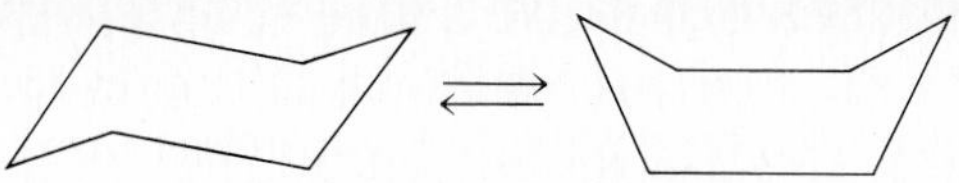

Conformers of cyclohexane

A second and important class of stereoisomers is that of the **configurational isomers.** The tetrahedral covalency of carbon leads to two different arrangements, or *configurations*, in space when four different atoms or groups of atoms are attached to a single carbon atom. Such a carbon represents a *chiral*[1] center in a molecule and results in the existence of two configurational isomers called **enantiomers,** that have the relationship of an object and its mirror image. Enantiomers have the unusual physical property of rotating the plane of polarized light in opposite directions, and for this reason they have been called "optical isomers." The enantiomers of 2-chlorobutane illustrate this type of configurational isomerism.

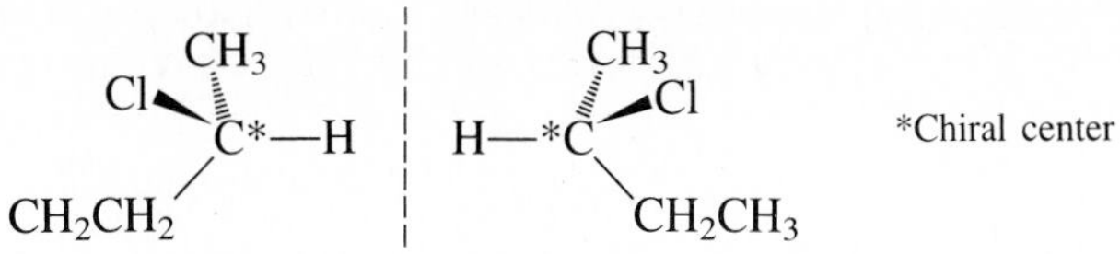

*Chiral center

Enantiomers of 2-chlorobutane

When a chiral center is created in a compound by a synthesis from an achiral starting material, as in the preparation of 1-phenylethanamine, **3,** from acetophenone, **1,** and ammonium formate, **2** (equation 1), the enantiomeric forms of **3** normally are formed in equal amounts, producing a **racemic modification,** which is *optically inactive.* The special technique required to separate the enantiomers, called *resolution,* is described in Section 7.6.

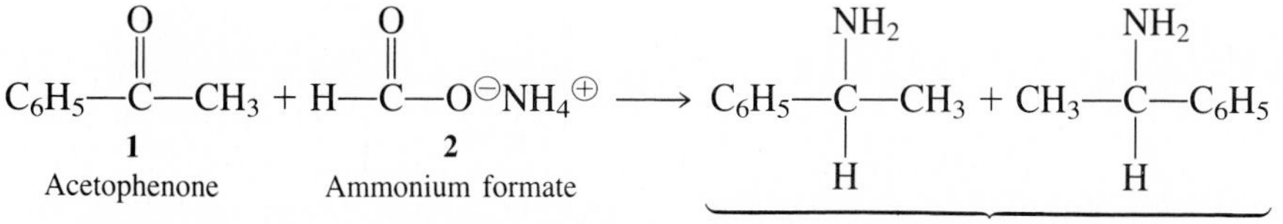

$$C_6H_5-\overset{\overset{O}{\|}}{C}-CH_3 + H-\overset{\overset{O}{\|}}{C}-O^{\ominus}NH_4^{\oplus} \longrightarrow C_6H_5-\underset{\underset{H}{|}}{\overset{\overset{NH_2}{|}}{C}}-CH_3 + CH_3-\underset{\underset{H}{|}}{\overset{\overset{NH_2}{|}}{C}}-C_6H_5 \qquad (1)$$

1   **2**
Acetophenone   Ammonium formate

3
1-Phenylethanamine

Many compounds containing chiral centers are found in nature in optically active forms. For example, both enantiomers of carvone, **4,** are found separately in the essential oils of certain plants.

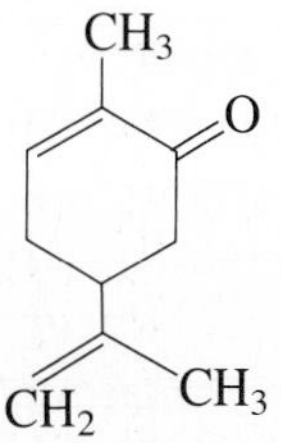

4
Carvone

[1]The term comes from the Greek word for "hand" and is appropriate because the hands are perhaps the best-known illustration of mirror-image dissymmetric objects.

The levorotatory (−) enantiomer is the major component of spearmint oil, and the dextrorotatory (+) enantiomer is the major component of both caraway seed oil and dill seed oil.

Achiral molecules may also have different configurations. In contrast to the case of chiral molecules, some of these may be represented adequately in two dimensions. Ethylenic *cis-trans* isomers belong to this category. They are sometimes called "geometric isomers," which is somewhat ambiguous because all stereoisomers differ in their geometry. Certain symmetrical cyclic compounds exhibit isomerism that is very similar to that of ethylenic *cis-trans* isomers, for example, *cis-* and *trans*-1,3-dimethylcyclobutane, **5** and **6**, respectively.

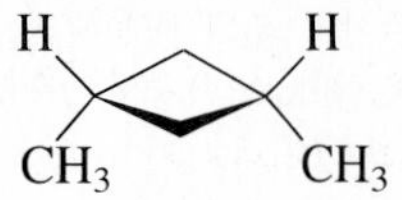

5

cis-1,3-Dimethylcyclobutane

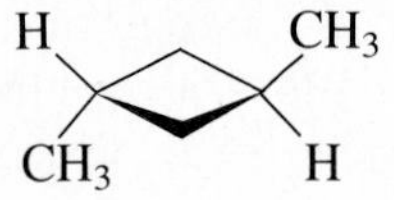

6

trans-1,3-Dimethylcyclobutane

Whether ethylenic or cyclic in structure, *stereoisomers* such as these *that are not enantiomers are classed as diastereomers*.

Compounds with two or more chiral centers may have not only enantiomers, but also diastereomers. Tartaric acid provides the classic example: There are three stereoisomers of tartaric acid, a pair of optically active enantiomers, **7** and **8**, and an achiral isomer, **9**, which is known as a *meso* form and is a diastereomer of the active forms.

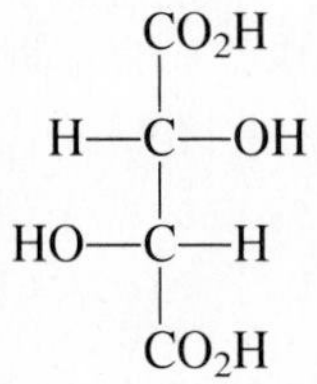

7

(+)-Tartaric acid

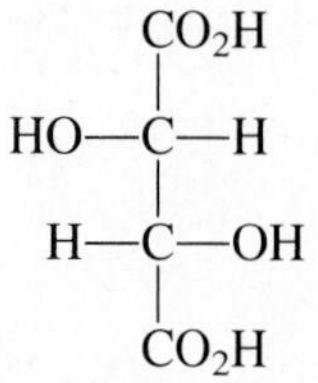

8

(−)-Tartaric acid

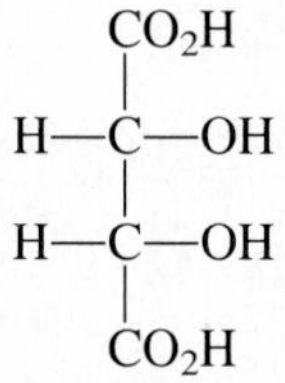

9

meso-Tartaric acid

Some cyclic compounds are chiral. For example, there are three isomers of 1,2-dimethylcyclobutane, **10, 11,** and **12.**

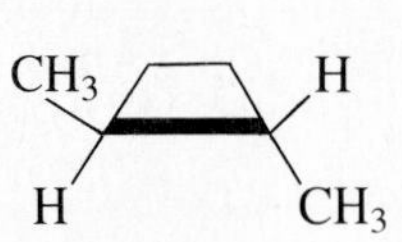

10

trans-1,2-Dimethylcyclobutane

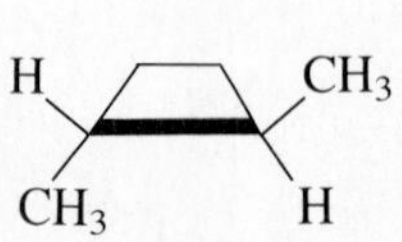

11

trans-1,2-Dimethylcyclobutane

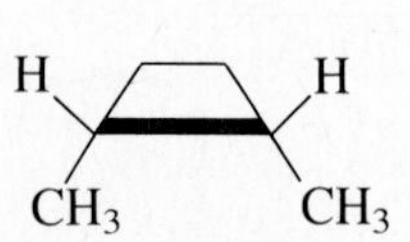

12

cis-1,2-Dimethylcyclobutane

Unlike 1,3-dimethylcyclobutane, the *trans* form of 1,2-dimethylcyclobutane is chiral and consists of a pair of optically active enantiomers, **10** and **11.** The *cis* form, **12,** is achiral and therefore optically inactive; it is a diastereomer of the optically active enantiomers. One may see from this example that there is an overlap between "*cis-trans* isomers" and "optical isomers." For this reason, one should use the term "optical isomers," if at all, only for those stereoisomers that *are* optically active. Thus, *meso* forms should not be included in such a class.

The experimental procedures in this chapter are chosen to illustrate (1) separation of diastereomers by chromatography (Section 7.2), (2) isomerization of one ethylenic diastereomer to another (Section 7.3), (3) the isolation of the enantiomers of carvone from their natural sources (Section 7.5), and (4) resolution of a racemate into its optically active enantiomers (Section 7.6). To facilitate the latter two studies, an introductory description of polarimetry is included (Section 7.4).

7.2 Separation of the Diastereomeric *cis-trans* Isomers of 1,2-Cyclohexanediol

The stereoisomers of 1,2-cyclohexanediol are analogous to those of 1,2-dimethyl-cyclobutane, **10, 11,** and **12,** consisting of a pair of *trans* enantiomers, **13** and **14,** and a *cis meso* diastereomer, **15.**

13	**14**	**15**
trans-1,2-Cyclohexanediol	*trans*-1,2-Cyclohexanediol	*cis*-1,2-Cyclohexanediol
		(*meso* isomer)

The stereochemistry of cyclic 1,2-diols is often described in organic textbooks because they can be produced stereoselectively from cycloalkenes. For example, a racemic mixture of **13** and **14** can be produced from cyclohexene by reaction with a peroxycarboxylic acid, followed by hydrolytic cleavage of the intermediate epoxide, **16** (equation 2).

Oxidation of cyclohexene by permanganate, on the other hand, produces **15,** presumably via a bicyclic ester, **17,** which is not isolated (equation 3).

In this experiment a commercial mixture of racemic *trans*-1,2-cyclohexanediol (**13, 14**) and *meso cis*-1,2-cyclohexanediol (**15**) is utilized. This material is readily available because it can be made by catalytic hydrogenation of catechol, **18** (equation 4), a phenolic compound that occurs in several plants and is synthesized industrially from other simple aromatic precursors.

$$\text{(cyclohexene)} + MnO_4^{\ominus} \longrightarrow \left[17 \right]^{\ominus} \xrightarrow[\text{H}^{\oplus}]{\text{H}_2\text{O}} \text{15} \qquad \textbf{(3)}$$

17

$$\text{(catechol)} + 3\ H_2 \xrightarrow[\text{heat, pressure}]{\text{catalyst}} \textbf{13} + \textbf{14} + \textbf{15} \qquad \textbf{(4)}$$

18
1,2-Dihydroxybenzene
(catechol)

One should remember that the physical and chemical properties of the enantiomers **13** and **14** are identical, except for their effect on plane-polarized light, but the physical and chemical properties of **15** are different from those of **13** and **14** since it is a diastereomer of them. The melting point of *trans*-1,2-cyclohexanediol is reported to be 104°C and that of *cis*-1,2-cyclohexanediol to be 98°C. Although these melting points are not very different, one can readily determine the identity of a sample of one of the diols, even though it may not melt sharply at either of the reported melting temperatures, assuming that one has available a pure sample of *one* of the isomers. This can be done by using the mixture melting-point technique (Section 3.2c).

The solubilities of the diastereomers are also different, but not *very* different, so that their separation by fractional crystallization is a tedious procedure.[2] In the experimental procedure presented here, the separation of these diastereomers by thin-layer chromatography (TLC) is described. (Consult Section 6.2 for a discussion of TLC technique.) The pure *trans* isomer is commercially available, but its high price reflects the difficulty of separating it from the *cis* isomer. However, since TLC requires such small amounts of material, it is feasible to use a tiny sample of the *trans* isomer in a side-by-side comparison of its retention factor (R_f) with those of the two isomers separated from the mixture by TLC. This allows assignment of each spot to an isomer and thus makes possible an assessment of the relative adsorption properties of the *cis* and *trans* isomers.

Separations of the diastereomers may also be accomplished by column chromatography (Section 6.1) and by gas chromatography (Section 6.5). Consult with your instructor for suggestions for performing these types of experiments.

[2]Fractional crystallization to separate a mixture of two components that do not differ markedly in their solubility properties requires many steps in which crystals are collected, the filtrate is cooled or evaporated to yield a second crop of crystals, the batches of crystals are combined and recrystallized, and the filtrates are evaporated to yield the more-soluble component. The more-soluble component must then be recrystallized repeatedly, often from a different solvent. Chromatographic techniques have in many cases superseded such tedious and time-consuming fractional crystallizations (see Chapter 6).

Pre-lab exercises for Section 7.2, Separation of the Diastereomeric *cis-trans* Isomers of 1,2-Cyclohexanediol, are found on page PL. 21.

EXPERIMENTAL PROCEDURE

DO IT SAFELY

1 Acetone and petroleum ether are highly flammable; do not have any flames in the vicinity of their use.

2 Do not allow iodine to come in contact with your skin and do not inhale its vapor, as it is corrosive and toxic.

Consult Chapter 6.2 for directions for performing a TLC separation.

Obtain or prepare *ca.* 10% solutions in acetone of a commercial mixture of *cis-* and *trans*-1,2-cyclohexanediol and of pure *trans*-1,2-cyclohexanediol. Place spots of each solution side by side on a 10-cm × 2-cm TLC silica gel plate. Develop the chromatogram using a mixture of 75% petroleum ether (bp 60–80°C) and 25% 2-propanol. Remove the plate from the solvent mixture when the solvent front approaches the top of the plate, air-dry it for a few minutes, and place it in a closed jar with a few iodine crystals to make the spots visible.

EXERCISES

1. By comparing the R_f values of the separated spots from the mixture of the *cis* and *trans* diols with the R_f value of the *trans* isomer, decide which isomer is adsorbed more strongly by the silica gel. Suggest a reason for the difference in adsorptivity of the two isomers.

2. It may happen that the separation of the isomers by TLC is not complete—one of the "spots" may be found to be smeared out to some extent. Considering this, which one of the isomers do you think could be obtained pure more easily by column chromatography?

3. What would be the consequence of using pure petroleum ether as the eluting solvent?

4. What would be the consequence of using pure 2-propanol as the eluting solvent?

5. Compare Figures 7.2 and 7.3. In what areas are the IR absorptions similar and in what areas are they different? Explain the similarities and the differences.

SPECTRA OF STARTING MATERIALS

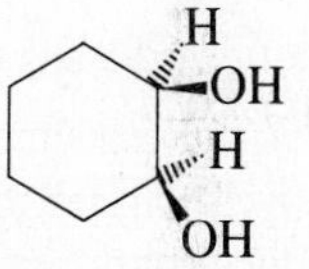

Chemical shifts: δ 21.6, 30.0, 70.4.

Figure 7.1 CMR data for *cis*-1,2-cyclohexanediol.

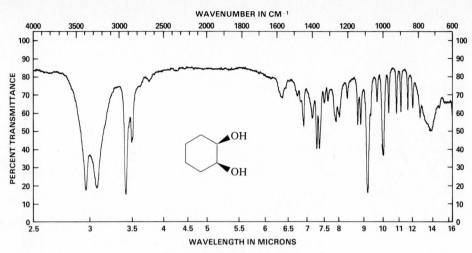

Figure 7.2 IR spectrum of *cis*-1,2-cyclohexanediol (KBr pellet).

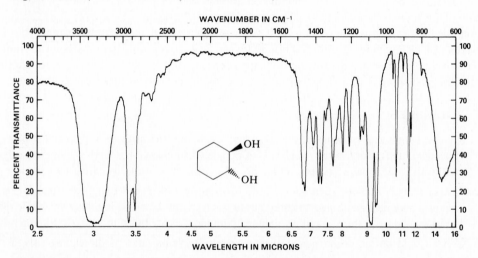

Figure 7.3 IR spectrum of *trans*-1,2-cyclohexanediol (KBr pellet).

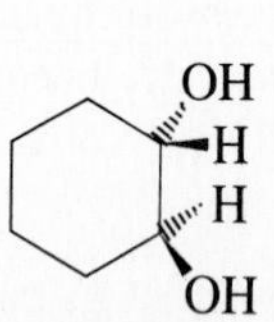

Chemical shifts: δ 24.2, 32.8, 74.7.

Figure 7.4 CMR data for *trans*-1,2-cyclohexanediol.

7.3 Isomerization of Dimethyl Maleate to Dimethyl Fumarate

Maleic and fumaric acids (**19** and **20**) are often cited as the classic examples of *cis-trans* stereoisomers, although they were given different names before they were known to be isomers.

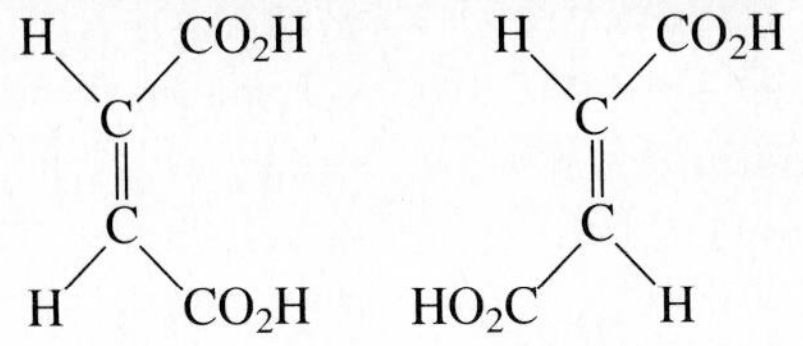

19
Maleic acid

20
Fumaric acid

They may be categorized as ethylenic-type diastereomers and as such may be expected to have different physical and chemical properties. This expectation is amply fulfilled, as is illustrated by their melting points, which are 130°C **(19)** and 287°C **(20),** and by the fact that fumaric acid is found in many plants and is a normal product of animal metabolism, whereas maleic acid is not found in nature and is, in fact, quite toxic.

Conversion of one configuration to another usually requires that bonds be broken and remade. In the case of ethylenic-type diastereomers, rupture of the pi bond in one isomer will allow rotation about the remaining C—C sigma bond, so that subsequent re-formation of the pi bond may result in the production of the other isomer. As a specific example, consider the possible consequences of addition of bromide ion to the dimethyl ester of maleic acid **(21),** a *cis* isomer (equation 5). This process would give the stabilized anion **22,** which could undergo rotation about the carbon-carbon bond to give **23.** Loss of bromide ion from **23** would provide the diastereomeric isomer of **21,** dimethyl fumarate **(24,** equation 5).

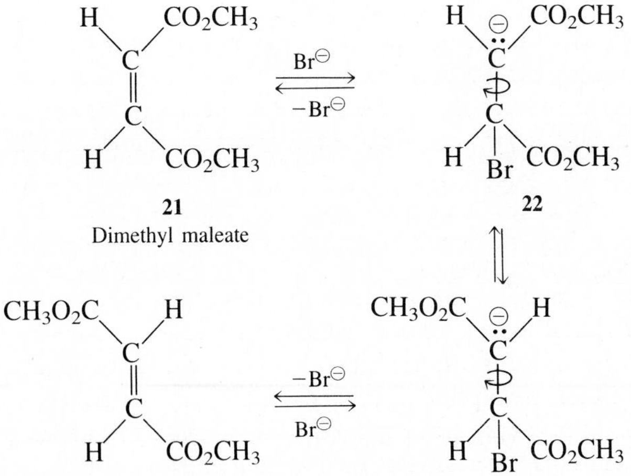

21
Dimethyl maleate

22

(5)

24
Dimethyl
fumarate

23

As described in the following Experimental Procedure, the conversion of dimethyl maleate to dimethyl fumarate is accomplished by treatment of the maleate isomer with a dilute solution of bromine in carbon tetrachloride. A potentially important reaction, the addition of bromine to the double bond of either the maleate or fumarate, is actually unlikely because the presence of the two electron-withdrawing carbomethoxy groups, —CO_2CH_3, on the carbons at the double bond decreases the electron density of the pi bond sufficiently to preclude facile attack of the electrophilic Br^+. However, this does not necessarily prevent the nucleophilic attack of Br⁻, which could possibly induce the isomerization as pictured in equation 5. Another possible mechanism can be envisioned in which a bromine *atom* adds to dimethyl maleate to afford a free radical intermediate analogous to the anion **22.** Therefore, the mechanism of the isomerization of dimethyl maleate to its diastereomer may be either nucleophilic or free radical in nature. Observations made in the experiment should provide a basis for a decision between these two mechanisms.

It should be noted that the properties of the dimethyl esters of maleic and fumaric acids are very different, as are those of the parent acids. Dimethyl maleate is a liquid at room temperature (melting point $-19°C$), whereas dimethyl fumarate is a solid (melting point $101-102°C$).

Pre-lab exercises for Section 7.3, Isomerization of Dimethyl Maleate to Dimethyl Fumarate, can be found on page PL. 23.

EXPERIMENTAL PROCEDURES

DO IT SAFELY

1 Bromine is a hazardous chemical. Do not breathe its vapors or allow it to come into contact with the skin because it may cause serious chemical burns. All operations involving the transfer of solutions of bromine should be carried out in a ventilation hood; it is prudent to wear rubber gloves. If you get bromine on your skin, wash the area quickly with soap and warm water and soak the skin in 0.6 *M* sodium thiosulfate solution (up to 3 h if the burn is particularly serious).

2 Do *not* use your mouth with a pipet to transfer the bromine; use a dropper with a rubber bulb.

3 Do not rinse glassware containing residues of bromine with acetone. See the Do It Safely note 2 in Section 12.2.

Place 1.5 mL of dimethyl maleate in each of three 150-mm test tubes and with a dropper add enough of a 0.6 *M* solution of bromine in carbon tetrachloride to *two* of the tubes so that an orange solution results. Add an equal volume of carbon tetrachloride to the third test tube. Stopper all the test tubes and place one of the tubes containing bromine in the dark and expose the other two tubes to strong light. If decoloration of a solution should occur, add an additional portion of the bromine solution. After 30 min cool all three solutions in ice water, observe in which test tube(s) crystals appear, and isolate the precipitate by vacuum filtration (Section 2.16). Wash the crystals free of bromine with a little *cold* carbon tetrachloride and press them as dry as possible on the filter disk. Recrystallize the product from ethanol (Section 3.1) and determine its melting point and weight. The reported melting range for dimethyl fumarate is $101-102°C$.

Add a few drops of cyclohexene as needed to any solutions or containers in which the color of bromine is evident in order to dispel that color, and then discard all solutions in the bottle labeled "organic liquid waste."

EXERCISES

1. Write a reasonable mechanism for the bromine-catalyzed isomerization of dimethyl maleate to dimethyl fumarate observed in this experiment.

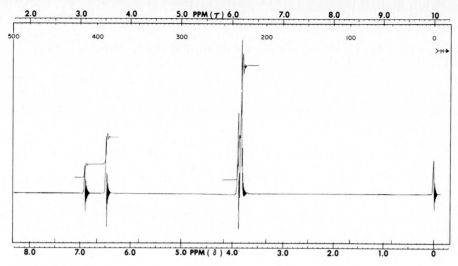

Figure 7.5 PMR spectrum of a mixture of dimethyl fumarate and dimethyl maleate.

2. What is the function of light in the isomerization reaction?

3. What is the purpose of exposing a sample of dimethyl maleate containing no bromine to light?

4. Suggest a reason why decoloration of the solution of bromine and dimethyl maleate is slow.

5. Figure 7.5 is a PMR spectrum of a mixture of dimethyl fumarate and dimethyl maleate. Calculate the percentage of each isomer in this mixture.

6. Write an equation for the reaction of cyclohexene with bromine.

7. Which of the isomeric esters is more thermodynamically stable? Suggest a reason.

SPECTRA OF STARTING MATERIALS AND PRODUCTS

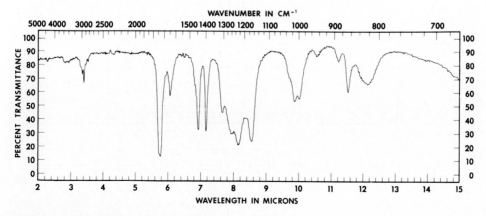

Figure 7.6 IR spectrum of dimethyl maleate.

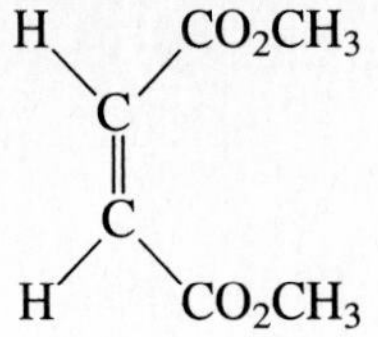

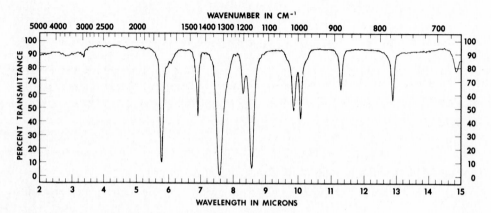

Chemical shifts: δ 52.1, 130.1, 165.8.

Figure 7.7 CMR data for dimethyl maleate.

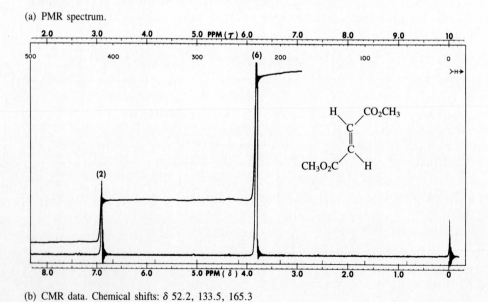

Figure 7.8 IR spectrum of dimethyl fumarate (KBr pellet).

(a) PMR spectrum.

(b) CMR data. Chemical shifts: δ 52.2, 133.5, 165.3

Figure 7.9 NMR data for dimethyl fumarate.

7.4 Polarimetry

As noted in Section 7.1, chiral molecules are optically active; that is, they have a rotational effect on the plane of polarized light[3] as it passes through the substance. The phenomenon may be explained as follows. Consider that a chiral molecule has a dissymmetric distribution of electron density. As the polarized light passes through the molecule, its electric component will necessarily be affected in a dissymmetric manner by the electrons of the molecule. The result is that the plane of the polarized light is twisted about the axis of propagation. Any substance having this effect on polarized light is termed optically active. An achiral molecule (one that is superimposable with its mirror image) will be optically inactive, since the light, as it passes through the substance, will on the average encounter a symmetric distribution of electrons and will thus be unaffected. It must be realized that such an experiment is macroscopic rather than microscopic, because even with a very small sample of the substance under consideration, a very large number of molecules are present. Any measurement must necessarily be an average observation of the effect on all rays within the polarized light by all molecules through which they pass.

In a symmetric environment enantiomers are identical with regard to all physical constants and chemical properties, with one exception: although each member of the enantiomeric pair rotates the plane of polarized light by angles of the same magnitude, they do so in *opposite* directions. By convention, when looking through the sample *into* the beam of light, a rotation observed to be clockwise, or to the right, is termed a *positive* rotation, and a counterclockwise rotation is termed a *negative* rotation. If the sample under consideration is a mixture of exactly equal amounts of the two enantiomers, the rotation of one of the components will be exactly offset by the equal but opposite rotation of the other. The net rotation will be observed as zero. Such a mixture is called a **racemic modification.** In order that there be any net observed rotation in a mixture of enantiomers, one member of the pair must be present in an amount greater than the other.

The observed angle of rotation depends on a number of factors: (1) the nature of the compound, (2) its *concentration* in solution or its *density* as a neat liquid, (3) the length of the sample through which the light must pass (the *path length*), (4) *temperature*, (5) *solvent*, and (6) the *wavelength* of light. The concentration, or density, and path length of the sample are important considerations because they determine the average number of chiral molecules through which a beam of light will pass. The magnitude of rotation is the result of a cumulative effect, and if, on the average, the light passes through a greater number of chiral molecules, the observed magnitude of rotation will be greater.

It is certainly desirable to present the optical rotation as a *physical constant* of the active compound. Therefore, it is common practice to report the **specific rotation** rather than the

---

[3]Electromagnetic radiation (including light) is composed of propagated electric and magnetic fields. From a wave-mechanical point of view, a ray of light is composed of two plane waves, one electric and one magnetic, oriented at right angles to each other. For simplicity, considering only the electric waves, ordinary light is completely unordered, with these waves oriented at all possible angles in the plane perpendicular to the direction of propagation. In plane-polarized light, each ray has the plane of its electric wave oriented parallel to the corresponding plane of all other rays.

observed angle of rotation. The specific rotation is calculated by correcting the observed rotation to *unit* concentration and path length using equation 6.

$$[\alpha] = \frac{\alpha}{1 \times c} \quad \text{or} \quad \frac{\alpha}{1 \times d} \tag{6}$$

$[\alpha]$ = specific rotation (degrees)

α = observed rotation (degrees)

l = path length (decimeters)

c = concentration (g/mL of solution)

d = density (g/mL, neat)

To specify the other variables on which the rotation depends, the temperature and wavelength (nm) of the light employed are presented as superscript and subscript, respectively, on the symbol for specific rotation; the solvent used and the concentration of solute in it are denoted in parentheses following the numerical value and sign of the specific rotation. For example, a specific rotation might be reported as $[\alpha]_{490}^{25°}$ +23.4° (2.1 CH_3OH). A sodium lamp emitting light at 589 nm, the sodium D line, is usually used as a light source. In this case it is common practice to use D in place of the numerical value of the wavelength in the expression, as for example, in the rotation $[\alpha]_D^{25°}$ −15.2° (1.0 H_2O).

Various types of commercial polarimetric apparatus are available. Some, such as the Rudolph polarimeter, are manually operated and require direct readings by the operator, whereas other, newer varieties are automatic and utilize photoelectric cells for measurements of greater accuracy and precision. A simply constructed student-type polarimeter has been described (see the reference at the end of this section) that may be used quite satisfactorily for the measurements of rotation in Sections 7.5 and 7.6.

The basic principles of operation of all polarimeters are illustrated in Figure 7.10. Ordinary light passes through a polarizing element (the polarizer in the figure) which in a precise instrument is a Nicol prism but in a simple instrument may be a piece of Polaroid film. The beam of polarized light then passes through the sample tube, where rotation of the plane of

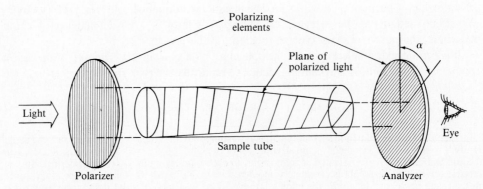

Figure 7.10 Schematic illustration of optical rotation (α = about −45°).

the light will be effected by a neat optically active liquid or a solution of an optically active substance. To determine the angle of rotation, the viewer turns the analyzer, another Nicol prism or piece of Polaroid film, until the light intensity observed is the same as that observed when the sample tube is empty or contains only pure solvent. The number of degrees through which the analyzer must be turned is the observed optical rotation, α, which can be converted to the specific rotation, $[\alpha]$, by applying equation 6.

REFERENCE

Shavitz, R., "An Easily Constructed Student Polarimeter," *Journal of Chemical Education*, **1978,** *55,* 682.

7.5   Isolation of the Enantiomers of Carvone from Essential Oils

Recall that all of the physical properties of enantiomers are identical, except with respect to plane-polarized light, and that all of their chemical properties are identical, except in an asymmetric environment. One of the simplest and most interesting examples of a difference of chemical properties in an asymmetric environment is the demonstration that the *odors* of the enantiomers of carvone, **4,** are different!

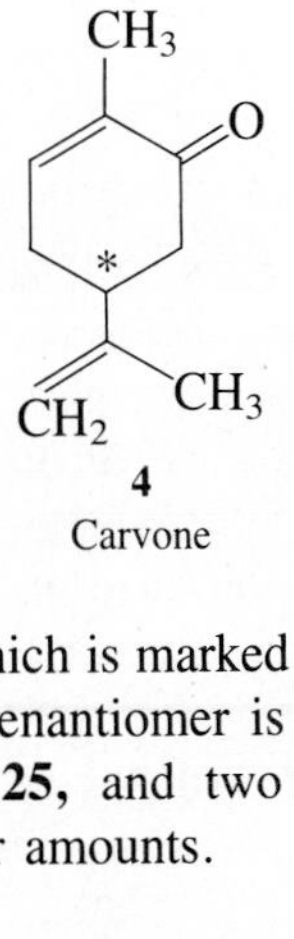

4

Carvone

Carvone contains one chiral center, which is marked with an asterisk in **4,** and thus it exists in two mirror-image forms. The $(-)$-enantiomer is the principal component of spearmint oil, which also contains limonene, **25,** and two other terpene hydrocarbons, α- and β-phellandrene, **26** and **27,** in smaller amounts.

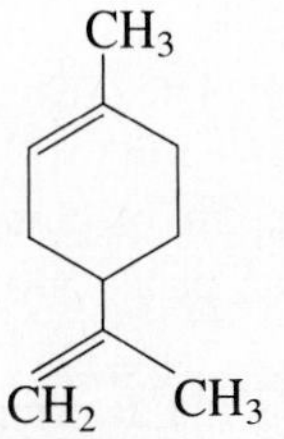

25	**26**	**27**
Limonene	α-Phellandrene	β-Phellandrene

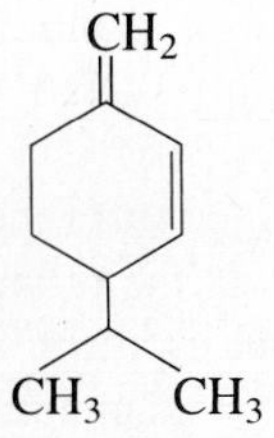

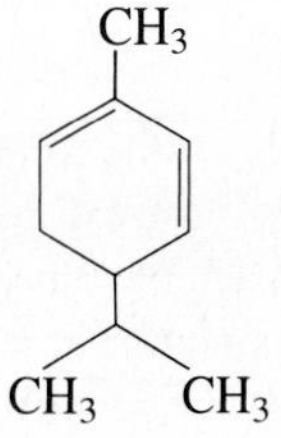

Caraway seed oil (and also dill seed oil) contains the (+)-enantiomer of carvone, as well as a larger amount of limonene than in spearmint oil. It is rather unusual for both enantiomers of a compound to occur in natural products, especially such readily accessible sources as spearmint oil and caraway seed oil.

Although these essential oils contain several aromatic (in the sense of odorous) compounds, as we have noted, it has been unequivocally proved that the marked difference in the odor of the two essential oils is attributable to the enantiomers of carvone! It should not be so surprising that enantiomers can have different odors if one considers that the odor receptors in the nose represent a chiral system, and hence should "react" differently with the two enantiomers of carvone, so that they produce diastereomeric olfactory sensations. Further information about the relationship between chirality and odor can be found in the references at the end of this section.

In the experimental procedures of this section, spearmint oil and caraway seed oil are subjected to fractional distillation in order to separate the carvone enantiomers from the other minor components of the oils. The boiling points of the carvone enantiomers are identical, of course, being 230°C at atmospheric pressure (760 torr); the boiling point of limonene is 177°C, which is sufficiently lower so that it can be separated easily by fractional distillation. The boiling points of the phellandrene isomers are near that of limonene, so they also can be separated easily from carvone in spearmint oil, along with the limonene. The fractional distillation should be carried out under reduced pressure in order to avoid any thermal decomposition of the carvone at higher temperatures. Consult Section 4.6 for details of this technique.

The pertinent tests that may be applied to the isomeric carvone products obtained from the two essential oils are numerous and varied, and the number performed will depend upon the time and facilities available to the students. In order to compare the properties of the carvone enantiomers, the instructor will probably wish to divide the laboratory class and assign spearmint oil to one group as starting material and caraway seed oil to the other. The cost of caraway seed oil is more than double the cost of spearmint oil, so the instructor may wish to assign more students to use the spearmint oil.

The simplest and most obvious observation of difference in properties to be made is certainly that of the odors. Although most people can detect a marked and characteristic difference in odor, it has been found that 8 to 10% of humans do not have this ability. (Achiral noses?) It should be interesting to check this statistic with each group of students and instructors.

If gas chromatography is available, samples of the original oils and of the distillation products should be analyzed and compared. Typical chromatograms are depicted in Figures 7.11–7.14.

The IR and PMR spectra of the carvone enantiomers are identical, of course, but if it is convenient, spectra of products from both essential oils may be obtained and compared with Figures 7.15 and 7.16.

A simple "wet test" can be made to confirm the identity of products from the enantiomers and an achiral reactant. One such is the production of a 2,4-dinitrophenylhydrazone derivative from the enantiomeric carvone isomers.

The ultimate test of the identity of the enantiomers is the determination of their optical rotation. Directions are given in the Experimental Procedure section for such determinations using a simple polarimeter and the procedure described in Section 7.4.

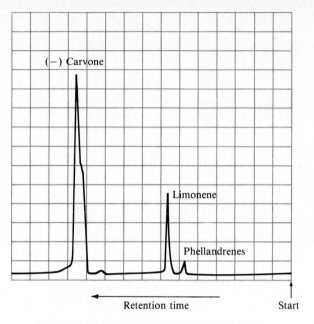

Figure 7.11 Gas chromatogram of spearmint oil.

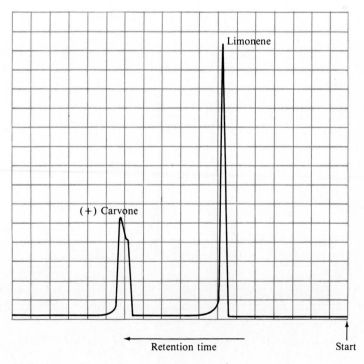

Figure 7.12 Gas chromatogram of caraway seed oil.

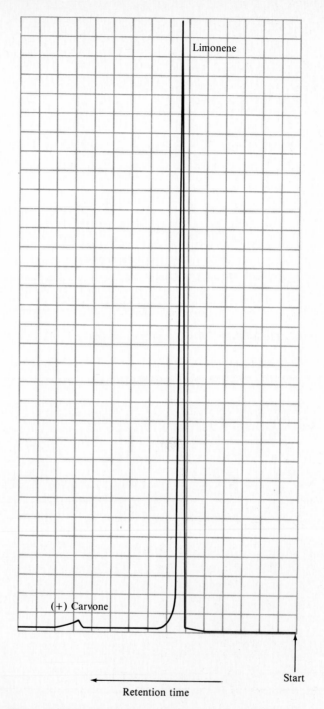

Figure 7.13 Gas chromatogram of Fraction 1 from distillation of caraway seed oil.

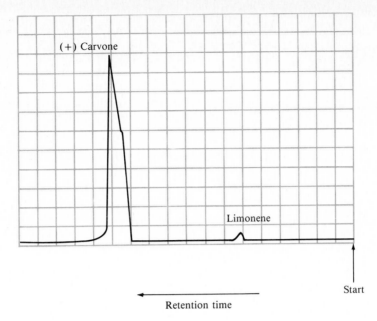

Figure 7.14 Gas chromatogram of Fraction 2 from distillation of caraway seed oil.

Pre-lab exercises for Section 7.5, Isolation of the Enantiomers of Carvone from Essential Oils, are found on page PL. 25.

EXPERIMENTAL PROCEDURE

DO IT SAFELY

1 If an oil bath is used for heating, be sure that it is supported well, and do not handle it when it is hot.

2 Examine all the glassware carefully, especially the stillpot, for cracks and stars, which might lead to implosion during the vacuum distillation.

A. Isolation of Carvone by Vacuum Fractional Distillation of Spearmint Oil or Caraway Seed Oil

The most important aspect of this experiment is the proper assembly of the distillation apparatus so that a satisfactory vacuum can be maintained. The following notes duplicate parts of the discussion in Chapter 4, but it may be helpful to repeat here some of the details that are most pertinent to this experiment.

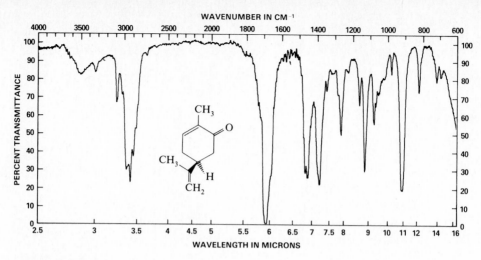

(a) (+)-Enantiomer from caraway seed oil.

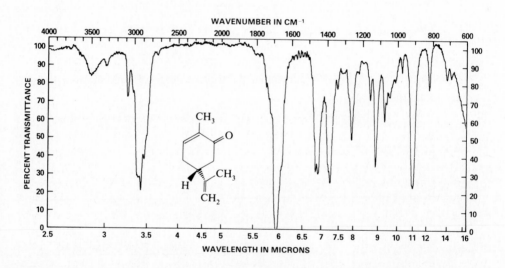

(b) (−)-Enantiomer from spearmint oil.

Figure 7.15 IR spectra of the enantiomeric carvones.

Assemble the apparatus for vacuum distillation using a 50-mL stillpot, a Claisen connecting tube, and a stillhead, as shown in Figure 2.17. Grease all the ground-glass joints lightly and make sure that all rubber stoppers fit well. Use rubber tubing that is thick-walled so that it will not collapse under vacuum. Bore the holes in the rubber stopper used for the safety

(a) PMR spectrum of carvone.

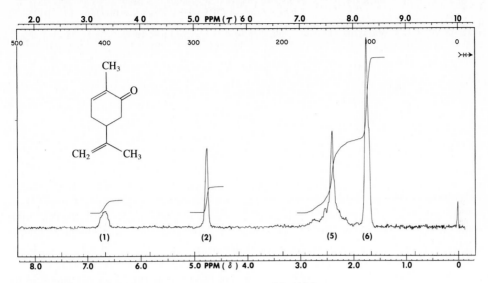

(b) CMR data. Chemical shifts: δ 110.4, 135.4, 144.1, 146.7, 198.8.

Figure 7.16 NMR data for carvone.

trap cleanly so that the glass tubes will seal well. Make sure the capillary ebullition tube is so fine that only a *very* small stream of bubbles is produced by blowing through the tube when the capillary end is immersed in acetone. Insulate the Claisen connecting tube and stillhead well by wrapping them with glasswool. If the stillpot is to be heated with an oil bath, use a liquid that has a flash point well over 200°C; silicone oil is ideal. The bath may be heated with a burner, but an electrical resistance coil connected to a variable transformer is preferable. An electric heating mantle may also be used in conjunction with a variable transformer, but an oil bath allows better control because a thermometer suspended in the oil will give a direct measure of the heat supplied to the stillpot. If no suitable oil bath is available, a heating mantle or other type of electric heating device may be used, but care should be taken not to overheat. Weigh the flasks that will be used as stillpot and receivers before assembling the apparatus.

Place 30 mL of spearmint oil or caraway seed oil in the 50-mL stillpot, attach a 25-mL round-bottomed flask as a receiver, make sure all connections are tight, and lower the pressure using an aspirator or whatever system is available in the laboratory. A pressure of about 25–30 torr is ideal for the distillation in which carvone, the major component of both spearmint and caraway seed oils, is separated from limonene and other terpenes present in smaller amounts. The following procedure is described for a distillation performed at a pressure of 25 torr. If you cannot achieve this pressure, use the accompanying Table 7.1 to interpolate the appropriate boiling temperatures corresponding to your pressure.

TABLE 7.1   Boiling Points of Carvone and Limonene at Different Pressures

Pressure (torr)	Carvone bp, °C	Limonene bp, °C
760	231	177
100	157	110
40	132	87
30	125	79
20	115	71
10	100	55

Begin the distillation by applying heat to the stillpot. If an oil bath is used, the first drops of distillate will appear when the oil bath is at about 150°C; the stillhead boiling temperature will be about 80°C. Collect the fraction that distils in the range 80°C to about 120°C. In the case of caraway seed oil, the temperature may remain constant in the range 80°C to 85°C for a short time, since this oil contains a larger amount of limonene than does spearmint oil. In any case, when the stillhead temperature rises and then levels off again at about 120°C, reduce the heat, open the stopcock that releases the vacuum and remove the receiver containing the first distillation fraction. Attach another 25- or 50-mL round-bottomed flask as receiver and close the stopcock at the trap so as to reduce the pressure again. When a stable vacuum has been reestablished, raise the heat, and collect the distillate (carvone) that boils at about 120–125°C. During this distillation, the temperature of the oil bath may be raised to about 190°C. When no more drops of distillate are obtained, discontinue the heating, release the vacuum, and remove the receiver containing the second distillation fraction, which should be almost pure carvone. Weigh the receivers and record the weights of the two distillation fractions. If time permits, allow the stillhead and Claisen connecting tube to drain into the stillpot (30 min should be sufficient) and then wipe the stillpot clean and weigh it to determine the amount of residual oil.

The amount of carvone obtained from 30 mL of the essential oils should be about 16 g from spearmint oil and about 12 g from caraway seed oil, but it may be more or less, depending on the source of the oils.

B. Studies of the Enantiomers of Carvone from Spearmint Oil and Caraway Seed Oil

1. Odor   Compare the odor of the carvone obtained from spearmint oil with that of the carvone from caraway seed oil. Smell the first fraction obtained from each oil and compare its odor with that of the carvone fraction.

2. Gas Chromatography   Gas chromatograms of representative samples of carvone from spearmint oil and from caraway seed oil should be obtained under identical conditions so that the retention times may be compared. Chromatograms of the original essential oils should also be obtained, as well as one from the first distillation fractions, so that an estimate of the success of the distillations can be made. A suitable column is a 2-m × 3-mm 15% QF-1 on Chromosorb W 80/100 support, programmed 100–220°C at 20°C/min.

3. IR and PMR Spectra   If possible, an IR and a PMR spectrum should be obtained from a sample of carvone from each of the essential oils, and the spectra should be compared with those of Figures 7.15 and 7.16.

4. Optical Rotation   The amount of carvone obtained from 30 mL of the essential oils will probably be enough to provide a 5-cm light path in a simple polarimeter such as that described in Section 7.4, so that the rotation of a neat sample can be determined. If some students do not obtain enough carvone for an individual determination of rotation, they should combine their products, making certain that the carvone samples were obtained from the same essential oil. The specific rotations can be calculated from the observed rotations using the formula of equation 6 and the description given in Section 7.4. The reported density of carvone is 0.960 at 20°C, and the specific rotation is ±62° at 20°C.

5. 2,4-Dinitrophenylhydrazone of Carvone   Prepare this derivative from a 0.5-g sample of your distilled carvone using the general procedure of Section 27.5, a2. The reported melting point of the crystalline product is 193–194°C. If recrystallization is necessary, a mixture of ethanol and ethyl acetate is a satisfactory solvent. Obtain a sample of a derivative from another student who used a carvone sample distilled from the other essential oil and make a mixture with your derivative. Determine the melting point of the mixture.

EXERCISES

1. What difficulty would result from using an ebullition tube having a capillary bore that is too large?
2. Why should a constant vacuum be established before heating is begun for the distillation?
3. If you observed a rotation of 180° for an optically active liquid in a simple polarimeter, how could you determine experimentally whether the sign of rotation is actually + or −?
4. What melting point should you expect from a mixture of the 2,4-dinitrophenylhydrazone from (+)-carvone and the 2,4-dinitrophenylhydrazone from (−)-carvone if both derivatives are pure?
5. Identify the chiral center in limonene, **25**.
6. The absolute configuration of (−)-carvone is shown in structure **4a**. Tell whether this configuration should be designated as R or S.

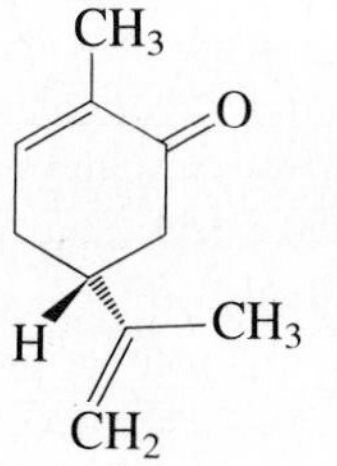

4a

(−)-Carvone

REFERENCES

1. Murov, S. L.; Pickering, M., *Journal of Chemical Education,* **1973,** *50,* 74.
2. Russell, G. F.; Hills, J. I., *Science,* **1971,** *172,* 1043.
3. Friedman, L.; Miller, J. G., *Science,* **1971,** *172,* 1044.

7.6 Resolution of Racemic 1-Phenylethanamine

1-Phenylethanamine (**3**) may be prepared from acetophenone (**1**) and ammonium formate (**2**) (see reference 1 at the end of this section). The product obtained is not optically active, even though it contains an asymmetric carbon atom (marked in formula **3** with an asterisk), because the enantiomers are produced in exactly equal amounts; that is, a racemic modification is formed.

$$
\underset{\substack{\textbf{1}\\ \text{Acetophenone}}}{C_6H_5-\overset{\overset{\textstyle O}{\|}}{C}-CH_3} + \underset{\substack{\textbf{2}\\ \text{Ammonium}\\ \text{formate}}}{H-\overset{\overset{\textstyle O}{\|}}{C}-O^{\ominus}NH_4^{\oplus}} \rightarrow \underset{\substack{\textbf{3}\\ \text{1-Phenylethanamine}}}{C_6H_5-\overset{\overset{\textstyle NH_2}{|}}{\underset{\underset{\textstyle H}{|}}{C^*}}-CH_3} + H_2O + CO_2 \quad \textbf{(1)}
$$

Enantiomers cannot be separated from one another by the standard methods of crystallization, distillation, or chromatography. This is because they have identical physical properties, with the exception of their rotation of polarized light in opposite directions, as noted in Section 7.4. However, **diastereomeric isomers (diastereomers)** do differ in physical properties such as solubility, boiling point, and chromatographic adsorption characteristics. The most generally useful procedure for *resolving a racemic modification* of enantiomers involves converting the enantiomers to diastereomers, separating the diastereomers by one of the standard experimental procedures, and then regenerating the enantiomers from the separated diastereomers. This procedure requires the use of an optically active resolving agent; fortunately, nature provides an abundant source of these. Most of the naturally occurring organic compounds that have chiral centers are found in optically active form and hence can be used to resolve racemic modifications produced by synthesis.

In this experiment optically active (+)-tartaric acid is used as the resolving agent for racemic 1-phenylethanamine (see reference 2 at the end of this section). The resolution scheme is outlined in Figure 7.17. By reaction with the optically active acid, the racemic amine is converted to a mixture of diastereomeric salts, which have different solubilities in methanol and hence can be separated by fractional crystallization. To obtain *both* enantiomeric isomers in a state of high optical purity may require a large number of careful and tedious crystallization steps, but it is usually possible to obtain by only one or two crystallization steps the enantiomer that gives the less soluble salt in reasonable optical purity. In some cases it is also possible to obtain the other enantiomer without resorting to many repeated crystallizations of the more soluble diastereomer. For example, in the research on which this

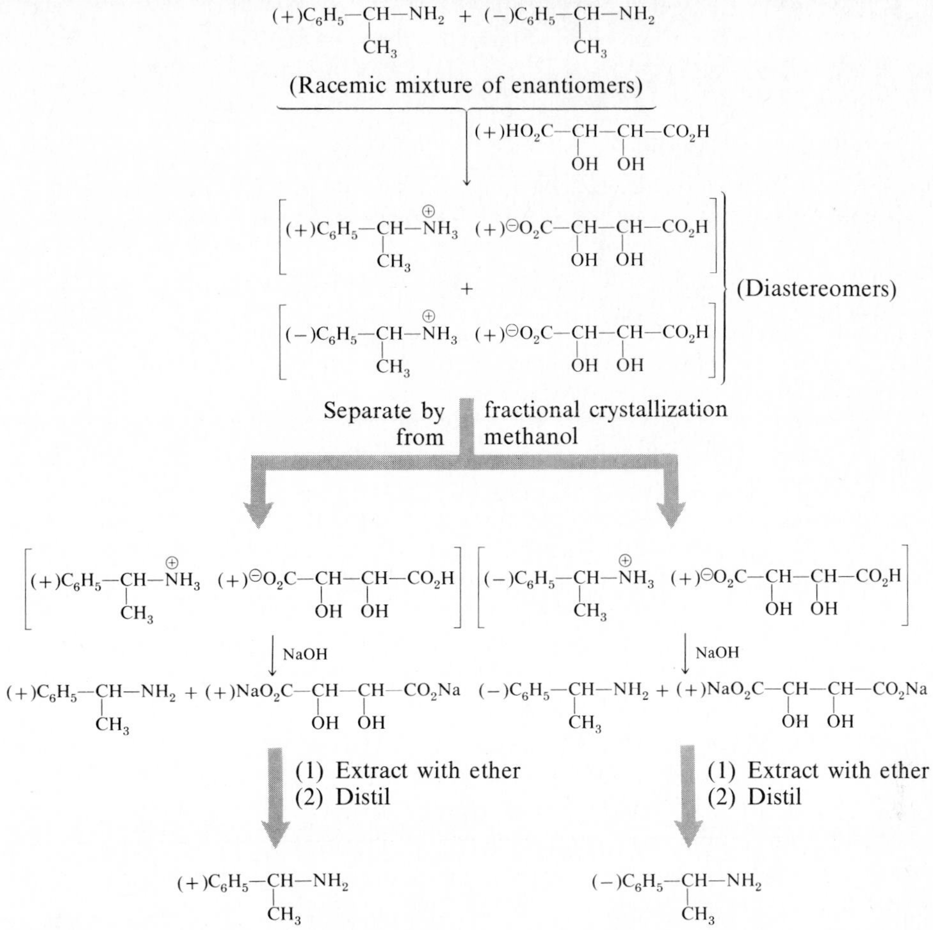

$$(+)C_6H_5-\underset{\underset{CH_3}{|}}{CH}-NH_2 \; + \; (-)C_6H_5-\underset{\underset{CH_3}{|}}{CH}-NH_2$$

(Racemic mixture of enantiomers)

$$(+)HO_2C-\underset{\underset{OH}{|}}{CH}-\underset{\underset{OH}{|}}{CH}-CO_2H$$

$$\left[(+)C_6H_5-\underset{\underset{CH_3}{|}}{CH}-\overset{\oplus}{NH_3} \quad (+)^{\ominus}O_2C-\underset{\underset{OH}{|}}{CH}-\underset{\underset{OH}{|}}{CH}-CO_2H\right]$$

+

$$\left[(-)C_6H_5-\underset{\underset{CH_3}{|}}{CH}-\overset{\oplus}{NH_3} \quad (+)^{\ominus}O_2C-\underset{\underset{OH}{|}}{CH}-\underset{\underset{OH}{|}}{CH}-CO_2H\right]$$

(Diastereomers)

Separate by fractional crystallization from methanol

$$\left[(+)C_6H_5-\underset{\underset{CH_3}{|}}{CH}-\overset{\oplus}{NH_3} \quad (+)^{\ominus}O_2C-\underset{\underset{OH}{|}}{CH}-\underset{\underset{OH}{|}}{CH}-CO_2H\right]$$

$$\left[(-)C_6H_5-\underset{\underset{CH_3}{|}}{CH}-\overset{\oplus}{NH_3} \quad (+)^{\ominus}O_2C-\underset{\underset{OH}{|}}{CH}-\underset{\underset{OH}{|}}{CH}-CO_2H\right]$$

NaOH

NaOH

$$(+)C_6H_5-\underset{\underset{CH_3}{|}}{CH}-NH_2 + (+)NaO_2C-\underset{\underset{OH}{|}}{CH}-\underset{\underset{OH}{|}}{CH}-CO_2Na$$

$$(-)C_6H_5-\underset{\underset{CH_3}{|}}{CH}-NH_2 + (+)NaO_2C-\underset{\underset{OH}{|}}{CH}-\underset{\underset{OH}{|}}{CH}-CO_2Na$$

(1) Extract with ether
(2) Distil

(1) Extract with ether
(2) Distil

$$(+)C_6H_5-\underset{\underset{CH_3}{|}}{CH}-NH_2$$

$$(-)C_6H_5-\underset{\underset{CH_3}{|}}{CH}-NH_2$$

Figure 7.17 Resolution of (±)-1-phenylethanamine by means of (+)-tartaric acid.

experiment is based (see reference 3 at the end of this section), it was found that the *hydrogen sulfate salt* of the enantiomer that gave the more soluble tartrate salt was *less soluble* than the hydrogen sulfate salt of the racemic amine, so that both enantiomers could be obtained in optical purity quite conveniently.

By measuring the optical rotation of the 1-phenylethanamine recovered from the less soluble salt in this experiment, it will be possible to determine not only whether it is the (+)- or (−)-enantiomer but also the extent of optical purity achieved in the resolution. If you are to do this part of the experiment, consult your instructor for special experimental directions.

Pre-lab Exercises for Section 7.6, Resolution of Racemic 1-Phenylethanamine, can be found on page PL. 27.

EXPERIMENTAL PROCEDURE

DO IT SAFELY

1 Methanol is flammable; do not heat it in a beaker and do not use a flame, but use a steam or hot-water bath.

2 Take care to avoid contact of the 14 M NaOH with the skin, as it is corrosive. If contact occurs, wash the area with copious amounts of water.

3 Take the usual care in the extractions with diethyl ether; have no flames in the vicinity.

Dissolve an accurately weighed sample of approximately 5 g of racemic 1-phenylethanamine in 35 mL of methanol, and determine its specific rotation, [α], using a polarimeter and equation 1. The amounts of 1-phenylethanamine and methanol specified here are those that give satisfactory results when a polarimeter sample tube with a light path of 2 dm or more and a volume of 35 mL or less are used. If a tube of shorter light path and/or larger volume is used, either the amount of 1-phenylethanamine used as starting material must be increased, or two or more students will have to combine their yields of resolved product to give a solution of high enough concentration so that the observed optical rotation will be large enough for accuracy. Students should check with their instructor about the type of polarimeter sample tube that will be used.

In a 1-L Erlenmeyer flask place 0.208 mol of (+)-tartaric acid and 415 mL of methanol and heat to boiling. To the hot solution add cautiously (to avoid foaming), with stirring, the 35 mL of solution recovered from the polarimeter and enough additional racemic 1-phenylethanamine to make a total of 0.206 mol. Allow the solution to cool slowly to room temperature and to stand undisturbed for 24 hr or until the next laboratory period.★ The amine hydrogen tartrate should separate in the form of white prismatic crystals. If the salt separates in the form of needlelike crystals, the mixture should be reheated until *all* the crystals have dissolved and then allowed to cool slowly. If any prismatic crystals of the salt are available, they should be used to seed the solution.

Collect the crystals of the amine hydrogen tartrate (17–19 g) on a filter, and wash them with a small volume of cold methanol. Do not discard the filtrate before you ask your instructor if you are to turn it in for recovery of the other enantiomer of 1-phenylethanamine. Dissolve the crystals in about four times their weight of water, and add 15 mL of 14 M sodium hydroxide solution. Extract the resulting mixture with four 75-mL portions of diethyl ether. Do not discard the aqueous solution; pour it into a bottle provided for recovery of tartaric acid. Wash the combined ether extracts with 50 mL of saturated sodium chloride solution, dry the ether solution over anhydrous magnesium sulfate, and then filter it from the desiccant.★

Remove the ether from the solution by distillation from a steam bath using an unpacked Hempel fractionating column, and then arrange to distil the residue (1-phenylethanamine) under aspirator pressure, using *no* fractionating column (see Section 4.6). It will be necessary to use a burner or an electric heater because the amine has a boiling point of 94–95°C (28 torr). The yield of optically active 1-phenylethanamine should be 5–6 g.

Weigh the distilled product accurately, and then transfer it quantitatively into about 35 mL of methanol. Measure the volume of the methanol *solution* accurately. It is the weight of the 1-phenylethanamine (in grams) divided by the volume of the *solution* (in mL) that gives you the concentration (c) for equation 6. Transfer the solution to the polarimeter sample tube and determine the specific rotation from the observed rotation by using equation 6. The reported specific rotation of optically pure 1-phenylethanamine is $[\alpha]_D^{25}$ 40.1° (neat).

EXERCISES

1. Report the specific rotation of your resolved product as (+) or (−). Suppose the observed rotation was found to be 180°. How could you determine whether the rotation was (+) or (−)?

2. How could you increase the optical purity of your product?

3. Describe clearly the point in the experimental procedure at which the major part of the other enantiomer was removed from your product.

4. How could the (+)-tartaric acid be recovered so that it can be used over again to resolve more racemic amine?

5. The absolute configuration of (+)-1-phenylethanamine has been shown to be R. Make a perspective drawing of this configuration.

6. Suppose you had prepared a racemic organic acid and wished to obtain an optically active form of it. How could you do this?

7. Suggest a possible procedure for resolving a racemic alcohol.

REFERENCES

 1. Ingersoll, A. W., in *Organic Syntheses,* Collective Vol. II, A. H. Blatt, editor, John Wiley & Sons, New York, 1943, p. 503.
 2. Ault, A., *Journal of Chemical Education,* **1965,** *42,* 269; *Organic Syntheses,* Collective Vol. V, H. E. Baumgarten, editor, John Wiley & Sons, New York, 1973, p. 932.
 3. Theilacker, W.; Winkler, H.–G., *Chemische Berichte,* **1954,** *87, 690.*

chapter **8**

introduction to spectroscopy; infrared spectroscopy

One important aspect of the science of organic chemistry is the determination of the exact structure of organic compounds, whether isolated from natural sources (*natural products*) or produced in chemical reactions. Formerly this was a very time-consuming, laborious, and often impossible task for the "classical" organic chemist because the primary method of structure determination usually involved converting an "unknown" compound into a "known" compound by a series of chemical reactions. This procedure is now known to have led to erroneous structures in some instances because of reactions involving structural rearrangements that went unrecognized by the investigator. The modern organic chemist has various spectroscopic techniques available that supplement and often replace the classical chemical methods of structure proof. Spectroscopic data are capable of providing structural information that may have taken years to elucidate in the not-too-distant past. The development of spectroscopy as a tool for structure determination has greatly contributed to the rapid growth of organic chemistry in the last 20–30 years.

Some of the contemporary spectroscopic methods of analysis are **nuclear magnetic resonance (NMR), infrared (IR), ultraviolet (UV),** and **visible spectroscopy** and **mass spectrometry (MS).** It is important to learn how to obtain, analyze, and interpret data provided by the spectroscopic methods discussed in this and the next three chapters. The greatest emphasis will be placed on NMR and IR spectroscopy, which are generally more available in the undergraduate organic laboratory course. It should be noted that the modern organic chemist usually uses some or all of these spectroscopic methods in structure determination. The use of a single type of spectroscopy is seldom sufficient to define a structure unequivocally. Chapter 10 shows how IR and NMR can be used together in structure determination. The comments above may imply that spectroscopy completely replaces experimental "bench" chemistry, which is not true. The chemist often combines experimental and spectral work in structure determination, as is illustrated in Chapter 27.

This chapter is devoted to IR, Chapter 9 to NMR, Chapter 10 to a combination of IR and NMR, and Chapter 11 to UV/visible and MS. An overview of these spectral methods is given below.

8.1 Overview of Spectral Methods

The various spectral methods (except for MS) discussed in this text depend upon molecular absorption of radiation to produce a specific type of molecular excitation. The type of excitation that is induced depends upon the energy of the radiation incident on the absorbing molecule. Ultraviolet and visible spectroscopy normally involve excitation of electrons from an orbital of the ground electronic state to an orbital of the next higher electronic state of the molecule, for example, the transitions associated with the energies E_1, E_2, and E_3 of Figure 8.1. Infrared spectroscopy, in contrast, depends upon transitions among different molecular vibrational-rotational levels with the *same* electronic state (usually the ground state), examples being the processes of energies E_4 and E_5 in Figure 8.1. Finally, NMR spectroscopy involves realignment of the spins of atomic nuclei in a magnetic field as a result of absorption of energy.

As indicated in Figure 8.1, electronic transitions are of higher energy than those involving vibrational-rotational changes and consequently require light of greater energy in order to occur. The energy (in kcal/mole) associated with radiation of a particular frequency (ν) or wavelength (λ) is readily determined by application of equations 1a and 1b, respectively, in

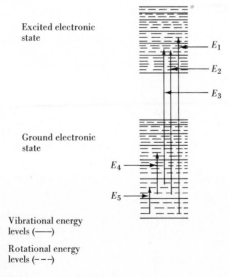

Figure 8.1 Rotational, vibrational, and electronic energy levels and transitions between them.

which N is Avogadro's number, h is Planck's constant $(6.6256 \times 10^{-27}$ erg $\cdot$ sec), c is the velocity of light $(2.998 \times 10^{10}$ cm/sec), and ν and λ would be measured, respectively, in Hertz (1.0 Hz = 1.0 cycle/sec) and nanometers (nm = 10^{-9} m).

$$E = Nh\nu \tag{1a}$$

$$E = Nhc/\lambda \tag{1b}$$

In summary, the various types of spectroscopy involve some sort of molecular excitation initiated by absorption of radiation; it is fundamentally the corresponding energy required to stimulate the transition from one molecular state to another that differentiates the methods. Table 8.1 presents the wavelengths of the various spectroscopic techniques of interest, the energies associated with them, and the basic molecular phenomena that are being induced. Reference to this chart as you learn more about the various methods should aid you in understanding the relationships among them.

Mass spectrometry, on the other hand, involves bombarding a molecule with high-energy particles (usually electrons) that cause it to *fragment* into smaller molecular ions, which can be identified on the basis of their mass. This technique also provides the molecular weight of the molecule with great accuracy. A brief introduction to MS is presented in Chapter 11.

Sometimes the terms *spectrometer* and *spectrophotometer* are used interchangeably and incorrectly. While both are instruments used to determine certain properties of molecules, a spectrophotometer makes use of some form of light in its operation, and a spectrometer does not. Hence, the terms "infrared or ultraviolet/visible spectrophotometer" and "mass spectrometer" represent correct usage.

The important types of structural information that can be obtained from the spectral methods presented in this text may be summarized as follows:

1. Infrared (IR) Spectroscopy IR is commonly used to ascertain the type(s) of *functional group(s)* in a molecule. For example, the presence of an —OH group, a carbonyl group (C=O), a C=C bond, a carbon-carbon triple bond or an aromatic ring can be deter-

TABLE 8.1 Relationships Between Wavelength, Energy, and Molecular Phenomena

Wavelength (nm)	Energy (kcal/mol)	Type of Spectroscopy	Molecular Phenomenon
200	143 ⎫		Excitation of valence electrons
		UV	from a filled to an unfilled
320	89 ⎭		orbital, for example, $\pi \rightarrow \pi^*$
400	71.5 ⎫		Same as above, except that
		Visible	absorption of light occurs in a
750	38.1 ⎭		region that is visible (colored) to the human eye
2000	14 ⎫	IR	Stretching and bending of
16000	2 ⎭		interatomic bonds
5×10^7	5.7×10^{-4} ⎫		Realignment of spins of atomic
		NMR	nuclei in an applied magnetic
2.1×10^{13}	1.3×10^{-9} ⎭		field

mined with infrared spectroscopy. IR can be just as important in providing information about which functional groups are absent. However, IR does not provide information about the number of hydrogens and other atoms contained in a molecule and usually does not allow a precise structural determination unless the IR spectrum of an unknown compound is found to be identical to that of a known compound.

2. Nuclear Magnetic Resonance (NMR) Spectroscopy NMR provides information about the number and arrangement of various atoms in a molecule, including but not limited to hydrogen. For example, in proton magnetic resonance (PMR) spectroscopy (Chapter 9), the presence of a —CH_3 or —CH_2CH_3 group can be determined, as can that of an aromatic ring and other groups that contain hydrogens. Some of the commonly encountered functional groups, such as a C—C, C=C or C=O, are not observed *directly* in a PMR spectrum, but their presence may be deduced indirectly when hydrogens are attached to carbon. On the other hand, carbon-13 magnetic resonance (CMR) spectroscopy (Chapter 9) does provide information concerning the presence or absence of most carbon-containing functional groups.

3. Ultraviolet (UV)/Visible Spectroscopy IR and NMR provide information about the functional group(s) and arrangements of atoms (structure) present in a molecule, respectively. On the other hand, UV/visible spectroscopy is useful for molecules that contain one or more double or triple bonds, as in alkenes or alkynes, or carbonyl groups. This technique is most useful for *conjugated* molecules, which have two or more double or triple bonds adjacent to each other. Compounds containing aromatic rings also can be analyzed by UV/visible spectroscopy.

4. Mass Spectrometry (MS) MS is useful for structural elucidation of most organic molecules. The molecule is broken apart into smaller fragments, many of which can be identified on the basis of their mass. This technique is currently the best method for determining the molecular weight of a molecule with great accuracy. However, considerable experience is required to analyze a mass spectrum because the original molecule is broken into many different small fragments. The great sensitivity of the instrument and the natural abundance of isotopes of a given element also provide potential complications in completely analyzing the spectrum.

8.2 Theory of Infrared Spectroscopy

The infrared spectra that are most useful and generally available for elucidating the structure of an organic compound are obtained by irradiation of the compound with light in the infrared (IR) region of the electromagnetic spectrum. The region of greatest interest is 4000 to 500 cm^{-1}, and the units are called either **reciprocal centimeters** or **wavenumbers.** The terms "reciprocal centimeters," "wavenumbers," and "frequency" (λ) are used interchangeably. As we shall see later, most infrared spectra are recorded in terms of these units. An alternative method for expressing the energy associated with this region is **wavelength,** λ, in microns, μ, with units of micrometers (10^{-6} m). The position of an IR absorption band can thus be expressed in either microns or wavenumbers; the latter tends to be preferred by organic chemists, but the spectra provided in this text contain both scales. Wavenumbers can be converted to microns, and vice versa, using equation 2.

$$\text{wavelength } (\lambda) = \frac{10{,}000}{\text{wavenumber (cm}^{-1})} \tag{2}$$

The interaction of IR radiation with an organic molecule can be explained qualitatively by imagining that the molecular covalent bonds between atoms are analogous to springs. These molecular springs are constantly vibrating and undergo stretching and bending motions (Figure 8.2) at frequencies that depend upon the masses of the atoms involved in the bond and upon the type of chemical bond joining the atoms. The frequencies of the various vibrations of the molecule correspond to those of IR radiation, so that absorption of radiation occurs and produces an increase in the *amplitude* of the molecular vibration modes. In other words, IR radiation is absorbed at certain characteristic frequencies by the molecule. No irreversible change occurs in the molecule, and the light energy it absorbs is quickly given off in the form of heat. *IR spectroscopy does not permanently destroy or change the structure of a molecule in any way*.

The analogy of bonds as springs can be extended to atoms by considering them to be balls of different mass. The vibrational energies associated with bonds (springs) and atoms (balls) are accompanied by various stretching and bending vibrations. Heavier atoms (balls) are moved apart with lower frequencies, and atoms (balls) that are held together by stronger bonds (stronger springs) are pulled together with higher frequencies. Classical physics provides a mathematical relationship between frequency of vibration, mass of the atoms attached to a bond, and bond strength (equation 3).

$$\nu = \frac{1}{2\pi c} \sqrt{\frac{k}{m^*}} \tag{3}$$

where ν = frequency of absorption in cm^{-1}
c = speed of light, 2.998×10^{10} cm/sec
k = force constant of bond

m^* = reduced mass of two atoms of a bond = $\dfrac{(m_1 + m_2)}{m_1 m_2}$

with m_1 and m_2 being the masses of the two atoms

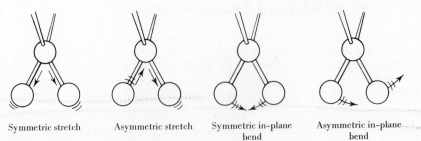

| Symmetric stretch | Asymmetric stretch | Symmetric in-plane bend | Asymmetric in-plane bend |

Figure 8.2 Some vibrational modes of a group of covalently bonded atoms.

This relationship indicates that the vibrational frequency is proportional to the square root of the force constant, k, which is unique to a given bond. The force constant is analogous to the strength of a spring. Equation 3 also shows that frequency is *inversely* proportional to the reduced mass of the atoms and supports the earlier statement that atoms of greater mass decrease the frequency of the absorption of energy. Also, stronger bonds (larger force constants) increase the frequency.

The frequency of bond movement (stretching and bending) depends on the bond strength, the atoms attached to a bond, and to some extent the environment of a particular bond in the molecule. In the ball and spring analogy, the motion of both the balls and the spring increases if they are hit. When IR radiation interacts with the bond in a molecule, the vibration of the bond also increases. However, a ball attached to a spring responds to all changes in energy that are supplied to it. On the other hand, a bond in a molecule is affected only when energy of a particular wavelength is absorbed. This results in absorption of energy and causes the energy level of the bond to be increased, as shown in Figure 8.1. Only certain energies affect certain bonds. When a bond is exposed to the characteristic energy that it absorbs, the amplitude of a particular vibration increases *suddenly,* an occurrence that is quite different from the *gradual* absorption of energy that is supplied to a ball-and-spring system. The IR instrument continuously changes the frequency of the IR radiation that strikes the sample. When the energy of the radiation corresponds exactly to the energy required to stretch or bend a particular bond, that energy is absorbed, and the amount of energy absorbed is recorded as a function of wavenumber on graph paper. The resulting plot of energy absorption versus wavenumber (energy) is called an **infrared (IR) spectrum.**

Suppose that an organic molecule is irradiated with light in the IR region and that it absorbs no energy at a particular frequency. At this frequency, all the energy is transmitted through the molecule because none of it is absorbed. Now suppose the same molecule is irradiated with light at a *different* frequency and that some of the energy is absorbed in a particular stretching or bending vibration. Some of the energy is absorbed and the rest is transmitted through the sample. The amount of energy absorbed or transmitted is viewed in one of two ways. If a sample absorbs 40% of the radiation at a given frequency, then 60% (100% − 40%) of the radiation is transmitted through it. The terms **percent transmittance** and **percent absorbance** are used to indicate the amount of energy transmitted through or absorbed by the sample. Percent transmittance, $\%T$, is defined as the ratio of the intensity of the light passing through a sample, I, to the intensity of the light striking the sample, I_0, multiplied by 100 (equation 4).

$$\%T = \frac{I}{I_0} \times 100 \tag{4}$$

Percent absorbance, $\%A$, is defined in equation 5.

$$\%A = 100 - \%T \tag{5}$$

For reasons discussed in Section 8.3, the IR instrument measures percent transmittance ($\%T$).

Figure 8.3 shows a typical IR spectrum, which contains scales giving both the wavenumber in cm^{-1} and wavelength in microns of the radiation as well as scales giving the percent

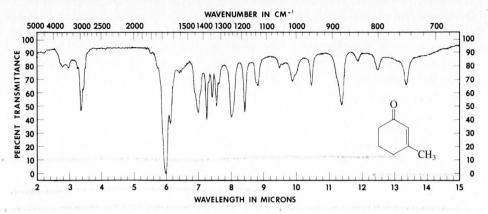

Figure 8.3 Infrared spectrum of 3-methyl-2-cyclohexenone.

transmittance of energy at each frequency. The IR spectra in this text and those obtained from most instruments contain these scales.

The significance and interpretation of the peaks in IR spectra are discussed in Section 8.7.

8.3 The Infrared Spectrophotometer

The operation of an infrared spectrophotometer will now be discussed briefly. A simplified diagram of a typical double-beam IR instrument is shown in Figure 8.4. Double-beam refers

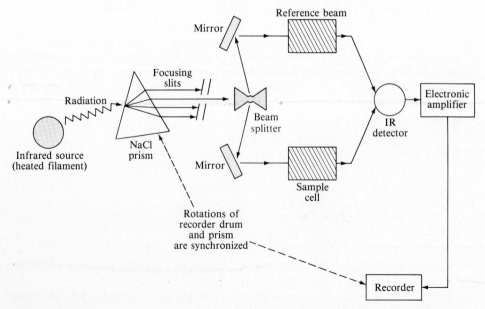

Figure 8.4 Simplified view of the recording double-beam infrared (IR) spectrophotometer. See text for discussion.

to the fact that the infrared radiation from the source is split into two beams of equal intensity. One beam passes through a cell containing the sample of the compound being studied and the other through a reference cell. The IR source (a heated filament) emits a continuum of wavelengths that pass through a prism or grating that separates the continuum into radiation of different wavelengths. Rotation of the prism or grating permits the different wavelengths to pass through the focusing slits to produce light of a single wavelength (*monochromatic light*), which strikes the beam splitter. The split beams are of equal intensity and are focused by the mirrors before passing through the sample and reference cells. The two beams then strike an IR detector, which compares their intensities and produces percent transmittance. The IR detector converts the transmittance into an electrical signal that is amplified and sent to a recorder that traces the IR spectrum. The rotation of the prism or grating is synchronized with the wavelength and wavenumber scales on the recorder to provide a spectrum that is a plot of the transmittance of light at different wavenumbers as it passes through the sample. The recorders are of two basic designs. Some instruments keep the paper stationary while the pen moves horizontally across it, and in others the paper moves while the pen is kept stationary. In either case, the pen moves vertically to record the transmittance as a function of wavenumber. The instrument is much more complex than this simplified view, and it should be emphasized that this discussion provides only an overview of the principles of its operation.

The determination of an IR spectrum involves the introduction of a sample into a cell such as is described below, which is then placed in the instrument. A sheet of preprinted, precalibrated IR paper is placed on the recorder, and the spectrum is recorded "automatically" after a few minor adjustments are made on the instrument. An IR spectrophotometer is usually a self-contained unit that includes all of the electronics and controls, the recorder, and a place for the sample. The recording of the spectrum is done nearly automatically.

The length of time required to record a spectrum depends on the instrument. A "routine" spectrum can be run in 5 to 10 minutes but does not show great detail. This type of spectrum is suitable for most purposes in organic chemistry, since it usually shows important absorptions that are useful in identifying functional groups. Some IR spectrophotometers, especially very expensive ones, are capable of producing "high resolution" spectra that may take hours to record. These show great detail and are used when small differences in wavenumber or absorption must be determined. Many technological advances in IR spectrophotometers have appeared in recent years, and the most expensive high-resolution units today are equipped with microprocessors that analyze the spectra and provide a vast amount of very accurate data.

The sample cells and various optical parts in the spectrophotometer, such as the prism, must be *transparent* to infrared radiation. They are made of various highly purified, fused inorganic salts such as NaCl (rock salt), KBr, CsBr, or other salts composed of Group I metals and halogens. IR radiation is absorbed by ordinary Pyrex glass, so it is unsatisfactory for use in cells. Sodium chloride and potassium bromide cells are used most frequently by organic chemists. Although one usually thinks of NaCl in the form of a white crystalline solid (table salt), sodium chloride and other IR cells are transparent because they are manufactured under high pressure.

IR cells must be handled with great care and must *not* come in contact with moisture. They are often stored in desiccators containing a drying agent such as calcium chloride. Even though they may be turned "off," IR spectrophotometers contain electric heaters that run constantly to keep the internal NaCl or KBr optics free from contact with moisture in the air.

Consequently, IR machines should be plugged in at all times. Turning them ''on'' activates the electronics, motors, and infrared source. Most machines take 10–15 minutes to warm up after being turned on.

8.4 Infrared Spectroscopy of Liquids, Solids, and Gases

Infrared spectra can be determined for liquids, solids, and gases, and the techniques for handling each will be discussed briefly in this section. Before this is done, the factors that determine the intensity of absorption in an IR spectrum should be considered. The amount of absorption, termed absorbance (A), at a given wavenumber is defined in equation 6.

$$A = \log_{10}\frac{I_0}{I} = kcl \tag{6}$$

where A = absorbance
I_0 = amount of radiation passing through the sample
I = amount of radiation originally striking sample
k = absorptivity of the sample
c = concentration, in g/L, of solute in solution and
l = path length, in cm, of cell containing sample

The absorptivity is a constant, but its magnitude is wavenumber-dependent, which results in the observation of absorption maxima and minima—that is, the peaks—in the spectrum. The two experimental variables that increase or decrease the intensities of *all* of the peaks in an IR spectrum are concentration and cell length.

Absorptivity is somewhat characteristic of the functional group absorbing the radiation. For example, the absorptivity of the stretching mode of a carbonyl group (C=O) is generally greater than that of a carbon-carbon double bond. This can be seen by comparing the intensities of the peaks in the spectrum shown in Figure 8.3—1630 cm^{-1} for C=C and 1690 cm^{-1} for C=O. The discussion of the interpretation of IR spectra (Section 8.7) will include generalizations regarding the relative intensities of various peaks associated with different functional groups.

The *general* techniques that are used to determine the IR spectra of liquids, solids, and gases are discussed in the following paragraphs.

1. Liquids IR spectra of liquids are usually run on the pure compound, often called a **neat liquid.** Peak intensities in the IR spectra of neat liquids can be controlled only by varying the cell length, since concentration is not a variable. The simplest but not necessarily the best technique involves placing a thin film of the pure liquid between two NaCl plates, which are sometimes called **NaCl windows.** The windows are placed in a cell holder, which is put into the sample compartment of the instrument. This type of *demountable* sample cell has several disadvantages, one being that the thickness of the film cannot be determined and is slightly different each time a sample is prepared. Another is that volatile samples evaporate to some extent after being placed in the sample compartment because of the heat generated by the infrared radiation. This causes a decrease in the intensity of the absorption peaks while the spectrum is being run because peak intensity is partially dependent on the amount of sample present in the cell. For example, the peaks may be very strong (low percent transmittance) at the beginning of the spectra (high wavenumbers) and be weak (high percent transmittance) toward the end of the spectrum as a result of sample evaporation. However, demountable IR cells are used frequently for neat (pure) liquid samples, the experimental details of which are discussed in Section 8.5b, part 1.

A better but more expensive technique utilizes a neat liquid cell that consists of two NaCl windows that are permanently mounted in a cell holder and separated by a Teflon spacer of known thickness, usually 0.020–0.030 mm. This **fixed-thickness cell** is filled with the liquid and sealed with Teflon stoppers, which avoids sample loss through evaporation and permits the sample thickness to be known with accuracy. The spectra obtained when this type of cell is used are better than those obtained with a demountable cell.

Whether the demountable or the fixed-thickness cell is used, the cell containing the sample is always placed in the sample compartment in the IR spectrophotometer. *Nothing is placed in the reference beam,* the reason being that the NaCl windows in the sample cell do not absorb IR radiation; thus, any observed absorption arises only from the sample contained in the cell.

2. Solids

2. Solids The IR spectra of solids can be obtained in one of several ways: the **solution method,** the **potassium bromide pellet method,** and the **mull method.** The solution method is the most desirable and the mull method the least in terms of the quality of spectra obtained.

(a) Solution Method Most organic solids are soluble in one or more organic solvents. The most commonly used solvents for IR spectra are carbon tetrachloride, chloroform, and carbon disulfide (CS_2). As a general rule, solutions containing 20 mole percent of solute give suitable IR spectra when the cell length is 0.2 mm. Shorter cells require a greater concentration and longer cells a lesser concentration, since the absorbance depends on both concentration and cell length (equation 6). Fixed-thickness cells must be used for solutions. However, if one attempted to determine the spectrum of a solid dissolved in a solvent using the technique described above for neat liquids, the spectrum would show peaks resulting from *both* the solid *and* the solvent. This problem is solved by placing one cell containing the solvent and the dissolved solute in the sample compartment and another cell of the same length containing *pure* solvent in the reference compartment. The cell containing pure solvent is called the **solvent cell.** Recall from the discussion of IR spectrophotometer operation (Section 8.3) that the IR detector compares the intensities of the two beams that strike it. The solvent in both the sample and the solvent cell absorbs radiation to the same extent at a given wavenumber, so that the IR detector ''sees'' the same intensities due to the solvent. In other words, the absorptions due to the solvent in both cells cancel one another. The absorption differences are therefore due only to the solid contained in the sample cell, and the IR spectrum recorded is that of the solid.

Since the sample cell also contains the solute, there is less solvent in this cell than in the solvent cell. The cancellation of solvent peaks is thus not complete, and the resulting IR spectrum usually contains small peaks due to the difference in solvent concentrations. However, the spectra obtained by this technique are usually satisfactory for most purposes, since the C—Cl bonds in CCl_4 and $CHCl_3$ absorb at low wavenumbers and do not obscure the peaks of greatest interest for most organic molecules. When CS_2 is used as the solvent, the C=S bond absorbs strongly at 1500 cm^{-1}, and care must be exercised in interpreting such spectra. With more elaborate, adjustable-path-length IR cells, the thickness of the sample cell can be increased slightly to *exactly* cancel the absorptions due to solvent.

Experimental details are given in Section 8.5c, part 1.

(b) Potassium Bromide (KBr) Pellet Method Another method of determining the infrared spectra of solids is the potassium bromide pellet method. Potassium bromide, like NaCl, is transparent in the infrared region. A potassium bromide pellet is a nearly transparent disk prepared by subjecting a finely ground mixture of dry, powdered spectral-grade KBr and the solid sample to a high pressure (*ca.* 8–20 tons per square inch). The heat generated by the high pressure causes the mixture to fuse. After being allowed to cool and resolidify, the disk is then placed in the spectrophotometer, and the IR spectrum is run.

Potassium bromide is extremely hygroscopic, and it ordinarily is impossible to prepare a disk that is completely free of water. Consequently, spectra of solids taken in KBr pellets

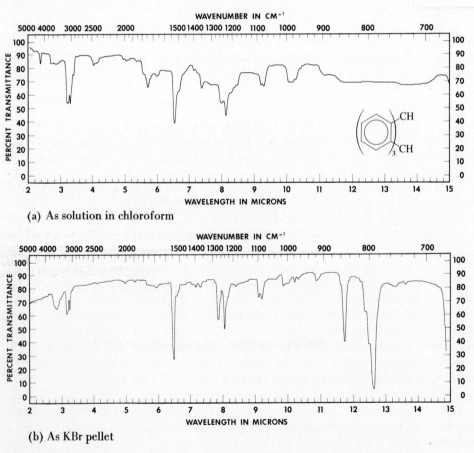

(a) As solution in chloroform

(b) As KBr pellet

Figure 8.5 Infrared spectra of triptycene as a solution in chloroform and as a KBr pellet.

almost invariably show absorption bands for water. The O—H peaks of water may partially or completely obscure N—H and O—H stretching in the $3200–3600$ cm^{-1} region, thus rendering this region of the spectrum difficult to interpret. This concept is illustrated by the two spectra shown in Figure 8.5.

Another limitation inherent in using the KBr pellet method for solids is that heat-sensitive compounds may decompose or react with KBr during the preparation of the pellet. This technique is not applicable to quantitative work because of the difficulty in preparing the disk reproducibly.

Experimental details are given in Section 8.5c, part 2.

(c) Mull Method An alternative to the solution method for IR spectra of solids is the mull method. A mull is a dispersion of finely divided solid in a carrier fluid, commonly mineral oil (Nujol) or perfluorokerosene. Nujol is a trade name for mineral oil and is obtained from crude oil by fractional distillation; it has the general formula $CH_3(CH_2)_nCH_3$, where $n = 18–22$. Perfluorokerosene is derived from kerosene, which comes from crude oil and is composed mostly of linear alkanes of 12–18 carbon atoms. Perfluorokerosene,

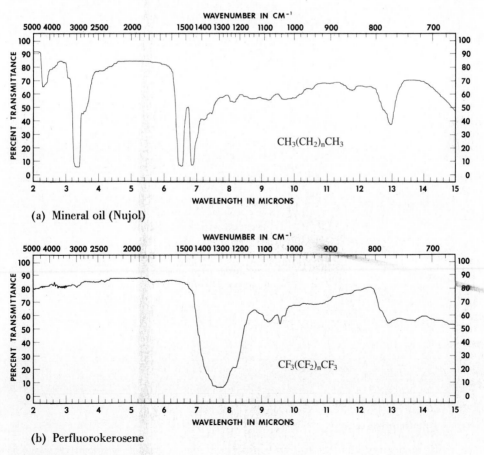

(a) Mineral oil (Nujol)

(b) Perfluorokerosene

Figure 8.6 Infrared spectra of the mulling fluids (a) mineral oil (Nujol) and (b) perfluorokerosene.

$CF_3(CF_2)_nCF_3$, $n = 10\text{--}16$, is obtained by subjecting kerosene to exhaustive fluorination. The prefix *per* in *perfluoro* means that the compound has all of its hydrogens replaced with fluorine. Another useful mulling fluid is hexachlorobutadiene, $CCl_2{=}CCl{-}CCl{=}CCl_2$, which has absorption peaks in the $500\text{--}1000$ cm^{-1} region.

These liquids have a reasonably high viscosity, which helps maintain the dispersion, and relatively few absorption bands in their IR spectra (Figure 8.6). The mull is placed between two cell windows in a demountable cell, which is positioned in the sample compartment of the instrument. Interpreting IR spectra obtained by the mull method requires that the peaks due to the mulling fluid be ignored, since the spectra contain peaks due to both the mulling fluid and the compound dispersed in it. As shown in Figure 8.6, the strong absorption peaks of mineral oil ($2800\text{--}3000$ cm^{-1} and $1300\text{--}1500$ cm^{-1}) correspond to locations where perfluorokerosene does *not* absorb. Essentially all portions of the IR spectrum of a solid can be observed if two spectra are obtained, one in Nujol and one in perfluorokerosene. The spectra shown in Figure 8.7 illustrate this technique; the peaks due to the mulling fluid are indicated in the spectra.

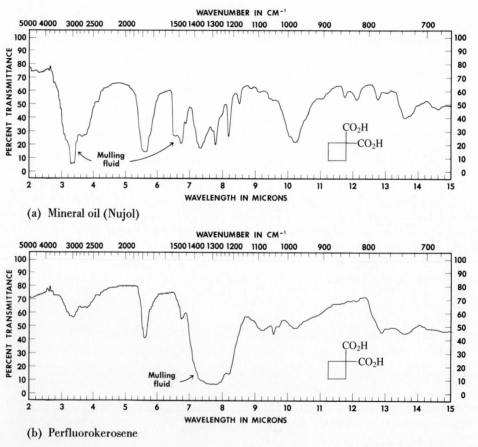

Figure 8.7 Infrared spectra of cyclobutane-1,1-dicarboxylic acid as mulls in (a) mineral oil (Nujol) and (b) perfluorokerosene.

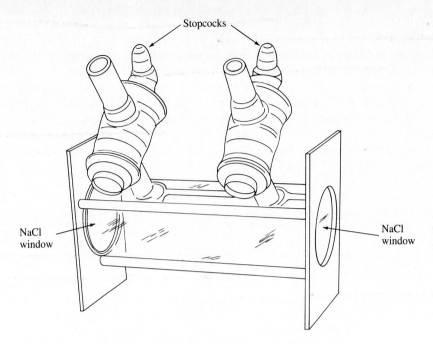

Stopcocks

NaCl window

NaCl window

Figure 8.8 An IR gas cell. The stopcock in the center of the cell permits introduction of the gaseous sample into the cell, and special high-vacuum techniques are normally required.

The mull method is not amenable to quantitative work because it is impossible to prepare the mull reproducibly for the same sample.

Experimental details are provided in Section 8.5c, part 3.

3. Gases Infrared spectra can be determined for gases. The path lengths of gas cells (Figure 8.8) are usually quite long (10–20 cm) in order to acquire a sufficiently high number of molecules to produce a decent spectrum. The windows at each end are usually made of NaCl. Since the apparatus required for trapping gaseous samples and transferring them to an IR gas cell is somewhat specialized for an introductory organic chemistry lab, this technique will not be discussed further.

8.5 Experimental Procedures for Spectrum Determination

(a) Handling IR Cells

The windows of most types of IR cells are composed of a clear fused salt such as sodium chloride or potassium bromide. Contact of the cells with moisture must be avoided at all times or the windows will become cloudy and transmit very little light. Breathing directly onto the windows, touching their faces with your fingers, or using the windows with samples that contain moisture should be avoided.

IR windows and cells should be cleaned with a *dry* solvent such as carbon tetrachloride or chloroform. *Absolute* ethanol can also be used, and it has the advantage of not subjecting the experimenter to chlorinated solvents, which are suspected of being carcinogens. It is good technique to wash the windows and cells at least three times. *Avoid inhaling the solvent vapors, and avoid letting the solvent come in contact with your skin.* Hydroxylic solvents, such as methanol or 95% ethanol, should *not* be used to clean cells, since they normally contain some water. Ideally, IR cells should be dried under a slow stream of dry nitrogen. If this facility is not available, the cells can be dried by *carefully* blotting the surfaces with laboratory tissue to remove the solvent. The window surfaces should never be rubbed. After use, cells should always be cleaned, dried, and stored in a desiccator.

(b) Liquid Samples (Neat Liquids)

Read Part a above for information about care and use of NaCl windows.

The infrared spectra of liquids can be determined using either a *demountable IR cell* or a *fixed-thickness cell*. Both techniques involve placing a thin *film* of sample between two sodium chloride IR windows.

(1) DEMOUNTABLE IR CELLS

A demountable IR cell is shown in Figure 8.9. It is good practice to clean and dry the sodium chloride plates before using them. One salt plate (window) is positioned in the holder

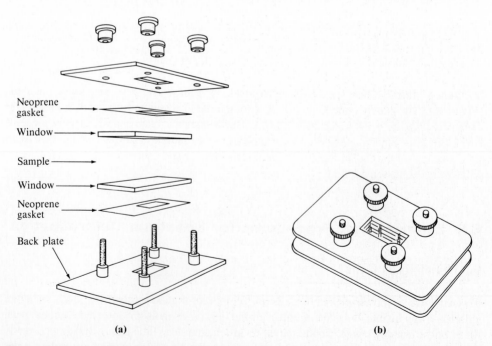

Neoprene gasket

Window

Sample

Window

Neoprene gasket

Back plate

(a)   (b)

Figure 8.9 A demountable IR cell. (a) Details showing correct assembly of cell. (b) Completely assembled cell.

on top of the gasket. A drop or two of the liquid whose spectrum is to be determined is placed in the center of the window, and the second window is placed on top of the first window containing the sample. The top of the holder is put in place and secured *gently* with the nuts. *Avoid applying undue pressure when securing the holders and windows together, as this may cause the salt plates to break. Apply only enough pressure to hold the windows in place.* The completely assembled demountable IR cell is also shown in Figure 8.9.

The assembled cell containing the sample is placed in the sample beam of the infrared spectrophotometer. Nothing is placed in the reference beam, since the only absorbing material in the cell is the liquid whose spectrum is being determined. Your instructor should provide information concerning the operation of the infrared spectrophotometer being used. After running the spectrum, three possible situations may be encountered: the peaks are too weak (for example, the strongest peak may be only 30–40% absorbance); the peaks are too strong (for example, the strongest peak may be more than 98% absorbance and the bottom of this peak appears flat); or the peaks are about right (for example, the strongest peak is about 90% absorbance). If the peaks are too weak, place more sample between the windows and rerun the spectrum. If the peaks are too strong, tighten the nuts on the cell holder slightly (*do not overtighten!*) to make the film thinner, and then rerun the spectrum. Alternatively, a "chopper" may be placed in the reference beam to reduce peak size; your instructor should be consulted if this alternative is to be used. After obtaining a suitable IR spectrum, remove, clean, and store the NaCl windows and cell holder in a desiccator. *Do not remove the spectrum until it has been calibrated (Section 8.5d).*

(2) FIXED-THICKNESS IR CELLS

A fixed-thickness IR cell is shown in Figure 8.10. It is already assembled, with a spacer whose thickness (usually 0.020–0.030 mm) should be indicated on the holder. This type of cell is cleaned by removing the Teflon plugs, placing a few drops of dry carbon tetrachloride or chloroform in one of the openings, and forcing the solvent out with a gentle stream of dry nitrogen. The clean, dry cell is filled by placing a drop or two of liquid in one of the openings until the window is filled. The caps are inserted in both openings, and the spectrum is run. Remove, clean, and store the cell in a desiccator. *Calibrate the spectrum before removing it from the spectrophotometer (Section 8.5d).*

(c) Solid Samples

The experimental procedures that can be used to prepare samples for the determination of the infrared spectra of solids are discussed below.

(1) SOLUTION METHOD

First, determine the solubility of the solid in a suitable IR solvent. Good spectra are obtained when the concentration of the solid in the solvent is approximately 15–20 mole percent, which means that about 20 mg of a sample whose molecular weight is about 100 should dissolve in about 1 mL of solvent. Test the solubility of the solid in carbon tetrachloride, chloroform, and carbon disulfide, in that order. The best choice for the solvent is CCl_4, followed by $CHCl_3$, and finally CS_2, the reason being that carbon tetrachloride causes fewer

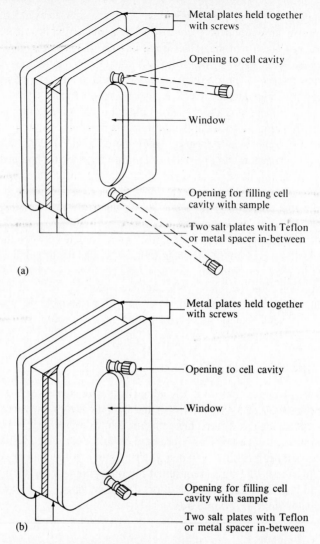

Figure 8.10 A fixed-thickness IR cell. (a) Cell with plugs removed for cleaning or filling. (b) Completely assembled cell.

absorption peaks and therefore obscures fewer peaks of the desired compound. For qualitative purposes, the samples are seldom weighed accurately. It is easier to estimate the amount of solid and then run the spectrum. If the peaks are too strong, the solution is too concentrated and is diluted before the spectrum is run again. If the peaks are too weak, more sample is added to the solvent before the spectrum is rerun.

The IR spectrum of the solution is determined in much the same way as that of a neat liquid. The solution cells look identical to the fixed-thickness cells (Figure 8.10) and differ only in the thickness of the spacer. Before using solution cells, clean and dry them according to the directions given above for fixed-thickness IR cells.

Two solution cells of *identical* thickness must be used, normally 0.20–0.25 mm. The solution of the solvent plus sample is placed in one cell, which should be marked "sample cell," and the pure solvent is placed in a cell that should be marked "solvent cell" or "reference cell." The solvent and sample cells are normally marked so that one cell is always used for the solvent and the other is always used for the sample. This helps prolong the life of at least one of the cells, since the sample cell may, on occasion, be filled with solutions that contain moisture. The solvent cell has less chance of having moisture introduced into it. The sample cell is then placed in the sample beam and the solvent cell in the reference beam of the infrared spectrophotometer. After obtaining a suitable spectrum, remove, clean, and store the cells in a desiccator. *Do not remove the recorded spectrum until it has been calibrated (Section 8.5d).*

(2) POTASSIUM BROMIDE (KBr) PELLET METHOD

The potassium bromide pellet is prepared by completely and intimately mixing 1–2 mg of solid sample with about 100–300 mg of anhydrous, spectral-grade powdered potassium bromide. The potassium bromide must be kept in a tightly closed bottle and stored in a desiccator at all times when not actually being used. The mixing is done by thorough grinding of the potassium bromide–sample mixture in a mortar and pestle made of agate. The mixture is pressed into a solid pellet (disk) under high pressure using a potassium bromide press. There are many different styles of presses, and your instructor should demonstrate the one available in your laboratory. With some presses, the pellet is formed in a die that can be placed directly in the sample beam. Other presses produce pellets that have to be mounted before placement in the sample beam. The pellet containing the sample is placed in the sample beam, and a "chopper" (Figure 8.11) is placed in the reference beam. The chopper compensates for the loss of sample beam intensity due to scattering, and the adjustable screen on the chopper should be set so that the recorder pen indicates 100% transmission (0% absorbance) at the wavenumber where the sample absorbs the least. This wavenumber is easily determined by running the spectrum quickly, and some instruments permit this to be

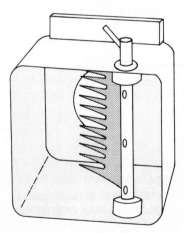

Figure 8.11 A beam attenuator, which is used to decrease the intensity of the infrared radiation in the reference beam.

done by manual scanning. Use of the chopper yields a better-looking spectrum, since it has the effect of moving the entire spectrum to the top of the paper and permits nearly full-scale pen movement for the strongest peak. If the peaks are too strong, the pellet must be redone using less sample, and if they are too weak, more sample must be added to the KBr when the pellet is remade. After obtaining an acceptable spectrum, clean the pellet holder and return it to a desiccator. *Calibrate the spectrum before removing it from the instrument (Section 8.5d).*

(3) MULL METHOD

A mull is first prepared by thoroughly grinding several milligrams of solid sample in a mortar and pestle made of agate. Then several drops of the mulling fluid, either Nujol (mineral oil) or perfluorokerosene, are added, and grinding is continued until a smooth paste is formed. This paste is used to prepare an IR cell using two NaCl windows in the same manner as for a neat liquid using a demountable cell (Section 8.4b, part 1). Failure to grind the sample finely will result in a poor-quality IR spectrum. If the absorption peaks are too weak, a new mull must be prepared using more sample, and if they are too strong, a new mull containing less sample must be prepared. After obtaining a good spectrum, remove and clean the cell windows and store them in a desiccator. *Calibrate the spectrum before removing the paper from the instrument (Section 8.5d).*

(d) Calibration of IR Spectra

IR spectrophotometers used for routine spectra may often have some mechanical slippage in the paper holder or drive gear. As a result, the absorption peaks in IR spectra often provide wavenumber readings that do *not* exactly correspond to the *correct* wavenumbers. This problem is circumvented by calibrating the spectra, using a small portion of the spectrum produced by a standard compound whose absorption peaks are known accurately. The most commonly used standard is a **polystyrene film,** which is a thin sheet of polystyrene mounted in a cardboard holder. The IR spectrum of polystyrene is shown in Figure 8.12.

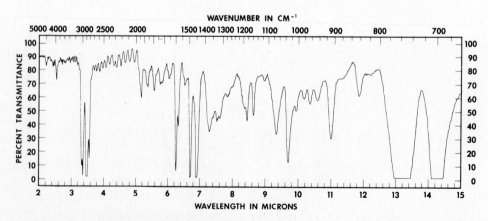

Figure 8.12 Infrared spectrum of polystyrene film.

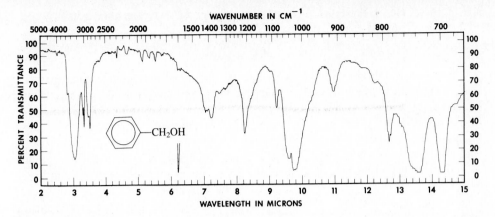

Figure 8.13 Infrared spectrum of benzyl alcohol (phenylmethanol) with the 1601 cm^{-1} peak of polystyrene inserted for calibration.

Calibration of IR spectra is accomplished as follows. After running the spectrum of a compound by any of the methods presented in this section, the cells containing the sample and the reference solvent (if used) are removed. *Do not remove the IR spectral copy from the recorder.* The polystyrene film is then placed in the sample beam, and a small portion of the polystyrene spectrum is recorded on the *same* sheet that contains the spectrum of the sample. For most purposes, it is adequate to use the polystyrene absorption peak occurring at 1601 cm^{-1}. This may be done either by manually scanning to this peak or by allowing the machine to scan to this point automatically. The recorder pen is lifted from the paper until just before this absorption peak occurs; the pen is then lowered, the one peak is printed, and the pen is raised again. Often the color of the recorder pen is changed when printing the calibration peak to avoid problems in interpreting the spectrum. Figure 8.13 shows an IR spectrum with the calibration peak recorded on it.

The calibration peaks are used in the following manner. If the 1601 cm^{-1} peak of polystyrene appears at 1601 cm^{-1} on the paper, then the wavenumbers of the peaks of the original sample are correct and can be read directly from the spectrum. However, if the 1601 cm^{-1} polystyrene peak appears at 1590 cm^{-1} on the spectrum, this means that all the peaks from the original compound are lower than they should be. Correct wavenumber readings for the peaks can be obtained by *adding* 11 cm^{-1} to the wavenumbers that are read on the original spectrum. On the other hand, if the 1601 cm^{-1} polystyrene peak appeared at 1611 cm^{-1} on the spectrum, this means that all the peaks of the original compound are higher than they should be. Correct wavenumber readings for the peaks are now obtained by *subtracting* 10 cm^{-1} from the wavenumbers that are read on the original spectrum. This correction factor can be used, assuming that all the peaks observed in the spectrum have been shifted to higher or lower wavenumbers by the same amount.

IR spectra should always be calibrated to ensure that correct wavenumbers for each absorption peak are obtained. For example, the carbon-oxygen double bond in aldehydes, ketones, carboxylic acids, and esters absorbs in the 1700–1760 cm^{-1} region. Failure to determine the exact wavenumber may lead to the incorrect identification of functional groups.

8.6 Uses of Infrared Spectroscopy: Identification of Functional Groups and Compounds

Infrared spectroscopy can be used to determine whether or not two compounds are identical. If they are the same, their IR spectra must be the same, provided that the spectra were obtained under identical conditions (same solvent, concentration, cell length, and so forth). If the compounds are different, their IR spectra will be different. Sometimes they may be of similar structure, so that their IR spectra may be very similar and only careful analysis reveals a difference. Sometimes a compound may be impure, and when its IR spectrum is compared with that of the pure compound, the impure compound will show additional peaks owing to the impurity. Certain stereochemical features in a molecule may cause differences in the IR spectra. For example, a pair of enantiomers exhibit identical spectra while a pair of diastereomers exhibit different spectra.

One possible method for identifying a compound hinges upon the successful identification of the functional group(s) it contains. As a general rule, absorptions between 5000 and 1250 cm^{-1} involve *vibrational* excitations of various functional groups present in a molecule. For example, the stretching mode of the carbon-oxygen double bond of a ketone usually appears in the region 1680–1760 cm^{-1} (see Figure 8.3). Careful analysis of the portion of the IR spectrum from 5000 to 1250 cm^{-1} provides a great deal of information concerning the presence *or absence* of different functional groups. The IR spectra of many known compounds have been compared, and the result has been the establishment of numerous fairly reliable correlations between the wavenumber of an absorption and the type of functional group(s) contained in a molecule. The intensity of the absorption peak (weak, moderate, or strong) also can be utilized in the functional group determination. The normal frequency range of absorption for a number of functional groups is presented in abbreviated form in Table 8.2, and a more detailed discussion of these absorptions is given in Table 8.3.

The presence of a particular absorption peak may indicate a certain functional group. However, the absence of certain absorption peaks also provides valuable information concerning functional groups that are *not* present in the molecule. For example, the absence of a strong absorption peak between 1680 and 1800 cm^{-1} indicates the absence of a C=O bond and thus eliminates the following functional groups: aldehyde, ketone, carboxylic acid, ester, amide, acid chloride, and anhydride. Similarly, the absence of a strong, broad absorption peak between 3200 and 3600 cm^{-1} implies that the compound does not contain these groups: —OH (alcohol, phenol or carboxylic acid) or —NH$_2$ (amine or amide). As you become more experienced with IR spectroscopy, you should find that a great deal of information, both positive and negative, can be obtained by correctly analyzing an IR spectrum.

The absorption maxima observed between 1250 and 500 cm^{-1} in the IR spectrum usually are not associated with vibrational excitation of a particular functional group, but instead are the result of a complex *vibrational-rotational* excitation of the *entire* molecule. The absorption spectrum from 1250 to 500 cm^{-1} typically is characteristic and *unique* for every compound; the apt description **fingerprint region** is often applied to this portion of the spectrum. It is highly unlikely that two different organic compounds will exhibit *identical* spectra in the fingerprint region, even though their spectra might be quite *similar* in the 5000–1250 cm^{-1} region. This concept is illustrated in Figure 8.14, which shows the spectra of two alkanes.

The apparently unique IR spectrum of each different organic compound has led to the

TABLE 8.2 Abbreviated Table of Infrared Absorption Ranges of Functional Groups

Bond	Type of Compound	Frequency Range, cm^{-1}	Intensity
C—H	Alkanes	2850–2970	Strong
		1340–1470	Strong
C—H	Alkenes (\C=C/ H)	3010–3095	Medium
		675–995	Strong
C—H	Alkynes (—C≡C—H)	3300	Strong
C—H	Aromatic rings	3010–3100	Medium
		690–900	Strong
O—H	Monomeric alcohols, phenols	3590–3650	Variable
	Hydrogen-bonded alcohols, phenols	3200–3600	Variable, sometimes broad
	Monomeric carboxylic acids	3500–3650	Medium
	Hydrogen-bonded carboxylic acids	2500–2700	Broad
N—H	Amines, amides	3300–3500	Medium
C=C	Alkenes	1610–1680	Variable
C=C	Aromatic rings	1500–1600	Variable
C≡C	Alkynes	2100–2260	Variable
C—N	Amines, amides	1180–1360	Strong
C≡N	Nitriles	2210–2280	Strong
C—O	Alcohols, ethers, carboxylic acids, esters	1050–1300	Strong
C=O	Aldehydes, ketones, carboxylic acids, esters	1690–1760	Strong
NO_2	Nitro compounds	1500–1570	Strong
		1300–1370	Strong

hypothesis that if two spectra are completely identical as to the wavenumber and intensity of the absorption maxima—that is, if two spectra are *completely superimposable*—the spectra must be of the same compound. However, the spectra must be determined on samples that have the same concentration of solute, and the same IR cells must have been used. Comparison of spectra of unknown compounds with those of known compounds is a very useful technique for structure proof. Organic chemists who undertake the total synthesis of a natural product obtain IR spectra of the natural product and the synthesized compound. If they are identical, the synthesis is deemed successful.

Infrared spectroscopy provides valuable information concerning the nature of the functional groups present in molecules, and it can be used to identify an unknown compound if reference spectra of suspected known compounds are available. Many IR spectra have been reproduced in books available in reference libraries (see references at the end of this chapter). Often an IR spectrum alone does not provide sufficient information to identify an unknown compound unambiguously. When this is the case, other spectroscopic techniques, such as NMR and UV/visible, must be used, often in combination with chemical analyses. The

Text continues on p. 215

TABLE 8.3 Detailed Table of Characteristic Infrared Absorption Frequencies

The hydrogen stretch region (3600–2500 cm⁻¹). Absorption in this region is associated with the stretching vibration of hydrogen atoms bonded to carbon, oxygen, and nitrogen. Care should be exercised in the interpretation of very weak bands because these may be overtones of strong bands occurring at frequencies one-half the value of the weak absorption, that is, 1800–1250 cm⁻¹. Overtones of bands near 1650 cm⁻¹ are particularly common.

$\bar{\nu}$ (cm⁻¹)	Functional Group	Comments
(1) 3600–3400	O—H stretching Intensity: variable	3600 cm⁻¹ (sharp) unassociated O—H, 3400 cm⁻¹ (broad) associated O—H; both bands frequently present in alcohol spectra; with strongly associated O—H (CO₂H or enolized β-dicarbonyl compound) band is very broad (about 500 cm⁻¹ with its center at 2900–3000 cm⁻¹).
(2) 3400–3200	N—H stretching Intensity: medium	3400 cm⁻¹ (sharp) unassociated N—H, 3200 cm⁻¹ (broad) associated N—H; an NH₂ group usually appears as a doublet (separation about 50 cm⁻¹); the N—H of a secondary amine is often very weak.
(3) 3300	C—H stretching of an alkyne Intensity: strong	The *complete* absence of absorption in this region, 3300–3000 cm⁻¹, indicates the absence of hydrogen atoms bonded to C=C or C≡C and *usually* indicates the lack of unsaturation in the molecule. Because this absorption may be very weak in large molecules, some care should be exercised in this interpretation. In addition to the absorption at about 3050 cm⁻¹, aromatic compounds will frequently show *sharp* bands of medium intensity at about 1500 *and* 1600 cm⁻¹.
(4) 3080–3010	C—H stretching of an alkene Intensity: strong to medium	
(5) 3050	C—H stretching of an aromatic compound Intensity: variable; usually medium to weak	
(6) 3000–2600	OH strongly hydrogen-bonded Intensity: medium	A very broad band in this region superimposed on the C—H stretching frequencies is characteristic of carboxylic acids (see 1).
(7) 2980–2900	C—H stretching of an aliphatic compound Intensity: strong	Just as in the previous C—H entries (3–5), *complete* absence of hydrogen atoms bonded to tetravalent carbon atoms. The tertiary C—H absorption is weak.
(8) 2850–2760	C—H stretching of an aldehyde Intensity: weak	Either one or two bands *may* be found in this region for a single aldehyde function in the molecule.

The triple-bond region (2300–2000 cm⁻¹). Absorption in this region is associated with the stretching vibration of triple bonds.

$\bar{\nu}$ (cm⁻¹)	Functional Group	Comments
(1) 2260–2215	C≡N Intensity: strong	Nitriles conjugated with double bonds absorb at lower end of frequency range; nonconjugated nitriles appear at upper end of range.

$\bar{\nu}$ (cm^{-1})	Functional Group	Comments
(2) 2150–2100	C≡C Intensity: strong in *terminal* alkynes, variable in others.	This band will be absent if the alkyne is symmetrical, and will be very weak or absent if the alkyne is nearly symmetrical.

The double-bond region (1900–1550 cm^{-1}). Absorption in this region is *usually* associated with the stretching vibration of carbon-carbon, carbon-oxygen, and carbon-nitrogen double bonds.

$\bar{\nu}$ (cm^{-1})	Functional Group	Comments
(1) 1815–1770	C=O stretching of an acid chloride Intensity: strong	Conjugated and nonconjugated carbonyls absorb at the lower and upper ends, respectively, of the range.
(2) 1870–1800 and 1790–1740	C=O stretching of an acid anhydride Intensity: strong	*Both bands* are present; *each band* is altered by ring size and conjugation to approximately the same extent noted for ketones (see 4).
(3) 1750–1735	C=O stretching of an ester or lactone Intensity: very strong	This band is subject to all the structural effects discussed in 4; thus, a conjugated ester absorbs at about 1710 cm^{-1} and a γ-lactone absorbs at about 1780 cm^{-1}.
(4) 1725–1705	C=O stretching of an aldehyde or ketone Intensity: very strong	This value refers to the carbonyl absorption frequency of an acyclic, nonconjugated aldehyde or ketone in which no electronegative groups, for example, halogens, are near the carbonyl group; because this frequency is altered in a predictable way by structural alterations, the following generalizations may be drawn: (a) *Effect of conjugation:* Conjugation of the carbonyl group with an aryl ring or carbon-carbon double or triple bond lowers the frequency by about 30 cm^{-1}. If the carbonyl group is part of a cross-conjugated system (unsaturation on each side of the carbonyl group), the frequency is lowered by about 50 cm^{-1}. (b) *Effect of ring size:* Carbonyl groups in six-membered and larger rings exhibit approximately the same absorption as acyclic ketones; carbonyl groups contained in rings smaller than six absorb at higher frequencies, for example, a cyclopentanone absorbs at about 1745 cm^{-1} and a cyclobutanone absorbs at about 1780 cm^{-1}. The effects of conjugation and ring size are additive, for example, a 2-cyclopentenone absorbs at about 1710 cm^{-1}. (c) *Effect of electronegative atoms:* An electronegative atom (especially oxygen or halogen) bonded to the α-carbon atom of an

$\bar{\nu}$ (cm^{-1})	Functional Group	Comments
		aldehyde or ketone may raise the position of the carbonyl absorption frequency by about 20 cm^{-1}.
(5) 1700	C=O stretching of an acid Intensity: strong	This absorption frequency is lowered by conjugation, as noted under entry 4.
(6) 1690–1650	C=O stretching of an amide or lactam Intensity: strong	This band is lowered in frequency by about 20 cm^{-1} by conjugation. The frequency of the band is raised about 35 cm^{-1} in γ-lactams and 70 cm^{-1} in β-lactams.
(7) 1660–1600	C=C stretching of an alkene Intensity: variable	Nonconjugated alkenes appear at upper end of range, and absorptions are usually weak; conjugated alkenes appear at lower end of range, and absorptions are medium to strong. The absorption frequencies of these bands are raised by ring strain but to a lesser extent than noted with carbonyl functions (see 4).
(8) 1680–1640	C=N stretching Intensity: variable	This band is usually weak and difficult to assign.

The hydrogen bending region (1600–1250 cm^{-1}). Absorption in this region is commonly due to bending vibration of hydrogen atoms attached to carbon and to nitrogen. These bands generally do not provide much useful structural information. In the listing, the bands that are most useful for structural assignment have been marked with an asterisk (*).

$\bar{\nu}$ (cm^{-1})	Functional Group	Comments
(1) 1600	—NH$_2$ bending Intensity: strong to medium	This band in conjunction with bands in the 3300 cm^{-1} region is often used to characterize primary amines and unsubstituted amides.
(2) 1540	—NH— bending Intensity: generally weak	This band in conjunction with bands in the 3300 cm^{-1} region is often used to characterize secondary amines and monosubstituted amines. In the case of secondary amines this band, like the N—H stretching band in the 3300 cm^{-1} region, may be very weak.
(3) *1520 and 1350	NO$_2$ coupled stretching bands Intensity: strong	This pair of bands is usually very intense.
(4) 1465	—CH$_2$— bending Intensity: variable	The intensity of this band varies according to the number of methylene groups present; the more such groups, the more intense the absorption.

$\bar{\nu}$ (cm^{-1})	Functional Group	Comments
(5) 1410	—CH$_2$— bending of carbonyl-containing component Intensity: variable	This absorption is characteristic of methylene groups adjacent to carbonyl functions; its intensity depends on the number of such groups present in the molecule.
(6) *1450 and 1375	—CH$_3$ Intensity: strong	The band of lower frequency (1375 cm^{-1}) is usually used to characterize a methyl group. If two methyl groups are bonded to one carbon atom, a characteristic doublet (1385 and 1365 cm^{-1}) will be present.
(7) 1325	| —CH bending Intensity: weak	This band is weak and often unreliable.

The fingerprint region (1250–600 cm^{-1}). The fingerprint region of the spectrum is generally rich in detail, with many bands appearing. This region is particularly diagnostic for determining whether an unknown substance is identical with a known substance, the IR spectrum of which is available. It is not practical to make assignments to all these bands because many of them represent combination frequencies and therefore are very sensitive to the total molecular structure; moreover, many single-bond stretching vibrations and a variety of bending vibrations also appear in this region. Suggested structural assignments in this region must be regarded as tentative and are generally taken as corroborative evidence in conjunction with assignments of bands at higher frequencies.

$\bar{\nu}$ (cm^{-1})	Functional Group	Comments
(1) 1200	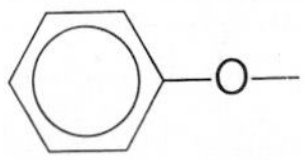 ⬡—O— Intensity: strong	It is not certain whether these strong bands arise from C—O bending or C—O stretching vibrations. One or more strong bands are found in this region in the spectra of alcohols, ethers, and esters. The relationship indicated between structure and band location is only approximate, and any structural assignment based on this relationship must be regarded as tentative. Esters often exhibit one or two strong bands between 1170 and 1270 cm^{-1}.
(2) 1150	| —C—O— | Intensity: strong	
(3) 1100	| —CH—O— Intensity: strong	
(4) 1050	—CH$_2$—O— Intensity: strong	

(5) 965	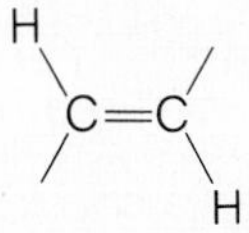 H
	\\ /
	  C=C
	/ \\
	       H
	C—H bending
	Intensity: strong
		This strong band is present in the spectra of *trans*-1,2-disubstituted ethylenes.

$\bar{\nu}$ (cm^{-1})	Functional Group	Comments
(6) 985 and 910	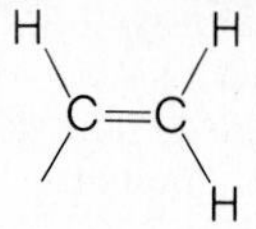 C—H bending Intensity: strong	The lower frequency band of these two strong bands is used to characterize a terminal vinyl group.
(7) 890	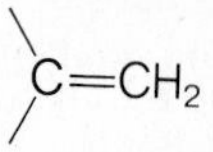 C=CH$_2$ C—H bending Intensity: strong	This strong band, used to characterize a methylene group, may be raised by 20–80 cm^{-1} if the methylene group is bonded to an electronegative group or atom.
(8) 810–840	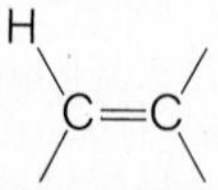 Intensity: strong	Very unreliable; this band is not always present and frequently seems to be outside this range, since substituents are varied.
(9) 700	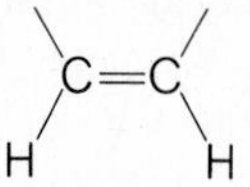 Intensity: variable	This band, attributable to a *cis*-1,2-disubstituted ethylene, is unreliable because it is frequently obscured by solvent absorption or other bands.
(10) 750 and 690	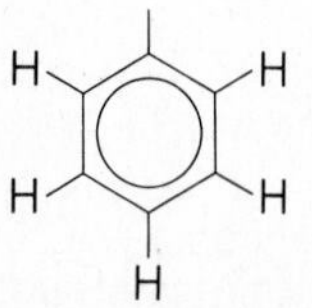 C—H bending Intensity: strong	These bands are of limited value because they are frequently obscured by solvent absorption or other bands. Their usefulness will be most important when independent evidence leads to a structural assignment complete except for position of aromatic substituents.
(11) 750	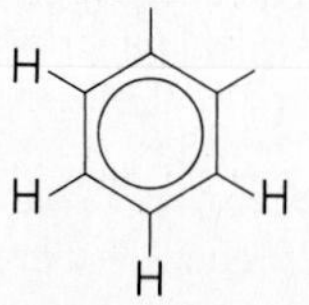 C—H bending Intensity: very strong	
(12) 780 and 700	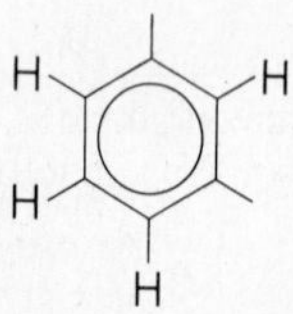 and 1, 2, 3 Intensity: very strong	

$\bar{\nu}$ (cm^{-1})	Functional Group	Comments
(13) 825	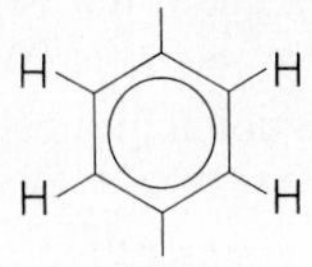 and 1, 2, 4 Intensity: very strong	
(14) 1400–1000	C—F Intensity: strong	The position of these bands is quite sensitive to structure. As a result, they are not particularly useful because the presence of halogen is more easily detected by chemical methods. The bands are usually strong.
(15) 800–600	C—Cl Intensity: strong	
(16) 700–500	C—Br Intensity: strong	
(17) 600–400	C—I Intensity: strong	

methodology for combining chemical and spectroscopic analyses of unknown compounds is presented in Chapter 27.

Infrared spectroscopy is useful in monitoring the progress of a chemical reaction in order to determine when it is complete. This can be done by withdrawing small portions of the reaction mixture, working them up, and taking an IR spectrum to see if any unchanged starting material remains. IR spectroscopy can also be used to monitor the purification of a compound by obtaining a spectrum on various fractions from a distillation or on different crops of crystals obtained during a recrystallization.

Infrared spectroscopy can be used for quantitative determinations. Solutions of different known concentrations of a pure compound are prepared and their spectra are determined, using the same IR cell. The absorbances, A, of a particular peak at each concentration are determined and plotted vs. concentration. Equation 6 indicates that there should be a linear relationship between concentration and absorbance at constant cell length. A solution of unknown concentration is then run, and the absorbance of the same peak is determined. From this absorbance value, the concentration of the sample can be determined from the "standardized" plot that has been constructed. Results obtained from this technique are usually accurate to at least ±5%.

8.7 Interpretation of Infrared Spectra

A good way to learn how to identify the functional groups in organic compounds on the basis of their infrared spectra is to examine spectra of known compounds. The organic chemist is usually interested in identifying *only* key absorption peaks that indicate which functional groups are present, as even the most experienced infrared spectroscopists are hard-pressed to identify each and every peak in an infrared spectrum. Thus, the beginning student should become familiar with the methodology of functional group identification. In this section, infrared spectra of compounds containing common functional groups encountered in organic chemistry are presented.

These spectra are presented together to provide the student the opportunity to compare easily the characteristic absorption peaks of many different functional groups. Comparing these spectra with ones obtained from "unknown" compounds should be useful, especially when performing qualitative organic analyses (Chapter 27).

(a) Alkanes

The IR spectra of alkanes exhibit strong absorptions at **2850–2970** cm^{-1} (stretching) and at **1340–1470** cm^{-1} (bending) due to the C—H bond (more correctly, hydrogen bonded to sp^3-hybridized carbon). Figure 8.14 shows the IR spectra of two alkanes. Note that the regions between 1250 and 5000 cm^{-1} are very similar, while the rest of the spectra (the fingerprint regions) are quite different. These spectra support the hypothesis that each compound has a unique IR spectrum, especially in the fingerprint region. The peaks of greatest interest are marked on the spectra.

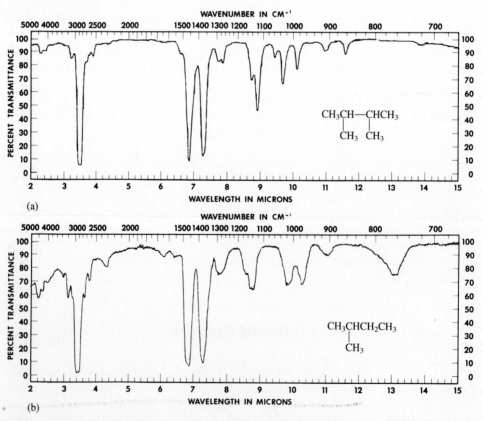

Figure 8.14 Infrared spectra of (a) 2,3-dimethylbutane and (b) 2-methylbutane.

(b) Cycloalkanes

The IR spectra of cycloalkanes are similar to those of alkanes, with the absorption maxima appearing at approximately the same places. Figure 8.15 shows the IR spectrum of cyclohexane, which should be compared with the spectra of the alkanes in Figure 8.14.

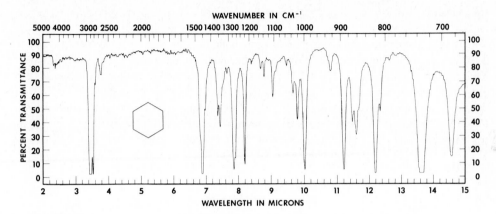

Figure 8.15 Infrared spectrum of cyclohexane.

(c) Alkenes

The IR spectra of alkenes exhibit peaks due to the stretching of hydrogen attached to the C=C (vinyl hydrogens) at **3010–3095** cm^{-1} of medium intensity and at **675–995** cm^{-1} of strong intensity. Here we see the effect of hydrogens attached to sp^2-hybridized carbon atoms; absorptions are at higher wavenumbers than those of C—H bonds in alkanes. Alkenes also exhibit absorptions at **1610–1680** cm^{-1} due to stretching of the C=C. These are of variable intensity, and symmetrically substituted double bonds tend to exhibit very weak peaks due to C=C stretching. Figure 8.16 shows the IR spectrum of an alkene; note that the vinyl hydrogens absorb at a higher wavenumber than the alkyl hydrogens.

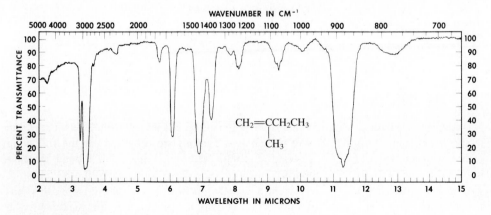

Figure 8.16 Infrared spectrum of 2-methyl-1-butene.

These comments about the C=C stretching absorptions are for an "isolated" double bond. When an alkene is conjugated with another C=C, a C=O, or an aromatic ring, the C=C stretching frequency is shifted to lower wavenumber.

(d) Alkynes

The hydrogen attached to a carbon-carbon triple bond (a *terminal* alkyne) absorbs strongly at **3300** cm^{-1}, and this is typical for hydrogen attached to *sp*-hybridized carbon. Additionally, alkynes absorb in the **2100–2260** cm^{-1} region owing to the carbon-carbon triple bond. There are several structural features that may cause a triple bond to go undetected in infrared spectroscopy; for example, alkynes with no hydrogens attached to the triple bond exhibit no C—H stretching peaks, and a symmetrical alkyne (two identical substituents attached to the triple bond) often exhibits no triple-bond stretching. Figure 8.17 shows the IR spectrum of an alkyne.

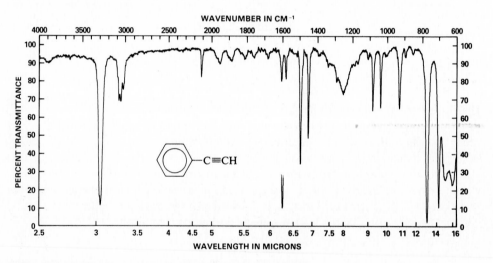

Figure 8.17 Infrared spectrum of phenylethyne (phenylacetylene).

(e) Alkyl Halides

Alkyl halides exhibit many infrared absorptions typical of alkanes, with additional peaks due to the carbon-halogen bond. Strong stretching vibrations are observed for the following carbon-halogen bonds: the C—F bond at **1000–1400** cm^{-1}, the C—Cl bond at **600–800** cm^{-1}, the C—Br bond at **500–600** cm^{-1}, and the C—I bond near **500** cm^{-1}. The IR spectrum of a typical alkyl halide, 2-chloro-2-methylbutane, is given in Figure 8.18. Compare this spectrum with that of 2-methylbutane (Figure 8.14b) and note the peak due to the halogen.

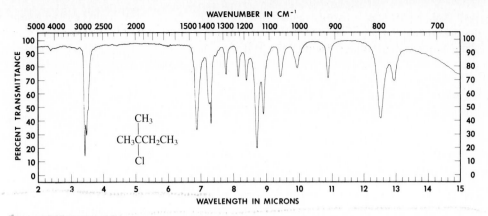

Figure 8.18 Infrared spectrum of 2-chloro-2-methylbutane.

(f) Aromatic Hydrocarbons

Aromatic hydrocarbons exhibit medium absorptions due to C—H stretching at **3010–3095** cm^{-1} and strong C—H out-of-plane bending vibrations at **690–900** cm^{-1} due to hydrogen attached to sp^2-hybridized carbon. The peaks in the **690–900** cm^{-1} region are useful in deducing the location of the substituents on the ring; see Table 8.3 for further discussion. Stretching absorptions of the carbon-carbon bonds in the aromatic ring are of variable intensity, occurring in the **1500–1600** cm^{-1} region. Figure 8.19 shows the IR spectrum of a monosubstituted aromatic hydrocarbon.

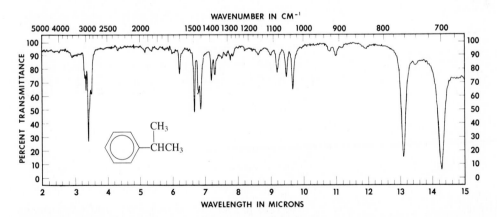

Figure 8.19 Infrared spectrum of isopropylbenzene (cumene).

(g) Aromatic and Aliphatic Compounds

Infrared spectroscopy is useful for distinguishing between aromatic and aliphatic compounds. Aromatics generally have a weak C—H stretching peak in the **3000** cm^{-1} region, sharp peaks of moderate intensity in the **1500–1600** cm^{-1} region, and several strong absorptions at wavenumbers less than **900** cm^{-1}. Aromatic peaks are generally quite sharp and rather symmetrical, while aliphatic peaks tend to be broad and often appear unsymmetrical. Aliphatic compounds absorb around 2950 cm^{-1} (C—H region) and seldom have peaks at wavenumbers less than 900 cm^{-1}. The spectrum of isopropylbenzene (Figure 8.19) should be compared with the spectra of alkanes, cycloalkanes, alkenes, alkynes, and alkyl halides (Figures 8.14–8.18).

(h) Alcohols

Alcohols show absorptions stemming from both the C—O and the O—H bonds. The O—H bond can be either monomeric or hydrogen-bonded. Alcohols are usually monomeric in very dilute solutions, but they are hydrogen-bonded in neat liquids and in more concentrated solutions. The O—H peak of an alcohol is usually obvious in an IR spectrum. For monomeric alcohols, the O—H peak appears at **3590–3650** cm^{-1} and is usually a sharp peak. The O—H peak for hydrogen-bonded alcohols is usually broad and appears between **3200** and **3600** cm^{-1}. The C—O absorption is strong and appears between **1050** and **1300** cm^{-1}. The IR spectrum of cyclohexanol is given in Figure 8.20.

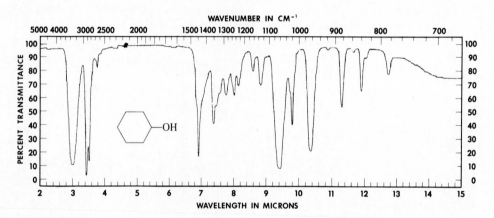

Figure 8.20 Infrared spectrum of cyclohexanol.

(i) Phenols

Phenols have C—O and O—H absorptions similar to those of alcohols. Monomeric and hydrogen-bonded O—H peaks are observed for both alcohols and phenols. However, phenols may be distinguished from alcohols because of the presence of the aromatic ring in a phenol (see part g above). The IR spectrum of a phenol is given in Figure 8.21.

(j) Aldehydes

The infrared spectra of aliphatic aldehydes have the characteristic C=O stretching at **1705–1725** cm^{-1} and the C—H stretching of the hydrogen *attached* to the C=O at **2760–2850** cm^{-1}. When the carbonyl group is conjugated with an aromatic ring or carbon-carbon double or triple bond, the wavenumber of the C=O peak is lowered by about 30 cm^{-1}. Aliphatic aldehydes can be distinguished from aromatic aldehydes in the manner discussed in g above. The IR spectra of an aliphatic and an aromatic aldehyde are provided in Figure 8.22.

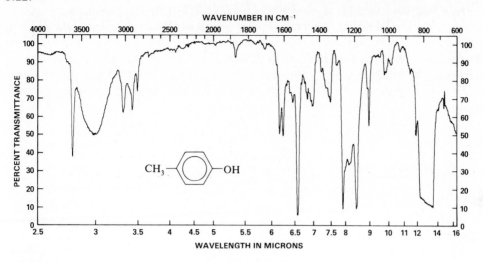

Figure 8.21 Infrared spectrum of p-cresol (in CCl$_4$ solution).

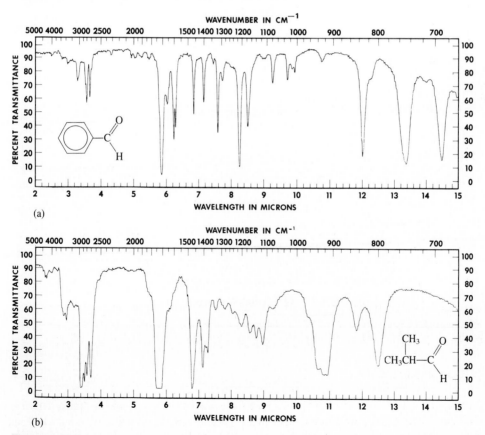

Figure 8.22 Infrared spectrum of (a) benzaldehyde and (b) 2-methylpropanal (isobutyraldehyde).

(k) Ketones

Ketones have the characteristic C=O stretching at **1705–1725** cm^{-1} in the same general region as that for aldehydes. Aliphatic ketones absorb at **1710** cm^{-1}, alkyl-aryl ketones at **1690** cm^{-1}, and diaryl ketones at **1670** cm^{-1}. The aromatic ring in the latter two causes the shift to lower wavenumbers. The C—H peak for the hydrogen attached to the carbonyl group is absent in the spectra of ketones. Aromatic ketones can be distinguished from aliphatic ones as discussed in g above. The IR spectra of an aliphatic and an aromatic ketone are shown in Figure 8.23.

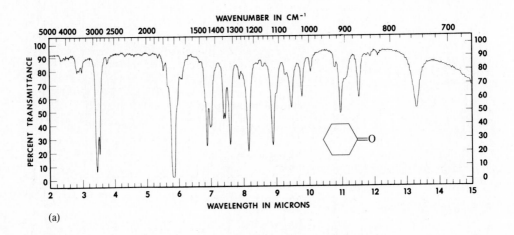

(a)

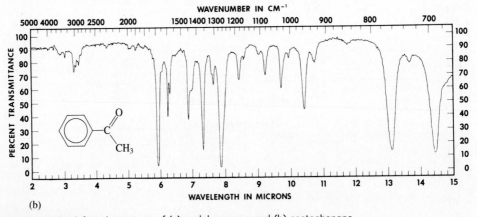

(b)

Figure 8.23 Infrared spectrum of (a) cyclohexanone and (b) acetophenone.

(l) Carboxylic Acids

Carboxylic acids exhibit two strong characteristic absorptions, one that is fairly sharp at **1700** cm^{-1} due to the C=O and another that is quite broad (about **500** cm^{-1}) and centered at **2900–3000** cm^{-1} due to the O—H group. Acids are usually hydrogen-bonded, which accounts for the broad, strong O—H band. The C=O absorption is lowered by conjugation with aromatic rings and other double bonds. Carboxylic acids also exhibit a strong peak in the **1100–1300** cm^{-1} region due to the C—OH bond, but the location of this absorption is variable and difficult to predict. Figure 8.24 provides IR spectra of an aliphatic and an aromatic carboxylic acid.

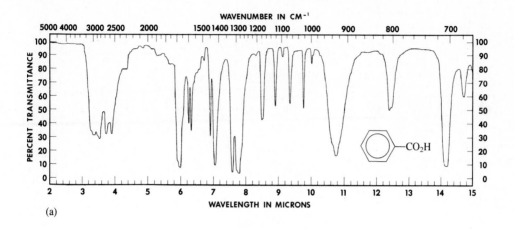

(a)

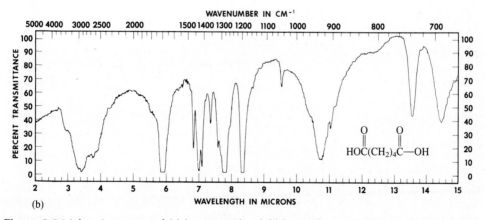

(b)

Figure 8.24 Infrared spectrum of (a) benzoic acid and (b) hexanedioic acid (adipic acid) (KBr pellets).

(m) Esters

Esters show a strong peak at **1735–1750** cm^{-1} due to the C=O; this band is lowered to about **1710** cm^{-1} in a conjugated ester. They also often show one or two strong bands between **1170** and **1270** cm^{-1}. The IR spectra of an aliphatic and an aromatic ester are provided in Figure 8.25.

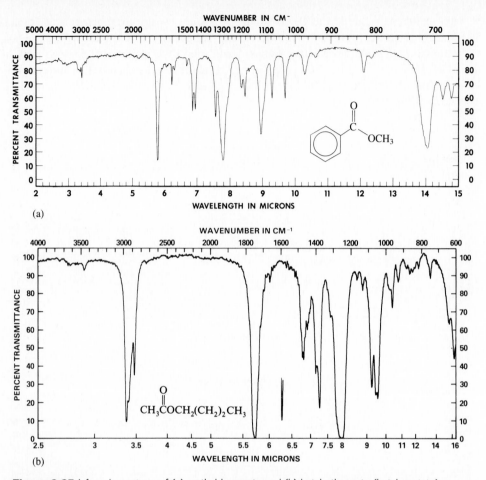

Figure 8.25 Infrared spectrum of (a) methyl benzoate and (b) butyl ethanoate (butyl acetate).

(n) Amides

The C=O bond of an amide absorbs at **1650–1690** cm^{-1}, and this band is lowered by about 20 cm^{-1} by conjugation. The cyclic amides, which are called lactams, absorb 35–70 cm^{-1} higher than 1650–1690 cm^{-1}. Amides bearing at least one hydrogen attached to the amide nitrogen, as in —NH_2 and —NHR, produce N—H stretching bands of medium intensity at **3200–3400** cm^{-1}. Unassociated N—H bonds absorb at about **3400** cm^{-1} (sharp peak), while associated N—H bond absorptions are broad and appear at about **3200** cm^{-1}. The —NH_2 group usually appears as two fairly sharp peaks separated by 50–100 cm^{-1}. The hydrogen stretching band associated with —NHR is usually very weak. An amide nitrogen containing *no* hydrogens, as in —NR_2, shows no absorption in the 3200–3400 cm^{-1} region. IR spectra of several amides are given in Figure 8.26.

(o) Acid Chlorides

Acid chlorides exhibit a strong C=O stretching peak within the range **1770–1815** cm^{-1}, with conjugated acid chlorides absorbing near **1770** cm^{-1} and unconjugated ones near

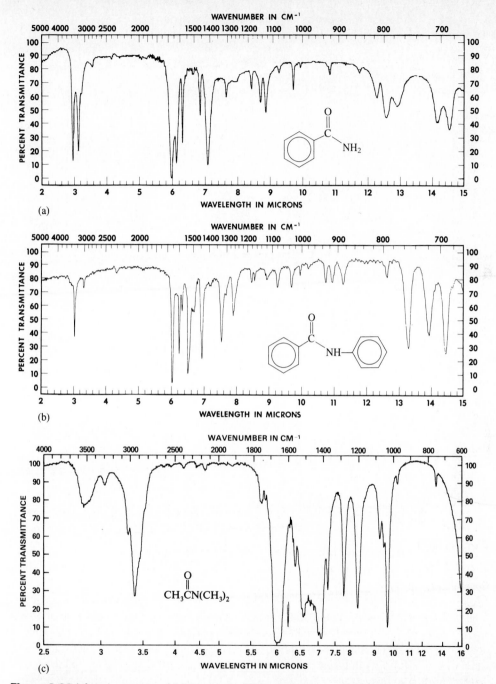

Figure 8.26 Infrared spectrum of (a) benzamide (KBr pellet), (b) benzanilide (KBr pellet), and (c) *N,N*-dimethylacetamide.

1815 cm^{-1}. Note that the electronegative chlorine attached to a C=O causes the carbonyl absorption to shift to higher wavenumbers; for example, aldehydes and ketones absorb at 1705–1725 cm^{-1}, and the attachment of chlorine to the C=O shifts its absorption to 1770–1815 cm^{-1}. The IR spectrum of an acid chloride is provided in Figure 8.27.

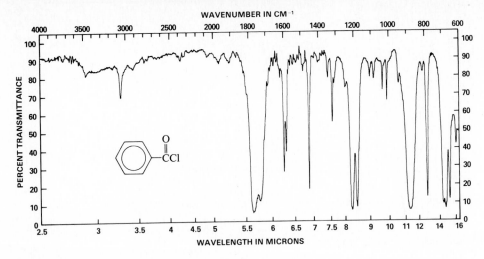

Figure 8.27 Infrared spectrum of benzoyl chloride.

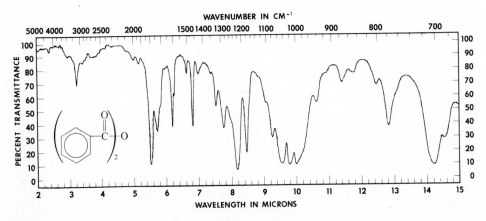

Figure 8.28 Infrared spectrum of benzoic anhydride (KBr pellet).

(p) Anhydrides

The infrared spectra of acid anhydrides contain *two* C$=$O stretching peaks, one at **1800–1870** cm^{-1} and the other at **1740–1790** cm^{-1}. The position of each peak is shifted, depending on ring size if the anhydride is cyclic, and also by conjugation. The C—O single-bond stretching mode results in a peak in the **1050–1200** cm^{-1} region. Figure 8.28 shows an IR spectrum of an anhydride.

(q) Ethers

Ethers produce strong peaks in the **1050–1200** cm^{-1} region due to stretching of the C—O—C bonds. Alkyl ethers usually absorb in the **1050–1150** cm^{-1} region, and aryl or vinyl ethers

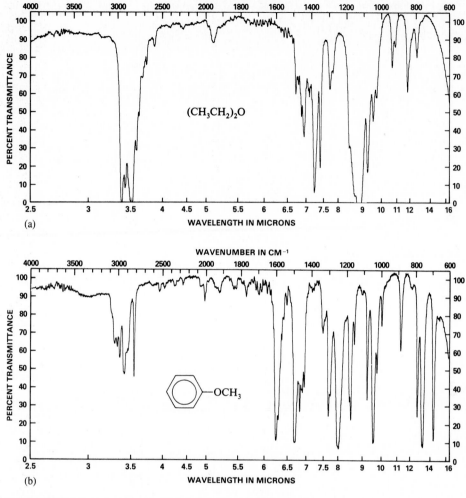

Figure 8.29 Infrared spectrum of (a) diethyl ether and (b) anisole.

give peaks at **1200–1275** cm^{-1}. Alcohols and ethers produce absorption peaks in approximately the same region for the C—O stretching modes. However, alcohols also exhibit a strong peak at 3200–3600 cm^{-1} because of the O—H group and thus are distinguished readily from ethers. Figure 8.29 shows the IR spectra of two typical ethers.

(r) Nitro Compounds

The nitro group, —NO_2, produces two strong, coupled stretching bands in the infrared, one at **1350** cm^{-1} and the other at **1520** cm^{-1}. Nitro groups are commonly encountered as substituents on aromatic rings but are seldom found in aliphatic compounds. The IR spectrum of an aromatic nitro compound is given in Figure 8.30.

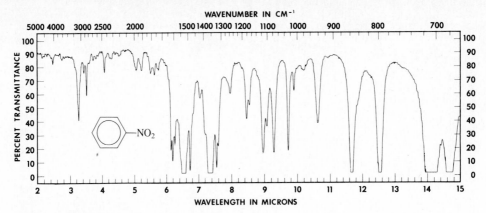

Figure 8.30 Infrared spectrum of nitrobenzene.

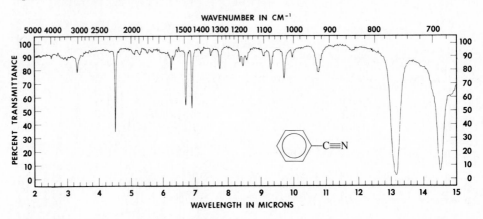

Figure 8.31 Infrared spectrum of benzonitrile.

(s) Nitriles

The carbon-nitrogen triple bond, —C≡N, produces a strong band in the **2215–2265** cm^{-1} range. Conjugated nitriles absorb near **2215** cm^{-1} and nonconjugated nitriles near **2260** cm^{-1}. Figure 8.31 provides the IR spectrum of a nitrile.

(t) Amines

Amines have absorption bands of medium intensity in the **3200–3400** cm^{-1} region due to N—H stretching when there is hydrogen attached to nitrogen as in RNH_2 (1° amines) and RR′NH (2° amines), but not in RR′R″N (3° amines). Unassociated amines give a sharp peak at **3400** cm^{-1} and associated amines a broad peak at **3200** cm^{-1}. The —NH_2 group usually appears as a doublet (two peaks) with a separation of about 50 cm^{-1}. The N—H stretching of a secondary amine is usually weak. The —NH_2 group produces a medium to strong peak at **1600** cm^{-1} due to bending, and this band in conjunction with bands in the 3300 cm^{-1} region can be used to characterize primary amines and unsubstituted amides. The —NH—

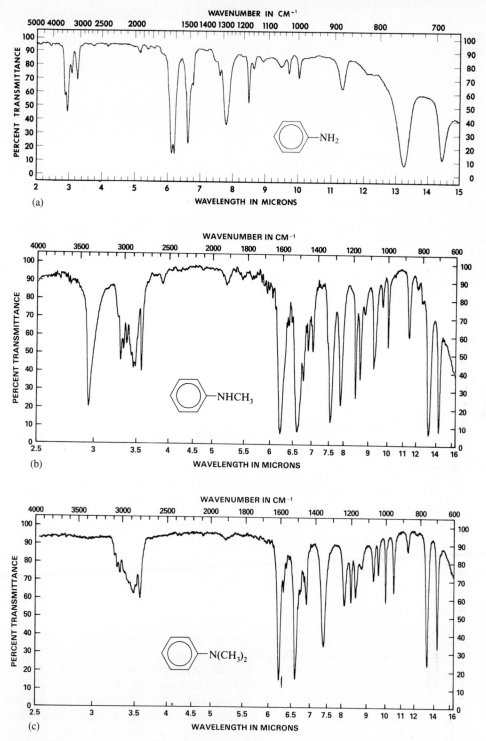

Figure 8.32 Infrared spectrum of (a) aniline, (b) N-methylaniline, and (c) N,N-dimethylaniline.

group produces a weak peak due to bending at **1540** cm^{-1}, and this band in conjunction with bands in the 3300 cm^{-1} region can be used to characterize secondary amines and monosubstituted amides. The peaks for secondary amines may be very weak at both 1540 and 3300 cm^{-1}. However, all amines and amides exhibit a strong peak at **1180–1360** cm^{-1} due to the C—N bond. Figure 8.32 contains IR spectra of primary, secondary, and tertiary amines.

EXERCISES

1. Define or explain the following terms:
   a. spectrometer
   b. spectrophotometer
   c. reciprocal centimeters
   d. wavenumbers
   e. wavelength
   f. percent transmittance
   g. percent absorbance
   h. neat liquid
   i. absorption maxima
   j. demountable sample cell
   k. fixed-thickness cell
   l. reference beam
   m. potassium bromide pellet
   n. mull method
   o. solvent cell
   p. calibration of spectrum
   q. solvent reference cell

2. Convert the following wavenumbers into microns:
   a. 1690 cm^{-1} b. 2990 cm^{-1} c. 680 cm^{-1}

3. Convert the following into wavenumbers:
   a. 6.05 microns   b. 15.2 microns   c. 3.1 microns

4. Why is it imperative that infrared windows be protected from moisture?

5. Why should excessive pressure be avoided when assembling a demountable infrared cell?

6. Discuss the disadvantages of using the following mulling fluids:
   a. Nujol   b. perfluorokerosene

7. Discuss the potential disadvantages of using the following solvents when determining the infrared spectra of solids:
   a. carbon tetrachloride   b. chloroform   c. carbon disulfide

8. Which infrared sampling technique(s) can be used to determine the infrared spectrum of
   a. a pure liquid?   b. a solid?   c. a gas?

9. What is the purpose of calibrating an infrared spectrum, and how is this done?

10. Why must an infrared cell be cleaned thoroughly before and after use? What solvents are suitable for this purpose?

11. The infrared spectrum of a certain substance was determined in a solvent cell whose length was 0.5 mm; the percent transmittance of a certain peak was 75% when the concentration of the solute was 10% in carbon tetrachloride. Assuming that the absorption coefficient does not vary with concentration, will the percent absorption increase, decrease, or remain unchanged when the spectrum is obtained under the following conditions? Briefly explain your answer in each case.
   a. The concentration is increased to 15% while the path length is unchanged.
   b. The concentration is decreased to 5% while the path length is unchanged.
   c. The path length is decreased to 0.3 mm while the concentration is unchanged.
   d. The path length is increased to 0.7 mm while the concentration is unchanged.
   e. Both the concentration and path length are doubled.

    f. The concentration is doubled and the path length is reduced to 0.25 mm.

    g. The concentration is cut in half and the path length is doubled.

12. What problems might be experienced if an infrared spectrum of a solid dissolved in chloroform is determined without using a reference cell?

13. Figures 8.33 and 8.34 show the infrared spectra of 4-methyl-2-pentanol and 4-methyl-1-pentene. Which spectrum belongs to which compound? Briefly explain.

$$(CH_3)_2CHCH_2CHCH_3 \qquad (CH_3)_2CHCH_2CH{=}CH_2$$
$$|$$
$$OH$$

                                    4-Methyl-1-pentene

4-Methyl-2-pentanol

14. Figures 8.35 and 8.36 show the infrared spectra of dimethyl fumarate and *cis*-cyclohexane-1,2-dicarboxylic acid. Which spectrum belongs to which compound? Briefly explain.

        CH$_3$O$_2$C   H   CO$_2$H

              C$=$C

        H   CO$_2$CH$_3$ CO$_2$H

        Dimethyl fumarate   *cis*-Cyclohexane-1,2-
                                      dicarboxylic acid

15. Figures 8.37 and 8.38 show the infrared spectra of acetanilide and styrene. Which spectrum belongs to which compound? Briefly explain.

        C$_6$H$_5$NHCOCH$_3$ C$_6$H$_5$CH$=$CH$_2$

          Acetanilide   Styrene

Spectra for Exercises 13 to 15 follow. References can be found on page 233.

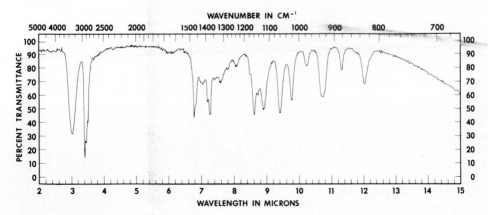

Figure 8.33 Infrared spectrum for Exercise 13.

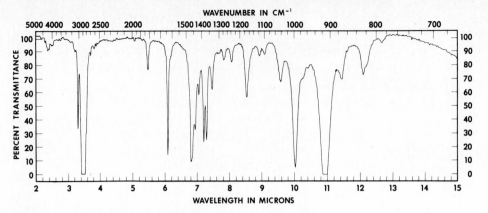

Figure 8.34 Infrared spectrum for Exercise 13.

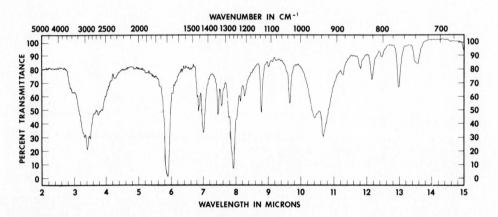

Figure 8.35 Infrared spectrum for Exercise 14.

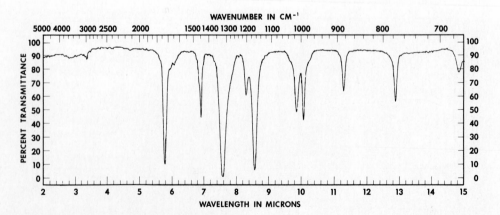

Figure 8.36 Infrared spectrum for Exercise 14.

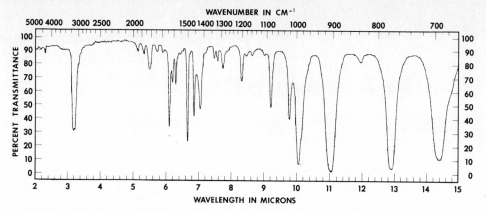

Figure 8.37 Infrared spectrum for Exercise 15.

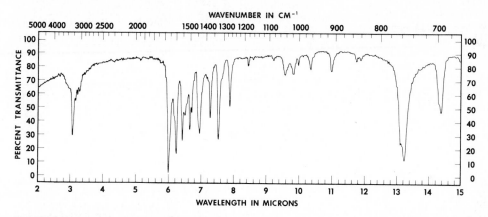

Figure 8.38 Infrared spectrum for Exercise 15.

REFERENCES

1. Silverstein, R. M.; Bassler, G. C.; Morrill, T. C. *Spectrometric Identification of Organic Compounds,* 4th edition, Wiley, New York, 1981, pp. 95–180.

2. Dyer, J. R. *Applications of Absorption Spectroscopy of Organic Compounds,* Prentice-Hall, Englewood Cliffs, N.J., 1965, pp. 22–57.

3. Pasto, D. J.; Johnson, C. R. *Organic Structure Determination,* Prentice-Hall, Englewood Cliffs, N.J., 1969, pp. 109–158.

4. Cross, A. D.; Jones, R. A. *Introduction to Practical Infrared Spectroscopy,* 3rd edition, Plenum Press, New York, 1969.

5. Conley, R. T. *Infrared Spectroscopy,* 2nd edition, Allyn and Bacon, Boston, 1972.

6. Lambert, J. B.; Shurvell, H. F.; Verbit, L.; Cooks, R. G.; Stout, G. H. *Organic Structural Analysis,* Macmillan, New York, 1976.

7. Nakanishi, K.; Solomon, P. H., *Infrared Absorption Spectroscopy,* 2nd edition. Holden-Day, San Francisco, 1977.

8. Bellamy, L. J. *The Infrared Spectra of Complex Molecules,* 3rd edition, Halsted, New York, 1975.

9. *Aldrich Library of Infrared Spectra,* 2nd edition, C. J. Pouchert, editor, Aldrich Chemical Co., Milwaukee, 1975.

10. *Sadtler Standard Spectra, Midget Edition,* IR vols. 1–26, Sadtler Research Laboratories, Philadelphia.

11. Swinehart, J. S. *Organic Chemistry,* Appleton-Century-Crofts, New York, 1969, Appendix III.

nuclear magnetic resonance spectroscopy

Nuclear magnetic resonance (NMR) spectroscopy was developed in the early 1950s and became readily available to organic chemists in the early 1960s. Today the combination of infrared spectroscopy (IR, Chapter 8) and NMR spectroscopy provides a powerful method of structure determination and compound identification. NMR spectroscopy is discussed in this chapter, and the next chapter shows how IR and NMR data can be used together in identifying a compound.

Nuclear magnetic resonance is an important experimental technique that is based on the interesting property of *nuclear spin* exhibited by various nuclei. Among organic compounds, the common elements that exhibit nuclear spin are hydrogen (^{1}H), fluorine (^{19}F), phosphorus (^{31}P), and the minor isotopes of carbon (^{13}C) and nitrogen (^{15}N). Atomic nuclei must have nuclear spin in order to be studied by NMR spectroscopy.

Nuclear magnetic resonance is a general term that applies to all nuclei that have magnetic properties. The most widely studied nucleus is hydrogen covalently bonded to other atoms in a molecule. In NMR discussions, this nucleus is often referred to as a *hydrogen* or a *proton,* although these terms are misnomers because there are no free hydrogen atoms or protons in most organic compounds. The term **PMR** (proton magnetic resonance) is often used to refer to the magnetic resonance of the hydrogen nucleus. Nuclear magnetic resonance due to the natural abundance of carbon-13, ^{13}C, has become increasingly more important in structure elucidation in the past few years and is called **CMR** (carbon magnetic resonance). PMR and CMR spectroscopy are discussed in this and the next chapter.

9.1 The Concept of Nuclear Magnetic Resonance Spectroscopy; the NMR Spectrometer

Although the hydrogen nucleus has been selected for the purpose of illustrating the concept of nuclear magnetic resonance spectroscopy, the following discussion applies equally well to other nuclei, such as ^{13}C, that exhibit this phenomenon. The hydrogen nucleus has a uniform distribution of charge, and it generates a small nuclear magnetic moment when it spins about an axis. This nucleus with its magnetic moment behaves as though it is a small bar magnet. The rotation of the hydrogen nucleus and its magnetic moment are shown in Figure 9.1a. When this nucleus is placed in an external magnetic field, the magnetic moment of the nucleus becomes aligned either *with* the applied field (same direction as the external field, Figure 9.1b) or *against* it (opposite direction, Figure 9.1c). There are two allowed energy levels, E_1 and E_2, of the hydrogen nucleus associated with these alignments, and these levels are said to be *quantized*.

Note that the magnetic dipole of the nucleus is not aligned parallel with the external field but traces out a circular path about the axis defined by the external magnetic field. This phenomenon is called *precession* and can be likened to a spinning gyroscope, the axis of which is not parallel to the gravitational field of the earth. The more stable state, E_1, is that in which the magnetic dipole of the nucleus is oriented *with* the external field (Figure 9.1b). Energy must be absorbed by the nucleus in order to "flip" the direction of the nuclear magnetic dipole, that is, to interconvert nuclei between the two spin states. Because the energy difference between these two states is quite small (approximately 0.01 to 1 *calorie*), there is statistically only a slight excess of nuclear magnetic dipoles oriented with the external field at room temperature.

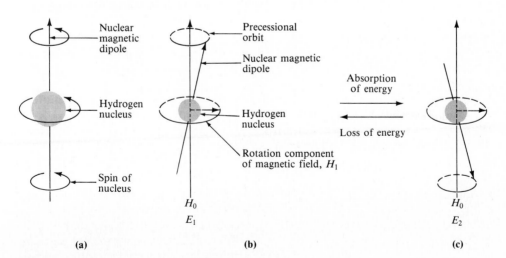

Figure 9.1 Spin properties of the hydrogen nucleus. (a) Rotation of hydrogen nucleus and its magnetic moment. (b) Magnetic moment of nucleus aligned with applied external magnetic field. (c) Magnetic moment of nucleus aligned against applied magnetic field.

The *frequency* of the precession, ω_0, of the nucleus is proportional to the strength of the applied external magnetic field, H_0, as given in equation 1. In this equation, γ_H is the *magnetogyric ratio* constant, which has a characteristic value for the particular nucleus being studied but is different for each nucleus. It should be emphasized that γ for a proton, γ_H, has a different value than that for carbon-13, γ_{13C}, but keep in mind that the theory of proton magnetic resonance is the focus of this discussion. This equation therefore requires that ω_0 increase as H_0 increases.

$$\omega_0 = \gamma_H H_0 \qquad\qquad (1)$$

Figure 9.1 shows the precession of protons about the axis of the nucleus in a strong, uniform magnetic field.

The transition between the two energy states, E_1 and E_2, can be induced if the nucleus, which is precessing at a frequency ω_0 in a external field H_0, is subjected to a second, oscillating magnetic field, H_1, having a rotation component *perpendicular* to H_0 (Figure 9.2). The oscillating magnetic field (H_1) can be generated by an oscillator coil placed at right angles to the applied magnetic field. The frequency of rotation of H_1 can be varied by changing the frequency, ν, of the oscillator used to produce it. The oscillating magnetic field is linear and is equivalent to two vectoral components that rotate in opposite directions. The component rotating in the same direction as the precession of the nuclear magnetic dipole of the proton[1] is the one that is important. When the angular velocity of H_1 is equal to the precessional frequency, ω_0, of the proton, the nucleus absorbs electromagnetic energy and flips to its higher energy state, E_2, as shown in Figure 9.2. This occurrence is referred to as

---

[1] This discussion focuses on the proton as the absorbing species, but it in fact can be applied to any nucleus that has the property of nuclear spin.

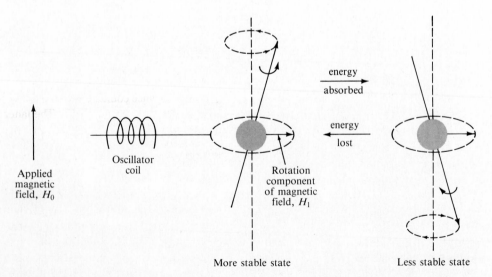

energy absorbed

energy lost

Oscillator coil

Applied magnetic field, H_0

Rotation component of magnetic field, H_1

More stable state   Less stable state

Figure 9.2 Interconversion of two states of hydrogen nucleus in applied magnetic field when subjected to second, oscillating magnetic field, H_1, generated by oscillator coil that is perpendicular to applied field, H_0.

the condition of *resonance*. Energy absorption is detected by the electronic detectors in the nuclear magnetic resonance spectrometer and recorded on paper as peaks to produce an NMR spectrum. The nuclear magnetic dipole of the proton then relaxes to its more stable, lower energy state (E_1). It should be pointed out that both excitation and de-excitation are occurring at resonance, but the *net* absorption of energy depends on unequal populations in E_1 and E_2. In the case when the external magnetic field is left on too long, there are equal populations in E_1 and E_2. In this situation, *saturation* occurs, and no absorption of energy would be observed.

There are two variables associated with achieving resonance and the resulting flipping of the nucleus: the applied external magnetic field (H_0) and the frequency of the oscillator (ν). The oscillator frequency required to produce the desired nuclear magnetic resonance transition depends upon the strength of the applied external magnetic field. Equation 2 gives the relationship between energy and the applied magnetic field, and equation 3 gives the energy associated with the frequency of the oscillator. When resonance occurs, the ΔE and $\Delta E'$ must be equal, so that equation 4 results.

$$\Delta E = \frac{h\gamma_H H_0}{2\pi} \tag{2}$$

$$\Delta E' = h\nu \tag{3}$$

$$h\nu = \frac{h\gamma_H H_0}{2\pi} \quad \text{or} \quad \nu = \frac{\gamma_H H_0}{2\pi} \tag{4}$$

where $\gamma_H H_0 = \omega_0$
ΔE = energy difference between two spin states
γ_H = magnetogyric ratio (a constant)
H_0 = applied external magnetic field strength at nucleus
h = Planck's constant
ν = frequency of oscillator required for resonance absorption

This equation indicates that at resonance, the frequency produced by the oscillator coil (ν) is directly proportional to the applied external magnetic field (H_0), so that the use of stronger external magnetic fields requires higher oscillator frequencies in order for resonance to occur. Moreover, equation 4 shows that the condition for resonance could be achieved either by holding H_0 constant and varying ν, or by keeping ν constant and changing H_0. The latter approach is more practical from the standpoint of instrument design and is used in many older NMR spectrometers. (Modern Fourier transform NMR spectrometers utilize a different technique of irradiating a sample to obtain an NMR spectrum, but the theory and operation of these instruments will not be discussed in this text.) It has been found that if the applied external magnetic field, H_0, is of an experimentally convenient magnitude of 14,092 gauss (G), then the frequency of the oscillator, ν, must be on the order of 60 megaHertz (MHz) in order to produce *proton* resonance. This oscillator frequency is in the radio-frequency (rf) region of the electromagnetic spectrum, so that an rf oscillator is required.[2] Note

---

[2]The external magnetic field, H_0, and/or the rf frequency, ν, must be changed when other nuclei that have nuclear spin are studied, because γ will be different for them. The values for H_0 and ν given here are for proton resonance only.

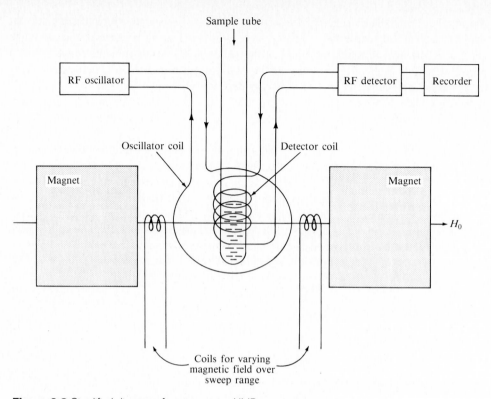

Figure 9.3 Simplified diagram of contemporary NMR spectrometer.

that the terms *Hertz (Hz)* and *cycles per second (cps)* are synonymous; one megaHertz (MHz) is equivalent to 1 million or 10^6 cps.

A schematic diagram of a contemporary nuclear magnetic resonance spectrometer is shown in Figure 9.3. This instrument consists of a large electromagnet or permanent magnet that produces the applied external magnetic field over the sweep range of the spectrum, and the rf oscillator coil and detector coil. Although not indicated in the figure, the sample tube must be spun while an NMR spectrum is recorded, the purpose of this being to ensure a homogeneous magnetic field in the sample tube.

9.2 Introduction to PMR Spectra

As mentioned earlier, the term "NMR" applies generally to the magnetic resonance spectra resulting from different elements having nuclear spin. The magnetic resonance of protons in organic molecules, which is called **proton magnetic resonance** spectroscopy or simply **PMR,** will now be discussed.

All the hydrogens in an organic molecule would absorb energy at the same place in a PMR spectrum *if* they were all equivalent (see equation 1). As a consequence, very little useful information would be obtained regarding the structure of the molecule. Fortunately, they are *not* all equivalent because the nuclei are in different magnetic environments owing to the three-dimensional electronic structure of the molecules. PMR spectroscopy provides information about the structure of the molecule being examined and is therefore a valuable tool in structure determination.

Consider the PMR spectrum of 1-nitropropane, shown in Figure 9.4. A number of significant features of this spectrum will be discussed, and some of the terminology associated with PMR spectroscopy will be introduced.

The strength of the applied magnetic field, H_0, is plotted along the horizontal axis and *increases* from left to right on the spectrum, which is called the *upfield* direction. Going from right to left is called the *downfield* direction. The *energy* required to produce the transition between the nuclear spin states, that is, E_1 and E_2, increases from left to right. In the spectrum, increasing intensity of absorption of energy is plotted in a vertical direction.

Several scales are given horizontally across the top and bottom of the spectrum. All of these scales can be used to indicate the location of the peaks on the spectrum. The scale from 0 to 500 at the top and running right to left represents cycles/sec in Hertz (Hz). The absorption of electromagnetic energy by most protons occurs within 600 Hz downfield from TMS (defined below) when a 60-MHz oscillator is used, thus accounting for the range printed on the paper. The scale from *ca.* 2 to 10 going from left to right at the top is given in ppm (parts per million) and is called the *tau* (τ) scale. At the bottom, the scale of *ca.* 8 to 0 from left to right is given in ppm (parts per million) and is called the *delta* (δ) scale. The significance of these various scales is discussed below.

A single, sharp absorption peak is located at a position defined as being at 0 Hz. The positions of the other peaks on the spectrum are measured relative to the position of this peak, which is produced by the protons of a standard reference compound, *tetramethylsilane,* $(CH_3)_4Si$, abbreviated *TMS*. Because this reference peak results from the introduction of *ca.* 1% of TMS, a liquid, into the sample whose spectrum is being determined, the TMS serves

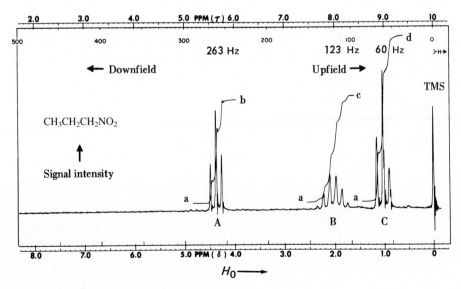

Figure 9.4 PMR spectrum of 1-nitropropane.

as an *internal standard*. In the spectrum given in Figure 9.4, three groups of peaks are found to be centered at 60, 123, and 263 Hz *downfield* from the TMS peak when the rf oscillator is set at 60 MHz. These values represent the **chemical shifts** of the three magnetically distinct types of hydrogens in the molecule.

These groups of peaks, labeled A, B, and C, result from absorption of energy by hydrogens. The amount of energy required to realign the nuclei in the magnetic field—that is, the energy necessary to produce *resonance* and thus absorption of energy—is clearly *not* identical for all the protons in the molecule. This may appear surprising, because equation 4 indicates that all hydrogen nuclei should undergo spin-flipping or resonance at the same H_0 when ν is held constant, and thus only *one single peak* would be expected.

The observation of more than one peak in the PMR spectrum can be explained *qualitatively* as follows. The three groups of peaks, A, B, and C, represent three "types" of hydrogens in 1-nitropropane, which are nonequivalent by virtue of the slightly different magnetic environments in which they exist. The magnetic environment of each proton in the molecule depends on two major factors: (1) the externally applied magnetic field, H_0, which is of uniform strength throughout the entire molecule and therefore *cannot* produce nonequivalence of the different hydrogens and (2) circulation of the electrons within the molecule, which provides an "electron shield" around the various atoms, including the hydrogens. This shield is *not* uniform because the electron density varies over the molecule. The existence of this internal electric field within the molecule induces a magnetic field about the protons and causes them to be magnetically nonequivalent because their nuclei are shielded to varying extents. The hydrogen nuclei that are immersed in a higher induced magnetic field will be more highly shielded than other, less-shielded nuclei when an external magnetic field is applied to the molecule. In other words, more highly shielded nuclei will experience an *effective* magnetic field, H_e, that is *less* than the applied external field, so that *higher* applied fields will be necessary to produce the desired nuclear spin transitions (resonance). In summary, the appearance of more than one group of peaks in the spectrum reflects the differences in the flux of the magnetic field induced within the molecule by circulation of electrons.

The *chemical shift* of a hydrogen—that is, its location relative to TMS—if measured in units of Hz is *dependent* on the frequency of the oscillator being used in the spectrometer (see equation 4). Rather than reporting chemical shifts in Hz, which would require specifying the rf frequency *and* the strength of the external magnetic field, it is useful to convert such shifts into units that are *independent* of the oscillator frequency. If the chemical shift (in Hz) is divided by the oscillator frequency (in Hz) and the result is multiplied by 10^6, a *frequency-independent* scale, called the **delta (δ) scale,** is obtained (equation 5).

$$\delta = \frac{\text{chemical shift in Hz}}{\text{oscillator frequency in Hz}} \times 10^6 \qquad \textbf{(5)}$$

Chemical shifts calculated in this manner are values on a standard scale, which has the TMS peak at δ 0.0, with peaks downfield from TMS having *positive* values ($+\delta$), and peaks upfield from TMS having *negative* values ($-\delta$).[3] This frequency-independent scale has many

---

[3] The sign convention should be opposite to the one given and should be $-\delta$ for peaks downfield from TMS. Common practice and usage correspond to the convention given here because most protons absorb downfield from TMS.

advantages today, since NMR spectrometers operate at higher frequencies such as 90 MHz, 200 MHz, 300 MHz, or 500 MHz, and the δ-values of chemical shift are the same regardless of the oscillator frequency.

The δ-values[4] for the groups of peaks observed for 1-nitropropane can be calculated as follows, using equation 5:

$$\text{Group A:} \quad [(60)/(60 \times 10^6)](10^6) \; = \delta_A \quad 1.00$$
$$\text{Group B:} \quad [(123)/(60 \times 10^6)](10^6) \; = \delta_B \quad 2.05$$
$$\text{Group C:} \quad [(263)/(60 \times 10^6)](10^6) \; = \delta_C \quad 4.38$$

Fortunately, these calculations seldom have to be performed, since most NMR spectrum paper provides scales that give directly the chemical shift in Hz and the corresponding δ-values.

In the early days of NMR spectroscopy, another frequency-independent scale, the *tau scale* (τ), was used to represent chemical shift. This scale is seldom used today but still appears in the earlier literature. The tau and delta scales are easily interconverted by using equation 6.

$$\text{tau}(\tau) = 10.0 - \text{delta}(\delta) \tag{6}$$

The delta scale is used exclusively in our discussions and tables, although some of the PMR spectra in this text have both scales.

The PMR spectrum of 1-nitropropane (Figure 9.4) merits further discussion regarding the location of the groups of peaks **(chemical shift),** the number of peaks in each group **(splitting pattern),** the area under each peak **(peak area),** and the distance between the peaks within each group **(coupling constant).** The next four sections discuss these features, which provide a great deal of information about the structure of this and other organic molecules.

9.3 Chemical Shift and Structure

The location of proton absorptions relative to TMS is called **chemical shift,** which is something of a misnomer since no chemicals are present other than the absorbing compound and solvent. The shift is due to the chemical environment in the molecule. Because the value of the chemical shift of a proton attached to a given type of functional group is nearly constant, the determination of the shifts of various protons in an unknown compound often provides valuable information about the functional groups present. The underlying question is *why* certain protons in a molecule appear at one place in the PMR spectrum while others appear at other places, a feature clearly illustrated by the three groups of peaks in 1-nitropropane (Figure 9.4). Our discussion in Section 9.2 suggests that different *molecular* environments produce different *electronic* environments around the hydrogens. The externally applied

---

[4]The method employed for expressing chemical shifts uses *parts per million (ppm)*. The peaks for Group A are centered at 60 Hz, and this can be converted into parts per million as follows: $(60 \text{ Hz})/(60 \times 10^6 \text{ Hz}) = 1.00 \times 10^{-6}$, which is 1 part per million (ppm).

magnetic field, H_0, is uniform over the entire molecule, so the magnetic field induced by the movement of the electrons and the proximity of one hydrogen to another cause the protons to be *nonequivalent*. The magnetic fields induced by the molecule cause the protons near these fields to be either *shielded* or *deshielded*. That is, each ''type'' of proton ''experiences'' an *effective* magnetic field, H_e, that is either greater or smaller than the applied magnetic field, H_0. A stronger magnetic field must be applied to a molecule containing shielded protons in order to cause their spins to flip so that the condition of resonance exists. Conversely, a weaker magnetic field is required for resonance to occur in a molecule containing deshielded protons. All nuclei are shielded to some extent, but the degree of shielding varies from proton to proton. The terms ''shielding'' and ''deshielding'' are used in a relative sense. Deshielded protons absorb further downfield from TMS than do shielded protons. The appearance of more than one group of peaks in a PMR spectrum reflects differences in shielding or deshielding and thus differences in the strengths of the induced magnetic field at different places in the molecule.

The chemical shifts observed for hydrogens in aliphatic hydrocarbons follow a predictable pattern. In a molecule containing methyl hydrogens, secondary (2°) hydrogens, and tertiary (3°) hydrogens, the methyl hydrogens absorb upfield (nearest TMS), the 2° appear downfield from the —CH_3 hydrogens, and the 3° absorb downfield from —CH_2— hydrogens. For example, the —CH_3 protons in propane, $CH_3CH_2CH_3$, appear at 0.9 ppm, whereas the —CH_2— hydrogens absorb downfield at 1.43 ppm.

As mentioned before, the *magnitude* of chemical shifts of protons attached to a particular functional group or type of carbon atom is nearly constant from compound to compound. Tables have been assembled that correlate chemical shifts with structural features, and the structural features of a molecule containing hydrogens can often be deduced on the basis of the PMR chemical shifts. Table 9.1 presents an abbreviated compilation, and Table 9.2 contains a more extensive listing of observed chemical shifts for a variety of specific types of hydrogens. The narrow range of 10 ppm (600 Hz) spans the proton chemical shifts for hydrogens found in nearly all organic compounds. This apparently low sensitivity of proton chemical shift often produces overlapping of absorption peaks of different types of hydrogens, which can make the interpretation of spectra difficult. This potential problem has been partially solved by the development of instruments operating at higher frequencies, for example, 90 MHz, 200 MHz, 300 MHz, or 500 MHz, which produce spectra with much greater resolution of peaks. Even so, the chemical shift range observed at higher frequencies is still δ 0–10, and the correlations provided in Tables 9.1 and 9.2 apply for all instruments, regardless of their operating frequency.

Some of the factors that influence the degree of shielding of protons and thus the chemical shift are discussed next.

(a) Effect of Electron-Withdrawing Groups on Chemical Shift

The PMR spectra of molecules that contain electron-withdrawing groups usually exhibit a downfield chemical shift of *some* of the protons. The inductive effect of these groups causes electrons to be withdrawn from nearby protons, which are therefore deshielded to some extent as a consequence. However, the inductive effect is greatly diminished as hydrogens

Text continues on p. 245

TABLE 9.1 Chemical Shifts of Hydrogens Attached to Various Functional Groups

Functional Group; Hydrogen Type Shown as **H**	Chemical Shift, ppm δ	Functional Group; Hydrogen Type Shown as **H**	Chemical Shift, ppm δ
TMS, $(CH_3)_4Si$	0	Alcohols, ethers	
Cyclopropane	0–0.4	$HO-\overset{\mid}{\underset{\mid}{C}}-H$	3.4–4
Alkanes			
RCH_3	0.9		
R_2CH_2	1.3	$RO-\overset{\mid}{\underset{\mid}{C}}-H$	3.3–4
R_3CH	1.5		
Alkenes		Acetals	5.3
$-\overset{\mid}{C}=\overset{\mid}{C}-H$ (vinyl)	4.6–5.9		
$-\overset{\mid}{C}=\overset{\mid}{C}-CH_3$ (allyl)	1.7	Esters	
Alkynes		$R-\overset{O}{\overset{\|}{C}}-O-\overset{\mid}{\underset{\mid}{C}}-H$	3.7–4.1
$-C\equiv C-H$	2–3		
$-C\equiv C-CH_3$	1.8	$RO-\overset{O}{\overset{\|}{C}}-\overset{\mid}{\underset{\mid}{C}}-H$	2–2.6
Aromatic			
$Ar-H$	6–8.5	Carboxylic acids	
$Ar-\overset{\mid}{\underset{\mid}{C}}-H$ (benzyl)	2.2–3	$HO-\overset{O}{\overset{\|}{C}}-\overset{\mid}{\underset{\mid}{C}}-H$	2–2.6
Fluorides, $F-\overset{\mid}{\underset{\mid}{C}}-H$	4–4.5	$R-\overset{O}{\overset{\|}{C}}-O-H$	10.5–12
Chlorides		Aldehydes, ketones	
$Cl-\overset{\mid}{\underset{\mid}{C}}-H$	3–4	$R-\overset{O}{\overset{\|}{C}}-\overset{\mid}{\underset{\mid}{C}}-H$	2–2.7
$Cl-\overset{Cl}{\underset{\mid}{C}}-H$	5.8	Aldehydes	
Bromides, $Br-\overset{\mid}{\underset{\mid}{C}}-H$	2.5–4	$R-\overset{O}{\overset{\|}{C}}-H$	9–10
Iodides, $I-\overset{\mid}{\underset{\mid}{C}}-H$	2–4	Amides	
		$R-\overset{O}{\overset{\|}{C}}-\overset{\mid}{N}-H$	5–8
Nitroalkanes, $O_2N-\overset{\mid}{\underset{\mid}{C}}-H$	4.2–4.6	Alcohols, $R-O-H$	4.5–9
		Phenols, $Ar-O-H$	4–12
		Amines, $R-NH_2$	1–5

TABLE 9.2 Compilation of PMR Absorptions for Various Molecules

Listed below are the PMR chemical shifts observed for the protons of a number of different organic compounds. The shifts are classified according to whether they are methyl, methylene, or methine types of hydrogen atoms. The hydrogen shown in **bold** is responsible for the absorptions listed.

METHYL ABSORPTIONS

Compound	Chemical Shift (ppm) δ	Compound	Chemical Shift (ppm) δ
CH_3NO_2	4.3	CH_3CHO	2.2
CH_3F	4.3	CH_3I	2.2
$(CH_3)_2SO_4$	3.9	$(CH_3)_3N$	2.1
$C_6H_5COOCH_3$	3.9	$CH_3CON(CH_3)_2$	2.1
$C_6H_5-O-CH_3$	3.7	$(CH_3)_2S$	2.1
CH_3COOCH_3	3.6	$CH_2=C(CN)CH_3$	2.0
CH_3OH	3.4	CH_3COOCH_3	2.0
$(CH_3)_2O$	3.2	CH_3CN	2.0
CH_3Cl	3.0	CH_3CH_2I	1.9
$C_6H_5N(CH_3)_2$	2.9	$CH_2=CH-C(CH_3)=CH_2$	1.8
$(CH_3)_2NCHO$	2.8	$(CH_3)_2C=CH_2$	1.7
CH_3Br	2.7	CH_3CH_2Br	1.6
CH_3COCl	2.7	$C_6H_5C(CH_3)_3$	1.3
CH_3SCN	2.6	$C_6H_5CH(CH_3)_2$	1.2
$C_6H_5COCH_3$	2.6	$(CH_3)_3COH$	1.2
$(CH_3)_2SO$	2.5	$C_6H_5CH_2CH_3$	1.2
$C_6H_5CH=CHCOCH_3$	2.3	CH_3CH_2OH	1.2
$C_6H_5CH_3$	2.3	$(CH_3CH_2)_2O$	1.2
$(CH_3CO)_2O$	2.2	$CH_3(CH_2)_3Cl$, Br, I	1.0
$C_6H_5OCOCH_3$	2.2	$CH_3(CH_2)_4CH_3$	0.9
$C_6H_5CH_2N(CH_3)_2$	2.2	$(CH_3)_3CH$	0.9

METHYLENE ABSORPTIONS

Compound	Chemical Shift (ppm) δ	Compound	Chemical Shift (ppm) δ
$EtOCOC(CH_3)=CH_2$	5.5	$EtCH_2Cl$	3.4
CH_2Cl_2	5.3	$(CH_3CH_2)_4N^+I^-$	3.4
CH_2Br_2	4.9	CH_3CH_2Br	3.4
$(CH_3)_2C=CH_2$	4.6	$C_6H_5CH_2N(CH_3)_2$	3.3
$CH_3COO(CH_3)C=CH_2$	4.6	$CH_3CH_2SO_2F$	3.3
$C_6H_5CH_2Cl$	4.5	CH_3CH_2I	3.1
$(CH_3O)_2CH_2$	4.5	$C_6H_5CH_2CH_3$	2.6
$C_6H_5CH_2OH$	4.4	CH_3CH_2SH	2.4
$CF_3COCH_2C_3H_7$	4.3	$(CH_3CH_2)_3N$	2.4
$Et_2C(COOCH_2CH_3)_2$	4.1	$(CH_3CH_2)_2CO$	2.4
$HC\equiv C-CH_2Cl$	4.1	$BrCH_2CH_2CH_2Br$	2.4
$CH_3COOCH_2CH_3$	4.0	Cyclopentanone (α CH_2)	2.0
$CH_2=CHCH_2Br$	3.8	Cyclohexene (α CH_2)	2.0
$HC\equiv CCH_2Br$	3.8	Cycloheptane	1.5
$BrCH_2COOCH_3$	3.7	Cyclopentane	1.5
CH_3CH_2NCS	3.6	Cyclohexane	1.4
CH_3CH_2OH	3.6	$CH_3(CH_2)_4CH_3$	1.4
		Cyclopropane	0.2

METHINE ABSORPTIONS

Compound	Chemical Shift (ppm) δ	Compound	Chemical Shift (ppm) δ
C_6H_5**CHO**	10.0	C_6H_5Cl	7.2
p-ClC$_6$H$_4$**CHO**	9.9	**CHCl$_3$**	7.2
p-CH$_3$OC$_6$H$_4$**CHO**	9.8	**CHBr$_3$**	6.8
CH$_3$**CHO**	9.7	p-Benzoquinone	6.8
Pyridine (α **H**)	8.5	C_6**H$_5$**NH$_2$	6.6
p-C$_6$**H$_4$**(NO$_2$)$_2$	8.4	Furan (β **H**)	6.3
C_6H_5**CH**=CHCOCH$_3$	7.9	CH$_3$**CH**=CHCOCH$_3$	5.8
C_6H_5**CHO**	7.6	Cyclohexene (vinyl **H**)	5.6
Furan (α **H**)	7.4	(CH$_3$)$_2$C=**CH**CH$_3$	5.2
Naphthalene (β **H**)	7.4	(CH$_3$)$_2$**CH**NO$_2$	4.4
p-C$_6$**H$_4$**I$_2$	7.4	Cyclopentyl bromide (**H** at C-1)	4.4
p-C$_6$**H$_4$**Br$_2$	7.3	(CH$_3$)$_2$**CH**Br	4.2
p-C$_6$**H$_4$**Cl$_2$	7.2	(CH$_3$)$_2$**CH**Cl	4.1
C_6**H$_6$**	7.3	C_6H_5C≡C—**H**	2.9
C_6**H$_5$**Br	7.3	(CH$_3$)$_3$C—**H**	1.6

are located farther from the withdrawing group. This effect is clearly seen by comparing the chemical shifts in 1-nitropropane (Figure 9.4) with those of propane:

$$CH_3\text{—}CH_2\text{—}CH_2\text{—}NO_2 \qquad CH_3\text{—}CH_2\text{—}CH_3$$

A↑ B↑ C↑   D↑ E↑ D↑

δ 1.0 δ 2.05 δ 4.38 δ 0.9 δ 1.43 δ 0.9

1-Nitropropane   Propane

The —CH$_3$ groups in both absorb at nearly the same δ-value. However, the —NO$_2$ group is strongly electron-withdrawing, so the —CH$_2$— group marked **C** is shifted far downfield from the —CH$_2$— group in propane. 1-Nitropropane contains *two* —CH$_2$— groups that absorb at very different chemical shifts from one another. This is because the electron-withdrawing effect of the —NO$_2$ group falls off in going to carbon atoms further removed from it, so that —CH$_2$— group **B** is not as far downfield. Note, however, that both of the —CH$_2$— groups in 1-nitropropane are shifted downfield from the —CH$_2$— group in propane as a result of the inductive effect of the —NO$_2$ group.

In general, the absorptions by the protons attached to a carbon atom bearing an electron-withdrawing group are shifted downfield the most, and the magnitude of the downfield shift decreases in going farther from the withdrawing group. Electron withdrawal by the inductive effect through σ-bonds has the effect of increasing the partial positive charge on the hydrogens, which in turn causes those protons to absorb further downfield. With functional groups such as the carboxylic acid group, —CO$_2$H, this effect is so pronounced that the absorption by the proton in —CO$_2$H appears very far downfield, typically at values greater than δ 10.

The halogens are strong electron-withdrawing groups and cause the signals of protons nearby to be shifted downfield. The following examples illustrate this effect:

$$CH_3\text{—}CH_2\text{—}CH_2\text{—}Br \qquad CH_3\text{—}CH_2\text{—}Br$$

A↑　　B↑　　C↑　　　　　D↑　　E↑

δ 1.05　δ 1.9　δ 3.4　　　　δ 1.7　δ 3.4

1-Bromopropane　　　　　　Bromoethane

Note that the —CH_2— group bearing the Br atom has the same chemical shift in both molecules. The —CH_3 group in bromoethane is upfield from the —CH_2— group marked **B** in 1-bromopropane. This is in accord with the earlier statement that signals from —CH_3 groups are upfield from those of —CH_2— groups, and the comparison here is valid since the CH_3 and CH_2 groups are both one carbon atom removed from the carbon bearing the electron-withdrawing bromine atom. The —CH_3 group in 1-bromopropane has nearly the same chemical shift as the —CH_3 group in 1-nitropropane. Comparison of the chemical shifts for 1-bromopropane with those for 1-nitropropane indicates that the NO_2 group causes a greater downfield shift of both of the CH_2 signals, thus indicating that the NO_2 group must be more strongly electron-withdrawing than Br.

Atoms of different electronegativity cause different chemical shifts. The effect of the four halogens is shown by the chemical shift of the methyl group in the methyl halides: CH_3F, δ 4.26; CH_3Cl, δ 3.0; CH_3Br, δ 2.82; CH_3I, δ 2.16. Here fluorine causes the greatest downfield chemical shift, since it is the most electronegative element and thus the most electron-withdrawing halogen.

Increasing the number of halogens attached to the same carbon atom causes increased downfield shift, as illustrated by these examples: $CHCl_3$, δ 7.24; CH_2Cl_2, δ 5.28; CH_3Cl, δ 3.0. These chemical shifts can be compared to that for methane (δ 0.9).

There are many other functional groups in organic chemistry that are electron-withdrawing and that cause downfield chemical shifts. The sample PMR spectra in Section 9.8 merit careful inspection to see these effects first-hand. In our discussion of chemical shifts we have presented the actual δ-values. The beginning student should be able to examine an organic molecule and determine the *relative* chemical shifts for the different types of protons it contains.

(b) Effect of Ring Currents on Chemical Shift

The chemical shifts of protons attached to a carbon-carbon triple bond (a *terminal alkyne*) or to an aromatic ring (**aromatic protons**) are different from what one might predict. The hydrogens of benzene absorb at δ 7.23, and those of acetylene appear at δ 2.9.

$$H\text{—}C\equiv C\text{—}H$$

Ethyne
(acetylene)
δ 2.9

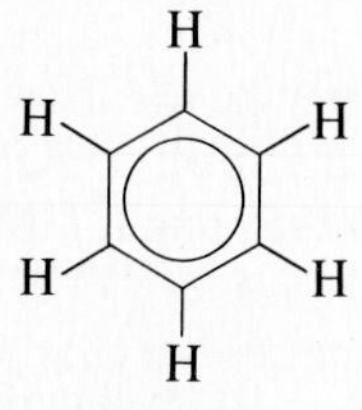

Benzene
δ 7.23

The chemical shifts of these two types of protons result from the degree of shielding that each experiences. The magnetic fields induced by the ring currents are responsible for the chemical shifts of aromatic protons and terminal alkyne protons. These compounds have the common feature of possessing π-electrons; benzene has 6 π-electrons that form a π-cloud, and acetylene contains 2 π-bonds. These π-electrons induce a magnetic field to occur when an external magnetic field is applied to these molecules, and this is called a **ring current.** The induced magnetic field is opposite to the external magnetic field (terminal alkyne) or in the same direction as the external magnetic field (aromatic hydrogens).

Protons are deshielded when the ring current is in the same direction as the external field. They experience a magnetic field from both the ring current and the external field. Less external magnetic field is required for resonance, and the protons appear farther downfield than might be expected. Such is the case for *aromatic protons*, which appear at δ 7.23. This ring current effect is shown in Figure 9.5a.

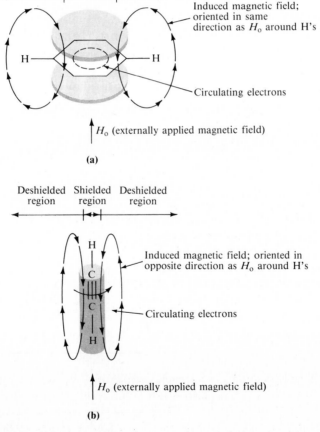

Figure 9.5 (a) Deshielding of protons in benzene due to induced ring currents. (b) Shielding of protons in ethyne (acetylene) due to induced ring currents.

On the other hand, protons are shielded when the ring current is in the opposite direction to the external field. The ring current cancels some of the applied magnetic field, and the protons experience an effective magnetic field that is less than the external field. Thus, a greater external magnetic field is required for resonance, and these protons appear further upfield than expected. This situation arises for hydrogens attached to a carbon-carbon triple bond (a terminal alkyne), which absorb at δ 2.9. This ring current effect is shown in Figure 9.5b.

The deshielding effect of the aromatic ring currents extends beyond the hydrogens attached directly to the aromatic ring (the aromatic hydrogens). For example, the absorptions of protons attached to the methyl group in toluene, $C_6H_5CH_3$, are shifted downfield to δ 2.32 as a result of the ring currents. Compare this chemical shift with that of a —CH_3 group in alkanes, which occurs at δ 0.9. The hydrogens attached to the methyl group in toluene are called **benzyl** hydrogens. The chemical shifts of different types of benzyl hydrogens follow the same pattern as aliphatic hydrogens, with secondary benzyl hydrogens, $ArCH_2$—, being downfield from primary benzyl hydrogens, $ArCH_3$, and tertiary benzyl hydrogens, $ArCH$—, being downfield from secondary benzyl hydrogens.

9.4 Spin-Spin Splitting Patterns and Coupling Constants

The PMR spectrum of 1-nitropropane (Figure 9.4) shows three groups of peaks, whose chemical shifts have already been rationalized. Note that each *group* of peaks consists of a set of peaks. These are caused by what is called **spin-spin coupling or splitting.** The magnetic interactions between nonequivalent hydrogen nuclei in the molecule produce this phenomenon, which is observed in most organic molecules. The coupling occurs between a hydrogen on one carbon atom and hydrogen attached to an *adjacent* carbon atom, provided that the hydrogens are *nonequivalent*. Coupling does not occur between magnetically equivalent hydrogens and is seldom observed when protons are separated from one another by *more* than two atoms. In general, coupling occurs between protons that have different chemical shifts and are therefore nonequivalent, and this is the topic of this discussion.

The three types of hydrogens in 1-nitropropane can be labeled as shown below:

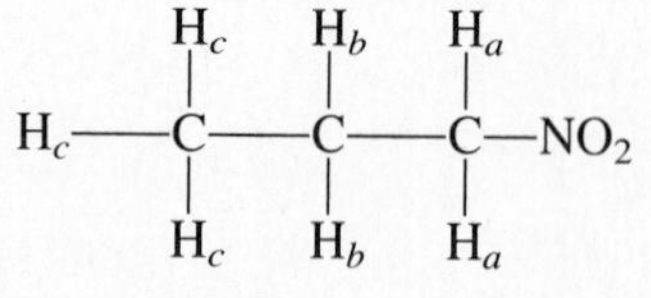

$$\begin{array}{cccc}
 & H_c & H_b & H_a \\
 & | & | & | \\
H_c\text{---} & C\text{---} & C\text{---} & C\text{---}NO_2 \\
 & | & | & | \\
 & H_c & H_b & H_a
\end{array}$$

We would expect the labeled protons to couple as follows: (1) H_a with H_b; (2) H_b with both H_a and H_c; and (3) H_c with H_b. Examination of the PMR spectrum reveals a different number of peaks for each type of hydrogen, which is called the **splitting pattern.** The terms "splitting" and "coupling" are often used interchangeably. For each type of hydrogen, there is a separation between the peaks, the magnitude of which is called the **coupling constant.** These topics are discussed below.

(a) Spin-Spin Splitting Patterns

The number of peaks into which the PMR absorption of a given type of hydrogen nucleus, H_i, is divided because of spin-spin coupling depends in a linear manner upon the number of protons coupling with H_i. The interaction occurs through the intervening bonds, and the strength of the interaction depends upon various factors, such as the stereochemical relationship between the atoms, the types of bonds that separate the atoms, and the number of adjacent atoms. About half of the neighboring protons have their nuclear spins aligned with the external field, and half are aligned against it. Thus, the magnetic field experienced by the absorbing hydrogens is influenced by the orientation of the nucleus of the hydrogen atoms on the neighboring carbon atom. This situation requires that slightly different amounts of energy be required for the transition between nuclear energy levels to take place, and these energy differences result in the observation of varying numbers of peaks for absorbing protons. The number of observed peaks depends on the number of adjacent hydrogens and can be predicted by considering all the possible combinations of spin states of those hydrogens. This concept, which will not be discussed further, can also be used to produce the relative heights of the observed peaks. The number of peaks that will be observed can be predicted as follows. If n protons, all of which have the same coupling constant with hydrogen H_i, are coupled with this nucleus, the number of peaks, N, into which a proton signal is split[5] can be calculated from the formula given in equation 7.

$$N = n + 1 \tag{7}$$

where N = number of peaks observed for absorbing
protons and
n = number of equivalent adjacent hydrogens
with the same coupling constant

If the coupling constants are not identical, the spectrum may be much more complicated, but a detailed discussion of such cases is not appropriate here.

Equation 7, often called the "$n + 1$ rule," is a good rule of thumb for relating the splitting pattern of a PMR signal to the number of adjacent hydrogens. *The number of neighboring hydrogens dictates the number of peaks into which an absorbing proton signal is split. Note should be made of the fact that "n" in equation 7 is NOT the number of absorbing protons.* The number of peaks observed for absorption by a proton or group of equivalent protons is called **multiplicity.** The terminology used to designate the number of peaks is simple. If there is one peak, it is called a **singlet,** two peaks are called a **doublet,** three peaks a **triplet,** four peaks a **quartet,** five peaks a **quintet,** six peaks a **sextet,** seven peaks a **septet,** eight peaks an **octet,** and so forth. Sometimes the peaks associated with absorbing protons are so complex that the number cannot be determined with accuracy, and in this case the PMR signal is called a **multiplet.**

The application of the $n + 1$ rule for predicting the number of peaks that will be observed in a PMR signal is illustrated by the following general examples.

[5]This statement is an oversimplification and is strictly true only in those cases in which the ratio of the chemical shift, in Hz, between coupled protons and their coupling constant is greater than 5 to 10. This point is further discussed in Section 9.7 in the interpretation of Figure 9.9. See also references 1 and 2 at the end of this chapter for detailed discussion of this point.

1. No Adjacent Hydrogens In the following general structures, assume that there are no adjacent hydrogens, so that $n = 0$ and $N = n + 1 = 0 + 1 = 1$. When the indicated protons absorb, they would appear as a singlet in each case.

$$R_3C-CH_3 \qquad R_3C-CH_2-CR_3 \qquad R_3C-\overset{\displaystyle |}{\underset{\displaystyle CR_3}{CH}}-CR_3$$

Even though a different number of hydrogens is contained in these molecules, each signal will appear as a singlet, although they have different chemical shifts.

2. One Adjacent Hydrogen Consider the following general structure where there is one hydrogen adjacent to the methyl group.

$$\underset{a}{R_2CH}-\underset{b}{CH_3}$$

The hydrogens labeled **a** and **b** absorb energy at different chemical shifts. Our interest is in how many peaks are observed for each type of proton. The signal of the $-CH_3$ group will appear as two peaks (a doublet), since there is one adjacent hydrogen. That is, $n = 1$, so that $N = n + 1 = 1 + 1 = 2$. The signal from the proton in the CH position marked **a** will appear as four peaks (a quartet) because there are three adjacent hydrogens (**b**). That is, $n = 3$, so that $N = n + 1 = 3 + 1 = 4$. Note again that the splitting pattern depends on the number of hydrogens attached to the carbon that is *adjacent* to the carbon bearing the protons providing the PMR signal.

3. Two Adjacent Hydrogens The following general structure contains two hydrogens adjacent to the methyl group, and the $n + 1$ rule allows us to predict the number of peaks that will be observed for both the CH_2 and the CH_3 groups.

$$\underset{a}{R-CH_2}-\underset{b}{CH_3}$$

When the CH_2 group (**a**) absorbs, the signal will appear as four peaks (a quartet), since there are three equivalent hydrogens (**b**) on the CH_3 group. Here, $n = 3$, so that $N = n + 1 = 3 + 1 = 4$. When the CH_3 group absorbs, the signal will contain three peaks (a triplet), since there are two equivalent hydrogens (**a**) on the CH_2 group. Verify this by application of the $n + 1$ rule.

4. More than Two Adjacent Hydrogens There are many examples of groups that contain more than two adjacent hydrogens. The following general formula is illustrative.

$$\underset{a}{R-CH_2}-\underset{b}{CH_2}-\underset{a}{CH_2}-R$$

Assume that the CH_2 groups marked **a** are chemically equivalent and have the same chemical shift and that the chemical shift of the groups marked **a** is different from that of the CH_2

group marked **b.** When the **a** groups absorb, three peaks (a triplet) will be observed, since these groups are equivalent and have two neighboring hydrogens (**b**). On the other hand, the signal from **b** will appear as five peaks (a quintet) since there is a *total* of four neighboring hydrogens—two on each of the CH_2 groups marked **a.** Verify that the $n + 1$ rule provides correct prediction for the splitting pattern in this example.

The key to predicting the splitting pattern depends on properly counting the *total* number of hydrogens, termed **vicinal hydrogens,** attached to all of the carbons adjacent to the carbon that bears the absorbing protons. Note again that the splitting pattern does not depend on the number of absorbing protons and that the $n + 1$ rule applies when the vicinal hydrogens are magnetically equivalent and have different chemical shifts from the absorbing proton(s).

Let us now apply the $n + 1$ rule to the PMR spectrum of 1-nitropropane, whose structure is redrawn for reference:

$$
\begin{array}{ccccccc}
 & & H_c & & H_b & & H_a \\
 & & | & & | & & | \\
H_c & \!\!-\!\! & C & \!\!-\!\! & C & \!\!-\!\! & C-NO_2 \\
 & & | & & | & & | \\
 & & H_c & & H_b & & H_a \\
\end{array}
$$

The signal from H_a protons should appear as a triplet, since they have two neighbors. The H_b protons have five neighboring hydrogens and should produce a sextet, *if* the coupling constant between H_a and H_b is equal to that between H_b and H_c. The H_c protons should yield a triplet, since they have two neighbors. The observed spectrum (Figure 9.4) confirms these predictions.

It should be apparent that the splitting pattern—the *multiplicity*—of the PMR absorption of a given type of proton can be utilized to determine the number of hydrogens on the carbon atom(s) adjacent to the carbon bearing the absorbing proton(s). This type of information is very useful for analyzing PMR spectra and ultimately for structure determination.

The $n + 1$ rule can be rearranged algebraically so that the number of neighboring protons, n, can be determined from the number of peaks, N. Simply rearrange equation 7 to give $n = N - 1$. For example, the appearance of a doublet in a PMR spectrum means that the absorbing protons have one neighboring hydrogen, a triplet indicates two adjacent hydrogens, and so forth. The additional PMR spectra presented later will illustrate this point.

Splitting patterns are not always easy to analyze. Peaks of two different groups of protons often overlap and add together. Other PMR techniques, such as spin decoupling, are available to aid in the analysis of spectra, but discussion of these methods is beyond the scope of this text. Instruments operating at higher frequencies, such as 200 MHz, 300 MHz, or 500 MHz, produce spectra with better resolution and peak separation. The references provided at the end of the chapter should be consulted for discussions about these other useful techniques.

(b) Spin-Spin Coupling Constants

The magnitude of the spin-spin splitting between hydrogen nuclei can be measured in Hz, and the values so obtained are called **coupling constants, *J*.** In contrast to chemical shifts, the *magnitude* of coupling constants is *independent* of the oscillator frequency, so units of Hz or cps are used with such values. Some typical hydrogen-hydrogen coupling constants, J_{HH}, are listed in Figure 9.6.

The previous discussion focused on coupling between hydrogens on adjacent carbons (*vicinal* hydrogens). However, the coupling constants in Figure 9.6 indicate that coupling can occur between hydrogens attached to the same carbon (*geminal* hydrogens). This phenomenon depends partly on the magnitude of the bond angle between the hydrogens and is observed only when the geminal protons are chemically nonequivalent.

The coupling constants observed in the PMR spectrum of 1-nitropropane are $J_{ab} = J_{bc} = 7$ Hz. These values are obtained by measuring the distance, in Hz, between the individual peaks that comprise each set of absorptions. These are indicated on the spectrum in Figure 9.4, but it is indeed difficult to determine the magnitude of these coupling constants from this figure because the original spectrum has been greatly reduced in size. Coupling constants are easier to measure on the original spectra as actually recorded, since they are much larger, typically $12'' \times 18''$.

Coupling constants provide valuable information regarding structure. If two sets of absorbing protons are on adjacent carbon atoms, the coupling constants *must* be identical because the proton on one carbon atom is coupling with that on the adjacent carbon, and vice versa. For example, the PMR spectrum of an ethyl group, CH_3CH_2—, consists of two sets of

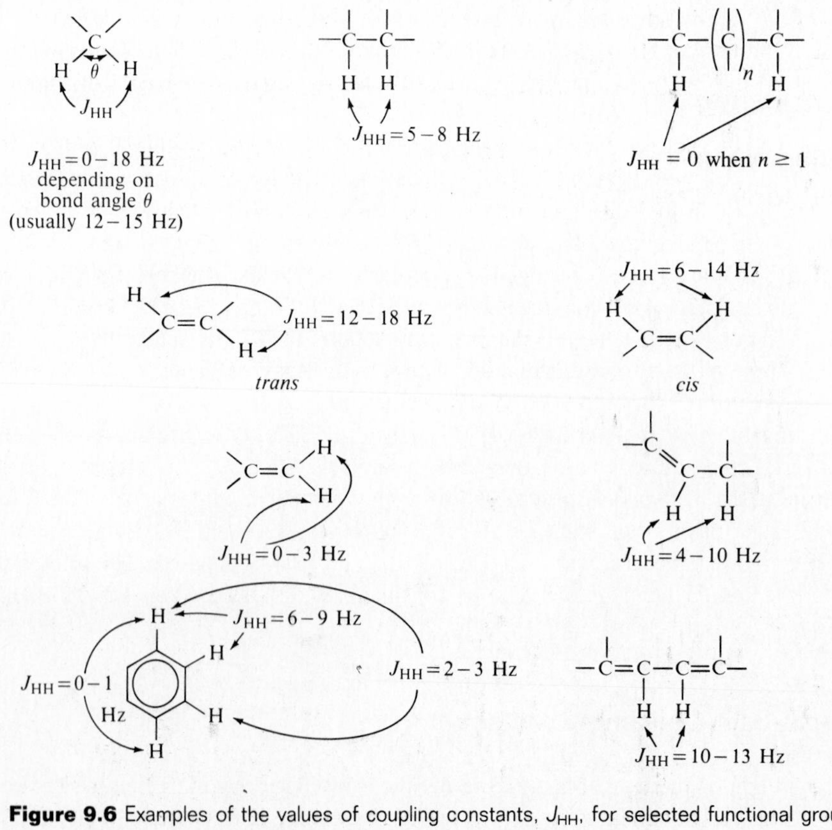

Figure 9.6 Examples of the values of coupling constants, J_{HH}, for selected functional groups.

peaks, one of which is a triplet and the other a quartet. The observed coupling constants in each set of peaks are identical and are approximately 7 Hz. On the other hand, suppose a PMR spectrum contained a triplet and a quartet where the coupling constants in the triplet were quite different from those in the quartet. This leads to the conclusion that the protons responsible for the triplet are *not* adjacent to the protons that cause the quartet. Coupling constants therefore provide additional information about the structure of a molecule.

9.5 Hydrogen Counting and Peak Area

The location of the absorption peaks is important in the analysis of IR spectra (Chapter 8), and little use is made of the intensity of the peaks other than to indicate which groups produce strong, medium, or weak absorption bands. In contrast, the areas of the peaks in a PMR spectrum can be used to determine the *relative* numbers of the different types of hydrogens in a molecule because the peak areas in a PMR spectrum are a linear function of the number of hydrogen nuclei producing the peaks. The areas can be determined by electronic integration. This is done by the same instrument that recorded the spectrum. Some NMR spectrometers are equipped with microprocessors that determine the peak areas and report them in printed form. Other spectrometers provide electronic integration as a second trace on the spectrum, as shown on the PMR spectrum of 1-nitropropane (Figure 9.4).

Experimentally, the PMR spectrum is recorded first, and then electronic integration is performed by readjusting the instrument and tracing the integration curves on the same spectrum. These are easily distinguished from the absorption peaks. The distance that the integration curve rises as it passes over a set of peaks is proportional to the area under the peaks. The height of each integration peak can be measured in millimeters or by counting vertically the number of squares on the spectrum chart. For 1-nitropropane, integration heights are obtained by measuring the distance between *a–b, a–c,* and *a–d;* for this spectrum, these heights are 28 mm for *a–b,* 28 mm for *a–c,* and 44 mm for *a–d.* These heights are directly proportional to the number of protons represented by the absorption peaks under them. The *relative* numbers of hydrogens are then computed by dividing the integration heights by the smaller or smallest height to produce small whole or fractional numbers. This computation is done below for 1-nitropropane:

Group A: 28/28 = 1.0
Group B: 28/28 = 1.0
Group C: 44/28 = 1.55

However, a molecule cannot contain fractional numbers of hydrogens, so the relative numbers obtained above are converted to whole numbers by multiplying *each* value by two:

Group A: 1.0   × 2 = 2.0
Group B: 1.0   × 2 = 2.0
Group C: 1.55 × 2 = 3.1

The accuracy of electronic integration is no better than 5–10%, and the relative abundance of the protons determined above (2.0:2.0:3.1) is within experimental error of the *absolute* ratio of 2:2:3 anticipated for 1-nitropropane.

Peak integration provides the "empirical formula" with respect to the hydrogens contained in a molecule. For example, one might obtain the ratio 1:2:2:3, which represents the *relative* numbers of protons. The actual molecule might contain this *absolute* number of protons or some other whole number ratio such as 2:4:4:6 or 3:6:6:9, both of which would be in the ratio of 1:2:2:3. Additional information, such as the molecular formula, must be available to deduce the structure represented by a PMR spectrum.

9.6 Procedure for Analyzing PMR Spectra

It should now be clear that the correct analysis of a PMR spectrum of an unknown compound can yield a wealth of information regarding its structure. Determination of the *chemical shift* of each peak or group of peaks allows one to speculate about the *type* of functional group to which the hydrogen nucleus producing the absorption is attached. The *spin-spin splitting pattern* and the *n + 1* rule provide information concerning the *number of nearest hydrogen neighbors* to the proton(s) producing a given absorption peak. *Integration* allows evaluation of the peak areas of each set of peaks and permits one to determine the *relative numbers of each type* of hydrogen present in the molecule.

In order to interpret a PMR spectrum, some additional type of information must be available about the compound whose spectrum is being analyzed. The molecular formula of the compound or the source of the compound (for example, its synthesis from a known compound) must be available. Even the most experienced spectroscopists could not deduce a complete and correct structure of a compound based *solely* upon a PMR spectrum unless some additional information were available.

The complete analysis of the PMR spectrum of an unknown compound is profitably carried out by the following sequence of steps.

1. Determine the relative numbers of the different types of hydrogens present by measuring the heights of the *integration* curve above each set of peaks. The heights can be measured with a ruler or by counting (vertically) the number of squares in the grid on the original spectrum. The spectra in this text have been reproduced without the grids, so a ruler must be used. The integration heights are then reduced to the simplest possible *whole number ratio* that represents the *relative numbers of protons in the molecule*. If the molecular formula of the unknown is available, the *relative* ratios can be converted into *absolute* numbers of the hydrogens present.

   The discussion above applies to the analysis of spectra that may be provided to you or that you may have run. *The relative abundances of each type of hydrogen have been provided above the peaks on most of the spectra contained in this text.*

2. Attempt to deduce the functional groups present in the molecule by examining the *chemical shifts* of the different sets of peaks. This can be done more easily if the molecular formula of the compound is available. Alternatively, the organic chemist might attempt to convert one compound into a new one and then determine the PMR spectrum of the product. Ideally, a tentative "guess" should be made about the type of hydrogens repre-

sented by *each* set of absorption peaks and thus the functional groups in which the hydrogens are present or to which they are adjacent.

3. Analyze the *spin-spin splitting* patterns in order to determine the molecular environments of each type of hydrogen. Correct application of the *n + 1* rule permits one to determine the number of nearest hydrogen neighbors.

After completing the above steps, it should be possible to identify the structure of the unknown compound. However, the structure assignment must be consistent with all of the information available from the spectrum: the chemical shifts, the number of hydrogens represented by each set of peaks as determined from the peak areas, and the splitting pattern. If the structure does not correctly explain *each* fact, then it is incorrect. A final check of this type helps ensure that the compound has been identified completely and correctly. In reality, sometimes it is possible to make a partial structural assignment that is consistent with the spectrum. This situation may arise if a spectrum contains many overlapping peaks so that the splitting patterns and peak areas for *each* type of hydrogen cannot be analyzed accurately.

The following example suggests the type of analysis involved in a partial structural assignment. Suppose a PMR spectrum contains a doublet at δ 0.90 with a peak area equivalent to 6 protons. The appearance of a doublet implies that there is *one* neighboring hydrogen, and therefore the doublet is probably due to *two* —CH_3 groups attached to a carbon bearing one hydrogen. This information strongly suggests the presence of an isopropyl group, —$CH(CH_3)_2$, in the molecule. Determination of the *number* of hydrogens represented by the other signals and analysis of the rest of the spectrum must support this assignment.

If an infrared spectrum of the unknown compound is also available, it can be analyzed as described in Chapter 8. This analysis should provide additional information concerning the functional groups present, and such information may be used to supplement or to confirm the structure derived from analysis of the PMR spectrum.

It is usually impossible to determine the presence of functional groups from PMR spectra, although the presence of some groups, such as —CO_2H, —CHO, ArOH, ROH, ArH, and carbon-carbon double bonds containing one or more hydrogens, can often be deduced in a PMR spectrum. Structural assignments based *solely* on interpretation of spectra should be confirmed by comparing the spectra of the unknown with those of the known compound, which has been synthesized in an unequivocal fashion. The structure could also be confirmed by using classical chemical methods to convert the unknown into a solid derivative, which can be identified on the basis of its melting point, as described in Chapter 27.

9.7 Examples of Structure Identification Using PMR

Additional examples of structure identification using PMR are illustrated in this section, in which the molecular formula of the compounds and the corresponding PMR spectra are given. These examples illustrate the approach outlined in Section 9.6 and involve reasonably simple compounds whose PMR spectra contain well-separated peaks. However, most compounds yield considerably more complex spectra, often with overlapping peaks, and their analysis is frequently more challenging and complex.

Example 1

Problem   Determine the structure of the compound having the molecular formula C_4H_9Cl whose PMR spectrum is given in Figure 9.7.

Solution

(a) Peak Areas: Hydrogen Counting   The PMR spectrum exhibits three absorption bands: a doublet centered at δ 1.04, a multiplet centered at δ 1.95, and a doublet centered at δ 3.35. Measurement of the height (in mm) of the integration peaks above each absorption band gives 34:5.5:11, respectively. Dividing each of the peaks by 5.5 yields 6:1:2. While the peak area measurements provide the *relative* number of protons represented by each peak, they must give the actual number of each type of proton for C_4H_9Cl because the molecular formula is given.

(b) Chemical Shift Identification   The doublet centered at δ 3.35 represents the protons attached to the carbon bearing the electron-withdrawing chlorine atom. The multiplet at δ 1.95 probably represents a methine hydrogen and is apparently shifted downfield slightly from the normal location of such a hydrogen peak owing to the presence of chlorine in the molecule. The doublet centered at δ 1.04 suggests the presence of methyl groups in the molecule because they typically absorb at this δ-value.

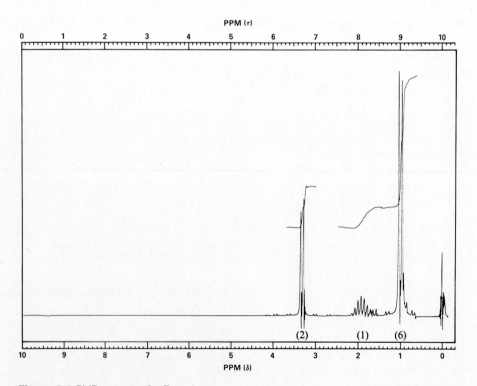

Figure 9.7 PMR spectrum for Example 1.

(c) Splitting Pattern The doublet at δ 3.35 represents two protons, and application of the $N = n + 1$ rule indicates they must have one neighboring hydrogen. The multiplet at δ 1.95 represents one proton; it is difficult to determine the number of neighboring protons because of the complexity of the signal. The doublet at δ 1.04 represents six protons and must be due to two methyl groups that have the same hydrogen as the vicinal neighbor.

(d) Determination of Structure From the information gained above, we can conclude that the molecule contains the following structural features:

    one CH_2—Cl group having one neighboring hydrogen
    two CH_3— groups having one neighboring hydrogen

At this point we have accounted for three carbon atoms, eight hydrogens, and one chlorine atom. Hence, we have only to account for one carbon and one hydrogen. Keeping in mind that the CH_2Cl group and the two methyl groups each have one neighboring hydrogen because of the splitting patterns, the following complete structure can be drawn:

$$
\begin{array}{c}
CH_3 \\
| \\
CH_3-CH-CH_2-Cl
\end{array}
$$

1-Chloro-2-methylpropane
(isobutyl chloride)

As a final check, the proposed structure should account for the observed chemical shifts, peak areas, and splitting pattern, which in fact it does.

Example 2

Problem The PMR spectrum given in Figure 9.8 was obtained for a compound having the molecular formula $C_{11}H_{16}$. On the basis of this information, deduce the structure of this compound.

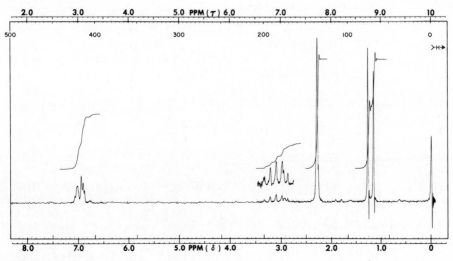

Figure 9.8 PMR spectrum for Example 2.

Solution

(a) Peak Areas: Hydrogen Counting The PMR spectrum exhibits four absorption bands: a multiplet centered at δ 7.0, a multiplet centered at δ 3.1, a singlet at δ 2.3, and a doublet centered at δ 1.2. Measurement of the height (in mm) of the integration peaks above each absorption band gives 15:5:30:30, respectively. Dividing each of the integration peak heights by 5 yields 3:1:6:6. While the peak-area measurements provide an indication of the *relative* number of protons represented by each peak, they must give the actual number of each type of proton for the molecule in question, since the molecular formula given contains 16 hydrogens because $3 + 1 + 6 + 6 = 16$.

(b) Chemical Shift Identification The peak centered at δ 7.0 probably represents hydrogen attached to an aromatic ring, as this type of hydrogen normally produces absorptions in this region. The multiplet centered at δ 3.1 is likely to result from benzylic hydrogens, which must have a number of adjacent protons. It should be pointed out that this absorption band consists of several small, ill-defined peaks on the baseline; the peaks shown above the baseline peaks represent an enlargement of the peaks contained on the baseline, thus providing a clearer indication of the number of peaks represented by this absorption. The singlet at δ 2.3 is also probably due to the presence of benzylic hydrogens. The doublet centered at δ 1.2 is likely to represent aliphatic hydrogens. These are tentative assignments, which are confirmed in the following paragraphs.

(c) Splitting Pattern The multiplet at δ 7.0 is complex. It is often difficult to determine the number of peaks contained in some aromatic proton absorptions. The multiplet centered at δ 3.1 appears to consist of five peaks, and application of the $N = n + 1$ rule suggests that the proton has 4 neighbors. However, absorptions of this complexity *usually* contain *more* than five peaks, some of which are indistinguishable. The singlet at δ 2.3 represents six hydrogens, none of which have any neighbors. The doublet at δ 1.2 also represents six hydrogens, which must have one neighboring hydrogen.

(d) Determination of Structure From the integration count, there are three aromatic hydrogens, and because the aromatic ring contains six carbon atoms, this accounts for six carbons and three hydrogens in the original molecule, $C_{11}H_{16}$, and leaves five carbons $(11 - 6)$ and 13 hydrogens $(16 - 3)$ in the substituents attached to the ring. The singlet at δ 2.3 represents six hydrogens that have no neighboring protons, and this peak must therefore result from the presence of *two* methyl groups, —CH_3, which are probably attached to the ring. We have now accounted for eight carbons and nine hydrogens in the molecule. The doublet at δ 1.2 represents six hydrogens that have one neighboring proton, so this absorption must be due to *two* aliphatic methyl groups attached to a carbon bearing one hydrogen. This leaves one carbon and one hydrogen unaccounted for, and we note that the multiplet at δ 3.1 represents one hydrogen that is likely to be benzylic because of the chemical shift.

The information gained thus far indicates that the original molecule contains these features:

one aromatic ring containing three substituents
two CH_3 groups attached to the ring
one CH group attached to the ring
two aliphatic CH_3 groups attached to a carbon bearing one hydrogen

The last two features are indicative of an isopropyl or 2-propyl group, $-CH(CH_3)_2$, which is attached to the ring. When put together, these facts indicate that one possible structure for the molecule might be the following:

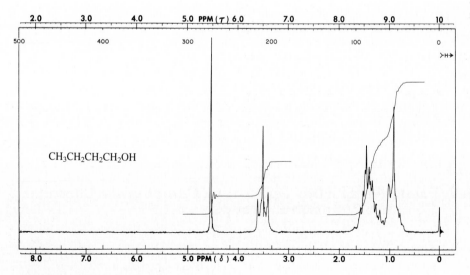

The PMR spectrum does not permit a clear-cut decision to be made regarding the exact location of these three substituents on the ring. For the six hydrogens of the two CH_3 groups to be *identical* and thus give one signal, the structure would have to be 1,3,5-substituted. However, the 1,2,4-orientation might not make the two CH_3 groups very different, probably not enough so to be seen in the PMR spectrum. In actuality, the PMR spectrum represented in Figure 9.8 is of the compound whose structure is shown above. It should be pointed out that infrared spectroscopy is frequently useful in determining the location of substituents on an aromatic ring.

Example 3

The PMR spectrum of 1-butanol, $CH_3CH_2CH_2CH_2OH$, is reproduced in Figure 9.9 and is presented as an example in which it is not so simple to integrate signals separately or to apply the *n + 1* rule, which worked so beautifully in the previous examples.

Figure 9.9 PMR spectrum of 1-butanol for Example 3.

Spectra in which the multiplicities of the various hydrogens can successfully be predicted by the $n + 1$ rule are commonly referred to as "first order," and the PMR spectra presented thus far fall in this category. The splitting patterns recorded in many of the spectra contained in this book, however, will be much more complex and essentially uninterpretable according to the rule. In many cases this will be the consequence of the presence in a compound of a variety of functionally similar hydrogens having very similar magnetic environments and therefore similar chemical shifts. This situation is exemplified by the methylene hydrogens of 1-butanol (Figure 9.9). Consideration of substituent effects allows assignment of the triplet centered at δ 3.5 to the methylene hydrogens at C-1 of the alcohol. However, it clearly is not possible to make separate assignments of the various bands in the region δ 1.1–1.7 to the two pairs of skeletally distinct methylene hydrogens at C-2 and C-3 responsible for these resonances. Rather, one simply assigns the multiplet centered at about δ 1.4 as representing both of the pairs at C-2 and C-3 and does not attempt to differentiate them.

Another complicating factor in applying the $n + 1$ rule is illustrated in Figure 9.9. The high-field resonance centered at about δ 0.9 can be assigned to the methyl group with confidence on the basis of the integration of this band and the anticipated chemical shift or methyl hydrogens. Nevertheless, the methyl group is not the uncomplicated triplet that would have been expected from application of the $n + 1$ rule. The theoretical basis for this departure from expectation is beyond the scope of this discussion, but in essence the failure of the rule to predict multiplicities arises whenever the difference in the chemical shift between coupled hydrogens is similar to, or not much larger in magnitude than, the value of the coupling constant between those hydrogens; in other words, if $\Delta\nu/J$ ($\Delta\nu$ is the difference in Hz between the chemical shifts of coupled hydrogens) is less than 5 to 10, that portion of the PMR spectrum will no longer be first order and have multiplicities predictable from the $n + 1$ rule. In this case of 1-butanol, the ratio for the methyl group and the methylene hydrogens with which it is coupled is about 5.

The remaining signal to be accounted for is the broad singlet at δ 4.6, which is due to the presence of the hydroxylic proton, that is, the proton attached to the OH group. It may seem strange that this proton appears as a singlet, albeit a broad singlet, since one would expect it to be coupled to the adjacent methylene hydrogens. The explanation for this phenomenon is presented in Section 9.8g.

Several additional solved examples involving the interpretation of PMR spectra are provided in Section 27.4. Numerous spectra are provided throughout this text, and it is recommended that as many as possible of them be examined. The following section provides PMR spectra of compounds containing many of the common functional groups.

9.8 Examples of PMR Spectra of Compounds Containing Common Functional Groups

The ability to interpret PMR spectra comes with experience in examining numerous spectra, and the general procedures for doing so are outlined in the preceding section. The PMR spectra of compounds containing most of the functional groups commonly encountered in organic chemistry are given in this section and interpreted with respect to *chemical shift*. These spectra are collected together to provide the student with the opportunity of comparing the absorption peaks that are characteristic of different types of organic molecules. This type of comparison is particularly useful in helping to solve "unknowns" in qualitative organic analysis (Chapter 27). Although the absorptions of many functional groups may not appear *directly* in a PMR spectrum, different groups usually have a profound effect upon the chemical shift of protons that may be attached to them. Hence, PMR spectroscopy provides considerable information about the structure of a molecule.

(a) Alkanes

The chemical shifts observed for alkanes depend on the *type* of hydrogen present in the molecule. In the absence of other nearby substituents, absorption occurs at these approximate values:

Primary	RCH_3	δ 0.9
Secondary	R_2CH_2	δ 1.3
Tertiary	R_3CH	δ 1.5

The PMR spectrum of pentane is shown in Figure 9.10. It shows a distorted triplet at δ 0.89 due to the terminal methyl groups. The integration curve shows an inflection point near δ 1.0, which divides the integration into two nearly equal parts. Six hydrogen absorptions are downfield from δ 1.0, and six are upfield from it. Alkanes are often difficult to analyze by PMR spectroscopy because the signals of similar types of protons overlap and produce complex, broad multiplets.

(b) Cycloalkanes

Most cycloalkanes exhibit the same chemical shift trends as alkanes, as illustrated by the PMR spectrum of methylcyclopentane (Figure 9.11). The nine ring hydrogens give the broad and poorly resolved peaks between δ 1.2 and 2.0, although the broad shoulder at δ 1.8 is due to the methine hydrogen (H_c). The methyl group appears as a doublet at δ 1.0, due to coupling with H_c with J_{H_a}-J_{H_c} = 5.6 Hz.

Cyclopropane is an exception, and its hydrogens absorb upfield from other aliphatic protons at δ 0.2. This unusually high upfield absorption is due to the anisotropy of the carbon-carbon bonds in the molecule, which causes the ring hydrogens to be *shielded*.

(c) Alkenes

The presence of a carbon-carbon double bond in a molecule affects the chemical shift of the hydrogens attached directly to the sp^2-hybridized carbon atoms (**vinylic hydrogens**) and also

Text continues on p. 263

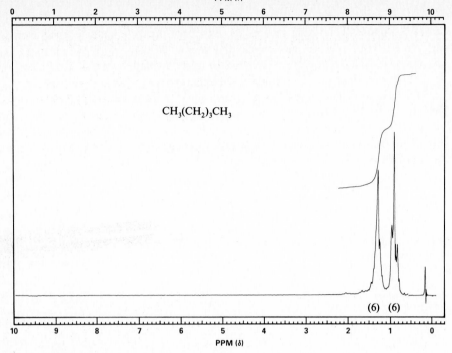

CH₃(CH₂)₃CH₃

(6)　(6)

Figure 9.10 PMR spectrum of pentane.

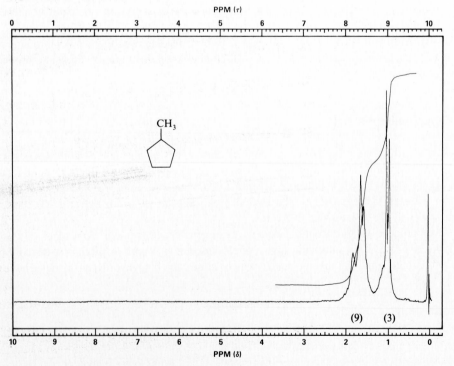

(9)　(3)

Figure 9.11 PMR spectrum of methylcyclopentane.

that of the hydrogens that may be present on an sp^3-hybridized carbon atom attached to the double bond **(allylic hydrogens).** These proton absorptions occur in the following ranges:

$$Vinylic \quad \text{C=C} \quad \delta\ \textbf{4.6–5.9}$$

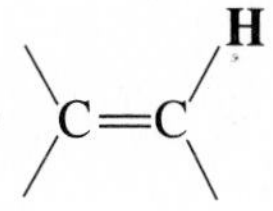

$$Allylic \quad \text{C=C} \quad \delta\ \textbf{1.7–2.5}$$

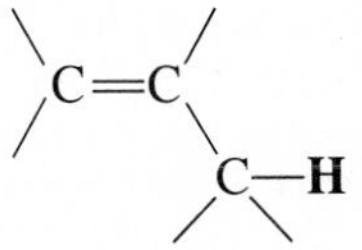

These downfield shifts are due to the presence of the π-cloud in the double bond, which deshields the vinylic hydrogens to a great extent and the allylic hydrogens to a lesser extent. These chemical shift values should be compared with those observed for alkynes (part d below). In comparing the shifts of different types of allylic protons, the following trends are observed:

3° allylic   2° allylic   1° allylic

←   **Downfield shift**

Note that this trend is the same as that observed for alkanes, but that the chemical shift range for allylic hydrogens is downfield from that of the hydrogens of alkanes. Other factors, such as the magnetic nonequivalence of vinylic protons, are discussed below.

In the PMR spectrum of 2-methyl-1-butene (Figure 9.12), the absorptions of the vinylic hydrogens, H_d, appear at δ 4.62. These protons have about the same chemical shift, but they are not magnetically equivalent because one of them is *cis* to a CH_3 group and the other is *cis* to a CH_2 group. Note that the δ 4.62 peak is not a sharp singlet but is somewhat broadened and contains several small spikes on the right-hand side of the peak. The signal of the hydrogens, H_b, on the methyl group attached to the double bond appears as a singlet at δ 1.7 because these hydrogens are deshielded, and thus the signal is shifted downfield from that of the other methyl group protons, H_a, in the molecule; the latter absorb at the expected chemical shift for a CH_3 group in an alkane. The distorted quartet centered near δ 2.0 is due to the methylene protons, H_c, and this signal is likewise shifted downfield from the δ 1.3 position of a CH_2 hydrogen signal in an alkane. The signals due to H_b and H_c are broadened due to long-range coupling, a topic that is beyond the scope of this text.

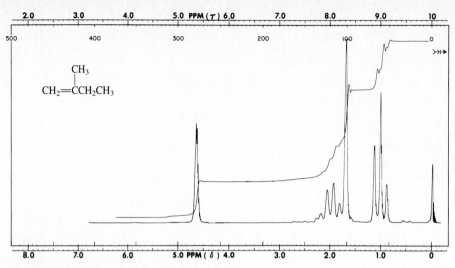

Figure 9.12 PMR spectrum of 2-methyl-1-butene.

(d) Alkynes

The signal from a hydrogen attached to a carbon-carbon triple bond in a *terminal* alkyne is shifted downfield as a result of deshielding by ring currents (Section 9.3a). As in alkenes, the signals of hydrogens that may be attached to carbon adjacent to the triple bond are also shifted downfield. The following absorption ranges are typically observed:

$$-C{\equiv}C-H \qquad \delta\ 2\text{--}3$$

$$-C{\equiv}C-\overset{|}{\underset{|}{C}}-H \quad \delta\ 1.8\text{--}2.5$$

The PMR spectrum of 2-methyl-3-butyn-2-ol (Figure 9.13) illustrates the chemical shifts for a terminal alkyne. In this spectrum, the absorption due to the CH_3 groups is shifted downfield to δ 1.5 owing to the presence of the electron-withdrawing oxygen atom on the carbon bearing them. Chemical shifts for alcohols will be discussed below. The acetylenic proton is not always observed as a singlet but is split as a result of long-range coupling when an alkyne has the following structural feature:

$$-\overset{|}{\underset{|}{C}}-C{\equiv}C-H$$
$$\phantom{-\overset{|}{\underset{|}{C}}}H$$

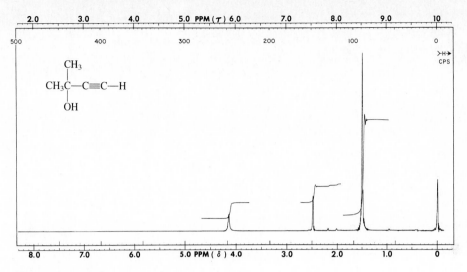

Figure 9.13 PMR spectrum of 2-methyl-3-butyn-2-ol.

(e) Aromatic Hydrocarbons

Hydrogens attached to an aromatic ring (*aromatic hydrogens*) have their absorptions shifted far downfield as a consequence of the ring currents associated with the π-cloud of the aromatic ring system (Section 9.3b). Additionally, the aromatic ring affects the chemical shifts of alpha-hydrogens (*benzylic hydrogens*) that may be present in aliphatic side-chains attached to the ring, owing to the fact that they are close enough to the ring to feel some effect of the π-cloud. Benzylic hydrogen signals are shifted downfield slightly compared with those of hydrogens contained in alkanes. The following summarizes these characteristic shifts:

$$\textit{Aromatic hydrogens} \quad \text{Ar—H} \qquad \delta\ 6\text{–}8.5$$

$$\textit{Benzylic hydrogens} \quad \text{Ar—}\overset{|}{\underset{|}{\text{C}}}\text{—H} \quad \delta\ 2.2\text{–}3$$

In later discussions about the chemical shifts of aromatic hydrogens present in different types of aromatic compounds, mention will be made of upfield and downfield shifts, and this will be in reference to shifts from the parent compound, benzene, which absorbs at δ 7.23. The chemical shift of the benzylic hydrogens depends on their type, with the following trends being observed.

$$\text{3° benzylic} \qquad \text{2° benzylic} \qquad \text{1° benzylic}$$
$$\longleftarrow \text{\textbf{Downfield shift}}$$

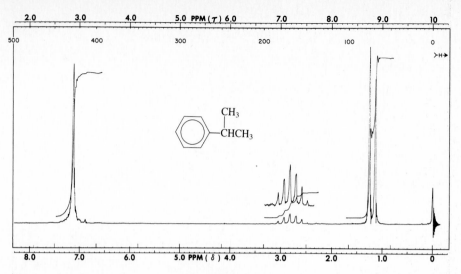

Figure 9.14 PMR spectrum of isopropylbenzene (cumene).

The PMR spectrum of isopropylbenzene (Figure 9.14) shows the signal of the hydrogens attached to the aromatic ring, which appears as a singlet at δ 7.2. In this case the *ortho, meta,* and *para* protons have nearly the same chemical shift, and this results in the observation of a singlet. However, this is not true for most aromatic compounds, a feature that is discussed in later parts of this section. The multiplet at δ 2.8 represents one proton and is due to the methine hydrogen in the isopropyl group; note that it is shifted downfield from the normal location of δ 1.5 for R_3CH in alkanes. This multiplet should appear as a septet because of the six equivalent neighboring protons, and it may be seen as such if one looks carefully and notices the *very* small peak on the left side of the enlarged portion of this absorption. The doublet centered at δ 1.2 represents six protons and thus the two methyl groups that are shifted downfield very slightly because of the ring currents associated with the aromatic system.

(f) Alkyl Halides

The presence of a halogen in an organic molecule causes the absorption of hydrogens attached to the carbon bearing the halogen to be shifted downfield owing to the electron-withdrawing effects of the halogen. The chemical shifts due to the presence of the different halogens are given below.

$$\textit{Alkyl fluorides} \quad \text{F——C——H} \quad \delta\ \textbf{4–4.5}$$

Alkyl chlorides $\mathrm{Cl{-}\overset{|}{\underset{|}{C}}{-}H}$ δ **3–4**

Alkyl bromides $\mathrm{Br{-}\overset{|}{\underset{|}{C}}{-}H}$ δ **2.5–4**

Alkyl iodides $\mathrm{I{-}\overset{|}{\underset{|}{C}}{-}H}$ δ **2–4**

In comparing different classes of protons attached to the carbon bearing the halogen, tertiary hydrogens appear farthest downfield, followed by secondary and then primary. If a proton is attached to a carbon atom that is once removed from the carbon bearing the halogen, that proton signal is shifted downfield *slightly* because of the inductive effect of the electronegative atom. The following examples illustrate these effects in alkyl chlorides.

CH_3Cl, δ **3.0** $\mathrm{CH_3{-}\overset{|}{\underset{|}{C}}{-}Cl}$, δ **1.5**

RCH_2Cl, δ **3.4** $\mathrm{R{-}CH_2{-}\overset{|}{\underset{|}{C}}{-}Cl}$, δ **1.7**

R_2CHCl, δ **4.0** $\mathrm{R_2CH{-}\overset{|}{\underset{|}{C}}{-}Cl}$, δ **1.6**

Increasing the number of halogens attached to a carbon bearing hydrogen also increases the extent of the downfield chemical shift of the proton, as illustrated below.

CH_3Cl, δ 3.0
CH_2Cl_2, δ 5.28
$CHCl_3$, δ 7.24

The PMR spectrum of 1-chloro-2-methylpropane (isobutyl chloride), given in Figure 9.7, has already been discussed and serves an an example of an alkyl halide.

(g) Alcohols

The electron-withdrawing oxygen atom in the OH group of an alcohol affects the chemical shifts of both the proton attached to oxygen and the hydrogen attached to the carbon atom that bears the OH group. As given below, downfield shifts are observed for both types of hydrogens.

$$-\overset{|}{\underset{|}{C}}-O-H \qquad \delta\ 1\text{--}5.5$$

$$H-\overset{|}{\underset{|}{C}}OH \qquad \delta\ 3.3\text{--}4$$

The chemical shift of the proton attached to oxygen is variable, and the value is highly dependent upon the concentration of the alcohol in the sample whose PMR spectrum is being determined. This concentration effect changes the extent of hydrogen-bonding between alcohol molecules and thus the chemical shift. At first glance, one might expect that the hydrogen attached to the OH group would cause spin-spin splitting with protons attached to the carbon bearing this group. This is rarely observed, however, because the weakly acidic alcohol hydrogen can undergo rapid exchange with the traces of water that are usually present: $RCH_2OH + H_2O \rightleftarrows RCH_2O^- + H_3O^+$. At one instant, the CH_2 group attached to oxygen "sees" the —OH proton, but the next instant it does not because the proton is gone. On the PMR time scale, this exchange is an equilibrium process that is too fast to be recorded by the instrument, and thus no splitting is observed. On the other hand, PMR spectra obtained from *pure, dry* alcohols indicate that the rate of exchange has been slowed down considerably, and coupling *may* be observed under these conditions. The PMR spectra presented in this text do *not* show this coupling, so that the OH proton signals appear as a single peak that is often broadened due to the type of exchange mentioned.

There is an easy experimental technique that can be used to ascertain which peak in a PMR spectrum is due to the proton on an OH group. The PMR spectrum of the alcohol is obtained, and then a small amount of deuterium oxide, D_2O, is added to the sample tube, which is shaken vigorously. This technique has the effect of replacing the acidic protons with deuterium, whereupon the equilibrium $ROH + D_2O \rightleftarrows ROD + HOD$ is established. The PMR spectrum is then obtained again, and because deuterium does not absorb in the PMR region, the peak that has disappeared is known to be due to the OH group. A new peak due to HOD appears at a different chemical shift. This same technique can be applied to determining which peak is due to the acidic hydrogen in a phenol, ArOH, or a carboxylic acid, RCO_2H.

The PMR spectrum of cyclohexanol (Figure 9.15) exhibits three major absorption regions: a singlet at δ 4.6 (one proton), a broad multiplet centered at δ 3.5 (one hydrogen) and a broad, complex peak between δ 1 and 2 (ten hydrogens). The δ 4.6 peak is due to the hydrogen attached to oxygen in the OH group, and the δ 3.5 peak results from the hydrogen attached to the carbon atom that bears the OH group. This peak is broad because of the equilibrium that exists between chair conformations in many cyclohexyl compounds; depending on the conformation, the absorbing hydrogen is either equatorial or axial, and it has

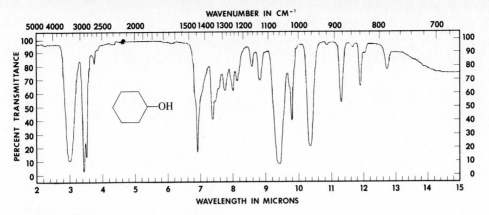

Figure 9.15 PMR spectrum of cyclohexanol.

been shown that equatorial protons absorb downfield from axial ones. The broad, complex absorption between δ 1 and 2 is due to the remainder of the hydrogens attached to the cyclohexane ring, and the two humps in this absorption are due to the equatorial and axial hydrogens of the cyclohexyl ring system.

(h) Ethers

The electron-withdrawing oxygen atom causes the absorption of the hydrogens that are attached to the adjacent carbon atom to be shifted downfield in a manner analogous to the effect in alcohols.

$$\text{H—C—O—R}\quad \delta\ \textbf{3.3–4}$$

The PMR spectrum of ethyl phenyl ether (phenetole), shown in Figure 9.16, provides an interesting example of how oxygen affects the chemical shift of the ethyl group and also the hydrogens attached to the aromatic ring. The quartet at δ 3.9 (two hydrogens) is shifted downfield considerably owing to the presence of the oxygen, and the triplet at δ 1.32 represents the methyl group, whose absorption appears slightly downfield as a result of the long-range electron-withdrawing effect of the oxygen. This spectrum shows the triplet-quartet splitting pattern typical of an ethyl group. Note should be made of the chemical shifts observed for the aromatic protons; the oxygen attached to the ring causes the *ortho* and *para* aromatic hydrogens to be more strongly shielded than the *meta* protons. The former signal, centered at δ 6.8, integrates to three hydrogens, whereas the signal from the *meta* hydrogens appears at about δ 7.15 as a distorted multiplet that integrates to two hydrogens. Recall that the aromatic hydrogens in benzene itself appear at δ 7.23, so that all the ring protons in ethyl phenyl ether are shifted upfield compared with those of benzene. This occurs because oxygen denotes electrons to the ring by *resonance,* and in general, electron-donating substituents on a ring cause an upfield shift, whereas electron-withdrawing groups cause a downfield shift.

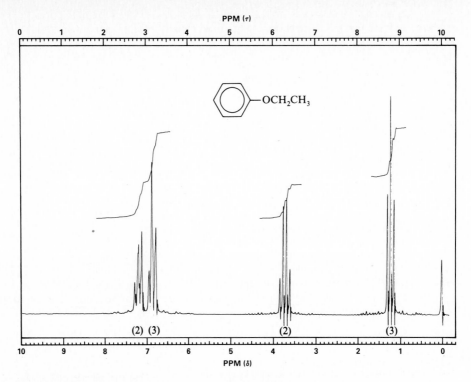

Figure 9.16 PMR spectrum of ethyl phenyl ether (phenetole).

Note that the aromatic hydrogen signals appear as a complex multiplet. It is often difficult to apply the $N = n + 1$ rule to the PMR signals of aromatic hydrogens.

(i) Aryl Halides and Nitro Compounds

Halides, —X, and nitro groups, —NO$_2$, are electron-withdrawing groups, and when attached to an aromatic ring, they cause the aromatic protons to be shifted downfield. The PMR spectrum of 4-nitrobromobenzene (Figure 9.17) illustrates the electron-withdrawing effects of these two substituents, with the doublet at δ 8.1 representing the two hydrogens *ortho* to the nitro group and the doublet at δ 7.7 the two protons *ortho* to bromine. This effect results from the fact that the nitro group is more strongly electron-withdrawing than bromine. Note that all of the aromatic hydrogen signals are shifted downfield from benzene. While it is often difficult to see well-defined splitting patterns for *ortho* and *meta* disubstituted aromatic compounds, the *para* isomers typically produce a doublet of doublets, such as that seen in Figure 9.17.

(j) Phenols

The OH group on phenols causes several interesting chemical shifts to be observed. Phenols

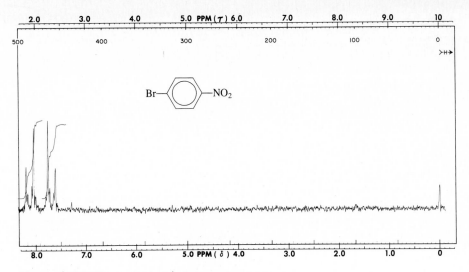

Figure 9.17 PMR spectrum of 4-nitrobromobenzene.

are weak acids but are more acidic than alcohols. As a result, the following chemical shift is observed:

$$\text{ArO—H} \quad \delta \; 4\text{–}12$$

This shift is variable; its value depends on the concentration of the phenol in solution and on the presence of electron-withdrawing or electron-donating substituents attached to the ring. As in alcohols, the peak due to the phenolic hydrogen can be identified by running the spectrum of the compound and then adding D_2O, which replaces the hydrogen with deuterium. When the spectrum is rerun, the peak that has disappeared is known to represent the OH group.

The PMR spectrum of phenol (Figure 9.18) illustrates the chemical shift of the OH group as a sharp singlet at δ 6.1 (one hydrogen) and the aromatic proton signals that appear as a complex set of peaks between δ 6.8 and 7.3. These hydrogens absorb at higher field than benzene (δ 7.23) because the oxygen attached to the ring strongly shields the *ortho* and *para* hydrogens, owing to electron-donation by resonance.

(k) Aldehydes

The presence of the carbon-oxygen double bond in an aldehyde produces two significant chemical shifts, as follows:

$$\text{R—C—H} \quad \delta \; 9\text{–}10$$

$$\text{H—C—C—H} \quad \delta \; 2\text{–}2.7$$

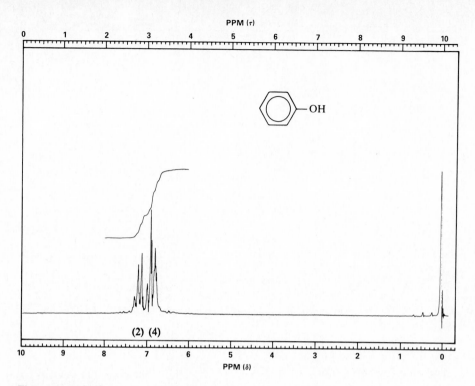

Figure 9.18 PMR spectrum of phenol.

The electron-withdrawing oxygen atom causes the aldehydic proton to be strongly deshielded, and its signal is shifted far downfield. The aldehyde group also causes the signals of the hydrogens attached to the carbon adjacent to the aldehyde group to be shifted downfield.

The PMR spectrum of butanal (butyraldehyde), shown in Figure 9.19, illustrates the type of chemical shifts typically observed for aldehydes. The absorption of the aldehydic hydrogen, H_d, appears as a triplet at δ 9.74, owing to the fact that it is split by the two hydrogens, H_c. The coupling constant is small, but the triplet is easily seen. The inset is used to show the peak of the aldehydic hydrogen. This method is frequently used to display absorptions that appear outside the normal spectrum format. The appearance of a downfield peak between δ 9 and 10 provides an *indication* of an aldehyde, although protons of carboxylic acids can also appear in this region. The H_c proton signals at δ 2.5 are shifted downfield as a result of these hydrogens being adjacent to the C=O, and they are split by H_b and H_d to give a complex triplet. The signals of the hydrogens at δ 1.8 (H_b) are shifted downfield slightly because of the inductive effect of the carbonyl group, and they appear as a distorted sextet as a result of the five neighboring hydrogens. The proton signals at δ 1.0 (H_a) appear as a triplet (two neighbors) at the expected chemical shift.

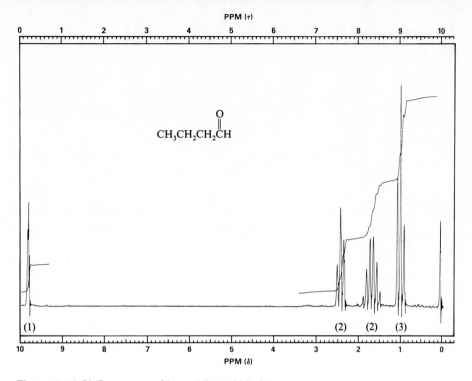

Figure 9.19 PMR spectrum of butanal (butyraldehyde).

(I) Ketones

As in aldehydes, the absorptions of hydrogens adjacent to the carbonyl group in ketones are shifted downfield.

$$R—\overset{\overset{\displaystyle O}{\|}}{C}—\overset{|}{\underset{|}{C}}—H \quad \delta\ 2\text{–}2.7$$

The PMR spectrum of 2-butanone (methyl ethyl ketone), shown in Figure 9.20, illustrates the downfield chemical shifts expected for a ketone. The signal of the three hydrogens on the methyl group adjacent to the carbonyl group, H_b, appears at δ 2.2 as a singlet because these hydrogens have no vicinal neighbors. The quartet at δ 2.5 results from the methylene hydrogens, H_c, which are split only by the hydrogens of the methyl group, H_a; the signal of the latter appears at δ 1.0 as a triplet owing to the two neighboring hydrogens. The $N = n + 1$ rule is obeyed in these splitting patterns.

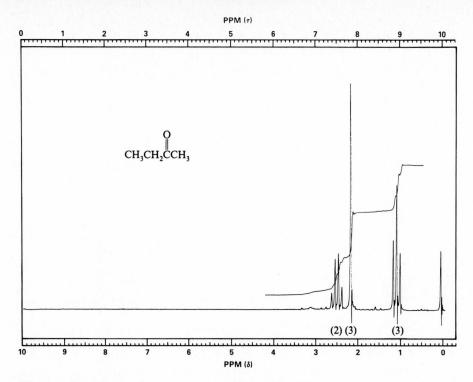

Figure 9.20 PMR spectrum of 2-butanone (methyl ethyl ketone).

(m) Carboxylic Acids

Owing to the presence of the carbon-oxygen double bond in carboxylic acids, the hydrogen attached to the —OH group and the hydrogens attached to the carbon atom adjacent to the C=O are shifted downfield.

$$
\begin{array}{c}
\text{O} \\
\parallel \\
\text{R—C—O—H} \quad \delta\ 10.5\text{–}12
\end{array}
$$

$$
\begin{array}{c}
\text{O} \qquad | \\
\parallel \qquad \\
\text{HOC—C—H} \quad \delta\ 2\text{–}2.7 \\
|
\end{array}
$$

In comparing the chemical shifts of hydrogen attached to oxygen in the —OH group in alcohols (δ 1–5.5), phenols (δ 4–12), and carboxylic acids (δ 10.5–12), note should be made of the downfield trend due to the increasing acidity of these functional groups. With each type of group, the hydroxylic proton may be shifted even farther downfield if strongly electron-withdrawing groups are present in the molecules. The hydrogens attached to the

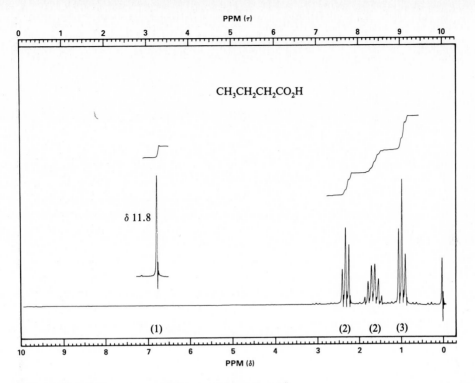

Figure 9.21 PMR spectrum of butanoic acid (butyric acid).

α-carbon atom in various types of molecules containing the carbon-oxygen double bond all appear at approximately the same chemical shift (δ 2–2.7).

The PMR spectrum of butanoic acid (butyric acid) is shown in Figure 9.21. The signal of the hydrogen of the —OH group occurs as a singlet at δ 11.97 and is unsplit because of the absence of neighboring hydrogens. Compare this value with those observed for an alcohol (Figure 9.15) and phenol (Figure 9.16). The splitting pattern and chemical shifts of the propyl chain are those expected. The triplet at δ 2.3 represents the two hydrogens α to the carboxyl group, H_c, which is shifted downfield because of the C=O; it is split by the two adjacent hydrogens. The complex multiplet centered at δ 2.7 is due to H_b and appears as a sextet owing to the five hydrogen neighbors. However, there is distortion in the peaks, which is probably due to the magnetic nonequivalence of H_a and H_c; higher-order splitting effects such as observed here are not uncommon. The triplet at δ 1.0 represents the three hydrogens of the methyl group, whose signal is not shifted because it is sufficiently far from the carboxyl group.

(n) Esters

The chemical shifts observed for esters result from the electron-withdrawing effects of both the acyl oxygen and the alkyl oxygen, as shown below.

$$RC\overset{\overset{\displaystyle O}{\|}}{}\!\!\!-O-\overset{|}{\underset{|}{C}}-H \quad \delta\ 3.7\text{–}4.1$$

$$RO\overset{\overset{\displaystyle O}{\|}}{C}\!\!-\overset{|}{\underset{|}{C}}-H \quad\quad \delta\ 2\text{–}2.6$$

Note that the chemical shifts of the hydrogens attached to the alkyl oxygen are shifted downfield somewhat compared with those of the alpha protons in alcohols and ethers (δ 3.3–4); this is due to the added electron-withdrawing effect of the acyl group. On the other hand, the absorptions of the hydrogens alpha to the acyl group appear at approximately the same chemical shift values as those in aldehydes, ketones, and carboxylic acids.

The PMR spectrum of ethyl ethanoate (ethyl acetate) is provided in Figure 9.22. The singlet at δ 2.0 represents three hydrogens that have no neighbors, and it must be due to the methyl group, H_b, attached to the acyl group. The quartet at δ 4.1, representing two protons that must have three adjacent hydrogens, is due to the methylene group attached to the alkyl oxygen atom. The triplet peak at δ 1.2 results from the hydrogens, H_a, of the methyl group and is shifted downfield slightly from that of alkane methyl groups because of the electron-withdrawing effect of the alkyl oxygen atom.

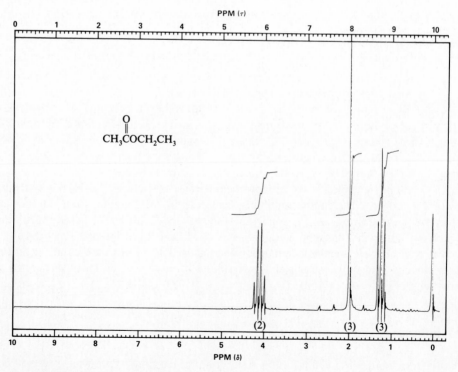

Figure 9.22 PMR spectrum of ethyl ethanoate (ethyl acetate).

(o) Amines

The chemical shift observed for the hydrogens attached to a nitrogen atom in an amine is given below.

$$R—NH_2 \quad \delta \; 1\text{–}5$$

This shift is variable and depends on the concentration of the amine in solution. Hydrogen-bonding between molecules plays a key role in dictating the shift value, as do the substituents contained in the alkyl or aryl group. Absorption peaks due to the amino hydrogens are usually singlets, even though the carbon to which the amino group is attached may contain hydrogens. The reason for this is analogous to that given previously for the lack of splitting observed for the hydroxyl hydrogen in an alcohol. It is due to quadrapole relaxation of nitrogen, a topic that will not be discussed here. In the case of a tertiary amine, R_3N, there is no amino hydrogen and thus no absorption.

The PMR spectrum of *N*-ethylaniline is shown in Figure 9.23. The absorption of the hydrogen attached to nitrogen appears as a broad peak at δ 3.5. The aromatic hydrogen absorptions appear as a complex set of peaks centered at δ 7.0. The peaks between δ 6.5 and 7.0 are due to the *ortho* and *para* ring hydrogens, which are shifted upfield as a result of the electron-releasing effect of nitrogen, which is a consequence of resonance between the lone

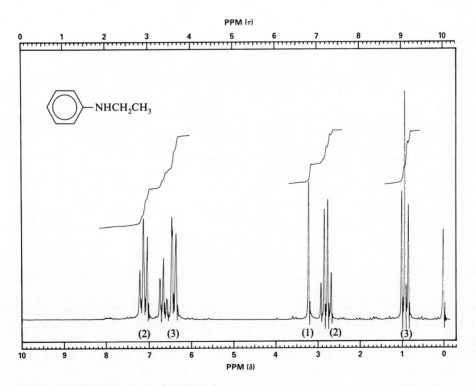

Figure 9.23 PMR spectrum of *N*-ethylaniline.

electron-pair on nitrogen and the ring. The peaks between δ 7.0 and 7.4 represent the two *meta* hydrogens. The quartet centered at δ 3.1 (two hydrogens) is due to the methylene hydrogens of the ethyl group, whose signal is shifted downfield because of the electron-withdrawing effect of the nitrogen. The triplet centered at δ 1.2 represents the three methyl hydrogens, and this peak is shifted downfield slightly by the presence of the nitrogen.

(p) Amides

The absorptions of protons attached to the amide nitrogen are shifted downfield considerably compared with those in an amine because of the presence of an additional electron-withdrawing group, the carbon-oxygen double bond. The amido hydrogens usually produce a broad peak, and their variable chemical shift is a consequence of factors such as concentration and solvent. There is extensive hydrogen-bonding in amides, and this also causes the amido hydrogen peak to be broadened considerably. Sometimes the absorption is so broad that a peak is not immediately evident in the spectrum, and the peak integration may be the only means by which it is detected. The hydrogens α to the acyl group have chemical shifts in the same general location as the α hydrogens in the other functional groups containing the C=O. The following summarizes the chemical shift regions for amides.

$$\underset{\text{RC—NH}_2}{\overset{\text{O}}{\|}} \text{ or } \underset{\overset{|}{\text{R}'}}{\underset{\text{RC—NH}}{\overset{\text{O}}{\|}}} \quad \delta \text{ 5–8}$$

$$\underset{\text{H}_2\text{NC—C—H}}{\overset{\text{O}}{\|}} \quad \delta \text{ 2–2.8}$$

The PMR spectrum of acetanilide, shown in Figure 9.24, shows the amido proton signal as a broad hump at δ 8.9. The signal of the methyl group appears as a singlet at δ 2.1 and is unsplit because there are no neighboring hydrogens. The aromatic hydrogens produce a complex multiplet between δ 7 and 7.6. The upfield aromatic ring hydrogen shift that was observed for the aromatic amine (Figure 9.23) does not occur for acetanilide, even though the latter compound has nitrogen with a nonbonded electron pair attached to the ring. This is due to the electron-withdrawing effect of the acyl group attached to nitrogen, which considerably reduces the electron-donating resonance effect that otherwise would cause the ring protons to be shielded and shifted upfield, as they are in aromatic amines.

9.9 Carbon-13 Magnetic Resonance (CMR) Spectroscopy

Carbon-13, ^{13}C, is an isotope of carbon and, like the hydrogen nucleus, has a nuclear spin. Carbon-13 has a *natural* abundance of about 1.1%, so that 1.1% of all the carbon atoms in an organic compound are ^{13}C. Thus, the nuclear magnetic resonance spectrum due to the

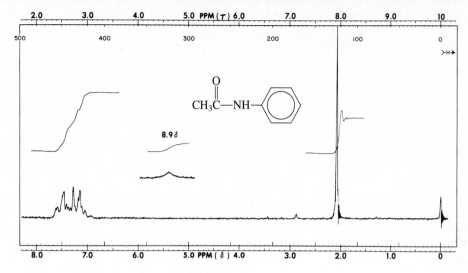

Figure 9.24 PMR spectrum of acetanilide.

carbon-13 atoms in a compound can be observed, and it is called a **carbon magnetic resonance (CMR)** spectrum. It should be pointed out that carbon-12 is not magnetic, so that ^{12}C does not absorb in nuclear magnetic resonance. The instrumentation required for CMR spectroscopy is much more expensive than that used for PMR spectroscopy. The CMR spectrometer requires the use of computer techniques to collect the weak positive signals that arise from the ^{13}C nucleus and add many spectra to get enough signal to be seen over the ''noise''. Unlike a PMR spectrum, which can be obtained in just a few minutes, the acquisition of a CMR spectrum usually requires several hours because of the signal-averaging techniques that are required. The reason for this is that the natural abundance of ^{1}H in hydrogen is greater than 99%, so that there are many absorbing atoms in a molecule, whereas the small number (1.1%) of ^{13}C atoms in carbon requires a much longer time for the acquisition of a CMR spectrum. Nevertheless, CMR spectroscopy today is becoming a routine tool for structure determination. Current electronic technology in the design of the CMR spectrometer permits a CMR spectrum to be obtained on as little as 10–20 mg of sample.

The principles of CMR spectroscopy are identical with those of PMR spectroscopy, which have been discussed in Section 9.1. When placed in an external magnetic field, the ^{13}C nuclei absorb energy in the radio-frequency region, causing their nuclear spins to go from one energy level to another. The carbon-13 nuclei in different magnetic environments may be either shielded or deshielded, a phenomenon that is also observed for the nuclei of hydrogen atoms in PMR spectroscopy. As in PMR, tetramethylsilane, $(CH_3)_4Si$, is used as the reference compound in CMR, and the chemical shifts observed for most carbon atoms are downfield from TMS; these shifts are reported in parts per million (ppm).

Modern CMR spectrometers are normally operated in a mode that *decouples* the ^{13}C signal from that of the proton(s) that may be attached to the carbon atom, a technique termed **proton decoupling.** Such CMR spectra, therefore, do not show the effects of the spin-spin coupling that is observed in PMR spectra. As a result, proton-decoupled CMR spectra are extremely simple and consist of a single sharp resonance line for each magnetically distinct carbon atom. On the other hand, *coupled* CMR spectra are made extremely complex because the magnitudes of proton-^{13}C coupling constants are large (120–230 Hz).

The interpretation of a CMR spectrum is accomplished on the basis of the observed ^{13}C chemical shifts, and the correlation between these values and structure is provided in Table 9.3. Note that most ^{13}C resonances are between 0 and 250 ppm downfield from TMS. In contrast to the chemical shifts observed in PMR spectroscopy (0 to 10 ppm), those found in CMR spectroscopy are quite large (0 to 250 ppm). Each value of the chemical shift listed in Table 9.3 depends on the environment of carbon in a molecule. The structural features of a molecule that cause downfield chemical shifts in CMR spectroscopy are similar to those prevalent in PMR. For example, the presence of electronegative elements causes a downfield chemical shift. Hybridization is also very important in CMR, as can be seen from the ^{13}C chemical shifts of sp^3-, sp^2-, and sp-hybridized carbon atoms. ^{13}C chemical shifts are also sensitive to stereochemistry and can be predicted very well from known, model compounds.

TABLE 9.3 **^{13}C Chemical Shifts in Carbon-13 Magnetic Resonance (CMR) Spectroscopy**

Absorbing Carbon Atom (shown as **C** or **Ar**)	Approximate Chemical Shift (ppm) δ
RCH$_2$**C**H$_3$	13–16
R**C**H$_2$CH$_3$	16–15
R$_3$**C**H	25–38
CH$_3\overset{\displaystyle O}{\overset{\displaystyle \|}{C}}$—R	30–32
CH$_3\overset{\displaystyle O}{\overset{\displaystyle \|}{C}}$—OR	20–22
R**C**H$_2$—Cl	40–45
R**C**H$_2$—Br	28–35
R**C**H$_2$—NH$_2$	37–45
R**C**H$_2$—OH	50–65
RC≡**C**H	67–70
R**C**≡CH	74–85
RCH=**C**H$_2$	115–120
R**C**H=CH$_2$	125–140
R**C**≡N	118–125
ArH	125–150
R$\overset{\displaystyle O}{\overset{\displaystyle \|}{C}}$—OR′	170–175
R$\overset{\displaystyle O}{\overset{\displaystyle \|}{C}}$—OH	175–185
R$\overset{\displaystyle O}{\overset{\displaystyle \|}{C}}$—H	190–200
R$\overset{\displaystyle O}{\overset{\displaystyle \|}{C}}$—CH$_3$	205–210

The CMR spectrum of 2-butanone, shown in Figure 9.25, illustrates the great simplicity that is typically observed in this type of NMR spectroscopy. The assignment of the various absorption peaks which correspond to the carbon atoms in 2-butanone is indicated on the spectrum. There are some interesting features that should be noted about the CMR spectrum of 2-butanone. As indicated previously, there is no observed spin-spin coupling, so each peak appears as a single, sharp absorption. Unlike PMR, the height of each peak is *not* directly related to the number of carbon atoms represented by each resonance, and there is no peak integration on a CMR spectrum. Note that the peaks are of unequal height, even though each one is due to the resonance of one carbon atom. However, there is a *rough* correlation between the intensity (peak heights) of the absorptions as follows: the more hydrogen atoms attached to a carbon, the more intense the resonance band of that carbon. For example, the ^{13}C absorption of carbon in a —CH_3 group is stronger (has a taller peak) than that in a carbonyl group. The assignment of the absorptions to various carbon atoms in 2-butanone becomes evident from the data provided in Table 9.3. The carbonyl carbon atom is the most *deshielded* carbon in the molecule, and its absorption appears at 207 ppm. The carbon atom that is most *shielded* is the methyl group farthest away from the carbonyl group, and its resonance is at 10 ppm. The methyl group adjacent to the C=O absorbs at 28 ppm, and the methylene carbon atom is responsible for the peak at 38 ppm. The methylene carbon atom is more deshielded than the methyl carbon attached to the carbonyl group, and the former appears downfield from it.

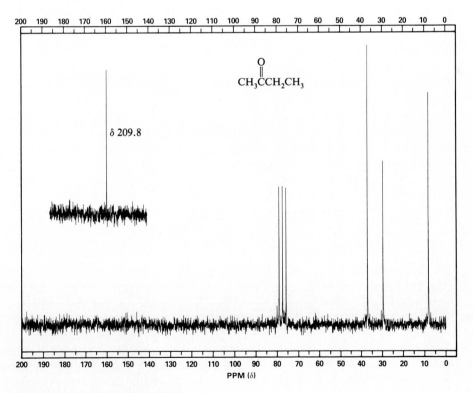

Figure 9.25 CMR spectrum of 2-butanone.

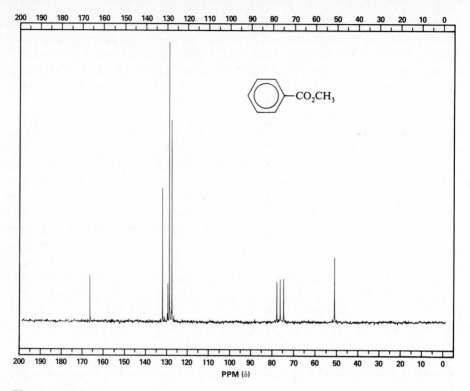

Figure 9.26 CMR spectrum of methyl benzoate.

The CMR spectrum of methyl benzoate is provided in Figure 9.26 and may be interpreted as follows. The carbonyl carbon atom of the ester is more shielded (165 ppm) than that of the carbon of the keto group in 2-butanone (207 ppm), and this shielding can be attributed to the presence of the two pairs of nonbonding electrons on the other oxygen atom of the ester group. The chemical shift observed for carbon in the methyl group in methyl benzoate (52 ppm) is more downfield than that of the carbon atom in the methyl group attached to the carbonyl group in 2-butanone, whose chemical shift is 30 ppm. The carbon atoms contained in the aromatic ring are also deshielded and are responsible for the absorptions observed between 128 and 132 ppm; the peak at 132 ppm is due to the ring carbon, which is attached to the ester group. The peaks observed between 128 and 130 ppm result from the remaining five carbon atoms of the aromatic ring.

9.10 Preparation of NMR Samples

The NMR spectra of greatest use to organic chemists are those obtained on liquids contained in special precision-ground glass tubes, 5 × 178 mm in size. Most commonly the substrate of interest, whether it is solid, liquid, or gaseous, is dissolved in an appropriate solvent, although spectra can also be obtained on neat liquids if their viscosities are not too high. With too-viscous liquids, broad rather than sharp bands are observed in the spectra, resulting in a loss of resolution between peaks. Approximately 0.3 mL of a 10–20% solution (by weight) is normally required. Sophisticated modern spectrometers, however, can provide usable NMR spectra on solutions containing a milligram or even less of solute. For PMR spectra, carbon tetrachloride is an excellent choice for the solvent, since it has no protons that would produce absorption peaks in addition to those of the solute. For samples that are insoluble in carbon tetrachloride, benzene, δ 7.23 (chemical shift), chloroform, δ 7.27, dichloromethane, δ 5.35, nitromethane, δ 4.33, acetonitrile, δ 2.00, and acetone, δ 2.17, may be suitable solvents if their proton resonances do not occur in a region of absorption of the solute. Many deuterated solvents, such as chloroform-d (CDCl$_3$), acetone-d_6 (CD$_3$COCD$_3$), and benzene-d_6 (C$_6$D$_6$), are available. Their use circumvents the problem of resonance absorptions of solvent protons in the PMR spectrum. Even though deuterium, ^{2}H, possesses the property of nuclear spin, the experimental conditions for achieving the resonance condition differ sufficiently from those required for hydrogen, so deuterium absorptions do not interfere with the PMR spectrum. The chemical shift of deuterium is vastly different from that of hydrogen. For CMR spectra, any of the solvents mentioned above for PMR spectra can be used, with carbon tetrachloride and deuterated chloroform the preferred solvents. However, any carbon-containing solvent will produce peak(s) in a CMR spectrum, so that care must be used in selecting a solvent that does not interfere with the peaks resulting from the compound whose spectrum is being determined.

Spectra of poor quality can result if a neat liquid sample or a solution contains undissolved solids such as the sample itself or contaminants (*even dust*). Moreover, trace amounts of ferromagnetic impurities picked up from contact of the sample or solvent with metals, for example, the iron of a "tin" can, cause havoc in the form of very broad and weak absorptions of the sample. If any solid material *at all* can be seen in the sample, it should be filtered. The simplest and most efficient method for filtering very small volumes is to insert a small plug of glasswool into a disposable pipet and to filter the sample directly into the NMR tube through the plug. After use the NMR tubes should be well cleaned and dried; they should be stored either in a closed container or with open end downward in order to avoid dust settling inside the tube.

REFERENCES

Books and Articles

1. Pasto, D. J.; Johnson, C. R. *Organic Structure Determination*, Prentice-Hall, Englewood Cliffs, N.J., 1969.

2. Silverstein, R. M.; Bassler, G. C.; Morrill, T. C. *Spectrometric Identification of Organic Compounds*, 4th edition, Wiley, New York, 1981.

3. Dyer, J. R. *Organic Spectral Problems*, Prentice-Hall, Englewood Cliffs, N.J., 1971.

4. Jackman, L. M.; Sternhell, S. *Applications of NMR Spectroscopy in Organic Chemistry*, 2nd ed., Pergamon, New York, 1969.

5. Levy, G. C.; Lichter, R. L.; Nelson, G. L. *Carbon-13 NMR*, 2nd ed., Wiley-Interscience, New York, 1980.

6. Tonelli, A. E.; Schilling, F. C. "^{13}C NMR Chemical Shifts and the Microstructure of Polymers," *Accounts of Chemical Research*, **1981**, *14*, 233.

7. Ault, A.; Ault, M. R. *An Introduction to Proton Nuclear Magnetic Resonance Spectroscopy*, Holden-Day, San Francisco, 1976.

8. Stothers, J. B. *Carbon-13 NMR Spectroscopy*, Academic Press, New York, 1972.

9. Abraham, R. J.; Loftus, P. *Proton and ^{13}C NMR Spectroscopy*, Heyden, London, 1978.

Catalogues of Spectra

10. Varian Associates, *High Resolution NMR Spectra Catalogue,* Vol. 1, 1962; Vol. 2, 1963.

11. *The Sadtler Collection of High Resolution Spectra,* Sadtler Research Laboratories, Philadelphia.

12. Pouchert, C. J.; Campbell, J. R. *Aldrich Library of NMR Spectra,* Vols. I–XI, Aldrich Chemical Company, Milwaukee, 1975.

13. *Nuclear Magnetic Resonance Spectra,* Sadtler Research Laboratories, Philadelphia.

14. Breitmaier, E.; Haas, G.; Voelter, W. *Atlas of Carbon-13 NMR Data,* 2 vols. Heyden, Philadelphia, 1979.

EXERCISES

1. Briefly define or explain the following:

a. tetramethylsilane (TMS)	**b.** downfield shift	**c.** upfield shift
d. delta (δ)	**e.** tau (τ)	**f.** parts per million
g. chemical shift	**h.** proton counting	**i.** $N = n + 1$
j. coupling constant	**k.** splitting pattern	**l.** singlet
m. doublet	**n.** triplet	**o.** quartet
p. multiplet	**q.** magnetically equivalent	**r.** chemically equivalent

2. Briefly explain the differences between PMR and CMR spectroscopy and indicate what structural features of a molecule are observed with these two types of spectroscopy.

3. Determine the number of peaks that would be observed in the PMR spectrum of the following molecules when each type of hydrogen absorbs energy:

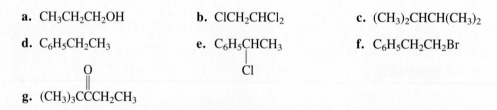

a. $CH_3CH_2CH_2OH$ **b.** $ClCH_2CHCl_2$ **c.** $(CH_3)_2CHCH(CH_3)_2$

d. $C_6H_5CH_2CH_3$ **e.** $C_6H_5CHCH_3$ **f.** $C_6H_5CH_2CH_2Br$
                                    |
                                    Cl

g. $(CH_3)_3C\overset{\displaystyle O}{\overset{\|}{C}}CH_2CH_3$

4. Determine the structure of the following compounds insofar as possible:
   a. Compound **1**, $C_{10}H_{14}$, whose PMR spectrum is provided in Figure 9.27.
   b. Compound **2**, C_3H_8O, whose PMR spectrum is provided in Figure 9.28.
   c. Compound **3**, C_8H_9Cl, whose PMR spectrum is provided in Figure 9.29.

5. The PMR chemical shifts, splitting patterns, and proton abundances for various molecules are given below. Deduce a structure or structures that are consistent with these data. (*Hint:* You may find it useful to sketch these spectra on paper.)
   a. C_4H_9Br: doublet at δ 1.04 (6 H), multiplet at δ 1.95 (1 H), doublet at δ 3.33 (2 H)
   b. $C_3H_6Cl_2$: quintet at δ 2.2 (4 H), triplet at δ 3.75 (2 H)
   c. $C_5H_{11}Br$: doublet at δ 0.9 (6 H), complex multiplet at δ 1.8 (3 H), triplet at δ 3.4 (2 H)

6. The CMR spectrum of 3-methyl-2-butanone **(4)** is given in Figure 9.30. Assign the observed peaks to the correct carbon atoms in the molecule.

$$CH_3-\underset{\underset{4}{\overset{\overset{\displaystyle O}{\|}}{C}}}{}-CH(CH_3)_2$$

3-Methyl-2-butanone

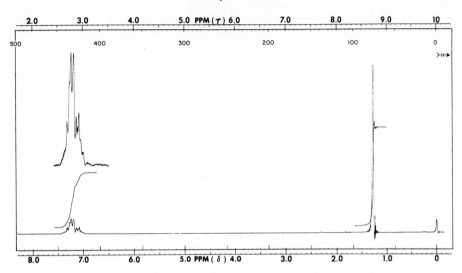

Figure 9.27 PMR spectrum for Exercise 4a.

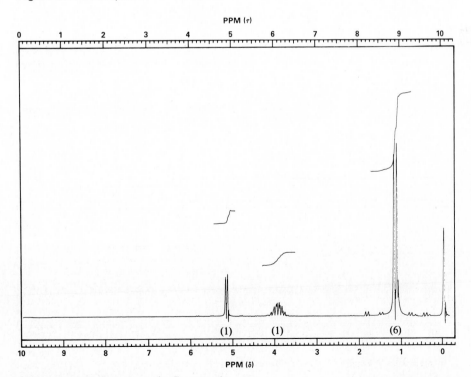

Figure 9.28 PMR spectrum for Exercise 4b.

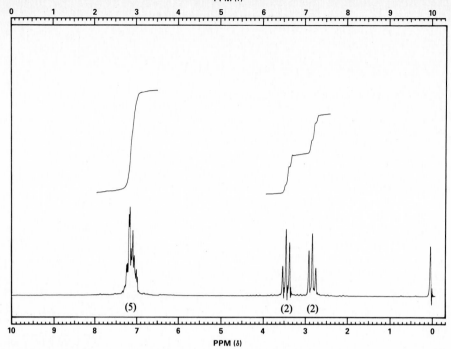

Figure 9.29 PMR spectrum for Exercise 4c.

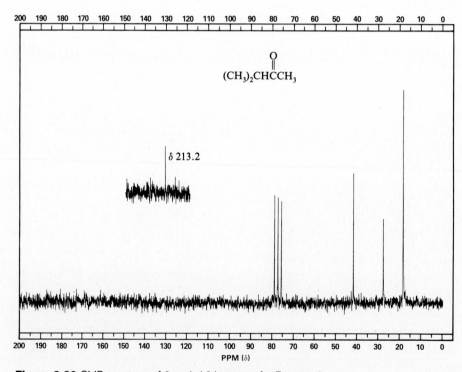

Figure 9.30 CMR spectrum of 3-methyl-2-butanone for Exercise 6.

spectroscopy: structure determination using IR and NMR

The previous two chapters dealt with infrared (IR) and nuclear magnetic resonance (NMR) spectroscopy as separate topics. Although each of these spectroscopic methods is useful alone in determining certain structural aspects of molecules, their greatest utility in the identification of unknown compounds is realized when they are used together. This chapter presents several examples of how IR and NMR can be used to determine the structure of compounds. As you study these examples, keep in mind that an IR spectrum provides information concerning the *functional groups* that may be contained in a substance. Proton magnetic resonance (PMR) spectroscopy gives information concerning the *number* and *type* of hydrogen atoms in a molecule, whereas carbon magnetic resonance (CMR) spectroscopy can be used to indicate the *type* of carbon atoms that are present in a substance and thus provide valuable information concerning its structure.

The structures of many unknowns whose molecular formulas are available can often be derived immediately by combining the information obtained from the IR and NMR spectra, with PMR spectra still being the most commonly encountered. However, proper interpretation of the spectra requires considerable practice on the part of the student. In order to provide an opportunity for such practice, the IR and PMR spectra of many of the organic compounds encountered as starting materials or products in the reactions described in this text have been reproduced. These spectra are labeled, so you will not be attempting to assign a structure to the substance producing them but will be interpreting the spectra in terms of the known structure.

The knowledge and experience you gain from analyzing the spectra of known compounds should provide a sound basis for interpreting the spectra of *unknown* substances and assigning a structure to them. Even though it may be possible to make structural assignments to unknowns based solely on the interpretation of spectra, these assignments must ultimately be confirmed either by comparing the spectra of the unknown with those of the known compound, which ideally has been synthesized in an unequivocal fashion, or by more classic means such as preparing one or more derivatives (Chapter 27) of the unknown and comparing their melting points to those of the known compound. Clearly, the classic methods have the severe limitation that the compound must have already been reported in the literature, along with physical properties of the derivatives.

This chapter provides examples of the interpretation of IR spectra in combination with PMR or CMR spectra of known and unknown compounds.

10.1 Analysis of IR and PMR Spectra of a Known Compound

The following example will serve to introduce the kind of analysis that can be done on the IR and PMR spectra of a known compound. Consider the spectra of 2-methyl-1-propanol **(1)**, which are given in Figure 10.1 and are analyzed below.

$$CH_3 \!-\! CH \!-\! CH_2 \!-\! OH$$
$$|$$
$$CH_3$$

1

2-Methyl-1-propanol
(isobutyl alcohol)

When the IR spectrum (Figure 10.1a) is examined, three prominent bands can be seen in the region 5000–1250 cm^{-1}. By referring to Tables 8.2 and 8.3, it is possible to assign the absorptions at about 2900 and 1465 cm^{-1} to the C—H bonds in the molecule. The broad and intense band at 3250 cm^{-1} undoubtedly is due to the O—H function in the alcohol, and the position and breadth of this band indicates that the alcohol is hydrogen-bonded. The very strong absorption at 1050 cm^{-1} is characteristic of a C—O single bond. As would be expected on the basis of the structure of **1**, no absorptions are seen for carbonyl functions, aromatic rings, or carbon-carbon multiple bonds.

Turning now to the PMR spectrum (Figure 10.1b), the first point of interest is the integration, which provides information concerning the areas under each set of peaks and, in turn, the *relative* number of each type of hydrogen in the molecule. Going *upfield,* the areas of the peaks are in the ratio, measured in mm, of 2.8:5.6:3.0:15. Dividing each number in this set by the smallest number, 2.8, the peak areas are in the simplest ratio of 1.0:2.0:1.1:5.4. Since the total number of hydrogens in **1** is 10, the latter ratio is within experimental error of the absolute ratio of 1:2:1:6.

It is now possible to analyze the spectrum by assigning the observed chemical shifts to specific types of hydrogens in the molecule. The peak areas and splitting patterns must also be in accord with the structure. The doublet at about δ 0.85 represents six hydrogens, and the chemical shift is consistent with the six hydrogens of the two methyl groups of the alcohol.

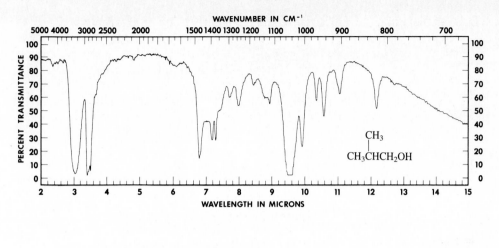

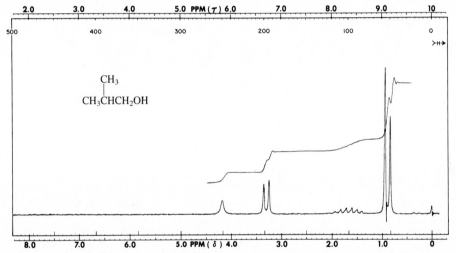

Figure 10.1 IR and PMR spectra of 2-methyl-1-propanol (isobutyl alcohol).

The splitting of the methyl absorptions is due to coupling with the lone methine (tertiary) hydrogen, $(CH_3)_2CH$. The multiplet from δ 1.4–2.0 can be assigned to the methine hydrogen because it represents an area of 1. Its chemical shift is consistent with the hydrogen being tertiary, and the splitting pattern is that expected of a hydrogen coupled with eight other neighboring hydrogens. Although the $n + 1$ rule predicts that this proton should appear as nine peaks, it is clearly impossible to see this number of peaks in the spectrum. The doublet at about δ 3.3 represents two hydrogens, and the chemical shift and splitting pattern are those anticipated for the CH_2 group in the alcohol. These protons are shifted downfield owing to the presence of the electron-withdrawing oxygen atom. This leaves the somewhat broad singlet of peak area 1 at δ 4.2 to be assigned to the hydroxylic hydrogen. For reasons discussed in Section 9.8g, this peak appears as a singlet and is not coupled with the adjacent CH_2 group.

10.2 Analysis of IR and CMR Spectra of a Known Compound

The example provided below will serve to illustrate the technique of analyzing the IR and CMR spectra of a known compound. Consider the spectra of methyl propanoate (methyl propionate), **(2),** which are provided in Figure 10.2.

$$CH_3CH_2\overset{\displaystyle O}{\overset{\displaystyle \|}{C}}OCH_3$$

2

Methyl propanoate
(methyl propionate)

Analysis of the IR spectrum of **2** reveals the presence of a carbon-oxygen double bond at 1740 cm^{-1}, which is indicative of the carbonyl group in the ester. The strong band at 1195 cm^{-1} results from absorption by the C—O—C bond of the methoxy group attached to the carbonyl carbon. As expected, no peaks are observed for other functional groups.

The CMR spectrum of **2** shows four absorptions, which is to be expected because there are four different types of carbon atoms in the molecule. The assignment of the four signals to the absorbing carbon atoms can be accomplished with the aid of Table 9.3 and are as follows:

$$CH_3CH_2CO_2CH_3$$
a   b   c   d

C_a is farthest removed from the carbon-oxygen double bond, and the peak at 9 ppm results from this carbon. The 28-ppm peak is attributed to C_b, and because of deshielding by the alkyl oxygen atom, the 51-ppm band is caused by C_d. This conclusion is reached by noting in Table 9.3 that the chemical shift is 50–65 ppm for carbon in RCH_2OH. The most deshielded carbon in the molecule is that contained in the carbonyl group, so the 175-ppm peak must be due to C_c. As this example illustrates, the interpretation of a CMR spectrum can be quite easy for simple molecules.

10.3 Analysis of IR and PMR Spectra of Unknown Compounds

The following two examples involve the identification of unknown compounds whose molecular formulas are provided.

Example 1

Problem The IR and PMR spectra of compound **3**, $C_{13}H_{11}NO$, are given in Figure 10.3. Deduce the structure of this compound.

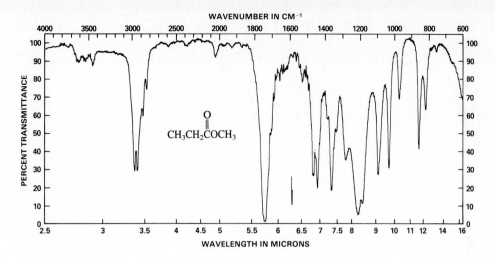

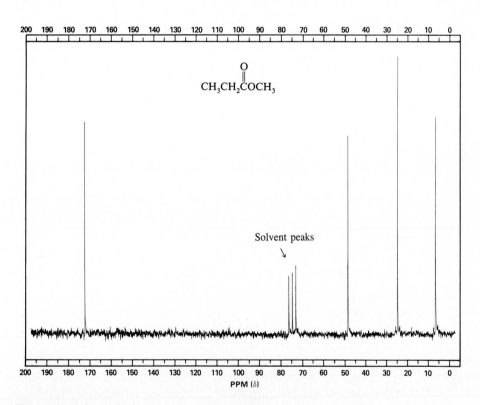

Figure 10.2 IR and CMR spectra of methyl propanoate (methyl propionate).

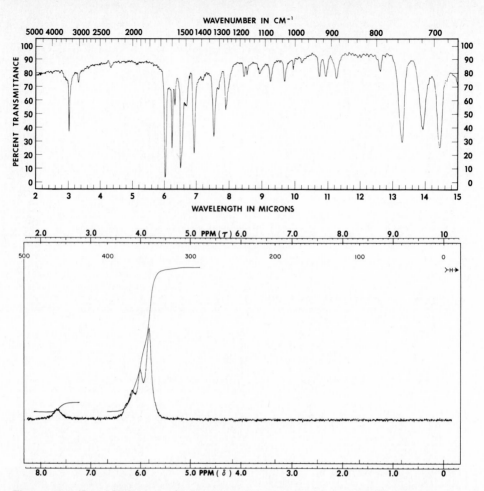

Figure 10.3 IR and PMR spectra for Example 1. Note that all peaks in the PMR spectrum are offset by 1.1 ppm upfield.

Solution Examination of the IR spectrum reveals several important absorption bands. The strong, sharp peak at 1670 cm^{-1} is possibly due to the C=O stretching of an amide. Note that this absorption is probably too low for the carbonyl of an ester, ketone, or aldehyde and too high for a carbon-carbon double bond. Reference to Tables 8.2 and 8.3 indicates that the C=O is likely to be conjugated. The PMR spectrum clearly shows that **3** contains at least one aromatic ring, owing to the absorptions between δ 7 and 8.1, and this should be kept in mind when accounting for the possible conjugation of the amide. Returning to the IR spectrum, the appearance of the sharp peak at 3350 cm^{-1} suggests the presence in **3** of an N—H bond of an amide and also indicates that the nitrogen contains just one hydrogen. The data in Tables 8.2 and 8.3 show that the absorption of an —NH$_2$ group usually appears as a doublet with 50 cm^{-1} separation, whereas that of a single N—H bond is usually sharp. The weak absorption at 3050 cm^{-1} reinforces the conclusion that an aromatic ring is present. The presence of the two strong bands at 690 and 750 cm^{-1} suggests that the aromatic ring is monosubstituted, although peaks in this region are not always reliable in indicating the location of substituents on a ring.

The PMR spectrum of **3** confirms the presence of the aromatic ring and the amide linkage. The complex multiplet between δ 7.0 and 7.7 is due to aromatic protons, and the broad "singlet" at δ 8.0 results from hydrogen attached to nitrogen in an amide. Integration of the peak areas yields a proton ratio of 10:1, respectively, which may be interpreted as follows. It may be concluded that **3** must contain *two* aromatic rings, each having five protons and six carbon atoms, and that the peak of area 1 must represent the amido hydrogen. Note that the relative peak areas correspond to the absolute number of protons contained in **3** because the given molecular formula indicates the presence of 11 hydrogens in the molecule. Recall also that the IR spectrum indicates that **3** is a secondary amide, a fact confirmed by the PMR spectrum.

This information may be put together to yield the structure of **3** as being *benzanilide*.

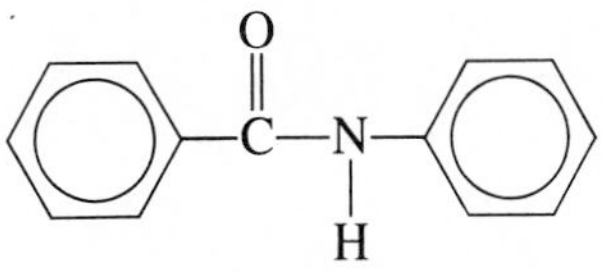

3
Benzanilide
(*N*-phenylbenzamide)

Example 2

Problem   The IR and PMR spectra of compound **4**, $C_9H_{10}O_2$, are given in Figure 10.4. What is the structure of this compound?

Solution   The IR spectrum contains two strong absorption bands, namely, the broad peak between 2800 and 3300 cm^{-1} and the peak at 1700 cm^{-1}, which give a strong indication that the molecule contains a carboxylic acid group, —CO_2H. This is further supported by the peak at δ 9.8 in the PMR spectrum, which is probably due to the carboxyl hydrogen. The peak at 1220 cm^{-1} in the IR spectrum is thus probably due to the C—O bond in the carboxylic acid group.

The PMR spectrum shows that **4** contains an aromatic ring on the basis of the band at δ 7.3. The area of the peaks, going *upfield,* gives the ratio of 1:5:4, and since the number of hydrogens contained in the unknown is 10, this ratio represents the absolute number of protons in the molecule. The four peaks between δ 2.6 and 3.0 may at first appear to be a quartet; if this were true, there would have to be another set of peaks in the spectrum with the same coupling constant, but we have already accounted for all the other peaks. Thus, these peaks must be due to the overlap of absorptions of hydrogens that are nearly equivalent. Before attempting to account for them, we should summarize information on the structure of **4** obtained thus far:

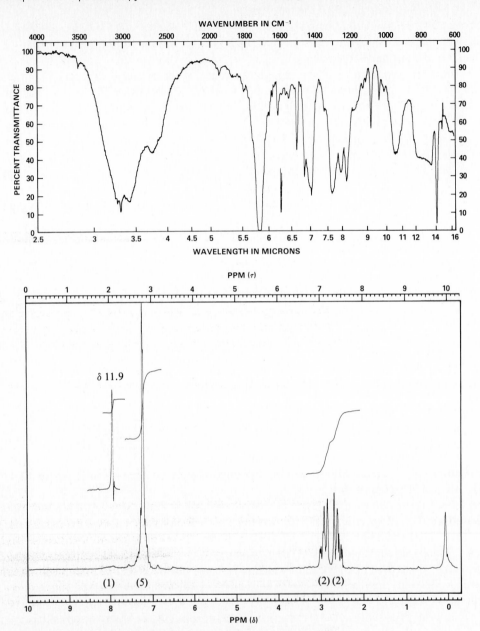

Figure 10.4 IR and PMR spectra for Example 2.

1 The aromatic ring is monosubstituted and accounts for six carbons and five hydrogens.

2 A carboxyl group, —CO_2H, is present.

With these facts we have now accounted for seven carbons, two oxygens, and six hydrogens; when these atoms are subtracted from the molecular formula, $C_9H_{10}O_2$, we are left with the job of accounting for two carbons and four hydrogens. The only possible structure that can be derived for these remaining atoms is —CH_2CH_2—, because we know there is a carboxyl group and a monosubstituted aromatic ring in **4.** When these various portions of the molecule are put together, we might conclude that the structure of the unknown is 3-phenylpropanoic acid (**4**).

$$\text{C}_6\text{H}_5\text{—CH}_2\text{CH}_2\text{—}\overset{\displaystyle \overset{\text{O}}{\|}}{\text{C}}\text{—OH}$$

4
3-Phenylpropanoic acid
(hydrocinnamic acid)

   Is the proposed structure in fact consistent with the only part of the molecule that was not easily determined by interpretation of the PMR spectrum, namely the ethano group, —CH_2CH_2—, which was assigned by process of elimination and assumed to be responsible for the peaks between δ 2.6 and 3.0? In structure **4,** it can be seen that the two hydrogens attached to each carbon atom are *not* chemically equivalent, because one of the CH_2 groups is adjacent to the electron-withdrawing carboxylic acid group while the other is deshielded by the aromatic ring. Each CH_2 group should appear as a triplet because each one has two neighboring hydrogens. Therefore, the four peaks at δ 2.6–3.0 apparently do not represent a quartet, but are actually two triplets of unequal chemical shift that are *superimposed* upon each other to give the appearance of a quartet. Thus, the proposed structure of **4** is consistent with the IR and PMR spectra, and this turns out to be the correct structure.

   The two examples discussed above illustrate the type of analysis that should be performed in deducing the structure of unknown compounds when the IR and PMR spectra and the molecular formula are given. It was possible to rationalize the hydrogens responsible for each set of peaks in the PMR spectrum provided in Example 1. However, as illustrated in Example 2, it may not be possible to account for *each* chemical shift, in which case as much information as possible should be deduced about the structure, often leaving only a few atoms unaccounted for.

   Many PMR spectra consist of a number of peaks, some of which may overlap one another. It is only with practice and careful thought that you will become proficient in using IR and PMR spectra in structure determination, and our goal in this section has been to introduce the thought processes involved in doing so. Note, however, that one often gives both the IR and PMR spectra a cursory initial examination to determine what functional groups may be present. The presence of some groups, such as alcohol or carbonyl, are immediately evident from the IR spectrum, while others, such as an aromatic ring, are unmistakable from the PMR spectrum. In other words, these two types of spectra do truly go hand-in-hand. Careful examination of the many IR and PMR spectra provided throughout this text should help give experience and insight into using spectroscopy as an important adjunct of structure determination.

10.4 Analysis of IR and CMR Spectra of an Unknown Compound

In this section we shall see how the analysis of an IR and a CMR spectrum can be used to identify the structure of an unknown compound. The spectra for compound **5**, $C_4H_{10}O$, are provided in Figure 10.5, and the structure of this compound can be deduced as follows.

Examination of the infrared spectrum of **5** reveals the presence of a strong, broad absorption peak between 3100 and 3600 cm^{-1}, which is indicative of an alcohol. This conclusion is further confirmed by the observation of the strong band at 1050 cm^{-1} due to the C—O bond. The molecular formula of **5** shows that it is a saturated compound, so it must contain no double bonds, a fact supported by the absence of any peaks in the IR spectrum for a carbon-oxygen or carbon-carbon double bond.

The CMR spectrum of **5** shows only two signals, which indicates that there are only two types of carbons in the unknown. The absorption at 69 ppm must be due to the carbon atom bearing the OH group, thus accounting for one of the four carbon atoms in **5**. The CMR spectrum indicates that there is only one other type of carbon atom in this molecule, but the band at 27 ppm is downfield from where the carbon atoms in CH_3 or CH_2 are observed (see Table 9.3). However, the electron-withdrawing oxygen atom of the OH group affects not only the chemical shift of the carbon that bears it (the 69-ppm band discussed above), but also the chemical shift of carbon atoms that may be attached to carbon in the COH group. Therefore, it is not unreasonable to assume that **5** contains CH_3 groups, whose carbon atoms are shifted downfield somewhat because of the presence of oxygen on the carbon to which they are attached. The only possible structure that fits the peaks observed in the CMR spectrum is 2-methyl-2-propanol (*tert*-butyl alcohol).

$$CH_3$$
$$|$$
$$CH_3—C—OH$$
$$|$$
$$CH_3$$

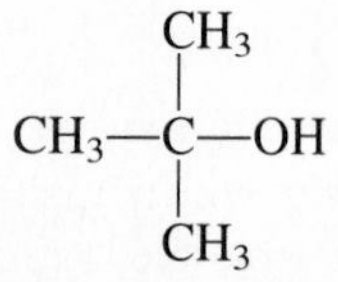

5

2-Methyl-2-propanol
(*tert*-butyl alcohol)

As an exercise, it is suggested that the structures of the other isomeric alcohols having the molecular formula $C_4H_{10}O$ be drawn; none of the other isomers have just two types of carbon atoms, and each of them would produce a CMR spectrum having more than two signals.

As mentioned previously, it is useful to obtain a PMR spectrum of an unknown compound in addition to the IR and CMR spectra of it. This example admittedly is rather simple, but it serves to illustrate the approach that can be used to apply CMR spectroscopy to structure identification.

EXERCISES

1. The IR and PMR spectra of various unknown compounds are given in the figures indicated below. Interpret both spectra in each part as completely as possible and attempt to assign a structure or structures to the unknowns on the basis of your interpretation.

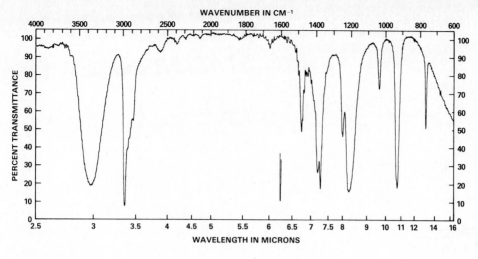

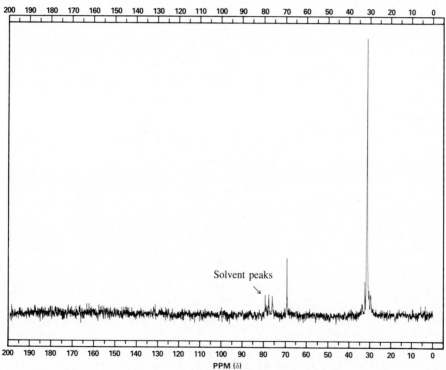

Figure 10.5 IR and CMR spectra for compound **5** in Section 10.4.

   a. Compound **6,** $C_{10}H_{14}O$, whose spectra are given in Figure 10.6.
   b. Compound **7,** $C_5H_{10}O_2$, whose spectra are given in Figure 10.7.
   c. Compound **8,** C_7H_9N, whose spectra are given in Figure 10.8.
   d. Compound **9,** $C_9H_{10}O$, whose spectra are given in Figure 10.9.
   e. Compound **10,** $C_{12}H_{18}$, whose spectra are given in Figure 10.10.

2. The IR and CMR spectra of compound **11**, C_3H_5Br, are provided in Figure 10.11. Deduce the structure of this compound, and assign ^{13}C absorptions to the correct carbon atoms in this unknown.

3. The IR and PMR spectra of citral are provided in Figure 10.12. Interpret these spectra as completely as possible.

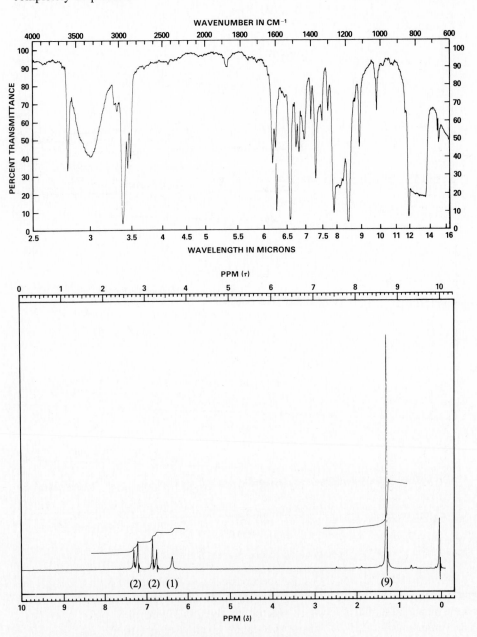

Figure 10.6 IR and PMR spectra for compound **6** in Exercise 1a.

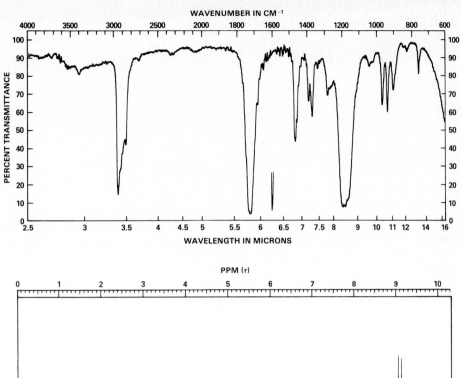

Figure 10.7 IR and PMR spectra for compound **7** in Exercise 1b.

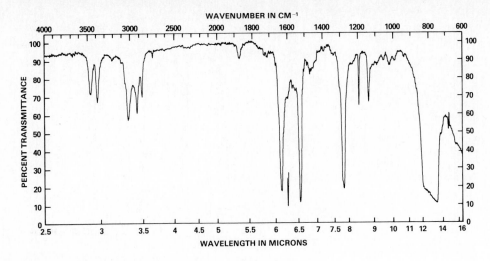

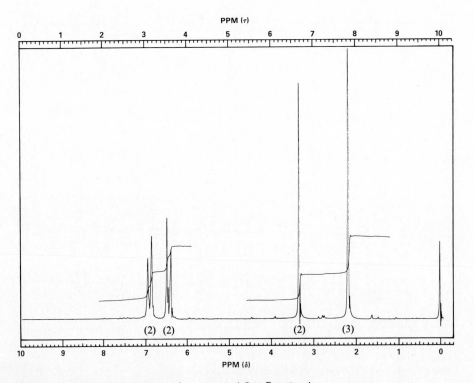

Figure 10.8 IR and PMR spectra for compound **8** in Exercise 1c.

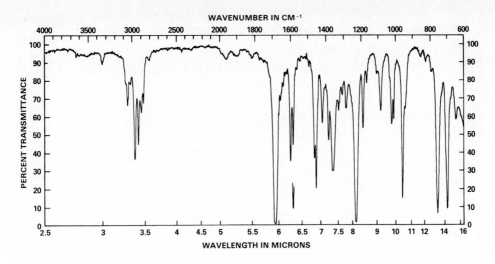

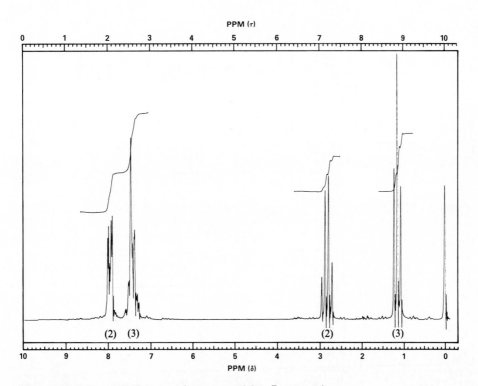

Figure 10.9 IR and PMR spectra for compound **9** in Exercise 1d.

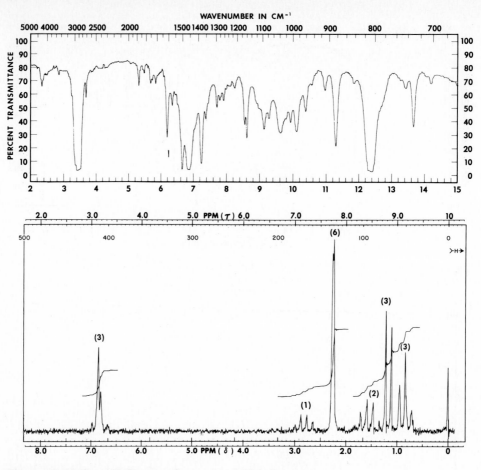

Figure 10.10 IR and PMR spectra for compound **10** in Exercise 1e.

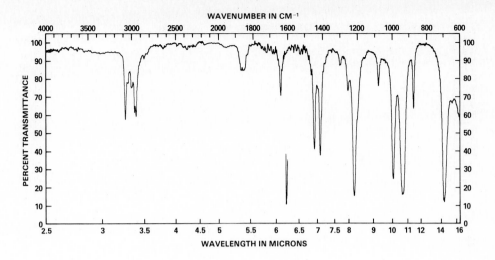

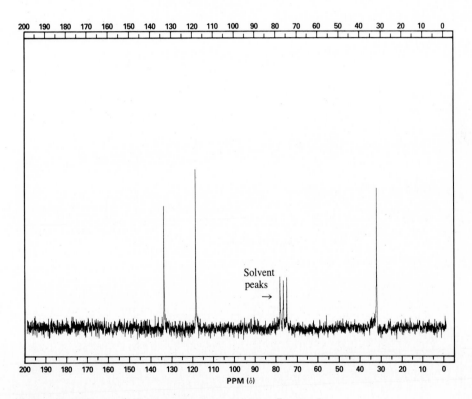

Figure 10.11 IR and CMR spectra for compound **11** in Exercise 2.

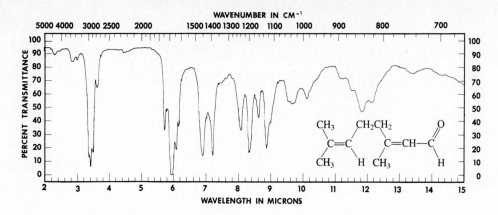

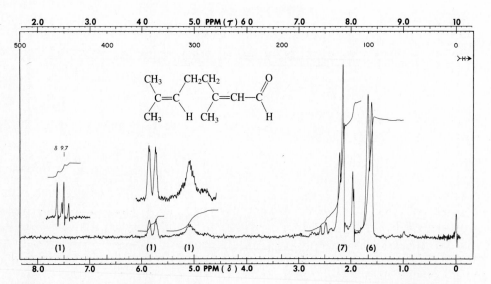

Figure 10.12 IR and PMR spectra of citral.

ultraviolet and visible spectroscopy; mass spectrometry

chapter **11**

11.1 Ultraviolet and Visible Spectroscopy

(a) Introduction

The technique of IR spectroscopy measures the stretching and bending of covalent bonds, as discussed in Chapter 8. On the other hand, ultraviolet (UV) spectroscopy and visible spectroscopy measure *electronic transitions* within molecules. Absorptions due to such transitions may occur in either or both the UV and visible regions, which, as Table 8.1 indicates, are adjacent to one another in the electromagnetic spectrum. Colored compounds absorb light in the visible region, whereas colorless ones may absorb energy in the UV region, a phenomenon that cannot be observed by the human eye.

Ultraviolet and visible spectroscopy both depend on the same fundamental molecular phenomenon, namely the excitation of an electron from a lower-energy to a higher-energy electronic state (see Figure 8.1). Table 8.1 indicates that the energy required for such an excitation ranges from about 38 to more than 100 kcal/mole and involves light in the wavelength range of 750 to 200 nm. These electronic changes are brought about by UV and visible light and generally involve either unshared pairs of electrons or electrons in π-bonds.

The two types of electronic excitations occurring in this range that are of greatest interest to an organic chemist are those involving promotion into an *antibonding* molecular orbital of an electron that originally occupied either a *nonbonding* molecular orbital (an n-electron) or a *bonding* molecular orbital (a σ-electron or a π-electron). However, the energy required to promote σ-type electrons to an antibonding molecular orbital is normally too high to be observed in the 750–200 nm region, so this type of molecular excitation is not a feature of

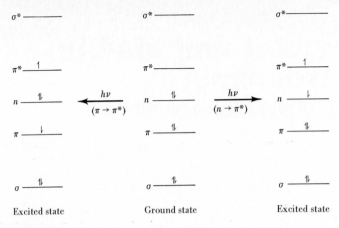

Figure 11.1 Energy diagram showing electronic transitions.

UV-visible spectra. Thus, UV-visible spectroscopy is limited to excitations of n- and π-electrons.

Because π-type antibonding molecular orbitals, designated π^*, are considerably lower in energy than the corresponding σ^* orbitals, the transitions normally stimulated by absorption of light having a wavelength in the UV-visible region involve populating the π^* state. The electronic transitions in which organic chemists generally are most interested are classified as $n \rightarrow \pi^*$ and $\pi \rightarrow \pi^*$, which are shown schematically in Figure 11.1.

(b) UV and Visible Spectra

Ultraviolet spectra are determined with a UV spectrophotometer, an instrument that is conceptually similar to an IR spectrophotometer (Section 8.3), except that the former measures absorption of light in the UV region, whereas the latter measures infrared radiation. A substance under study is dissolved in a suitable solvent, and the solution is placed in a cell. The same pure solvent is contained in another cell to serve as a reference. Methanol, ethanol, hexane, and water are common UV solvents because they do not absorb UV radiation; for this same reason, the optics and the cells that are used are made of quartz. The light source is usually a hydrogen lamp. The light passes through both cells, and the net amount of energy absorbed is measured and recorded. The spectrum that results is due entirely to the absorption of energy by the solute that is present, since dispersion of light by the sample cell and the solvent are compensated for by the reference cell.

The experimentally observed spectrum is subject to a number of variables, among which are solvent, concentration of the solution being examined, and the path length of the cell through which the light must pass. The amount of light absorbed by a particular solution is quantitatively defined by the **Beer-Lambert** law (equation 1).

$$A = \log \frac{I_0}{I} = \epsilon l c \qquad (1)$$

where A = absorbance or optical density

I_0 = intensity of incident light at a certain wavelength

I = intensity of light transmitted through sample at the same wavelength

ϵ = molar extinction coefficient

l = length of cell in cm and

c = concentration of sample in moles/liter

The UV spectrophotometer plots *absorbance*, A, as a function of the wavelength, λ, of the incident radiation, and this type of plot constitutes the UV spectrum of a compound. Since the concentration (c) and cell length (l) are known and the absorption (A) can be determined from the spectrum at each wavelength, the molar extinction coefficient (ϵ) can be calculated from equation 1. These coefficients typically range from 10 to 100,000. The intensity of the peaks in such a spectrum is affected by the experimental variables of concentration and path length, and it is common practice to redraw the spectrum by plotting ϵ or log ϵ *versus* wavelength. This type of plot is called the *standard UV spectrum*. However, the UV spectra provided in this text will be those that are actually observed experimentally, that is, plots of *absorbance* (also called *optical density*) *versus* wavelength. The concentration of the solution and the path length of the cell will be shown on the spectra, and in some instances a single spectrum will show traces made at more than one concentration so that both strong and weak absorbances can be discerned, as in the UV spectrum of 4-methyl-3-penten-2-one (**1**), shown in Figure 11.2.

$$:\!O\!:$$
$$\|$$
$$CH_3—C—CH\!=\!C—CH_3$$
$$|$$
$$CH_3$$

1

4-Methyl-3-penten-2-one

The entire UV-visible spectrum of a compound is often not reported, but rather only the wavelength, λ_{max}, and intensity, ϵ or log ϵ, of any maxima are given, along with the solvent in which the measurement was made. The crucial information contained in Figure 11.2 might thus be expressed in the following way:

$$\lambda_{max}^{95\%EtOH} \text{ 237 nm, log } \epsilon \text{ 4.1; 305 nm, log } \epsilon \text{ 1.8}$$

The solvent should be stated, since the values of both λ_{max} and ϵ are solvent-dependent, for reasons not discussed here. A discussion of solvent effects is contained in reference 4, and a recent interpretation of the origin of such effects is presented in reference 7.

4-Methyl-3-penten-2-one (**1**) contains both n- and π-electrons, the former being the nonbonding electron pairs on oxygen. Note that the absorption bands in its UV spectrum (Figure 11.2) are rather broad, with relatively poorly defined maxima. The diffuseness of the spectrum is a consequence of the fact that electronic transitions can occur from a variety of vibrational and rotational levels of the ground electronic state into a number of different such levels of the excited electronic state. Although the transitions themselves are quantized—that

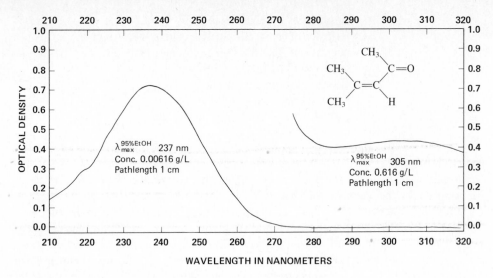

Figure 11.2 UV spectrum of 4-methyl-3-penten-2-one.

is, only certain of them are theoretically permissible—and should therefore appear as sharp "lines," the fact that closely spaced vibrational-rotational levels give rise to closely spaced lines causes coalescence of the discrete absorptions into a band *envelope* to produce the broad bands that are experimentally observed in the spectrum.

The most important characteristics of a UV spectrum are the locations of any maxima and their corresponding intensities. Figure 11.2 shows two maxima of different intensity at 238 and 314 nm. The location and intensity of a maximum are characteristic for a particular type of electronic excitation. For our purposes, they can be used to identify a specific type of **chromophore,** a term that refers to the particular arrangement of atoms responsible for absorption of the incident light. In the case of **1** and its UV spectrum, the chromophore is the α,β-unsaturated carbonyl moiety, with the longer-wavelength, less-intense maximum at 314 nm being assigned to the $n \to \pi^*$ excitation and the other maximum at 238 nm to the $\pi \to \pi^*$ process. Thus it is possible to detect the presence of various functional groups if they serve as chromophores in the UV-visible region of the electromagnetic spectrum. Correlations between UV absorption and structure are discussed in part c below.

Citral (**2**) is a natural product that can be obtained by steam distilling lemon grass oil (Section 5.7b), and its UV spectrum is shown in Figure 11.3.

2
Citral

This compound exhibits absorption maxima at 230 nm and 320 nm that are due to the α,β-unsaturated carbonyl chromophore; note that both **1** and **2** contain this general type of

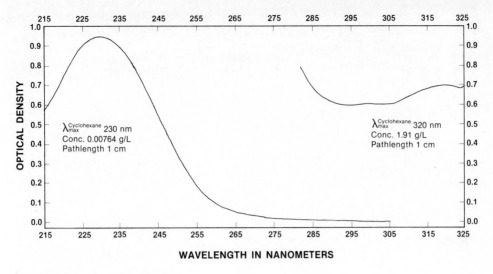

Figure 11.3 UV spectrum of citral.

chromophore, but **2** is an aldehyde, whereas **1** is a ketone. The observed λ_{max} for **2** are of the same order of magnitude and intensity as those for **1,** with the differences being attributed to the type of carbonyl group (aldehyde versus ketone) and the nature of the substituents attached to the α,β-unsaturated carbonyl moieties. Note should be made that the second carbon-carbon double bond in **2** is not part of the conjugated α,β-unsaturated carbonyl chromophore, owing to the fact that it is isolated from the conjugated system by two saturated carbon atoms (CH_2 groups). Isolated π-bonds do not produce $\pi \rightarrow \pi^*$ maxima at wavelengths greater than 200 nm.

The preceding discussion assumes that the molecular weight of the compound is known, but if it is unknown, the molar extinction coefficient cannot be calculated; that is, one would not be able to express the concentration c in *moles/liter* of solution. In this instance, the intensity of an absorption must be evaluated in a manner different from that described by equation 1. One way of doing so is given in equation 2, in which A and l have their usual meaning, but c is measured in g/100 mL of solution. Like ϵ, E is a characteristic measure of the absorptivity of the unknown substance.

$$E_{1\,cm}^{1\%} = \frac{A}{lc} \tag{2}$$

(c) Correlation Between Structure and UV Spectra

A compilation of a few of the chromophores commonly encountered by organic chemists, along with the wavelengths and intensities of their absorption maxima, is provided in Table 11.1. The measure of intensity, ϵ or log ϵ, is defined as the **molar absorptivity,** which is a measure of the probability that a particular electronic transition will occur. Because these

TABLE 11.1   Some Common Ultraviolet-active Chromophores and Their Properties

Chromophore	Wavelength of λ_{max} (nm)	Type of Excitation	log ϵ_{max}
(diene structure, R)	217–230	$\pi \rightarrow \pi^*$	4.0–4.3
(CH₂)ₙ ring polyene structure	240–280	$\pi \rightarrow \pi^*$	3.5–4.0
ketone, R—C(=O)—R′	270–300	$n \rightarrow \pi^*$	1.1–1.3
ester/acid, R—C(=O)—O—R′ (R′ = H, alkyl)	200–235	$n \rightarrow \pi^*$	1.0–1.7
α,β-unsaturated ketone	215–250	$\pi \rightarrow \pi^*$	4.0–4.3
	310–330	$n \rightarrow \pi^*$	1.3–1.5
unsaturated ester, O—R (R = H, alkyl)	200–240	$\pi \rightarrow \pi^*$	4.0–4.1
benzene-G (G = one or more alkyl groups)	256–272	$\pi \rightarrow \pi^*$	2.3–2.5
	200–210	$\pi \rightarrow \pi^*$	3.9–4.1
R—Het— benzene (Het = heteroatom)	265–290	$\pi \rightarrow \pi^*$	2.3–3.4
	210–230	$\pi \rightarrow \pi^*$	3.8–4.0

transitions are quantized, some are theoretically "allowed" and others are "forbidden," but nevertheless the latter can be experimentally observed in some cases, for reasons that will not be discussed in this text. When they do occur, these forbidden transitions do so with lower probability than the allowed transitions and are therefore observed as bands of lower intensity. Reference to Figure 11.2 and Table 11.1 suggests that $\pi \rightarrow \pi^*$ are allowed, whereas $n \rightarrow \pi^*$ are not, and this in fact corresponds to the prediction from theory. The $\pi \rightarrow \pi^*$ transition of the carbonyl group is normally observed only in *conjugated* carbonyl compounds. The carbonyl group has both nonbonding n-electrons and π-electrons, but when it is not conjugated, such as in simple ketones, the $\pi \rightarrow \pi^*$ excitation generally occurs at a wavelength of less than 200 nm, which is out of the usual UV range. On the other hand, simple ketones do exhibit an absorption maximum at 270–300 nm due to the $n \rightarrow \pi^*$ excitation, but it appears as a weak band with log $\epsilon = 1.0$–1.7.

Note that many of the UV-active chromophores given in Table 11.1 contain carbon-carbon double bonds, carbon-oxygen double bonds, conjugated systems such as dienes and α,β-unsaturated carbonyl compounds, or aromatic ring systems. It is compounds containing these chromophores that are amenable to examination by UV-visible spectroscopy. Increasing the extent of conjugation decreases the gap between the occupied and unoccupied energy levels of the molecule owing to delocalization, and this results in λ_{max} being shifted to longer wavelengths. If this shift is great enough, light is absorbed in the visible region, and a substance appears colored to the eye. For example, lycopene (**3**), which contains a highly conjugated chromophore, is found in watermelons and tomatoes and is responsible for their red color. Many pH indicators, such as Congo Red (**4**), textile fabric dyes, such as Fast Red A (**5**), and food colorings owe their color to the conjugated systems that they contain.

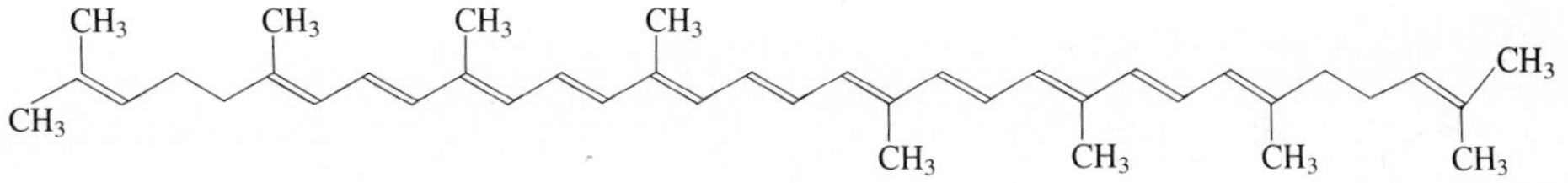

3
Lycopene

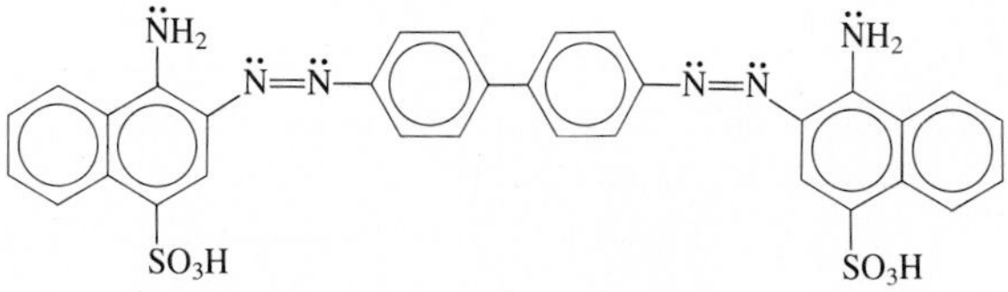

4
Congo Red

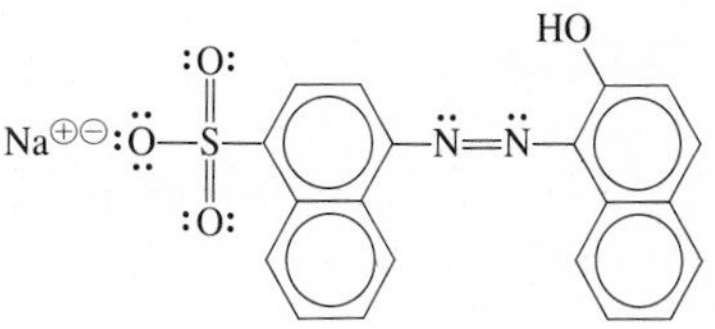

5
Fast Red A

A number of empirical generalizations have been developed that permit λ_{max} to be predicted as a function of the nature of the substituents and, in some instances, their location on the chromophore of interest. Although these rules are not provided here, they permit prediction of the relative locations of λ_{max} in compounds as closely related as **6** and **7**. These types of correlations are discussed in reference 4.

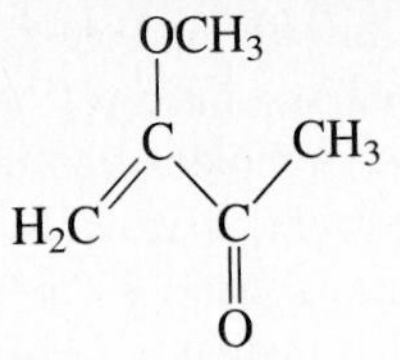

(d) Applications of UV–Visible Spectroscopy

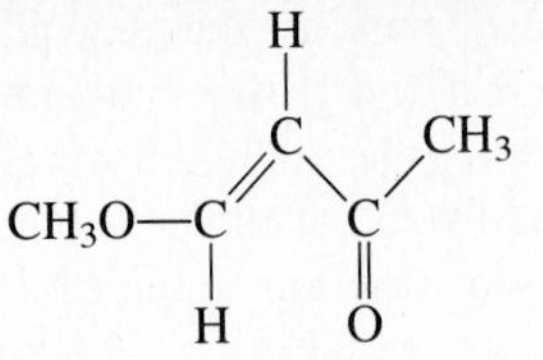

Examination of equation 1 reveals that if the quantities ϵ, l, and A are all known for a particular sample, it is possible to determine c, the concentration of the absorbing species in solution. The magnitude of ϵ can be determined experimentally by preparing solutions of the sample of known concentration and then determining A for each of these solutions in a cell of known path length. Given, then, that the various extinction coefficients and the path length, l, are known and are constant values, any change in A for a given sample as a function of time *must* result from a variation in c. Monitoring A for a particular solution over time will permit determination of the time-dependence of the concentration of the absorbing species. In other words, UV measurements permit the rate of appearance or disappearance of the species to be evaluated as a function of time, thus allowing reaction rates to be determined. Measurements of this type are typically done at one or two wavelengths, whose values are determined by the absorption maximum of the species whose appearance or disappearance is being studied. If the absorption curves of reactant and product overlap, the determination of the concentration *versus* time of each species may be somewhat more complex but can still be determined.

An example of this particular use of UV-visible spectroscopy is found in the quantitative determination of the relative rates of electrophilic bromination of substituted benzenes (equation 3), as described in Section 17.4. The rates of reaction that are determined in this experiment depend on following the disappearance of the absorbance due to molecular bromine, whose maximum is at 400 nm in the visible region. The UV-visible spectrophotometers used for these types of measurements are nonrecording and less expensive than the scanning ones used to determine the complete UV-visible spectrum. Their use is discussed in Section 17.4.

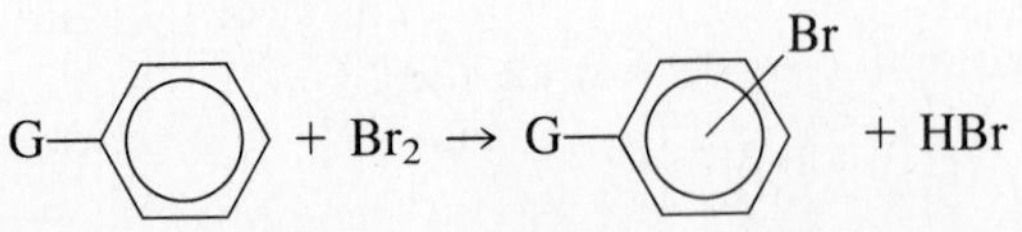

As mentioned earlier, UV-visible spectroscopy can also be used to help determine the structure of an unknown compound, provided that it contains a chromophore that absorbs in the UV-visible region.

(e) Sample Preparation for UV–Visible Spectroscopy

It is relatively easy to prepare samples for analysis by UV-visible spectroscopy. Most spectra are obtained in solutions, although cells appropriate for gaseous samples are available. The cells for UV work are constructed of quartz and commonly have a path length of 1 cm, whereas those for visible spectroscopy are made of the less expensive borosilicate glass; this latter type of glass is opaque to light in the UV region and is therefore not suitable for UV cells. Both UV and visible cells typically require about 3 mL of solution.

A variety of organic solvents, as well as water, can be used for UV-visible spectroscopy. They all share the common property of not absorbing significantly at wavelengths greater than about 220 nm, as indicated in Table 11.2. The wavelengths given in the table constitute the so-called "cutoff point" for the respective solvents. Below the given wavelength the solvent begins to absorb appreciably, so that this solvent may not be used for a solute that absorbs at wavelengths below that point. "Technical" and "reagent-grade" solvents often contain light-absorbing impurities and must usually be purified before use, but more expensive "spectral-grade" solvents, which can be used without being purified, are available for UV-visible spectroscopy. The choice of which solvent to use for a particular solute will depend on its solubility properties and the absence of chemical reaction between solvent and solute.

TABLE 11.2   Solvents for UV–Visible Spectroscopy

Solvent	Useful Spectral Range (Lower Limit)
Acetonitrile	<200 nm
Chloroform	245
Cyclohexane	205
95% Ethanol	205
Hexane	200
Methanol	205
Water	200

The concentration of the solution should be such that the observed value of A is in the range of about 0.3–1.5, since this permits the greatest accuracy in making the measurement. The approximate concentration should be estimated from the value of ϵ for any chromophores present (Table 11.1), from which c can be determined using equation 1. As a rough rule of thumb, 0.01–0.001 M solutions will give absorbances of appropriate magnitude for excitations of low intensity (log ϵ of about 1.0); dilution of this solution will then permit more intense absorptions to be observed in the desired range for A.

The solution must be weighed carefully, quantitatively transferred to a volumetric flask, and diluted accurately in order to achieve precise measurements of A and therefore of ϵ. Furthermore, the accidental introduction of even a minute amount of an intensively absorbing impurity into the solution can have a dramatic effect on the observed UV spectrum, so great care should be exercised in the handling and cleaning of all apparatus associated with the preparation of the solution. The cells should be rinsed thoroughly with the solvent that is being used, both before and after use, in order to minimize the development of contamination in the sample. Quartz cells to be used for UV work should never be rinsed with acetone for cleaning and drying purposes. Trace residues of acid or base catalysts on the surface of the quartz may, through an acid- or base-catalyzed aldol condensation, lead to the formation of trace quantities of 4-methyl-3-penten-2-one (Figure 11.2), whose presence may invalidate precise measurements of ϵ, when such work must be done. The outside optical surfaces of the cells must be clean and free of fingerprints. The cells are best cleaned using the same pure solvent that is being used for the solute, and they must also be rinsed thoroughly when changing from one concentration to another.

EXERCISES

1. For the UV spectra of Figures 11.2 and 11.3, assign the chromophores responsible for each maximum and calculate the value(s) of λ_{max} and of log ϵ.

2. Replot Figures 11.2 and 11.3 using values of log ϵ rather than optical density (absorbance, A) along the ordinate.

3. Repeat Exercise 1 for the following UV spectra, which are reproduced in this text:
   a. Figure 16.5   **b.** Figure 21.7   **c.** Figure 21.14
   d. Figure 21.39   **e.** Figure 24.4

REFERENCES

1. Dyer, J. R. *Applications of Absorption Spectroscopy of Organic Compounds,* Prentice-Hall, Englewood Cliffs, N.J., 1965.

2. Conley, R. T. *Infrared Spectroscopy,* Allyn and Bacon, Boston, 1966.

3. Pasto, D. J.; Johnson, C. R. *Organic Structure Determination,* Prentice-Hall, Englewood Cliffs, N.J., 1969. Describes applications of IR, NMR, and UV-visible spectroscopy and mass spectrometry.

4. Silverstein, R. M.; Bassler, G. C.; Morrill, T. C. *Spectrometric Identification of Organic Compounds,* 3d ed., John Wiley & Sons, New York, 1974. Includes IR, NMR, and UV spectroscopy and mass spectrometry.

5. Fleming, I.; Williams, D. H. *Spectroscopic Methods in Organic Chemistry,* McGraw-Hill Book Company, New York, 1966.

6. Haberfield, P. et al., *Journal of the American Chemical Society,* **1977,** 99, 6828.

11.2 Mass Spectrometry

The characterization and identification of the structures of organic molecules is greatly facilitated by the utilization of various spectroscopic methods such as infrared, ultraviolet-visible, and nuclear magnetic resonance, and these methods have the virtue of not destroying the substance being investigated. Another highly useful technique to organic chemists is **mass spectrometry (MS).** Although this technique has a variety of applications, two of its most important are in the determination of molecular weights and elemental compositions. It is *faster* and much more *accurate* and requires a much smaller amount of sample (*ca.* 10^{-6} g) than the classic methods, which it has almost completely replaced. Mass spectrometers are generally unavailable in the undergraduate organic laboratory, but mass spectrometry is sufficiently important that a brief introduction to it and the interpretation of mass spectra is presented here. Further information on this topic can be obtained by consulting the references at the end of the section or one of the contemporary lecture textbooks.

(a) The Concept of Mass Spectrometry

In MS, a sample of the substance is vaporized under high vacuum in the mass spectrometer, and a minute amount of the vapor is bombarded with electrons having energies in excess of 70 electron volts ($>$1400 kcal/mole). This can cause an electron within the molecule to be ejected to yield a charged species, which is positively charged and is called the **molecular ion, P^+** (equation 4). Although in reality the molecular ion is a **cation radical, $P^{\ddagger}$,** and contains an unpaired electron, it is usually abbreviated as P^+. Because of the high energy of the bombarding electrons, the molecule can also be broken into smaller parts, called *fragments* or daughter ions, which can also be positively charged. Bombardment can also give negatively charged ions, but in the usual mass spectrometer only the positively charged ions are detected.

$$M \quad + e^- \rightarrow \quad P^{\ddagger} \quad + 2\,e^- \qquad \textbf{(4)}$$

Molecule   Parent cation
radical

The mass-to-charge ratio, m/e, of the molecular ion and of each of the fragments is measured, but if each ion bears a single positive charge, the mass-to-charge ratio will be numerically equal to the mass of the fragment or molecular ion. However, polycharged cations are produced on occasion, thus making identification of the molecular ion and its fragments more difficult. Fragmentation of the molecular ion may occur very quickly to give one charged and one uncharged fragment, and the fragmentation products may break apart further to give even smaller charged and neutral parts (equation 5).

$$P^+ \rightarrow A^+ + B \quad \text{or} \quad A + B^+ \qquad \textbf{(5a)}$$

and then

$$A^+ \rightarrow C^+ + D \quad \text{or} \quad C + D^+ \qquad \textbf{(5b)}$$

The relative abundance of each fragment produced from the substance in the mass spectrometer is recorded as a function of the m/e ratio to provide the mass spectrum (see, for example, Figure 11.4). Because of the design of the instrument, only positively charged ions appear in the spectrum; neutral molecules and negatively charged ions are not observed. The mass spectrometer is a complex instrument whose construction and operation is described in reference 1 at the end of this section.

(b) Interpretation of Mass Spectra

Mass spectrometry is useful for measuring the molecular weight and molecular formula of a substance and also finds application in determination of its structure. The molecular ion peak, P^+, provides the molecular weight of the substance, and this peak can usually be detected. However, if the molecular ion undergoes rapid fragmentation, the number of them reaching the detector may be insufficient to produce a peak strong enough to be seen in the spectrum. Fortunately, this is usually not the case but is a possibility that exists for certain substances and must be recognized.

TABLE 11.3 Natural Abundance of Heavy Isotopes for Common Elements

Element	Relative Abundance (%) of Lowest-Mass Isotope	Relative Abundance (%) of Other Isotopes	
Carbon	100 ^{12}C	1.08 ^{13}C	
Hydrogen	100 ^{1}H	0.016 ^{2}H	
Nitrogen	100 ^{14}N	0.38 ^{15}N	
Oxygen	100 ^{16}O	0.04 ^{17}O	0.20 ^{18}O
Sulfur	100 ^{32}S	0.78 ^{33}S	4.40 ^{34}S
Chlorine	100 ^{35}Cl	32.5 ^{37}Cl	
Bromine	100 ^{79}Br	98.0 ^{81}Br	

Another problem encountered in mass spectroscopy arises from the existence of stable isotopes for many elements. Table 11.3 lists the stable isotopic forms of some of the common elements contained in organic molecules. The abundances are derived by setting the isotope of lowest atomic mass equal to 100% abundance and calculating the abundances of the other isotopes relative to it. Mass spectra often contain peaks of reasonable intensity that result from the presence of isotopes, such as peaks called $P + 1$ and $P + 2$ in which P is the mass of the molecular ion. These higher-mass peaks represent increases in mass of 1 or 2 due to heavier isotopes. For example, the mass spectrum of carbon monoxide shows a peak at $m/e = 28$ that is due to the parent peak, P^{+}, and corresponds to the molecular weight of CO. However, the spectrum also exhibits another peak at $m/e = 29$, which is about 1.12% of the intensity of P^{+}; this is the $P + 1$ peak, which is due to the natural abundance of stable isotopes of carbon and oxygen in the sample. To elaborate further, naturally occurring carbon contains ^{13}C in an amount equal to 1.08% of the amount of ^{12}C, and the amount of ^{17}O is 0.04% of the amount of ^{16}O in naturally occurring oxygen. The m/e peak of 28 in the mass spectrum of CO is due to $^{12}C^{16}O$, and the m/e peak at 29 is due to the presence of $^{12}C^{17}O$ (0.04%) and $^{13}C^{16}O$ (1.08%) in the sample; the sum of 1.08% + 0.04% gives 1.12%, which represents the observed abundance of the mixture of isotopes contained in the sample. Thus, the molecular weight of a substance is usually determined from the molecular ion peak observed in the spectrum, and the small peaks at slightly higher m/e are due to the presence of the heavier isotopes in the sample, that is, the $P + 1$ and $P + 2$ peaks, and are ignored. However, it is not always desirable to ignore these peaks. They may be useful if one is trying to distinguish between two compounds having the same molecular weight. For example, CO and N_2 have the same molecular weight of 28, and here the relative intensities of the $P + 1$ and $P + 2$ peaks allow the distinction to be made.

Tables containing the calculated intensities of the $P + 1$ and $P + 2$ peaks have been prepared for many combinations of carbon, hydrogen, oxygen, and nitrogen, and they permit one to limit the number of possible molecular formulas corresponding to a particular value of P^{+} (see reference 1). The formula given in equation 6 can be used to determine the intensity of the $P + 1$ peak for molecules containing more than one of each type of atom.

$$\text{Intensity of } P + 1 = (\text{no. of carbon atoms} \times \text{natural abundance of } ^{13}C) \qquad (6)$$
$$+ (\text{no. of hydrogen atoms} \times \text{natural abundance of } ^{2}H)$$
$$+ (\text{no. of nitrogen atoms} \times \text{natural abundance}$$
$$\text{of } ^{15}N) + \cdots$$

To summarize, the P^{+} peak provides the molecular weight of the original compound. The intensities of the $P + 1$ and the $P + 2$ peaks, relative to that of the P^{+} peak, allow one to deduce, with the aid of tables, possible molecular formulas for an unknown compound.

The large number of fragments that can form from the molecular ion normally make it difficult to interpret a mass spectrum completely. Nevertheless, application of theories of organic reaction mechanisms allows understanding of many aspects of the fragmentation patterns for a given molecular ion. For example, preferential bond breaking occurs if relatively stable allyl, benzyl, or tertiary carbocations result. There is also a tendency for stable neutral molecules, such as water, ammonia, nitrogen, carbon monoxide, carbon dioxide, ethylene, and acetylene, to result from fragmentation. Although these are not observed because only cations can be detected in the mass spectrum, their loss from a molecular ion to give the corresponding fragments gives clues as to the nature of functional groups that are present in the molecule. Consequently, examination of the *m/e* values for some of the fragments may also provide valuable information about the structural features present in an unknown compound.

The mass spectrum of 2-methylbutane **(8),** shown in Figure 11.4, contains several peaks that reveal a wealth of information in support of its structure.

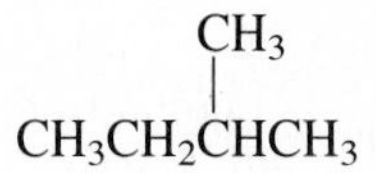

CH3
CH3CH2CHCH3

8

2-Methylbutane

When atomic mass units of 12.000 for carbon and 1.000 for hydrogen are used, the molecular weight of **8** is 72, and a peak at $m/e = 72$ is observed that is the molecular ion peak, P^+. The smaller peak at $m/e = 73$ is the $P + 1$ peak, which results from the presence of ^2H and ^{13}C in the compound. The peak at $m/e = 43$ is the strongest in the spectrum; its relative intensity is arbitrarily set as 100%, and the intensities of all other peaks in the spectrum are measured relative to it. The most intense peak in a mass spectrum is defined as the *base peak*. Other peaks of fairly high intensity appear at *m/e* of 15, 29, and 57, all of which can be accounted for by considering the possible ways in which the 2-methylbutane cation radical can fragment. The following scheme illustrates the origin of these peaks.

CH3
CH3—CH—CH2—CH3
1 2 3 4
Cation radical of
2-methylbutane, m/e = 72 (P)
Removal of CH3
from C-2
Break bond between C-2
and C-3
CH3CHCH2CH3 + CH3
m/e = 57
and
CH3CHCH2CH3 + CH3
m/e = 15
CH3CHCH3 + CH3CH2
m/e = 43
and
CH3CHCH3 + CH3CH2
m/e = 29

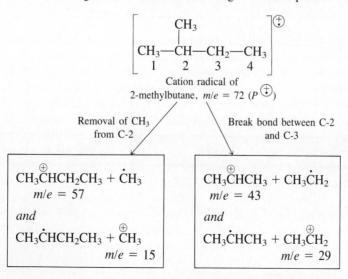

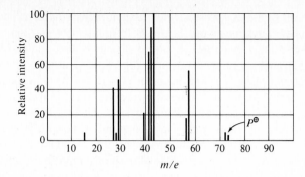

Figure 11.4 Mass spectrum of 2-methylbutane.

2-Methylbutane is first converted to the 2-methylbutane cation radical, which is the molecular ion. This ion can lose a methyl group from C-2 in two ways, to form either the 2-butyl cation radical and the methyl radical or the 2-butyl radical and the methyl cation radical. However, based on the known stability of carbocations, we can reason that the formation of the 2-butyl cation radical, a secondary carbocation, is greatly favored over the formation of the methyl cation radical, a methyl carbocation. The small peak at m/e 15 is due to the methyl ion, and the much more intense peak at m/e 57 can be attributed to the 2-butyl cation radical. The original cation radical can also fragment by breaking the bond between C-2 and C-3 to give either the 2-propyl cation radical and the ethyl radical or the 2-propyl radical and the ethyl cation radical. The formation of the 2-propyl cation radical, m/e 43, is favored over that of the ethyl cation radical, m/e 29, because the former is a secondary carbocation, which is more stable than the ethyl cation radical, a primary carbocation. Of all the cation radicals, the 2-propyl one is produced in the greatest amount and is the base peak (relative intensity = 100%). The other peaks in the spectrum arise from factors that we will not discuss, but some of them are due to $P + 1$ and $P + 2$ peaks of the various cation radicals.

The previous example illustrates the approach to applying mass spectrometry to structure identification or structure confirmation. A more practical use of this technique involves the determination of the mass spectrum of an unknown compound and, with practice, learning to deduce the structure of various fragments. For example, a peak at m/e 15 is often due to methyl, one at 29 is ethyl, and one at 43 is propyl. Once the fragments are identified, it is often possible to put them together to deduce the structure of the unknown compound.

The discussion so far assumes that the mass spectrometer can distinguish between ions that differ in mass by one atomic mass unit (amu), where the m/e values are measured to the nearest whole number. Most laboratories are equipped with these *low-resolution* mass spectrometers. More expensive instruments, called *high-resolution* mass spectrometers, can measure the m/e ratios to ±0.001 or ±0.0001 amu and provide very accurate measurements of molecular weights. This alleviates the need to calculate the relative intensities of the $P + 1$ and $P + 2$ peaks in order to distinguish between two possible molecules. For example, high-resolution mass spectrometry permits the distinction between CO (molecular weight

27.9949) and N_2 (molecular weight 28.0062) or between N_2H_4 (molecular weight 32.0375) and CH_4O (molecular weight 32.0262). In essence, then, mass spectrometry can be used to determine elemental composition.

EXERCISES

1. Determine the relative intensities of the P, $P + 1$, and $P + 2$ peaks in the following molecules:

   a. N_2 **b.** CH_4 **c.** CH_3NH_2 **d.** CH_3Br

2. In the mass spectrum of 2,2-dimethylpropane, the base peak occurs at m/e 57, peaks of moderate intensity are found at m/e 29 and 43, and a weak peak occurs at m/e 15.
   a. Account for the fragments that produce these peaks.
   b. What is m/e for P^+?

3. The mass spectrum of 3,3-dimethylheptane consists of the following m/e values, with relative intensities being given in parentheses: 43 (100%), 57 (100%), 71 (90%), 29 (40%), 99 (15%), and 113 (very low). The parent peak is missing.
   a. Indicate what fragments are responsible for each m/e value given.
   b. Suggest why the peak at 113 is of very low intensity.
   c. What is the m/e value of the parent peak? Suggest why it is not observed.

REFERENCES

1. Silverstein, R. M.; Bassler, G. C.; Morrill, T. C. *Spectrometric Identification of Organic Compounds,* 4th ed., John Wiley & Sons, New York, 1981, p. 3–93.
2. Beynon, J. H. *Mass Spectrometry and Its Application to Organic Chemistry,* Elsevier, Amsterdam, 1960.
3. Pasto, D. J.; Johnson, C. R. *Organic Structure Determination,* Prentice-Hall, Englewood Cliffs, N.J., 1969, pp. 243–294.
4. McLafferty, F. W. *Interpretation of Mass Spectrometry,* 2d ed., Benjamin/Cummins, Menlo Park, Calif., 1973.
5. Fleming, I.; Williams, D. H. *Spectroscopic Methods in Organic Chemistry,* McGraw-Hill Book Company, New York, 1966.
6. Lambert, J. B.; Shurvell, H. F.; Verbit, L.; Cooks, R. G.; Stout, G. H. *Organic Structural Analysis,* Macmillan, New York, 1976.

alkanes

12.1 Chlorination by Means of Sulfuryl Chloride

Alkanes (saturated hydrocarbons) are generally chemically inert even toward strong acids and bases. This characteristic greatly limits the chemistry of this class of substances under normal laboratory conditions. Some chemical reactions have been developed, however, that allow introduction of a limited number of functional groups into an alkane under relatively mild conditions. Subsequent reactions then can result in overall incorporation of a much wider variety of functional groups onto the original alkyl chain. For example, a saturated hydrocarbon, RH, can be transformed into a nitroalkane, RNO_2, or a hydroperoxide, ROOH, on reaction with nitrogen tetroxide or molecular oxygen, respectively (equations 1 and 2); reduction of the nitroalkane would give an amine, RNH_2, and of the hydroperoxide would yield an alcohol, ROH. Note that although introduction of the amino and hydroxyl groups into RH is *not* possible directly, this is precisely the net outcome of the two-step sequences described. A common characteristic of the types of chemical reactions that allow incorporation of functionality into alkanes is that **free-radical chain** processes are involved.

$$ R\text{---}H + HNO_3 \xrightarrow{400°C} \underset{\substack{\text{Nitroalkane} \\ \text{(mixture)}}}{R\text{---}NO_2} \xrightarrow[\text{catalyst}]{H_2} \underset{\substack{\text{Amine} \\ \text{(mixture)}}}{R\text{---}NH_2} \qquad \textbf{(1)} $$

$$ R\text{---}H + O_2 \xrightarrow{\text{Initiator}} \underset{\substack{\text{Hydroperoxide} \\ \text{(mixture)}}}{R\text{---}OOH} \xrightarrow[\text{catalyst}]{H_2} \underset{\substack{\text{Alcohol} \\ \text{(mixture)}}}{R\text{---}OH} \qquad \textbf{(2)} $$

The experiments described in this section allow conversion of alkanes to chloroalkanes (alkyl chlorides), RCl, by way of a free-radical substitution reaction. No reaction occurs when chlorine gas and a saturated hydrocarbon are mixed at room temperature, but irradiation with ultraviolet light ($h\nu$) or heating of the mixture at 200–400°C affords chloroalkanes and hydrogen chloride (equation 3). The photochemical or thermal activation of the mixture is required to convert some molecular chlorine into chlorine atoms, which themselves are free radicals. The generation of chlorine atoms is essential to the initiation of the substitution reaction between molecular chlorine and the alkane.

$$RH + Cl_2 \xrightarrow[h\nu]{\text{heat or}} RCl + HCl \qquad\qquad \textbf{(3)}$$

It is somewhat easier experimentally to generate chlorine atoms by use of an **initiator,** a substance that will decompose to free radicals under relatively mild reaction conditions (equation 4). The free radicals, In·, derived from the initiator molecule can react with molecular chlorine to produce InCl and the all-important chlorine atoms (equation 5). These in turn can react with the substrate hydrocarbon to give the desired chloroalkanes. In this experiment the usual procedure for chlorination of alkanes by use of molecular chlorine has been modified, for purposes of safety and convenience, by use of sulfuryl chloride, SO_2Cl_2, as the source of chlorine.

$$In\!-\!In \xrightarrow[h\nu]{\text{heat or}} In\cdot + In\cdot \qquad\qquad \textbf{(4)}$$

$$In\cdot + Cl\!-\!Cl \rightarrow InCl + Cl\cdot \qquad\qquad \textbf{(5)}$$

The mechanism of the free-radical chlorination of alkanes is fundamentally the same whether sulfuryl chloride or molecular chlorine is employed and can be divided into three discrete stages: **initiation, propagation,** and **termination.** In the following experiments the first step of initiation is the *homolytic* cleavage of azobisisobutyronitrile (IUPAC name: 2,2′-azobis[2-methylpropionitrile], **1**), commonly abbreviated as AIBN, into nitrogen and the free radical, $(CH_3)_2C(CN)\cdot$ (equation 6); this reaction occurs at a kinetically acceptable rate at 80–100°C. The initiator radical then attacks sulfuryl chloride to generate chlorine atoms, along with SO_2 (equations 7 and 8).

Initiation

$$
\underset{\substack{\textbf{1}\\ \text{Azobisisobutyronitrile}}}{
CH_3\!-\!\overset{\displaystyle CN}{\underset{\displaystyle CH_3}{\overset{|}{\underset{|}{C}}}}\!-\!N\!=\!N\!-\!\overset{\displaystyle CN}{\underset{\displaystyle CH_3}{\overset{|}{\underset{|}{C}}}}\!-\!CH_3}
\xrightarrow{80\text{–}100°C} N_2 + 2\ CH_3\!-\!\overset{\displaystyle CN}{\underset{\displaystyle CH_3}{\overset{|}{\underset{|}{C}}}}\!\cdot \qquad \textbf{(6)}
$$

$$\left(CH_3\!-\!\overset{\displaystyle CN}{\underset{\displaystyle CH_3}{\overset{|}{\underset{|}{C}}}}\!\cdot = In\cdot\right)$$

$$\text{In}\cdot + \text{Cl}-\overset{\overset{\displaystyle O}{\|}}{\underset{\underset{\displaystyle O}{\|}}{S}}-\text{Cl} \rightarrow \text{InCl} + \cdot\overset{\overset{\displaystyle O}{\|}}{\underset{\underset{\displaystyle O}{\|}}{S}}-\text{Cl} \tag{7}$$

$$\cdot\overset{\overset{\displaystyle O}{\|}}{\underset{\underset{\displaystyle O}{\|}}{S}}-\text{Cl} \rightarrow \overset{\overset{\displaystyle O}{\|}}{\underset{\underset{\displaystyle O}{\|}}{S}} + \text{Cl}\cdot \tag{8}$$

Propagation
$$\text{Cl}\cdot + \text{RH} \rightarrow \text{R}\cdot + \text{HCl} \tag{9}$$

$$\text{R}\cdot + \text{ClSO}_2\text{Cl} \rightarrow \text{RCl} + \cdot\text{SO}_2\text{Cl} \tag{10}$$

$$\cdot\text{SO}_2\text{Cl} \rightarrow \text{SO}_2 + \text{Cl}\cdot \tag{11}$$

Termination
$$\text{Cl}\cdot + \text{Cl}\cdot \rightarrow \text{Cl}_2 \tag{12}$$

$$\text{R}\cdot + \text{Cl}\cdot \rightarrow \text{RCl} \tag{13}$$

$$\text{R}\cdot + \text{R}\cdot \rightarrow \text{R}-\text{R} \tag{14}$$

The propagation steps include abstraction of a hydrogen atom from the hydrocarbon by chlorine atom to produce a new free radical, R· (equation 9), and attack of this radical upon sulfuryl chloride to yield an alkyl chloride and the radical ·SO$_2$Cl (equation 10). The latter, as noted in equation 7 and repeated in equation 8, functions as the precursor to chlorine atoms and is the final propagation step. A **chain reaction** is thus produced.

Once initiated, this chain reaction could, in principle, continue until either sulfuryl chloride or the alkane, depending upon which is the *limiting reagent* (see Section 1.7 for the definition of this term), is completely consumed. In actuality, termination reactions (equations 12–14, for example) interrupt the chain, so the process of initiation must be continued throughout the course of the reaction.

In summary, the various steps in a free-radical chain mechanism are described as follows. An **initiation step** generally involves the formation of a *low* concentration of free radicals from molecules and results in an *increase* in the concentration of free radicals present in the system. A **propagation step** produces *no net change* in the concentration of radicals present, and a **termination step** gives a *net decrease* in their concentration.

Free-radical halogenation of saturated hydrocarbons generally produces mixtures of several monohalogenated isomers as well as polyhalogenated products. Separation of such a mixture into its pure components is usually somewhat difficult, although such separations are performed in some industrial processes. Moreover, the reaction can be useful industrially in cases where *pure* individual haloalkanes are *not* required. For example, *n*-dodecane can be chlorinated to give a mixture of monochlorododecanes (equation 15). These chloroalkanes can then be converted by way of a Friedel-Crafts reaction (Section 17.2) into a mixture of phenyldodecanes, which are useful intermediates in the manufacture of one type of biodegradable detergent.

Methods other than halogenation of alkanes are commonly used for the preparation of pure haloalkanes. The reaction of a hydrogen halide with an alkene (equation 16) or an alcohol (equation 17) provides the corresponding haloalkane. Alcohols can also be readily

$$R-CH_2-CH_2-CH_2-CH_3 \xrightarrow{Cl_2} R-CH_2-CH_2-CH_2-CH_2Cl$$

$$+ R-CH_2-CH_2-\underset{\underset{Cl}{|}}{CH}-CH_3 + R-CH_2-\underset{\underset{Cl}{|}}{CH}-CH_2-CH_3$$

$(R = n\text{–}C_8H_{17})$

$$+ R-\underset{\underset{Cl}{|}}{CH}-CH_2-CH_2-CH_3 + \cdots \qquad \textbf{(15)}$$

converted to chloroalkanes by reaction with thionyl chloride, $SOCl_2$ (equation 18) or with phosphorus trichloride (equation 19, $X = Cl$); the corresponding bromides are produced when alcohols are treated with phosphorus tribromide (equation 19, $X = Br$). Procedures for some of these alternate ways to produce haloalkanes are given in Sections 13.3 and 18.2.

$$\underset{R}{\overset{R}{\diagdown}}C=C\underset{R}{\overset{R}{\diagup}} + HX \rightarrow R-\underset{\underset{X}{|}}{\overset{\overset{R}{|}}{C}}-\underset{\underset{H}{|}}{\overset{\overset{R}{|}}{C}}-R \qquad \textbf{(16)}$$

$$R-OH + HX \rightarrow R-X + H_2O \qquad \textbf{(17)}$$

$$RO-H + Cl-\underset{\underset{O}{\|}}{S}-Cl \xrightarrow{-HCl} R-O-\underset{\underset{O}{\|}}{S}-Cl \xrightarrow{heat} R-Cl + SO_2 \qquad \textbf{(18)}$$

$$3\ RO-H + PX_3 \rightarrow 3\ R-X + H_3PO_3 \qquad \textbf{(19)}$$

The Pre-lab exercises for Section 12.1, Chlorination by Means of Sulfuryl Chloride, can be found on page PL. 29.

EXPERIMENTAL PROCEDURE

DO IT SAFELY

1 The sulfuryl chloride used in this experiment reacts violently with water. Be sure that your glassware is *dry*. Take special care to avoid getting sulfuryl chloride on your skin, and do not breathe its vapors. This chemical should be weighed out in a ventilation hood, and we recommend that rubber gloves be worn when transferring it.

2 The hydrocarbons used in this experiment are *flammable*, and the use of burners should be avoided if possible.

3 Be certain that *all* connections in your apparatus are tight *prior* to heating of the reaction mixture.

4 When the reaction mixture is washed with aqueous sodium carbonate, carbon dioxide is generated in the separatory funnel. Be sure to vent the funnel frequently when shaking it (see Section 2.18) in order to relieve any gas pressure that develops.

This experiment provides for the chlorination of three alkanes: cyclohexane, heptane, and 1-chlorobutane. Optimally, each student should do all three parts because the number, type, and proportion of products from each of the three compounds illustrate important principles. Alternatively, the compounds may be divided among three groups of students and the results of each group compared.

$$\begin{array}{c} CH_2 \\ CH_2 \quad CH_2 \\ | \qquad | \\ CH_2 \quad CH_2 \\ CH_2 \end{array}$$

Cyclohexane $CH_3(CH_2)_5CH_3$ $CH_3CH_2CH_2CH_2Cl$

         Heptane   1-Chlorobutane

A. Cyclohexane

Fit a 100-mL round-bottomed flask with a water-cooled reflux condenser that is equipped at its top with a vacuum adapter connected as shown in Figure 12.1 to serve as a trap for the SO_2 and HCl produced in the reaction. *Either* fill the upper section of the adapter with calcium chloride, as in Figure 12.1, *or* attach a drying tube filled with calcium chloride to the female joint of the adapter. Place 33.6 g (43.3 mL, 0.400 mol) of cyclohexane,[1] 27.0 g (16.2 mL, 0.200 mol) of sulfuryl chloride, and 0.1 g of azobisisobutylonitrile (AIBN, **1**) in the flask, then weigh the flask and its contents. Connect the condenser and trap, turn on the vacuum so as to produce a *gentle* flow of air through the trap, and heat the mixture to a gentle reflux for 20 min, using either a heating mantle or an oil bath. Cool the reaction mixture, disconnect the flask from the condenser, and weigh it and its contents.★ If the loss of weight

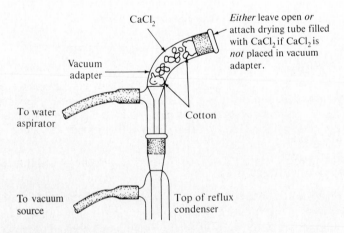

Figure 12.1 Details of gas trap arrangement.

[1] If the starting material has been stored in a metal container, it may be necessary to perform a simple distillation to remove traces of metal ions, which can inhibit the reaction.

is less than theoretical, add another small portion of the AIBN (**1**) and heat the mixture to reflux for 10 min. The theoretical loss of weight is calculated on the basis that 1 mole each of HCl and of SO_2 is evolved for each mole of sulfuryl chloride consumed. This calculation should be made *before* coming to laboratory.

After the theoretical amount of weight has been lost, cool the reaction mixture and *cautiously* pour it into 50 mL of *ice* water. Transfer the resulting two-phase solution to a separatory funnel and separate the layers.* If separation into two layers does not occur readily, add some sodium chloride to the separatory funnel and shake it. Wash the organic layer in the funnel with 0.5 *M* sodium carbonate solution until the aqueous washes are basic to litmus paper. Wash it once again with water and then dry it over about 3 g of *anhydrous* calcium chloride. (See Section 2.21 for a discussion of drying agents.) Filter the dried solution into a 100-mL flask fitted with a glass- or stainless steel–packed column for fractional distillation (Section 2.8) and carefully distil in order to separate the chlorinated products from unchanged starting material. After the unchanged starting material has been distilled (fraction 1), remove the column from the apparatus and replace it with a simple distillation head for the distillation of the chlorinated products. Suggested boiling ranges are as follows: fraction 1, ambient–85°C; fraction 2, 85–145°C.

Weigh each of the distillation cuts and either analyze them yourself by gas chromatography or submit samples for such analysis. Estimate the approximate yields of the chlorinated products, taking into account the amount of unchanged starting material that was recovered.

B. Heptane

Follow the directions given in part A, using 40.0 g (58.5 mL, 0.400 mol) of heptane (see footnote 1) in place of cyclohexane. Suggested boiling ranges are as follows: fraction 1, ambient–105°C; fraction 2, 105–160°C.

C. 1-Chlorobutane

Follow the directions given in part A, using 37.0 g (41.8 mL, 0.400 mol) of 1-chlorobutane (see footnote 1) in place of cyclohexane. Suggested boiling ranges are as follows: fraction 1, ambient–82°C; fraction 2, 82–165°C.

EXERCISES

1. Suggest at least one termination process for halogenation with sulfuryl chloride that has no counterpart in halogenation with molecular chlorine.

2. What is the reason for using less than the amount of sulfuryl chloride theoretically required to convert all the starting materials to monochlorinated products?

3. From which of the three starting materials should it be easiest to prepare a single pure monochlorinated product? Why?

4. What factors determine the proportion of monochlorinated isomers of heptane? of 1-chlorobutane?

5. Would it be proper to refer to azobisisobutyronitrile as a catalyst in the free-radical chlorination reaction? Why or why not?

6. What problems might attend the use of larger quantities of initiator for the free-radical chlorination reactions described in this section?

7. Why is the potential contamination of the chlorinated hydrocarbons with product(s) derived from azobisisobutyronitrile (1), for example, $(CH_3)_2C(CN)Cl$, of little concern in the halogenation experiments of this section?

8. How might a greater than theoretical loss of weight be "achieved" in a chlorination reaction performed according to the procedures of this section?

9. Why does addition of sodium chloride facilitate separation into layers of an emulsion consisting of water, chloroalkane(s), and starting alkane?

10. Why is it prudent to use caution when pouring the crude reaction mixture into ice water? Why is ice water rather than tap water specified?

11. Calculate the percent of each monochlorination product expected from heptane based on a relative reactivity of primary (1°):secondary (2°):tertiary (3°) hydrogens of 1.0:3.3:4.4. Refer to Figure 12.2 and calculate the observed ratio of 1°:2° chloroheptanes; compare the result with the anticipated theoretical ratio.

12. Refer to Figure 12.3 and calculate the percentage of each dichlorobutane present in fraction 2 (do not include the area of the peak due to 1-chlorobutane when making the calculation). Why does the observed ratio of products not agree with that predicted by use of the relative reactivities given in Exercise 11?

*13. Figure 12.4 gives the PMR spectrum of a mixture of dichlorobutanes obtained by free-radical chlorination of 1-chlorobutane.

a. Assign the multiplet at δ 5.80, the doublet at δ 1.55, and the triplet at δ 1.07, to the hydrogen nuclei of the three different isomers from which they arise. (*Hint*: Write the structures of the isomeric 1,x-dichlorobutanes and predict the multiplicity and approximate chemical shift of each group of hydrogens; see Section 9.3 for data needed to estimate chemical shifts.)

b. Calculate the approximate percentage of each isomer present in the mixture.

14. Calculate the heat of reaction for the reaction between cyclohexane and chlorine to yield chlorocyclohexane and hydrogen chloride, using the following bond dissociation energies and bond energies (in kcal/mol): C–H, 98.7; C–Cl, 81.0; Cl–Cl, 58.0; H–Cl, 103.2.

15. Chlorination of propene leads to high yields of 3-chloro-1-propene to the exclusion of products of substitution of the vinyl hydrogens. Why is this so?

16. Write the expected monochlorinated products of free-radical chlorination of methylcyclohexane with SO_2Cl_2. Predict which chlorinated isomer would be formed in highest yield, using the data on relative reactivities given in Exercise 11.

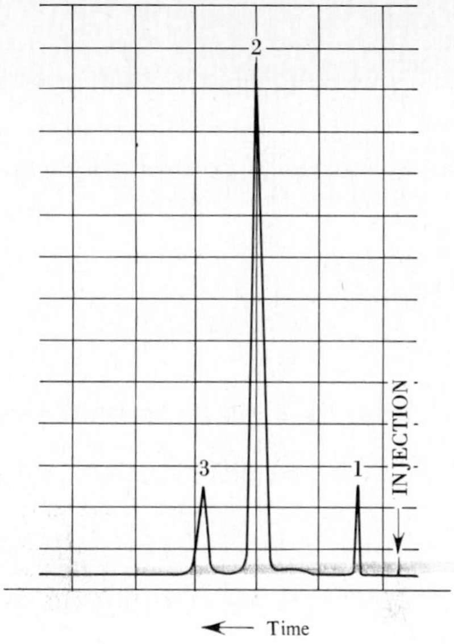

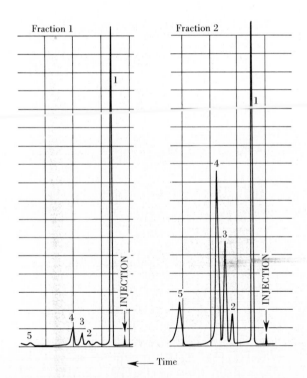

Figure 12.2 Gas chromatogram of fraction 2 resulting from chlorination of heptane. Peak 1: heptane; peak 2: 2-, 3-, 4-chloroheptanes; (bp 150°C, 151°C, and 152°C, respectively); peak 3: 1-chloroheptane (bp 159°C); column and conditions: 1.5 m, 5% silicone elastomer on Chromosorb W; 80°C, 40 mL/min.

← Time

Fraction 1

Fraction 2

← Time

Figure 12.3 Gas chromatograms of fractions 1 and 2 resulting from chlorination of 1-chlorobutane. Peak 1: 1-chlorobutane; peak 2: 1,1-dichlorobutane (bp 114–115°C); peak 3: 1,2-dichlorobutane (bp 121–123°C); peak 4: 1,3-dichlorobutane (bp 131–133°C); peak 5: 1,4-dichlorobutane (bp 161–163°C); column and conditions: 1.5 m, 5% silicone elastomer on Chromosorb W; 60°C, 40 mL/min.

REFERENCES

1. Kharasch, M. S.; Brown, H. C. *Journal of the American Chemical Society,* **1939,** *61,* 2142.
2. Reeves, P. C. *Journal of Chemical Education,* **1971,** *48,* 636.
3. Davies, D. I.; Parrott, M. J. *Free Radicals in Organic Synthesis,* Springer-Verlag, New York, 1978.
4. Pryor, W. A. *Introduction to Free Radical Chemistry,* Prentice-Hall, Englewood Cliffs, N.J., 1966.

SPECTRA OF STARTING MATERIALS AND PRODUCTS

The IR spectrum of cyclohexane is given in Figure 8.15.

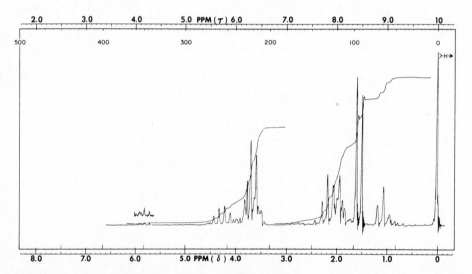

Figure 12.4 PMR spectrum of mixture of 1,*x*-dichlorobutanes.

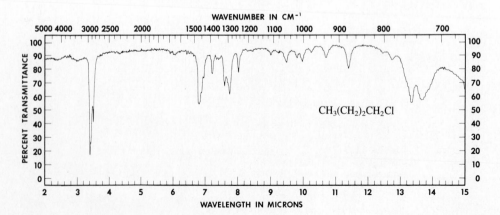

Figure 12.5 IR spectrum of 1-chlorobutane.

(a) PMR spectrum.

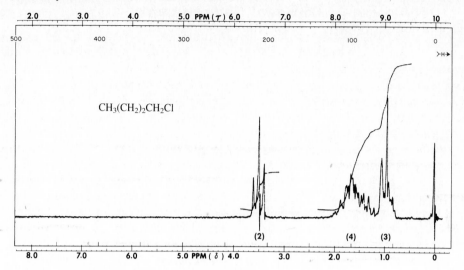

(b) CMR data. Chemical shifts: δ 13.4, 20.3, 35.0.

Figure 12.6 NMR data for 1-chlorobutane.

Chemical shift: δ 27.3.

Figure 12.7 CMR datum for cyclohexane.

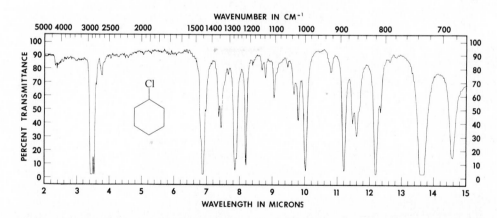

Figure 12.8 IR spectrum of chlorocyclohexane.

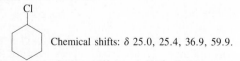

Chemical shifts: δ 25.0, 25.4, 36.9, 59.9.

Figure 12.9 CMR data for chlorocyclohexane.

12.2 Bromination: Relative Ease of Substitution of Hydrogen in Different Environments

The preceding experiment allows a comparison of the reactivity of hydrogen atoms bonded to primary (1°) and secondary (2°) carbon atoms[2] toward substitution by chlorine atoms on the basis of the relative amounts of primary and secondary chloroheptanes produced from heptane (Section 12.1, Exercise 11). In the present experiment the relative rates of reaction of bromine with several different hydrocarbons containing primary, secondary, and tertiary hydrogens will be determined.

Hydrogen atoms can also be classified according to the hybridization of the carbon atom to which they are attached. Specifically, hydrogens bound to sp^3-hybridized carbons are described as **aliphatic,** those attached to sp^2-hybridized carbons as either **vinylic** or **aromatic,** depending upon whether or not the carbon atom is part of an aromatic ring, and those bonded to sp-hybridized carbon as **acetylenic.** Examples of each of these categories are shown on the structures below. A further specification is used if the hydrogen atom is attached to a carbon atom that itself is bound to a vinylic or aromatic carbon. Such hydrogens are called **allylic** and **benzylic,** respectively, and examples of these are also given below.

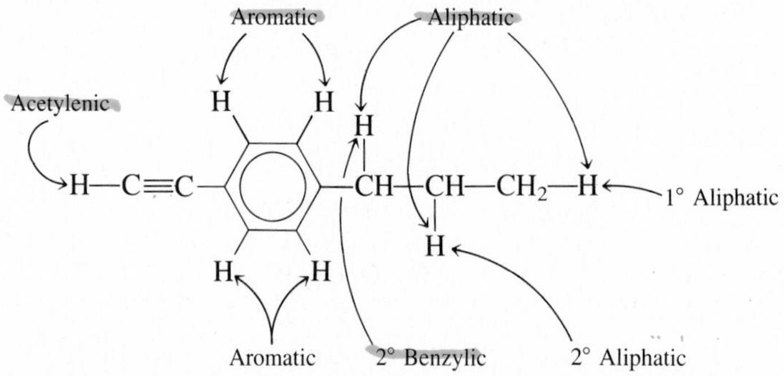

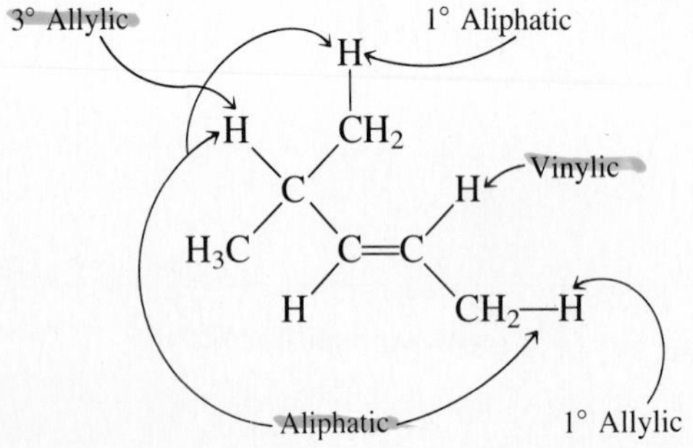

[2]In this discussion the slightly inaccurate but convenient practice will be followed of referring to a hydrogen atom according to the "degree" of the carbon atom to which it is bound; for example, a hydrogen atom bonded to a primary carbon atom will be called a "primary hydrogen."

The various classifications described above can be combined so that we speak of primary, secondary, and tertiary aliphatic, allylic, or benzylic hydrogens. Only one type of aromatic hydrogen atom and two types of vinylic hydrogens are possible. By careful consideration of your results from this experiment, you should be able to deduce an order of reactivity toward bromine atom for *seven* different types of hydrogens.

The experimental determination of relative reactivities is conceptually simple. An *excess* of hydrocarbon is allowed to react with molecular bromine, and the time required for the disappearance of the color due to bromine, as it is consumed in the reaction, is measured. The lengths of time required for decoloration of bromine are a function of the relative reactivities of the various hydrocarbons and of the reaction conditions used, and the *ratios* of these times represent the **relative** rates of the bromination reaction. Determination of the **absolute** rates of bromination of these same hydrocarbons could be done in a similar way, except that the concentration of bromine as a function of time would have to be determined much more precisely and the temperature of the reaction mixture more exactly controlled.

The measurement of relative reactivities is easier for bromination than for chlorination. Bromine is a less reactive, more selective reagent than is chlorine, so the rates of bromination of the various hydrocarbons are sufficiently different to allow the desired qualitative order of reactivity to be determined quite readily. Other advantages associated with the use of bromine are that it is a liquid rather than a gas at room temperature, a fact that makes it easier to handle in the laboratory, and that it is more highly colored than chlorine, so that decoloration is more readily discerned.

The reactions are to be carried out in carbon tetrachloride solution using a sequence of three different conditions: (1) at room temperature without special illumination, (2) at 50°C without special illumination, and (3) at 50°C with strong illumination. Under all these conditions, substitution of hydrogen by bromine occurs by a **free-radical chain mechanism** analogous to that of the chlorination reactions of Section 12.1. In this case the **initiation** step is the thermally or photochemically promoted dissociation of molecular bromine into bromine atoms (radicals), as shown in equation 20. The **propagation** steps are those of equations 21 and 22. The **termination** steps are analogous to those of equations 12–14 (Section 12.1).

$$Br_2 \xrightarrow[\text{or } h\nu]{\text{heat}} 2\ Br\cdot \tag{20}$$

$$Br\cdot + RH \rightarrow HBr + R\cdot \tag{21}$$

$$R\cdot + Br_2 \rightarrow RBr + Br\cdot \tag{22}$$

Pre-lab exercises for Section 12.2, Bromination: Relative Ease of Substitution of Hydrogen in Different Environments, are found on page PL. 31.

EXPERIMENTAL PROCEDURE

DO IT SAFELY

1 Bromine is a hazardous chemical. Do not breathe its vapors or allow it to come into contact with the skin because it may cause serious chemical burns. All operations involving the transfer of the pure liquid or its solutions should be done in a ventilation hood, and rubber gloves should be worn. If you get bromine on your skin, wash the area *immediately* with soap and warm water and soak the affected area in 0.6 *M* sodium thiosulfate solution (for up to 3 hr if the burn is particularly serious).

2 Bromine reacts with acetone to produce the powerful lachrymator, α-bromoacetone, $BrCH_2COCH_3$. Do *not* rinse glassware containing residual bromine with acetone! At the end of the experiment add small quantities of cyclohexene to any test tubes containing residual bromine, as indicated by coloration. Discard the solutions in an organic liquid-waste container *only* after the characteristic color of bromine has been discharged.

3 Avoid excessive inhalation of the vapors of any of the materials being used in this experiment. You should consider using an inverted funnel attached to a vacuum source and placed over the test tubes in order to lower the concentration of vapors in your area.

(*Note to the Instructor:* We recommend that the 1 *M* bromine in carbon tetrachloride solution be dispensed from burets [with Teflon stopcocks] placed in ventilation hoods. This will simplify the precise measurement of 1-mL portions of the solution and provide for greater safety in transferring the bromine.)

Construct a table in your notebook with the following four main headings: *Hydrocarbon*; *Types of Hydrogen,* entries under which will include the terms 1° aliphatic, 2° benzylic, *etc.*; *Conditions,* with the subheadings 25°C, 50°C, 50°C-hν); and *Elapsed Time,* entries under which will be the time required for reaction as measured by decoloration. In each of six 18- × 100-mm test tubes place a solution of 1 mL of hydrocarbon in 5 mL of carbon tetrachloride, using each of the following hydrocarbons: toluene, ethylbenzene, isopropylbenzene (cumene), *t*-butylbenzene, cyclohexane, and methylcyclohexane. Carefully label these test tubes to avoid confusing them and jeopardizing the interpretation of the experimental results. Place 1 mL of a 1 *M* solution of bromine in carbon tetrachloride in each of seven other test tubes and to *one* of them add 6 mL of carbon tetrachloride. Label this tube "Control" and the others according to which hydrocarbon is to be placed in each tube. Now add the solutions of hydrocarbons *to* those containing bromine in rapid succession. This addition should be done with agitation to ensure good mixing. Note and record the time of mixing *and* the elapsed times required for the reddish color of bromine to be discharged in each reaction mixture. Because the rates of bromination in some cases are quite fast, it is advisable to repeat the determination in order to be confident that the *relative* orders of decoloration are known.

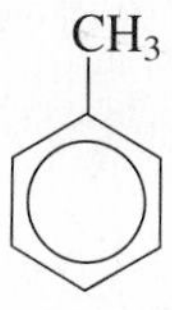

Toluene 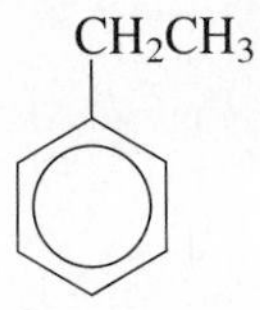   Ethylbenzene 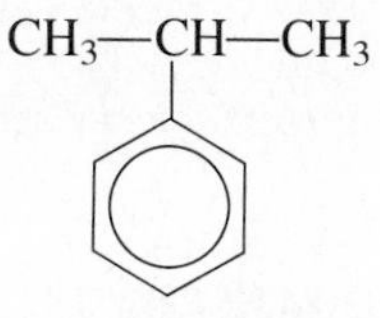   Isopropylbenzene 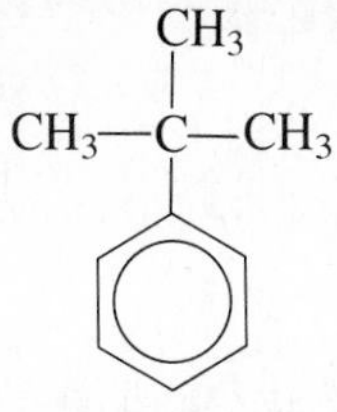   *t*-Butylbenzene

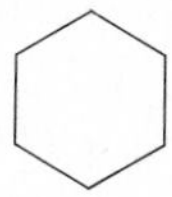

Cyclohexane 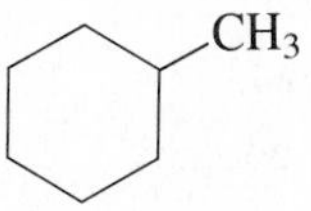   Methylcyclohexane

After watching for decoloration at room temperature for 20 min, place the "Control" and those reaction mixtures in which color remains in a beaker of water that is kept at 50°C on a steam bath or hot plate. Do not heat the water much above this temperature because carbon tetrachloride has a boiling point of 77°C. After 20–30 min at this temperature, suspend an *unfrosted* 100- or 150-watt light bulb over the test tubes at a distance of 10–13 cm (4–5 in.). Continue to observe and record the times at which the color is discharged in any tubes still containing bromine. The experiment can be discontinued when only one colored solution, other than the "Control," remains because it is obvious that this hydrocarbon reacts most slowly.

On the basis of your results, answer each of the exercises below.

EXERCISES

1. Arrange the six hydrocarbons in increasing order of reactivity toward bromination.
2. On the basis of the order of reactivity of the hydrocarbons, deduce the order of reactivity of the seven different types of hydrogens found in these compounds, that is, (1) primary aliphatic, (2) secondary aliphatic, (3) tertiary aliphatic, (4) primary benzylic, (5) secondary benzylic, (6) tertiary benzylic, and (7) aromatic.
3. Clearly explain how you arrived at your sequence in Exercise 2.
4. Draw the structure of the *major* monobromination product expected from each of the hydrocarbons used in this experiment.
5. Comment on the need for heat and/or light in order for bromination to occur with some of the hydrocarbons.
6. Why is a "Control" sample needed in this experiment?
7. Perform the calculations necessary to demonstrate that bromine is indeed the *limiting reagent* (Section 1.7). The densities and molecular weights needed to complete the calculations can be found in various handbooks of chemistry (Section 28.1c).

REFERENCES

1. Davies, D. I.; Parrott, M. J. *Free Radicals in Organic Synthesis,* Springer-Verlag, New York, 1978.
2. Pryor, W. A. *Introduction to Free Radical Chemistry,* Prentice-Hall, Englewood Cliffs, N.J., 1966.

SPECTRA OF STARTING MATERIALS

The IR spectrum of cyclohexane is given in Figure 8.15. The IR and PMR spectra of isopropylbenzene are given in Figures 8.19 and 9.14, respectively.

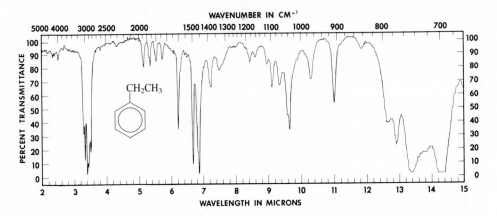

Figure 12.10 IR spectrum of ethylbenzene.

(a) PMR spectrum.

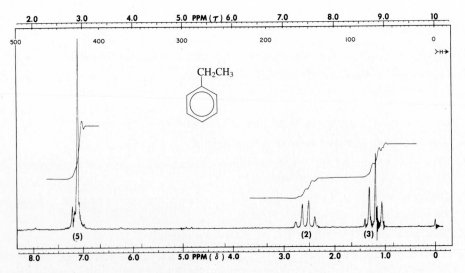

(b) CMR data. Chemical shifts: δ 15.6, 29.1, 125.7, 127.9, 128.4, 144.2.

Figure 12.11 NMR data for ethylbenzene.

CH(CH$_3$)$_2$

Chemical shifts: δ 24.1, 34.2, 125.8, 126.4, 128.4, 148.8.

Figure 12.12 CMR data for isopropylbenzene (cumene).

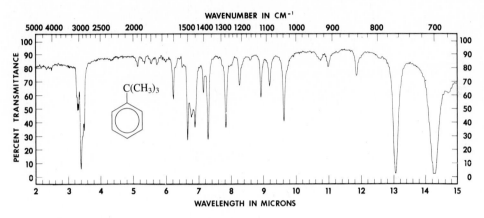

Figure 12.13 IR spectrum of *t*-butylbenzene.

(a) PMR spectrum.

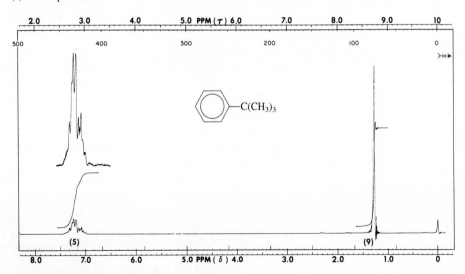

(b) CMR data. Chemical shifts: δ 31.3, 34.5, 125.1, 125.3, 128.0, 150.8.

Figure 12.14 NMR data for *t*-butylbenzene.

alkenes
preparations and reactions

Alkanes are extremely limited in the types of chemical reactions that they undergo, as was noted in the preceding chapter. Fortunately, most other families of organic compounds undergo a much greater variety of reactions than do the alkanes. This increase in chemical diversity can be attributed to the presence of functional groups, for example, heteroatom-containing substituents such as hydroxyl (OH) and various types of π-bonds such as the carbon-carbon double bond of alkenes and the carbon-oxygen double bond of aldehydes and ketones. These substituents impart enhanced reactivity to the substance containing them.

The importance of the role of a heteroatom in increasing chemical reactivity is nicely illustrated by the haloalkanes, **1.** These compounds are more reactive than alkanes because the greater *electronegativity* of the halogen atom, relative to hydrogen and carbon, serves to *polarize* the carbon-halogen bond and generate a partial positive charge on the carbon atom bearing this substituent. This polarization has important consequences for molecular reactivity, primarily for two reasons: (1) The *acidity* of hydrogen atoms near the carbon-halogen bond is enhanced, making the molecule susceptible to elimination reactions promoted by transfer of protons to strong bases; reactions of this type are introduced in Section 13.1. (2) The substituted carbon atom is subject to attack by *nucleophiles,* which are electron-rich and frequently anionic species, leading to displacement of the halogen with the formation of substitution products; this type of reaction is examined in some detail in Section 18.1.

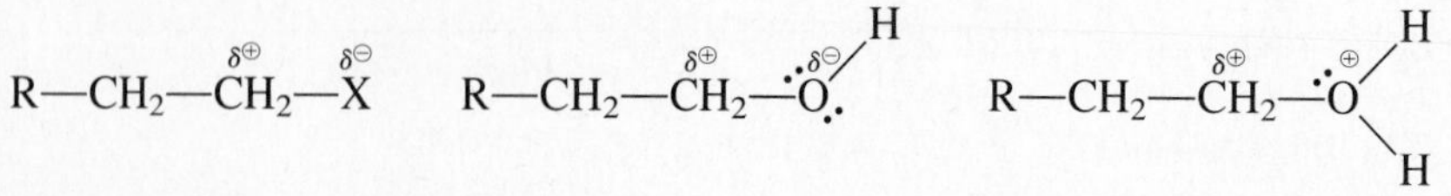

$$R-CH_2-\overset{\delta\oplus}{C}H_2-\overset{\delta\ominus}{X} \qquad R-CH_2-\overset{\delta\oplus}{C}H_2-\overset{\delta\ominus}{\ddot{O}}{:}^{H} \qquad R-CH_2-\overset{\delta\oplus}{C}H_2-\overset{\oplus}{\ddot{O}}\overset{H}{\underset{H}{<}}$$

1   **2**   **3**

Furthermore, the partial negative charge and the availability of nonbonding electrons on the halogen atom of **1** produce significant interactions with reagent and *solvent* molecules by way of electrostatic attractions and direct complexation. For example, the complexation of nonbonding electrons of a halogen substituent with Lewis acids, with subsequent production of a carbocation, plays a central role in the Friedel-Crafts alkylation of aromatic hydrocarbons by haloalkanes, as discussed in Section 17.2, and illustrates in an extreme way the increase in positive charge at the halogen-bearing carbon atom as a result of complexation. In summary, interactions of a covalently bound halogen atom with polar species serve to weaken the carbon-halogen bond, facilitating the elimination and substitution reactions mentioned above. None of these features influences the reactivity of the alkanes because carbon and hydrogen have similar electronegativities, so there is little or no polarization within the molecules.

Alcohols, **2**, are more reactive than alkanes, mainly because of the nonbonding electron pairs present on the oxygen atom. These electron pairs complex with Lewis acids, as do those of halogens, and with other electron-deficient species, known collectively as *electrophiles*. The simplest such interaction is the reaction of an alcohol, **2**, with a proton-donating acid to produce an oxonium ion, **3**. The positively charged oxygen atom in **3** plays much the same role as the halogen atom in **1**. A positive charge on the adjacent carbon atom is induced, enhancing the acidity of nearby hydrogens and leading to elimination (dehydration) reactions, the subject of Section 13.2, and making the carbon atom susceptible to substitution reactions, as discussed in Chapter 18.

The ways in which alkenes react as a result of the presence of highly polarizable π-bonding electrons are described in Section 13.3.

13.1 Dehydrohalogenation of Haloalkanes

As noted in the preceding section, the halogen atom of a haloalkane enhances the acidity of nearby hydrogen atoms by virtue of its effect on molecular polarity. This effect decreases the farther the hydrogen atom is from the carbon-halogen bond, so that the α-hydrogens of **1**, those on the carbon atom bearing the halogen, are more acidic than those at the adjacent β-position. One might therefore predict that the α-hydrogens would react more readily with bases. However, this prediction is generally not borne out experimentally, probably because this pathway does not lead directly to stable products and consequently is of relatively high energy. In contrast, the abstraction by base of a proton from the β-position, with simultaneous departure of halide ion from the α-position, leads smoothly in a *concerted (one-step) reaction* to the formation of the carbon-carbon π-bond of the alkene **4** (equation 1). The transition state for the reaction is shown in equation 2, in which the curved arrows symbolize the flow of electrons required to produce the double bond. This important transformation of haloalkanes to alkenes is called **dehydrohalogenation** and represents an example of an *elimination* reaction, with halide ion as the *leaving group*.

$$\overset{\beta}{R}\overset{\alpha}{CH_2}CH_2{-}X + B\colon^{\ominus} \rightarrow RCH{=}CH_2 + BH + X^{\ominus} \qquad \textbf{(1)}$$
$$\quad \mathbf{1} \qquad\qquad\qquad\qquad \mathbf{4}$$

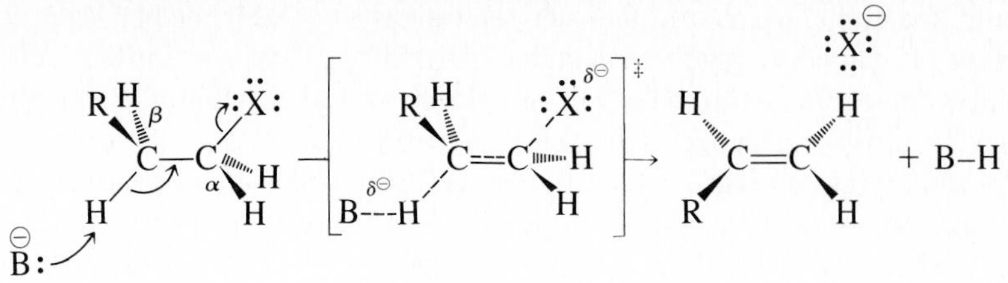

1   Transition state   4   (2)

This type of reaction is classified as an *E2 process,* where E stands for elimination and 2 refers to the molecularity of the rate-determining step of the reaction. Two species, the haloalkane and the base, B:, must collide in order to pass through the transition state of the rate-determining step of the reaction, and the molecularity is therefore two. The rate of the reaction is equal to a proportionality constant k_2, the rate constant, times the product of the concentrations of each of these species (equation 3).

$$\text{Rate} = k_2[\text{haloalkane}][\text{B}\!:] \qquad (3)$$

A second type of mechanism for an elimination reaction of haloalkanes is shown in equation 4. It involves generation of a *carbocation,* **5** (often referred to as a *carbonium ion,* although *carbenium ion* is a more systematic name), by rate-determining heterolysis of the carbon-halogen bond; base then rapidly abstracts a β-proton to give the alkene, **4.** This reaction has a rate that is dependent *only* on the concentration of the haloalkane (equation 5). Molecules of solvent are also involved in the rate-determining step because they aid in the departure of halogen through the process of solvation of the developing ions. However, although some of the solvent may be consumed by side reactions in which it functions as a nucleophile (equation 6), it is present in such a large excess relative to the reactants that its concentration remains essentially unchanged throughout the course of the reaction. Its concentration is therefore incorporated as an arithmetic constant into the value of the rate constant, k_1. Elimination reactions of this sort are designated as E1, and examples of them are encountered in Section 13.2. The overall result of reactions occurring by this mechanism is the same as those proceeding by an E2 pathway; namely, conversion of haloalkanes to alkenes.

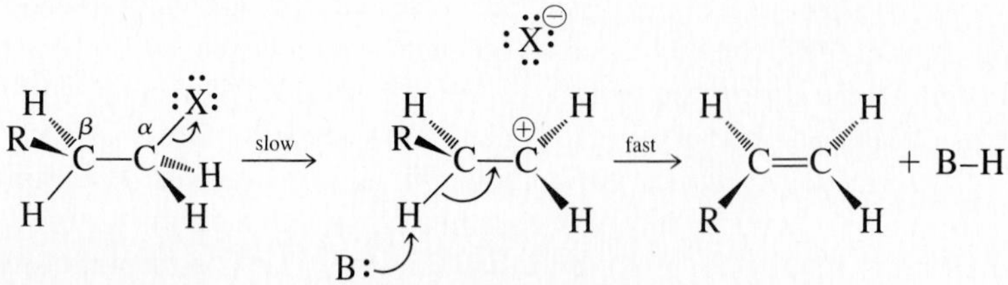

1   5   4   (4)

$$\text{Rate} = k_1[\text{haloalkane}] \qquad (5)$$

$$\text{RCH}_2\text{CH}_2\!-\!\text{X} \xrightarrow{-\text{X}^{\ominus}} \text{RCH}_2\overset{\oplus}{\text{CH}}_2 \xrightarrow[\text{(Solvent)}]{\text{HOS}} \text{RCH}_2\text{CH}_2\!-\!\text{OS} + \text{H}^{\oplus} \qquad (6)$$

Comparison of equations 2 and 5 allows an important conclusion: If E1 and E2 processes are proceeding simultaneously, the latter pathway could be favored by using an excess of base, thereby making it kinetically more competitive. This is because the rate of the E1 reaction will be unchanged by the increase in base concentration, whereas the rate of the E2 process will be increased. Conversely, the absence of strong bases and the use of more polar solvents tends to favor the E1 mechanism at the expense of the E2 elimination.

A variety of functional groups *other* than halide may serve as leaving groups in eliminations reactions. Among these are the sulfonate group, **6,** dimethylsulfide, **7,** trimethylamine, **8,** and water. All of these are relatively *weak* bases. This is an important characteristic of all so-called ''good'' leaving groups because the ability of a functional group to leave is related to its propensity for accepting the electron pair that bound it in the original molecule. This tendency for accepting an electron pair varies *inversely* with the base strength of the species formed.

| **6** | **7** | **8** |
| Sulfonate | Dimethyl sulfide | Trimethyl-amine |

Substitution reactions (equations 6 and 7) often compete with elimination reactions and decrease the yield of the desired alkene.

$$+ \; X:^{\ominus} \qquad\qquad (7)$$

The degree to which this side reaction is important depends on the nature of the substrate and of the base being used. As might be expected from steric considerations, competition with substitution is most important when the carbon atom bearing the leaving group is primary, becomes less so when it is secondary, and is usually unimportant when it is tertiary. The use of strong bases favors elimination over substitution, so base-solvent combinations such as alkoxides (RO^-) in alcohols, amide ion (NH_2^-) in benzene or diethyl ether, or potassium hydroxide in ethanol are very effective in promoting the elimination reaction. The competition between substitution and elimination is discussed in more detail in Section 18.1.

Elimination may give a mixture of products if the leaving group is unsymmetrically located on the carbon skeleton. For example, 2-chloro-2-methylbutane (equation 8) yields both 2-methyl-2-butene **(9)** and 2-methyl-1-butene **(10).** Elimination reactions are normally *irreversible* under the experimental conditions used so that the alkenes **9** and **10** are the products of two competing elimination reactions rather than of a single elimination process followed by an equilibrium that converts **9** to **10,** or *vice versa.* Consequently, the ratio of alkenes formed is subject to the *relative rates* of those two reactions and is therefore determined by the relative free energies of their respective transition states (**kinetic control**) rather than by the relative free energies of the alkenes themselves (**equilibrium control**; see Chapters 16 and 18).

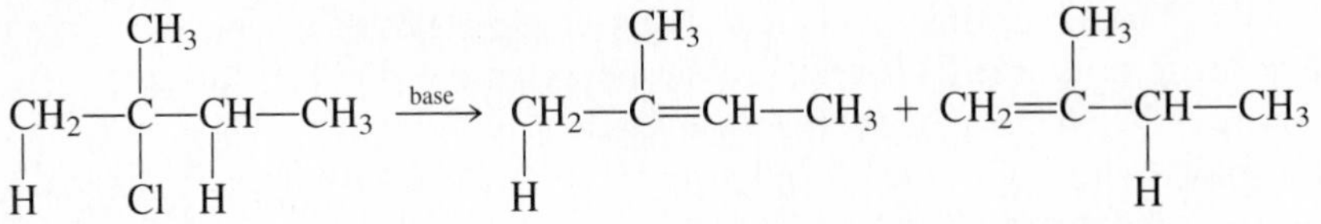

$$CH_2\!\!-\!\!\underset{\underset{H}{|}}{\overset{\overset{CH_3}{|}}{C}}\!\!-\!\!\underset{\underset{Cl}{|}}{\overset{}{CH}}\!\!-\!\!\underset{\underset{H}{|}}{\overset{}{CH_3}} \xrightarrow{\text{base}} CH_2\!\!-\!\!\underset{\underset{H}{|}}{\overset{\overset{CH_3}{|}}{C}}\!\!=\!\!CH\!\!-\!\!CH_3 + CH_2\!\!=\!\!\underset{}{\overset{\overset{CH_3}{|}}{C}}\!\!-\!\!\underset{\underset{H}{|}}{\overset{}{CH}}\!\!-\!\!CH_3 \qquad \textbf{(8)}$$

$$\qquad\qquad\qquad\qquad\qquad\qquad\qquad\qquad\quad \mathbf{9} \qquad\qquad\qquad\qquad\qquad \mathbf{10}$$

2-Chloro-2-methylbutane $\qquad\qquad$ 2-Methyl-2-butene $\qquad\qquad$ 2-Methyl-1-butene

The transition states for elimination in each direction for 2-chloro-2-methylbutane are shown below (**9**‡ and **10**‡). The H—C—C—Cl grouping is pictured as having the hydrogen and chlorine atoms at a *dihedral angle* of 180°, a relationship that is variously called *trans*-coplanar and *anti*-periplanar. It is from this geometry that E2 reactions occur most readily. If the dihedral angle between the β-hydrogen atom and the leaving group is much different from 180°, the free energy of activation, $\Delta G^{\ddagger}$, for elimination increases substantially.

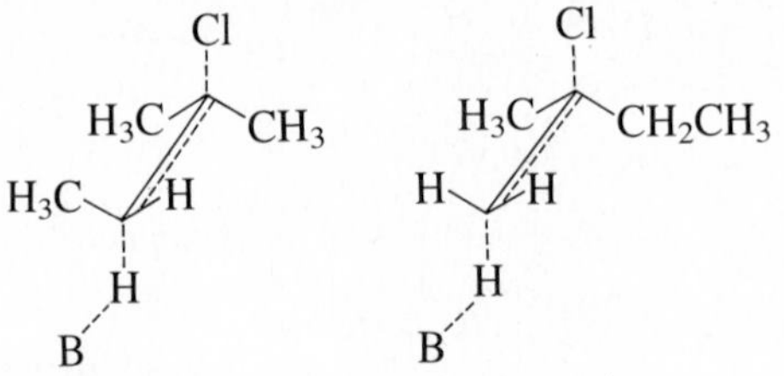

$\mathbf{9}^{\ddagger}$ $\qquad\qquad\qquad$ $\mathbf{10}^{\ddagger}$

The predominant product in an E2 elimination is the *more highly substituted alkene,* in the absence of complicating factors such as the steric factors discussed below. Because an increase in the number of alkyl substituents on the double bond almost always *increases* the stability of alkenes, that is, decreases their free energy, this outcome means that the elimination is resulting in preferential formation of the *thermodynamically more stable alkene.* The relative free energies of the transition states **9**‡ and **10**‡ reflect the relative energies of the products, **9** and **10**, respectively, formed from them. This is interpreted to mean that the formation of a *partial* carbon-carbon double bond is a major contributor to the total free energies of the two transition states. Thus, the free energy of activation for the formation of **9** is less than that for **10**, and **9** would be the preferred product in the example of equation 8. Be aware that prediction of the major product is not always so simple because of other considerations, such as those discussed below.

Relative free energies of transition states of competing elimination reactions may also be influenced by steric crowding. This will tend to increase the energies of the transition states and thus the overall energy requirements of the reactions. If steric factors are more important along the pathway to one product than to another, the proportion of products formed may be dramatically affected relative to that predicted on the basis of the degree of substitution of the double bond. This is particularly true when the reaction path to the more highly substituted alkene suffers from the greater steric crowding, in which case it may well be the *minor* component of the mixture. An example of this is shown in equation 9.

$$
\begin{array}{ccc}
\underset{\substack{|\\ CH_3}}{\overset{\substack{CH_3 \\ |}}{CH_3C}}-CH_2CHCH_3 & \rightarrow & \underset{\substack{|\\ CH_3}}{\overset{\substack{CH_3 \\ |}}{CH_3C}}-CH=CHCH_3 + \underset{\substack{|\\ CH_3}}{\overset{\substack{CH_3 \\ |}}{CH_3CCH_2CH}}=CH_2 \quad \textbf{(9)}\\
\underset{Cl}{|}
\end{array}
$$

| 4-Chloro-2,2-dimethylpentane | *Minor* 4,4-Dimethyl-2-pentene | *Major* 4,4-Dimethyl-1-pentene |

Unfavorable steric interactions in the transition state may result from interference of the substituents on the H—C—C—L grouping (L represents a leaving group) to the approach of the base, B:. This is particularly important when the base and/or the substituents are bulky. Bulky substituents bound to the carbon atom from which the proton is removed, not surprisingly, cause greater hindrance to the approaching base than do substituents on the carbon atom bearing the leaving group. With reference to the 2-chloro-2-methylbutyl system (equation 8), the presence of the methyl group at carbon 3 should cause the energy of transition state **9**‡ to be more adversely affected than **10**‡, the more sterically demanding is the base used to effect the elimination. A sufficiently bulky base might even result in the less substituted alkene becoming the major product.

Similar trends toward favoring the less-substituted alkene would also result from the use of a larger leaving group L or from a change in the substrate so that the size of the group(s) crowding the base is increased. Steric effects caused by increasing the size of the leaving group make it more difficult to attain the *trans*-coplanar arrangement of HCCL leading to the more substituted alkene. When dimethylsulfide (**7**) or trimethylamine (**8**) is the leaving group, the less-substituted alkene is usually the major product, regardless of the structure of the base.

The following series of experiments demonstrates the steric effect as a molecular parameter of the E2 reaction and the general techniques for performing base-promoted elimination reactions.[1]

Pre-lab exercises for Section 13.1, Dehydrohalogenation of Haloalkanes, are given on page PL. 33.

EXPERIMENTAL PROCEDURE

For the elimination reaction, it is suggested that each student be assigned a different base. The results may then be collected, averaged for each base, and presented to the class in order to define the trend in product ratio as a function of the size of the base used.

[1]These experiments are based in part on the following reference: Brown, H. C.; Moritani, I. *Journal of the American Chemical Society*, **1953**, *75*, 4112.

DO IT SAFELY

1 The majority of materials used in this experiment are highly flammable. **Use no flames.**

2 Refer to Sections 2.7 and 4.4 for precautions regarding simple distillations. Pay particular attention to those concerning the assembly and integrity of your glassware.

3 The solutions used in this experiment are *highly caustic.* Take care not to allow them to come into contact with your skin. If this should happen, flood the affected area with water and then thoroughly rinse the area with a solution of *dilute* (about 1%) acetic acid. We recommend that you wear rubber gloves while preparing and transferring solutions in this experiment.

4 If it is necessary for you to handle sodium *metal* during the experiment, remember that this metal reacts *violently* with water, with the formation, and possible *explosive* combustion, of hydrogen gas. Use only *dry* containers, forceps, and so forth. Do *not* handle pieces of sodium metal with your bare fingers.

5 If you are to handle solid sodium methoxide or potassium *t*-butoxide, avoid spilling these during the weighing process. Although they will be hydrolyzed rather rapidly in moist air, the resulting solution will be strongly alkaline. Clean up spillages with a water-soaked paper towel, and then wash your hands.

A. Elimination with Alcoholic Potassium Hydroxide

In this procedure, a Hempel column will serve two purposes. It will first function as a *reflux condenser* and then as a *fractional distillation column,* its more usual purpose in experimental procedures. This dual-purpose use minimizes the amount of glassware assembly required.

Place 0.075 mol of potassium hydroxide (5.0 g, correcting for the fact that commercial potassium hydroxide contains approximately 15% by weight of water; see Exercise 9) and 50 mL of *absolute* ethanol in a *dry* 100-mL round-bottomed flask. Attach a calcium chloride tube (Section 2.23) to the flask and warm the mixture on a hot-water bath until the potassium hydroxide has dissolved. Cool the flask to room temperature, using an ice-water bath, and add 0.050 mol (5.3 g, 6.2 mL) of 2-chloro-2-methylbutane and a few boiling chips to the flask.

Continuation Equip the flask for fractional distillation, as shown in Figure 2.16. If you are using glassware equipped with ground-glass joints, be sure to lubricate the joint connecting the Hempel column to the flask with a hydrocarbon or silicone grease. Lubrication of these joints is particularly important in this experiment because the strong bases being used may cause the joints to "freeze." In order to increase the efficiency of the Hempel column as a reflux condenser, fill it with Raschig rings, coarsely broken glass tubing, or other packing. Using a short piece of tubing, fit the vacuum adapter holding the 25-mL receiving flask with a calcium chloride drying tube and immerse the receiving flask in an ice-water bath. Attach water hoses to the *Hempel column* and circulate water through the jacket of the column during the period of reflux for this reaction. Heat the reaction mixture with a heating mantle or hot-water bath (see Section 2.5) for a period of 2 hr at a *gentle* reflux (see Section 2.19). This should be sufficient time to allow the reaction to go to about 95% completion. By the end of this period some solid should have precipitated and may cause some bumping.★

At the end of the reflux period, cool the flask containing the reaction mixture with an ice-water bath, turn off the cooling water, and remove the water hoses from the Hempel column, allowing the water to drain from the jacket. Reconnect the water hoses to the condenser so that the apparatus is now set for fractional distillation. If any low-boiling distillate has condensed in the receiving flask during the reflux period, allow it to remain, and continue to cool this flask in an ice-water bath. Distil the product mixture, collecting all distillate boiling below 45°C (2-methyl-1-butene, bp 31°C; 2-methyl-2-butene, bp 38°C). Transfer the product to a preweighed sample bottle with a *tight-fitting* stopper or cap and determine the yield. Perform qualitative tests that will demonstrate the presence of alkenes in the distillate (Section 27.5b). Analyze your product for GC analysis or submit a sample of it for such analysis. After obtaining the results, calculate the relative percentages of the two isomeric alkenes formed (see Section 6.6). Typical GC tracings of the products from this elimination and from one in which potassium *t*-butoxide was used as the base are shown in Figure 13.1. Figure 13.6 shows the PMR spectrum of the product mixture from a representative experiment.

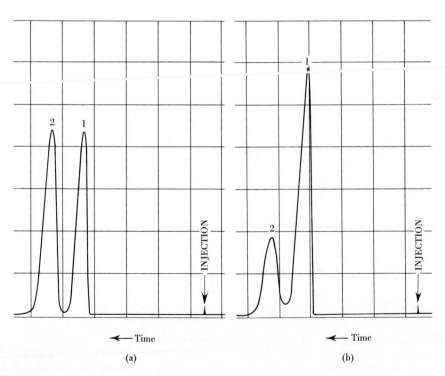

Figure 13.1 Typical GC traces of the products of elimination of 2-chloro-2-methylbutane. Assignments: peak 1: 2-methyl-1-butene; peak 2: 2-methyl-2-butene. (a) Elimination with KOH, showing approximately 44% 2-methyl-1-butene. (b) Elimination with KOC(CH₃)₃, showing approximately 76% 2-methyl-1-butene. Analyses were performed at 45°C on a 3-m column of 30% silicone gum rubber supported on Chromosorb P.

B. Elimination with Sodium Methoxide

Prepare a solution of sodium methoxide in *anhydrous* methanol by either of the following methods.

Method 1: *Use of sodium methoxide.* Dissolve 0.075 mol (4.1 g) of sodium methoxide powder in 50 mL of *anhydrous* methanol contained in a *dry* 100-mL round-bottomed flask. Use a clean, *dry* spatula for transferring the methoxide, taking care to perform the weighing operation as rapidly as possible to minimize the reaction between the methoxide and atmospheric water vapor (equation 10). The bottle from which the sodium methoxide is taken should be kept *tightly* closed.

$$NaOCH_3 + H_2O \rightarrow HOCH_3 + NaOH \qquad \text{(10)}$$

Method 2: *Use of sodium metal.* Prepare a solution of 0.075 mol (1.7 g) of sodium metal in 50 mL of *anhydrous* methanol in the following way. In each of two 50-mL beakers place 15 mL of *dry* toluene and weigh one of the two. To the *weighed* beaker must now be added the amount of sodium metal required for the experiment. This metal is normally stored under mineral oil for protection from water and oxygen. Using a small *dry* knife or *dry* spatula, cut a piece of sodium metal about the size of a pea, stick the tip of the knife or spatula into the metal, and rinse the mineral oil off the metal by swirling it in the *unweighed* beaker of toluene. Briefly blot the piece of sodium with a *dry* paper towel, and transfer the piece of metal to the *weighed* beaker. Repeat the sequence until the correct amount of sodium has been placed in the weighed beaker.

Place 50 mL of *anhydrous* methanol in a *dry* 100-mL round-bottomed flask fitted with a calcium chloride tube (Section 2.23). Remove a piece of sodium from the weighed beaker with the aid of the knife or spatula used previously, briefly blotting it as before, and add it to the flask containing the methanol. **Caution: No flames in the vicinity; hydrogen gas is evolved!** Replace the drying tube on the flask and wait until the initially vigorous gas evolution subsides. Then continue adding the sodium one piece at a time until all of it has been added. Allow the metal to react completely before continuing.

Complete the experiment as follows. Cool the methanolic sodium methoxide solution to room temperature with the aid of an ice-water bath and add *to* this solution 0.050 mol (5.3 g, 6.2 mL) of 2-chloro-2-methylbutane and a few boiling chips. Then follow the directions in the paragraph headed "Continuation" in part A.

C. Elimination with Potassium *t*-Butoxide

Place 0.075 mol (8.4 g) of solid potassium *t*-butoxide and 50 mL of *anhydrous* *t*-butyl alcohol in a *dry* 100-mL round-bottom flask. Use the same precautions for handling potassium *t*-butoxide as described for sodium methoxide in part B. Attach a calcium chloride tube (Section 2.23) and warm the flask with a water bath to assist dissolution of the base. If all of the solid has not dissolved in 5 min, continue the experiment with the heterogeneous mixture. Cool the flask to room temperature with an ice-water bath, and add 0.050 mol (5.3 g, 6.2 mL) of 2-chloro-2-methylbutane and a few boiling chips to the flask. Complete the experiment from this point by following the directions in the ''Continuation'' section of part A. A typical GC tracing and a typical PMR spectrum of the products of this elimination are shown in Figures 13.1 and 13.7, respectively.

EXERCISES

1. What would be the expected results of the alkoxide-promoted eliminations if the alcohols used as solvents contained water?

2. What is the solid material that precipitates as the eliminations proceed?

3. Why does the *excess* of base used in these eliminations favor the E2 over the E1 elimination?

4. If all the elimination reactions in the Experimental Procedure section had proceeded by the E1 mechanism, would the results be expected to be different from those actually obtained? Why?

5. From the results of the experiments that were performed and/or from the data in Figure 13.1, what conclusions can be drawn concerning the effect of *relative* base size upon product distribution?

6. What differences in product distributions would be expected for the eliminations of 2-chloro-2-methylbutane and 2-chloro-2,3-dimethylbutane with excess methanolic sodium methoxide?

7. If the leaving group in the 2-methyl-2-butyl system were larger than a methyl group, why would 2-methyl-1-butene be expected to be formed in greater amounts than if the leaving group were smaller than methyl, regardless of which base is used? Use ''sawhorse'' structural formulas to support your explanation.

8. Refer to Figures 9.12, 13.4a, 13.6, and 13.7, and calculate the percentage compositions of the mixtures of isomeric methylbutenes obtained from reaction of 2-chloro-2-methylbutane with potassium hydroxide and potassium *t*-butoxide. Note that Figures 9.12 and 13.4a are PMR spectra of the *pure* alkenes and that Figures 13.6 and 13.7 are PMR spectra of the *mixtures*. In the latter two spectra, the integration of the resonances in the region of δ 5 has been electronically amplified in the upper ''stepped'' line so that the relative areas of the two multiplets in that region can be more accurately measured.

9. Commercial potassium hydroxide contains approximately 15% by weight of water. Verify that to obtain 0.075 mol of potassium hydroxide, 5.0 g of the commercial material must be used.

SPECTRA OF STARTING MATERIALS AND PRODUCTS

The IR and PMR spectra of 2-methyl-1-butene are contained in Figures 8.16 and 9.12, respectively.

(a) PMR spectrum.

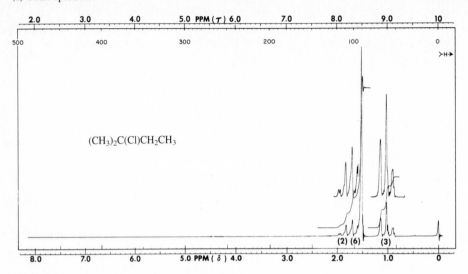

$(CH_3)_2C(Cl)CH_2CH_3$

(2) (6)   (3)

(b) CMR data. Chemical shifts: δ 9.5, 32.0, 39.0, 70.9.

Figure 13.2 NMR data for 2-chloro-2-methylbutane.

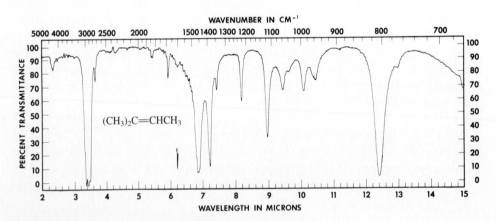

$(CH_3)_2C{=}CHCH_3$

Figure 13.3 IR spectrum of 2-methyl-2-butene.

(a) PMR spectrum.

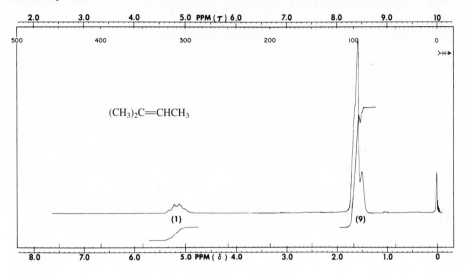

(b) CMR data. Chemical shifts: δ 13.4, 17.3, 25.6, 118.8, 132.0.

Figure 13.4 NMR data for 2-methyl-2-butene.

$CH_2{=}C(CH_3)CH_2CH_3$

Chemical shifts: δ 12.5, 22.3, 31.0, 108.8, 147.5.

Figure 13.5 CMR data for 2-methyl-1-butene.

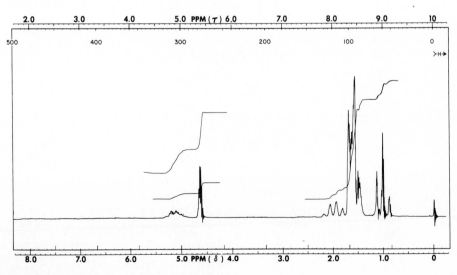

Figure 13.6 PMR spectrum of the product mixture from the elimination of 2-chloro-2-methylbutane with potassium hydroxide.

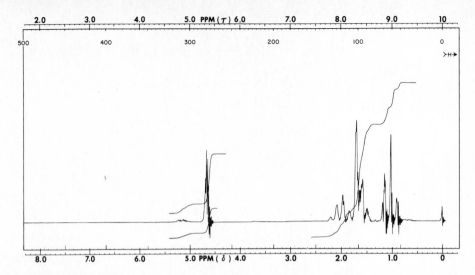

Figure 13.7 PMR spectrum of the product mixture from the elimination of 2-chloro-2-methylbutane with potassium *t*-butoxide.

13.2 Dehydration of Alcohols

The acid-catalyzed dehydration of alcohols is an extremely important method for the preparation of alkenes (equation 11). Mechanistically, the elimination involves an initial rapid protonation of the oxygen atom to form an *oxonium ion* **(11),** a process that transforms a poor leaving group, hydroxide, into an excellent one, water. The next step, at least with secondary and tertiary alcohols, requires *endothermic* decomposition of the oxonium ion into a carbocation **(12)** and water, and it is kinetically a first-order, unimolecular process. Subsequently, the carbocation quickly loses a proton from an adjacent carbon atom to give the alkene. This step is undoubtedly promoted by a molecule of water or alcohol in the reaction medium acting as a weak base to remove the proton, as shown in equation 11. With primary alcohols, ionization to a carbocation is too endothermic, so that elimination of water in this case is a second-order, bimolecular process involving attack by a molecule of water or alcohol, serving as a base, on a β-hydrogen atom of the oxonium ion, **11** (equation 12). This class of alcohols, therefore, undergoes the E2-type of elimination (see Section 13.1).

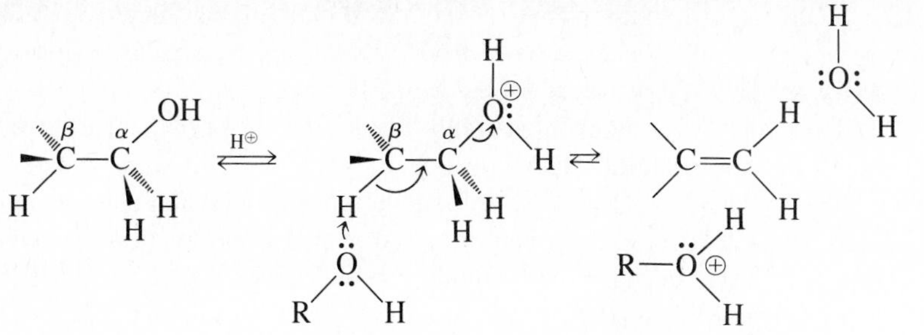

(12)

11

For secondary and tertiary alcohols the decomposition of **11** to **12** is the rate-determining step for the overall process and therefore controls the rate of dehydration. A representation of the activated complex for this step of the reaction is shown here as **13.**

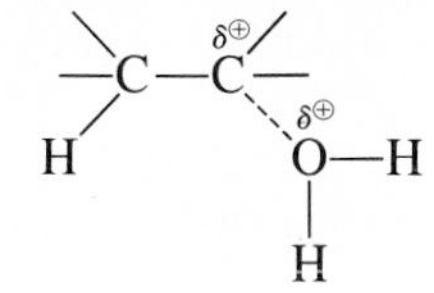

13

The free energy of this activated complex is predominantly influenced by the developing positive charge on the carbon atom as the carbon-oxygen bond breaks. Therefore the energies of activation for dehydration of various types of alcohols parallel the order of stability of the carbocations formed in each case. Since the relative stabilities of carbocations increase with an increasing number of alkyl or aryl substituents, tertiary carbocations are more stable than secondary, which in turn are more stable than primary. This directly relates to the observation that tertiary alcohols undergo dehydration more rapidly and at lower temperatures than the other classes of alcohols. To reiterate, this is because the most stable of the three classes of carbocations is formed in the rate-determining step in the dehydration of tertiary alcohols. As noted above, the energy of primary carbocations is so high that an alternate mechanism for elimination intervenes.

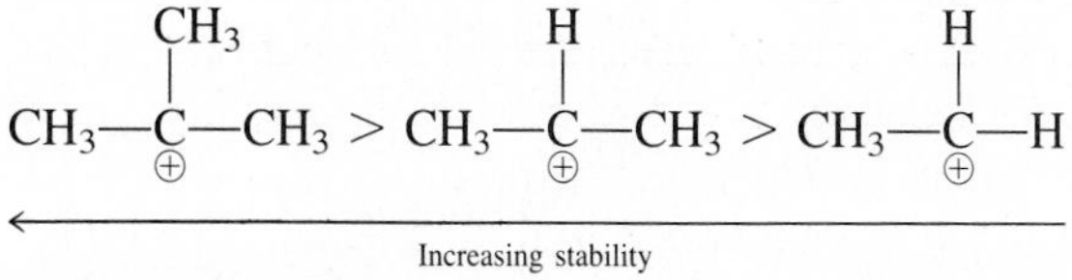

Increasing stability

Each of the steps along the reaction pathway is reversible, so that under the conditions of an elimination experiment, the alkene may be partially rehydrated to alcohol. In order to carry an elimination to completion (ideally, 100% conversion, although this is seldom achieved in practice), the alkene may be selectively distilled from the reaction mixture as it is formed, since its boiling point will be lower than that of the parent alcohol. (Why?) This has the effect of shifting the equilibrium to the right and maximizing the yield of alkene. *The constant removal from a reaction mixture of products as they are formed is a technique often used to afford high yields of products from reversible reactions.*

The same overall outcome could potentially result from development of a method for continuous removal of water rather than alkene from the reaction mixture. However, in the experiments included at the end of this section, it is more advantageous to remove the alkenes. This is not only because the alkenes formed in each experiment have lower boiling points than water but also because they would be produced in reduced yield if left in the presence of the sulfuric acid catalyst, which promotes the formation of polymeric products.

An E1 elimination is usually accompanied to some extent by substitution reactions. These competing reactions involve combination of the intermediate carbocation with the anion of the acid used for catalysis or with some other nucleophile present, such as the solvent. For example, if hydrochloric acid were chosen as the catalyst for dehydration, the chloride ion would trap the carbocation to give some chloroalkane (equation 13). This reaction produces an undesired by-product and, more important, involves *consumption* of the acid catalyst. Because the formation of chloroalkane is irreversible under the reaction conditions normally used for dehydration, the consequence is that the desired dehydration process is terminated altogether!

$$ROH \xrightarrow[-H_2O]{H^\oplus \text{ (HCl)}} R^\oplus \xrightarrow{Cl^\ominus} R\text{---}Cl \qquad \textbf{(13)}$$

The extent to which substitution products will be formed depends on the amount and nature of the acid catalyst used. However, if the formation of these products is reversible under the reaction conditions, these substitution reactions will often not be of significance in affecting the yield of an elimination if the alkene is distilled as it is formed, thereby driving all equilibria toward this type of product (equation 14). The catalysts recommended in this experiment are sulfuric or phosphoric acid. The formation of substitution products by reaction of a carbocation with anions derived from these acids (equation 14; Nu = bisulfate, HSO_4^-, and dihydrogen phosphate, $H_2PO_4^-$, respectively) *is* reversible under the reaction conditions.

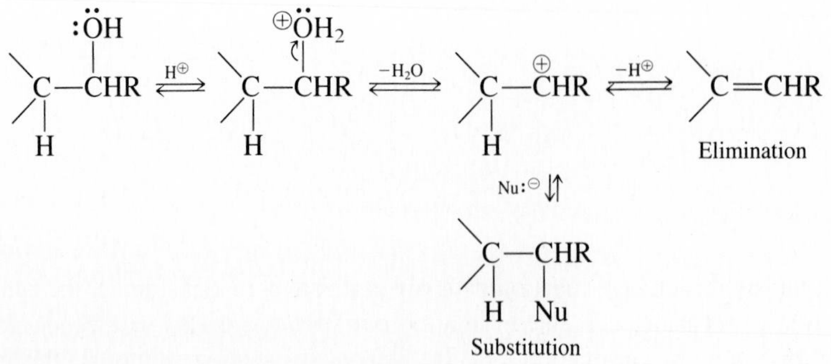

Just as with the E2 reaction (Section 13.1), it is common for two or more isomeric alkenes to be formed in E1 dehydration reactions. If the carbocationic center is unsymmetrically located on the carbon skeleton and if hydrogen atoms are attached to two of the adjacent carbon atoms, different products may be formed, depending on which proton is lost. This possibility is illustrated in equation 15.

$$
\begin{array}{c}
\underset{H_A}{\overset{H}{R-C}}-\overset{OH}{CH}-\underset{H_B}{\overset{R''}{C}}-R' \\
\downarrow {\scriptstyle H^{\oplus},\ -H_2O}
\end{array}
$$

$$
\underset{\textbf{15}}{\overset{H}{\underset{R}{>}}C=CH-\underset{H_B}{\overset{R''}{C}}-R'} \xleftarrow{-H_A^\oplus} \underset{\textbf{14}}{R-\underset{H_A}{\overset{H}{C}}-\overset{\oplus}{CH}-\underset{H_B}{\overset{R''}{C}}-R'} \xrightarrow{-H_B^\oplus} \underset{\textbf{16}}{R-\underset{H_A}{\overset{H}{C}}-CH=\overset{R''}{\underset{R'}{C}<}} \qquad \textbf{(15)}
$$

The proportion of the two alkenes, **15** and **16**, formed is dependent upon the relative free energies of the transition states for the two competing deprotonations of the carbocation **14**. In these transition states the C—H bond presumably is partially broken and the C—C double bond is partially formed, as shown in a general way in **17**. The presence of partial double bond character in the activated complexes suggests that their relative energies will parallel those of the corresponding alkenes, which of course have full double bonds. As a consequence, the lower-energy transition state will be that one leading to the more stable alkene. Alkene stability usually increases as the number of alkyl substituents on the double bond increases. Thus, the more highly substituted alkene normally predominates when competing E1 eliminations are possible. A similar argument was made in Section 13.1 regarding competing E2 eliminations.

17

The mixture of alkenes produced in E1 reactions in many cases cannot be entirely explained by the orientational factors described above. This is because a carbocation is highly prone to rearrangement if a more stable ion can be produced. Indeed, some rearrangement may occur even when the new carbocation is less stable than the first, although the extent of this may be small. Such rearrangements are accomplished by the migration of either an alkyl anion (R:$^-$) or a hydride ion (H:$^-$) from a carbon atom adjacent to the carbocationic center. Loss of a proton from the new cationic intermediate may then lead to other isomeric alkenes.

Rearrangements of carbocations can be illustrated using 3-methyl-2-pentyl ion (**18**) as an example. Recalling that the relative stability of carbocations decreases from tertiary to secondary to primary, primary ion **19** should *not* be produced by hydride migration from C-1 because this transformation would be endothermic, whereas hydride migration from C-3 to give the tertiary ion **20** should be exothermic and therefore a favorable process. An alternate path involving rearrangement of a methyl group from C-3 to give **21** is also possible but is expected to be less important than that giving **20**. There is less of a thermodynamic "driving force," *i.e.,* tendency to minimize free energy, in this case, since it represents an essentially *isothermal* conversion of one secondary ion, **18**, into another, **21**. In other words, such a rearrangement is neither exothermic nor endothermic.

$$
\begin{array}{c}
\text{CH}_3 \\
| \; ③ \; ② \quad ① \\
\text{CH}_3\text{CH}_2\!-\!\overset{\oplus}{\text{C}}\!-\!\text{CH}\!-\!\text{CH}_2 \\
| \qquad | \\
\text{H} \qquad \text{H} \\
\textbf{18}
\end{array}
\qquad\qquad
\begin{array}{c}
\text{CH}_3 \\
| \\
\text{CH}_3\text{CH}_2\text{C}\!-\!\text{CH}\!-\!\overset{\oplus}{\text{CH}_2} \\
| \quad | \\
\text{H} \quad \text{H} \\
\textbf{19}
\end{array}
$$

$$
\begin{array}{c}
\text{CH}_3 \\
| \\
\text{CH}_3\text{CH}_2\underset{\oplus}{\text{C}}\!-\!\text{CH}\!-\!\text{CH}_3 \\
| \\
\text{H} \\
\textbf{20}
\end{array}
\qquad\qquad
\begin{array}{c}
\text{CH}_3 \\
| \\
\text{CH}_3\text{CH}_2\underset{\oplus}{\text{CH}}\!-\!\text{CHCH}_3 \\
\\
\textbf{21}
\end{array}
$$

Experimental procedures for the dehydration of 4-methyl-2-pentanol and cyclohexanol are given below. 4-Methyl-2-pentanol yields a mixture of isomeric alkenes, including 4-methyl-1-pentene (**22**), *trans*-4-methyl-2-pentene (**23**), *cis*-4-methyl-2-pentene (**24**), 2-methyl-2-pentene (**25**), and 2-methyl-1-pentene (**26**), as shown in equation 16.

$$
\underset{\text{4-Methyl-2-pentanol}}{(\text{CH}_3)_2\text{CHCH}_2\overset{\text{OH}}{\overset{|}{\text{C}}}\text{HCH}_3} \xrightarrow[\Delta]{\text{H}^{\oplus}} \underset{\textbf{22} \atop \text{4-Methyl-1-pentene}}{(\text{CH}_3)_2\text{CHCH}_2\text{CH}\!=\!\text{CH}_2} + \underset{\textbf{23} \atop \textit{trans}\text{-4-Methyl-2-pentene}}{(\text{CH}_3)_2\text{CH}\overset{\displaystyle H}{\underset{\displaystyle H}{\diagdown\!\!\!\diagup\;\text{C}=\text{C}\;\diagup\!\!\!\diagdown}}\text{CH}_3} +
$$

(16)

$$
\underset{\textbf{24} \atop \textit{cis}\text{-4-Methyl-2-pentene}}{(\text{CH}_3)_2\text{CH}\diagup\!\!\!\text{C}=\text{C}\!\diagdown\text{CH}_3} \; + \; \underset{\textbf{25} \atop \text{2-Methyl-2-pentene}}{\text{CH}_3\diagdown\!\text{C}=\text{C}\!\diagup\text{CH}_2\text{CH}_3} \; + \; \underset{\textbf{26} \atop \text{2-Methyl-1-pentene}}{\text{CH}_2\!=\!\text{C}\diagup\!\!\!\overset{\text{CH}_3}{\diagdown\text{CH}_2\text{CH}_2\text{CH}_3}}
$$

Products **25** and **26** are typically found to make up approximately 45% of the product mixture, showing that rearrangement is an important phenomenon. These products arise by loss of a proton from the carbocation formed by two successive hydride migrations (equation 17).

$$
\begin{array}{c}
\text{CH}_3 \\
| \\
\text{CH}_3\!-\!\text{C}\!-\!\text{CH}\!-\!\underset{\oplus}{\text{CH}}\!-\!\text{CH}_3 \\
| \\
\text{H}
\end{array}
\xrightarrow{\sim\text{H}:^{\ominus}}
\begin{array}{c}
\text{CH}_3 \\
| \\
\text{CH}_3\!-\!\text{C}\!-\!\underset{\oplus}{\text{CH}}\!-\!\text{CH}\!-\!\text{CH}_3 \\
| \qquad\quad | \\
\text{H} \qquad\quad \text{H}
\end{array}
\xrightarrow{\sim\text{H}:^{\ominus}}
$$

(17)

$$
\begin{array}{c}
\text{CH}_3 \\
| \\
\text{CH}_3\!-\!\underset{\oplus}{\text{C}}\!-\!\text{CH}\!-\!\text{CH}\!-\!\text{CH}_3 \\
| \qquad | \\
\text{H} \qquad \text{H}
\end{array}
$$

The rather complex reaction mixture may be analyzed by gas chromatography (see Section 6.6). If retention times of authentic samples of the methylpentenes are compared with those of the components of the mixture, the peaks in the gas chromatogram of the mixture can be assigned to the different methylpentenes. If the areas under the peaks are measured, the percentage of each component of the mixture may be obtained.

Cyclohexanol undergoes acid-catalyzed dehydration without rearrangement to yield a single product, cyclohexene (equation 18). After purification, the reaction product may be identified by comparing its IR spectrum with that of pure cyclohexene.

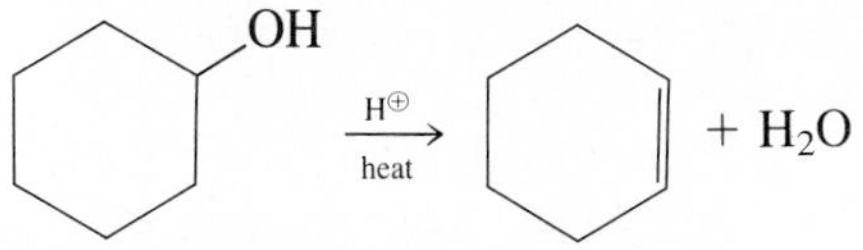

| Cyclohexanol | Cyclohexene |

$$\xrightarrow[\text{heat}]{H^{\oplus}} \quad + H_2O \qquad (18)$$

Pre-lab exercises for Section 13.2, Dehydration of Alcohols, can be found on page PL. 35.

EXPERIMENTAL PROCEDURE

DO IT SAFELY

1 The majority of materials used in this experiment are highly flammable. This is particularly true of the product alkenes. **Use no flames.**

2 Refer to Sections 2.7 and 4.4 for precautions regarding simple distillations. Pay particular attention to those concerning the assembly and integrity of your glassware.

3 There are several operations within the experiment that require you to pour, transfer, and weigh chemicals and reagents that you should *not* get on your skin. We recommend wearing rubber gloves during these operations, particularly when dealing with the strongly acidic catalyst and during the work-up and washing steps in which a separatory funnel is used. Should the catalyst or acidic solutions come in contact with the skin, *immediately* flood the affected area with water and then wash it with 5% sodium bicarbonate solution.

4 If you do the tests for unsaturation on the products of these experiments, read the caution regarding *bromine as a hazardous chemical* in the "Do It Safely" part of Section 12.2.

A. Dehydration of 4-Methyl-2-pentanol

Place 20.0 g (21.2 mL, 0.200 mol) of 4-methyl-2-pentanol in a 100-mL round-bottomed flask and to this add 10 mL of 9 *M* sulfuric acid. (85% Phosphoric acid can also be used as the catalyst but is somewhat less satisfactory. The reaction is slower and the yields are lower. Use 5 mL of this acid if you are instructed to use it.) Thoroughly mix the contents of the flask by swirling it. Add two or three boiling chips, and assemble the flask for fractional distillation according to Figure 2.15. The receiving flask should be immersed in an ice-water bath. Heat the reaction flask with a heating mantle or an oil bath *(no flames!),* collecting all distillates, *but keeping the head temperature below* 90°C. If the reaction mixture is not heated too strongly, the head temperature will remain below 80°C for most of the period of reaction. When about 10 mL of liquid remains in the reaction flask, discontinue heating.

Transfer the organic distillate to a small separatory funnel, add 5–10 mL of 3 *M* aqueous sodium hydroxide solution, and shake the funnel well, being cautious to vent the funnel from time to time (see Section 2.18). Allow the layers to separate and remove the aqueous layer. If you are uncertain about which is the aqueous layer or if an emulsion has formed so that separation of layers is slow, refer to Section 2.18.

Transfer the organic layer to a *dry* 50-mL Erlenmeyer flask and add 1–2 g of *anhydrous* calcium chloride (see Section 2.21).★ Occasionally swirl the mixture during a period of 10 min and then decant the dried organic mixture into a *dry* 50-mL distilling flask. Add boiling stones and distil *(no flames!)* the mixture through a pre-dried simple distillation apparatus (see Section 2.24 and Figure 2.14). Collect the fraction boiling between 53 and 69°C in a preweighed *dry* 25-mL receiving flask or *dry* sample bottle.★ If the bottle is used, its top should be flush with the bottom of the ground-glass joint of the vacuum adapter. Some product may be lost since an ice-water bath is not being used to cool the receiver, but such cooling tends to cause condensation of atmospheric water vapor along with the alkenes. This problem can be avoided if a receiving flask is used and a calcium chloride tube is attached to the nipple of the vacuum adapter. Consult your instructor about whether such an adaptation is to be used.

Calculate the yield of alkenes obtained. If the procedure is properly performed, a yield of 75–85% may be anticipated. The expected products of the reaction and their boiling points are 4-methyl-1-pentene (53.9°C), *cis*-4-methyl-2-pentene (56.4°C), *trans*-4-methyl-2-pentene (58.6°C), 2-methyl-1-pentene (61°C), and 2-methyl-2-pentene (67.3°C).

Test the distillate for unsaturation, using both the bromine in carbon tetrachloride and the Baeyer tests (see Section 27.5b). Either analyze your product by GC or submit a sample of it for such analysis. Calculate the percentage composition of your own product mixture. Figure 13.8 shows a typical GC tracing of the products of this reaction.

You should also obtain an IR spectrum of your product or submit a sample of it for such an analysis. By comparison of the absorption peaks from the IR spectrum that results with those from the spectrum of each of the *pure* expected products, identify as many of the components of your mixture as you can (see Figures 13.11, 13.13, 13.15, 13.17, and 13.19).

B. Dehydration of Cyclohexanol

Place 20 g (25 mL, 0.20 mol) of cyclohexanol in a 100-mL round-bottomed flask and to this add 10 mL of 9 *M* sulfuric acid. (85% Phosphoric acid can also be used as the catalyst but is somewhat less satisfactory. The reaction is slower and the yields are lower. Use 5 mL of this acid if you are instructed to use it.) Thoroughly mix the contents of the flask by swirling it. Add two or three boiling chips, and assemble the flask for fractional distillation according to Figure 2.15. The receiving flask should be immersed in an ice-water bath. Heat the reaction

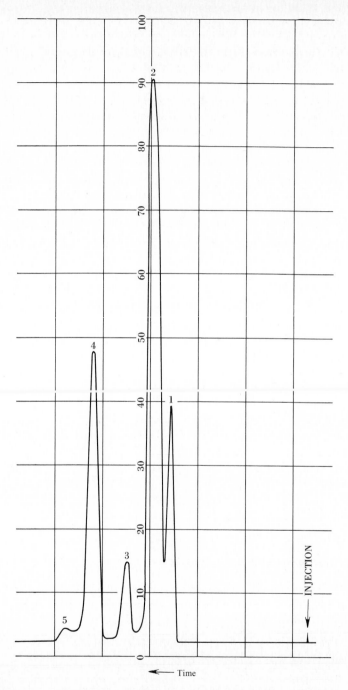

← Time

Figure 13.8 GC tracing of the product mixture from dehydration of 4-methyl-2-pentanol. The peaks have been assigned as follows: (1) 4-methyl-1-pentene (14.4%), (2) *cis*- and *trans*-4-methyl-2-pentene (combined 52.1%), (3) 2-methyl-1-pentene (6.5%), (4) 2-methyl-2-pentene (26.9%), and (5) unidentified. This analysis was made with a 2.4-m column packed with 5% SF-96 on 60/80 Chromosorb W at 35°C.

flask with a heating mantle or an oil bath *(no flames!)*, collecting all distillates, *but keeping the head temperature below* 90°C. If the reaction mixture is not heated too strongly, the head temperature will remain below about 60–70°C for most of the period of reaction. When about 10 mL of liquid remains in the reaction flask, discontinue heating.

Transfer the organic distillate to a small separatory funnel, add 5–10 mL of 3 *M* aqueous sodium hydroxide solution, and shake the funnel well, being careful to vent the funnel from time to time (see Section 2.18). Allow the layers to separate and remove the aqueous layer. If you are uncertain about which is the aqueous layer or if an emulsion has formed so that separation of layers is slow, refer to Section 2.18.

Transfer the organic layer to a *dry* 50-mL Erlenmeyer flask and add 1–2 g of *anhydrous* sodium sulfate (see Section 2.21).★ Occasionally swirl the mixture during a period of 5–10 min; then decant the dried organic mixture into another *dry* 50-mL Erlenmeyer flask, and add a fresh 1–2 g portion of *anhydrous* sodium sulfate. Swirl the flask occasionally during the next 5 min or so and then filter the liquid into a *dry* 50-mL distilling flask. The product *must* be dry at this stage in order to obtain pure cyclohexene, since water and cyclohexene form a minimum-boiling azeotrope (see Section 4.5). Add boiling stones and distil *(no flames!)* the mixture through a pre-dried simple distillation apparatus (see Section 2.24 and Figure 2.14). Collect the fraction boiling between 80 and 85°C in a preweighed *dry* 25-mL receiving flask or *dry* sample bottle.★ If the bottle is used, its top should be flush with the bottom of the ground-glass joint of the vacuum adapter. Some product may be lost since an ice-water bath is not being used to cool the receiver, but such cooling tends to cause condensation of atmospheric water vapor along with the alkenes. This problem can be avoided if a receiving flask is used and a calcium chloride tube is attached to the nipple of the vacuum adapter. Consult your instructor about whether such an adaptation is to be used.

Calculate the yield of distillate isolated, on the assumption that it is pure cyclohexene, and obtain both IR and GLC analyses of the reaction product. Compare your IR spectrum with that of authentic cyclohexene (Figure 13.21). Adjust your calculation of the yield of cyclohexene produced by using the results of the GLC analysis. Perform both the bromine in carbon tetrachloride and the Baeyer tests for unsaturation on samples of your product (see Section 27.5b).

EXERCISES

General Questions for Parts A and B

1. What would be the consequence of distilling the slurry of alkenes and the drying agent? In other words, why is the organic solution separated from the drying agent by filtration *prior* to the final distillation?

2. Which of the following primary alcohols would be most likely to dehydrate by the E1 mechanism? by the E2 mechanism? Explain.

$$CH_3CH_2OH \qquad (CH_3)_3CCH_2OH$$

3. Give structures for the products of dehydration of each of the following alcohols. For each, order the products with respect to preference of formation.

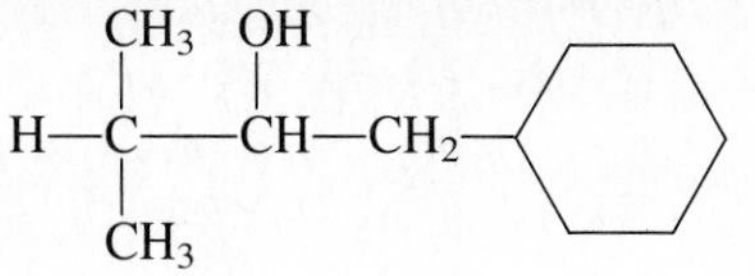

(a)

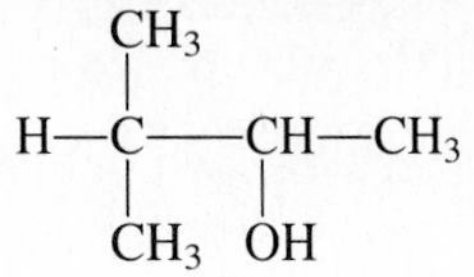

$$CH_3 \quad OH$$
$$H-\overset{\displaystyle |}{\underset{\displaystyle |}{C}}-\overset{\displaystyle |}{CH}-CH_2-\text{(cyclohexyl)}$$
$$CH_3$$

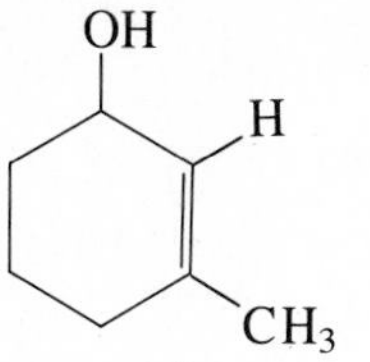

(b)

$$CH_3$$
$$H-\overset{\displaystyle |}{\underset{\displaystyle |}{C}}-\overset{\displaystyle |}{CH}-CH_3$$
$$CH_3 \quad OH$$

(c) cyclohexene ring with OH, H, and CH₃ substituents

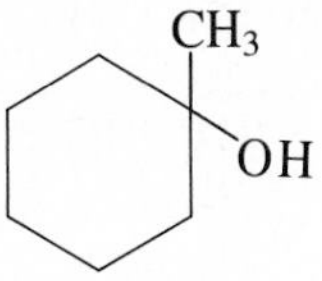

(d) cyclohexane ring with CH₃ and OH substituents

4. What consequences would be expected if the dehydration step of this procedure were conducted under reflux rather than with distillation? Assume that the work-up of the reaction would be unchanged.

Questions for Part A

5. Give a detailed mechanism explaining each of the products obtained from the dehydration of 4-methyl-2-pentanol.

6. Near the end of the dehydration of 4-methyl-2-pentanol, a white solid may precipitate from the reaction mixture. What is the solid likely to be?

7. Give the structure, including stereochemistry, of the product of addition of bromine to *cis*-4-methyl-2-pentene and that of addition to *trans*-4-methyl-2-pentene (see Section 13.3). Are these products identical, enantiomeric, or diastereomeric, relative to one another? Explain.

Questions for Part B

8. Give a detailed mechanism for the acid-catalyzed dehydration of cyclohexanol to cyclohexene.

9. Define a "minimum-boiling azeotrope."

10. Give the structure, including stereochemistry, of the product of addition of bromine to cyclohexene (see Section 13.3). Is this dibromide, at least in principle, separable into enantiomers? Explain.

SPECTRA OF STARTING MATERIALS AND PRODUCTS

The IR and PMR spectra of cyclohexanol are given in Figures 8.20 and 9.15, respectively.

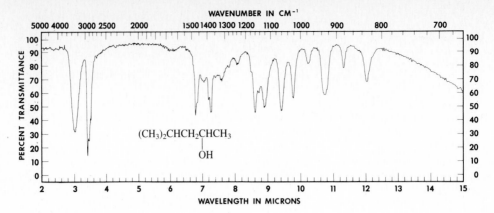

Figure 13.9 IR spectrum of 4-methyl-2-pentanol.

(a) PMR spectrum.

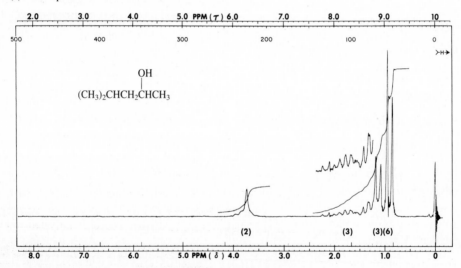

(b) CMR data. Chemical shifts: δ 22.5, 23.2, 23.9, 24.9, 48.8, 65.8.

Figure 13.10 NMR data for 4-methyl-2-pentanol.

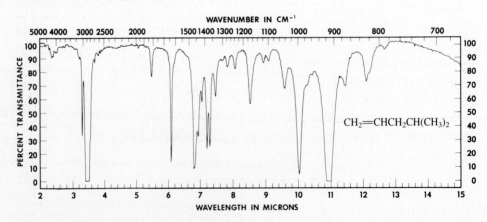

358   **Figure 13.11** IR spectrum of 4-methyl-1-pentene.

(a) PMR spectrum.

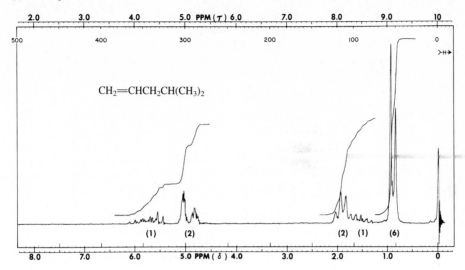

$CH_2{=}CHCH_2CH(CH_3)_2$

(1) (2) (2) (1) (6)

(b) CMR data. Chemical shifts: δ 22.3, 28.1, 43.7, 115.5, 137.8.

Figure 13.12 NMR data for 4-methyl-1-pentene.

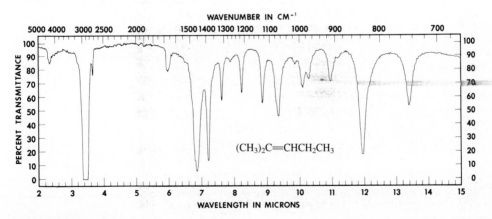

$(CH_3)_2C{=}CHCH_2CH_3$

Figure 13.13 IR spectrum of 2-methyl-2-pentene.

(a) PMR spectrum.

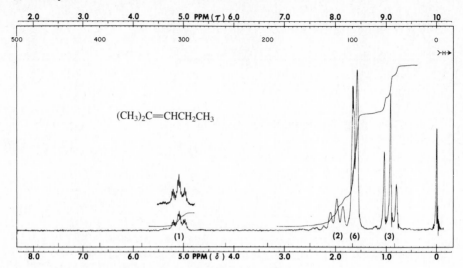

(b) CMR data. Chemical shifts: δ 14.5, 17.5, 21.6, 25.7, 126.9, 130.6.

Figure 13.14 NMR data for 2-methyl-2-pentene.

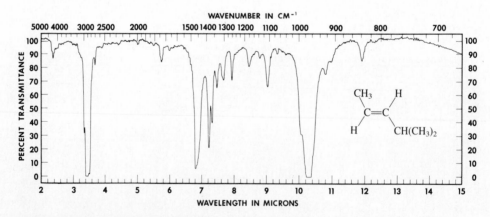

Figure 13.15 IR spectrum of *trans*-4-methyl-2-pentene.

(a) PMR spectrum.

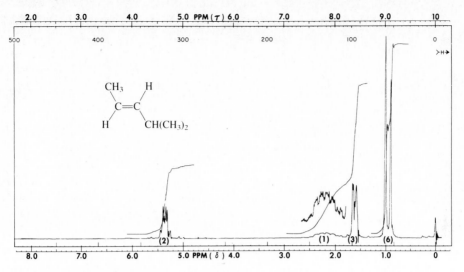

(b) CMR data. Chemical shifts: δ 17.6, 22.7, 31.5, 121.6, 139.4.

Figure 13.16 NMR data for *trans*-4-methyl-2-pentene.

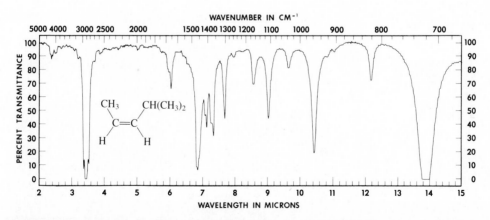

Figure 13.17 IR spectrum of *cis*-4-methyl-2-pentene.

(a) PMR spectrum.

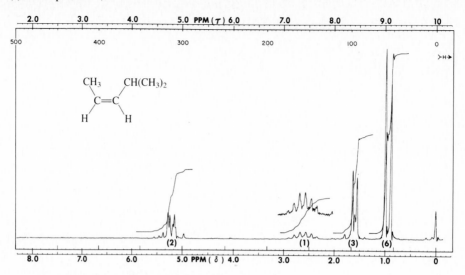

(b) CMR data. Chemical shifts: δ 12.7, 23.1, 26.4, 121.4, 138.6.

Figure 13.18 NMR data for *cis*-4-methyl-2-pentene.

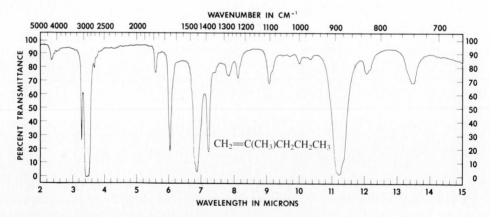

Figure 13.19 IR spectrum of 2-methyl-1-pentene.

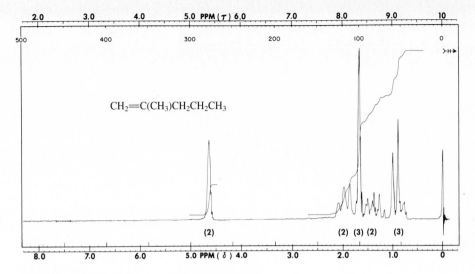

$CH_2=C(CH_3)CH_2CH_2CH_3$

(2)

(2) (3) (2) (3)

Figure 13.20 PMR spectrum of 2-methyl-1-pentene.

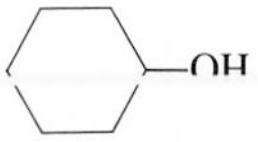

OH   Chemical shifts: δ 24.5, 25.9, 35.5, 70.1.

Figure 13.21 CMR data for cyclohexanol.

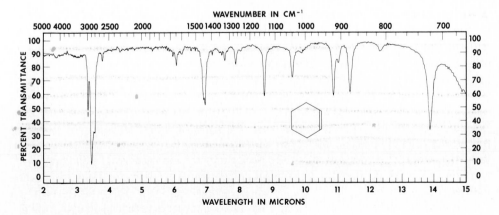

Figure 13.22 IR spectrum of cyclohexene.

(a) PMR spectrum.

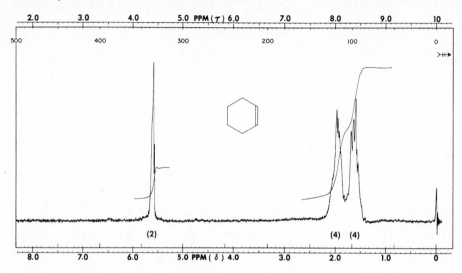

(b) CMR data. Chemical shifts: δ 22.9, 25.4, 127.3.

Figure 13.23 NMR data for cyclohexene.

13.3 Addition Reactions of Alkenes

Alkenes are very useful starting materials for organic syntheses, since these compounds undergo a large variety of reactions involving the carbon-carbon π-bond. In most instances the chemistry of this bond is characterized by the *addition* of a reagent, X—Y, across the π-bond so that saturation occurs (equation 19).

$$R_2C{=}CR_2 + X{-}Y \rightarrow R_2C{-}CR_2 \qquad\qquad \textbf{(19)}$$
$$\underset{X\quad Y}{\vert\qquad\vert}$$

The rich chemistry of alkenes is heavily dependent on contributions from two factors: (1) The strength of the carbon-carbon π-bond is no more than 65 kcal/mol. This value is substantially below those of typical σ-bonds involving carbon, which have strengths in the range of 80–100 kcal/mol. Addition of a reagent across the double bond is therefore usually exothermic because *one* π-bond is being replaced by *two* σ-bonds. (2) The π-bonding electrons are more loosely held to the bonded carbon atoms than are σ-electrons, making them much more polarizable. The π-electron "cloud" (molecular orbital) consequently is readily distorted through electrostatic interaction with an approaching reagent, which enhances the reactivity of the alkene toward attack. These features are not necessarily important in *all* reactions of alkenes, but they are a significant influence in many of the reactions that alkenes undergo.

The detailed mechanism through which the addition proceeds is dependent upon both the nature of X—Y and the conditions under which the reaction is performed. One of the more common mechanisms of addition of a reagent to an alkene involves the attack of an electrophile, E^+, an electron-deficient species, on the double bond to give an intermediate onium ion that may be acyclic (**27a**) or cyclic (**27b**), which then reacts with a nucleophilic reagent, Nu^-, to give the product **28** (equation 20).

$$R_2C{=}CR_2 + E^{\oplus} + Nu{:}^{\ominus} \leftrightarrows \left[\begin{array}{c} R_2C{-}CR_2 \\ \mid \quad {\oplus} \\ E \end{array}\right] \text{ or } \left[\begin{array}{c} R_2C{-}\!\!-\!\!-CR_2 \\ \diagdown \; E \; \diagup \\ {\oplus} \end{array}\right] \xrightarrow{Nu^{\ominus}} \begin{array}{c} R_2C{-}CR_2 \\ \mid \quad \mid \\ E \quad Nu \end{array} \quad \textbf{(20)}$$

$$E{-}Nu$$

$$\qquad\qquad\qquad\quad \textbf{27a} \qquad\qquad \textbf{27b} \qquad\qquad\qquad \textbf{28}$$

For unsymmetrical X—Y compounds, **Nu is the more electronegative of the two atoms**. The bromination of alkenes, a reaction used both as a qualitative test for unsaturation, as described in Chapter 27, and as a quantitative measure of the amount of unsaturation present in a compound, generally proceeds by this mechanism. The *orientation* of addition of E—Nu to an unsymmetrical alkene is such as to result in bond formation between the electrophile, E^+, and the carbon atom bearing the larger number of hydrogen atoms (Markovnikov's rule; equation 21, for example). The *stereochemistry* of the addition is such that E and Nu are *anti* to one another, as shown by the observation that *trans*-1,2-dibromocyclopentane (**30**) is obtained from the addition of bromine to cyclopentene.

$$R_2C{=}CH_2 + HBr \xrightarrow{CCl_4} \begin{array}{c} R_2C{-}CH_2 \\ \mid \quad \mid \\ Br \quad H \end{array} \qquad \textbf{(21)}$$

$$\textbf{29}$$

30

A related stepwise mechanism, which is observed less commonly, involves initial attack on the alkene by a nucleophile, Nu^-, to produce the carbanion **31**, which then reacts with E^+ to give **32** (equation 22).

$$R_2C{=}CR_2 + Nu{:}^{\ominus} \rightleftarrows \begin{array}{c} R_2C{-}\overset{\ominus}{C}R_2 \\ \mid \\ Nu \end{array} \xrightarrow[\text{(or E{-}Nu)}]{E^{\oplus}} \begin{array}{c} R_2C{-}CR_2 \\ \mid \quad \mid \\ Nu \quad E \end{array} \qquad \textbf{(22)}$$

$$\textbf{31} \qquad\qquad\qquad\qquad \textbf{32}$$

In fact, addition by this mechanism occurs easily only if the double bond bears substituents such as cyano (—C≡N), nitro (—NO$_2$), or carbonyl (C=O), which are capable of stabilizing a negative charge. The 1,4-addition of 2-methylpropanal to 3-buten-2-one (equation 23), an experiment described in Chapter 21, represents a mechanism of this type.

$$(CH_3)_2CHCHO + CH_2 = CH - \overset{\overset{\displaystyle O}{\|}}{C}CH_3 \xrightarrow{H^\oplus} \left[(CH_3)_2\overset{\overset{\displaystyle}{|}}{\underset{\underset{\displaystyle CHO}{|}}{C}} - CH_2 - CH = \overset{\overset{\displaystyle OH}{|}}{C}CH_3 \right]$$

$$\downarrow \qquad\qquad\qquad \textbf{(23)}$$

$$(CH_3)_2\overset{\overset{\displaystyle}{|}}{\underset{\underset{\displaystyle CHO}{|}}{C}}CH_2CH_2\overset{\overset{\displaystyle O}{\|}}{C}CH_3$$

A reagent, X—Y, can also often be added to an alkene stepwise by a free-radical process if free-radical initiators are present or if the reaction mixture is exposed to light of the proper wavelength. In such reactions the radical, X·, adds to the double bond to yield an intermediate radical **(33)**. The ensuing reaction between **33** and X—Y provides the product and regenerates X· (equation 24). With unsymmetrical alkenes, X becomes bonded to that carbon atom bearing the larger number of hydrogens to yield the more stable of the two possible radicals as an intermediate (equation 25).

$$R_2C = CR_2 + X\cdot \rightarrow R_2\overset{\overset{\displaystyle}{|}}{\underset{\underset{\displaystyle X}{|}}{C}} - \overset{\displaystyle\cdot}{C}R_2 \xrightarrow{X-Y} R_2\overset{\overset{\displaystyle}{|}}{\underset{\underset{\displaystyle X}{|}}{C}} - \overset{\overset{\displaystyle}{|}}{\underset{\underset{\displaystyle Y}{|}}{C}}R_2 + X\cdot \qquad \textbf{(24)}$$

$$\textbf{33}$$

$$R_2C = CH_2 + HBr \xrightarrow[\text{(initiator)}]{\text{Peroxides}} R_2\overset{\overset{\displaystyle}{|}}{\underset{\underset{\displaystyle H}{|}}{C}} - \overset{\overset{\displaystyle}{|}}{\underset{\underset{\displaystyle Br}{|}}{C}}H_2 \qquad \textbf{(25)}$$

$$\textbf{34}$$

The products of polar and of free-radical addition of hydrobromic acid to an alkene, **29** and **34** (see equations 21 and 25, respectively) are different, since bromine atom, Br·, is the chain carrier in the latter process. One must therefore carefully control reaction conditions in order to obtain pure products when free-radical and polar addition can produce different isomers. Such control may still be important even in those instances in which the *direction* of addition is of no importance, since the *stereoselectivity* of free-radical addition is normally low, and products resulting from both *anti-* and *syn-* addition of X—Y to the double bond can be obtained.

Reaction conditions that tend to favor *electrophilic* addition are low temperatures, the presence of ionic salts like sodium or potassium bromide in the case of the bromination reaction, and the absence of light and peroxides. In contrast, *free-radical* addition is favored by performing the reaction in the gas phase or in nonpolar solvents and in the presence of strong light or peroxides.

An experimental procedure for the polar addition of HBr to 1-hexene to yield 2-bromohexane (equation 26) is given below.

$$CH_2 = CH(CH_2)_3CH_3 + HBr \rightarrow \overset{\overset{\displaystyle}{|}}{\underset{\underset{\displaystyle H}{|}}{C}}H_2 - \overset{\overset{\displaystyle}{|}}{\underset{\underset{\displaystyle Br}{|}}{C}}H(CH_2)_3CH_3 \qquad \textbf{(26)}$$

This type of reaction is normally rather difficult to accomplish in the undergraduate laboratory, owing both to the low solubility in alkenes of the concentrated aqueous solutions of the hydrohalic acids and to the extensive protonation of water by these acids to give hydronium ion (equation 27), which is a *weaker* acid than the *undissociated* hydrohalic acid and is unable to protonate the alkene rapidly under mild reaction conditions. The presence of water along with a hydrohalic acid also introduces the possibility of competing acid-catalyzed addition of water to the alkene (equation 28), the reverse of acid-catalyzed dehydration (Section 13.2). These problems can be alleviated by the use of the acids in their *anhydrous* form, but this introduces difficulties in handling procedures because both HCl and HBr are highly corrosive gases at room temperature.

$$H-X + H_2O \rightleftharpoons H_3O^{\oplus} + X^{\ominus} \qquad \textbf{(27)}$$

$$R_2C{=}CR_2 + H_3O^{\oplus} \rightleftharpoons R_2C{-}CR_2$$
$$\underset{\text{OH} \quad \text{H}}{\vert \qquad \vert} \qquad \textbf{(28)}$$

A solution to these experimental difficulties is found by addition of a catalytic amount of a quaternary ammonium salt, such as **35** or **36,** to the heterogeneous mixture of the *aqueous* acid and the alkene. This salt facilitates transfer of HBr *from* the aqueous phase *into* the organic phase so that the addition becomes possible. The manner in which it does so is discussed below.

$$[CH_3(CH_2)_6CH_2]_3\overset{\oplus}{N}CH_3 \ Cl^{\ominus} \qquad CH_3(CH_2)_{14}CH_2\overset{\oplus}{N}(CH_3)_3 \ \overset{}{B}r^{\ominus}$$

<table>
<tr><td style="text-align:center">**35**</td><td style="text-align:center">**36**</td></tr>
<tr><td style="text-align:center">Trioctylmethylammonium chloride
(Trade name: Aliquat 336)</td><td style="text-align:center">Hexadecyltrimethylammonium bromide
(or cetyltrimethylammonium bromide)</td></tr>
</table>

The ammonium salt itself is partially soluble in both phases; that is, it partitions between both of them because of its combined **lipophilic** (nonpolar-loving) and **hydrophilic** (polar-loving) character. These properties are due to the alkyl groups and the ammonium function, respectively. By forming a complex, **37,** with the acid, the quaternary ammonium salt, **35,** for example, can "drag" HBr out of the aqueous phase and bring it, in only a *partially* hydrated form, into the presence of the alkene (see below). It can then repartition into the aqueous phase to regenerate **35,** and a type of catalytic cycle results, the salt being the catalyst. The net result is that the transfer of the HBr into the organic phase essentially "dehydrates" the acid, making it much more reactive toward the alkene.

$$HBr + R_4\overset{\oplus}{N}X^{\ominus} \rightleftharpoons R_4\overset{\oplus}{N}X^{\ominus}{-}{-}{-}HBr$$

37

Aqueous Phase

Phase Interface

Organic Phase

$$R_4\overset{\oplus}{N}X^{\ominus} + CH_3\overset{\underset{\displaystyle |}{\displaystyle Br}}{C}H(CH_2)_3CH_3 \xleftarrow{\ CH_2{=}CH(CH_2)_3CH_3\ } R_4\overset{\oplus}{N}X^{\ominus}{-}{-}{-}HBr$$

Compounds that promote the transport of reagents between immiscible layers by means of ion pairs like **35** are called **phase transfer catalysts.** Their presence can have dramatic effects on the rates of bimolecular reactions between reagents contained in immiscible phases; rate accelerations of 10^4 to 10^9 (!) are not uncommon when a phase transfer catalyst is added to a heterogeneous liquid mixture of reagents. Note that the technique currently emphasizes the use of *cationic* organic catalysts that form ion pairs with electron-pair donors (the Br of HBr in our case); it is much more difficult and synthetically less useful to prepare stable *anionic* phase transfer catalysts that would complex with electron-pair acceptors such as anions. A sampling of the wide applications of this method of catalyzing organic reactions is given in equations 29–31.

Substitution: $\quad C_7H_{15}CH_2Br$ (in CH_2Cl_2) + NaCN (in H_2O) $\xrightarrow[\substack{\text{transfer} \\ \text{catalyst}}]{\text{Phase}}$ $C_7H_{15}CH_2CN$ **(29)**

$\qquad\qquad\qquad$ (Organic phase) $\qquad\qquad$ (Aqueous phase)

Oxidation: $\quad C_6H_5CH_2OH$ (in benzene) + $KMnO_4$ (in H_2O) $\xrightarrow[\text{catalyst}]{\text{Phase transfer}}$ $C_6H_5CO_2H$ **(30)**

$\qquad\qquad\quad$ (Organic phase) $\qquad\qquad\quad$ (Aqueous phase)

Elimination: $C_6H_5CH_2CH_2Br$ (in CH_2Cl_2) + NaOH (in H_2O) $\xrightarrow[\substack{\text{transfer} \\ \text{catalyst}}]{\text{Phase}}$ $C_6H_5CH{=}CH_2$ **(31)**

$\qquad\qquad\qquad$ (Organic phase) $\qquad\qquad\quad$ (Aqueous phase)

Given the general principles that form the basis for phase transfer catalysis, it is clear that one factor that determines the overall rate of reaction will be the efficiency of partitioning of reagents between phases. This will be a function of, among other things, the total surface area of the two immiscible reagents in contact with each other. To increase this area, the reaction mixture must be agitated vigorously in order to produce emulsification, wherein tiny droplets of the immiscible layers develop. This is normally accomplished by rapid mechanical or magnetic stirring. However, if the appropriate apparatus is unavailable, vigorous shaking will sometimes produce the desired result, as in the optional experimental procedure given below.

A final point concerning additions to alkenes is noteworthy. So far, three general *stepwise* mechanisms of addition of a reagent, X—Y, to an alkene have been discussed, mechanisms that are distinguished by the type of species—electrophile, nucleophile, or free radical—involved in the initial step of the reaction. This step, at least in principle, could be reversible, and in the case of appropriately substituted alkenes could lead to geometric isomerization (equation 32). An example of just such a process is described in Section 7.3.

$(* = \odot, \oplus, \ominus)$ $\qquad\qquad\qquad\qquad\qquad\qquad\qquad\qquad$ **(32)**

A fourth type of mechanism for the reaction of the reagent X—Y with an alkene is nonstepwise in nature and consequently is called *concerted* addition (equation 33). As might be expected, the stereochemistry of the reaction is *syn*- addition of the reagent across the π-bond. Some reactions following this general mechanism are hydroxylation, hydroboration, ozonolysis, hydrogenation, and the Diels-Alder reaction. All these reactions are of considerable synthetic importance, and examples of the last two are described in Sections 20.2 and Chapter 15, respectively. The decoloration of potassium permanganate by alkenes, a qualitative test for the presence of carbon-carbon double bonds (Section 27.5b), involves a mechanism of this sort, in that the initial step of the process involves *syn*- addition of permanganate across the π-bond, which causes decoloration of the solution. Subsequent decomposition of the intermediate **38** results in formation of a 1,2-diol and manganese dioxide, the brown precipitate that forms in the reaction (equation 34). The 1,2-diol usually undergoes further oxidation by permanganate, a fact that makes the attempted synthesis of such diols by permanganate-promoted oxidation of alkenes a low-yield reaction.

$$R_2C{=}CR_2 + X{-}Y \rightarrow \left[\begin{array}{c} R_2C{=}CR_2 \\ | \quad\quad | \\ X\,{-}\,{-}\,{-}\,Y \end{array} \right] \rightarrow \begin{array}{c} R_2C{-}CR_2 \\ | \quad\quad | \\ X \quad\quad Y \end{array} \tag{33}$$

$$R_2C{=}CR_2 + KMnO_4 \xrightarrow{\;H_2O\;} \left[\begin{array}{c} R_2C{-}CR_2 \\ O \quad\quad O \\ \diagdown\!\!\!\!\diagup \\ Mn \\ \diagup\;\diagdown \\ O \quad\quad O \end{array} K^{\oplus} \right]^{\ominus} \xrightarrow{\;H_2O\;} \begin{array}{c} R_2C{-}CR_2 \\ | \quad\quad | \\ OH\;\;OH \end{array} + MnO_2\downarrow \tag{34}$$

Purple   *Brown*

38

Pre-lab exercises for Section 13.3, Addition Reactions of Alkenes, are found on page PL. 37.

EXPERIMENTAL PROCEDURE

DO IT SAFELY

1 Concentrated hydrobromic acid (47–49%) is a corrosive and toxic material. Measure out the amount required inside a fume hood and avoid inhalation of its vapors. Use rubber gloves when transferring the acid between containers. If the acid should come in contact with your skin, flood the affected area immediately and thoroughly with water and rinse it with 5% sodium bicarbonate solution.

2 Quaternary ammonium salts are toxic substances and can be absorbed through the skin. Should they accidentally come in contact with your skin, wash the affected area immediately with copious amounts of water.

3 If a flame is used in the distillation step, be certain that all joints in the apparatus are well lubricated and tightly mated because 1-hexene and the solvents used, particularly petroleum ether, are highly flammable.

A. Procedure Using Magnetic Stirrers

In a 100-mL round-bottomed flask combine 4.2 g (6.2 mL, 0.050 mmol) of 1-hexene, 27.8 mL (0.25 mol) of 47–49% aqueous hydrobromic acid, and *either* 2 g (2.3 mL, 0.005 mol) of trioctylmethylammonium chloride **(35)** *or* 3 g (0.005 mol) of hexadecyl-trimethylammonium bromide **(36)**. Equip the flask with a stirring bar and a reflux condenser, and bring the rapidly stirred (see Section 2.6) heterogeneous reaction mixture to a gentle reflux (see Section 2.19) with the aid of a heating mantle or oil bath (see Section 2.5). Continue heating and stirring for at least 2 hr and then allow the mixture to cool to room temperature.★

Purification

1. *Trioctylmethylammonium chloride* as catalyst. Transfer the two-phase mixture to a separatory funnel, rinse the reaction flask with 30 mL of petroleum ether (bp 60–80°C)[2] and add the rinse to the separatory funnel. Shake the contents of the funnel thoroughly in order to effect extraction (see Section 2.18) and allow the layers to separate. Three layers should form! Ascertain that the lowest one is an aqueous phase and remove it. Wash the two phases remaining in the funnel with two 25-mL portions of 1.2 *M* sodium bicarbonate solution. Vent the funnel *frequently* because gas is evolved in this step and excessive pressure *must not* build up in the funnel. This time the mixture will separate into only two layers. Transfer the organic layer to a 125-mL Erlenmeyer flask and dry it with swirling over 1–3 g of *anhydrous* magnesium sulfate (see Section 2.21) for at least 0.5 hr.★ Decant (see Section 2.15) or filter (see Section 2.12) the dried solution into a 50-mL round-bottomed flask, equip the flask for simple distillation (see Section 2.7), and distil the liquid, using a heating mantle, oil bath, or flame (see Section 2.5). Care with respect to the *rate* of heating must be taken throughout the course of this distillation because severe foaming can occur. Do not attempt to remove the solvent too rapidly, because this may result in excessive loss of product due to foaming. After solvent and unchanged 1-hexene have been removed,★ the bromohexane should be collected as a single fraction, bp 130–140°C. The product is a colorless, mobile liquid, and its actual yield should be in the range of 45–60% of theoretical.

That the product is 2-bromohexane rather than 1-bromohexane, or a mixture of the two, can be ascertained by spectroscopic methods (IR and PMR) and/or by GC, assuming authentic samples of the two isomers are available. It is also possible to demonstrate whether or not Markovnikov addition has occurred by subjecting the product to the sodium iodide in acetone and the silver nitrate tests for classification of haloalkanes (Section 27.5c) and by determining the boiling point of the product more precisely with the aid of a micro boiling-point apparatus (see Section 2.4).

2. *Hexadecyltrimethylammonium bromide* as catalyst. This salt tends to promote the formation of emulsions, so the workup of the reaction mixture must be modified slightly to minimize this problem. Transfer the two-phase mixture to a separatory funnel, rinse the reaction flask with 50 mL of dichloromethane, and add the rinse to the separatory funnel. Shake the funnel gently but thoroughly in order to effect extraction (see Section 2.18) and allow the layers to separate. If an emulsion has formed because of too-vigorous shaking, see Section

---

[2]Petroleum ethers are mixtures of *alkanes* boiling over various temperature ranges; they do *not* contain the grouping, —C—O—C—, that is present in the class of compounds known as ethers.

2.18. Ascertain which of the two phases is the aqueous phase (see Section 2.18) and remove it. Wash the organic phase with one 25-mL portion of 1.2 *M* sodium bicarbonate solution. Vent the funnel *frequently,* because gas is evolved in this step and excessive pressure *must not* build up in the funnel. This time the mixture will separate into only two layers. Transfer the organic layer to a 125-mL Erlenmeyer flask and dry it with swirling over 1–3 g of *anhydrous* magnesium sulfate (see Section 2.21) for at least 0.5 h.★ Decant (see Section 2.15) or filter (see Section 2.12) the dried solution into a 50-mL round-bottomed flask, equip the flask for simple distillation (see Section 2.7), and distil the liquid, using a heating mantle, oil bath, or flame (see Section 2.5). Care with respect to the *rate* of heating must be taken throughout the course of this distillation because severe foaming can occur. Do not attempt to remove the solvent too rapidly, because this may result in excessive loss of product due to foaming. After solvent and unchanged 1-hexene have been removed,★ the bromohexane should be collected as a single fraction, bp 130–140°C. The product is a colorless, mobile liquid, and its actual yield should be in the range of 45–60% of theoretical. Analyze the product according to the methods described for trioctylmethylammonium chloride.

B. Procedure Not Using Magnetic Stirrers

If magnetic stirring is not available, **36** must be used as the phase transfer catalyst, because **35** fails to promote the reaction in unstirred media. The following modified procedure is used:

Combine 6.3 g (9.3 mL, 0.075 mol) of 1-hexene, 40 mL (0.375 mol) of 47–49% aqueous hydrobromic acid, and 4 g (0.0075 mol) of hexadecyltrimethylammonium bromide (**36**) in a 100-mL round-bottomed flask. Swirl the heterogeneous mixture to effect mixing, equip the flask with a reflux condenser, and bring the two-phase mixture to reflux (see Section 2.19) using a heating mantle, oil bath, or flame (see Section 2.5). It is useless to add boiling chips to minimize bumping of the boiling mixture because the organic phase, which boils at the lower temperature, is the *upper* layer and thus is unaffected by the presence of the chips. Nevertheless, gentle heating should produce a smooth reflux of the organic layer. Swirl the flask gently every 15 min or so to effect mixing. This can be done by loosening or removing the clamp to the condenser and then loosening the clamp holding the flask *at the point where this clamp is attached to the ring stand. Do not loosen this clamp where it is affixed to the neck of the reaction flask!* Continue reflux and occasional swirling for at least 2 hr, and then allow the reaction mixture to cool to room temperature. The remainder of the workup is as described above under Part A.2. The yield should be in the range of 35–45%. Analyze the product according to the methods described above under Part A.1.

REFERENCES

   1. Landini, D.; Rolla, F. *Journal of Organic Chemistry,* **1980,** *45,* 3527.
   2. Starks, C. M.; Liotta, C. *Phase Transfer Catalysis: Principles and Techniques,* Academic Press, New York, 1978.
   3. Weber, W. P.; Gokel, G. W. *Phase Transfer Catalysis in Organic Synthesis,* Springer-Verlag, New York, 1977.
   4. Dehmlow, E. V.; Dehmlow, S. S. *Phase Transfer Catalysis,* Verlag Chemie, Weinheim, 1980.

EXERCISES

1. In general, the yield of addition product is higher when the reaction mixture is stirred than when it is not. Why is this?

2. Look up the densities of 47–49% aqueous hydrobromic acid, dichloromethane, and chloroform. Why might dichloromethane be better than chloroform for extraction of the aqueous acid layer obtained in the initial separation of the reaction mixture into two phases (Parts A.2 and B)?

3. Devise an experiment that would demonstrate that a phase transfer catalyst accelerates the rate of reaction between 1-hexene and aqueous hydrobromic acid.

4. When trioctylmethylammonium *chloride* is used in place of **36** as the phase transfer catalyst for this reaction, the possibility arises that some 1- or 2-chlorohexane will be obtained as a by-product. Of course the same possibility exists with **35** as the catalyst. Why would contamination of the desired addition product by such a by-product be of minor concern in this experimental procedure?

5. Why are **35** and **36** only partially soluble in 1-hexene? in aqueous hydrobromic acid?

6. Outline a procedure that would allow monitoring of the course of this reaction as a function of time. In other words, how might the reaction mixture be analyzed periodically so that you could determine when the addition reaction was complete?

7. Why would it be difficult to perform the polar addition of HBr to 1-pentene under the conditions used in this procedure?

8. Refer to Figure 13.30a and determine whether the free fatty acid polymer (FFAP) stationary phase used for the GC analysis shown there separates the bromohexanes on the basis of their relative boiling points or some other molecular property. Explain your answer.

SPECTRA OF STARTING MATERIALS AND PRODUCTS

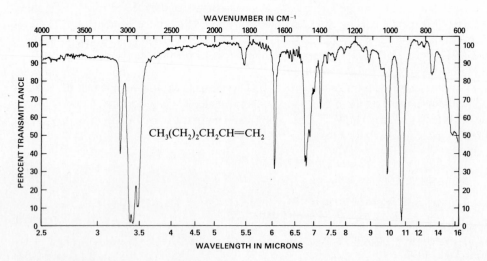

Figure 13.24 IR spectrum of 1-hexene.

(a) PMR spectrum.

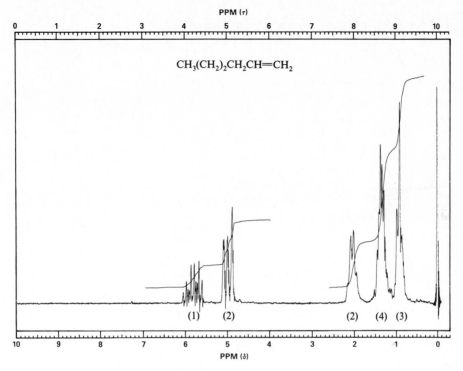

(b) CMR data. Chemical shifts: δ 14.0, 22.5, 31.6, 33.8, 114.3, 139.2.

Figure 13.25 NMR data for 1-hexene.

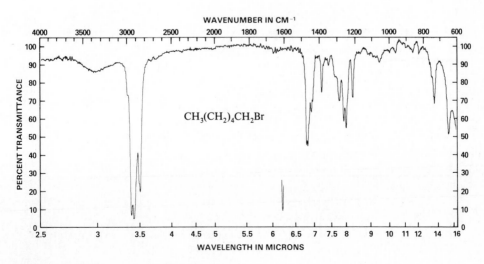

Figure 13.26 IR spectrum of 1-bromohexane.

(a) PMR spectrum.

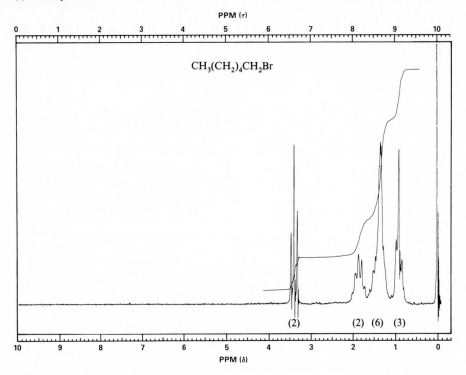

(b) CMR data. Chemical shifts: δ 14.0, 22.7, 28.1, 31.2, 33.1, 33.4.

Figure 13.27 NMR data for 1-bromohexane.

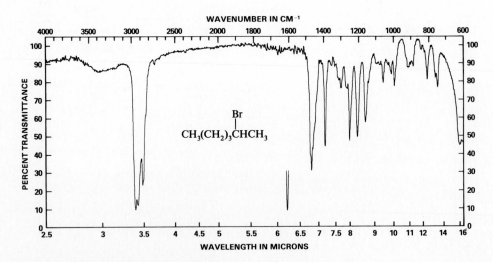

Figure 13.28 IR spectrum of 2-bromohexane.

(a) PMR spectrum.

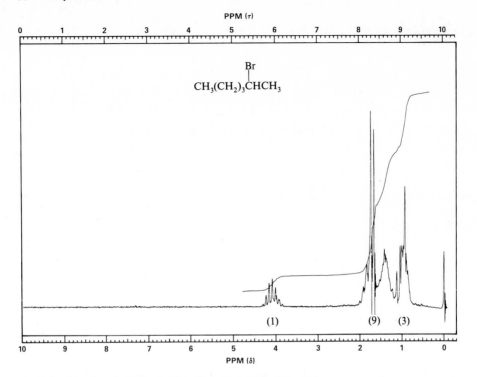

(b) CMR data. Chemical shifts: δ 13.9, 22.1, 26.5, 29.9, 41.0, 51.0.

Figure 13.29 NMR data for 2-bromohexane.

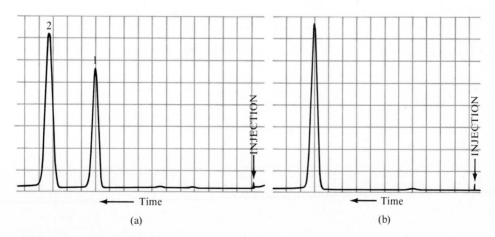

Figure 13.30 GC traces of the bromohexanes. (a) Mixture of 1-bromohexane (peak 2) and 2-bromohexane (peak 1). (b) Distillate from addition of HBr to 1-hexene. Analysis on 3-m 15% FFAP on 60/80 Chromosorb P/A at 130°C, flow rate of 60mL/min.

alkynes

Alkynes are similar to alkenes (see Chapter 13) insofar as their reactions are concerned. In particular, alkynes undergo electrophilic *addition* reactions with reagents such as E—Nu (equation 1). Using carefully controlled reaction conditions, it is possible to stop the addition reaction at the alkene stage in some cases.

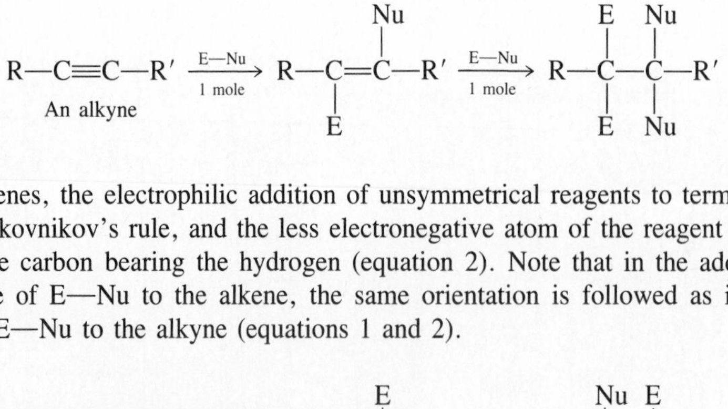

$$R-C\equiv C-R' \xrightarrow[\text{1 mole}]{\text{E---Nu}} R-\overset{\displaystyle}{\underset{\displaystyle E}{C}}=\overset{\displaystyle Nu}{\underset{\displaystyle}{C}}-R' \xrightarrow[\text{1 mole}]{\text{E---Nu}} R-\overset{\displaystyle E}{\underset{\displaystyle E}{C}}-\overset{\displaystyle Nu}{\underset{\displaystyle Nu}{C}}-R' \qquad (1)$$

An alkyne

As with alkenes, the electrophilic addition of unsymmetrical reagents to terminal alkynes follows Markovnikov's rule, and the less electronegative atom of the reagent becomes attached to the carbon bearing the hydrogen (equation 2). Note that in the addition of the second mole of E—Nu to the alkene, the same orientation is followed as in the initial addition of E—Nu to the alkyne (equations 1 and 2).

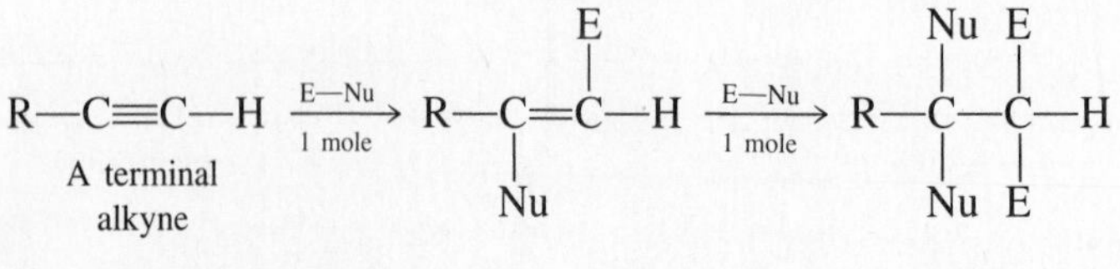

$$R-C\equiv C-H \xrightarrow[\text{1 mole}]{\text{E---Nu}} R-\overset{\displaystyle E}{\underset{\displaystyle Nu}{C}}=\overset{\displaystyle}{\underset{\displaystyle}{C}}-H \xrightarrow[\text{1 mole}]{\text{E---Nu}} R-\overset{\displaystyle Nu}{\underset{\displaystyle Nu}{C}}-\overset{\displaystyle E}{\underset{\displaystyle E}{C}}-H \qquad (2)$$

A terminal alkyne

Ethyne (acetylene), the simplest alkyne, is a widely used commercial organic compound, but because it is a gas at room temperature, it is not experimentally suitable for investigating the properties and reactions characteristic of the carbon-carbon triple bond. The alkyne used in our study, 2-methyl-3-butyn-2-ol **(1)**, has a terminal triple bond and therefore has chemical properties quite similar to those of ethyne. Alkyne **1** has the advantages of being inexpensive and readily available commercially, as well as being easily handled. The presence of a hydroxyl group in **1** has little effect on the chemical properties of the carbon-carbon triple bond. The main effect of this group is on the physical properties of the molecule, such as melting and boiling points. Acetylenic *hydrocarbons* having the same molecular weight would melt and boil much lower.

$$CH_3-\underset{\underset{CH_3}{|}}{\overset{\overset{OH}{|}}{C}}-C\equiv C-H$$

1
2-Methyl-3-butyn-2-ol

Alkynes respond to some of the same tests for unsaturation used on alkenes. For example, both alkenes and alkynes decolorize a solution of bromine in carbon tetrachloride. Either one or two moles of bromine may react per mole of alkyne, with two moles being required for complete reaction. Alkynes also decolorize potassium permanganate solutions, as do alkenes, in the Baeyer test.

The acetylenic hydrogen of a terminal alkyne ($-C\equiv C-H$) is relatively acidic compared with the hydrogens in alkanes and alkenes. The acidity of hydrocarbons has been studied extensively, and it has been found that the order of decreasing acidity is as follows:

$$H-C\equiv C- \;>\; \underset{H}{\overset{}{>}}C=C\overset{}{<} \;>\; H-\overset{|}{\underset{|}{C}}-\overset{|}{\underset{|}{C}}-$$

$K_a \sim 10^{-25}$	$K_a \sim 10^{-45}$	$K_a < 10^{-50}$
(Most acidic)		(Least acidic)

This acidity is an important property of terminal alkynes, and a simple test for this group is the ready formation of a precipitate on addition of a solution containing silver ammonia complex (silver nitrate dissolved in ammonia). Terminal alkynes give a solid silver salt, which results when the terminal hydrogen is removed by ammonia acting as a base and is replaced by silver ion. This reaction is shown in equation 3 in general terms; compound **1** is a representative example of a compound that would so react.

$$\underset{\text{A terminal alkyne}}{R-C\equiv C-H} + Ag(NH_3)_2^{\oplus} \rightarrow \underset{\text{A silver acetylide}}{\underline{R-C\equiv C-Ag}} + NH_3 + NH_4^{\oplus} \qquad \textbf{(3)}$$

This unique reaction of the terminal alkynes provides an easy method for separating them from nonterminal alkynes. The former form insoluble silver salts, whereas the latter do not and remain in solution. This test also differentiates terminal alkynes from alkenes. The silver acetylide salts can be reconverted to the terminal alkynes on treatment with hydrochloric acid, as shown in equation 4. *Dry* silver salts of this type are quite sensitive to shock and in this condition tend to decompose *explosively*. Care should be taken to ensure that they never dry out before decomposition by hydrochloric acid.

$$R—C≡C—Ag + HCl → R—C≡C—H + AgCl \qquad \textbf{(4)}$$

A useful reaction involving the carbon-carbon triple bond is hydration to give an aldehyde or ketone. For example, the hydration of acetylene was formerly used commercially for the preparation of acetaldehyde. (The high cost of acetylene and the development of newer processes based on ethylene has made this method obsolete as an industrial process.) The addition of water is catalyzed by mercuric sulfate in the presence of sulfuric acid. The reaction proceeds via the initial formation of vinyl alcohol, which is unstable and tautomerizes to acetaldehyde, as shown by equation 5.

$$H—C≡C—H + H_2O \xrightarrow[\text{HgSO}_4]{\text{H}_2\text{SO}_4} \left[CH_2{=}C\overset{\displaystyle OH}{\underset{\displaystyle H}{\big<}} \right] \rightleftharpoons CH_3—C\overset{\displaystyle O}{\underset{\displaystyle H}{\big\|}} \qquad \textbf{(5)}$$

Ethyne (acetylene)	Vinyl alcohol (unstable; not isolated)	Ethanal (acetaldehyde)

With substituted acetylenes, hydration occurs in accordance with Markovnikov's rule, the proton going to the carbon atom already bearing the greater number of hydrogens, and the product is a ketone. 2-Methyl-3-butyn-2-ol (**1**) is a monosubstituted acetylene that, under typical hydration conditions, will give a ketone, 3-hydroxy-3-methyl-2-butanone (**2**), as shown by equation 6.

$$
\underset{\textbf{1}}{\overset{\displaystyle HO}{\underset{\displaystyle CH_3}{CH_3{-}\overset{|}{\underset{|}{C}}{-}C≡C{-}H}}} + H_2O \xrightarrow[\text{HgSO}_4]{\text{H}_2\text{SO}_4}
\underset{\textit{(Unstable; not isolated)}}{\left[\overset{\displaystyle HO\quad OH}{\underset{\displaystyle CH_3}{CH_3{-}\overset{|}{\underset{|}{C}}{-}\overset{|}{C}{=}CH_2}} \right]} \rightleftharpoons
$$

$$
\underset{\substack{\textbf{2}\\ \text{3-Hydroxy-3-methyl-2-}\\ \text{butanone}}}{\overset{\displaystyle HO\quad O}{\underset{\displaystyle CH_3}{CH_3{-}\overset{|}{\underset{|}{C}}{-}\overset{\|}{C}{-}CH_3}}} \qquad \textbf{(6)}
$$

The hydration reactions often require specific catalysts, some of which are Cu(I), Hg(II), and Ni(II) ions, but of these mercuric ion is used most often. The role these catalysts play in the reaction is not completely understood, but it is felt that they may form a π-complex with the triple bond and thus render the alkynes more soluble in the polar aqueous solvent. It is also possible that these metal ions actually *add* directly to the triple bond and are then removed in a subsequent displacement reaction. The hydration of **1** is carried out using mercuric sulfate as the catalyst in the presence of sulfuric acid. To avoid the potential difficulty inherent in a highly exothermic reaction, dilute acid is used, and the alkyne is added in two portions.

It has been asserted above that the addition of water (an unsymmetrical reagent) to **1** (an unsymmetrical alkyne) will occur according to Markovnikov's rule to give **2** as the final product. This can readily be verified by identifying the product of the hydration experiment, since if the water were added in an anti-Markovnikov fashion, 3-methyl-3-hydroxybutanal (**3**) would be formed rather than the ketone **2**.

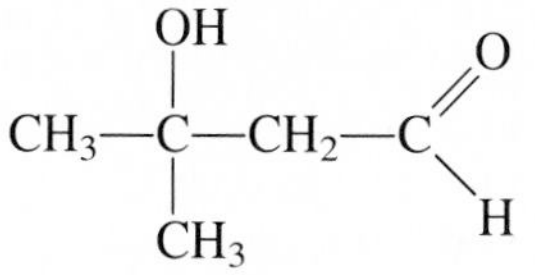

$$CH_3-\overset{\displaystyle OH}{\underset{\displaystyle CH_3}{C}}-CH_2-\overset{\displaystyle O}{\underset{\displaystyle H}{C}}$$

3
3-Hydroxybutanal

In order to distinguish between the two possible products, a solid derivative of the product can be prepared for identification purposes. (See Chapter 27 for additional information concerning identification of organic compounds through preparation of derivatives.) One suitable solid derivative for carbonyl-containing compounds is the semicarbazone (Section 27.5a.2). Compounds **2** and **3** have been prepared by independent methods, and their semicarbazone derivatives have been made: The semicarbazone (**4**) of ketone **2** melts at 162.5°C, whereas aldehyde **3** first dehydrates under the conditions of semicarbazone formation, and the resulting unsaturated aldehyde, $(CH_3)_2C=CHCHO$, then reacts to form a semicarbazone (**5**), which melts at 222–223°C.

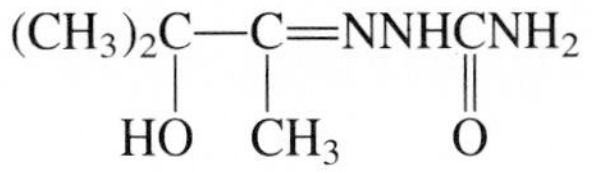

$$(CH_3)_2\underset{\displaystyle HO}{C}-\underset{\displaystyle CH_3}{C}=\overset{}{\underset{\displaystyle O}{NNHCNH_2}} \qquad (CH_3)_2C=\underset{\displaystyle H}{CH}-\underset{\displaystyle O}{C}=NNHCNH_2$$

4 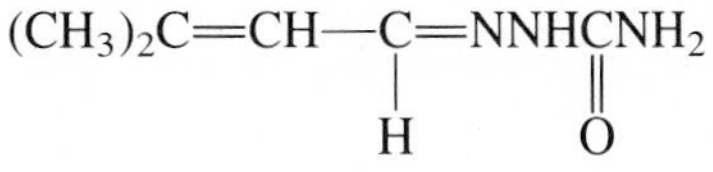   **5**

Thus, if the semicarbazone of the product is prepared and its melting point determined, the principal mode of hydration may be deduced.

Pre-lab exercises for Chapter 14, Alkynes, can be found on pages PL. 39 and 41.

EXPERIMENTAL PROCEDURE

DO IT SAFELY

1 The concentrated sulfuric acid must be handled carefully. Be sure that you add the acid slowly *to* water, and not in the reverse order.

2 The residual solution from the steam distillation, containing the mercury salts, must be discarded in a special waste container, *not* poured down the drain.

3 Take special care to destroy *all* of the silver salt with hydrochloric acid before discarding it.

A. Qualitative Tests for the Triple Bond

Reaction with Bromine in Carbon Tetrachloride See the experimental procedure given in Section 27.5b. Follow the same procedure for the qualitative test for the carbon-carbon triple bond in 2-methyl-3-butyn-2-ol.

Baeyer Test for Unsaturation See the experimental procedure given in Section 27.5b and follow it for the reaction of 2-methyl-3-butyn-2-ol with potassium permanganate.

Reaction with Silver Ammonia Complex; Formation of a Silver Acetylide and its Decomposition Prepare a solution of silver ammonia complex from 5 ml of 0.1 *M* silver nitrate solution by adding ammonium hydroxide solution dropwise. Brown silver oxide forms first; add *just enough* ammonium hydroxide to dissolve the silver oxide. Dilute the solution by adding 3 ml of water. Add 3 ml of the diluted silver ammonia complex solution to about 0.1 ml of 2-methyl-3-butyn-2-ol.

Note the formation of the silver acetylide salt. If you have at your disposal a disubstituted acetylene, treat it similarly and note any reactions that occur. Filter the silver salt from the aqueous solution; be careful not to let it dry, for the dry salt is explosive. Treat the silver salt with a small amount of dilute hydrochloric acid, and observe what changes occur, especially in the color and form of the precipitate. Ultimately, destroy *all* solid salt by treatment with hydrochloric acid.

B. Preparation of 3-Methyl-3-hydroxy-2-butanone; the Hydration of 2-Methyl-3-butyn-2-ol[1]

Add 18 mL of *concentrated* sulfuric acid *carefully* to 115 mL of water contained in a 500-mL round-bottomed flask. Dissolve 1 g of mercuric oxide in the resulting warm solution[2] and then cool the flask to about 50°C. Attach a reflux condenser to the flask, and then add *in one portion* 12.6 g (0.15 mol) of 2-methyl-3-butyn-2-ol through the condenser. The precipitate that forms immediately is presumably the mercury complex of the alkyne. Shake the reaction flask to mix the contents. When this is done, an exothermic reaction ensues as the precipitate dissolves, and the solution turns light brown. Allow the reaction to proceed by itself for about two minutes, and then heat the mixture to reflux. It may be that the exothermic reaction will not begin on its own; if it does not, heat the mixture until it starts to reflux. As soon as the reflux begins, remove the heat and cool the reaction mixture to 50°C. Add a second 12.6-g portion of alkyne through the condenser. A precipitate will form again. Mix the contents of the flask by shaking vigorously but carefully, and then heat the mixture to reflux for 15 to 20 min.★ While the reaction mixture is cooling to room temperature, prepare apparatus for steam distillation *with external steam supply,* as in Figure 2.19, taking care that all connections of the rubber tubing are tight. A water trap such as that shown in Figure

---

[1]The experimental procedure is a modification of that reported by Rose, N. C. *Journal of Chemical Education,* **1966,** *43,* 324.

[2]It has been found that the purity of the mercuric oxide greatly affects the yield of the hydration reaction. Use a good grade of mercuric oxide.

2.21(b) should be placed between the source of the steam (either a house line or an individual generator) and the reaction flask. If a house steam line is used, take care to blow the water out of the line before connecting it to the trap. If the flow of steam is interrupted at any time, the inlet tube must be removed immediately from the reaction flask to prevent its contents from being sucked back into the trap. Do not heat the reaction flask with a burner, but allow the steam distillation to proceed until about 150 mL of distillate has been collected, or until the reaction flask becomes almost full because of the condensation of some of the steam.★

Transfer the distillate to a separatory funnel, add about 25 g of potassium carbonate sesquihydrate to the distillate, and then saturate the solution with sodium chloride. A second layer may form; if so, do not separate it but continue with the extraction as described. Extract the mixture with three 35-ml portions of dichloromethane and combine the extracts. Dry the organic extracts over 8 g of *anhydrous* potassium carbonate, filter, and distil. After removal of the dichloromethane, collect the product boiling between 138 and 141°C. The average yield is between 10 and 15 g; the product should be a colorless liquid.

C. Identification of the Product: Spectroscopic Methods

The IR and NMR spectra of both the starting material **(1)** and the product **(2)** are provided at the end of this chapter. Although it is possible to identify the product by converting it to a known solid derivative (see below), it is also quite easy to prove the structure using spectral methods. An examination of the infrared spectra provided will indicate the absence of the C=O group in **1,** whereas an intense absorption for this group is clearly present in **2.** As an exercise and with the aid of Tables 8.2 and 8.3, identify the absorption attributed to this functional group. The PMR spectra will also show a terminal acetylenic hydrogen for **1,** whereas the hydrogens of the newly formed CH_3 adjacent to the C=O will be observed for **2.** Identify the peaks in the PMR spectra, using Tables 9.1 and 9.2.

D. Identification of the Product: Semicarbazone Formation

Add 1 ml of the product to a solution prepared from 1 g of semicarbazide hydrochloride and 1.5 g of sodium acetate dissolved in 5 ml of water. Shake or stir vigorously. The solid that forms is the crude semicarbazone. Recrystallize about one-third of the crude solid as follows. Add to it about 5 ml of toluene, heat the mixture on a steam bath, and add 95% ethanol dropwise until all of the solid has dissolved. Cool and collect the crystals by filtration. Determine the melting point and use this information to confirm the identity of the product obtained by hydration of 2-methyl-3-butyn-2-ol.

EXERCISES

1. It was stated that 2-methyl-3-butyn-2-ol is commercially available and is inexpensive. Suggest a method of preparation of this compound from two simple inexpensive organic compounds.

2. How might you separate, and ultimately obtain in pure form, 1-octyne and 2-octyne from a mixture containing both of them?

3. Give the structures for the products that you would expect to obtain on hydration of 1-octyne and of 2-octyne.

4. It is well known that alcohols can be oxidized to ketones by oxidizing agents such as potassium permanganate. An examination of alkyne **1** will show that it has a hydroxyl group as well as a triple bond. Therefore, a student might conclude that the positive permanganate test is due to the oxidation of the hydroxyl group and not to reaction of the triple bond. Could this be the case? If so, what additional "control" experiments might you suggest to eliminate this possibility from further consideration?

5. Consider the compounds hexane, 1-hexene, and 1-hexyne. What similarities and what differences would you expect in the reactions of these compounds with (a) bromine in carbon tetrachloride, (b) an aqueous solution of potassium permanganate, (c) an aqueous solution of sulfuric acid, (d) a solution of silver ammonia complex? Give the structures of the products, if any, that would be obtained from these reactions.

SPECTRA OF STARTING MATERIAL AND PRODUCT

The PMR spectrum of 2-methyl-3-butyn-2-ol is given in Figure 9.13.

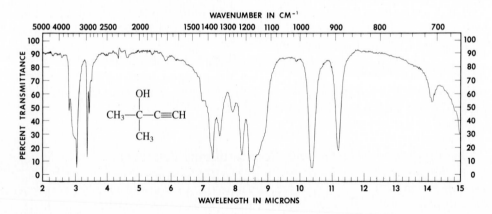

Figure 14.1 IR spectrum of 2-methyl-3-butyn-2-ol.

$$CH_3-\underset{\underset{CH_3}{|}}{\overset{\overset{OH}{|}}{C}}-C\equiv CH$$

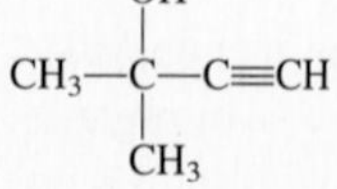

Chemical shifts: δ 31.3, 64.8, 70.4, 89.1.

Figure 14.2 CMR data for 2-methyl-3-butyn-2-ol.

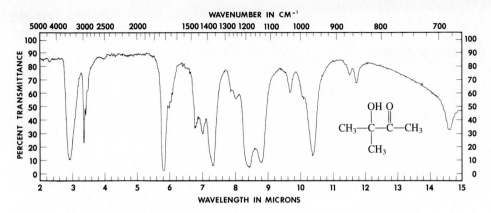

Figure 14.3 IR spectrum of 3-methyl-3-hydroxy-2-butanone.

(a) PMR spectrum.

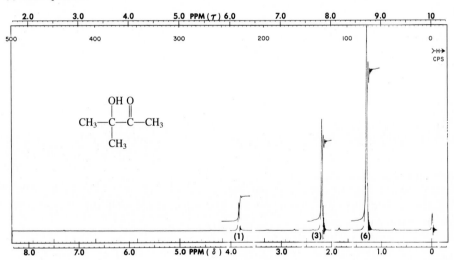

(b) CMR data. Chemical shifts: δ 23.8, 26.4, 76.7, 213.2.

Figure 14.4 NMR data for spectrum of 3-methyl-3-hydroxy-2-butanone.

dienes
the Diels-Alder reaction

The construction of carbocyclic rings from acyclic precursors is of great importance in organic chemistry. One of the most useful methods available to the organic chemist for accomplishing this involves the reaction between a 1,3-diene and an alkene, often referred to as a **dienophile** (Gr., *philos*, loving), which affords a derivative of cyclohexene (equation 1). If an alkyne is used instead of an alkene as the dienophile, a derivative of 1,4-cyclohexadiene is formed (equation 2). This ring-forming process is an example of a **cycloaddition** reaction and is called the Diels-Alder reaction in recognition of Otto Diels and Kurt Alder, two German chemists who discovered it and were responsible for its initial development. Because the reaction results in the formation of new carbon-carbon bonds between the two π-bonded carbon atoms of the dienophile and the 1- and 4-carbon atoms of the diene, there is a net 1,4- addition of the dienophile to the diene.

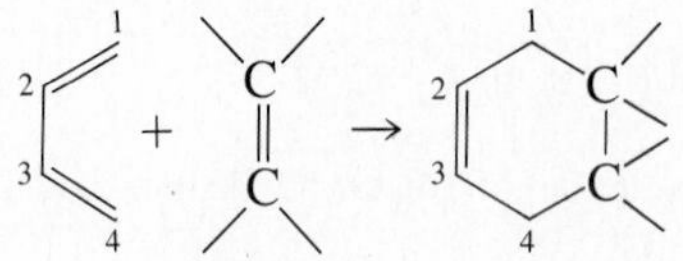

$$\text{(1)}$$

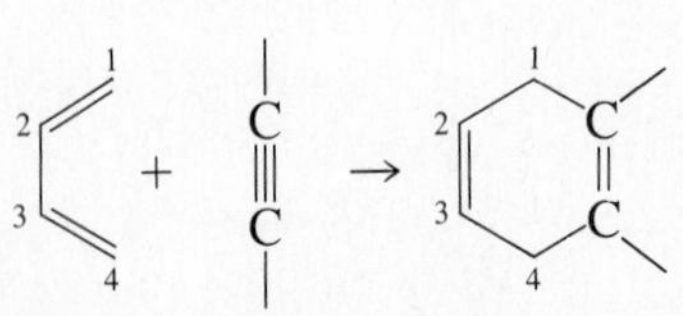

$$\text{(2)}$$

The Diels-Alder reaction is of quite general utility, because many types of dienes and dienophiles have been found to give good yields of adducts, and this broad scope accounts in large part for its significance in organic synthesis. The presence of electron-releasing substituents such as alkyl and alkoxy (—OR) on the diene and electron-withdrawing groups like

cyano (—C≡N) and carbonyl ($\diagdown$ C=O) on the dienophile increase the yield obtained in the

reaction. Put in another way, the Diels-Alder process is generally most efficient when an electron-rich diene and an electron-poor dienophile are used, although the reaction can be made to occur when the diene is electron-poor and the dienophile electron-rich.

Much research has been directed toward understanding the mechanism of the Diels-Alder reaction, and the process is now fairly well understood. In deriving a mechanism for this cycloaddition, the following facts must be considered. The reaction exhibits first-order dependency upon the concentration of both the diene and the dienophile, making the process second-order overall. Kinetic measurements show that the rate of the reaction is normally *not* significantly affected by (1) addition of catalysts such as bases or free-radical chain initiators, (2) irradiation with light of various wavelengths, (3) the phase (gas or liquid), or (4) the polarity of the solvent in which the reaction is performed.

As might be expected, there are exceptions to these generalizations. For example, Lewis acids catalyze certain types of Diels-Alder reactions (see reference 3 at the end of this chapter), and use of water rather than 2-methylheptane (isooctane) as the solvent for the Diels-Alder reaction between 1,3-cyclopentadiene and butenone (equation 3) increases the rate of the reaction by a factor of about 400, presumably for reasons of hydrophobicity (students interested in more information on this type of solvent effect should consult reference 5 at the end of the chapter). The accumulated evidence favors the conclusions that the transition state of the reaction consists of a single molecule of each of the reactants and that neither highly polar intermediates, such as carbocations, nor free radicals are involved in the mechanism. The generally accepted mechanism is thus one in which π-bond breaking and σ-bond making occur in a more or less synchronous (concerted) fashion in the transition state, so that little charge or free-radical character is developed (equation 3).

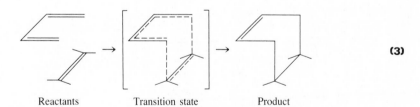

| Reactants | Transition state | Product |

(3)

Theoreticians have recently shown that the symmetries and electronic populations of the molecular orbitals of a diene and dienophile and of the cycloadduct derived from reaction between them are such that a quantum-mechanically continuous transformation of the populated orbitals of reactants into those of the Diels-Alder product is possible. Thus the reaction is said to be allowed as a concerted process and to be controlled by **orbital symmetry.** Most modern organic lecture textbooks contain a discussion of orbital symmetry and its ramifications with respect to many types of reactions. These as well as many supplemental sources, such as those of references 7 and 8 at the end of this chapter, can be consulted for details.

There is an extremely important limitation to the Diels-Alder reaction that is a consequence of the concerted nature of the cycloaddition. This is the requirement that the diene *must* be capable of attaining a conformation in which the two conjugated double bonds of the diene are on the same "side" of the single bond joining them; that is, the dihedral angle between the double bonds must be 0°. Such a conformation is designated as "*s-cis*" (*cis* about the *s*ingle bond) and is the geometry needed to give a *cis* double bond in the cycloadduct. A concerted reaction of a dienophile with a diene in its *s-trans* conformation would lead to a *trans* double bond, which is prohibited in a six-membered ring for geometric reasons. An attempt to construct a molecular model of *trans*-cyclohexene should convince you on this point.

s-cis   s-trans

An implication of the necessity of the *s-cis* conformation in order for the Diels-Alder reaction to occur is that dienes that have difficulty achieving this relationship should react more slowly or not at all. A test of this hypothesis is possible by determining the structure(s) of the cycloadduct(s) obtained from the reaction between a *mixture* of *cis*- and *trans*-1,3-pentadiene and maleic anhydride (equation 4), as is described in the Experimental Procedures (Part D).

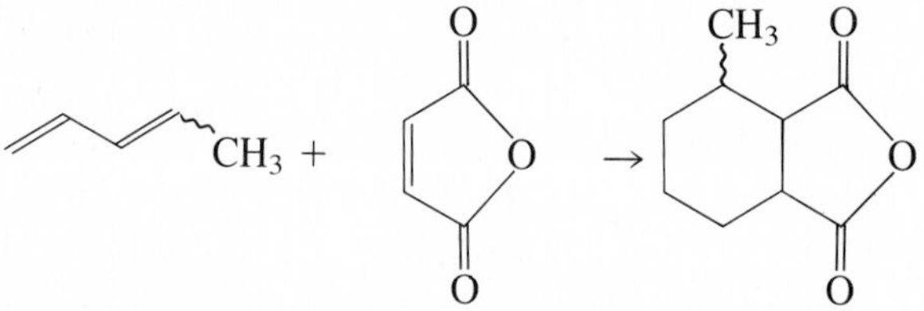

(4)

1,3-Pentadiene   Maleic anhydride
(*cis* and *trans*)

The highly ordered relationship between the diene and dienophile in the transition state depicted in equation 3 might lead one to anticipate high stereoselectivity in the Diels-Alder reaction, and this expectation is amply fulfilled. As an example, the cycloadduct resulting from reaction of *trans,trans*-2,4-hexadiene and a dienophile is exclusively *cis*-3,6-dimethylcyclohexene (equation 5), whereas the product obtained from *cis,trans*-2,4-hexadiene is *trans*-3,6-dimethylcyclohexene. Similarly, reaction between 1,3-butadiene and the *trans*-ester, dimethyl fumarate, results solely in formation of the *trans*-diester **1** (equation 6), whereas the *cis* isomer of **1** results when the *cis*-ester, dimethyl maleate, is used.

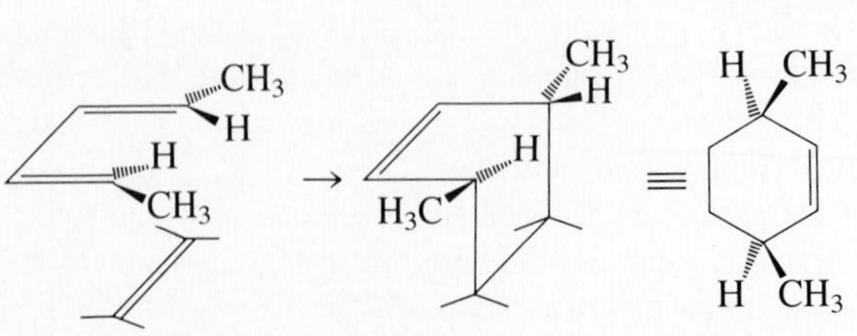

(5)

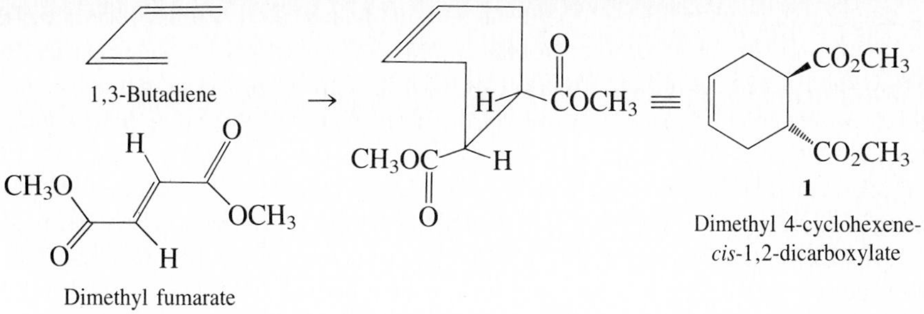

1,3-Butadiene

Dimethyl fumarate

Dimethyl 4-cyclohexene-
cis-1,2-dicarboxylate

(6)

1

A second type of stereoselectivity is observed for the Diels-Alder reaction. The addition of a dienophile such as maleic anhydride to a *cyclic* diene like 1,3-cyclopentadiene could in principle provide two products, **2** and **3** (equation 7). In fact, only a single cycloadduct, **2,** is observed experimentally. The explanation for this stereoselectivity is not entirely clear, but one attractive theory is that stabilization, that is, a lowering in energy, of the transition state **4** leading to formation of **2** is provided by through-space interaction between the *p*-orbitals of the diene and those of the dienophile, including those at the carbonyl carbon atoms. Analogous stabilization is not possible in the transition state **5** required for generation of **3**. The transition state **4** is often characterized as the one having "maximum accumulation of double bonds," which is simply an alternate way of expressing the fact that stabilization results from development of the through-space overlap of the *p*-orbitals. The expectation of a preference for formation of products resulting from this type of transition state is called the "Alder rule." The reaction between maleic anhydride and 1,3-cyclopentadiene is described in the Experimental Procedures (Part A).

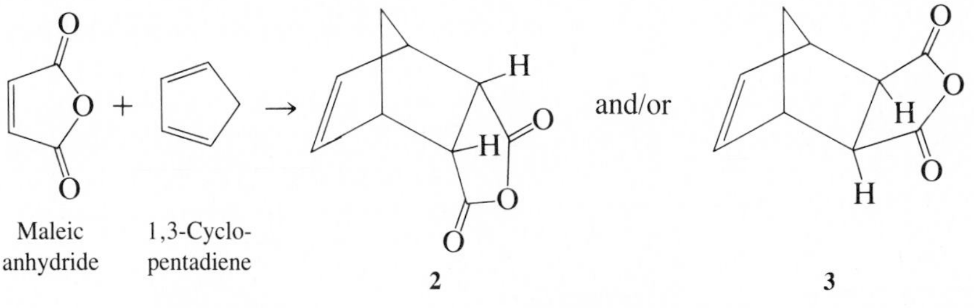

Maleic anhydride

1,3-Cyclo-pentadiene

and/or

(7)

2

3

→ 2

→ 3

4

5

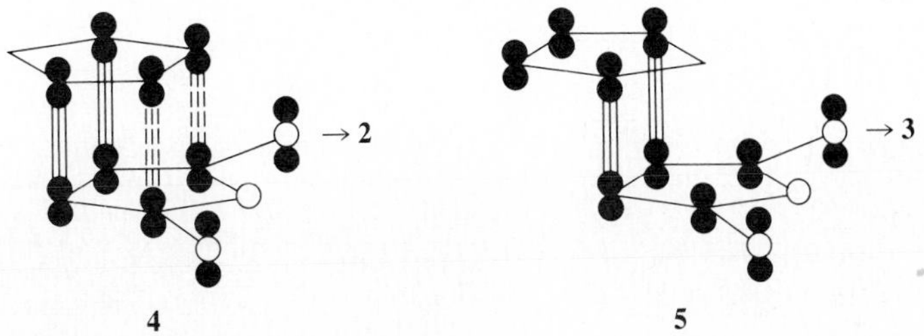

All Diels-Alder reactions are not so stereoselective as the one between cyclopentadiene and maleic anhydride, and mixtures of products are sometimes obtained. However, the product resulting from interaction of the diene and the dienophile in a transition state analogous to **4** generally predominates over that arising from a transition state similar to **5.**

This synthetic method for construction of six-membered rings is remarkably free of complicating side reactions, and the yields of the desired product are often nearly quantitative. Probably the single most important side reaction observed is dimerization or polymerization of the diene that is used. In the former process, one molecule of the diene can serve as the dienophile and another as the diene (equation 8); this is important mainly in those instances in which a relatively electron-rich dienophile such as ethylene is involved. Polymerization tends to dominate with those dienes for which the *s-cis* conformation is difficult to achieve for steric reasons.

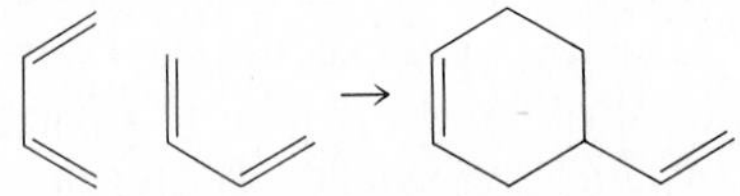

4-Vinylcyclohexene

(8)

Pre-lab exercises for the Diels-Alder reaction are found on pages PL. 43–49.

EXPERIMENTAL PROCEDURES

A. Reaction of 1,3-Cyclopentadiene with Maleic Anhydride

Monomeric 1,3-cyclopentadiene is not commercially available because it readily dimerizes by way of a Diels-Alder reaction at room temperature to give dicyclopentadiene (equation 9), which can be purchased. Fortunately, the equilibrium between the monomer and the dimer can be established at the boiling point of the dimer (170°C), a process commonly called "cracking," and the lower-boiling 1,3-cyclopentadiene can then be isolated by fractional distillation. The diene must be kept cold in order to prevent its redimerization prior to use in the Diels-Alder reaction.

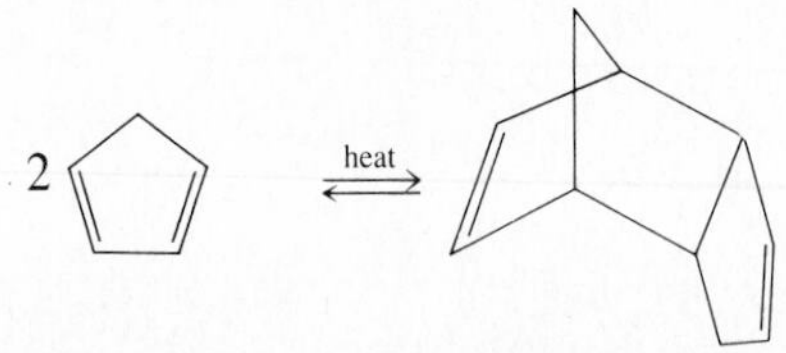

1,3-Cyclopentadiene   Dicyclopentadiene

(9)

DO IT SAFELY

1 1,3-Cyclopentadiene is a mildly toxic and rather volatile substance. Prepare and use it in a hood, if possible. Keep the diene cold at all times to minimize vaporization and possible inhalation of its vapors.

2 Use open flames *only* in those steps in which you are directed to do so. The organic solvents and the 1,3-cyclopentadiene used in this experiment are highly flammable. Be certain that all joints in the apparatus are tight before heating the dicyclopentadiene to produce the monomer.

Place 20 mL of dicyclopentadiene in a 100-mL round-bottomed flask, and attach the flask to a fractional distillation apparatus that is equipped with an *ice-cooled* receiver and has an unpacked fractionating column. Using either a small burner or a heating mantle, gently heat the dimer until brisk refluxing occurs (*Caution:* occasional foaming) and the monomer begins to distil in the range 40–42°C. Distil the monomer as rapidly as possible, but do not permit the head temperature to exceed 43–45°C. About 6 g of monomer should be obtained after distillation for about 30 min. Terminate the distillation when approximately 5 g of residue remains in the flask. If the distilled 1,3-cyclopentadiene is cloudy because of condensation of moisture in the cold receiver, add about 1 g of *anhydrous* calcium chloride to dry it.

While the distillation is in progress, place 6.0 g (0.061 mol) of maleic anhydride in a 125-mL Erlenmeyer flask, and dissolve it in 20 mL of ethyl acetate by heating on a steam or hot-water bath. Add 20 mL of petroleum ether (bp 60–80°C), and cool the solution thoroughly in an ice-water bath. Be sure that the solution is still homogeneous prior to execution of the next step. To this cooled solution add 4.8 g (6.0 mL, 0.073 mol) of dry 1,3-cyclopentadiene, and swirl the resulting solution gently until the exothermic reaction subsides and the adduct separates as a white solid.★ Heat the mixture on a steam or hot-water bath until the solid has redissolved, and then allow the solution to cool slowly to room temperature.★ Filter the solution and determine the yield and melting point of the solid anhydride obtained. The reported melting point is 164–165°C. The yield should be about 80% of theoretical. Test the product for unsaturation (Section 27.5b).

The anhydride can be converted to the corresponding diacid in the following manner. Place 4 g of anhydride and 25 mL of distilled water in a 125-mL Erlenmeyer flask and heat the mixture over a Bunsen burner until boiling occurs and all the oil that initially forms has dissolved. Allow the solution to cool to room temperature and then induce crystallization by scratching the flask at the air-liquid interface.★ After crystallization has begun, cool the flask to 0°C to complete the process, and filter the solution. Determine the yield and melting point of the product. The reported melting point of the expected dicarboxylic acid is 180–182°C. Perform appropriate tests to show whether hydrolysis has destroyed the carbon-carbon double bond (see Section 27.5b). Also test a saturated aqueous solution of the diacid with litmus or pHydrion paper and record the result.

B. Reaction of 1,3-Cyclopentadiene with *p*-Benzoquinone

p-Benzoquinone (**6**) has two carbon-carbon double bonds that can serve as dienophiles. In this experiment the Diels-Alder adduct resulting from reaction of 2 mol of 1,3-cyclopentadiene with 1 mol of *p*-benzoquinone will be prepared (equation 10).

(10)

1,3-Cyclopentadiene   *p*-Benzoquinone

> **DO IT SAFELY**
>
> **1** Read the comments regarding the preparation of 1,3-cyclopentadiene in the Do It Safely section for Part A.
>
> **2** p-Benzoquinone is a toxic substance. Take care in handling it to avoid contact with skin. Should contact occur, wash the area thoroughly with soap and warm water, not with solvents such as acetone or alcohol.

Prepare dry 1,3-cyclopentadiene as described in Part A. Add a solution of 3.2 g (4.0 mL, 0.049 mol) of this diene and 10 mL of toluene to 2.7 g (0.025 mol) of p-benzoquinone[1] dissolved in 20 mL of toluene, and swirl the resulting solution until the mildly exothermic reaction has subsided.★ After 1 hr cool the reaction mixture to 0°C and filter to isolate the solid adduct.★

In order to collect a second crop of product, concentrate the mother liquor to one-third of its original volume by distillation, cool the pot residue,★ collect the precipitate, and wash it with a few milliliters of ice-cold toluene.★ Combine the two crops of adduct and recrystallize the product by dissolving it in a minimum amount of boiling acetone (*Caution:* no flames) and then adding water until turbidity begins to appear in the boiling mixture. Clarify the solution by addition of a *small* amount of acetone and allow the solution to cool slowly to room temperature. Complete the recrystallization by cooling the solution to 0°C.

The pure adduct forms beautiful iridescent white needles or platelets and has a reported melting point of 157–158°C. The expected yield is 60–70%.

C. Reaction of 1,3-Butadiene and Maleic Anhydride

1,3-Butadiene is a gas at room temperature (bp −4.4°C). Diels-Alder reactions involving this diene are therefore normally performed in closed steel pressure vessels called autoclaves into which the diene is introduced under pressure. However, 1,3-butadiene can be generated conveniently by the thermal decomposition of 3-sulfolene (equation 11) and will then react with a dienophile that is present. Such an *in situ* preparation of 1,3-butadiene will be used in this experiment to prepare 4-cyclohexene-*cis*-1,2-dicarboxylic anhydride (equation 12).

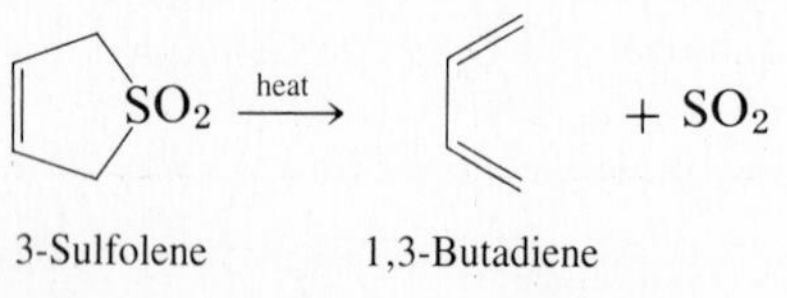

$$\text{3-Sulfolene} \xrightarrow{\text{heat}} \text{1,3-Butadiene} + SO_2 \qquad (11)$$

---

[1]Commercial p-benzoquinone is generally rather impure and may require purification prior to use in this experiment. However, a grade of this chemical having mp 113–115°C can be used without further purification. The quinone can be purified by sublimation in an apparatus such as that pictured in Figure 2.20. Consult your instructor for details of this method of purification.

1,3-Butadiene	Maleic	4-Cyclohexene-*cis*-1,2-
	anhydride	dicarboxylic anhydride

(12)

DO IT SAFELY

Be certain that all joints in the apparatus are tight and well lubricated before the reaction mixture is heated. The organic solvents used are flammable, and the sulfur dioxide that is evolved is toxic and ill-smelling. It is important to be sure that the gas trap is functioning properly before heating of the reaction mixture is started in order to avoid emission of gaseous sulfur dioxide into the laboratory.

Place 10.0 g (0.085 mol) of 3-sulfolene, 6.0 g (0.061 mol) of finely pulverized maleic anhydride, and 4 mL of dry xylene in a 100-mL round-bottomed flask equipped with a water-cooled reflux condenser. Fit the condenser with the gas trap described in Section 12.1 (Figure 12.1). Warm the flask gently while swirling to effect solution, then heat the mixture to a gentle reflux, using a small burner, heating mantle, or oil bath; continue heating for about 30 min.★ Cool the solution, add an additional 30 mL of xylene or toluene and about 1 g of decolorizing carbon, and then heat the mixture on a steam bath with constant swirling for about 5 min. During this period the condenser should be equipped with a calcium chloride drying tube rather than the gas trap to prevent introduction of water into the reaction mixture. Filter the hot mixture (*Caution:* check for flames in your vicinity) using a short-stemmed funnel and a fluted filter (see Figure 2.16), and carefully add petroleum ether (60–80°C) to the filtrate until cloudiness develops. Set the solution aside to cool to room temperature.★ Collect the crystals by filtration and dry them thoroughly. Record their weight and determine the melting point of the product. The reported melting point is 103–104°C. Perform qualitative tests for unsaturation, and hydrolyze 4 g of the anhydride to the diacid according to the procedure described in Part A. Determine the yield and melting point of the diacid and test it for unsaturation. The reported melting point of 4-cyclohexene-*cis*-1,2-dicarboxylic acid is 164–166°C.

If isolation of the anhydride is not desired, direct hydrolysis of it to the diacid can be accomplished in the following way. After the initial 30-min period of reflux, pour the reaction mixture into about 25 mL of water, and heat the resulting initially heterogeneous mixture for at least 30 min. Allow the mixture to stand until the next laboratory period. The resulting crude diacid has mp 160–164°C and can be purified by recrystallization from water.

D. Reaction of Maleic Anhydride with *cis*- and *trans*-1,3-Pentadiene

A mixture of *cis*- and *trans*-1,3-pentadiene is commercially available in a stated purity of 65–70%, although the GC trace in Figure 15.1 indicates that the two isomers constitute

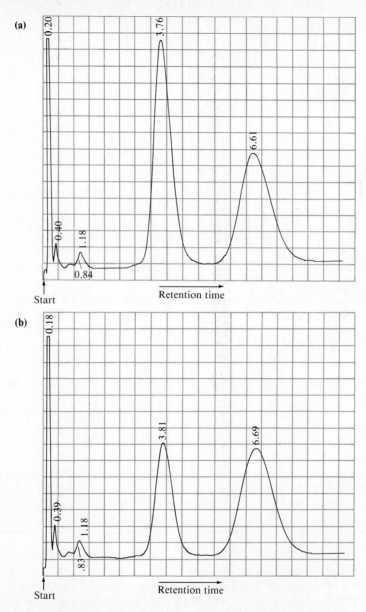

Figure 15.1 GC analyses of *cis*- and *trans*-1,3-pentadienes. (a) Commercial sample. (b) Distillate following reaction with maleic anhydride. Analysis on 1 m 30% AgNO₃-ethylene glycol on 60/80 Chromosorb P, at room temperature. Numbers above peaks are retention times, measured in min.

about 80% of the particular sample analyzed. The goal of this experiment is to determine the relative ease of Diels-Alder reaction of these two dienes with maleic anhydride (equation 4). Assuming the rule of ''maximum accumulation of double bonds'' is applicable, the possible products are cycloadducts **7** and **8,** the melting points of which are 40–41°C and 62–63°C, respectively. Determination of which product is formed preferentially is thus a matter of isolating the solid and determining its melting point. As successful completion of exercises 6 and 7 will show, it is then possible to assess the relative reactivities of the two dienes toward maleic anhydride.

7   8

DO IT SAFELY

1 Use open flames *only* in those steps in which you are directed to do so. The organic solvents and the 1,3-pentadiene used in this experiment are highly flammable. Be certain that all joints in the apparatus are well lubricated and tight before the reaction mixture is heated to reflux and before distillation of solvent.

2 1,3-Pentadiene is a low-boiling, unpleasant-smelling liquid. Avoid release of its vapor into the laboratory by keeping vessels containing it closed as much as possible.

Fit a 100-mL round-bottomed flask with a reflux condenser and a drying tube containing calcium chloride, and dry the apparatus by heating it with a Bunsen burner (see Section 2.24). Place 4 g (0.04 mol) of powdered maleic anhydride in the flask and add *to* it a solution containing 10 g of a mixture of *cis-* and *trans-*1,3-pentadiene of 65–70% purity and 20 mL of toluene. Warm this mixture *gently* with a Bunsen burner, oil bath, or heating mantle, and swirl the flask to effect solution. Then heat the resulting solution, which may be turbid, at a *gentle* reflux for 30 min. Be prepared to remove the heating source if the reaction becomes exothermic, in which case refluxing will continue in the absence of additional heating. After allowing the mixture to cool, equip the reaction flask for simple distillation and distil about 25 mL of volatiles from the reaction mixture. Allow the turbid distillation residue to cool to room temperature and add about 10 mL of technical grade ether to effect coagulation of any gummy polymer that may have formed. Filter the solution through a pad of filter-aid (Celite, see Section 2.12) and concentrate the filtrate to a volume of about 15 mL on the steam cone in a ventilation hood or by simple distillation. *(No flames!)* Cool the residue in an ice-water bath to effect crystallization, and collect the slightly off-white product by filtration. About 3.5–4 g of crude product should be obtained.

Reserve a small portion of the product for the purpose of determining a melting point of crude material, and recrystallize the remainder by dissolving it in about 4 mL of warm toluene and then adding petroleum ether (bp 60–80°C) until turbidity appears. Be certain that the temperature of the solution is not above the melting point of either of the possible anhydrides at the time turbidity is achieved. Determine the weight and melting point of the purified material and use the latter as a basis for deciding which of the two possible Diels-Alder adducts has been obtained. The reported melting points of **7** and **8** are 40–41°C and 62–63°C, respectively.

Hydrolysis of the anhydride to the diacid can be accomplished according to the procedure given in Part A. The reported melting points of the diacids derived from **7** and **8** are 161–162°C and 156–158°C, respectively.

EXERCISES

General Questions

1. Why does a 1,3-diene like 1,3-cyclopentadiene dimerize much more readily than does one like *trans*-1,3-pentadiene?

2. Write the structure of the product resulting from the reaction of *cis,trans*-2,4-hexadiene and 2,3-dimethyl-2-butene.

3. Write structures for the products expected in the following possible Diels-Alder reactions. If no reaction is anticipated, write "N.R."

(a) 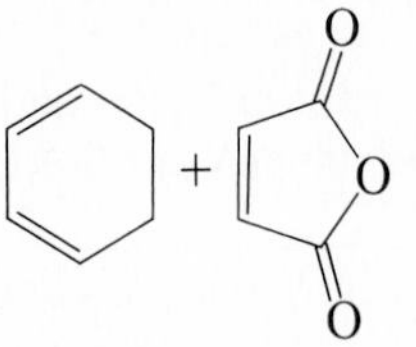 +

(b) 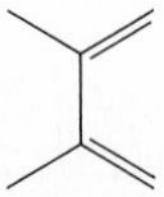 + CH_2=CHCHO

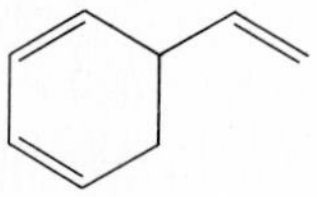

(c)  + CH_2=CH_2

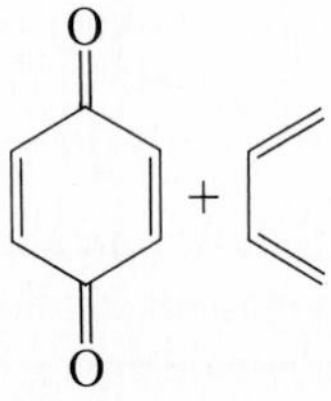

(d) + CH_3O_2C—C≡C—CO_2CH_3

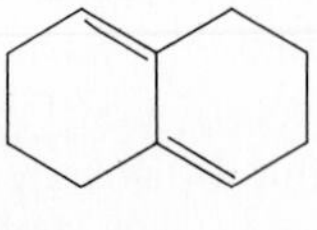

(e) → C_8H_{10}

(f) + (2 moles)

(g) + CH_3O_2C—C≡C—CO_2CH_3

Questions for Part A

4. What problems might arise if the 1,3-cyclopentadiene were not dry prior to its reaction with maleic anhydride?

5. Why is it good technique to induce crystallization of the diacid derived from the Diels-Alder adduct *before* cooling the solution to 0°C?

6. The "cracking" of dicyclopentadiene to two moles of 1,3-cyclopentadiene (equation 9)

is an example of a reverse *(retro)* Diels-Alder reaction. Give the products to be anticipated from an analogous reaction with the compounds shown below.

(a)

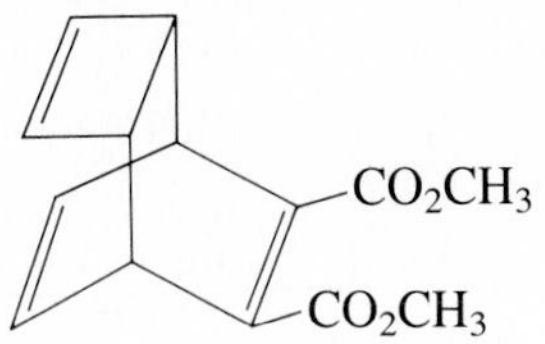

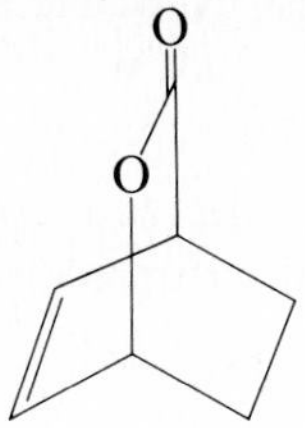

(b)

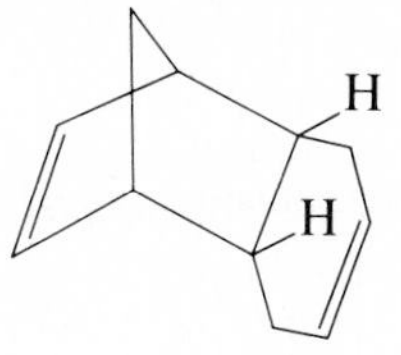

(c)

(d) 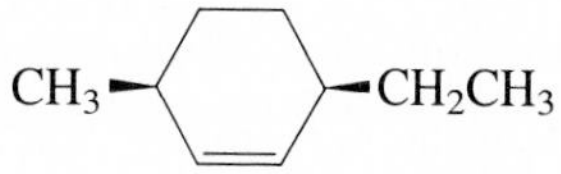 CH₃—⟨⟩—CH₂CH₃

(e)

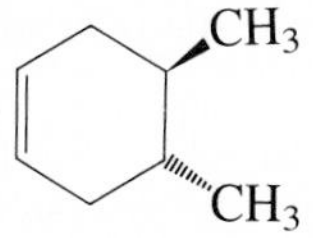

Questions for Part B

7. The "cracking" of dicyclopentadiene to two moles of 1,3-cyclopentadiene (equation 9)
is an example of a reverse *(retro)* Diels-Alder reaction. Predict the products to be
anticipated from an analogous reaction with the compounds shown below.

(a)

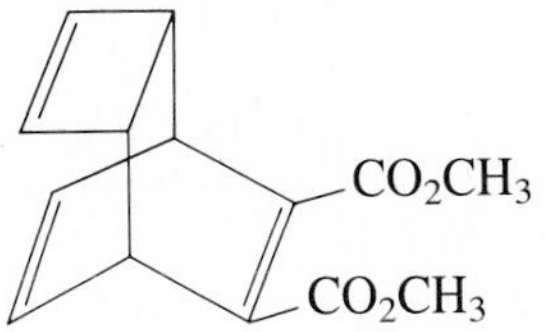

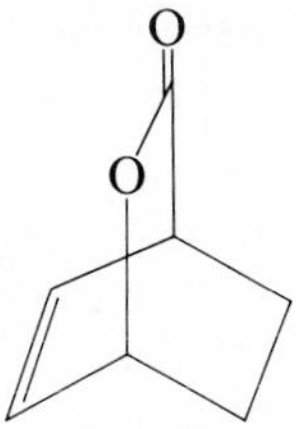

(b)

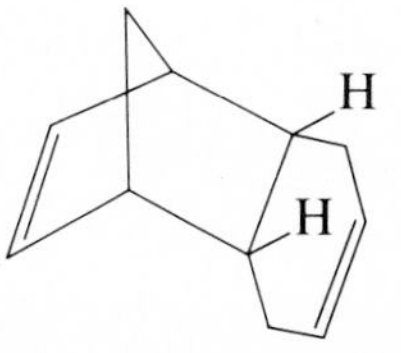

(c)

(d) 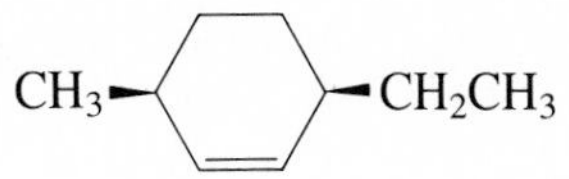 CH₃—⟨⟩—CH₂CH₃

(e)

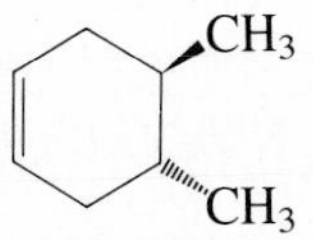

8. Pure *p*-benzoquinone is yellow, but the Diels-Alder adduct with 1,3-cyclopentadiene
made in this experiment is colorless. Why is this?

Questions for Part C

9. In the reaction of 1,3-butadiene with maleic anhydride, a small amount (4 mL) of xylene was used initially, but an additional and larger amount (30 mL) of xylene or toluene was subsequently required. Why was the total amount of organic solvent needed not added at the beginning of the reaction? Why was an additional quantity required in the procedure?

10. Why was it important to prevent introduction of water into the reaction mixture in the decolorizing step that precedes the isolation of the anhydride?

11. Write the structure, including stereochemistry, of the expected addition product of bromine to the Diels-Alder adduct obtained by this procedure.

Questions for Part D

12. a. Which cycloadduct, **7** or **8**, is formed by the reaction of the mixture of *cis*- and *trans*-1,3-pentadiene with maleic anhydride?

   b. Draw the transition states required for formation of this product from the *cis*-diene and from the *trans*-diene.

   c. Based on the Alder rule, as illustrated by the stereochemical outcome in the reaction of 1,3-cyclopentadiene and maleic anhydride (equation 7), which of the acyclic dienes was responsible for formation of the observed product?

13. Why does one geometric isomer of 1,3-pentadiene react faster than the other in a Diels-Alder reaction?

14. In Figure 15.1 are given GC traces of (a) a commercial sample of ''65–70%'' *cis*- and *trans*-1,3-pentadiene and (b) the distillate boiling below 60°C obtained after reaction of the mixture of dienes with maleic anhydride, as described in this chapter.

   a. Calculate the ratio of the peaks at 3.8 min and 6.6 min relative to one another in the two traces. These two components constitute 82.5 and 86.2% of the total volatiles in the sample before and after reaction, respectively.

   b. Assign the peaks to *cis*- and to *trans*-1,3-pentadiene and defend your assignment. For your answer, assume that the polymer by-product that was produced is constituted of *equal* proportions of both of the isomeric dienes.

   c. Why is it necessary in part b that consideration be taken of the composition of the polymeric by-product?

   d. Do the results of the analysis of GC correspond to what you would have expected based on your answer in question 12?

   e. Calculate the theoretical yield for the reaction of maleic anhydride and the reactive isomer of 1,3-pentadiene, given the quantities of materials used in the experiment and the GC trace of the mixture of dienes used.

   f. Look up the boiling points of *cis*- and of *trans*-1,3-pentadiene.

   (1) Do the two isomers emerge from the GC column in the order of their increasing boiling points?

   *(2) If a GC column that has a stationary phase that allows separation of compounds on the basis of their relative boiling points is used to analyze mixtures of *cis*- and *trans*-1,3-pentadiene, a specific example being a 3-m column containing 10%

SE-30 on Chromosorb P and operating at room temperature, only a *single* peak rather than two is observed. This is because the boiling points of the isomers are too similar. How can you account for the fact that a 1-m GC column containing silver nitrate in ethylene glycol on Chromosorb P and operating at room temperature readily separates the two isomers?

15. What function does the filter-aid serve in the initial filtration of the reaction mixture? Should this filtration be done by gravity or vacuum?

REFERENCES

1. Sauer, J. *Angewandte Chemie, International Edition,* **1966,** *5*, 211. A general review of the Diels-Alder reaction.

2. Beltrame, P. *Comprehensive Chemical Kinetics,* **1973,** *9*, 94. A review emphasizing mechanistic aspects of the Diels-Alder reaction.

3. Carruthers, W. *Some Modern Methods of Organic Synthesis,* 2nd ed., Cambridge University Press, Cambridge, 1978. A review emphasizing synthetic applications of the Diels-Alder reaction.

4. Sample, T. E.; Hatch, L. F. *Journal of Chemical Education,* **1968,** *45*, 55.

5. Breslow, R.; Martra, U. *Tetrahedron Letters,* **1984,** *25*, 1239, and references cited therein.

6. Brocksom, T. J.; Constantino, M. G. *Journal of Organic Chemistry,* **1982,** *47*, 3450.

7. Woodward, R. B.; Hoffmann, R. *The Conservation of Orbital Symmetry,* Academic Press, New York, 1970.

8. Lehr, R.; Marchand, A. *Orbital Symmetry,* Academic Press, New York, 1972.

SPECTRA OF STARTING MATERIALS AND PRODUCTS

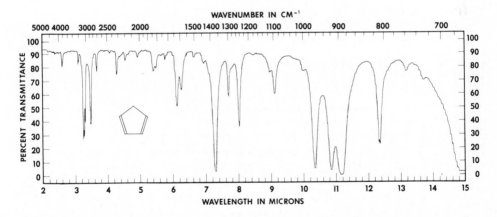

Figure 15.2 IR spectrum of 1,3-cyclopentadiene.

(a) PMR spectrum.

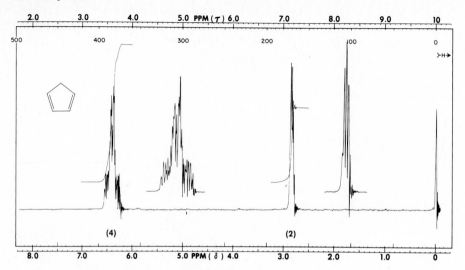

(b) CMR data. Chemical shifts: δ 42.2, 133.0, 133.4.

Figure 15.3 NMR data for 1,3-cyclopentadiene.

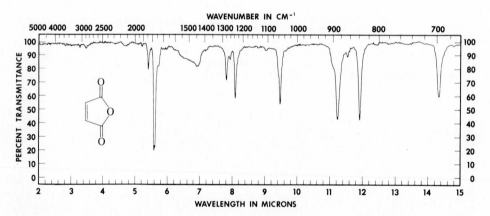

Figure 15.4 IR spectrum of maleic anhydride (in CCl_4 solution).

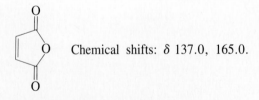 Chemical shifts: δ 137.0, 165.0.

Figure 15.5 CMR data for maleic anhydride.

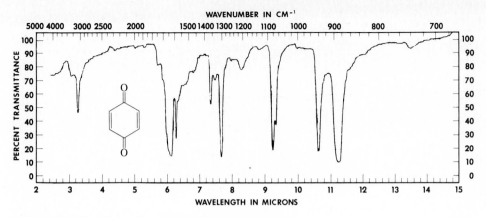

Figure 15.6 IR spectrum of *p*-benzoquinone (KBr pellet).

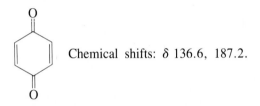

Chemical shifts: δ 136.6, 187.2.

Figure 15.7 CMR data for *p*-benzoquinone.

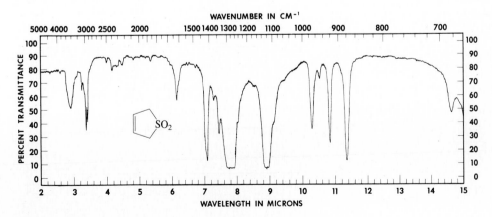

Figure 15.8 IR spectrum of 3-sulfolene (in CCl₄ solution).

(a) PMR spectrum.

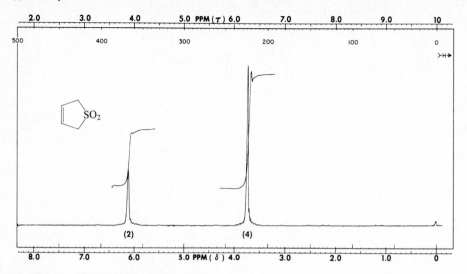

(b) CMR data. Chemical shifts: δ 55.7, 124.7.

Figure 15.9 NMR data for 3-sulfolene.

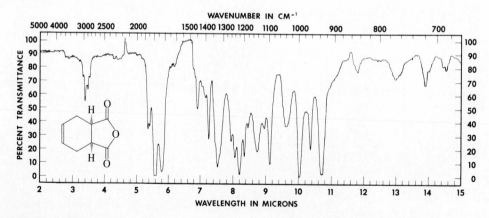

Figure 15.10 IR spectrum of 4-cyclohexene-*cis*-1,2-dicarboxylic anhydride (KBr pellet).

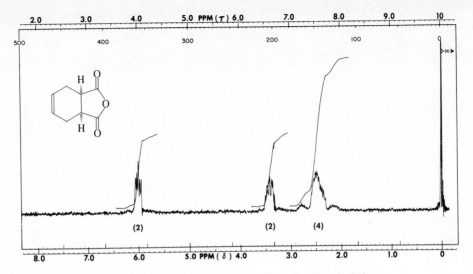

Figure 15.11 PMR spectrum of 4-cyclohexene-*cis*-1,2-dicarboxylic anhydride.

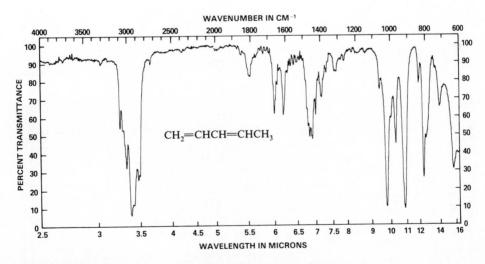

$$CH_2{=}CHCH{=}CHCH_3$$

Figure 15.12 IR spectrum of commercial mixture of 1,3-pentadienes.

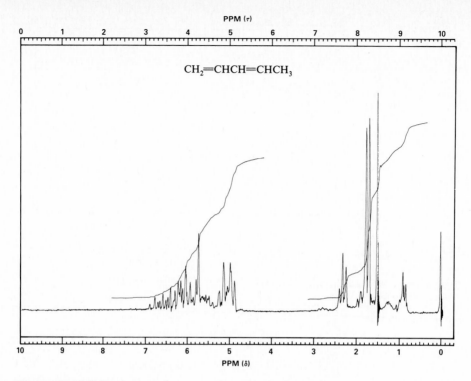

Figure 15.13 PMR spectrum of commercial mixture of 1,3-pentadienes.

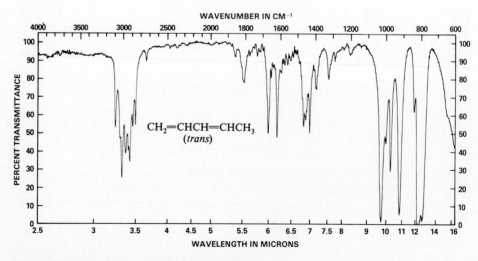

Figure 15.14 IR spectrum of *trans*-1,3-pentadiene.

(a) PMR spectrum.

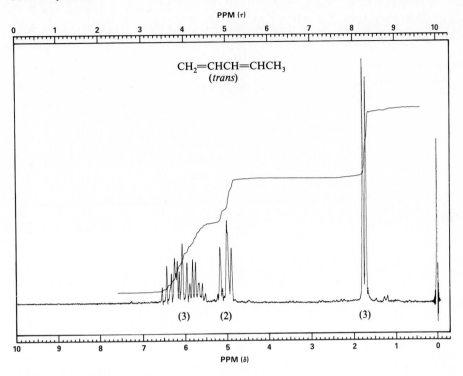

(b) CMR data. Chemical shifts: δ 17.6, 113.9, 129.0, 132.9, 137.5.

Figure 15.15 NMR data for *trans*-1,3-pentadiene.

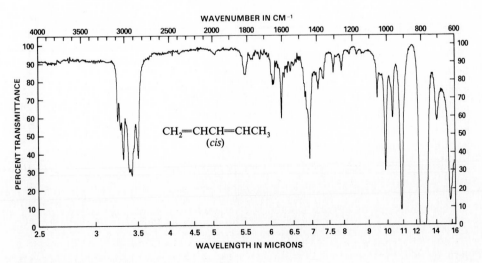

Figure 15.16 IR spectrum of *cis*-1,3-pentadiene.

(a) PMR spectrum.

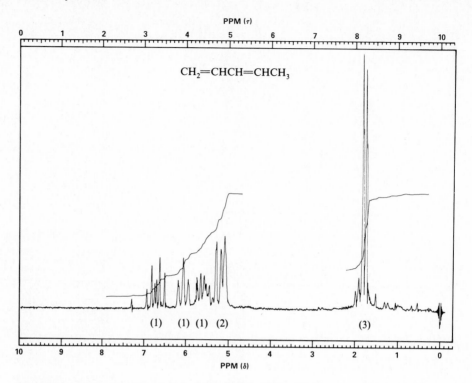

(b) CMR data. Chemical shifts: δ 12.9, 116.1, 126.1, 130.6, 132.2.

Figure 15.17 NMR data for *cis*-1,3-pentadiene.

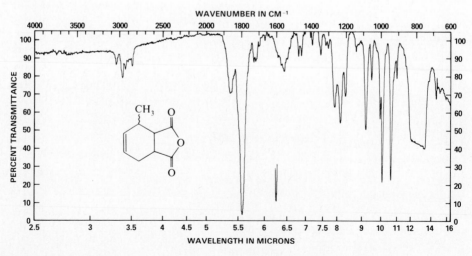

Figure 15.18 IR spectrum of Diels-Alder adduct of maleic anhydride and a 1,3-pentadiene (in CCl$_4$ solution).

(a) PMR spectrum.

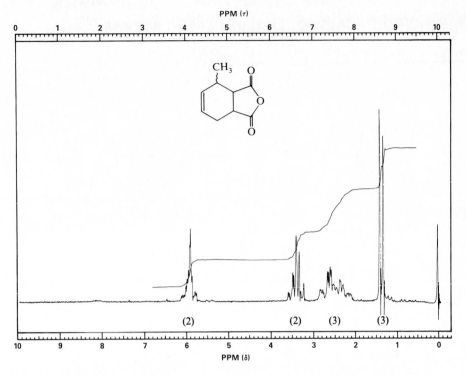

(b) CMR data. Chemical shifts: δ 17.1, 24.1, 30.3, 41.3, 45.5, 127.7, 135.3, 172.2, 175.0.

Figure 15.19 NMR data for Diels-Alder adduct of maleic anhydride and a 1,3-pentadiene.

kinetic and equilibrium control of a reaction

In predicting which of two competing reactions will predominate, a usefule rule of thumb is to choose the reaction that is more exothermic. This rule is based on the fact that *most commonly* the more exothermic reaction will have the lower energy of activation and thus will have the faster rate of reaction. For example, Figure 16.1 depicts the potential energy changes involved in the conversion of compound **X** into the products **Y** and **Z** by two competing reactions (equations 1 and 2).

$$\mathbf{X} \rightleftarrows \mathbf{Y} \tag{1}$$

$$\mathbf{X} \rightleftarrows \mathbf{Z} \tag{2}$$

The more exothermic reaction (equation 2) produces the more stable product, **Z** (the one lying at the lower energy level). Note that the energy of the transition state leading to **Z** ($X_Z^{\ddagger}$) is lower than the energy of the transition state leading to **Y** ($X_Y^{\ddagger}$); that is, the relative energies of the transition states are the same as the relative energies of the products to which they lead. Hence the energy of activation required for production of **Z** [$E_a(Z)$] is lower than the energy of activation required for production of **Y** [$E_a(Y)$], so that the rate of the reaction producing **Z** is faster than the rate of the reaction producing **Y**.

Most organic reactions either are practically irreversible or are carried out under conditions such that equilibrium between products and starting materials is not attained, so the yields of products are determined by the relative rates of competing reactions, as described in the preceding paragraph. The major product of such a reaction is said to be the product of **kinetic control.** However, when the experimental conditions are favorable for equilibrium to be reached between starting materials and products, the product that predominates initially because of kinetic control *may not* be the major product after equilibrium has been attained.

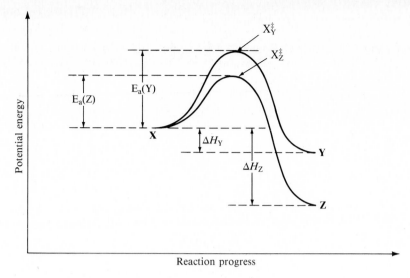

Figure 16.1 Typical reaction profile for competing reactions.

In some reactions the product of kinetic control is less stable than another product formed at a slower rate. An example of such a system (equations 3 and 4) is represented by Figure 16.2, in which product **B** is shown to be more thermodynamically stable than product **A**.

$$\mathbf{X} \rightleftarrows \mathbf{A} \qquad\qquad\qquad (3)$$

$$\mathbf{X} \rightleftarrows \mathbf{B} \qquad\qquad\qquad (4)$$

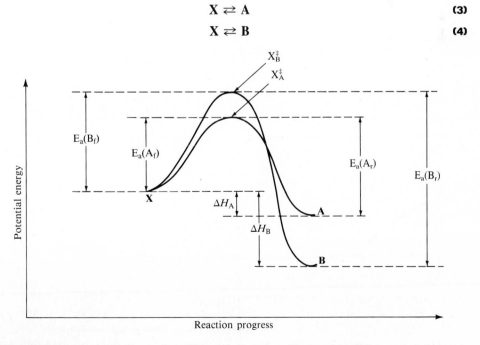

Figure 16.2 Reaction profile that predicts different products from kinetic and equilibrium control of competing reactions.

It should be noted that in contrast to the more usual relationships depicted in Figure 16.1, the relative energies of the transition states are not in the same order as the relative energies of the products to which they lead. Thus, although product **A** will predominate initially in the reaction mixture, if the reactions are allowed to come to equilibrium, the more stable product **B** will be found to be the major product. For this reason **B** is called the product of **equilibrium control** (or thermodynamic control), whereas **A** is called the product of *kinetic control*.

It is interesting to note that although **A** is produced initially more rapidly than **B** because of the relative energies of activation for the competing forward reactions $[E_a(A_f) < E_a(B_f)]$, **A** is also reconverted to starting material (**X**) more rapidly than **B**, because the energies of activation for the reverse reactions have the same relationship $[E_a(A_r) < E_a(B_r)]$, owing to the greater stability of **B** than **A**. This is not true for competing reactions that have the more usual energy relationships represented by Figure 16.1, so that even when equilibrium conditions are attained in these systems, the same product (**Z**) will predominate. The product of kinetic control is also the product of equilibrium control, although the proportion of products is not necessarily the same under all experimental conditions.

There are many well-known reactions of various mechanistic types in which kinetic and equilibrium control lead to different major products under different experimental conditions. For example, the addition of hydrogen bromide to 1,3-butadiene at low temperatures gives the 1,2-adduct, 3-bromo-1-butene, **1,** as the major product (equation 5a), whereas at room temperature the major product is the 1,4-adduct, 1-bromo-2-butene (**2,** equation 5b).

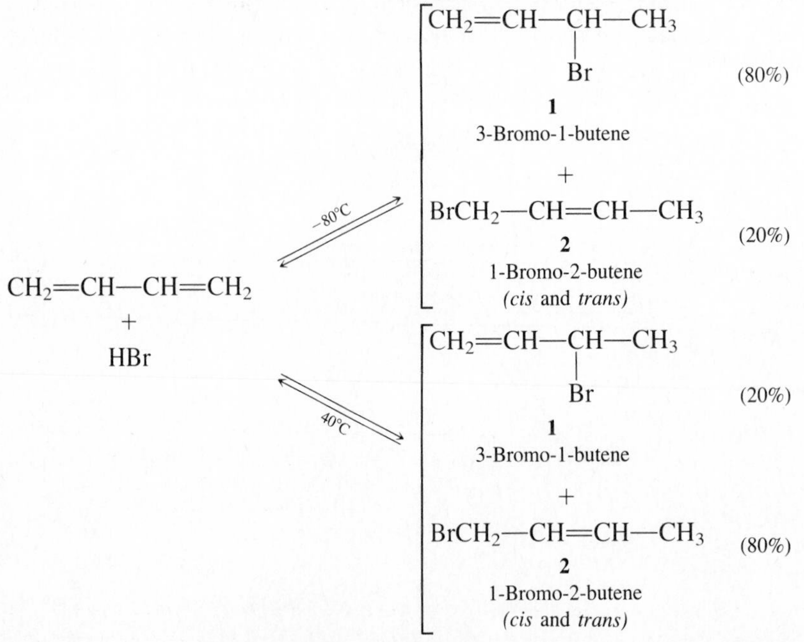

$$CH_2{=}CH{-}CH{=}CH_2$$
$$+$$
$$HBr$$

$-80°C$

$$CH_2{=}CH{-}CH{-}CH_3$$
$$|$$
$$Br$$
(80%)

1
3-Bromo-1-butene

(5a)

$+$

$$BrCH_2{-}CH{=}CH{-}CH_3$$
2
(20%)

1-Bromo-2-butene
(*cis* and *trans*)

$40°C$

$$CH_2{=}CH{-}CH{-}CH_3$$
$$|$$
$$Br$$
(20%)

1
3-Bromo-1-butene

(5b)

$+$

$$BrCH_2{-}CH{=}CH{-}CH_3$$
(80%)
2
1-Bromo-2-butene
(*cis* and *trans*)

The 1,2-adduct is the product of kinetic control, since equilibrium is not attained at low temperatures. The 1,4-adduct, although formed more slowly, is the more stable product, so it predominates if equilibrium has been established at higher temperatures. The greater stability of the 1,4-adduct is an example of the general effect of a higher degree of substitution on sp^2 carbon atoms.

When this reaction is analyzed in terms of Figure 16.2, the 1,2-adduct may be identified with **A** and the 1,4-adduct with **B.** The first step in the mechanism of the reaction is the addition of a proton to the diene to produce the resonance-stabilized carbocation **3** (equation 6), and **3** may be identified with **X** in Figure 16.2. The two transitions states, XB‡ and XA‡, correspond to those formed when bromide ion reacts at the two partially positive charged carbon atoms of **3** to give the products **1** and **2** (equation 7).

$$CH_2{=}CH{-}CH{=}CH_2 + H^{\oplus} \rightarrow \left[CH_3{-}\overset{\oplus}{CH}{-}CH{=}CH_2 \leftrightarrow CH_3{-}CH{=}CH{-}\overset{\oplus}{CH_2} \right] \quad \text{(6)}$$
$$\mathbf{3}$$

$$\mathbf{3} + Br^{\ominus} \rightarrow \mathbf{1} + \mathbf{2} \quad \text{(7)}$$

Although the reaction of 1,3-butadiene with hydrogen bromide is the example most given in textbooks to illustrate the principle of kinetic and equilibrium control of a reaction, it does not lend itself readily to a simple laboratory experiment. 1,3-Butadiene is a gas at room temperature, and the analysis of the product mixture containing the two isomeric liquid bromobutenes is not convenient.

The example chosen for an experiment to illustrate the principle of kinetic and equilibrium control of products involves the competing reactions of semicarbazide (**4**) with the two carbonyl compounds, cyclohexanone (**5**, equation 8) and 2-furaldehyde (**6**, equation 9) to give two different products. In this case one compound, semicarbazide, reacts with two different compounds, cyclohexanone and 2-furaldehyde, whereas in the previous example a reagent (Br⁻) reacts at two different positions in a single carbocation to produce two different compounds.

$$\text{(8)}$$

4	**5**	**7**
Semicarbazide	Cyclohexanone	

$$\text{(9)}$$

4	**6**	**8**
Semicarbazide	2-Furaldehyde	

The principle is the same, nevertheless. The organic products of these reactions (**7** and **8**), which are known as the semicarbazones of the respective carbonyl compounds, are crystalline solids and have distinctive melting points by which they may be identified easily. This allows a simple experimental determination of which compound is the kinetically controlled product and which is the equilibrium-controlled product.

The reactions of carbonyl compounds with nucleophiles such as semicarbazide are subject to significant effects of pH on rates and equilibrium constants. This is largely because of the different way in which the reactants are affected by reaction with hydrogen ions.[1] The addition of a hydrogen ion to a carbonyl compound (equation 10) makes the carbonyl carbon atom *more* electrophilic because of its partial positive charge, as shown in the resonance hybrid of the conjugate acid **9.** On the other hand, addition of a hydrogen ion to the nucleophilic semicarbazide (**4,** equation 11) converts it to its conjugate acid **(10),** which is *not* nucleophilic, so that the desired reaction (equation 12) is disfavored.

$$R_2C\!\!=\!\!\overset{..}{\underset{..}{O}} + H^{\oplus} \rightleftharpoons \left[R_2C\!\!=\!\!\overset{\oplus}{\underset{..}{O}}H \leftrightarrow R_2\overset{\oplus}{C}\!\!-\!\!\overset{..}{\underset{..}{O}}H \right] \qquad \textbf{(10)}$$
$$\underset{\textbf{9}}{}$$

$$\underset{\textbf{4}}{H_2N\overset{\overset{\displaystyle O}{\|}}{C}NH\overset{..}{N}H_2} + H^{\oplus} \rightleftharpoons \underset{\textbf{10}}{H_2N\overset{\overset{\displaystyle O}{\|}}{C}NH\overset{\oplus}{N}H_3} \qquad \textbf{(11)}$$

$$\textbf{4 + 9} \rightarrow \left[H_2N\overset{\overset{\displaystyle O}{\|}}{C}NH\overset{\oplus}{N}H_2\overset{\overset{\displaystyle OH}{|}}{C}R_2 \right] \rightarrow H_2N\overset{\overset{\displaystyle O}{\|}}{C}NHN\!\!=\!\!CR_2 + H_2O + H^{\oplus} \qquad \textbf{(12)}$$

At very low pH (high H^+ concentration) the concentration of the nucleophile **(4)** is reduced, and at high pH (low H^+ concentration) the concentration of the activated electrophile **(9)** is reduced. Accordingly, for the reaction of each carbonyl compound with each nucleophilic reagent there is an optimum pH at which the product of the concentrations of the conjugate acid of the carbonyl compound and the nucleophilic reagent is maximized. In order to produce and maintain this optimum pH for reactions of aldehydes and ketones with reagents such as semicarbazide, phenylhydrazine, and hydroxylamine, these reactions are carried out in *buffered* solutions.

A buffered solution is one that resists changes in pH. In general, this requires the presence of a pair of substances in the solution, one of which neutralizes hydroxide ions and the other of which neutralizes hydrogen ions. For example, the $H_2PO_4^-/HPO_4^{2-}$ **buffer system** is produced by the addition of dibasic potassium phosphate (K_2HPO_4) to semicarbazide hydrochloride. The $H_2PO_4^-$ component of the buffer system is produced as shown in equation 13.

$$\underset{\substack{\text{Semicarbazide}\\\text{hydrochloride}}}{H_2N\overset{\overset{\displaystyle O}{\|}}{C}NH\overset{\oplus}{N}H_3\ Cl^{\ominus}} + HPO_4^{2\ominus} \rightarrow H_2N\overset{\overset{\displaystyle O}{\|}}{C}NHNH_2 + H_2PO_4^{\ominus} + Cl^{\ominus} \qquad \textbf{(13)}$$

The function of the two ions in neutralizing hydroxide and hydrogen ions is illustrated in equations 14 and 15. Each different buffer system functions to maintain the pH within a

---

[1]Actually, these reactions are affected not only by the concentration of hydrogen ions (of which pH is a measure) but also by the concentration and nature of any weak acids that may be present. A discussion of these effects is beyond the scope of this book; further information may be found in textbooks on physical-organic chemistry, in connection with specific and general acid catalysis.

rather narrow range characteristic of the weak acid and weak base of that particular system. In the case of the phosphate system, the range is *ca*. pH 6.1–6.2. A carbonic acid/bicarbonate buffer system (H_2CO_3/HCO_3^-) maintains a pH of *ca*. 7.1–7.2.

$$H_2PO_4^{\ominus} + OH^{\ominus} \rightarrow HPO_4^{2\ominus} + H_2O \qquad \textbf{(14)}$$

$$HPO_4^{2\ominus} + H^{\oplus} \rightarrow H_2PO_4^{\ominus} \qquad \textbf{(15)}$$

The maximum rates of the reactions of most aldehydes and ketones with semicarbazide occur in the pH range 4.5–5.0; the rates decrease at higher or lower pH. For the purpose of making derivatives of carbonyl compounds (Chapter 27), semicarbazide is best used in an acetate buffer ($HC_2H_3O_2/C_2H_3O_2^-$) solution, which maintains a pH in the maximum rate range, 4.5–5.0. However, for the purpose of demonstrating the principle of kinetic and equilibrium control of reactions, buffers that maintain higher pHs, and thus produce lower rates, are desirable. Parts A–C of the Experimental Procedure involve a phosphate buffer system, whereas part D employs the bicarbonate system. This allows a comparison of the way in which the difference in rates in the two buffer systems affects the product ratio. Analysis of the products from the various parts of these experiments should provide strong clues as to which of the semicarbazones is the product of kinetic control and which is the product of equilibrium control.

It is implicit in the theory of kinetic and equilibrium control, as illustrated by Figure 16.2, that the "kinetic product" that is produced more rapidly is also reconverted to starting material more rapidly. The "equilibrium product," on the other hand, being the more stable product, is not so readily reconverted to starting material. Experimental Procedure E provides tests of the relative stabilities of the two semicarbazone products toward the reverse reaction. The results of these experiments should provide additional evidence of the identity of one of the semicarbazones as the product of kinetic control and the other as the product of equilibrium control.

Pre-lab exercises for Chapter 16, Kinetic and Equilibrium Control of a Reaction, can be found on page PL. 51.

EXPERIMENTAL PROCEDURE

DO IT SAFELY

A metal block-type melting point apparatus is appropriate for the determinations of the melting points of the semicarbazones in this experiment. A liquid-filled apparatus such as a Thiele tube should not be used, because the melting point of one of the compounds is over 200°C, and heating the liquid to this temperature may cause it to smoke or even catch fire.

Note: The instructor may wish to divide the class and assign some of the students to only certain parts of these procedures rather than having all students do all parts. If this is done, the results of the different groups may be collected and made available to the whole class, so that each student may then record the results, draw his or her own conclusions, and write them up individually.

A. Preparation of Cyclohexanone Semicarbazone

In a 50-mL Erlenmeyer flask, dissolve 1.0 g of semicarbazide hydrochloride and 2.0 g of dibasic potassium phosphate (K_2HPO_4) in 25 mL of water. Using a 1-mL graduated pipet, deliver 1.0 mL of cyclohexanone into a test tube and dissolve it in 5 mL of 95% ethanol. Pour the ethanolic solution into the aqueous semicarbazide solution, and swirl or stir the mixture immediately. Allow 5 or 10 min for crystallization of the semicarbazone to reach completion, then collect the crystals by vacuum filtration and wash them on the filter with a little cold water. Dry the crystals in air and determine their weight and melting point. The reported melting point of cyclohexanone semicarbazone is 166°C.★

B. Preparation of 2-Furaldehyde Semicarbazone

For best results the 2-furaldehyde should be redistilled just before use. Prepare the semicarbazone of 2-furaldehyde by following the procedure of part A exactly, except use 0.8 mL of 2-furaldehyde instead of 1.0 mL of cyclohexanone. The reported melting point of 2-furaldehyde semicarbazone is 202°C.★

C. Reactions of Semicarbazide with Cyclohexanone and 2-Furaldehyde in Phosphate Buffer Solution

Dissolve 3.0 g of semicarbazide hydrochloride and 6.0 g of dibasic potassium phosphate in 75 mL of water. This solution will be referred to as *solution W*.

Prepare a solution of 3.0 mL of cyclohexanone and 2.5 mL of 2-furaldehyde in 15 mL of 95% ethanol. This solution will be referred to as *solution E*.★

1 Cool a 25-mL portion of solution W and a 5-mL portion of solution E *separately* in an ice bath to 0–2°C, then add solution E to solution W and swirl them together; crystals should form almost immediately. Place the mixture in an ice bath for about 3–5 min, and then collect the crystals by vacuum filtration and wash them on the filter with a little cold water. Dry the crystals and determine their weight and melting point.

2 Add a 5-mL portion of solution E to a 25-mL portion of solution W at room temperature; crystals should be observed in about 1–2 min. Allow the mixture to stand at room temperature for 5 min, cool it in an ice bath for about 5 min, and then collect the crystals by vacuum filtration and wash them on the filter with a little cold water. Dry the crystals and determine their weight and melting point.

3 Warm a 25-mL portion of solution W and a 5-mL portion of solution E *separately* on a steam bath or in a water bath to 80–85°C, then add solution E to solution W and swirl them together. Continue to heat the mixture for 10–15 min, cool it to room tempera-

ture, and then place it in an ice bath for about 5–10 min. Collect the crystals by vacuum filtration, and wash them on the filter with a little cold water. Dry the crystals and determine their weight and melting point.★

D. Reactions of Semicarbazide with Cyclohexanone and 2-Furaldehyde in Bicarbonate Buffer Solution

Dissolve 2.0 g of semicarbazide hydrochloride and 4.0 g of sodium bicarbonate in 50 mL of water. Prepare a solution of 2.0 mL of cyclohexanone and 1.6 mL of 2-furaldehyde in 10 mL of 95% ethanol. Divide each of these solutions into two equal portions.

1 Mix one-half the aqueous solution and one-half the ethanolic solution at room temperature. Allow the mixture to stand at room temperature for 5 min, cool it in an ice bath for 5 min, and then collect the crystals by vacuum filtration and wash them on the filter with a little cold water. Dry the crystals and determine their weight and melting point.

2 Warm the other portions of the aqueous and ethanolic solutions *separately* on a steam bath or in a water bath to 80–85°C, then combine them and continue heating the mixture for 10–15 min. Cool the solution to room temperature, and then place it in an ice bath for 5 to 10 min. Collect the crystals by vacuum filtration, and wash them on the filter with a little cold water. Dry the crystals and determine their weight and melting point.

E. Tests of Reversibility of Semicarbazone Formation

1 In a 25-mL Erlenmeyer flask place 0.3 g of cyclohexanone semicarbazone (prepared in part A), 0.3 mL of 2-furaldehyde, 2 mL of 95% ethanol, and 10 mL of water. Warm the mixture until a homogeneous solution is obtained (about 1 or 2 min should suffice) and then continue warming an additional 3 min. Cool the mixture to room temperature and then in an ice bath. Collect the crystals on a filter, and wash them with a little cold water. Dry the crystals and determine their melting point.★

2 Repeat the preceding experiment, but use 0.3 g of 2-furaldehyde semicarbazone (prepared in part B) and 0.3 mL of cyclohexanone in place of the cyclohexanone semicarbazone and 2-furaldehyde.★

On the basis of your results from experiments C, D, and E, deduce which semicarbazone is the product of kinetic control and which is the product of equilibrium control. To do this, you must first determine from the melting point of the crystals produced in parts C1, C2, and C3 whether the product in each part is the semicarbazone of cyclohexanone, 2-furaldehyde, or a mixture of the two. Note that in C1 the crystals of product separate almost immediately, in C2 after 1 or 2 min, and in C3 only after 10–15 min at a higher temperature, so that the reaction time is shortest in C1, intermediate in C2, and longest in C3.

In considering the results of the experiments in part D, compare the properties of the products with those obtained in part C under the same conditions; i.e., compare the product of D1 with that of C2, and the product of D2 with that of C3.

In considering the results of the experiments in part E, remember that the equilibrium product, being the thermodynamically more stable, is not easily converted into the less stable kinetic product. However, the kinetic product can be converted more easily into the more stable equilibrium product. This follows because the reverse reaction of the less stable kinetic product has a lower activation energy than the reverse reaction of the equilibrium product.

Your completed laboratory report should include the diagram called for in Exercise 1 and answers to Exercises 2 and 3 as well, unless you were instructed to omit some parts of the experiment.

EXERCISES

1. On the basis of the results from experiments C, D, and E, draw a diagram similar to Figure 16.2, and clearly label the products corresponding to **A** and **B.**

2. On the basis of the results from the experiments of part D, explain the effect of the higher pH on the reactions between semicarbazide and the two carbonyl compounds.

3. What results might be expected if sodium acetate buffer (which provides a pH of approximately 5) were used in experiments analogous to those of part C? Explain.

4. Calculate the exact pH of a buffer solution prepared by adding equimolar amounts of glacial acetic acid and sodium acetate to distilled water, given pK_a for acetic acid $= 4.75$.

5. What different result, if any, would you expect to find if, in experiment D2, the heating period at 80–85°C was extended to one hour?

6. There are two NH_2 groups in semicarbazide **(4),** yet only one of them reacts with the carbonyl group in forming a semicarbazone. Explain.

7. Figure 16.5 is the UV spectrum of 2-furaldehyde.
   a. Calculate ϵ_{max} for the absorption bands present in the spectrum.
   b. Assuming that the UV spectrum of 2-furaldehyde is the same in 95% ethanol as it is in methanol, and that cyclohexanone does not absorb significantly in the UV spectrum, show how the concentration of 2-furaldehyde in solution E could accurately be determined.

SPECTRA OF STARTING MATERIALS

The IR spectrum of cyclohexanone is given in Figure 8.23a.

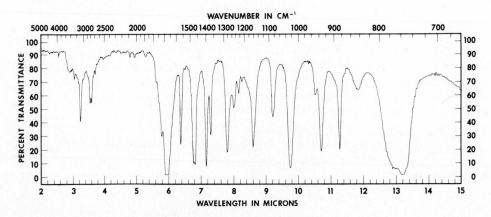

Figure 16.3 IR spectrum of 2-furaldehyde.

(a) PMR spectrum.

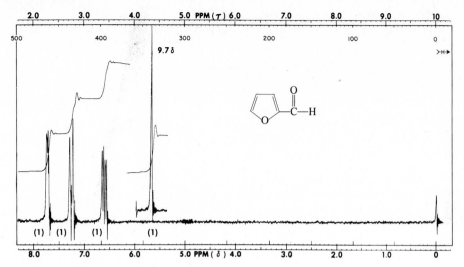

(b) CMR data. Chemical shifts: δ 112.9, 121.9, 148.6, 153.3, 178.1.

Figure 16.4 NMR data for 2-furaldehyde.

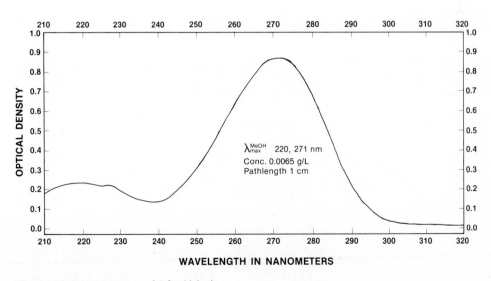

Figure 16.5 UV spectrum of 2-furaldehyde.

(a) PMR spectrum.

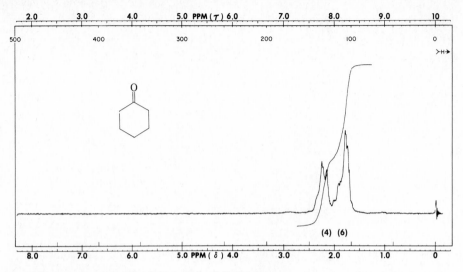

(b) CMR data. Chemical shifts: δ 25.1, 27.2, 41.9, 211.2.

Figure 16.6 NMR data for cyclohexanone.

electrophilic aromatic substitution

17.1 Introduction

Electrophilic aromatic substitution is a very important part of organic chemistry, because the introduction of many functional groups onto an aromatic ring is accomplished in this way. The general form of this reaction may be represented by equation 1, in which Ar—H is an aromatic compound (an ''arene'') and E^+ is any one of a number of different electrophiles that may replace H.

$$Ar—H + E^{\oplus} \rightarrow Ar—E + H^{\oplus} \qquad \text{(1)}$$

Although equation 1 represents the overall net reaction correctly, it is greatly oversimplified. For example, the electrophile must usually be generated from the starting materials during the course of the reaction, and various bases assist in the removal of H^+.

The chemical kinetics of electrophilic aromatic substitution reactions have been studied extensively. For many but not all cases the rate of substitution has been found to be first order in the aromatic compound and first order in the electrophile (equation 2):

$$\text{rate} = k_2[Ar—H][E^{\oplus}] \qquad \text{(2)}$$

Because of its bimolecular nature it is often termed an S_E2 reaction (S = substitution, E = electrophilic, and 2 = bimolecular). A general mechanism for the S_E2 reaction is given below. This mechanism has been substantiated by considerable evidence, especially that coming from studies of isotope effects.

Step 1: Formation of electrophile (an equilibrium reaction):

$$E—Nu \xrightleftharpoons{\text{catalyst}} E^{\oplus} + Nu^{\ominus}$$ **(3)**

Step 2: Reaction of electrophile with arene (slow step):

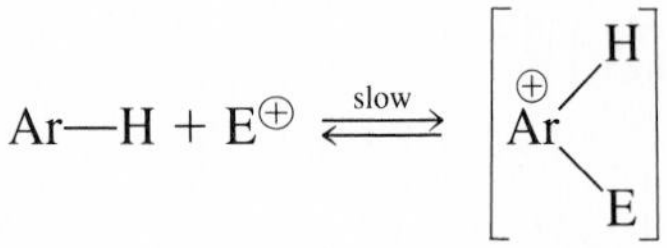

$$Ar—H + E^{\oplus} \xrightleftharpoons{\text{slow}} \left[\overset{\oplus}{Ar} \begin{smallmatrix} \diagup H \\ \diagdown E \end{smallmatrix} \right]$$ **(4)**

Intermediate

Step 3: Loss of proton to give product (fast step):

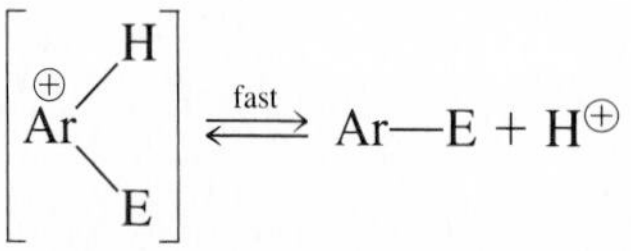

$$\left[\overset{\oplus}{Ar} \begin{smallmatrix} \diagup H \\ \diagdown E \end{smallmatrix} \right] \xrightleftharpoons{\text{fast}} Ar—E + H^{\oplus}$$ **(5)**

The electrophile is most often produced by the reaction between a catalyst and a compound that contains a potential electrophile (equation 3). The second-order nature of the reaction may be seen to be related to equation 4, in which *two* moles react to give the intermediate. If indeed the kinetics are second order, then this must be the slow, rate-limiting step in the overall reaction; the loss of a proton (equation 5) that follows must be fast relative to the reaction of equation 4.

In this chapter, experiments illustrating three types of electrophilic aromatic substitution reactions are presented: Friedel-Crafts alkylation in Section 17.2, nitration in Section 17.3, and bromination in Section 17.4.

17.2 Friedel-Crafts Alkylation of *p*-Xylene with 1-Bromopropane

The Friedel-Crafts alkylation reaction is one of the classic types of electrophilic aromatic substitution and as such has been subjected to extensive mechanistic study. It is also of great industrial importance as the most versatile method for attaching alkyl side chains to aromatic rings. The two main limitations to this reaction as a synthetic tool are (1) the difficulty of preventing the introduction of more than one alkyl group onto an aromatic ring, owing to the activating effect of the first group introduced, and (2) the occurrence of rearrangements of the alkyl group. The first difficulty can often be largely resolved by using a large excess of the arene (note the proportion of *p*-xylene to 1-bromopropane used in this experiment). High-speed stirring also helps to prevent polyalkylation. In the case of the alkylation of benzene by ethylene to give ethylbenzene, an intermediate in the industrial production of styrene (see Chapter 24), efficient agitation is essential, for otherwise the ethylbenzene is selectively dissolved in the catalyst layer, where it is further alkylated. There has been considerable confusion about the second limitation, the occurrence and extent of alkyl group rearrangement; this has sometimes been exaggerated and sometimes overlooked, especially in early work. This aspect of the reaction is discussed further in the following paragraphs.

A generally accepted mechanism of alkylation is outlined in equations 6, 7, and 8. The active electrophile, R^+, is produced from an alkyl halide by reaction with a Lewis acid, commonly $AlCl_3$ (equation 6). The electrophile reacts with the arene, shown in equation 7 as benzene, to produce the intermediate resonance-stabilized sigma complex, which then ejects a proton to produce HCl, re-form the aromatic system, and regenerate the catalyst (equation 8).

$$R\text{---}Cl + AlCl_3 \rightleftharpoons R^{\oplus} + AlCl_4^{\ominus} \qquad \textbf{(6)}$$

$$R^{\oplus} + \text{[benzene]} \xrightarrow{\text{slow}} \left[\text{resonance structures of sigma complex} \right] \qquad \textbf{(7)}$$

This resonance-stabilized intermediate may be represented by the single symbol

[structure: sigma complex with H and R, ⊕ in ring]

$$\text{[sigma complex]} + AlCl_4^{\ominus} \xrightarrow{\text{fast}} \text{[R-substituted benzene]} + HCl + AlCl_3 \qquad \textbf{(8)}$$

In the case of secondary and tertiary alkyl halides, the electrophile may reasonably be represented as a carbocation, R^+. However, the electrophile produced from a primary alkyl halide under typical Friedel-Crafts reaction conditions is probably not a true primary carbocation, but is more likely a polarized complex such as $R^{\delta+}\text{----}X\text{----}Al^{\delta-}Cl_3$. Nevertheless, as will be seen in the following discussion, whether it is a carbocation or a polarized complex, the electrophile from a primary alkyl halide is capable of rearrangements by means of hydride and methide shifts. *For simplicity,* the electrophiles from all three types of alkyl halides are represented here as carbocations, R^+.

When a carbocation is involved as an intermediate, rearrangement is to be expected in some but not necessarily all cases. (For rearrangements of carbocations that accompany elimination reactions, see Section 13.2.) For example, rearrangement of an unstable carbocation to a more stable one is to be expected, but a rearrangement that involves little gain in stability (decrease in energy) or that requires a higher-energy intermediate is not necessarily to be expected.

These generalizations can be illustrated by the reactions of the isomeric butyl halides in Friedel-Crafts alkylations. An *n*-butyl halide (**1**) reacts with benzene to give a mixture of *n*-butylbenzene (**3**) and *sec*-butylbenzene (**6**) in a ratio of about 1:2. The *n*-butyl carbocation (**2**), or its equivalent polarized complex, undergoes a 1,2-hydride shift to give the more stable secondary carbocation (**5**). The proportion of **3** and **6** produced is determined by the competition between the direct reaction of **2** with benzene and its rearrangement to **5** before reaction with benzene. However, alkylation with a secondary butyl halide **4** (Figure 17.1), which involves the secondary carbocation **5** as an intermediate, gives only *sec*-butylbenzene (**6**); no rearrangement to a *t*-butyl derivative (**12**) occurs because of the high energy of the primary isobutyl carbocation **8,** which is required as an intermediate between **5** and **11.** On the other hand, an isobutyl halide (**7**) gives only *t*-butylbenzene (**12**) and *no* "unrearranged" isobutylbenzene (**9**). The intermediate produced from **7** and AlCl$_3$, although it actually may

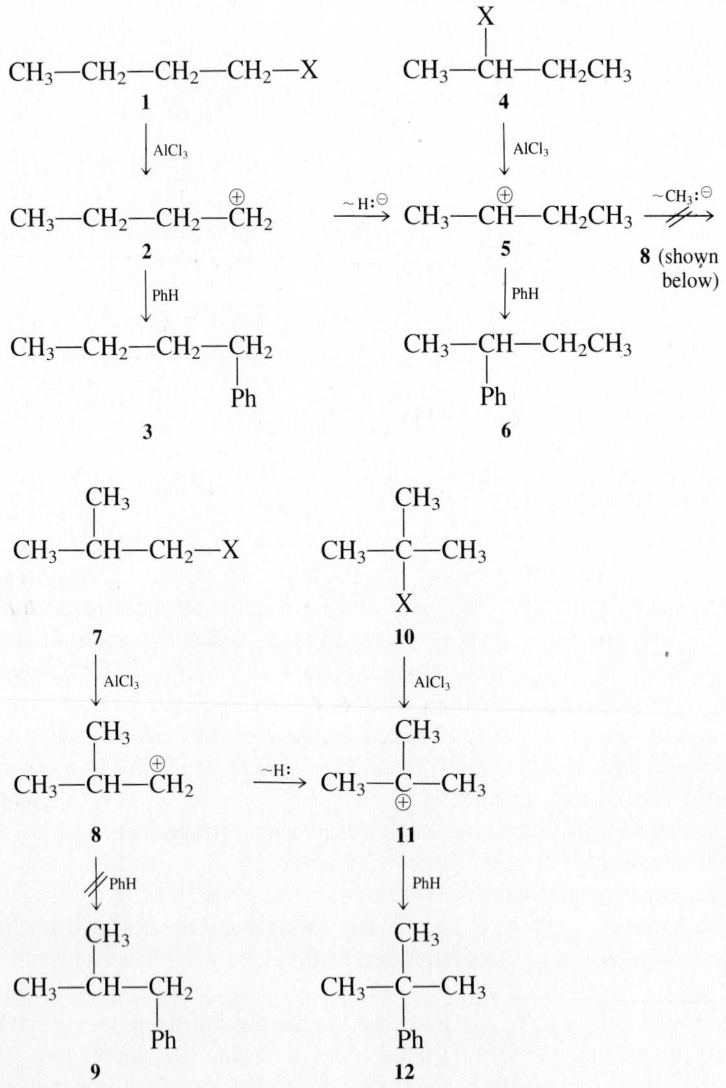

Figure 17.1 Alkylations of benzene with butyl halides.

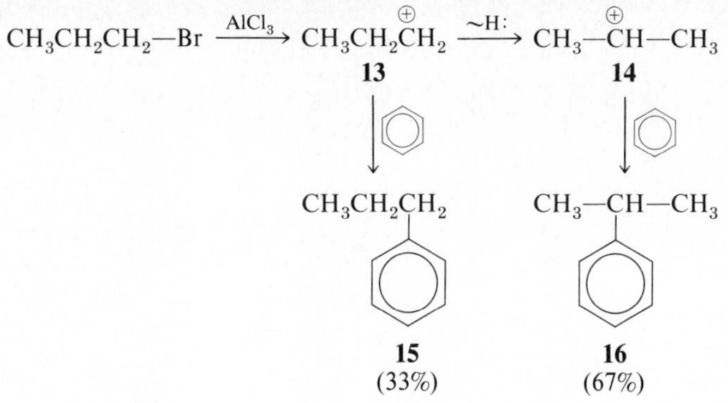

$$CH_3CH_2CH_2{-}Br \xrightarrow{AlCl_3} CH_3CH_2\overset{\oplus}{C}H_2 \xrightarrow{\sim H:} CH_3{-}\overset{\oplus}{C}H{-}CH_3$$

13   **14**

CH₃CH₂CH₂

15
(33%)

CH₃—CH—CH₃

16
(67%)

Figure 17.2 Alkylation of benzene with 1-bromopropane.

not be a true primary carbocation (**8**), but rather a polarized complex, is nonetheless capable of rearrangement to the stable tertiary carbocation so rapidly that direct alkylation by **8** does not compete successfully, and none of the unrearranged product is formed. As might be expected, *t*-butyl halide (**10**) gives only *t*-butylbenzene (**12**).

The extent of rearrangement accompanying alkylation is also dependent on factors such as the nature of the arene (when other than benzene), the temperature, the solvent (if any), and the nature and concentration of the catalyst. The present experiment relates to the first of these factors, the effect of using different arenes as substrates in alkylation by the same alkyl halide.

The reaction of 1-bromopropane with benzene and aluminum chloride gives a mixture of *n*-propylbenzene (**15**) and isopropylbenzene (**16**, cumene), with only a small variation in the proportion of the isomers at different temperatures (Figure 17.2). Apparently, the rate of rearrangement of the primary propyl carbocation **13** to the secondary propyl ion **14** is not as rapid as the rate of conversion of the primary isobutyl carbocation **8** to the *tertiary* butyl ion **11,** so that direct reaction of **13** with benzene can compete with some success with the rearrangement of **13** to **14.** One might expect that an arene which is more reactive than benzene toward electrophilic substitution (that is, a more nucleophilic arene) would compete better than benzene for reaction with the primary carbocation **13** before it rearranges to **14.** Methyl groups are known to activate the benzene ring toward electrophilic substitution; hence the reaction of 1-bromopropane with toluene, *p*-xylene, and mesitylene might be expected to give increasing proportions of *n*-propylarene:isopropylarene.

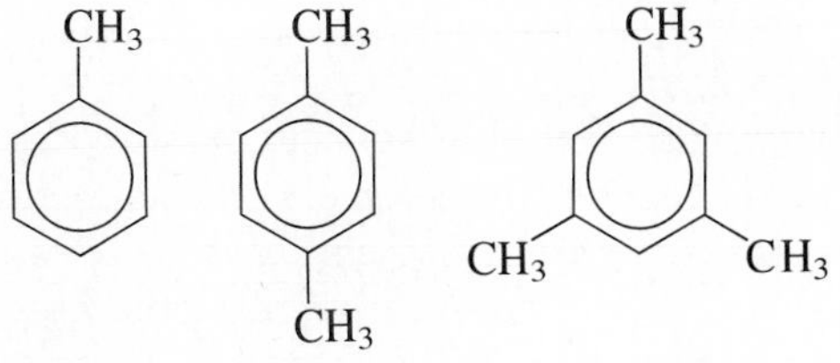

Toluene   *p*-Xylene   Mesitylene

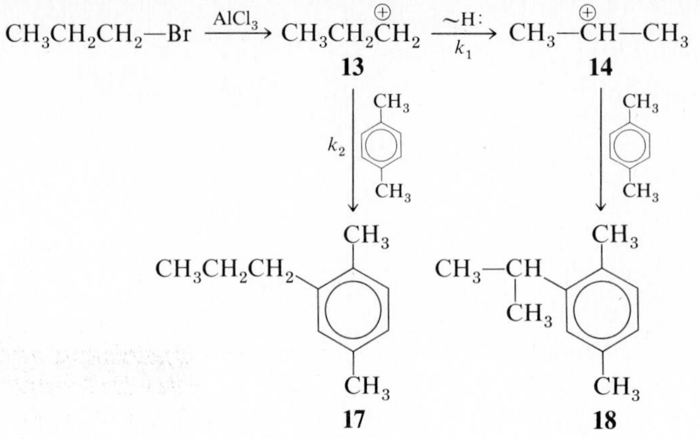

$$CH_3CH_2CH_2-Br \xrightarrow{AlCl_3} CH_3CH_2\overset{\oplus}{C}H_2 \xrightarrow[k_1]{\sim H:} CH_3-\overset{\oplus}{C}H-CH_3$$

13   **14**

$$k_2 \downarrow \text{(p-xylene)} \qquad \qquad \downarrow \text{(p-xylene)}$$

17 1,4-dimethyl-2-n-propylbenzene (CH$_3$CH$_2$CH$_2$ substituent)

18 1,4-dimethyl-2-isopropylbenzene ((CH$_3$)$_2$CH substituent)

Figure 17.3 Alkylation of *p*-xylene with 1-bromopropane.

This prediction is based on the reasonable assumption that the rate of rearrangement of intermediate **13** to **14** (k_1 in Figure 17.3) will be little different in the presence of benzene, *p*-xylene, or mesitylene, whereas the rate of reaction of **13** with *p*-xylene (k_2) or mesitylene will be significantly faster than with benzene.

In this experiment the reaction of 1-bromopropane with *p*-xylene has been chosen for several practical reasons. *p*-Xylene is used rather than toluene because it is more nucleophilic than toluene and because alkylation of *p*-xylene does not give rise to orientational isomers. Note that propylation of toluene would yield *ortho-*, *meta-*, and *para*-propyltoluenes, which would complicate the determination of the ratio of *n*-propyltoluenes to isopropyltoluenes. Mesitylene is less satisfactory than *p*-xylene because alkylation between two methyl groups on the ring involves considerable steric hindrance, which introduces an additional factor. The choice of 1-bromopropane rather than 1-chloropropane is based solely on economics, owing to the rather unusual circumstance that this particular bromide is less expensive than the corresponding chloride.

Mixtures of *n*-propyl-*p*-xylene (**17**, 1,4-dimethyl-2-*n*-propylbenzene) and isopropyl-*p*-xylene (**18**, 1,4-dimethyl-2-isopropylbenzene) can be analyzed conveniently by gas chromatography and by IR and PMR spectroscopy. Students may wish to compare their results with those reported in the reference given at the end of this section.

Pre-lab exercises for Section 17.2, Friedel-Crafts Alkylation of *p*-Xylene with 1-Bromopropane, can be found on page PL. 53.

EXPERIMENTAL PROCEDURE

Alkylation of *p*-Xylene with 1-Bromopropane

DO IT SAFELY

1 Anhydrous aluminum chloride reacts vigorously with water, even the moisture on your hands, producing fumes of hydrogen chloride. Do not allow it to touch your skin. The reagent bottle and a scale should be placed in a hood so that you can quickly weigh out the required amount and place it in your round-bottomed flask *in the hood.*

2 *p*-Xylene is flammable. Make sure that your apparatus is assembled correctly and have your instructor inspect it before you begin the distillation. Since the boiling temperature of *p*-xylene and the propylxylenes is higher than 100°C, steam heating is not adequate and electric heating is desirable. If a gas burner is used, take care to keep the flame away from the distillate.

Note: The *p*-xylene used in this experiment should be dry. It can be dried by azeotropic distillation, discarding the first distillate, which may be cloudy if the *p*-xylene did contain some water.

Equip a 250-mL round-bottomed flask with a Claisen connecting tube, a water-cooled condenser, a gas trap, and an addition funnel, as shown in Figure 17.4. Place 2.7 g (0.020 mol) of anhydrous powdered aluminum chloride in the flask and immediately cover it with 50 mL of dry *p*-xylene. Measure 18.2 mL (24.6 g, 0.200 mol) of 1-bromopropane in a 100-mL graduated cylinder and pour it into the separatory funnel (stopcock closed). Prepare an ice-water bath so that it will be ready should it become necessary to cool the reaction mixture. Turn on the water to the reflux condenser and to the aspirator (or connect the trap to the house vacuum source) and then begin adding the 1-bromopropane dropwise to the mixture of *p*-xylene and aluminum chloride. Loosen the clamp holder of the lower clamp so that the reaction mixture can be swirled gently from time to time as the 1-bromopropane is being added. If magnetic stirring is available, of course it will not be necessary to swirl the reaction mixture manually. If the evolution of hydrogen bromide becomes too vigorous, raise the ice-water bath so as to cool the reaction mixture, and reduce the rate at which the 1-bromopropane is being added. The addition should take about 15 min. After all the 1-bromopropane has been added, allow another 45 min for reaction, swirling the reaction mixture every few minutes. Pour the mixture into a 250-mL beaker containing 40 g of crushed ice. After stirring the mixture until all the ice has melted, pour it into a 125-mL separatory funnel and drain off the lower (aqueous) layer and discard it. Pour the organic layer into a 125-mL Erlenmeyer flask containing about 3 g of *anhydrous* calcium chloride (4–6 mesh) and swirl it for about 5 min.★ Filter the dried solution into a 100-mL round-bottomed flask and equip the flask for fractional distillation with either a Vigreux or an unpacked Hempel column (Figure 2.15). Using either electrical or gas heating and after making sure that all connections are tight, distil the excess *p*-xylene and forerun into one receiver (labeled A) until the distillation temperature reaches 180°C. Discontinue heating and allow

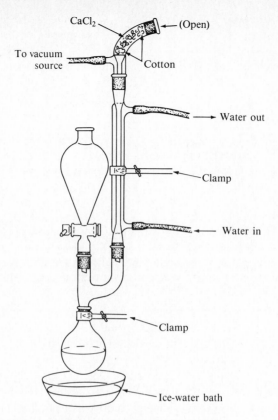

Figure 17.4 Apparatus for Friedel-Crafts alkylation.

any liquid in the column to drain into the flask.★ Replace the column with a simple distilling head (Figure 2.14) and resume distillation, collecting the fraction boiling between 180 and 207°C in a second receiver (labeled B). If any material boiling below 180°C is obtained in the second distillation, add it to the material in the first receiver.

Note: If the column and distillation head are insulated well with glasswool and/or aluminum foil, it will not be necessary to discontinue the distillation through the column and change to a simple distilling head. The distillation may be continued through the column, collecting the fraction boiling between 180°C and 207°C in a second receiver, labeled B. When the distillation temperature begins to rise above 207°C, stop the heating, allow the residue to cool, and pour it into an organic waste receptacle provided by your instructor.

Record the weights of fractions A and B. Submit or save samples of each fraction for GC, IR, and/or PMR analysis (see Figures 17.5 to 17.8). Using the weights of fractions A and B and the analytical results, calculate the yield of propylxylenes, and estimate the proportion of the isomers.

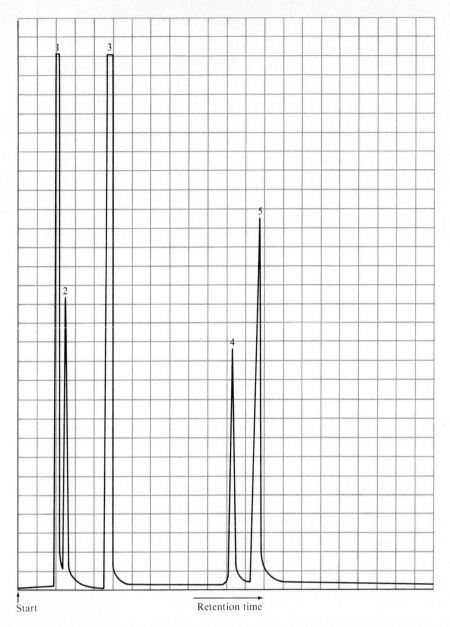

Start

Retention time

Figure 17.5 GC trace of reaction mixture from alkylation of *p*-xylene with 1-bromopropane. The peaks are identified as (1) diethyl ether (solvent), (2) benzene and/or toluene (present as impurity in *p*-xylene), (3) *p*-xylene, (4) isopropyl-*p*-xylene, and (5) *n*-propyl-*p*-xylene. Column: 15-m, 2% silicone oil on DC-550 HiPak.

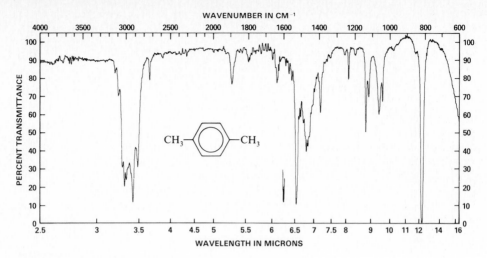

Figure 17.6 IR spectrum of *p*-xylene.

(a) PMR spectrum.

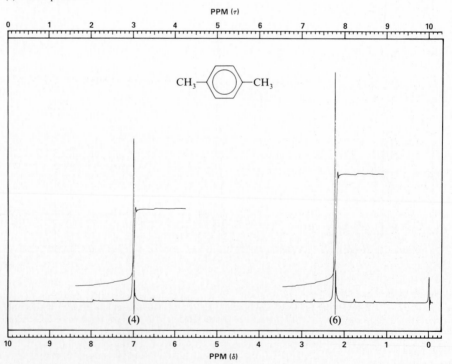

(b) CMR data. Chemical shifts; δ 20.9, 129.0, 134.6.

Figure 17.7 NMR data for *p*-xylene.

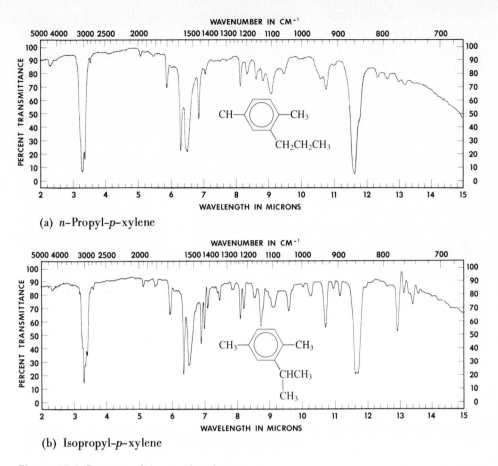

(a) *n*–Propyl–*p*–xylene

(b) Isopropyl–*p*–xylene

Figure 17.8 IR spectra of the propyl-*p*-xylenes.

EXERCISES

1. What compound(s) do you think might be present in the higher-boiling residue remaining in the distillation flask after collection of the propylxylene isomers?

2. Calculate the molar ratio of 1-bromopropane to *p*-xylene used in the experiment. What would be the effect on the experimental results of using a ratio twice that actually used? Consider this in connection with your answer to Exercise 1.

3. Using Figure 17.5, calculate the approximate proportion of isopropyl-*p*-xylene and *n*-propyl-*p*-xylene in the mixture analyzed by GC.

4. Alkylation of toluene with 1-bromopropane gives a mixture of four isomeric propyl-toluenes. Write formulas for these compounds, and on the basis of the results of your alkylation of *p*-xylene predict the relative amounts in which they would be formed.

5. Suggest a procedure for preparing pure *n*-propyl-*p*-xylene, that is, one that does not require separating it from isopropyl-*p*-xylene.

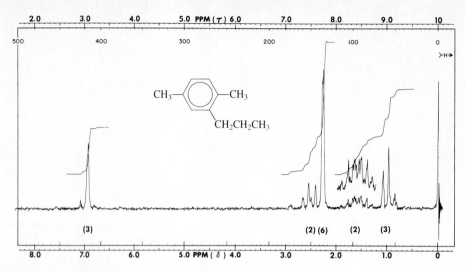

(a) n–Propyl–p–xylene

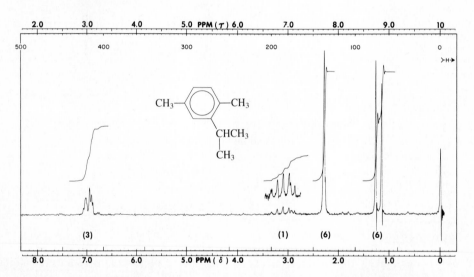

(b) Isopropyl–p–xylene

Figure 17.9 PMR spectra of the propyl-p-xylenes.

6. What products are expected from alkylation of benzene with each of the following alkyl halides, using aluminum chloride as a catalyst?

    a. 1-chlorobutane   **c.** 1-chloro-2-methylpropane

    b. 2-chloropentane   **d.** 2-chloro-2-methylpropane

REFERENCE

Roberts, R. M.; Shiengthong, D. *Journal of the American Chemical Society,* **1964,** *86,* 2851.

17.3 Nitration of Bromobenzene

Reaction of an aromatic compound such as benzene with a mixture of concentrated sulfuric and nitric acids leads to the introduction of a nitro group onto the aromatic ring.

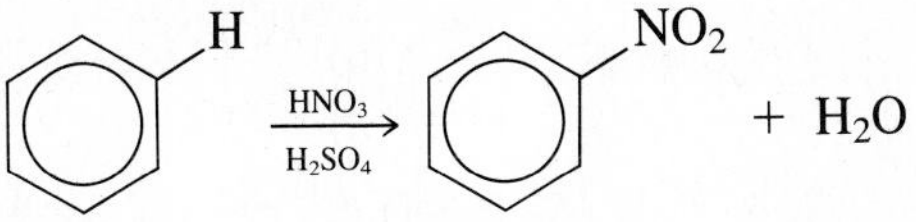

$$\text{(9)}$$

This transformation is an example of an electrophilic aromatic substitution reaction (Section 17.1) in which the electrophilic species is the nitronium ion, $NO_2^{\oplus}$. This cation is produced by reaction of sulfuric acid with the weaker nitric acid, as shown in equation 10, in a process reminiscent of the rate-determining step of dehydration of alcohols (Section 13.2) in which an oxonium ion fragments with loss of water to produce a carbocation. Notice that with respect to sulfuric acid, nitric acid is acting as a *base,* not an acid!

$$\text{(10)}$$

The rate-determining step in the nitration reaction is that in which the nitronium ion reacts with the aromatic ring, localizing a pair of pi electrons in the formation of a sigma bond to form the resonance-stabilized intermediate **19.**

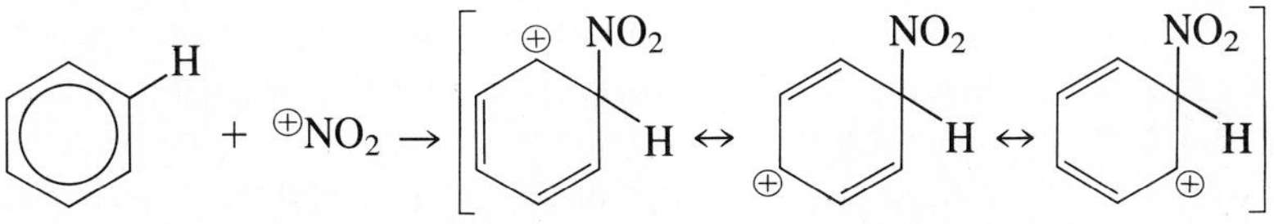

$$\text{(11)}$$

In the final step the intermediate rapidly loses a proton, probably to a molecule of water, to regenerate the aromaticity of the ring.

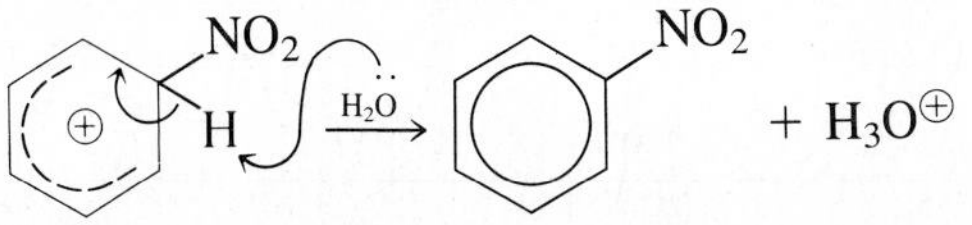

$$\text{(12)}$$

The reaction of the nitronium ion with a substituted benzene such as bromobenzene **(20),** the substrate used in this experiment, is similar in its mechanism to that of benzene, but more complex in its detailed analysis. Owing to the presence of the bromine atom on the ring, the sixfold symmetry of benzene is destroyed so that there are three different sites for the attack of the nitronium ion, each leading to a different product (equation 13).

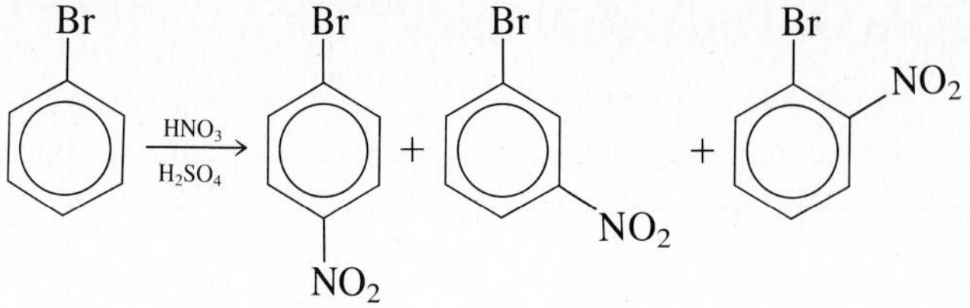

(13)

$$\underset{20}{} \qquad \underset{21}{} \qquad \underset{22}{}$$

For electronic reasons, attack at either the *ortho*- or the *para*-position occurs with a lower energy of activation than for attack at the *meta*-position. This is because an electron pair on the bromine atom in the intermediate produced by attack at either of these positions may be delocalized, providing *additional* resonance stabilization of the intermediate; note the additional resonance structure in **23**, as compared to **19**. This additional stabilization by bromine is *not* provided in the intermediate resulting from *meta*-attack.

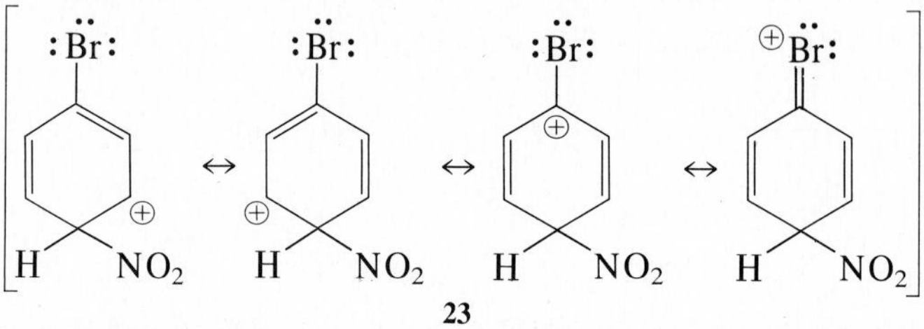

$$\underset{23}{}$$

Thus bromine is an example of an *ortho,para*-directing group. Typically, the *para*-substituted product **21** predominates in the reaction mixture, owing to the steric inhibition by the large bromine atom to the approach of the electrophile to the *ortho*-position.

The introduction of a second nitro group to provide 2,4-dinitrobromobenzene (**24**, equation 14) is possible in principle. However, under the conditions in which the reaction is conducted in our experiment, very little dinitration is observed. Owing to the strong *deactivating* effect of the first nitro group introduced onto the ring, the energy of activation for the introduction of the second is raised. Thus, by keeping the temperature of the reaction mixture below 60°C, dinitration is largely avoided. In addition, the major product **21** precipitates from the reaction mixture, which effectively removes it from the reaction and further reduces the extent of dinitration.

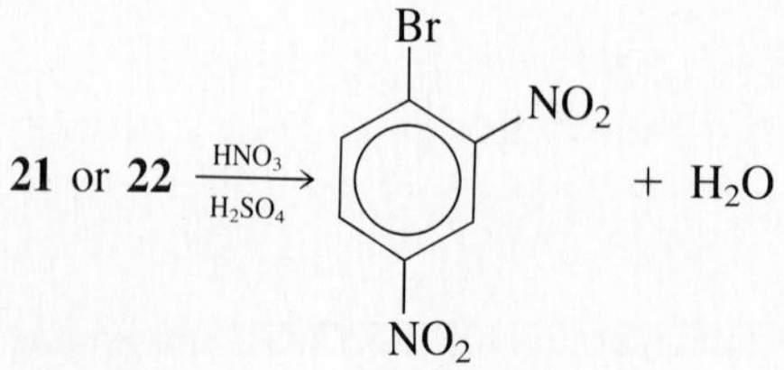

(14)

$$\underset{24}{}$$

4-Nitrobromobenzene (**21**) has a more symmetrical structure than 2-nitrobromobenzene (**22**). Consequently it is somewhat less polar and therefore less soluble than the *ortho*-isomer.

This is dramatically demonstrated in ethanol, in which the *ortho*-isomer is very soluble at room temperature but in which the *para*-isomer is soluble only to the extent of 1.2 g per 100 mL. Advantage may be taken of this difference in solubility in order to separate these isomers by the technique of *fractional crystallization*. This is accomplished by dissolution of the product mixture from nitration of bromobenzene into hot 95% ethanol, followed by cooling of the hot solution. The less soluble *para*-isomer selectively crystallizes from the cooling solution, from which it may then be filtered. By concentrating the mother liquors, a second crop of 4-nitrobromobenzene may be obtained. Ordinarily, in a fractional crystallization procedure, once the less-soluble component of the mixture has been mostly removed from the solution, the other component (such as the *ortho*-isomer in this experiment) is induced to crystallize. In this experiment, however, owing to the very low melting point (mp 40–41°C) of 2-nitrobromobenzene, it is very difficult to induce its crystallization in the presence of impurities. We have therefore elected the option of isolating 2-nitrobromobenzene by column chromatography.

EXPERIMENTAL PROCEDURE

DO IT SAFELY

1 Concentrated sulfuric and nitric acids may each cause severe chemical burns if they are allowed to come into contact with your skin. Take proper precautions not to allow this to happen. Watch carefully for drips and runs on the outside surface of reagent bottles and graduated cylinders when you pick them up. Wash any affected area immediately with cold water, and apply 0.6 M sodium bicarbonate solution to the area.

2 The nitrobromobenzenes produced in the experiment are irritating to sensitive skin areas. If you should have these materials on your hands and then accidentally rub or touch your face, you may notice a slight stinging and sensitivity in that area. Apply clean mineral oil to a soft paper or cloth towel, and gently swab the area.

Pre-lab exercises for Section 17.3, Nitration of Bromobenzene (Part A) can be found on page PL. 55.

A. Nitration of Bromobenzene

Prepare a mixture of 28.5 g (20.0 mL) of concentrated nitric acid and 37.0 g (20.0 mL) of concentrated sulfuric acid in a 250-mL round-bottomed flask, and cool it to room temperature by means of a water bath. Equip the flask with a Claisen connecting tube fitted with a water-cooled condenser and a thermometer that extends into the flask, as shown in Figure 17.10. Through the top of the condenser add 15.7 g (10.5 mL, 0.100 mol) of bromobenzene to the flask in 2–3 mL portions over a period of about 15 min. Loosen the clamp to the flask and shake the flask vigorously (but carefully) and frequently during the addition. Do

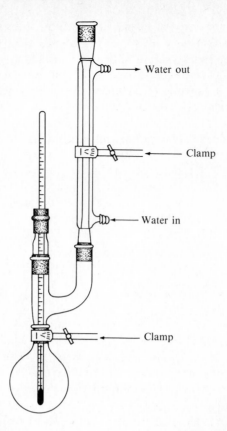

Figure 17.10 Apparatus for nitration of bromobenzene.

not allow the temperature of the reaction mixture to rise above 50–55°C during the addition. The temperature may be controlled by allowing more time between the addition of successive portions of bromobenzene and by cooling the reaction flask as necessary with an ice-water bath.

After the addition is complete and the exothermic reaction has subsided, heat the flask with a steam bath for 30 min, keeping the temperature of the reaction mixture below 60°C. Cool the flask to room temperature and then pour the reaction mixture into 200 mL of cold water in a 500-mL beaker. Isolate the crude nitrobromobenzene by vacuum filtration. Wash the filter cake thoroughly with cold water and allow the crystals to drain under vacuum until nearly dry.★

Transfer the crystals to a 250-mL Erlenmeyer flask along with approximately 5 mL of 95% ethyl alcohol per gram of crystals. Heat this mixture to boiling in order to dissolve the crude product. Set the flask aside and allow the contents to cool slowly to room temperature. Isolate the nearly pure crystals of 4-nitrobromobenzene by vacuum filtration. Wash the crystals with a little *ice-cold* alcohol, allowing the washes to drain into the filter flask with the mother liquors. When the crystals have dried, determine their weight and melting point. About 8–10 g of 4-nitrobromobenzene (**21**) may be expected.★ The reported melting point is 125–126°C.

Evaporate the mother liquors to a volume of 45–50 mL on a steam bath, preferably in the hood, and allow the solution to cool to room temperature. A second crop of 1–2 g of 4-nitrobromobenzene may be obtained in this way. It may either be combined with the first crop and the whole batch recrystallized, or it may be submitted separately.

Further concentrate the mother liquors from the second crop to a volume of 15–20 mL. Cooling will result in the formation of an oil that contains 2-nitrobromobenzene (**22**). Separate the oil from the two-phase mixture by means of a pipet. The 2-nitrobromobenzene may be isolated and purified by column chromatography, as described in part B.

Pre-lab exercises for Section 17.3, Nitration of Bromobenzene (Part B), can be found on page PL. 57.

B. Thin-Layer and Column Chromatography

In two small vials prepare solutions of 4-nitrobromobenzene and of the oil containing 2-nitrobromobenzene in about 0.5 mL of chloroform. Obtain a 3- × 8-cm strip of silica gel chromatogram sheet (without fluorescent indicator). Following the procedure of Section 6.2, apply spots of each of these two solutions. Allow the spots to dry and develop the chromatogram in a TLC chamber with 9:1 (v:v) hexane:chloroform as the solvent.

When the solvent has climbed to within about 2 cm of the top of the plate, remove the developed chromatogram from the chamber, quickly mark the solvent front with a pencil, and allow the plate to dry. The spots may be visualized by placing the dry plate in a chamber whose atmosphere is saturated with iodine vapor. Calculate the R_f values of the spots observed and identify them as either 2- or 4-nitrobromobenzene. (A small orange spot may be observed very near the origin for the oil; this spot is identified as 2,4-dinitrobromobenzene.)

Following the procedure given in Section 6.1, prepare a column using a 50-mL buret and 15–20 g of silica gel. (Unless a large column is available to you, it will not be possible to submit the *entire* sample of oil obtained in part A to column chromatography.) Apply a 0.5-g sample of the oil containing the 2-nitrobromobenzene to the head of the column and rinse the inside of the buret with 1 mL of chloroform. Open the stopcock and allow the liquid to drain just to the top of the sand. Fill the buret with 9:1 (v:v) hexane:chloroform and elute the column until a total of 100 mL of the solvent has passed through the column. Do not allow the level of liquid to drain below the sand at the top of the column. Collect the eluent in 20- to 25-mL fractions in a series of numbered 50-mL Erlenmeyer flasks. After 100 mL of the 9:1 solvent has passed through the column, continue by passing 50 mL of a 4:1 (v:v) hexane:chloroform solution through the column and collecting 20- to 25-mL fractions.

Evaporate each of the fractions obtained to dryness on a steam bath in the hood. Characterize any solid residues obtained as either 2- or 4-nitrobromobenzene through either melting point determinations or TLC analysis. 2-Nitrobromobenzene has a reported melting point of 40–41°C; the melting point of 4-nitrobromobenzene is 125–126°C.

EXERCISES

1. The *o*:*p* ratio in the mononitration of bromobenzene has been reported to be 38:62. Using this ratio and the amount of 4-nitrobromobenzene you obtained, calculate the experimental yield of your reaction.

2. Why is the bromobenzene added in portions over a 15-min period rather than all at once?

3. Explain why 4-nitrobromobenzene **(21)** predominates in the product mixture over 2-nitrobromobenzene **(22).**

4. Using resonance structures, show why no detectable amount of 3-nitrobromobenzene is found in the product mixture.

5. Explain how TLC may be used to select the most appropriate solvent for use in a column chromatographic separation.

6. Explain why 4-nitrobromobenzene **(21)** has a larger R_f than the 2-isomer, **22,** even though it is the less soluble of the two.

SPECTRA OF STARTING MATERIALS AND PRODUCTS

The IR spectrum of bromobenzene is included in Figure 22.3; the PMR spectrum for 4-nitrobromobenzene is given as Figure 9.17.

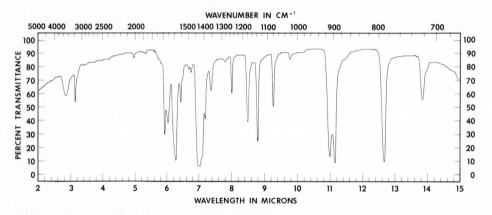

Figure 17.11 IR spectrum of 4-nitrobromobenzene (KBr pellet).

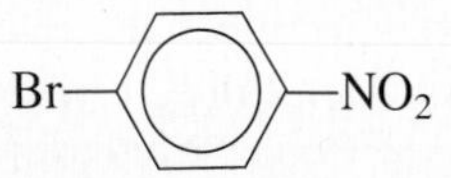

Br—⟨○⟩—NO$_2$ Chemical shifts: δ 124.9, 129.8, 132.5, 147.0.

Figure 17.12 CMR data for 4-nitrobromobenzene.

17.4 Relative Rates of Electrophilic Aromatic Substitution

Electrophilic aromatic substitution reactions are among the best understood of all organic reactions from the mechanistic standpoint, and most of the information obtained about aromatic substitution is internally consistent. The qualitative principles of the reaction that are usually discussed in organic chemistry textbooks include the effect of a substituent already present in an aromatic compound on the reactivity of that compound toward a new, entering substituent and on the orientation of the entering substituent (*ortho, meta,* or *para*). Normally, however, the subject of reactivity is discussed in terms of activation or deactivation of the molecule with respect to benzene, and little if any information is given about the *quantitative* differences in rates and reactivities of substituted aromatic compounds. The purpose of this discussion and accompanying experimental work is to give the student some feeling for the magnitude of the actual differences in reactivity.

A number of studies of electrophilic substitutions on arenes have been reported in which the experimental conditions allowed a direct comparison of the relative rates. For example, the relative rates of benzene and toluene toward halogenation, acetylation, sulfonation, nitration, and methylation have been determined. In all cases the reaction of toluene was more rapid, which has led to the well-known generalization in textbooks that a methyl group activates an aromatic ring toward reaction with electrophiles. For example, bromination of toluene is 605 times faster than bromination of benzene.

The effect that different substituents on the benzene ring have on the rate of a single type of reaction has been shown to be enormous. The difference in the rate of bromination of phenol and of nitrobenzene is about 10^{17}; the hydroxyl group produces an *activation* amounting to a factor of 10^{12} and the nitro group a *deactivation* by a factor of 10^5. The relative rates of bromination of several monosubstituted benzenes are studied in the experiments of this section.

In preparation for the measurement of reaction rates, it will be helpful to review some of the factors that influence them (see Section 18.4 for a brief summary). These factors include temperature, concentration of reactants, solvent composition, and nature and concentration of catalysts. It is of course possible to change any one of them while keeping all other factors constant, so that the effect of each of the variables can be determined.

The rate expression for the bromination of an aromatic compound is given by equation 15, in which Br_2 has been inserted in place of the E^+ of equation 2 in Section 17.1, since bromine is the precursor to the actual electrophile.

$$\text{Rate} = k_2[\text{ArH}][\text{Br}_2] \tag{15}$$

Note that the rate expression could be more complex under certain conditions. Bromination of less reactive aromatic compounds requires the use of a Lewis acid catalyst for reaction to occur at a reasonable rate, and water, in *trace* amounts, has been found to accelerate the rate of these electrophilic reactions. However, by avoiding the use of a catalyst and by performing the bromination in 15 *M* acetic acid, a mixed solvent composed of 90% acetic acid and 10% water, both of these complications are excluded from the rate studies. This much water in the solvent ensures a concentration that remains essentially constant during the course of the reaction and hence does not enter into the rate expression. The rate expression of equation 15 states that the rate of the reaction is proportional to the concentration of the aromatic substrate and to the concentration of bromine, and as written it is a second-order rate expression. Although one can, in principle, work with this rate expression, the calculations are much more difficult and time-consuming than are those for first-order reactions. Thus the experimental procedure is designed to provide a "pseudo-first-order" reaction to simplify the calculations. This is accomplished by maintaining the concentration of the aromatic substrate at a high level so that it does not change appreciably throughout the course of the reaction. Under these conditions the rate expression becomes

$$\text{rate} = k_1[\text{Br}_2] \tag{16}$$

where k_1 is a new rate constant that will contain all the factors of constancy for the reagents and solvents. Rate constant k_1 is called a *pseudo*-first-order rate constant. Pseudo-rate constants in general refer to reactions in which the concentration of one or more reactants that appear in the rate expression are chosen so that they do not appreciably change as the reaction proceeds. Thus they remain constant and are automatically incorporated in the pseudo-rate constant.

The *integrated* form of equation 16 is given in equation 17,

$$[\text{Br}_2]_t = [\text{Br}_2]_0 e^{-k_1 t} \tag{17}$$

and can be rearranged to the expressions shown in equations 18 and 19,

$$\ln \frac{[\text{Br}_2]_0}{[\text{Br}_2]_t} = k_1 t \tag{18}$$

$$2.303 \log \frac{[\text{Br}_2]_0}{[\text{Br}_2]_t} = k_1 t \tag{19}$$

where $[\text{Br}_2]_0$ = initial concentration of bromine and $[\text{Br}_2]_t$ = concentration of bromine at any time t.[1] From the experimental data, one can prepare a graph of log $[\text{Br}_2]_0/[\text{Br}_2]_t$ versus time, t. If the reaction is first order, the graph should show a reasonably straight line with a slope of $k_1/2.303$ if common logarithms are used. It is important to note that $[\text{Br}_2]_t$ is the concentration of bromine *remaining* at time t, and not the amount of bromine that has reacted.

The preparation of the reagents must be done carefully so that the initial concentrations are known accurately, and a large water bath must be used to ensure a constant temperature throughout the reaction.

One of the main reasons for choosing bromination reactions for carrying out rate studies is that it is easy to follow the rate of disappearance of the bromine color as the reagent reacts with the aromatic substrate. The value of choosing reaction conditions so that the rate of the reaction is proportional only to the concentration of the bromine thus becomes quite apparent.

An important limitation must be placed on the types of aromatic compounds to be studied. Because no Lewis acid catalyst is present, the study is necessarily limited to compounds that are much more reactive than benzene. Benzene and compounds of similar or lower reactivity would react so slowly with bromine in the absence of a catalyst that it would not be possible to obtain good kinetic data in a single laboratory period. This limitation presents no serious problem, since it will still be possible to determine the relative reactivities of a series of compounds and to illustrate the principles involved in these sorts of measurements.

Two types of experiments are presented here. The first describes a procedure for determining qualitative differences in the rates of bromination of several aromatic compounds by

[1]See Section 18.4 for a more detailed discussion of these equations for first-order rate constants. The equations are very similar to those presented for the hydrolysis of tertiary halides.

simply observing how long it takes them to decolorize bromine solutions. The second procedure, which provides more accurate, or quantitative, measurement of differences in reaction rates, requires the use of an inexpensive visible/ultraviolet spectrophotometer. A nonrecording instrument can be used, and the rate of disappearance of bromine can be determined by reading absorbance values on the meter.[2] If a compound strongly absorbs light at a given wavelength, then the ability of light to pass through a sample of that compound is diminished. The spectrophotometer measures the amount of light that has been absorbed, this value being called absorbance. Absorbance is defined as in equation 20, and is often called *optical density*.[3]

$$\text{absorbance } (A) = \log \frac{I_0}{I} = \epsilon c l \qquad \textbf{(20)}$$

The absorbing species in these experiments is bromine in all instances. The same cell will be used for all measurements for a given aromatic substrate, so that for a given kinetic run the values of ϵ and l in equation 20 remain constant, and equation 19 may be combined with equation 20 to give equation 21.

$$2.303 \log \frac{A_0}{A_t} = k_1 t \qquad \textbf{(21)}$$

$$\text{where } A_0 = \epsilon[\text{Br}_2]_0 l$$
$$A_t = \epsilon[\text{Br}_2]_t l$$

This form of equation 19 is useful in that it relates a series of measurements of absorbance to the rate constant k_1. By measuring the rate constants for several substrates one may quantitatively evaluate the reactivities of these substrates toward bromination. Additional instructions for analyzing the data collected are provided within the Experimental Procedure.

EXPERIMENTAL PROCEDURE

DO IT SAFELY

1 Bromine is a hazardous chemical that may produce severe chemical burns if allowed to come in contact with the skin. Even though the solutions used in the experiments are dilute, take proper precautions. Wash your hands and soak any affected area for a few minutes in a 0.6 *M* sodium thiosulfate solution.

2 Do *not* use pipets filled by mouth to transfer *any* of the solutions in this experiment, either the bromine solutions or the acetic acid solutions of the substrates.

[2]Several relatively inexpensive instruments are available, such as the Bausch and Lomb Spectronic 70. A discussion of UV-visible spectroscopy is provided in Section 11.2.
[3]See Section 11.2 for a discussion of equation 20 and a definition of its terms.

In the following experiments it would be convenient for two students to work together.

Pre-lab exercises for Section 17.4, Relative Rates of Electrophilic Aromatic Substitution, Part A, Qualitative Rate Comparisons, can be found on page PL. 59.

A. Qualitative Rate Comparisons

Stock solutions of the aromatic substrates should be prepared containing 0.2 mol of each substrate per liter of 15 M (90%) acetic acid. Prepare a water bath using a 1-L beaker and adjust and maintain the temperature at 35 $\pm$ 2°C. Place 2 mL of the stock solutions of each of the following substrates in separate small test tubes: (a) phenol, (b) anisole, (c) diphenyl ether, (d) acetanilide, (e) p-bromophenol, and (f) α-naphthol. If stock solutions are provided, use special care to ensure that one solution is not contaminated by another. Suspend the carefully labeled test tubes partially in the water bath by looping a piece of copper wire around the neck of the test tube and over the rim of the beaker.

Prepare several disposable pipets with a capacity of 2.0 mL and equip them with rubber bulbs. Calibrate these by comparison with known volumes, and mark the droppers by scratching the glass with a file. These will be used to introduce the bromine solution into the test tubes containing the substrates. It is essential that the addition be done as rapidly as possible and that the volumes be fairly accurate. Transfer about 45 mL of a solution containing 0.05 M bromine in 15 M acetic acid to an Erlenmeyer flask and allow the solution to equilibrate in the water bath for a few minutes. Add 2.0 mL of the bromine solution to one of the test tubes containing a substrate. Make the addition rapidly, mix the solution quickly, and note the exact time of addition. Observe the reaction mixture and note how long it takes for the bromine color to become faint yellow or to disappear. Repeat this procedure with each of the substrates, making sure that you use the same end-point color for each one.

When the reaction is slow—that is, no decolorization occurs within 5 min—go on to another compound while waiting for the end-point to be reached. Record the reaction times, and on the basis of these observations arrange the compounds in order of increasing reactivity toward bromine.

If it is impossible to determine the relative rates accurately at 35°C, repeat the experiment with the compounds in doubt at 0°C, using an ice-water bath.

Pre-lab exercises for Section 17.4, Relative Rates of Electrophilic Aromatic Substitution, Part B, Quantitative Measurements, can be found on page PL. 61.

B. Quantitative Measurements

Obtain three sample tubes for use with the spectrophotometer. These tubes, called *cuvettes,* resemble small test tubes, but the dimensions and glass thickness are carefully controlled in manufacture.

Have at your disposal stock solutions (0.5 M) of anisole, diphenyl ether, and acetanilide in 15 M acetic acid, as well as a 0.02 M solution of bromine in 15 M acetic acid. You will also need a supply of two or three 2-mL disposable pipets for accurate sample measurement. *Caution:* Do not use your mouth with these pipets; use rubber bulbs.

Since all solutions absorb light to some extent, and although they may appear transparent to the naked eye, it is necessary to calibrate the spectrophotometer to zero absorbance using a solution containing all components of the solution to be measured *except the absorbing species* (bromine in this case). Into a clean and dry cuvette place 2 mL of 15 M acetic acid and 2 mL of a 0.5 M solution in 15 M acetic acid of the substrate whose rate of substitution is to be measured. Use 2-mL pipets to ensure accurate measurement of these small volumes. Stir the solution to make sure that it is homogeneous and place the cuvette in the spectrophotometer. Note that the cuvette has an alignment mark to ensure that it is always placed in the spectrophotometer in the same orientation. Align this mark with the corresponding engraved mark on the front of the sample holder. Close the door to the sample holder and adjust the absorbance reading to zero. This calibration and all future readings are done at 400 nm, the wavelength at which elemental bromine absorbs light.

In preparing for and executing the following experiments, you must be well organized so that you can work quickly. It may take several tries before you are able to work rapidly enough to get acceptable results. Once you have zeroed the spectrophotometer for a given substrate, clean and dry the *same* cuvette used in that step, and continue with the following operations: (1) to the cuvette add 2 mL of 0.02 M bromine solution, (2) add 2 mL of the 0.5 M solution of substrate, (3) stir quickly, one time only, (4) have a second person record the exact time of mixing, (5) place the cuvette in the spectrophotometer with the correct orientation and close the door to the sample compartment, and (6) record a series of absorbance readings and the times at which these readings were taken. Obtain as many readings as possible before the absorbance drops below about 0.5, although readings may be taken below 0.5. It is easier to record the time at which the needle on the meter crosses a line on the absorbance scale rather than to attempt interpolation between those lines. Record the laboratory temperature.

Repeat the above for each of the other substrates. For those runs that are particularly fast it is advisable to perform duplicate runs.

Treatment of Data

1 The data recorded above constitute a series of values A_t (see equation 21). To obtain A_0, it is necessary to extrapolate these data to the initial time of mixing, t_0. Since the absorbance decreases exponentially during the course of the run, it is easiest to do this by plotting log A_t (vertically) versus time (seconds, horizontally). To determine log A_0 extrapolate the straight line obtained to zero time. Take the antilog of this intercept to obtain A_0. This must be done separately for each substrate.

2 For each substrate, using the value of A_0 just obtained and the recorded values of A_t, calculate a series of values of 2.303 log A_0/A_t. On graph paper plot these values (vertically) versus time (seconds, horizontally). Draw the best straight line possible through these points. Calculate the slope of that line to obtain k_1 (sec^{-1}) for that substrate.

3 After obtaining k_1 for each substrate, divide all values of k_1 by the *smallest* value to obtain the *relative* reactivities of the substrates toward electrophilic aromatic bromination.

EXERCISES

Questions for Part A

1. Write structures for the major monobromination products that would be formed from each of the substrates used. Explain your predictions.

2. Based on the results obtained in this experiment, arrange the substrates in order of *decreasing* order of reactivity toward bromine. Explain the reactivity order in terms of inductive and/or resonance effects.

3. **a.** Compute the initial concentrations of bromine and of substrate in the reaction mixture, assuming that no reaction had occurred.
   b. Which reagent, bromine or substrate, was present in excess?
   c. Suggest a reason why the experimental procedure calls for one reagent to be present in excess over the other.
   d. Would you be able to determine the rates of reaction if the concentrations of substrate and of bromine had been reversed over what was called for in the experimental procedure? Explain your answer.

Questions for Part B

1. Write structures for the major monobromination products that would be formed from each of the substrates used. Explain your predictions.

2. Based on the results obtained in this experiment, arrange the substrates in order of *decreasing* order of reactivity toward bromine. Explain the reactivity order in terms of inductive and/or resonance effects.

3. **a.** Compute the initial concentrations of bromine and of substrate in the reaction mixture, assuming that no reaction had occurred.
   b. Which reagent, bromine or substrate, was present in excess?
   c. Suggest a reason why the experimental procedure calls for one reagent to be present in excess over the other.
   d. Would you be able to determine the rates of reaction if the concentrations of substrate and of bromine had been reversed over what was called for in the experimental procedure? Explain your answer.

4. Show that equation 21 derived from combining equations 19 and 20. On the basis of this derivation, explain why it is not necessary to use either the extinction coefficient, ϵ, or the cell length, l, in the Treatment of Data.

5. Explain why the various rate constants, k_1, can be used to deduce the relative reactivities of the various substrates toward bromination.

chapter **18**

nucleophilic
aliphatic substitution
preparation of haloalkanes and
chemical kinetics

18.1 Theory of Nucleophilic Aliphatic Substitution; Competition with Elimination

Nucleophilic aliphatic substitution reactions have widespread use in organic syntheses. Many types of compounds containing different functional groups can be prepared by varying the reactants. This chapter deals with two aspects of nucleophilic substitution at saturated, sp^3-hybridized carbon, namely, the conversion of different alcohols into the corresponding haloalkanes and the evidence from chemical kinetics and rate studies to support one of two possible mechanisms by which substitution can occur.

(a) The Concept of Nucleophilic Aliphatic Substitution

The substitution of one group for another at a saturated carbon atom is a commonly utilized process in effecting organic molecular transformations. One form of this reaction is exemplified in equation 1, where *Nu:* is a symbol for a generalized *nucleophile* and represents a molecule or ion that has *nucleophilic* character, and *L* represents a *leaving* group. Nucleophiles have in common the property of bearing *at least one nonbonding pair of electrons* and being either *neutral* or *negatively charged;* the nonbonding pair of electrons is donated to carbon with the concomitant formation of a new covalent bond. Examples of nucleophiles include $:\!\ddot{C}l\!:^-$, $:\!\ddot{B}r\!:^-$, $H\ddot{O}\!:^-$, $:\!N\!\equiv\!C\!:^-$, $H_2\ddot{O}\!:$, and $:\!NH_3$. Note that each of these molecules or ions fits the general definition above for nucleophiles.

The leaving group, L, must have the ability to accept the pair of bonding electrons from the alkyl group, R, as the R—L bond breaks. The best leaving groups are those that are *conjugate bases* of strong acids. For example, a leaving group such as Cl^- is the conjugate base of the strong acid, HCl, and Cl^- is said to be a moderately good leaving group. On the other hand, HO^- is a poor leaving group since it is the conjugate base of the weak acid water. The ease with which a group leaves can be determined experimentally by studying rates of reaction for various leaving groups.

$$R—L + Nu\text{:} \rightarrow R—Nu + L\text{:} \tag{1}$$

(b) Types of Nucleophilic Substitution

Although nucleophilic substitution is a very general reaction for aliphatic compounds, R—L, the mechanism followed in a given transformation is strongly dependent upon the structure of the alkyl group, R. There are two distinctly different mechanisms about which an enormous amount of discussion and moderate agreement has accumulated over the years, and these are designated by the symbols S_N1 (S for substitution, N for nucleophilic, and 1 for unimolecular) and S_N2 (2 stands for bimolecular). The two types of mechanistic pathways are depicted in equations 2 and 3, respectively.

S_N1 mechanism:

$$R_3C—L \overset{slow}{\longleftrightarrow} R_3C^{\oplus} + L\text{:}^{\ominus} \tag{2a}$$

$$R_3C^{\oplus} + Nu\text{:}^{\ominus} \overset{fast}{\longrightarrow} R_3C—Nu \tag{2b}$$

S_N2 mechanism:

$$Nu\text{:}^{\ominus} \quad \overset{R}{\underset{H}{\overset{|}{C}}}—L \rightarrow \left[Nu \overset{\delta\ominus}{---} \overset{R}{\underset{H\ \ H}{\overset{|}{C}}} \overset{\delta\ominus}{---} L \right] \rightarrow Nu—\overset{R}{\underset{H}{C}}_{H} + L\text{:}^{\ominus} \tag{3}$$

Transition state

The S_N1 pathway involves two successive steps, the first of which is the heterolytic cleavage of the leaving group from the remainder of the molecule; the leaving group takes the pair of bonding electrons with it. Equation 2a shows this process in a greatly simplified fashion, in that what is not shown is that this step is aided considerably by polar interactions between solvent molecules and the incipient cationic and anionic centers. This phenomenon is investigated kinetically in the experiments of Section 18.4. The carbon atom acquires a positive charge and becomes a *carbocation*. In the second step the nucleophile combines with the positively charged carbon atom to form the product. Normally, the concentration of the nucleophile, Nu:, is high compared with that of the L:⁻ that has been produced, so that the reverse of reaction 2a is relatively unimportant. One would anticipate that the first of the two steps will be slower, since it involves net bond breaking, which is an endothermic process. The second step involves bond formation, which is an exothermic process that should be favored energetically. The experimental observation that the rates of many substitution reactions depend only on the concentration of the substrate, R—L, and are independent of the nucleophile concentration is consistent with this type of mechanism. This is stated mathematically by equation 4.

$$S_N1 \text{ reaction: Rate} = k_1[R—L] \tag{4}$$
$$\text{where } k_1 \text{ is the rate constant}$$

The S_N2 mechanism, on the other hand, is represented as a direct attack of the nucleophile at an angle of 180° to the C—L bond. Because the C—L bond is being broken concurrently with the formation of the C—Nu bond, both the substrate, RCH_2—L, and the nucleophile are involved in the transition state of the rate-determining step, and this is a bimolecular reaction. Therefore the rate of an S_N2 reaction depends on the nature and concentration of the substrate *and* the nucleophile, as shown in equation 5.

$$S_N2 \text{ reaction: Rate} = k_2[RCH_2L][Nu\!:\!] \tag{5}$$
$$\text{where } k_2 \text{ is the rate constant}$$

In accordance with the S_N1 and S_N2 mechanisms given above for nucleophilic aliphatic substitution, the following two factors have been found to play an important role in dictating the preferred mode of reaction for a particular substrate.

1 In going from CH_3—L to R_3C—L, the number of alkyl groups attached to the central carbon atom is increased, thus making it more difficult for effective backside attack by the nucleophile. This decreases the ease with which the S_N2 process can occur.

2 As the number of alkyl groups attached to the central carbon atom increases, the incipient carbocation in the S_N1 reaction becomes more stable, and this facilitates its formation along the S_N1 pathway. The relative stabilities of substituted carbocations are discussed in Section 13.2.

These two effects reinforce one another and yield the following general trends:

$$\xleftarrow{\text{Increasing ease of } S_N2 \text{ reaction}}$$
$$CH_3—L \quad RCH_2—L \quad R_2CH—L \quad R_3C—L$$
$$\xrightarrow{\text{Increasing ease of } S_N1 \text{ reaction}}$$

The possibility exists that a given nucleophilic substitution reaction can occur by either or both of the mechanisms presented above. Only one of the two paths is observed for many substrates, especially those that are primary (1°, RCH_2—L) or tertiary (3°, R_3C—L), with the former occurring predominantly by S_N2 and the latter by S_N1. Secondary substrates (2°, R_2CH—L) are prone to react by both mechanisms, and the pathway is dictated by factors such as solvent, reaction conditions, and nature of the nucleophile. The mechanism by which a particular reaction occurs must usually be confirmed *experimentally*. For example, if the structures of the starting substrate and the product(s) are determined, it may be possible to deduce the mechanism by which the process occurred. To elaborate, we learned in Sections 13.2 and 17.2 that carbocations are prone to molecular rearrangement if a more stable cation can be formed. Since carbocations are involved in the S_N1 mechanism, the formation of a rearranged product points to this mechanism. On the other hand, the lack of rearrangement does not necessarily exclude an S_N1 process. A specific example of such a possibility is discussed in Section 18.2, where one of the many experimental techniques used to elucidate reaction mechanisms is presented.

(c) Competition Between Substitution and Elimination

In all reactions in which nucleophilic aliphatic substitution occurs, there is the potential of competing elimination reactions that produce alkenes. For the most part, bimolecular elimination reactions, E2 (E stands for elimination and 2 for bimolecular), compete with the S_N2 process, and unimolecular elimination reactions, E1, compete with S_N1 substitution. E1 reactions are discussed in detail in Section 13.2 and E2 reactions in Section 13.1.

The competition between substitution and elimination is shown in equations 6 and 7. In general, *elimination is favored* when strongly basic and only slightly polarizable nucleophiles are used, examples being $R\ddot{O}:^-$, $H_2\ddot{N}:^-$, $H:^-$, and $H\ddot{O}:^-$. *Substitution is favored* with weakly basic and highly polarizable nucleophiles such as I^-, Br^-, Cl^-, H_2O, and $CH_3CO_2^-$. Polarizability is a measure of the ease of distortion of the electron cloud of the species by a nearby charged center, which is a full or partial positive charge in substitution reactions.

S_N2 *versus* E2:

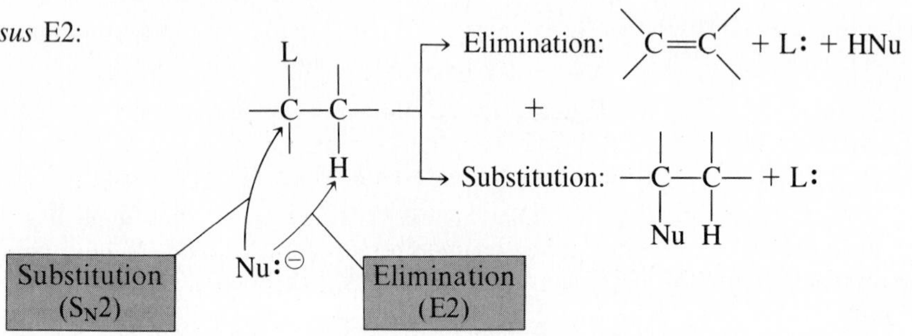

$$ \text{(6)} $$

S_N1 *versus* E1:

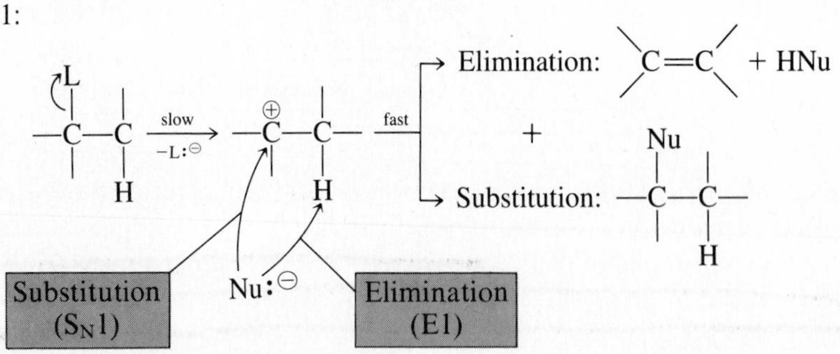

$$ \text{(7)} $$

18.2 Preparation of 1-Bromobutane: An S_N2 Reaction

A common way of converting a primary alcohol to a haloalkane is shown in equation 8. Since this method involves the *reversible* reaction between an alcohol and a mineral acid (HX = HCl, HBr, HI), the best yield of haloalkane is obtained when the *position of equilibrium* lies far to the right.

$$HX + R\text{—}OH \underset{\text{heat}}{\overset{H_2SO_4}{\rightleftharpoons}} R\text{—}X + H_2O \qquad \textbf{(8)}$$

The preparation of 1-bromobutane may be accomplished by heating 1-butanol with concentrated hydrobromic acid in the presence of concentrated sulfuric acid (equation 9).

$$\underset{\substack{\text{1-Butanol} \\ (n\text{-butyl alcohol})}}{CH_3CH_2CH_2CH_2\text{—}OH} + HBr \underset{\text{heat}}{\overset{H_2SO_4}{\rightleftharpoons}} \underset{\substack{\text{1-Bromobutane} \\ (n\text{-butyl bromide})}}{CH_3CH_2CH_2CH_2\text{—}Br} + H_2O \qquad \textbf{(9)}$$

The mechanism for this reaction has been shown to occur in two steps (equation 10).

$$n\text{–}C_3H_7CH_2\overset{\cdot\cdot}{\underset{\cdot\cdot}{O}}H + H^{\oplus} \rightleftharpoons n\text{–}C_3H_7CH_2\text{—}\overset{H}{\underset{H}{\overset{\oplus}{O}}}{:} \rightleftharpoons n\text{–}C_3H_7CH_2\text{—}Br + H_2\overset{\cdot\cdot}{\underset{\cdot\cdot}{O}}{:} \qquad \textbf{(10)}$$

1

The first step involves the protonation of the alcohol to give oxonium ion **(1),** which is an example of Lewis acid–Lewis base complex formation. The oxonium ion then undergoes displacement by the bromide ion to form the bromoalkane and water. In this S_N2 reaction *water is the leaving group and bromide ion is the nucleophile.* The sulfuric acid serves two distinct and important purposes: (1) it is a dehydrating agent that reduces the activity of water and shifts the position of equilibrium to the right, and (2) it provides an added source of hydrogen ions to increase the concentration of oxonium ion **(1).** The use of *concentrated* hydrobromic acid also helps to establish a favorable equilibrium.

The fact that acid is required for this reaction can be gleaned from the experimental observation that *no* reaction occurs between 1-butanol and NaBr. Note that the leaving group in equation 7 is the weakly basic water molecule. The discussion in Section 18.1 indicates that the leaving groups in nucleophilic substitution reactions must be only *weakly basic.* Thus, consideration of the possible reaction between 1-butanol and NaBr (equation 11) reveals that the leaving group would be the *strongly basic* hydroxide ion if the reaction did occur.

$$CH_3CH_2CH_2CH_2\text{—}OH + NaBr \xrightarrow{\quad\times\quad} \underset{\textit{Not formed}}{CH_3CH_2CH_2CH_2\text{—}Br} + NaOH \qquad \textbf{(11)}$$

Hence, theory predicts that reaction 11 should *not* occur, and, in fact, it is the reverse reaction between a bromoalkane and hydroxide ion that does occur. A smooth conversion of 1-butanol to 1-bromobutane is observed in the presence of acid (equation 9), which results in the weakly basic water molecule being the leaving group.

The S_N2 mechanism shown in equation 10 can be confirmed experimentally by the use of gas chromatography to determine that the product of the reaction is 1-bromobutane. A *possible* alternate mechanism leading to its formation is an S_N1 process, but this would involve formation of the 1-butyl cation; if formed, this very unstable cation would be expected to undergo some molecular rearrangement so that two isomeric products, 1-bromobutane and 2-bromobutane, would result (equation 12).

$$CH_3CH_2CH_2CH_2-\overset{..}{\underset{..}{O}}H + H^{\oplus} \rightleftarrows CH_3CH_2CH_2CH_2-\overset{\overset{H}{|}}{\underset{\underset{H}{}}{\overset{\oplus}{O}:}}$$

$$CH_3CH_2\overset{\oplus}{C}HCH_3 \xleftarrow{\ \sim H:^{\ominus}\ } CH_3CH_2CH_2\overset{\oplus}{C}H_2 + H_2O \qquad (12)$$

$$\Big\downarrow :\overset{..}{\underset{..}{Br}}:^{\ominus} \qquad\qquad\qquad \Big\downarrow :\overset{..}{\underset{..}{Br}}:^{\ominus}$$

$$\underset{\underset{Br}{|}}{CH_3CH_2CHCH_3} \qquad\qquad\qquad CH_3CH_2CH_2CH_2Br$$

2-Bromobutane   1-Bromobutane

The concept of product identification is exceedingly important in elucidating mechanisms, and the experimental procedures below allow you to decide whether or not the major product of the reaction between 1-butanol and hydrobromic acid is actually only 1-bromobutane, in which case this reaction must therefore involve the S_N2 pathway. On the other hand, if two major products, 1-bromobutane and 2-bromobutane, were observed, one might conclude that the reaction occurred at least in part by an S_N1 mechanism. 2-Bromobutane can also be formed by a different route involving the dehydration of 1-butanol to give a mixture of 1-butene and 2-butene, followed by addition of HBr across the carbon-carbon double bond. However, the reaction conditions are such that very little dehydration occurs.

   The mixture of hydrobromic acid and sulfuric acid may be prepared in two ways. One method is to add concentrated sulfuric acid to concentrated hydrobromic acid, and the second is to generate the hydrobromic acid *in situ* by adding concentrated sulfuric acid to sodium bromide (equation 13).

$$NaBr + H_2SO_4 \rightleftarrows HBr + NaHSO_4 \qquad (13)$$

Both of these methods work well and give good yields of the bromoalkane when low molecular weight alcohols are used. With higher molecular weight alcohols, the *in situ* generation of HBr is not effective because of the low solubility of these alcohols in concentrated salt solutions. In these instances, concentrated hydrobromic acid is used. With very high molecular weight alcohols, the method of choice is to add hydrogen bromide gas to the alcohol at elevated temperatures or to select a different reactant, such as PBr_3.

   Although the addition of concentrated sulfuric acid is desirable for bromoalkane preparation, its presence may cause two important side reactions to occur. The alcohol may react with sulfuric acid to form an ester, the alkyl hydrogen sulfate (**2**, equation 14).

$$ROH + H_2SO_4 \rightleftarrows RO-SO_3H + H_2O \qquad (14)$$

2

An alkyl
hydrogen
sulfate

This reaction is reversible, and the alcohol is regenerated because its concentration is reduced as the bromoalkane is formed; that is, the position of equilibrium in equation 14 is shifted to the left. The formation of the alkyl hydrogen sulfate **(2)** would not directly decrease the yield of bromoalkane if it were not for the fact that it can undergo two other reactions to give undesired side products: (1) On heating, **2** can undergo elimination to give a mixture of alkenes (equation 15). (2) It can react with another molecule of alcohol to give a dialkyl ether by an S$_N$2 reaction in which the nucleophile is ROH (equation 16).

$$RO\text{---}SO_3H \xrightarrow{\text{heat}} \text{alkenes} + H_2SO_4 \qquad \textbf{(15)}$$

$$RO\text{---}SO_3H + ROH \xrightarrow{\text{heat}} R\text{---}O\text{---}R + H_2SO_4 \qquad \textbf{(16)}$$

An ether

Both of these side reactions consume alcohol and result in a decreased yield of bromoalkane. The temperatures used for the substitution reaction on *primary* alcohols are generally not high enough to cause these side reactions to be of great importance.

Concentrated sulfuric acid cannot be used in the preparation of secondary bromoalkanes from *secondary* alcohols because these alcohols are far more easily dehydrated than primary alcohols to give alkenes (equation 15). In fact, this type of dehydration reaction is a method for synthesizing alkenes (Section 13.2). One means of circumventing this problem is to use concentrated (48%) hydrobromic acid without the added sulfuric acid. However, haloalkanes are generally better prepared from secondary alcohols by use of phosphorus trihalides (PX$_3$, X = Br or I, equation 17) or thionyl halides (SOX$_2$, X = Cl or Br, equation 18). Thionyl halides must be used when the alcohol is susceptible to *carbocationic rearrangements* (Sections 13.2 and 17.2) so that the presence of acid must be avoided. In these cases, pyridine is used as the solvent, because it is a weak base and neutralizes the acid (HX) as it is formed (equation 18).

$$3\ ROH + \underset{\substack{\text{Phosphorus} \\ \text{trihalide}}}{PX_3} \rightarrow 3\ RX + H_3PO_3 \qquad \textbf{(17)}$$

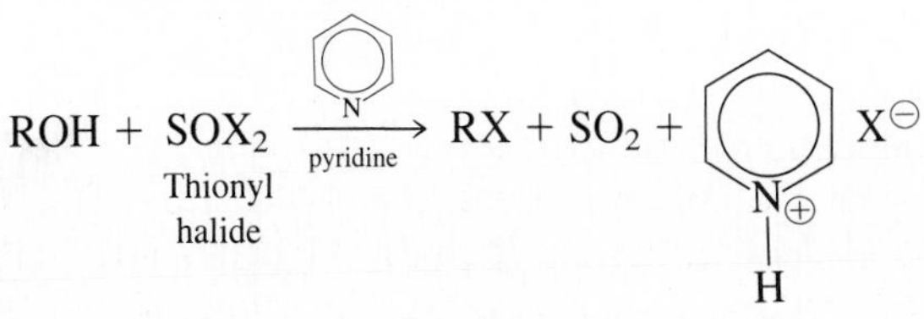

$$ROH + \underset{\substack{\text{Thionyl} \\ \text{halide}}}{SOX_2} \xrightarrow[\text{pyridine}]{} RX + SO_2 + \quad X^{\ominus} \qquad \textbf{(18)}$$

Pre-lab exercises for Section 18.2, Preparation of 1-Bromobutane, can be found on page PL. 63.

EXPERIMENTAL PROCEDURE

DO IT SAFELY

1 Examine your glassware for cracks and chips. This experiment involves heating concentrated acids, and defective glassware could break under these conditions and spill hot chemicals on you and those working around you.

2 Concentrated sulfuric acid and water mix with the evolution of substantial quantities of heat. *Always add the acid to the water* in order to disperse the heat through warming of the water. Add the acid slowly and with swirling to ensure continuous and thorough mixing.

3 It is advisable to wear rubber gloves in this experiment. Be very careful when handling concentrated sulfuric acid. When it is poured from the reagent bottle, some may run down the outside of the bottle. In the event that any concentrated sulfuric acid comes in contact with your skin, immediately wash it off with copious amounts of cold water and then with dilute sodium bicarbonate solution to help neutralize any residual acid.

Place 0.30 mol of sodium bromide in a 250-mL round-bottomed flask, and add about 35 mL of water and 0.30 mol of 1-butanol. Mix thoroughly and cool the flask in an ice bath. *Slowly* add 35 mL of *concentrated* sulfuric acid to the cold mixture with swirling and continuous cooling. Remove the flask from the ice bath, add boiling chips, attach a reflux condenser, warm the flask gently until most of the salts have dissolved, and heat the mixture to gentle reflux. A noticeable reaction occurs and two layers form, the upper layer being the alkyl bromide; the inorganic layer has a high concentration of salts and thus is much more dense than 1-bromobutane. Continue heating under reflux for 45 min.★

Equip the flask for simple distillation. Distil the mixture rapidly, and collect the distillate in an ice-cooled receiver. Codistillation of 1-bromobutane and water occurs. Continue the distillation until the distillate is clear. The head temperature should be around 115°C at this point. The increased boiling point is caused by codistillation of sulfuric acid and hydrobromic acid with water.★

Transfer the distillate to a separatory funnel. Add about 30 mL of water and shake the funnel gently with venting. Note that two layers are formed; decide which of these is the organic layer (Section 2.18). Separate the layers and discard the *aqueous* layer, but be sure you know which layer is which. Wash the organic layer with 10 mL of 2 *M* aqueous sodium hydroxide solution and then with about 20 mL of water.

Transfer the cloudy 1-bromobutane layer to a small Erlenmeyer flask, and dry it over a little *anhydrous* magnesium sulfate.★ Swirl occasionally for a period of 10–15 min. Filter the mixture by gravity into a clean, dry 50-mL round-bottomed flask. To avoid the loss of any 1-bromobutane that is adsorbed by the drying agent on the filter paper, pour about 5 mL of dichloromethane through the filter and allow it to drain into the flask along with the remainder of the 1-bromobutane. Add two or three boiling chips, and equip the flask for simple distillation. Carefully remove the dichloromethane as a forerun and collect the product, boiling at 90–103°C. Compute the yield of the product isolated. Yields of 70–80% may be obtained in this experiment. Analyze your product by GC on a nonpolar column (SF-96 or SE-30 is satisfactory). Obtain the retention times of standard samples of 1-bromobutane and of 2-bromobutane. In your report discuss the relative percentages of 1- and 2-bromobutanes in your product in terms of the mechanism(s) of the reaction. Obtain IR spectra of your product and of 1-butanol and discuss the differences observed.

EXERCISES

1. In the procedure some water was added to the initial reaction mixture. How might the yield of 1-bromobutane be affected by the failure on the part of the student to add the water, and what product(s) would be favored? How might the yield of product be affected by adding, for example, twice as much water as is called for while keeping the quantities of the other reagents the same?

2. In the purification process, the organic layer is washed with 2 M NaOH and then with water. What is the purpose of these washes?

3. After the washes described in Exercise 2, the 1-bromobutane is treated with anhydrous magnesium sulfate. Why is this done, and what does it remove?

4. The final step of the purification process involves adding some dichloromethane, followed by a simple distillation.
   a. Why is simple rather than fractional distillation used?
   b. Why is this final distillation performed? Suggest what is removed by this distillation.

5. The following reaction was carried out and the indicated products were isolated.

$$(CH_3)_2CHCHCH_3 \xrightarrow{\text{HBr}} (CH_3)_2CHCHCH_3 + (CH_3)_2CCH_2CH_3$$
$$\qquad\quad | \qquad\qquad\qquad\qquad\quad | \qquad\qquad\qquad |$$
$$\qquad\quad OH \qquad\qquad\qquad\qquad Br \qquad\qquad\qquad Br$$
$$\qquad\qquad\qquad\qquad\qquad\qquad\qquad 10\% \qquad\qquad\qquad 90\%$$

Suggest reasonable mechanisms for this reaction, and indicate whether each is S_N1 or S_N2.

6. A student desired to prepare 1,3-dibromopropane, $BrCH_2CH_2CH_2Br$, and used 0.3 moles of 1,3-propanediol, $HOCH_2CH_2CH_2OH$, in place of 1-butanol. If the same quantities of the other reagents called for in the experiment are used, would 1,3-dibromopropane be formed in good yield? Why or why not?

SPECTRA OF STARTING MATERIALS AND PRODUCTS

The IR and PMR spectra of 1-chlorobutane (Figures 12.5 and 12.6, respectively) are similar to those of 1-bromobutane, so that the spectra of the latter compound are not included. The PMR spectrum of 1-butanol is given in Figure 9.9.

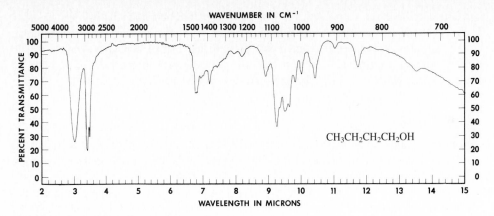

Figure 18.1 IR spectrum of 1-butanol.

$CH_3CH_2CH_2CH_2OH$ Chemical shifts: δ 13.9, 19.2, 35.0, 62.2.

Figure 18.2 CMR data for 1-butanol.

18.3 Preparation of Tertiary Chloroalkanes: An S_N1 Reaction

Several different reagents, such as HX, PX_3 and SOX_2, can be used to prepare haloalkanes from alcohols, as discussed in Section 18.2. However, only one reagent gives acceptable yields for the preparation of tertiary haloalkanes, namely the reaction of a mineral acid with the alcohol, the mechanism of which is typically S_N1. Phosphorus trihalides or thionyl halides are not applicable for the preparation of *tert*-haloalkanes from alcohols because elimination reactions predominate.

The reaction of 2-methyl-2-propanol (*tert*-butyl alcohol) with hydrochloric acid (equation 19) produces 2-chloro-2-methylpropane (*tert*-butyl chloride) and is illustrative of the S_N1 reaction.

$$
\begin{array}{c}
CH_3 \\
| \\
CH_3{-}\underset{\underset{\displaystyle CH_3}{|}}{C}{-}OH \;+\; HCl
\end{array}
\;\rightleftharpoons\;
\begin{array}{c}
CH_3 \\
| \\
CH_3{-}\underset{\underset{\displaystyle CH_3}{|}}{C}{-}Cl
\end{array}
\;+\; H_2O
\qquad \text{(19)}
$$

2-Methyl-2-propanol   2-Chloro-2-methylpropane
(*tert*-butyl alcohol)   (*tert*-butyl chloride)

The mechanism, shown in equation 20, involves three steps, the first of which is the protonation of the alcohol (equation 20a). The second step (equation 20b) is the slowest, rate-determining step in this sequence, although it is faster than the corresponding S_N2 attack of Cl^- on oxonium ion (**3**).

$$CH_3-\overset{\overset{\displaystyle CH_3}{|}}{\underset{\underset{\displaystyle CH_3}{|}}{C}}-\ddot{O}H + H^{\oplus} \;\rightleftharpoons\; CH_3-\overset{\overset{\displaystyle CH_3}{|}}{\underset{\underset{\displaystyle CH_3}{|}}{C}}-\overset{\oplus}{\ddot{O}}\overset{\displaystyle H}{\underset{\displaystyle H}{\big\langle}} \qquad \textbf{(20a)}$$

3

$$CH_3-\overset{\overset{\displaystyle CH_3}{|}}{\underset{\underset{\displaystyle CH_3}{|}}{C}}-\overset{\oplus}{\ddot{O}}\overset{\displaystyle H}{\underset{\displaystyle H}{\big\langle}} \;\overset{slow}{\rightleftharpoons}\; CH_3-\overset{\overset{\displaystyle CH_3}{|}}{\underset{\underset{\displaystyle CH_3}{|}}{\overset{\oplus}{C}}} \;+\; H_2\ddot{O}\colon \qquad \textbf{(20b)}$$

Relatively stable
tertiary carbocation

$$CH_3-\overset{\overset{\displaystyle CH_3}{|}}{\underset{\underset{\displaystyle CH_3}{|}}{\overset{\oplus}{C}}} \;+\; \colon\!\overset{\cdot\cdot}{\underset{\cdot\cdot}{Cl}}\!\colon^{\ominus} \;\overset{fast}{\rightleftharpoons}\; CH_3-\overset{\overset{\displaystyle CH_3}{|}}{\underset{\underset{\displaystyle CH_3}{|}}{C}}-Cl \qquad \textbf{(20c)}$$

The reverse is true for the reaction of the oxonium ions of primary and secondary alcohols. This reactivity difference between tertiary oxonium ions and primary and secondary oxonium ions is a reflection of the relative stabilities of the three types of carbocations that result from the loss of a molecule of water. Although *relatively* slow compared with the other steps in the sequence, reaction 20b is quite rapid from an *absolute* viewpoint. Thus, with concentrated hydrochloric acid a favorable equilibrium is established in a few minutes at room temperature and gives high yields of 2-chloro-2-methylpropane. The equilibrium is less favorable with dilute hydrochloric acid.

The principal side reaction is E1 elimination, resulting from the loss of a proton from the incipient carbocation to give 2-methylpropene (equation 21). Under the reaction conditions, however, the elimination reaction is reversible through Markovnikov addition of HCl to the alkene to give the desired product, 2-chloro-2-methylpropane (equation 22). Therefore, elimination is not a serious complication.

$$CH_3-\overset{\overset{\displaystyle CH_2-H}{|}}{\underset{\underset{\displaystyle CH_3}{|}}{\overset{\oplus}{C}}} \;\rightarrow\; CH_3-\overset{\overset{\displaystyle CH_2}{\|}}{\underset{\underset{\displaystyle CH_3}{|}}{C}} \;+\; H^{\oplus} \qquad \textbf{(21)}$$

$$CH_3-\overset{\overset{\displaystyle CH_2}{\|}}{\underset{\underset{\displaystyle CH_3}{|}}{C}} \;+\; H^{\oplus} \;\rightarrow\; CH_3-\overset{\overset{\displaystyle CH_3}{|}}{\underset{\underset{\displaystyle CH_3}{|}}{\overset{\oplus}{C}}} \;\overset{Cl^{\ominus}}{\longrightarrow}\; CH_3-\overset{\overset{\displaystyle CH_3}{|}}{\underset{\underset{\displaystyle CH_3}{|}}{C}}-Cl \qquad \textbf{(22)}$$

Reactions involving the addition of HX to an alkene (equation 22) are well known and are discussed in Section 13.3.

Pre-lab exercises for Section 18.3, Preparation of Tertiary Chloroalkanes, are found on page PL. 65.

EXPERIMENTAL PROCEDURE

DO IT SAFELY

1 This experimental procedure requires the use of a separatory funnel *with venting* because of the potential liberation of gases. Be careful when using it.

2 It is recommended that rubber gloves be worn throughout the experiment because concentrated hydrochloric acid is being used. If any acid spills on your skin, wash it off with large volumes of water and then with dilute sodium bicarbonate solution to neutralize the residual acid.

A. 2-Chloro-2-methylpropane

Place 0.50 mol of 2-methyl-2-propanol and 1.5 mol of hydrogen chloride provided by reagent-grade *concentrated* (12 *M*) hydrochloric acid in a 250-mL separatory funnel. After mixing, swirl the contents of the separatory funnel gently *without* the stopper on the funnel. After swirling about 1 min, stopper and invert the funnel; after inverting, vent to release excess pressure by carefully opening the stopcock. Avoid shaking the funnel until this is done. Shake the funnel for several minutes, with intermittent venting. Allow the contents to stand until the mixture has separated into two distinct layers, which should be completely clear.

Separate the layers and wash the organic layer with 50 mL of *saturated* sodium chloride solution and then with 50 mL of *saturated* sodium bicarbonate solution. On initial addition of the bicarbonate solution, vigorous gas evolution will normally occur; gently swirl the *unstoppered* separatory funnel until the vigorous gas effervescence ceases. Stopper the funnel and invert it gently; vent the funnel immediately to release gas pressure. Shake the separatory funnel gently, with *frequent* venting; then shake it vigorously, again with frequent venting. Separate the organic layer (the density of a saturated sodium bicarbonate solution is about 1.1), wash it with about 40 mL of water, and then with 40 mL of *saturated* sodium chloride solution. Carefully remove the aqueous layer.

Transfer the 2-chloro-2-methylpropane to a small Erlenmeyer flask, and dry with *anhydrous* magnesium sulfate.★ Swirl occasionally for a period of 10–15 min. Filter the mixture into a small round-bottomed flask, and equip it for simple distillation. Collect the fraction boiling from 49–52°C in a receiver cooled with ice water. Determine the percent yield of the product. The yield of 2-chloro-2-methylpropane should be 75–85%.

B. 2-Chloro-2-methylbutane

2-Chloro-2-methylbutane may be prepared from 0.50 mol of 2-methyl-2-butanol, using the same general procedure as that described for 2-chloro-2-methylpropane. The 2-chloro-2-methylbutane will be obtained as the product on final distillation; material boiling from 83–85°C should be collected. Determine the percent yield of isolated product. The yield should be about 75%.

EXERCISES

1. Draw the structures of all the other alcohols that are isomeric with 2-methyl-2-propanol. Arrange these alcohols in order of *increasing* reactivity toward concentrated hydrochloric acid. Which, if any, of these alcohols would you expect to give a reasonable yield of the corresponding chloroalkane under such reaction conditions?

2. The work-up procedures for the reactions in this section call for washing the "crude" chloroalkanes with sodium bicarbonate solution.
   a. What purpose does this wash serve?
   b. This washing procedure is accompanied by vigorous gas evolution, which increases the difficulty of handling and requires considerable caution. Alternatively, one might consider using a dilute solution of sodium hydroxide instead of the sodium bicarbonate. Discuss the relative advantages and disadvantages of using these two basic solutions in the work-up. On the basis of these considerations, why were you instructed to use sodium bicarbonate, even though it is more difficult to handle?

3. The *rate* of reaction of alcohols with concentrated hydrochloric acid is increased when anhydrous zinc chloride is added to the acid. Primary alcohols are, for all practical purposes, unreactive toward pure concentrated hydrochloric acid, but they react at a reasonable rate when zinc chloride is added. Suggest an explanation for this fact. (*Hint:* Zinc chloride is a *Lewis acid.*)

4. On prolonged heating with concentrated hydrochloric acid, 2,2-dimethyl-1-propanol (neopentyl alcohol), a primary alcohol, reacts according to the equation

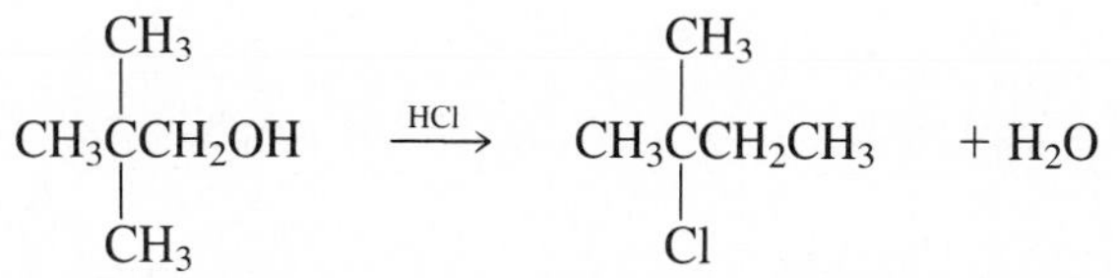

2,2-Dimethyl-1-propanol
(neopentyl alcohol)

2-Chloro-2-methylbutane
(*tert*-pentyl chloride)

   Suggest a mechanism for this reaction, and account for the fact that the observed product is a tertiary chloroalkane even though the starting alcohol is primary.

5. Why is *saturated* sodium chloride solution, rather than dilute solution, specified for the final washing of the product?

6. The procedures call for drying the chloroalkanes with anhydrous magnesium sulfate. Could solid sodium hydroxide or potassium hydroxide be used for this purpose? Explain.

SPECTRA OF STARTING MATERIALS AND PRODUCTS

The IR and PMR spectra of 2-chloro-2-methylbutane are provided in Figures 13.2 and 13.3, respectively.

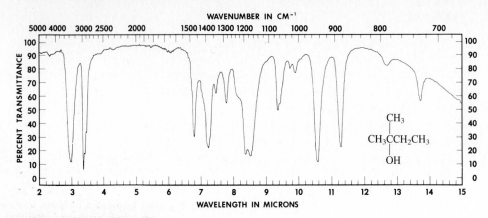

Figure 18.3 IR spectrum of 2-methyl-2-butanol.

(a) PMR spectrum.

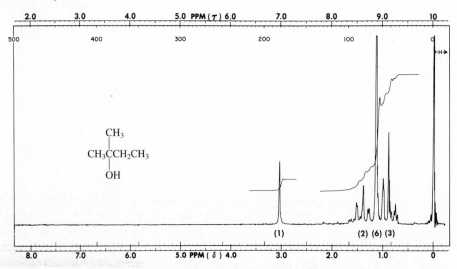

(b) CMR data. Chemical shifts: δ 8.7, 28.6, 36.5, 71.1.

Figure 18.4 NMR data for 2-methyl-2-butanol.

18.4 Chemical Kinetics: Evidence for Nucleophilic Substitution Mechanisms

(a) Dependence of Concentration upon Rate of Reaction

Continuous changes in the concentrations of reactants and products are both expected and observed during the course of chemical reactions. Thus, as a reaction proceeds, the concentration of each reactant decreases until that of the limiting reagent becomes zero, at which point the reaction stops. This assumes that the reaction is an irreversible rather than an equilibrium process. The concentration of the product concomitantly increases from zero at the beginning of the reaction to its maximum value when the reaction stops. The rate or velocity of a reaction refers to how fast these concentrations change *as a function of time*. The concept of relative reactivity is also relevant here because the more reactive of two compounds will reach complete reaction in a shorter time and will consequently exhibit a greater *rate* of change of concentration.

The study of chemical reactions that involves the investigation of the interplay of factors and variables that influence the rate of reaction is called *chemical kinetics*. It would be difficult to overstate the importance of this area of research, since it provides great insight into the nature and details of the mechanism for the reaction being studied.

Some of the most important evidence in support of the S_N1 and S_N2 mechanisms has been obtained from kinetic studies. The *rate* of nucleophilic substitution at primary and most secondary carbon atoms bearing leaving groups has been found to be proportional to the concentrations of the substrate *and* the nucleophile and is represented by the *rate law* of equation 23.

$$\text{rate of reaction} = k_2[\text{R—L}][\text{Nu:}] \qquad \textbf{(23)}$$

The proportionality constant, k_2, is known as the *rate constant* and relates the rate of reaction with the concentrations of reactants. For example, doubling the concentration either of R—L or of Nu: causes the rate of reaction to double, and doubling the concentrations of both causes a fourfold increase in the reaction rate. Equation 23 is consistent with the proposal that the transition state for the rate-determining step involves both the substrate and the nucleophile, as is the case for an S_N2 reaction (equation 3).

On the other hand, the rates of substitution at tertiary and some secondary carbon atoms susceptible to nucleophilic attack are found to be proportional to the concentration of substrate but *independent* of the concentration of the nucleophile. Equation 24 gives the rate law for this type of reaction, where k_1 is the rate constant.

$$\text{rate of reaction} = k_1[\text{R—L}] \qquad \textbf{(24)}$$

Therefore, the transition state of the rate-determining step involves only the substrate and not the nucleophile. This is interpreted as indicating an initial slow ionization of the substrate to yield a carbocation, as represented in the S_N1 mechanism (equation 2). Molecules of solvent are involved in these reactions through solvation of ions. However, the concentration of the solvent remains essentially *constant* within experimental error during the course of the reaction so that it is included as part of the rate constant, k_1.

Both k_1 and k_2 are rate constants, and the subscripts "1" and "2" are included to indicate that they are for monomolecular and bimolecular reactions, respectively. Rate constants can have different units; the unit for k_1 is $(\text{time})^{-1}$; k_2 has the units of $(\text{concentration})^{-1}(\text{time})^{-1}$.

Both S_N1 and S_N2 reactions can be studied from a kinetic standpoint, but bimolecular reactions are experimentally more difficult, and the calculations of their rate constants are more tedious. Therefore, the present studies are limited to a detailed examination of the kinetics of S_N1 reactions, as discussed below.

(b) The S_N1 Reaction and Its Rate Law

An S_N1 reaction is a first-order reaction, in that the rate of reaction is proportional to the first power of the concentration of the substrate (equation 24), so that doubling its concentration doubles the rate of reaction. A graph of *rate* versus *concentration* in such a case yields a straight line of slope k_1, where rate refers to the amount of product formed in some finite amount of time. Although the rate of reaction depends on the concentration of one or more of the reactants, the rate *constant* is independent of their concentration.

When a substance is being consumed in a first-order reaction, its concentration decreases exponentially with time. If C_0 is the *initial* concentration of the substrate at time $t = 0$, which is called t_0, and C_t is its concentration at any elapsed time t, where t is measured in any unit of time (for example, sec, min, or hr) after the reaction is started, the relationship between these variables is given in equation 25.

$$C_t = C_0 e^{-k_1 t} \tag{25}$$

The rate constant, k_1, has the units of $(\text{time})^{-1}$, as for example, sec^{-1}. Equation 25 may be rewritten as equation 26 or 27.

$$\ln(C_0/C_t) = k_1 t \tag{26}$$

$$2.303 \log(C_0/C_t) = k_1 t \tag{27}$$

If the initial concentration of the reactant, C_0, is known and if its concentration, C_t, is measured at various known time intervals, t, while the reaction is proceeding, the rate constant may be determined in the following ways.

1 The values of C_0, C_t and t measured at each point during the reaction may be substituted into equation 26 or 27, which is then solved for k_1. This produces several values of k_1, which can be averaged. The *correct* rate constant is not easily obtained by this method because the average value will be affected without bias, that is, without compensation for any measurements that may be incorrect due to experimental error.

2 A better method for calculating k_1 from experimental data is to plot $\log(C_0/C_t)$ versus time t. Assuming that the reaction is first order and the data are fairly accurate, a reasonably straight line should be obtained if it is drawn so that it lies closest to the largest number of points on the graph. This line is drawn *with bias* and purposely gives more "weight" to the majority of points that will lie close to the line; those points that appear to be in error are ignored or given less credence. The slope of this "best" straight line is the rate constant k_1 if natural logarithms (equation 26) are used or $k_1/2.303$ if logs to the base 10 (equation 27) are used. In the latter case, the slope must be multiplied by 2.303 to obtain k_1.[1]

The discussion and experiments presented below illustrate methods of studying chemical kinetics and determining the effects of structure on reactivity, as exemplified by the solvolysis of tertiary haloalkanes.

---

[1] An alternative method of obtaining the best straight line is the application of a least-squares treatment to the data. This mathematical treatment will also yield a "biased" rate constant but will do so without the subjective influence of human judgment. Consult the following reference for details: Brown, G. H; Sallee, C. M. *Quantitative Chemistry*, Prentice-Hall, Englewood Cliffs, N.J., 1963, pp. 123ff.

Solvolysis refers to a substitution reaction in which the solvent, HOS, functions as the nucleophile (equation 28). In principle, solvolysis reactions can be carried out in any nucleophilic solvent, such as water (hydrolysis), alcohols (alcoholysis), carboxylic acids (acetolysis with acetic acid and formolysis with formic acid), or liquid ammonia (ammonolysis).

$$R—X + HOS \rightarrow R—OS + H^{\oplus} + X^{\ominus} \qquad \text{(28)}$$

A major limitation in choosing a solvent is the solubility of the substrate in the solvent. It is important that the reaction mixture be homogeneous, for if it is not, surface effects at the interface of the phases will make the kinetic results difficult to interpret and probably nonreproducible as well. In the experiment described here, the solvolyses will be carried out in mixed solvents consisting of 2-propanol (isopropyl alcohol) and water.

The rate expression given in equation 26 or 27 must be used to determine the rate constants, k_1, for these solvolysis reactions, and the quantities C_0, C_t, and t must be measured. Because C_t is the concentration of substrate *remaining* at time t and *not* the concentration of substrate that has reacted, a procedure must be used that permits C_t to be measured. The procedure we use to measure C_t is based on the fact that for every molecule of haloalkane that reacts, one molecule of HX is produced (equation 28). Thus, the progress of the reaction may be followed by determining the concentration of hydrogen ion, $[H^+]$, produced as a function of time. Note, however, that $[H^+]$ represents the amount of haloalkane that has *reacted* and not the amount that remains. At any time t, $[H^+]_t = C_0 - C_t$, so that C_t is determined by subtracting $[H^+]_t$ from C_0 (equation 29).

$$C_t = C_0 - [H^+]_t \qquad \text{(29)}$$

The value of $[H^+]_t$ is determined experimentally by withdrawing an accurately measured sample of material from the reaction mixture with a volumetric pipet, this sample being called an *aliquot,* and adding it to a quantity of 98% 2-propanol sufficient to "quench" the reaction so that it no longer proceeds at a measurable rate. The elapsed time, t, is calculated by subtracting the time that the aliquot was removed from the time, t_0, that the reaction was started. The quenched sample is titrated with standardized dilute sodium hydroxide solution whose concentration is *ca.* 0.04 M. This titration yields $[H^+]$, and if C_0 is known, C_t can be calculated from equation 29.

The value of C_0 can be determined in two different ways. (1) The haloalkane being studied can be weighed accurately and then its weight converted into moles. The volume of 2-propanol–water solvent can be measured accurately, and the molarity of the haloalkane, which is its initial concentration, C_0, is obtained by dividing the moles of haloalkane by the volume of solvent in L. (2) A more reliable and easier procedure for determining C_0 is to allow the reaction to go to completion, at which time all the haloalkane will have reacted. An aliquot of the reaction mixture is then withdrawn and titrated with the standardized sodium hydroxide solution to yield the "infinity point." The concentration of hydrogen ion, $[H^+]$, obtained at this time must be equal to the initial concentration of the haloalkane, C_0; that is, $[H^+]_\infty = C_0$. Substitution of this value of C_0 into equation 29 yields equation 30.

$$C_t = [H^+]_\infty - [H^+]_t \qquad \text{(30)}$$

As long as the *same* sodium hydroxide solution is used for *all* titrations, including the infinity titration, it is *not* necessary to use standardized base in the titrations. Furthermore, it will not be necessary actually to calculate the $[H^+]$ concentrations because they are directly proportional to the *volume* of sodium hydroxide solution required to titrate the acid in the various aliquots. With this consideration and using equation 30, equation 27 becomes equation 31.

$$2.303 \log \frac{(\text{mL of NaOH})_\infty}{(\text{mL of NaOH})_\infty - (\text{mL of NaOH})_t} = k_1 t \qquad \textbf{(31)}$$

It must be emphasized that this equation can be used only if the volumes of all aliquots, including that of the infinity titration, are the same and that the *same* solution of sodium hydroxide must be used for all titrations. Then the rate constant can be determined using *only* the volumes of NaOH required for the titrations at various times *t*, including the infinity point titration.

(c) Factors Influencing the Rate of S$_N$1 Reactions

Some of the factors that influence the rate of S$_N$1 reactions are described in the following paragraphs, and the experimental procedures provided allow some of them to be investigated.

1. Solvent Composition The nature of the solvent greatly influences the rate of S$_N$1 reactions. The solvent effect may be considered as a dependence of the rate on the *polarity* of the solvent, which influences the rate of the initial ionization (equation 2a), the slow step of the S$_N$1 reaction. The more polar the solvent, the more rapid the ionization. This is because of the greater ability of the more polar solvent to stabilize charged species, both positive and negative, through solvation.

The experiments that are provided make use of mixed solvents consisting of 2-propanol and water. The effects of solvent composition on the rate can be studied using mixtures of 60:40, 55:45, and 50:50 (volume:volume) 2-propanol:water.

2. Effect of Concentration Equation 24 indicates that the *rate* of solvolysis is proportional to the concentration of haloalkane, and this assumes that the *rate constant* is independent of haloalkane concentration.

3. Effect of Temperature Temperature changes also influence the rate of a chemical reaction. A *rough* rule of thumb is that the rate doubles for each 10°C rise in temperature. For example, increasing the temperature from 25°C to 45°C, a 20°C rise, increases the rate of reaction by a factor of *about* 4. The effect of temperature on solvolysis reactions can be conveniently examined in the laboratory by determining rates at room temperature and at 0°C.

4. Effect of Alkyl Group Structure Tertiary haloalkanes (alkyl halides) may be used in solvolysis reactions with complete confidence that the substitutions are proceeding by the

S_N1 mechanism. Two suitable compounds that can be used to illustrate the effect of alkyl group structure are 2-chloro-2-methylpropane (*tert*-butyl chloride) and 2-chloro-2-methylbutane (*tert*-pentyl chloride). Determination of the rate of solvolysis of these compounds will provide data that illustrate the reactivity difference between two tertiary alkyl groups, the *tert*-butyl group, $(CH_3)_3C$—, and the *tert*-pentyl group, $(CH_3)_2(C_2H_5)C$—.

5. Effect of Leaving Group Although no experiments are included here to show the effect of different leaving groups on the rate of a solvolysis reaction, such effects should be observed because the bond to the leaving group is being broken in the rate-determining step of the reaction. For example, bromoalkanes are more reactive than the corresponding chloroalkanes, since bromide is a better leaving group than chloride. These reactivity differences are reflected in the values of the rate constants.

The composition of the final product mixture in an S_N1 reaction has no influence on the rate of the reaction or the value of the rate constant. This follows since the rate-determining step in an S_N1 reaction is the *first* step, namely carbocation formation, and the product(s) are rapidly formed from reaction of various nucleophiles with the carbocation. For example, three products are possible from the solvolysis of 2-chloro-2-methylpropane in aqueous 2-propanol (equations 32–35), and each one is formed from the incipient carbocation. All three reactions share the same rate-determining step (equation 32).

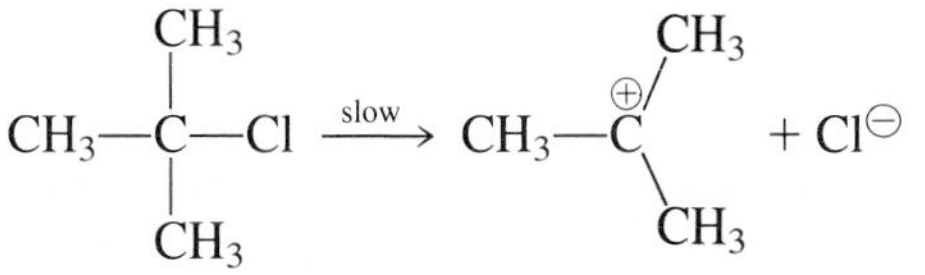

$$\underset{\underset{CH_3}{|}}{\overset{\overset{CH_3}{|}}{CH_3-C-Cl}} \xrightarrow{\text{slow}} \underset{CH_3}{\overset{CH_3}{CH_3-\overset{\oplus}{C}}} + Cl^{\ominus} \qquad (32)$$

$$\underset{CH_3}{\overset{CH_3}{CH_3-\overset{\oplus}{C}}} \xrightarrow{\text{fast}} \underset{CH_3}{\overset{CH_2}{CH_3-C}} + H^{\oplus} \qquad (33)$$

$$\underset{CH_3}{\overset{CH_3}{CH_3-\overset{\oplus}{C}}} + H-OH \xrightarrow{\text{fast}} \underset{\underset{CH_3}{|}}{\overset{\overset{CH_3}{|}}{CH_3-C-OH}} + H^{\oplus} \qquad (34)$$

$$\underset{CH_3}{\overset{CH_3}{CH_3-\overset{\oplus}{C}}} + (CH_3)_2CHOH \xrightarrow{\text{fast}} \underset{\underset{CH_3}{|}}{\overset{\overset{CH_3}{|}}{CH_3-C-OCH(CH_3)_2}} + H^{\oplus} \qquad (35)$$

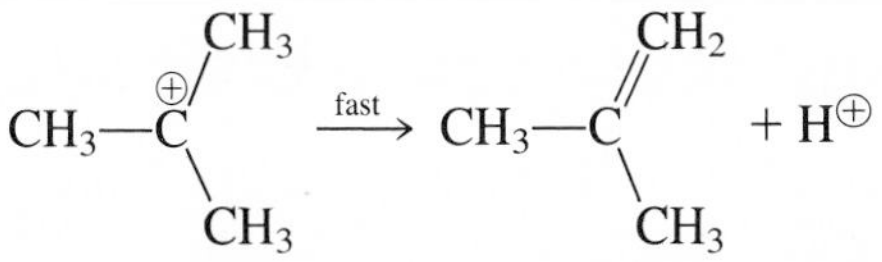

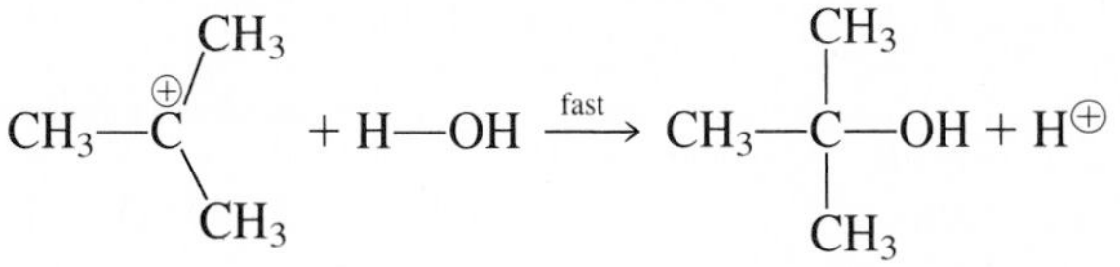

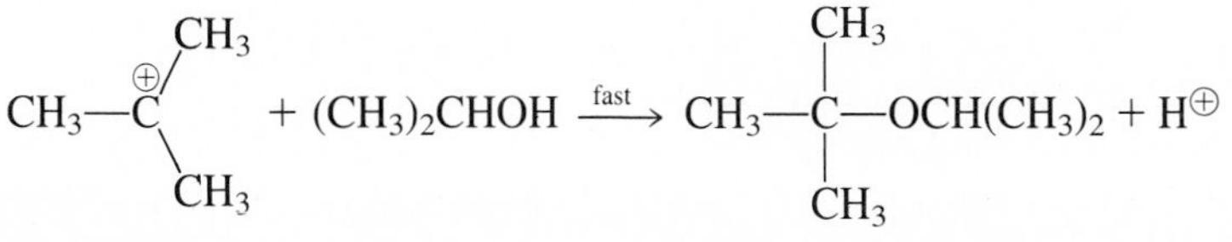

Pre-lab exercises for Section 18.4, Chemical Kinetics, are found on page PL. 67.

EXPERIMENTAL PROCEDURE

Note: The experimental procedure, down to the "Treatment of Data" section, should be read *completely* before you start the experiment.

A. General Comments

In experiments of the type given here, you should strive for consistency and accuracy. Prepare yourself ahead of time, and establish a systematic approach to the collecting and recording of necessary data. Before coming to class, prepare in your notebook a table for recording the following: (1) which solvent you are using and the volume used in preparing the reaction mixture, (2) the weight or volume of haloalkane used in preparing the reaction mixture, (3) the time, t_0, at which the kinetic run is initiated, (4) the temperature of the reaction mixture, (5) the results of a "blank" titrimetric determination, (6) a series of times at which aliquots are withdrawn, and (7) the initial and final buret readings observed in the titration of each aliquot.

This experiment involves a series of quantitative measurements carried out in a relatively short time. You should be prepared to work *rapidly* and *carefully* in order to maintain a high standard of accuracy. Buret readings should be made to the nearest 0.02 mL if possible, although precision within 0.05 mL will normally be satisfactory. Time measurements should be made *at least* to the nearest minute.

B. General Kinetic Procedure

Throughout this experiment, use the same pipet and buret, as well as the same sodium hydroxide solution.

Using a graduated cylinder, accurately measure into a 250-mL Erlenmeyer flask equipped with a well-fitting rubber stopper 100 mL of the solvent that has been assigned to you. Using a thermometer, measure and record the temperature of this solution. In a second flask obtain 80 mL of 98% 2-propanol to use for quenching purposes.

Obtain 125–150 mL of approximately 0.04 M sodium hydroxide solution in a third flask. Stopper the flask with a well-fitting rubber stopper. Set up a 50-mL buret, rinse it with a small amount of the sodium hydroxide solution, fill it, see that all air bubbles are out of the tip, and cover the top of the buret with a test tube or small beaker to minimize absorption of carbon dioxide from the air.

Put about 2 mL of phenolphthalein indicator solution in a test tube and have it available, with a dropper, for use in each titration. Connect a short length of rubber tubing to the nearest aspirator or vacuum line for use in drawing air through the pipet for a minute or two after each sampling in order to dry it before taking the next sample. Have available a watch that may be read at least to the nearest minute.

To initiate a kinetic run, add the haloalkane to the solvent mixture. A sample size of approximately 1 g is satisfactory, either weighed out or measured with a 1-mL pipet. Swirl the mixture gently to obtain homogeneity. Note and record as t_0 the time of addition. Keep the flask tightly stoppered to avoid evaporation and as a consequence a change of concentration.

While waiting to make the first measurement, determine a "blank" correction for the solvent. Using a graduated cylinder, measure into a 125-mL Erlenmeyer flask a separate 10-mL portion of the 2-propanol–water mixture being used. (*Note:* Do *not* use the mixture containing the haloalkane.) Next add 10-mL of 98% 2-propanol and 4–5 drops of phenolphthalein to the blank, and titrate with base to a faint pink color that persists for 30 sec. In all titrations use a white background (paper or a towel) below the titration flask, and accentuate the lower edge of the meniscus in the buret by holding dark paper or some other dark object just below it to make it easier to read. The blank correction will probably be no more than 0.05–0.15 mL.

At regular intervals take a 10-mL sample from the reaction mixture with a 10-mL pipet, and add it to 10-mL (measured with a graduated cylinder) of 98% 2-propanol contained in a 125-mL flask. Be sure to note the time of addition of the aliquot, probably best taken as the time at which one-half of it has been added to the alcohol used to quench the reaction. Titrate with base to the phenolphthalein end point, as in the blank determination.

The suggested *approximate* times of taking aliquots using various solvents under various conditions are listed below:

1 50% 2-propanol–water and 2-chloro-2-methylpropane: 10, 20, 35, 50, 75, and 100 min

2 55% 2-propanol–water and 2-chloro-2-methylpropane: 15, 30, 50, 75, 100, and 135 min

3 60% 2-propanol–water and 2-chloro-2-methylpropane: 20, 40, 70, 100, 130, and 170 min

4 50% 2-propanol–water and 2-chloro-2-methylbutane: 10, 20, 30, 40, 50, and 60 min

5 55% 2-propanol–water and 2-chloro-2-methylbutane: 15, 30, 45, 60, 80, 110, and 140 min

6 60% 2-propanol–water and 2-chloro-2-methylbutane: 20, 40, 60, 80, 100, and 120 min

The *fastest* of the above reactions will require about 4 hr to reach 99.5% completion, and the slowest requires over 12 hr. Therefore it is most convenient to wait until the next laboratory period to perform the infinity titration necessary to obtain C_0. Stopper the reaction flask *tightly* to avoid evaporation, and store the flask in your desk. Be sure also to save at least 30 mL of the sodium hydroxide solution in a *tightly* stoppered flask so that it will be available for the infinity titration in the following laboratory period.

C. Treatment of Data

1 Using the buret readings, determine by difference the number of mL of sodium hydroxide solution used in each titration. Apply the blank correction to all values by subtracting it from each volume. Use the corrected volumes in your calculations. Using the recorded times for each aliquot, determine in each case the elapsed time from t_0. Apply equation 31 by calculating the log term, multiplying that value by 2.303, and *plotting* the result (vertically) versus the elapsed time t (in hours) for each kinetic point. Draw the best straight line through the points. Determine the slope of the line; this slope is the rate constant k_1. It should be about 0.2 hr^{-1} for 60% 2-propanol-water and 2-chloro-2-methylpropane and about 0.8 hr^{-1} for 50% 2-propanol-water and 2-chloro-2-methylpropane.

2 Using the same data, calculate the value of k_1 separately for each kinetic point using equation 31. Compare the *average* of these values with the rate constant obtained graphically. Also compare this average with each of the values that were averaged.

Which procedure, graphical or averaging, allows you most easily to spot a point that is likely in error?

3 The half-life, $t_{1/2}$, is the time necessary for one-half of the original alkyl halide to react and is given by equation 36.

$$t_{1/2} = \frac{0.69}{k_1} \tag{36}$$

Calculate the half-life of your reaction using the value of k_1 obtained from the graph. Go back and examine your experimental data. About one-half of the total volume of NaOH used in the infinity titration should have been consumed in a titration done at this time. If not, an error has been made in the calculations, and they should be rechecked.

4 Consider the magnitudes and types of errors involved in the various measurements made in this experiment. Using the unavoidable errors in measurement that would have been expected in volume, time, and titration measurements, calculate the *maximum* error in k_1 that can be expected, for example, $k_1 = 0.80 \text{ hr}^{-1} \pm 0.05 \text{ hr}^{-1}$.

EXERCISES

1. Give as many advantages as possible for using an infinity titration and equation 31 instead of calculating haloalkane concentrations and using equation 27. Be sure to consider what the alternatives are for obtaining C_0.

2. Show how equation 31 can be derived from equation 27. (*Hint:* Set up the concentrations C_0 and C_t in terms of moles/liter for the volume of NaOH used in each titration, and recognize the fact that $[OH^-] = [H^+]$. Substitute these values into equation 27, and cancel out constants that appear in the new equation.)

3. Equation 36 defines *half-life*, which is the time required for a reaction to reach 50% completion, and it applies to any first-order reaction. Show how equation 36 can be derived from equation 27. (*Hint:* When the reaction is 50% complete, $C_t = \frac{1}{2} C_0$.)

4. Suppose the flask had not been stoppered until the following laboratory period when the infinity titration was done. Would the calculated rate constant have been larger or smaller than the "correct" value if some evaporation had occurred?

5. Why does the titration end-point color fade after 30–60 sec?

6. List the possible errors involved in the determination of rate constants by the procedure you used. State the relative importance of each.

chapter **19**

oxidation reactions of alcohols, carbonyl compounds, and arenes

19.1 Introduction

One of the most important types of reactions of organic compounds is their oxidation, which results not only in simple functional group transformations, such as conversion of alkenes to epoxides (equation 1), but also in degradation of molecules by cleavage of carbon-carbon bonds (equation 2). Oxidation processes are vital to the maintenance of life. For example, metabolic energy is derived from the overall oxidation of carbohydrates, fats, and proteins to carbon dioxide and water, among other products (equation 2, for instance), and detoxification of potential poisons commonly involves their oxidation in the liver to more benign substances. Similarly, the single most important source of energy for industry and society in general is the oxidation (burning) of fuels such as gas, coal, and petroleum.

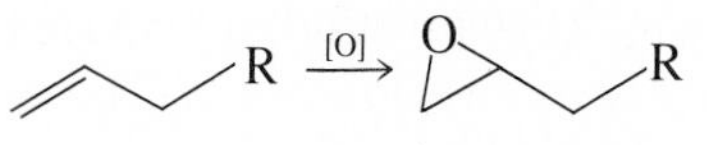

$$\text{An epoxide} \qquad (1)$$

$$\text{HOCH}_2(\text{CHOH})_4\text{CHO} \xrightarrow{[O]} \text{CO}_2 + \text{H}_2\text{O} + \text{energy} \qquad (2)$$
A carbohydrate

Oxidation reactions also are exceedingly important processes in the organic laboratory, for they open the way for introducing various functional groups onto hydrocarbon chains. Thus, primary and secondary alcohols can be converted to aldehydes and ketones, and these may in turn be transformed into carboxylic acids, as shown (equations 3 and 4, respectively).

$$RCH_3 \xrightarrow{[O]} RCH_2OH \rightarrow \underset{\text{Aldehyde}}{R\overset{\displaystyle O}{\overset{\|}{C}}H} \rightarrow \underset{\text{Carboxylic acid}}{R\overset{\displaystyle O}{\overset{\|}{C}}{-}OH} \tag{3}$$

1° Alcohol

$$R_2CH_2 \xrightarrow{[O]} \underset{\text{2° Alcohol}}{R_2CHOH} \rightarrow \underset{\text{Ketone}}{R_2C{=}O} \rightarrow \underset{\text{Carboxylic acids}}{RCOOH + R'COOH} \tag{4}$$

At least in principle, the alcohols could be derived by partial oxidation of hydrocarbons, as shown in equations 3 and 4, although this is generally done in the laboratory only on unsaturated hydrocarbons (equations 5 and 6), since it is very difficult to find reaction conditions that allow controlled oxidation of the fully saturated compounds.

$$\diagdown\!\!\diagup\!\!\diagdown\!\!R \xrightarrow[(O_2)]{[O]} \diagdown\!\!\diagup\underset{\text{OH}}{\diagdown}R \tag{5}$$

$$\diagdown\!\!\diagup\!\!\diagdown\!\!R \xrightarrow[(OsO_4)]{[O]} HO\diagdown\!\!\diagup\underset{\text{OH}}{\diagdown}\diagup R \tag{6}$$

One process for the preparation of methanol by partial oxidation of a mixture of propane and butane (equation 7) is of *industrial* importance, however, but it also involves destruction of carbon-carbon bonds and produces a number of other oxidation products. Consequently, a substantial research effort is under way at present to increase the number and selectivity of industrially useful processes based on controlled oxidation of saturated hydrocarbons.

$$CH_3CH_2CH_3 + CH_3CH_2CH_2CH_3 \xrightarrow{[O]} CH_3OH + CH_2O + CH_3CHO + \text{other products} \tag{7}$$

An oxidation reaction need *not* result in the introduction of oxygen into an organic molecule. In the most general sense and in analogy to concepts of oxidation as applied to inorganic compounds, oxidation of an organic substance simply requires an increase in the oxidation state of carbon. Several systems exist for defining the oxidation states of carbon atoms, but the one provided here involves the following steps:

1. Select the carbon atom whose oxidation state is to be defined.
2. Assign oxidation numbers to the atoms attached to this carbon atom using the following values:
    a. +1 for hydrogen;
    b. −1 for halogen, nitrogen, oxygen and sulfur;
    c. 0 for carbon.
3. Add the oxidation numbers of these atoms. If a heteroatom (2b) is bound to carbon by more than one bond, multiply its oxidation number by the number of bonds linking it to the carbon atom involved.
4. The sum of the number from step 3 and the oxidation number of the carbon atom under consideration must equal the charge on the carbon atom, which is zero unless a positive or negative charge is present on it.

Application of this technique shows that the dehydrogenation of an alkane to an alkene (equation 8), a key industrial process, and of an alkene to an alkyne (equation 9) are oxidation reactions even though oxygen is not incorporated in these reactions. Determination of the changes in the oxidation number of the various carbon atoms during the conversion of acetic acid to carbon dioxide (equation 10) illustrates how the method is used when heteroatoms are bound to carbon by more than one bond.

$$RCH_2CH_2R \rightarrow RCH{=}CHR$$
$$-2 -2 -1 -1$$

$$RCH{=}CHR \rightarrow R{-}C{\equiv}C{-}R$$
$$-1 -1 0 0$$

Net change: loss of $2e^{\ominus}$ (oxidation of carbon)

(8)

(9)

$$\underset{-3\ +3}{CH_3\overset{\displaystyle O}{\overset{\|}{C}}{-}OH} \rightarrow 2\ \underset{+4}{O{=}C{=}O}$$

(10)

The oxidation reactions described in this chapter all involve an increase in the oxygen content of products relative to reactants. The source of oxygen in our experiments is either water or an oxidized form of a metal rather than molecular oxygen. A number of different metals having a variety of oxidation states are used most commonly as laboratory oxidizing agents.

19.2 Preparation of Aldehydes and Ketones by Oxidation of Alcohols

Aldehydes and ketones play a central role in organic synthesis, and efficient methods for their preparation are very important. These substances can be synthesized from alkynes by acid-catalyzed hydration (equation 11; see also Chapter 14) or hydroboration followed by oxidation (equation 12) and from carboxylic acids or their derivatives by reaction with organometallics or a variety of reducing agents (equations 13–16).

$$R{-}C{\equiv}C{-}R(H) \xrightarrow[\substack{H_2SO_4 \\ H_2O}]{HgSO_4} R{-}\overset{\displaystyle O}{\overset{\|}{C}}{-}CH_2{-}R(H)$$

(11)

$$R{-}C{\equiv}C{-}R(H) \xrightarrow[\substack{(2)\ H_2O_2/HO^{\ominus}}]{(1)\ B_2H_6} R{-}CH_2{-}\overset{\displaystyle O}{\overset{\|}{C}}{-}R(H)$$

(12)

$$R{-}\overset{\displaystyle O}{\overset{\|}{C}}{-}OH \xrightarrow[\substack{(2)\ H_3O^{\oplus}}]{(1)\ R'Li} R{-}\overset{\displaystyle O}{\overset{\|}{C}}{-}R'$$

(13)

$$R{-}\overset{\displaystyle O}{\overset{\|}{C}}{-}Cl + R_2'Cd \quad or \quad R_2'CuLi \rightarrow R{-}\overset{\displaystyle O}{\overset{\|}{C}}{-}R'$$

(14)

$$R-C\equiv N \xrightarrow[\text{(2) } H_3O^{\oplus}]{\text{(1) } R'MgX} R-\overset{\overset{\displaystyle O}{\parallel}}{C}-R' \qquad \textbf{(15)}$$

$$R-C\equiv N \xrightarrow[\text{(2) } H_3O^{\oplus}]{\text{(1) } LiAlH_4/-80°C} R-\overset{\overset{\displaystyle O}{\parallel}}{C}-H \qquad \textbf{(16)}$$

One of the most common synthetic methods for the preparation of aldehydes and ketones is the oxidation of primary and secondary alcohols with chromic acid, H_2CrO_4, or potassium permanganate (equation 17), a reaction involving a two-electron change in the oxidation number of the functionalized carbon atom. A description of the use of the former oxidizing agent for conversion of alcohols to aldehydes and ketones follows, and directions for the use of both of these reagents are included in the Experimental Procedures.

$$R-\overset{\overset{\displaystyle OH}{|}}{CH}-R'(H) \xrightarrow[\text{or } KMnO_4]{H_2CrO_4} R-\overset{\overset{\displaystyle O}{\parallel}}{C}-R'(H) \qquad \textbf{(17)}$$

Oxidation no.: 0(2°)   +2 (ketone)
            −1(1°)   +1 (aldehyde)

Chromic acid is not stable for long periods and is therefore produced when required by the reaction of sodium or potassium dichromate with an excess of an acid, such as sulfuric or acetic acid (equation 18), or by dissolution of chromic anhydride in water (equation 19). In the latter preparation either sulfuric or acetic acid is also added because the rate of oxidation of alcohols by chromic acid is much faster in acidic solutions. For the preparation or oxidation of substances that would decompose under strongly acidic conditions, either chromic anhydride complexed with pyridine or basic solutions of potassium permanganate can be used as the oxidizing agent.

$$Na_2Cr_2O_7 + 2\ H_2SO_4 \rightarrow \left[H_2Cr_2O_7\right] \xrightarrow{H_2O} 2\ H_2CrO_4 + 2\ NaHSO_4 \qquad \textbf{(18)}$$

$$CrO_3 + H_2O \rightarrow H_2CrO_4 \qquad \textbf{(19)}$$

The general mechanism of oxidation is well understood. Alcohols form chromate esters (**1**) in the presence of chromic acid, just as they do when allowed to react with carboxylic acids (see Section 22.2c). The ester **1** is relatively unstable and decomposes by an E2 elimination (see Section 13.1) to produce the carbonyl compound (equation 20); the elimination is normally the rate-determining step in the reaction.

$$\underset{R}{\overset{R'(H)}{\diagdown}}CH-OH + HO\overset{\overset{\displaystyle O}{\parallel}}{\underset{\underset{\displaystyle O}{\parallel}}{Cr}}OH \xrightarrow{-H_2O} \left[\underset{R}{\overset{R'(H)}{\diagdown}}CH-O\overset{\overset{\displaystyle O}{\parallel}}{\underset{\underset{\displaystyle O}{\parallel}}{Cr}}OH\right] \rightarrow \underset{R}{\overset{R'(H)}{\diagdown}}C=O + H_2CrO_3 \qquad \textbf{(20)}$$

1
Chromate ester

The stoichiometry of the oxidation of alcohols by chromic acid can be determined by evaluating the changes in oxidation numbers of the oxidizing agent and substrate alcohol. During the course of the oxidation of the alcohol, as represented in equation 20, chromium undergoes a two-electron *reduction* from the $+6$ to the *unstable* $+4$ valence state, and the carbon atom experiences a two-electron *oxidation*. If this corresponded to the *overall* course of the reaction, a one-to-one ratio of chromic acid to alcohol would be required, as is shown in equation 20. However, because $+4$ is an unstable oxidation state of this chromium, a disproportionation reaction occurs between chromium(IV) and chromium(VI) to produce chromium(V), a new oxidizing agent written as HCr^VO_3 (equation 21). The two moles of chromium(V) thus formed will oxidize two more moles of alcohol to the carbonyl compound, and chromium(III), a *stable* oxidation state and written here as H_3CrO_3, will result (equation 22). In the presence of sulfuric acid, the latter species is converted to Cr_2SO_4 (equation 23). Summation of equations 20–23 results in equation 24, which shows that in the chromic acid oxidation of alcohols only *two* equivalents of chromic acid are required to oxidize *three* equivalents of an alcohol to the corresponding carbonyl compound.

$$H_2Cr^{IV}O_3 + H_2Cr^{VI}O_4 \rightarrow 2\ HCr^VO_3 + H_2O \tag{21}$$

$$2\ HCr^VO_3 + 2\ R\!\!-\!\!\overset{\overset{\displaystyle R'(H)}{|}}{C}HOH + \rightarrow 2\ \overset{\overset{\displaystyle R'(H)}{\diagup}}{\underset{\underset{\displaystyle R}{\diagup}}{C}}{=}O + 2\ H_3CrO_3 \tag{22}$$

$$2\ H_3CrO_3 + 3\ H_2SO_4 \rightarrow Cr_2(SO_4)_3 + 6\ H_2O \tag{23}$$

$$3\ R\overset{\overset{\displaystyle OH}{|}}{C}HR'(H) + 2\ H_2CrO_4 + 3\ H_2SO_4 \rightarrow 3\ R\overset{\overset{\displaystyle O}{||}}{C}R'(H) + Cr_2(SO_4)_3 + 8\ H_2O \tag{24}$$

A less-detailed approach to determination of the stoichiometry of this oxidation is also useful. As noted earlier, the carbon atom of the alcohol undergoes a two-electron oxidation. Chromium, on the other hand, has a *net* change involving a *three*-electron reduction ($+6$ to $+3$). To balance the total electronic change, then, requires three equivalents of alcohol for each two equivalents of chromium(VI) (equation 25), the same result as obtained above.

$$\underset{\substack{\text{Oxidation no. 0 (2°)} \\ -1\ (1°)}}{R\!\!-\!\!\overset{\overset{\displaystyle OH}{|}}{C}H\!\!-\!\!R'(H)} + \underset{+6}{Cr(VI)} \rightarrow \underset{\substack{+2\ \text{(ketone)} \\ +1\ \text{(aldehyde)}}}{R\!\!-\!\!\overset{\overset{\displaystyle O}{||}}{C}\!\!-\!\!R'} + \underset{+3}{Cr(III)} \tag{25}$$

As noted earlier, chromic acid is prepared from other sources, and it is necessary to determine how much of these substances must be used. Examination of equation 18 shows that *two* equivalents of chromic acid are produced from *one* equivalent of dichromate. Thus, if either sodium or potassium dichromate is used to generate chromic acid, only *one* equivalent of dichromate is required to oxidize *three* equivalents of alcohol. In the case of chromic anhydride, however, only *one* equivalent of chromic acid is provided per equivalent of the anhydride (equation 19), and therefore *two* equivalents of chromic anhydride per *three* equivalents of alcohol are necessary to achieve the stoichiometry shown in equations 24 and 25.

Some major side reactions complicate the chromic acid oxidation of a primary alcohol to an aldehyde, the most important of which is the further conversion of the aldehyde to a carboxylic acid (equation 26). This undesired overoxidation can be minimized by adding the chromic acid *to* the primary alcohol, so that an excess of the oxidizing agent is *not* present in the reaction mixture, and by distilling the aldehyde from the reaction mixture as it is formed. The need to distil the aldehyde away from the oxidant means that the aldehyde must have a boiling point of less than about 150°C if it is to be prepared in high yield using chromic acid as the oxidant.

$$3 \ \underset{\substack{\| \\ O}}{R-C-H} + 2 \ H_2CrO_4 + 3 \ H_2SO_4 \rightarrow 3 \ \underset{\substack{\| \\ O}}{R-C-OH} + Cr_2(SO_4)_3 + 5 \ H_2O \quad \textbf{(26)}$$

Even when the aldehyde is volatile, a poor yield of product is sometimes obtained. This is due to the facile conversion of the aldehyde to a hemiacetal, **2,** which is subsequently oxidized to the ester of a carboxylic acid (equation 27). Occasionally this "side" reaction can be turned into a useful synthesis of the ester, as is the case in the preparation of 1-butyl butanoate by the chromic acid oxidation of 1-butanol [equation 27; $R = CH_3(CH_2)_2$].

$$RCH_2OH \xrightarrow{H_2CrO_4} \underset{\substack{\| \\ O}}{R-C-H} \xrightarrow[RCH_2OH]{H^\oplus} \underset{\substack{ \\ OCH_2R}}{R-\overset{OH}{\underset{|}{C}}-H} \xrightarrow{H_2CrO_4} \underset{\substack{\| \\ O}}{R-C-OCH_2R} \quad \textbf{(27)}$$

$$\underset{\text{A hemiacetal}}{\mathbf{2}}$$

Ketones are much more stable toward oxidizing agents in mildly acidic media than are aldehydes, and the side reactions mentioned for the oxidation of primary alcohols do not occur to a significant extent in the oxidation of secondary alcohols. Under alkaline or *strongly* acidic conditions, *enolizable* ketones will undergo oxidation with cleavage of a carbon-carbon bond to give two carbonyl fragments. For example, cyclohexanone, which can be obtained in good yield by the chromic acid oxidation of cyclohexanol, can be converted to hexanedioic acid on treatment with potassium permanganate under mild alkaline conditions (equation 28). The reaction undoubtedly requires initial conversion of the ketone to the enol, **3,** the double bond of which is then hydroxylated and cleaved by the oxidant (equation 29). The aldehyde-carboxylate that results is further oxidized to the dicarboxylate, which gives the diacid upon acidification (equation 30).

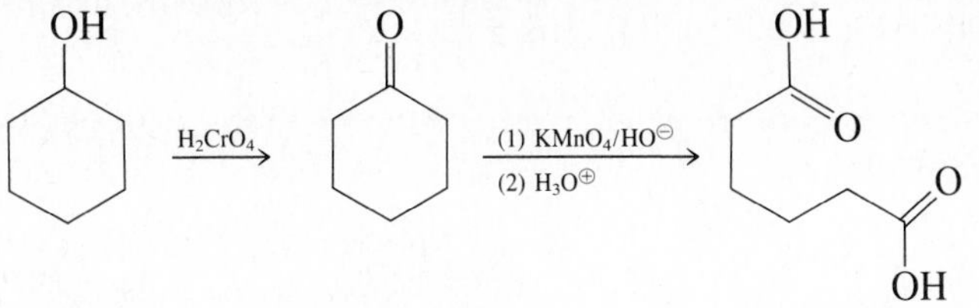

Cyclohexanol Cyclohexanone Hexanedioic acid

(28)

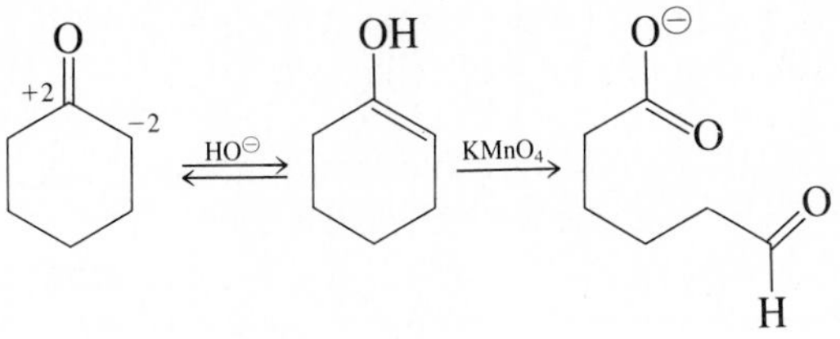

3 **4**

(29)

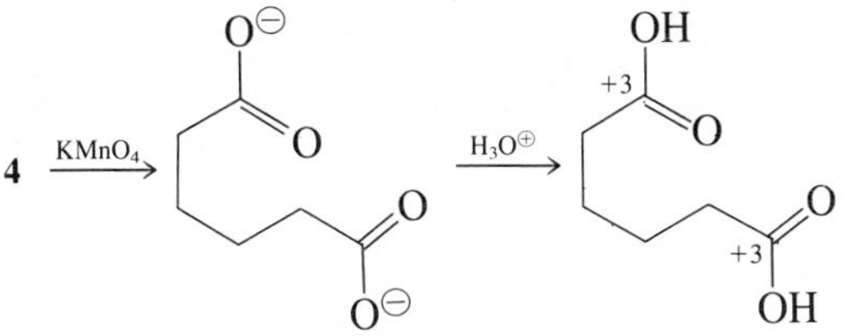

(30)

The stoichiometry of the oxidation of cyclohexanone can be determined by consideration of the changes in oxidation states that occur. The two carbon atoms of interest undergo a *net* six-electron oxidation, as indicated on the structures of cyclohexanone (equation 29) and hexanedioic acid (equation 30). Reference to the half reaction for the oxidant (equation 31) shows that permanganate is reduced to manganese dioxide, a three-electron reduction. Thus, *two* moles of the oxidant are required per mole of ketone to effect the reaction of equation 28.

$$\mathrm{Mn^{VII}O_4^{\ominus}} + \tfrac{1}{2}\,\mathrm{H_2O} \rightarrow \mathrm{Mn^{IV}O_2} + \mathrm{HO^{\ominus}} + \tfrac{3}{2}\,:\ddot{\underset{\cdot\cdot}{\mathrm{O}}}\cdot \qquad \textbf{(31)}$$

Cyclohexanone is a *symmetrical* ketone and can give only the single enol, **3.** If a ketone is *not* symmetrical, it can produce two different enols (equation 32), each of which will be oxidized by permanganate to different products. The formation of complex mixtures of products when unsymmetrical ketones are oxidized is a complication that detracts from the synthetic utility of this reaction.

$$\mathrm{RCH_2{-}\overset{\overset{\displaystyle O}{\|}}{C}{-}CH_2R'} \xrightleftharpoons{\mathrm{HO^{\ominus}}} \mathrm{RCH{=}\overset{\overset{\displaystyle O{-}H}{|}}{C}{-}CH_2R'} + \mathrm{R{-}CH_2{-}\overset{\overset{\displaystyle O{-}H}{|}}{C}{=}CHR'} \qquad \textbf{(32)}$$

EXPERIMENTAL PROCEDURE

DO IT SAFELY

1 When diluting sulfuric acid with water, **always** add the acid *to* the water and swirl the container to ensure continuous mixing. The dissolution of sulfuric acid in water generates heat, and when the acid is added to water, the heat is dispersed through warming of the water. Swirling prevents the layering of the denser sulfuric acid at the bottom of the flask and the attendant possibility that **hot** acid will be splattered by the steam generated when the two layers are suddenly mixed later by agitation.

2 When preparing and handling solutions of chromic acid or potassium permanganate, it is advisable to wear rubber gloves to avoid contact of these reagents with your skin. Either solution will cause unsightly stains on your hands for several days, and the chromic acid–sulfuric acid solution may cause severe chemical burns. If these oxidants come in contact with the skin, wash the affected area thoroughly with soap and warm water and, in the case of chromic acid, rinse the area with 5% sodium bicarbonate solution.

3 Residues of manganese dioxide can be difficult to remove from glassware. Rinse stained glassware with 10% sodium bisulfite solution, but if this fails, add 6 *M* HCl to the flask and heat it in a ventilation hood until the stain has disappeared.

4 Be certain that all joints in the apparatus are lubricated and tightly mated before the alcohol is heated to boiling.

Pre-lab exercises for Section 19.2, Part A, Preparation of 2-Methylpropanal, can be found on page PL. 69.

A. Preparation of 2-Methylpropanal

Prepare a solution of chromic acid by dissolving 0.10 mol of potassium dichromate in 175 mL of water and then *slowly* adding, with swirling, 22 mL of *concentrated* sulfuric acid. Allow this solution to cool to room temperature, using an ice-water bath if desired to hasten the process. If a stock solution of chromic acid has been prepared for your use, obtain 175 mL of the solution. Fit a 250-mL round-bottomed flask with a Claisen adapter, and equip the adapter with a dropping funnel directly above the flask and an unpacked Hempel column on the parallel side arm. To the top of the Hempel column attach a stillhead bearing a thermometer and a water-cooled condenser. Position a 50- or 100-mL graduated cylinder to collect the distillate that will be produced. Place 0.30 mol of 2-methyl-1-propanol and 25 mL of water in the flask along with two or three boiling chips, and heat the mixture until *gentle* boiling begins. Discontinue heating and immediately begin to add the red-orange solution of chromic acid from the dropping funnel to the hot mixture of alcohol and water at a rate such that all of the acid will have been added in about 15 min. This rate will cause the reaction mixture to boil vigorously, and a mixture of alcohol, aldehyde, and water will steam-distil (see Section 2.10), giving a head temperature of 80–85°C. This temperature should be maintained by appropriate adjustment of the rate of addition of acid. After all the chromic acid has been added and the mixture has stopped distilling on its own, heat and distil the dark green [chromium(III)] reaction mixture for an additional 15 min.★ Note the amount of water contained in the two-phase distillate, transfer the mixture to a separatory funnel, and add 0.5 g of sodium carbonate. Shake the funnel thoroughly, with venting to relieve any pressure, and then saturate the aqueous layer by adding about 0.2 g of sodium chloride for each milliliter of water to salt out any dissolved product. Shake the funnel again to effect solution of the salt, remove the organic layer, and dry it over anhydrous sodium sulfate.★ Filter the crude organic product into a 50-mL distilling flask equipped for fractional distillation, and isolate the 2-methylpropanal, bp 63–66°C. If the fractionation is continued after the aldehyde has distilled, unchanged alcohol, bp 107–108°C, and some isobutyl isobutyrate, bp 148–149°C, can be isolated.

Prepare and isolate the 2,4-dinitrophenylhydrazone of 2-methylpropanal, mp 186–188°C, by the procedure given in Section 27.5a, Aldehydes and Ketones, 1.A. Apply the chromic acid in acetone test described in Section 27.5a, Aldehydes and Ketones, 1.C, to 2-methylpropanal.

The preparation of the sodium bisulfite addition product of the aldehyde (equation 33) can be accomplished by adding 6 mL of *saturated* sodium bisulfite solution to 1 mL of the aldehyde contained in a 125-mL Erlenmeyer flask. Swirl the flask and then allow the mixture to stand for 10 min. Reaction will be indicated by warming of the solution. Add 25 mL of 95% ethanol, swirl the mixture well, and cool the resulting solution in an ice-salt bath. Collect the sodium bisulfite addition product by vacuum filtration, washing the filter cake once with 95% ethanol and once with diethyl ether. You should *not* attempt to determine the melting point of this solid. The 2-methylpropanal can be regenerated by adding 5–10 mL of either 1 M sodium carbonate solution or dilute hydrochloric acid to the addition compound and gently warming the mixture.

$$(CH_3)_2CHCHO + NaHSO_3 \rightarrow (CH_3)_2CH{-}\overset{\overset{\displaystyle O{-}H}{|}}{\underset{\underset{\displaystyle SO_3^{\ominus}Na^{\oplus}}{|}}{C}}{-}H \qquad \textbf{(33)}$$

2-Methylpropanal	Sodium bisulfite	Sodium bisulfite addition product of 2-methylpropanal

Although sodium bisulfite addition compounds are generally not good derivatives for characterizing carbonyl compounds, they are often used to purify aldehydes, methyl ketones, and cyclic ketones such as cyclohexanone and cyclopentanone.

Pre-lab exercises for Section 19.2, Part B, Preparation of Cyclohexanone, can be found on page PL. 71.

B. Preparation of Cyclohexanone

Prepare a solution of chromic acid by dissolving 0.081 mol of potassium dichromate in 125 mL of water and then *slowly* adding, with swirling, 19 mL of *concentrated* sulfuric acid. Allow this solution to cool to room temperature, using an ice-water bath if desired to hasten the process. If a stock solution of chromic acid has been prepared for your use, obtain 150 mL of the solution. Add the solution of chromic acid in *one* portion *to* a mixture of 0.20 mol of cyclohexanol and 75 mL of water in a 500-mL Erlenmeyer flask. Thoroughly mix the solutions by swirling and determine the temperature of the reaction mixture. Do *not* use a thermometer to stir this solution. The mixture should quickly become warm. When the temperature reaches 55°C, cool the flask in an ice-water bath or a pan of cold water so that a temperature of 55–60°C is maintained. When the temperature of the solution no longer exceeds 60°C upon removal of the flask from the cooling bath, allow the flask to stand for 1 hr with occasional swirling.★

Transfer the reaction mixture to a 500-mL round-bottom flask, add 100 mL of water and two or three boiling chips, and equip the flask for fractional distillation through an unpacked column. Distil the mixture until approximately 100 mL of distillate, which consists of an aqueous and an organic layer, has been collected.★ Place the distillate in a separatory funnel, saturate the aqueous layer by adding sodium chloride (about 0.2 g of salt per milliliter of water will be required) and swirling the mixture to effect solution. Separate the layers, and extract the aqueous layer with 15 mL of dichloromethane. Combine this extract with the organic layer, and dry the solution over anhydrous magnesium sulfate.★ Filter the solution into a 50- or 100-mL flask, and equip the flask for simple distillation. Wrap the flask and stillhead with some insulating material (towel or glasswool) for optimum efficiency in this distillation. Remove the low-boiling dichloromethane and then continue the distillation, collecting cyclohexanone as a colorless liquid in the boiling range 152–155°C. Calculate the percent yield of product obtained.

Following the procedure given in Section 27.5(a), Aldehydes and Ketones, A.1, prepare and isolate the 2,4-dinitrophenylhydrazone of cyclohexanone, mp 162–163°C. Also apply the chromic acid test described in Section 27.5(a), Aldehydes and Ketones, A.3, to your product.

Pre-lab exercises for Section 19.2, Part C, Oxidation of Cyclohexanone to Hexanedioic Acid, can be found on page PL. 73.

C. Oxidation of Cyclohexanone to Hexanedioic Acid (Adipic Acid)

In a 500-mL Erlenmeyer flask combine 0.10 mol of cyclohexanone and a solution of 0.20 mol of potassium permanganate in 250 mL of water. Add 2 mL of 3 *M* sodium hydroxide solution to this mixture and note the temperature of the resulting solution. Allow the mixture to warm to 45°C and maintain that temperature by intermittent cooling with an ice- or cold-water bath. When the temperature no longer rises above 45°C upon removal of the flask from the cooling bath, allow the solution to stand for an additional 5–10 min; its temperature should drop during this time. Complete the oxidation by heating the mixture to boiling for a few minutes with a hot plate or a burner. Test for the presence of permanganate by placing a drop of the reaction mixture on a piece of filter paper; any unchanged permanganate will appear as a purple ring around the spot of brown manganese dioxide. If permanganate remains, add *small* portions of *solid* sodium bisulfite to the reaction mixture until the spot test is negative.★ Vacuum-filter the mixture through a pad of filter-aid (see Section 2.16), thoroughly wash the brown filter cake with water,★ and then concentrate the filtrate to a volume of about 65 mL by heating it with a hot plate or a burner.★ If the concentrate is colored, add decolorizing carbon, reheat the solution to boiling for a few minutes, and then filter by gravity. Carefully add concentrated hydrochloric acid to the filtrate until the solution tests acidic to litmus paper, and then add an additional 15 mL of acid.★ Allow the solution to cool to room temperature and isolate the precipitated hexanedioic acid by vacuum filtration. The acid is a white solid, mp 152–153°C, and can be recrystallized from ethanol-water if necessary.

EXERCISES

Questions for Part A

1. Write the structure of the 2,4-dinitrophenylhydrazone of 2-methylpropanal. Is it possible for geometric isomers to be produced in the formation of this product? Explain.

2. Why is the two-phase distillate obtained in the oxidation of 2-methyl-1-propanol first saturated with sodium chloride *before* separation of the organic layer?

3. Write a mechanism for the formation of isobutyl isobutyrate in this experiment.

*4. What modifications in the procedure for oxidation of 2-methyl-1-propanol with chromic acid might be made to *maximize* the yield of isobutyl isobutyrate?

5. The costs of sodium dichromate *dihydrate,* potassium dichromate, and chromium anhydride (CrO_3) are $15.00 lb, $4.10/lb, and $6.70/lb, respectively. Determine which of these reagents would be the most economical source of chromic acid to use for oxidation of 2-methyl-1-propanol to 2-methylpropanal.

Questions for Part B

6. Why is the chromic acid solution added *to* the alcohol, rather than the reverse, when the preparation of the ketone is being attempted?

7. Why is the two-phase distillate obtained in the oxidation of cyclohexanol first saturated with sodium chloride *before* separation of the organic layer?

8. Write the structure of the 2,4-dinitrophenylhydrazone of cyclohexanone. Is it possible for geometric isomers to be produced in the formation of this product? Explain.

9. The costs of sodium dichromate *dihydrate,* potassium dichromate, and chromium anhydride (CrO_3) are $15.00/lb, $4.10/lb, and $6.70/lb, respectively. Determine which of these reagents would be the most economical source of chromic acid to use for oxidation of cyclohexanol to cyclohexanone.

Questions for Part C

10. Why is the reaction mixture made alkaline in the oxidation of cyclohexanone to hexanedioic acid?

11. Why does acidification of the concentrated reaction mixture cause precipitation of the hexanedioic acid?

12. Why is filter-aid used in the workup procedure?

13. Give the products to be expected on oxidation of 2-methylcyclohexanone with alkaline potassium permanganate.

REFERENCES

1. *Oxidation in Organic Chemistry,* K. B. Wiberg, editor, Academic Press, New York, 1965, Vol. 5-A, Chapters 1 and 2.
2. *Oxidation,* R. L. Augustine, editor, M. Dekker, New York, 1969, Vol. 1, Chapters 1 and 2.

SPECTRA OF STARTING MATERIAL AND PRODUCTS

The IR and PMR spectra of 2-methyl-1-propanol are provided in Figure 10.1; the IR spectrum of 2-methylpropanal is given in Figure 8.22(b). The IR and PMR spectra of cyclohexanol are contained in Figures 8.19 and 9.15, respectively, and the IR spectrum of cyclohexanone is presented in Figure 8.23(a).

$(CH_3)_2CHCH_2OH$ Chemical shifts: δ 19.1, 30.9, 69.4.

Figure 19.1 CMR data for 2-methyl-1-propanol.

(a) PMR spectrum.

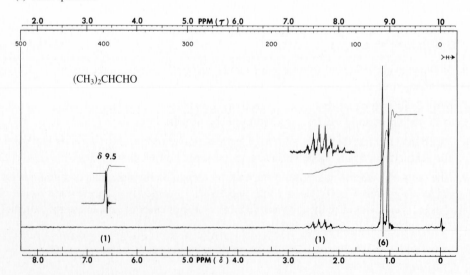

(b) CMR data. Chemical shifts: δ 15.5, 41.2, 204.7.

Figure 19.2 NMR data for 2-methylpropanal.

$$\overset{\displaystyle O}{\overset{\|}{HOC}}(CH_2)_4\overset{\displaystyle O}{\overset{\|}{COH}} \quad \text{Chemical shifts: } \delta\ 24.3,\ 33.6,\ 174.8.$$

Figure 19.3 CMR data for hexanedioic acid.

19.3 Base-Catalyzed Oxidation-Reduction of Aldehydes: The Cannizzaro Reaction

Aldehydes that have *no* hydrogen atoms on the carbon atom adjacent to the carbonyl group (the α-carbon atom) undergo mutual oxidation and reduction in the presence of *strong* alkali (equation 34). Those aldehydes having hydrogens on the α-carbon atom preferentially undergo other types of base-promoted reactions such as the aldol condensation (equation 35), as described in Section 21.2. The occurrence of the mutual oxidation-reduction, called the Cannizzaro reaction, is a consequence of the fact that an aldehyde is intermediate in oxidation state between an alcohol and a carboxylic acid and can be converted into either one by a decrease or gain, respectively, of two in the oxidation number of its carbonyl carbon atom, as indicated in equation 34.

$$2\ R_3\overset{\displaystyle O}{\overset{\|}{CCH}} \xrightarrow[\text{(2) } H_3O^\oplus]{\text{(1) } HO^\ominus} R_3CCH_2OH\ +\ R_3\overset{\displaystyle O}{\overset{\|}{CCOH}} \qquad \textbf{(34)}$$

$$\quad +1 \qquad\qquad\qquad -1 \qquad\qquad +3$$

$$2\ RCH_2\overset{\displaystyle O}{\overset{\|}{CH}} \xrightarrow[\text{(2) } H_3O^\oplus]{\text{(1) } HO^\ominus} RCH_2\overset{\displaystyle O-H}{\overset{|}{CH}}-\underset{\displaystyle R}{\overset{\displaystyle O}{\overset{\|}{CHCH}}} \qquad \textbf{(35)}$$

The mechanism of the reaction, which has considerable experimental support, follows quite logically from the ease with which nucleophiles add to the carbonyl group, particularly that of an aldehyde. The first step of the reaction, then, is attack of the basic catalyst on the carbonyl group to give a tetrahedral intermediate (equation 36), followed by transfer of hydride from this intermediate to the carbonyl group of another aldehyde function (equation 37). This second step is the one in which oxidation and reduction occur; the remaining steps (equations 38 and 39) simply illustrate the types of acid-base chemistry expected to occur in the strongly basic medium in which the reaction is performed. Summation of equations 36–39 gives equation 34.

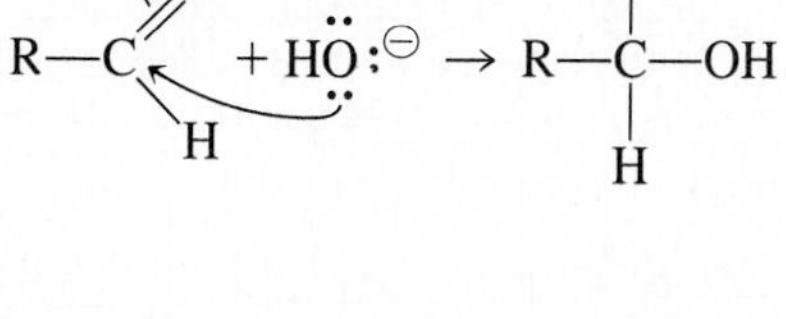

$$R-\overset{\displaystyle \ddot{O}:}{\underset{\displaystyle H}{C}}\ +\ H\ddot{O}:^\ominus\ \rightarrow\ R-\overset{\displaystyle :\ddot{O}:^\ominus}{\underset{\displaystyle H}{C}}-OH \qquad \textbf{(36)}$$

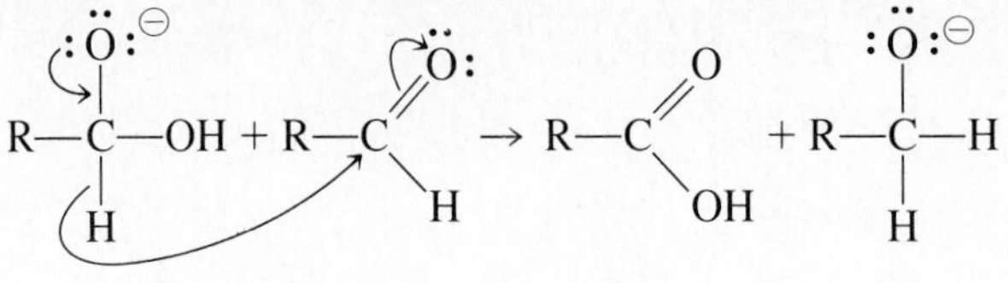

$$R-\underset{\underset{H}{|}}{\overset{\overset{:\ddot{O}:^{\ominus}}{|}}{C}}-OH + R-\underset{\underset{H}{|}}{\overset{\ddot{O}:}{C}} \longrightarrow R-\overset{\overset{O}{\|}}{C}-OH + R-\underset{\underset{H}{|}}{\overset{:\ddot{O}:^{\ominus}}{C}}-H \qquad \textbf{(37)}$$

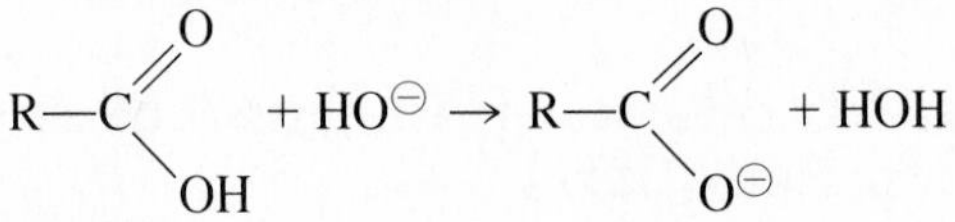

$$R-\overset{\overset{O}{\|}}{C}-OH + HO^{\ominus} \longrightarrow R-\overset{\overset{O}{\|}}{C}-O^{\ominus} + HOH \qquad \textbf{(38)}$$

$$R-\underset{\underset{H}{|}}{\overset{\overset{:\ddot{O}:^{\ominus}}{|}}{C}}-H + HOH \longrightarrow R-\underset{\underset{H}{|}}{\overset{\overset{OH}{|}}{C}}-H + HO^{\ominus} \qquad \textbf{(39)}$$

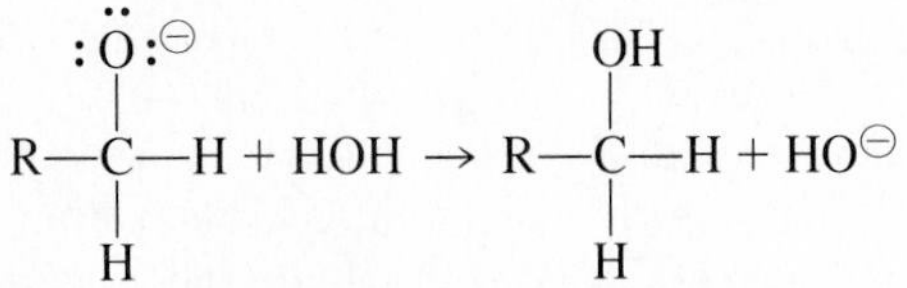

Aromatic aldehydes, formaldehyde, and trisubstituted acetaldehydes all undergo the Cannizzaro reaction. *Crossed Cannizzaro reactions* are also known in which two *different* aldehydes react to produce the acids and alcohols corresponding to each of the two aldehydes (equation 40). In this experiment, benzaldehyde, an aromatic aldehyde, will be converted to benzyl alcohol and potassium benzoate, which gives benzoic acid upon acidification of the reaction mixture (equation 41).

$$\text{RCHO} + \text{R}'\text{CHO} \xrightarrow{\text{HO}^{\ominus}} \text{RCO}_2\text{H} + \text{R}'\text{CO}_2\text{H} + \text{RCH}_2\text{OH} + \text{R}'\text{CH}_2\text{OH} \qquad \textbf{(40)}$$

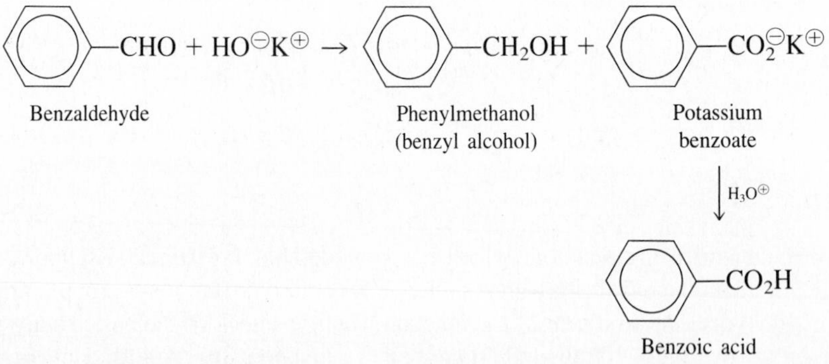

Benzaldehyde   Phenylmethanol   Potassium
              (benzyl alcohol)   benzoate

(41)

Benzoic acid

Pre-lab exercises for Section 19.3, Base-Catalyzed Oxidation-Reduction of Aldehydes: The Cannizzaro Reaction, can be found on page PL. 75.

EXPERIMENTAL PROCEDURE

DO IT SAFELY

The solution of potassium hydroxide is corrosive. Should it come in contact with the skin, flood the affected area immediately with water and then rinse the area with 1% acetic acid.

Dissolve 2 g of solid potassium hydroxide in 2 mL of distilled water by swirling it in a 100-mL beaker, then cool the mixture to room temperature. Put 2 mL of benzaldehyde in an 18- × 150-mm test tube and add the concentrated potassium hydroxide solution to it. Cork the tube securely and shake the tube vigorously until an emulsion is formed. Allow the stoppered tube to stand in your desk until the next laboratory period. Crystallization should occur in the interim.

Add about 1 mL of water to the mixture, stopper the tube, and shake it. If all the crystals originally present do not dissolve, add a little more water, break up the solid mass with a glass rod, and stopper and shake the tube again. Repeat this procedure until all of the solid is in solution. Pour the solution into a separatory funnel, and extract it three times with 10-mL portions of diethyl ether; shake the mixture *gently* to avoid formation of an emulsion. The ethereal solution may be dried over magnesium sulfate and examined by GC[1] to determine the proportion of benzyl alcohol and unchanged benzaldehyde present.

Following the extraction with ether, pour the alkaline aqueous solution into a mixture of 5 mL of concentrated hydrochloric acid, 4 mL of water, and about 5 g of crushed ice. Stir vigorously during the addition. Cool the resultant mixture by placing its container in an ice-water bath, and collect the crystalline benzoic acid.★ Dry this product and determine its melting point. Save a sample of the benzoic acid for comparison with the product of the experiment of Section 19.4.

If the reaction is carried out on a larger scale, the benzyl alcohol can be isolated in the following way. Shake the ethereal solution with aqueous sodium bisulfite solution to remove the benzaldehyde by formation of its sodium bisulfite addition product (equation 32, Section 19.2) and then wash it with water. Dry the organic solution over magnesium sulfate, filter it into a flask equipped for simple distillation, and distil. After the ether has been removed, drain the water from the condenser of the apparatus, and distil the benzyl alcohol, bp 205°C. Alternatively, a crystalline derivative of the alcohol may be prepared, after removal of the ether from the solution by distillation. The phenylurethan (mp 78°C) or the 3,5-dinitrobenzoate (mp 113°C) is suitable (see Section 27.5(a), Alcohols, B.1 and B.2).

EXERCISES

1. When the Cannizzaro reaction is carried out on benzaldehyde in D_2O solution, no deuterium becomes attached to carbon in the benzyl alcohol formed. How does this support the mechanism given in equations 36–39?

2. By what means would aqueous sodium bisulfite remove unchanged benzaldehyde from the reaction mixture?

3. Write an equation for the reaction of a mixture of benzaldehyde and formaldehyde with concentrated potassium hydroxide solution, followed by acidification of the reaction mixture. Show all organic products.

4. The Cannizzaro reaction occurs much more slowly in *dilute* than in *concentrated* potassium hydroxide solution. Why is this?

5. How would propanal react with concentrated potassium hydroxide solution under the

---

[1]A 1.5-m column packed with 5% silicone gum rubber as the stationary phase is satisfactory for the analysis. In one experiment about 20% of the original benzaldehyde was found to be present after 24 hours at room temperature.

conditions of this experiment? Also answer this same question for the case of 2,2-dimethylpropanal.

SPECTRA OF STARTING MATERIALS AND PRODUCTS

The IR spectra of benzyl alcohol, benzaldehyde, and benzoic acid are given in Figures 8.13, 8.22a, and 8.24a, respectively.

(a) PMR spectrum.

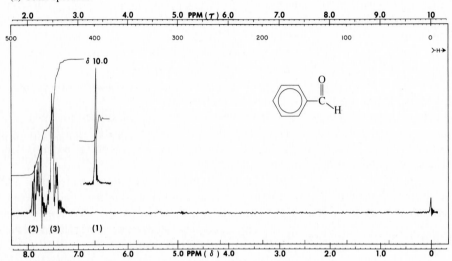

(b) CMR data. Chemical shifts: δ 129.0, 129.7, 134.4, 136.6, 192.0.
Figure 19.4 NMR data for benzaldehyde.

(a) PMR spectrum.

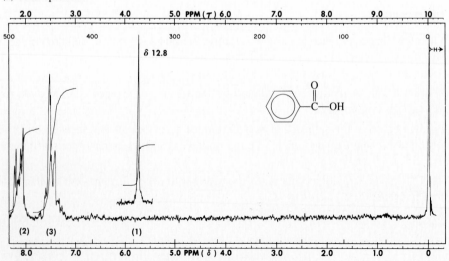

(b) CMR data. Chemical shifts: δ 128.5, 129.5, 130.3, 133.8, 172.7.
Figure 19.5 NMR data for benzoic acid.

(a) PMR spectrum.

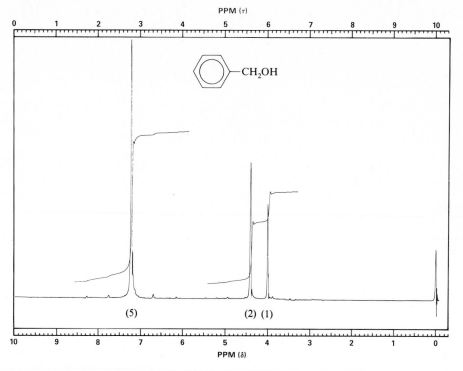

(b) CMR data. Chemical shifts: δ 64.5, 126.9, 127.2, 128.3, 141.0.

Figure 19.6 NMR data for benzyl alcohol.

19.4 Carboxylic Acids

Carboxylic acids can be synthesized in a variety of ways, including the hydrolysis of nitriles or of any derivative of the acid and oxidation of an aldehyde or ketone (Section 19.2). The carbonation of a Grignard reagent (Section 12.2) is another useful synthetic route to such acids.

A special method available for the preparation of aryl carboxylic acids involves the oxidation of the side-chain R group of an alkyl-substituted benzene with alkaline permanganate or with chromic acid solution (equation 42). The carbon atom bonded to the aromatic ring must bear *at least one* hydrogen for this method of preparation to be efficient. If more than one alkyl side chain is present, each will be oxidized to a carboxyl group, and a polycarboxylic acid will result (equation 43). Identification of the mono- or polycarboxylic acid obtained from this side-chain oxidation allows determination of the positions of attachment of the side chains on the aromatic ring, if this is not previously known.

$$ArCH_2R \xrightarrow[\text{or } H_2CrO_4]{KMnO_4/base} ArCO_2H \qquad (42)$$

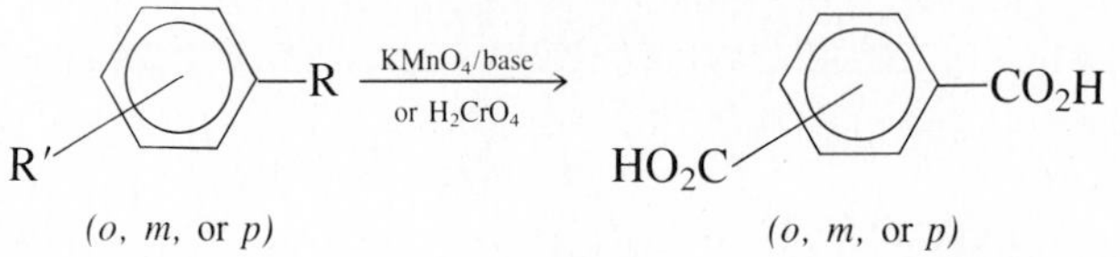

$$(o, m, \text{or } p) \qquad (o, m, \text{or } p) \qquad (43)$$

In the side-chain oxidation described in this experiment, ethylbenzene, a commercially available compound made by Friedel-Crafts alkylation (see Section 17.2) of benzene with ethene, is converted to benzoic acid by reaction with alkaline permanganate (equation 44). The sodium carbonate shown in the equation is produced by reaction of carbon dioxide with aqueous sodium hydroxide, the carbon dioxide itself being derived from the methyl group that is lost in the oxidation.

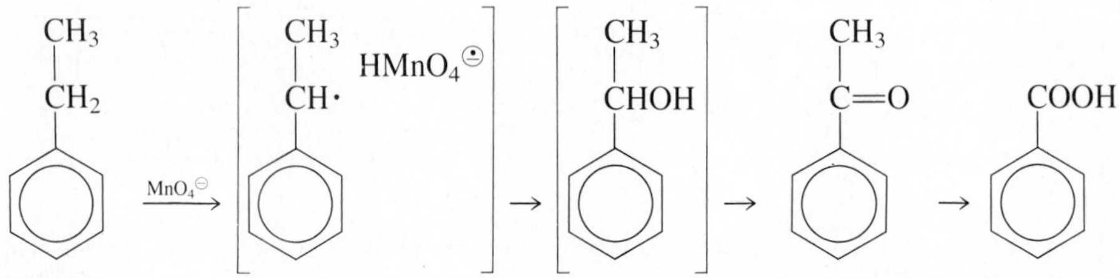

The mechanism of this type of oxidation is not well understood but may involve removal of a benzylic hydrogen atom in a free-radical process initiated by permanganate ion. Subsequent reactions lead first to the alcohol and then to the ketone, which is further oxidized to the carboxylic acid (equation 44). The mechanism of this final oxidation step parallels that given for the oxidation of cyclohexanone to hexanedioic acid (Section 19.2). The stoichiometry of the reaction can be determined by using the techniques developed in Section 19.1 and by recalling equation 31 (see exercise 1).

Pre-lab exercises for Section 19.4, Carboxylic Acids, can be found on page PL. 77.

EXPERIMENTAL PROCEDURE

DO IT SAFELY

1 When preparing and handling solutions of potassium permanganate, it is advisable to wear rubber gloves to avoid contact of the solution with your skin; it will cause unsightly stains on your hands for several days. If this oxidant comes in contact with the skin, wash the affected area thoroughly with soap and warm water.

2 Residues of manganese dioxide on glassware can be difficult to remove. Rinsing the glassware with 10% sodium bisulfite solution will sometimes work. In more difficult cases, 6 *M* HCl should be placed in the glassware and heated **in a ventilation hood.**

Place 0.071 mol of potassium permanganate, 120 mL of water, 1.5 mL of 3 *M* aqueous sodium hydroxide, and 0.016 mol of ethylbenzene in a 500-mL round-bottomed flask fitted with a reflux condenser, and gently heat this mixture at reflux for 2 hr (or up to 3.5 hr if time permits). *Severe bumping will occur if strong heating is used.*★ Test the hot solution for unchanged permanganate by placing a drop of the reaction mixture on a piece of filter paper; if a purple ring appears around the brown spot of manganese dioxide, permanganate remains. Destroy any excess permanganate by adding small amounts of solid sodium bisulfite to the mixture until the spot test is negative. Do not add a large excess of bisulfite. Add 2 g of filter-aid (Celite) to the hot mixture to promote its rapid filtration (see Section 2.16). Filter the mixture by vacuum and rinse the reaction flask and filter cake with two 10-mL portions of hot water.★ Concentrate the filtrate to about 30 mL by performing a simple distillation. If the concentrate is turbid, filter it by gravity; the filtrate should be clear.★ Acidify the aqueous residue with concentrated hydrochloric acid, adding acid until no more benzoic acid separates; about 5–10 mL of acid will be required.★ Cool this mixture in an ice-water bath, and isolate the benzoic acid by vacuum filtration. The crude acid can be purified by recrystallization from hot water or by sublimation. If the latter procedure is used, the crude acid must be *dry*. Determine the yield and melting point of the pure benzoic acid obtained.

EXERCISES

1. Write a balanced equation for conversion of ethylbenzene and permanganate ion to benzoate and carbonate ions and manganese dioxide in the presence of hydroxide ion. Refer to equation 31 for the half-reaction involving reduction of permanganate.

2. Would it be feasible to attempt the preparation of benzaldehyde by permanganate oxidation of toluene? Explain.

3. Would you expect a carboxylic acid to be more or less soluble in pure water than in water containing strong mineral acids? Explain.

4. *t*-Butylbenzene is *not* oxidized to benzoic acid by permanganate. Why not?

5. What effect on yield of isolated product is expected if the concentration step is omitted prior to acidification? Explain.

SPECTRA OF STARTING MATERIAL AND PRODUCTS

IR and NMR data for ethylbenzene can be found in Figures 12.10 and 12.11. The IR spectrum of benzoic acid is given in Figure 8.22a, and NMR data for this compound is contained in Figure 19.5.

reduction reactions of double bonds

alkenes, carbonyl compounds, and imines

20.1 Introduction

Many different functional groups containing double and triple bonds can be reduced, a process that involves the addition of the elements of one or two molecules of hydrogen across the π-bond(s). Reductions can be accomplished in many different ways, including catalytic hydrogenation and chemical reduction, both of which can be done in the laboratory with reasonable ease. Reduction is also commonly encountered in biological systems where enzymes play a central role, but this chapter concerns nonbiological processes.

Equations 1–4 illustrate some of the types of reduction reactions commmonly performed by organic chemists, with the symbol [H] representing a variety of chemical and catalytic methods of reduction. Catalytic reduction refers to reactions involving the addition of hydrogen gas to various substances in the presence of metal catalysts such as platinum, nickel and palladium.

Alkenes:

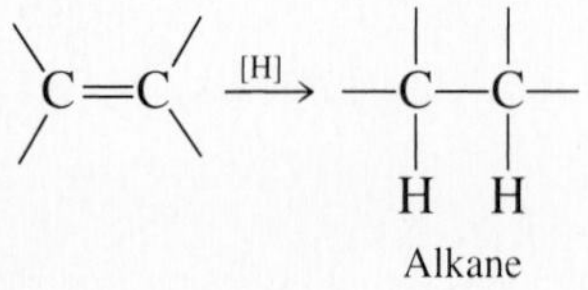

(1)

Alkynes:

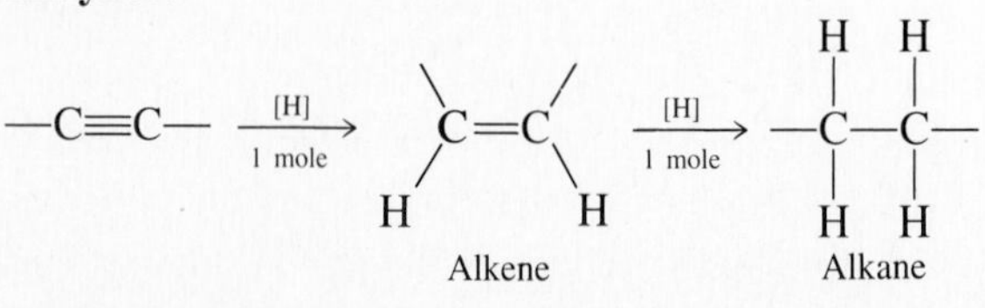

(2)

Carbonyl compounds:

$$\backslash C=O \xrightarrow{[H]} -\overset{|}{\underset{|}{C}}-O$$

$$\overset{}{H} \quad \overset{}{H}$$

(3)

Alcohol

Imines:

$$\backslash C=N- \xrightarrow{[H]} -\overset{|}{\underset{|}{C}}-\overset{|}{N}-$$

$$\overset{}{H} \quad \overset{}{H}$$

(4)

Amine

Many other functional groups can also be reduced, among which are those of carboxylic acids and derivatives of carboxylic acids (esters, amides, anhydrides), nitro groups, aromatic rings, and nitriles. The general picture emerges that any organic compound containing one or more π-bonds is, in principle, capable of adding the elements of hydrogen to form a product containing only single bonds. The starting compounds are said to be **unsaturated** and the products to be **saturated**, which means saturation with respect to hydrogen; hence, they cannot react further under conditions of reduction.

The oxidation of organic compounds involves an increase in the oxidation state of carbon, that is, a net loss of electrons. A procedure for determining the oxidation states of carbon in organic molecules is discussed in Section 19.1. On the other hand, reduction involves a decrease in the oxidation state of carbon and hence a net gain of electrons. For example, the oxidation number of each carbon atom in ethylene, $CH_2=CH_2$, is -2 and in ethane, CH_3CH_3, is -3, so if ethylene is converted to ethane by, for example, catalytic hydrogenation, the reaction involves a net gain of $2\ e^-$ by carbon. Exercise 7 at the end of Section 20.2 provides further experience in determining the oxidation numbers of carbon compounds that undergo reduction.

Catalytic hydrogenation is the most important *industrial* method used for reduction, and hydrogen gas in the presence of a variety of different catalysts is required for this purpose. Some of the commonly exploited terms that appear in advertising for margarines and cooking oils have become part of the household vocabulary. For example, vegetable oils are often called ''polyunsaturated'' because they consist of glyceride molecules which contain many carbon-carbon double bonds per molecule. When subjected to catalytic hydrogenation, these oils react with hydrogen gas and some or all of these double bonds are converted into single bonds. Hydrogenation of vegetable oils produces solid substances, called fats, that are used in margarine and in cooking. The term ''partially hydrogenated'' means that some but not all of the double bonds in the oil were reduced by hydrogen, while ''fully hydrogenated'' or ''saturated'' means that all the carbon-carbon double bonds have been reduced. Other commercial applications of catalytic hydrogenation are commonly utilized, as in the reduction of low molecular weight alkenes to give high-octane gasoline, for example.

Catalytic hydrogenation is useful for small-scale reductions in the research laboratory, and applications of this process include the syntheses of new compounds and the determination of the degree of unsaturation, that is, the number of double and/or triple bonds, in a compound whose structure is to be determined. This process requires that the hydrogenation be done quantitatively, and techniques have been developed for doing so on a very small amount of compound and with great accuracy. One major problem facing the experimenter who desires to carry out a catalytic hydrogenation is deciding which catalyst must be used. This decision may be reached by analogy with similar reactions that have been reported in the literature or by trial and error. Section 20.2 contains the discussion of and experimental procedure for the catalytic hydrogenation of a compound containing a carbon-carbon double bond.

A variety of more convenient and versatile "chemical reducing agents" have been developed, among which are the metal hydride compounds such as lithium aluminum hydride, $LiAlH_4$, and sodium borohydride, $NaBH_4$. A major consideration in selecting a chemical reducing agent involves its possible reaction with other functional groups that may be present in a molecule, that is, the selectivity of the reagent. Many different types of selective reducing agents have been developed in recent years that will react with only one type of functional group. For example, lithium aluminum hydride reduces aldehydes, ketones, and esters, whereas sodium borohydride will reduce aldehydes and ketones but *not* esters. Furthermore, experimental ease and safety often play a role in the selection of a metal hydride reducing agent; for example, if both lithium aluminum hydride and sodium borohydride can be used, the latter is preferred because it is safer and easier to use. Section 20.3 provides several chemical reductions that utilize sodium borohydride.

The nitro group, $—NO_2$, can be reduced to the amino group, $—NH_2$, by catalytic hydrogenation or by using lithium aluminum hydride. However, this conversion is more conveniently accomplished by using finely divided metals, such as iron or tin with hydrochloric acid, or stannous chloride. Section 23.2b describes the reduction of nitrobenzene, $C_6H_5NO_2$, to aniline, $C_6H_5NH_2$, using iron metal and HCl, and Section 23.4A gives a procedure for the reduction of the nitro group in 2,6-dimethylnitrobenzene into an amino group using stannous chloride, $SnCl_2$.

20.2 Catalytic Hydrogenation of the Carbon-Carbon Double Bond: Hydrogenation of 4-Cyclohexene-*cis*-1,2-dicarboxylic Acid

The addition reactions of alkenes, discussed in Section 13.3, can be extended to catalytic hydrogenation in which one mole of hydrogen is added (equation 1). A wealth of information has been acquired about the mechanism of this reaction, which is known to occur by the concerted *syn* addition of two atoms of hydrogen. For example, the only product obtained from the catalytic hydrogenation of 1,2-dimethylcyclopentene (**1**) is *cis*-1,2-dimethylcyclopentane (**2**), and this supports the view that both of the hydrogen atoms add to the *same* side of the double bond (*syn* addition, equation 5).

(5)

1,2-Dimethylcyclopentene   *cis*-1,2-Dimethylcyclopentane

The stereochemistry of this reaction has been rationalized by the suggestion that hydrogen gas is adsorbed on the surface of the finely divided platinum catalyst to produce hydrogen atoms. The π-electrons of the double bond are believed to complex with the catalyst surface; the hydrogen atoms are then transferred from the surface to the carbon atoms to form the saturated product. Other catalysts such as palladium and nickel can be used equally well.

The experiment that follows involves the conversion of 4-cyclohexene-*cis*-1,2-dicarboxylic acid (**3**) to cyclohexane-*cis*-1,2-dicarboxylic acid (**4**) using catalytic hydrogenation (equation 6).

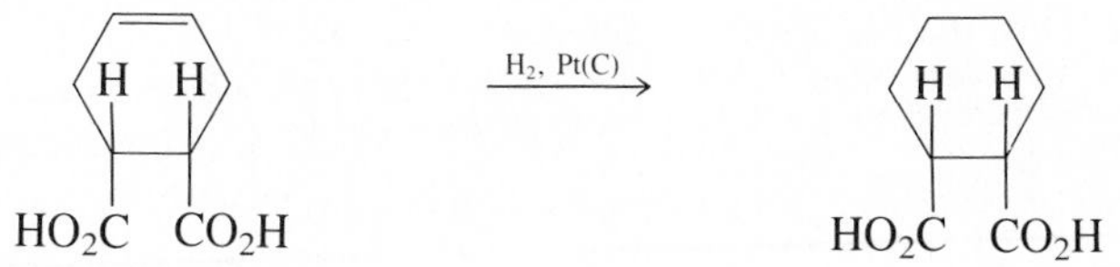

3	**4**
4-Cyclohexene-*cis*-1,2-dicarboxylic acid	Cyclohexane-*cis*-1,2-dicarboxylic acid

The synthesis of **3** is accomplished by the Diels-Alder reaction between butadiene and maleic anhydride, followed by hydrolysis of the adduct, as discussed in the experimental procedures of Chapter 15, part C.

The reduction can be accomplished in several ways, but generally it involves the use of hydrogen gas either at or slightly above atmospheric pressure and a noble metal such as palladium, platinum, or nickel as the catalyst. For our purposes it will be convenient to utilize sodium borohydride as the *in situ* source of hydrogen by allowing it to react with concentrated hydrochloric acid, a reaction that involves the union of H^+ from HCl with the hydride ions, $H:^-$, contained in $NaBH_4$ (equation 7). The metal catalyst, platinum, is formed concomitantly by the reduction of chloroplatinic acid, H_2PtCl_6, with hydrogen in the presence of decolorizing carbon, which serves as a solid support for the finely divided metallic platinum produced (equation 8).

$$NaBH_4 + HCl + 3\ H_2O \rightarrow 4\ H_2 + B(OH)_3 + NaCl \qquad \textbf{(7)}$$

$$H_2PtCl_6 + 2\ H_2 \rightarrow Pt^0 + 6\ HCl \qquad \textbf{(8)}$$

Note should be made of the selectivity of the hydrogenation reaction depicted in equation 6. The carboxylic acid groups, $-CO_2H$, are unaffected by the process, and their stereochemistry is unchanged even though the *trans* diacid (**4**) is more stable.

Pre-lab exercises for Section 20.2, Hydrogenation of 4-Cyclohexene-*cis*-1,2-dicarboxylic Acid, can be found on page PL. 79.

EXPERIMENTAL PROCEDURE

DO IT SAFELY

1 Hydrogen gas is extremely flammable. Use no flames in this experiment, and make sure that there are no open flames nearby.

2 When removing the diethyl ether by simple distillation, use a steam bath or hot-water bath, as directed. **Do not use a hot plate.**

3 Avoid skin contamination by the solutions you are using, preferably by wearing rubber gloves throughout the experiment. If you do get these solutions on your hands, wash them thoroughly with soap and water. In the case of contact with the sodium borohydride solution, rinse the affected areas with 1% acetic acid solution. In the case of acid burns, apply a paste of sodium bicarbonate to the area for a few minutes.

Prepare a reaction vessel for hydrogenation by tying a heavy-walled balloon to the sidearm of a 125-mL filter flask. Also prepare a 1 M aqueous solution of sodium borohydride by dissolving 0.4 g (0.01 mol) of sodium borohydride in 10 mL of water and adding 0.1 g of sodium hydroxide as a stabilizer. Place 10 mL of water, 1 mL of a 5% solution of chloroplatinic acid ($H_2PtCl_6 \cdot 6H_2O$), and 0.5 g of decolorizing carbon in the reaction flask and add, with swirling, 3 mL of the 1 M sodium borohydride solution. Allow the resulting slurry to stand for 5 min to permit formation of the catalyst. During this time dissolve 1 g (0.006 mol) of 4-cyclohexene-cis-1,2-dicarboxylic acid (see Chapter 15, part C) by heating it with 10 mL of water.

Pour 4 mL of *concentrated* hydrochloric acid into the catalyst-containing reaction flask, and then add the hot aqueous solution of the diacid to the flask. Seal the flask with a serum cap, and wire the cap securely in place. Draw 1.5 mL of the 1 M sodium borohydride solution into a plastic syringe, push the needle of the syringe through the serum cap, and inject the solution dropwise while swirling the flask. If the balloon on the sidearm of the flask becomes inflated, stop the addition of sodium borohydride solution until deflation occurs. When the syringe is empty, remove it from the flask, and refill it with an additional 1.5 mL of solution. The dropwise addition of this further quantity of sodium borohydride should cause the balloon to inflate and to remain inflated, indicating a positive pressure of hydrogen in the system. Remove the syringe from the serum cap and allow the flask to stand, occasionally swirling for 5 min. Heat and swirl the flask on the steam cone until the balloon deflates to a constant size and then heat for an additional 5 min. A total of no more than 20 min of heating should be required.

Release the pressure from the reaction flask by pushing through the serum cap the needle of a syringe from which the barrel has been removed, filter the hot reaction mixture by suction, and place the filter paper in a container reserved for ''recovered catalyst.''★ Cool the filtrate and extract it with three 25-mL portions of *technical* ether. Combine the extracts, wash them with 10 mL of saturated sodium chloride solution, and then filter the organic solution through a piece of fluted filter paper containing solid anhydrous sodium sulfate into a tared round-bottomed flask. Remove the ether by simple distillation, discarding the distillate in the organic liquid-waste bottle. The product will remain as a solid residue. Allow the crude product to air-dry, and then determine its weight and melting point.

Transfer the bulk of the crude diacid by scraping it into a 25-mL Erlenmeyer flask. Add about 2 mL of water to the larger flask, heat to dissolve any residual diacid, and pour the hot solution into the smaller flask. Bring all the diacid into solution at the boiling point by adding no more than two additional mL of water. After solution has been accomplished, add 3 drops of *concentrated* hydrochloric acid to decrease the solubility of the diacid in the solution, allow the mixture to cool to room temperature, and then place the flask in an ice-water bath to effect more complete crystallization of product. Isolate the product and determine its melting point and yield. The reported melting point is 192°C. Ascertain that the product is saturated by performing the qualitative tests for unsaturation described in Section 27.5b. Also determine whether hydrogenation has affected the acidic nature of the molecule by testing an aqueous solution of the diacid with litmus paper.

EXERCISES

1. Draw the structure of the product of catalytic hydrogenation of
   a. 1,2-dimethylcyclohexene   **b.** *cis*-2,3-dideuterio-2-butene
   c. *trans*-2,3-dideuterio-2-butene   **d.** *cis*-3,4-dimethyl-3-hexene

2. Why does the addition of hydrochloric acid to an aqueous solution of a carboxylic acid decrease its solubility in water?

3. Aqueous solutions of sodium borohydride can be prepared and kept for short periods of time, as in this experiment. On the other hand, lithium aluminum hydride reacts explosively with water. Suggest a possible reason for these great differences in reactivity with water.

4. *Technical* diethyl ether contains small traces of water and was used to extract the product from the filtrate in this experiment. Why were you instructed to use the technical rather than the anhydrous grade for this purpose?

5. What is the purpose of washing the ethereal extracts with saturated sodium chloride solution rather than with water?

6. What is the purpose of pouring the ether extracts through sodium sulfate?

7. Determine the oxidation numbers of the carbon atoms undergoing change in the following reactions, and indicate the net gain or loss of electrons in each.
   a. $CH_3Cl \rightarrow CH_4$
   b. $CH_3CHO \rightarrow CH_3CH_2OH$
   c. acetylene $\rightarrow$ ethylene
   d. $CH_3CO_2H \rightarrow CH_3CHO$
   e. $CH_3CH{=}NH \rightarrow CH_3CH_2NH_2$

SPECTRA OF STARTING MATERIAL AND PRODUCT

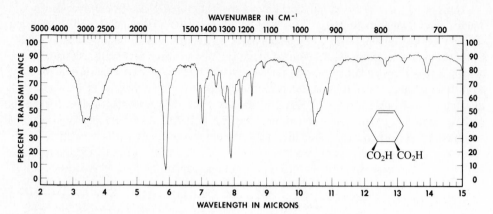

Figure 20.1 IR spectrum of 4-cyclohexene-cis-1,2-dicarboxylic acid.

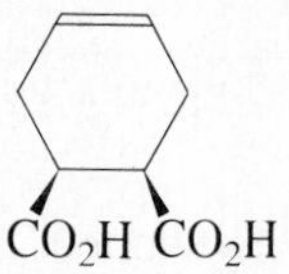

CO$_2$H CO$_2$H Chemical shifts: δ 25.9, 39.2, 125.2, 174.9.

Figure 20.2 CMR data for 4-cyclohexene-cis-1,2-dicarboxylic acid.

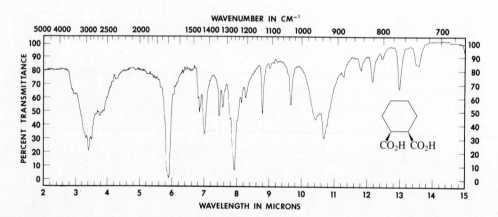

Figure 20.3 IR spectrum of cyclohexane-cis-1,2-carboxylic acid.

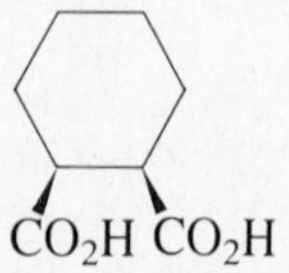

CO$_2$H CO$_2$H Chemical shifts: δ 23.8, 26.3, 42.3, 175.6.

Figure 20.4 CMR data for cyclohexane-cis-1,2-dicarboxylic acid.

20.3 Reduction of Imines; Preparation of Amines

Imines can be prepared from primary amines and carbonyl compounds such as aldehydes and ketones, as discussed in Section 21.1A. Although aliphatic carbonyl compounds and primary aliphatic amines give imines that are unstable and polymerize or hydrolyze to regenerate the starting compounds, stable imines can be isolated if either an aromatic carbonyl compound or an aromatic primary amine is used.

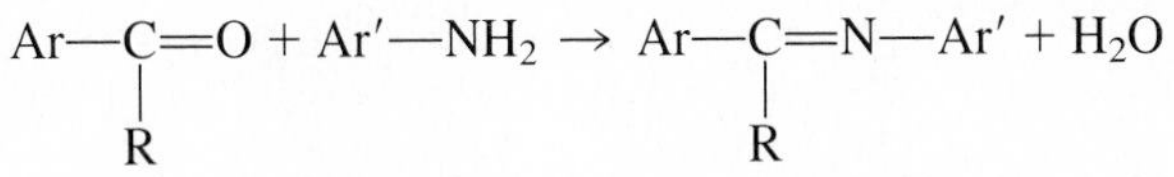

$$\text{Ar—C}{=}\text{O} + \text{Ar}'\text{—NH}_2 \rightarrow \text{Ar—C}{=}\text{N—Ar}' + \text{H}_2\text{O} \qquad \textbf{(9)}$$
$$\underset{\text{R}}{|} \qquad\qquad\qquad \underset{\text{R}}{|}$$

An imine

Imines may undergo addition of one mole of hydrogen in the presence of a catalyst such as nickel to produce secondary amines (equation 10), a process that is analogous to the hydrogenation of alkenes to alkanes and of carbonyl compounds to alcohols.

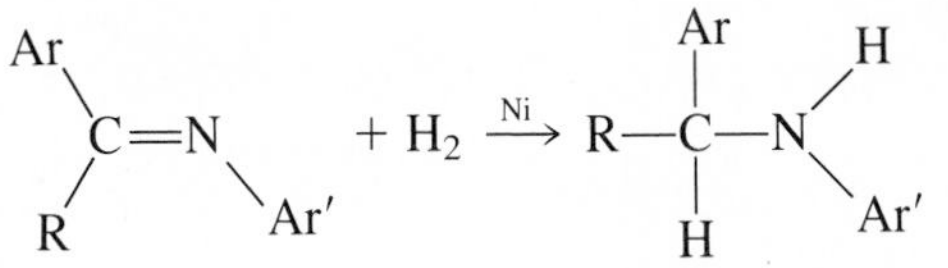

$$\underset{\underset{\text{R}}{}}{\overset{\text{Ar}}{}}\text{C}{=}\text{N} \underset{\text{Ar}'}{} + \text{H}_2 \xrightarrow{\text{Ni}} \text{R—C—N} \qquad \textbf{(10)}$$

Imine   Secondary amine

If the desired compound is a secondary amine, it is not necessary to isolate the imine. For example, reaction mixtures from an aliphatic or aromatic carbonyl compound and an aliphatic or aromatic amine or even ammonia may be subjected directly to catalytic hydrogenation to produce primary or secondary amines in good yields (equations 11 and 12). Presumably, imines are transitory intermediates in these reactions, which are called *reductive aminations*.

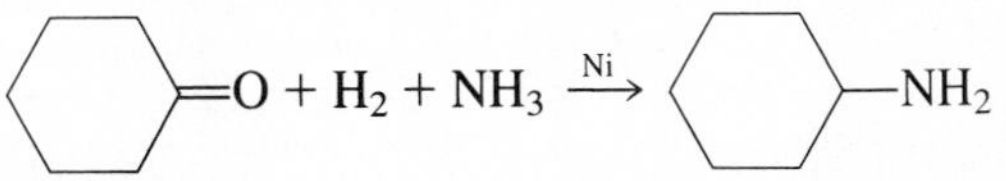

$$\bigcirc{=}\text{O} + \text{H}_2 + \text{NH}_3 \xrightarrow{\text{Ni}} \bigcirc\text{—NH}_2 \qquad \textbf{(11)}$$

Cyclohexanone   Cyclohexylamine

$$\text{C}_6\text{H}_5\text{C}\overset{\nearrow\text{O}}{\underset{\searrow\text{H}}{}} + \text{H}_2 + \text{C}_6\text{H}_5\text{NH}_2 \xrightarrow{\text{Ni}} \text{C}_6\text{H}_5\text{—NH—CH}_2\text{C}_6\text{H}_5 \qquad \textbf{(12)}$$

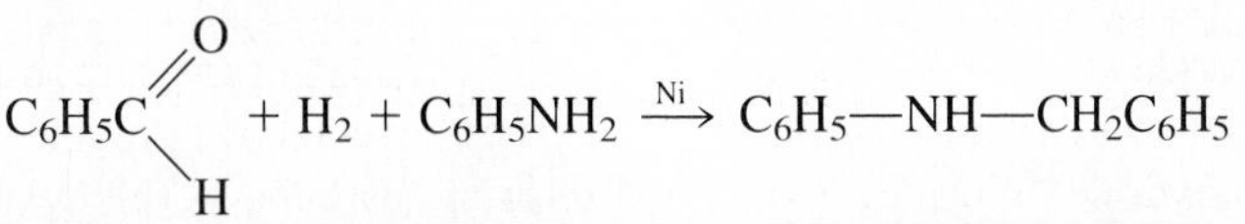

Benzaldehyde   Aniline   Benzylphenylamine

One of the limitations to this procedure for the synthesis of amines is that any other functional groups in the starting compounds that are sensitive to catalytic hydrogenation, such as $C{=}C$, $N{=}O$ (as in NO_2), or $C{—}X$, may also be reduced under the conditions of the reaction. This limitation can be overcome if a *selective* metal hydride, such as sodium borohydride, is used rather than catalytic hydrogenation.

This section describes the preparation of a stable imine and its subsequent reduction with sodium borohydride. *N*-Cinnamyl-*m*-nitroaniline (**9**) is synthesized starting with cinnamaldehyde (**5**) and *m*-nitroaniline (**6**). This experiment illustrates the selectivity of the reducing agent, sodium borohydride, which reduces the intermediate imine (**7**) but not the carbon-carbon double bond, the nitro group, or benzene rings that are present in it (equation 15).

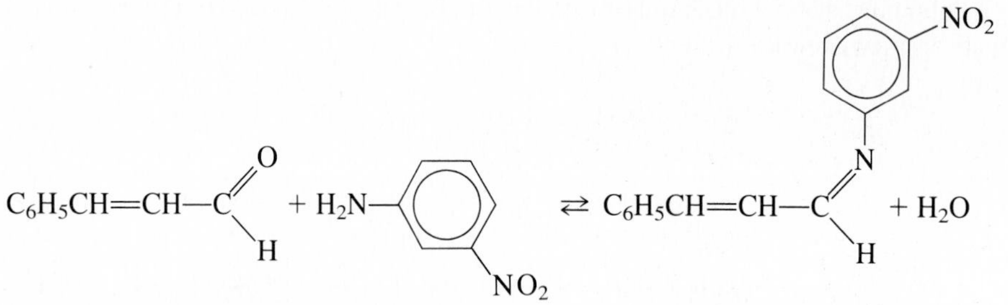

$$C_6H_5CH=CH-\overset{\displaystyle O}{\underset{\displaystyle H}{C}} \ + \ H_2N-\!\!\!\bigcirc\!\!\!-NO_2 \ \rightleftharpoons \ C_6H_5CH=CH-\overset{\displaystyle N}{\underset{\displaystyle H}{C}} \ + \ H_2O \qquad \textbf{(13)}$$

5	**6**	**7**
Cinnamaldehyde	*m*-Nitroaniline	*N*-Cinnamylidene-*m*-nitroaniline

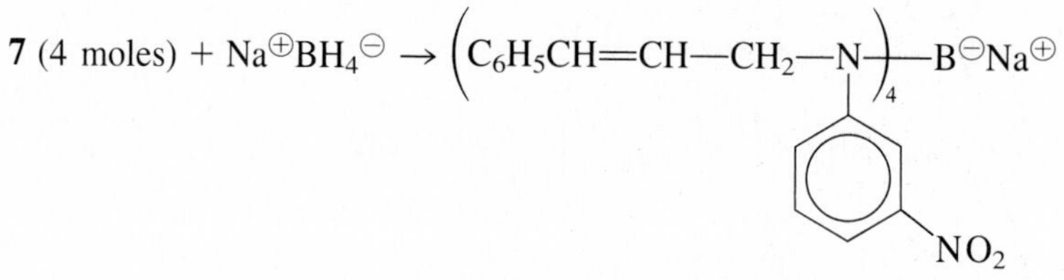

$$\textbf{7} \ (4 \text{ moles}) + Na^{\oplus}BH_4^{\ominus} \rightarrow \left(C_6H_5CH=CH-CH_2-N \right)_4 B^{\ominus}Na^{\oplus} \qquad \textbf{(14)}$$

8

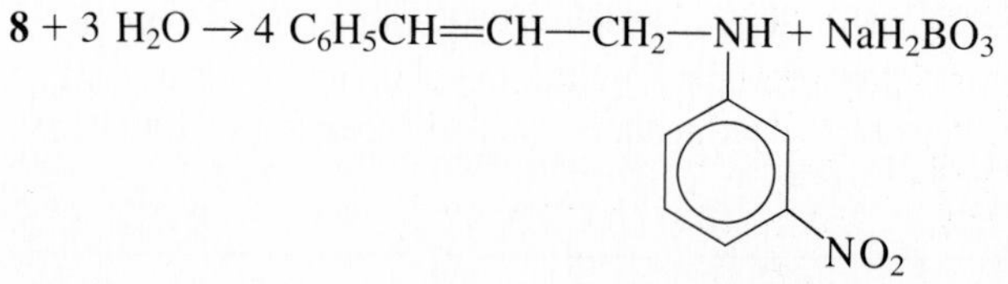

$$\textbf{8} + 3 \ H_2O \rightarrow 4 \ C_6H_5CH=CH-CH_2-NH + NaH_2BO_3 \qquad \textbf{(15)}$$

9

N-Cinnamyl-*m*-nitroaniline

The reaction of sodium borohydride with the imine (equation 14) is analogous to the addition of a Grignard reagent to a carbonyl compound. A hydride ion, H:$^-$, is transferred from the borohydride anion, BH_4^-, to the electrophilic carbon of the carbon-nitrogen double bond, and the electron-deficient boron atom becomes attached to nitrogen. All four hydrogens of the borohydride anion are transferred to imine carbons in this way, producing an organoboron anion, **8**, which is subsequently decomposed with water to yield the desired secondary amine (**9**, equation 15).

Although the formation of the imine is reversible (equation 13), it can be brought to completion by removing the water from the reaction mixture as it is produced. A convenient way to do this is to heat the carbonyl compound and primary amine in cyclohexane and then allow the cyclohexane to distil continuously during the time that reaction is occurring. Cyclohexane forms a minimum-boiling azeotrope with the water and removes it as it is formed.

Sodium borohydride is used in methanol solution, and although it reacts slowly with the solvent, its rate of reaction with the amine is much faster. In small-scale preparations it is more convenient to use a sufficient excess of the NaBH$_4$ to allow for its reaction with the solvent than to use a less-reactive solvent in which it is less soluble.

Pre-lab exercises for Section 20.3, Reduction of Imines, can be found on page PL. 81.

EXPERIMENTAL PROCEDURE

DO IT SAFELY

1 Methanolic solutions of sodium borohydride are highly caustic, and contact with the skin should be avoided. It is recommended that rubber gloves be worn when handling such solutions. If contact occurs, wash the area with copious quantities of cold water.

2 Flasks that contain methanolic sodium borohydride must **not** be stoppered. The solution evolves hydrogen gas, and dangerous buildup of pressure could occur in a stoppered flask.

Note: For best results the cinnamaldehyde must be freshly distilled before use in this experiment.

In a 100-mL round-bottomed flask place 0.022 mol of cinnamaldehyde, 0.020 mol of *m*-nitroaniline, 25 mL of cyclohexane, and three or four boiling stones. Set up the apparatus for simple distillation using a graduated cylinder as a receiver, and heat the reaction mixture on a steam bath or with a heating mantle. If a burner must be used, take the proper precautions against the flammability of cyclohexane. Distil until most of the cyclohexane has been removed; 20–22 mL of distillate should be obtained in 25–30 min. *Discontinue* heating when the distillation rate is seen to decrease. Add another 25 mL of cyclohexane and repeat the distillation, discontinuing the heating as before when the rate of distillation decreases.

During the distillation prepare a solution of 0.020 mol of sodium borohydride in 15 mL of methanol. This solution should be prepared in an *unstoppered* vessel no more than 5 to 10 min before it is used because of the reaction between sodium borohydride and methanol. When the distillation has been completed, remove the heat source, and either pour out about 0.5 mL of the residual liquid or, preferably, insert a micropipet and take a 0.5-mL sample of the liquid. Add 3–4 mL of methanol to this sample in a test tube, swirl the tube to effect solution, and then place the test tube in ice. (Read the directions of the next paragraph while the solution is cooling.) Collect any crystals that separate, dry them, and determine their melting point. The reported melting point of the imine, *N*-cinnamylidene-*m*-nitroaniline, is 92–93°C. The imine may be purified by recrystallization from methanol.

Add 20 mL of methanol to the remainder of the residue from the distillation (crude imine). Attach a Claisen connecting tube equipped with a water-cooled reflux condenser and an addition funnel.[1] Pour the methanolic solution of sodium borohydride into the addition

---

[1] If a Claisen adapter is not available, the condenser may be attached directly to the flask and an addition funnel inserted in the top of the condenser through a *slotted* cork or rubber stopper. The slot is to prevent a closed system.

funnel (*stopcock closed*), and then add this solution to the imine solution dropwise, or in several portions, at a rate such that the addition is completed within 5 min. Swirl the reaction mixture while the addition is being made. After all of the borohydride solution has been added, heat the reaction mixture at reflux for 15 min.

Cool the reaction mixture to room temperature, and pour it into 50 mL of water. Stir the mixture, and allow it to stand with occasional stirring for 10–15 min. Collect the orange crystals and wash them with water.★ Dry the product, and determine its yield. Determine the melting point of the product, N-cinnamyl-m-nitroaniline. It may be recrystallized from 95% ethanol; the melting point of the pure secondary amine is 106–107°C.

EXERCISES

1. What causes the turbidity in the distillate collected during the heating of the carbonyl compound and primary amine in cyclohexane solution?

2. Why was the use of *dry* cyclohexane not specified?

3. Although 95% ethanol is a satisfactory solvent for the recrystallization of N-cinnamyl-m-nitroaniline, it is not suitable for recrystallization of the corresponding imine from which it was produced. Why?

4. Determine the molar ratio of NaBH$_4$ to imine that you actually used in the experiment. Why do you think that it is necessary to use a greater molar ratio than theoretical?

5. Explain how sodium borohydride is "stabilized" toward reaction with solvent methanol by the addition of sodium hydroxide.

6. In Figures 20.7 and 20.8, assign as many of the IR absorption peaks to the components of the molecules as you can. Which peaks provide evidence of the conversion of the imine to the secondary amine? Which peaks show the retention of the carbon-carbon double bond and of the nitro group?

7. After the reaction between sodium borohydride and the imine is complete, the reaction mixture is treated with water to produce the desired secondary amine. Explain this reaction by indicating the source of the hydrogen that ends up on the amino nitrogen.

8. Although sodium borohydride is fairly unreactive toward methanol, the addition of a mineral acid to this solution results in the rapid destruction of the NaBH$_4$. Explain.

9. Suppose that each of the following pairs of compounds were subjected to the reactions provided in the experimental procedures. Draw the structures of the imines that would be formed, as well as those of the final amine product.
   a. cinnamaldehyde and aniline
   b. benzaldehyde and m-nitroaniline
   c. benzaldehyde and aniline

10. Propose a mechanism to show how imine **7** can be hydrolyzed by water to produce **5** and **6**.

11. Explain why aliphatic imines are much less stable than those that have at least one aromatic group attached to the carbon-nitrogen double bond.

SPECTRA OF STARTING MATERIAL AND PRODUCT

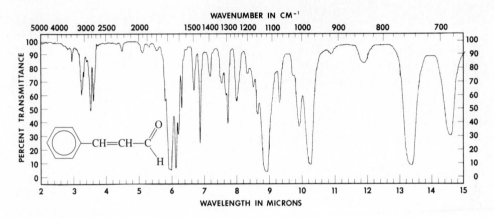

Figure 20.5 IR spectrum of cinnamaldehyde.

(a) PMR spectrum.

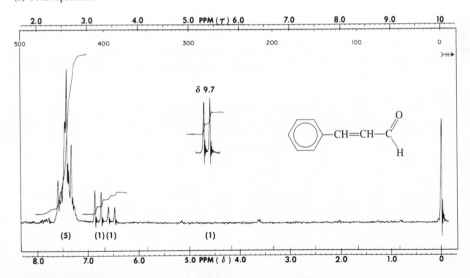

(b) CMR data. Chemical shifts: δ 128.5, 129.0, 131.1, 134.1, 152.3, 193.2.

Figure 20.6 NMR data for cinnamaldehyde.

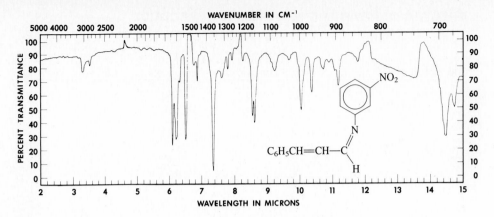

Figure 20.7 IR spectrum of N-cinnamylidene-m-nitroaniline (in CCl₄ solution).

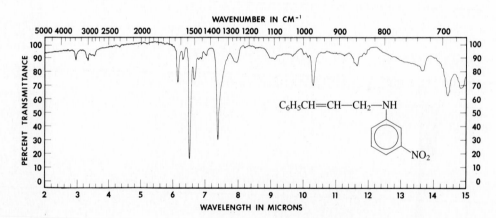

Figure 20.8 IR spectrum of N-cinnamyl-m-nitroaniline (in CCl₄ solution).

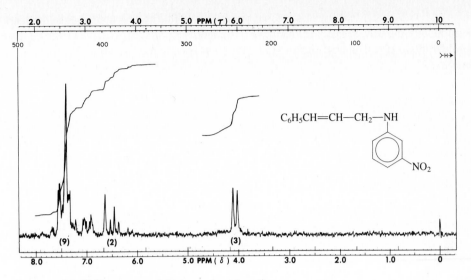

Figure 20.9 PMR spectrum of *N*-cinnamyl-*m*-nitroaniline.

20.4 Reduction of Carbonyl Compounds; Preparation of Alcohols

Carbonyl compounds, specifically aldehydes and ketones, can be reduced by a variety of procedures, including catalytic hydrogenation and metal hydride reduction. The reduction of aldehydes provides a convenient synthetic route to primary alcohols (equation 16), and secondary alcohols can be obtained from ketones (equation 17).

$$R-\underset{\underset{H}{|}}{\overset{\overset{O}{\|}}{C}} \xrightarrow{[H]} RCH_2OH \qquad \text{(16)}$$

$$R = \text{alkyl or aryl}$$

$$R-\underset{\underset{R'}{\diagdown}}{\overset{\overset{O}{\|}}{C}} \xrightarrow{[H]} \underset{\underset{R'}{|}}{RCHOH} \qquad \text{(17)}$$

$$R \text{ and } R' = \text{alkyl or aryl}$$

As with the reduction of other functional groups containing a double or triple bond, catalytic hydrogenation of carbonyl groups is the method of choice in industrial reactions, but metal hydrides are used most frequently for this purpose in the research laboratory. Although both lithium aluminum hydride and sodium borohydride reduce aldehydes and ketones to alcohols, a virtue of the latter is that it may be used in alcoholic and even aqueous solutions. In contrast, lithium aluminum hydride requires the use of anhydrous ethereal solvents such as diethyl ether and tetrahydrofuran. Sodium borohydride does react slowly with alcohol solvents and with water, but it reacts much more rapidly with carbonyl compounds, so that its reaction with the solvent is not troublesome. The reduction of acetophenone (**10**) by sodium borohydride to give 1-phenylethanol (**11**) after hydrolysis (equation 18) is described in this procedure.

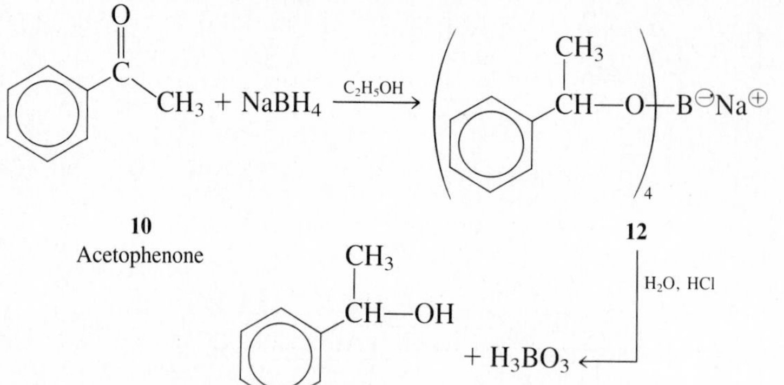

(18a)

10
Acetophenone

12

(18b)

11
1-Phenylethanol

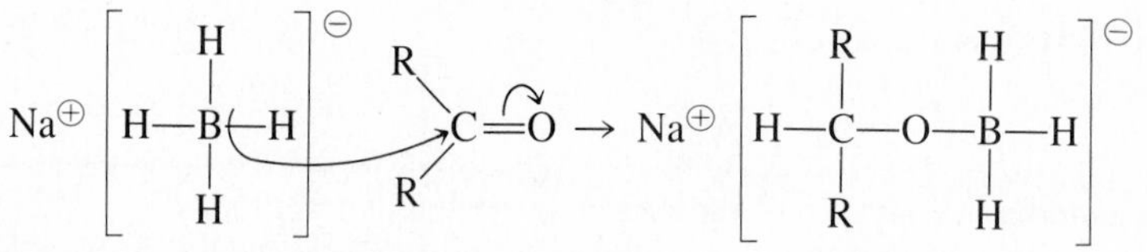

(19)

This reaction involves the transfer of a hydride ion, $H:^-$, from $NaBH_4$ to the carbonyl carbon as the first step in the mechanism (equation 19). All of the hydrogens attached to boron are transferred in this way to produce the intermediate borate salt (**12,** equation 18a), which is decomposed upon addition of water and acid to yield 1-phenylethanol (**11,** equation 18b). To purify 1-phenylethanol, it must be separated from water, ethanol, inorganic acids, and salts; this is accomplished by evaporating the ethanol on a steam bath and then extracting the 1-phenylethanol with ether. The ethereal solution is dried and the solvent is distilled. The desired product may be isolated in pure form by distillation under reduced pressure. Because the alcohol is prone to dehydration to give styrene, which readily polymerizes (equation 20), some anhydrous potassium carbonate is added to the distillation flask in order to neutralize traces of acid that may be present.

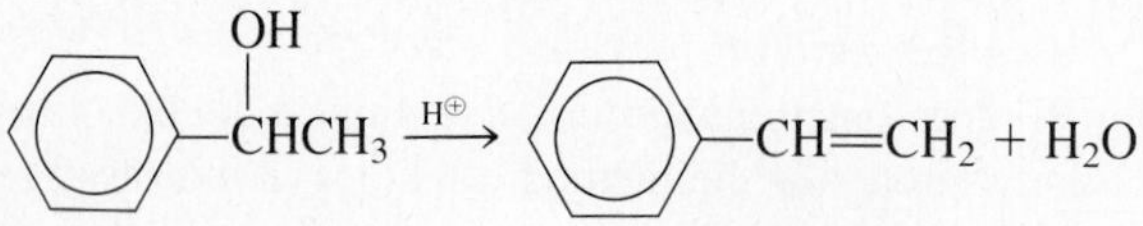

(20)

Styrene

Pre-lab exercises for Section 20.4, Reduction of Carbonyl Compounds, can be found on page PL 83.

EXPERIMENTAL PROCEDURE

> ### DO IT SAFELY
>
> **1** Methanolic solutions of sodium borohydride are highly caustic; contact with the skin should be avoided. It is recommended that rubber gloves be worn when handling such solutions. If contact occurs, wash the area with copious quantities of cold water.
>
> **2** Flasks that contain methanolic sodium borohydride must **not** be stoppered. The solution evolves hydrogen gas, and a dangerous buildup of pressure could occur in a stoppered flask.

In a 150-mL beaker dissolve 0.032 mol of sodium borohydride in 25 mL of 95% ethanol, and add 0.100 mol of acetophenone dropwise. Stir the mixture with a stirring rod during the addition, and keep the temperature below 50°C by decreasing the rate of addition and by external cooling with an ice bath, if necessary. After the addition of the acetophenone is complete, allow the reaction mixture, which contains a white precipitate, to stand for 15 min at room temperature. Add about 10 mL of 3 M hydrochloric acid solution dropwise and with stirring; hydrogen will be evolved and most of the white solid will dissolve. Place the beaker containing the reaction mixture on a steam bath, and concentrate the solution by evaporation of the ethanol until two liquid layers separate.

Add 20 mL of diethyl ether, and transfer the mixture to a 125-mL separatory funnel. Shake the mixture gently, allow the layers to separate, and remove the ether layer.★ Extract the aqueous layer with a 10-mL portion of diethyl ether, and add this to the other ether solution. Dry the combined ether solutions over anhydrous magnesium sulfate; briefly swirl the solution with the drying agent, and then decant the solution into a dry flask. Add another portion of drying agent and repeat the swirling; filter the solution into a tared 100-mL round-bottomed flask.★

If pure 1-phenylethanol is to be isolated, add about 1 g of anhydrous potassium carbonate, and arrange for vacuum distillation (Section 2.8). Remove the diethyl ether at atmospheric pressure (steam or hot-water bath), then reduce the pressure to below 20 torr, if possible, and distil using an oil bath; the boiling point of 1-phenylethanol is 102.5–103.5°C (19 torr), 97°C (13 torr). Calculate the yield of 1-phenylethanol.

EXERCISES

1. Determine the molar ratio of NaBH$_4$ to acetophenone that you actually used in the experiment. Why do you think that it is necessary to use a greater molar ratio than theoretical?

2. After the reaction between sodium borohydride and the carbonyl compound is complete, the reaction mixture is treated with water and acid to produce the desired secondary alcohol. Explain this reaction by indicating the source of the hydrogen that ends up on oxygen.

3. Although sodium borohydride is fairly unreactive toward methanol, the addition of a mineral acid to this solution results in the rapid destruction of the NaBH$_4$. Explain.

4. In the vacuum distillation to give pure 1-phenylethanol, anhydrous potassium carbonate was added to neutralize small traces of remaining acid in order to prevent dehydration. Why is this alcohol more prone to acid-catalyzed dehydration than is 2-phenylethanol?

5. Suggest the structure of the white precipitate formed in the reaction of acetophenone with sodium borohydride, and write an equation for its reaction with water and hydrochloric acid.

6. Distillation under reduced pressure is used to purify 1-phenylethanol. Explain why distillation at atmospheric pressure is undesirable. (*Hint:* A side product is formed to a great extent with atmospheric distillation. What is it likely to be?)

7. Besides $NaBH_4$ and $LiAlH_4$, what other reagents might be expected to reduce acetophenone to 1-phenylethanol?

8. Draw the structure of the product that results from complete reduction of the following compounds by $NaBH_4$:
   a. cyclohexanone   **b.** 3-cyclohexen-1-one   **c.** 1,4-butanedial   **d.** 4-oxohexanal

9. Draw the structure of the product that would be formed from allowing each of the compounds in Exercise 8 to react with excess hydrogen gas in the presence of nickel catalyst.

10. Why does concentration of the ethanolic reaction mixture, followed by addition of hydrochloric acid, result in the formation of two layers?

11. What is the expected consequence of working up the reaction *without* acidification?

SPECTRA OF STARTING MATERIAL AND PRODUCT

The IR spectrum of acetophenone is given in Figure 8.23a.

(a) PMR spectrum.

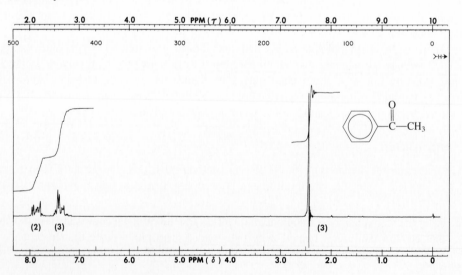

(b) CMR data. Chemical shifts: δ 26.3, 128.6, 128.3, 133.0, 137.3, 197.4.

Figure 20.10 NMR data for acetophenone.

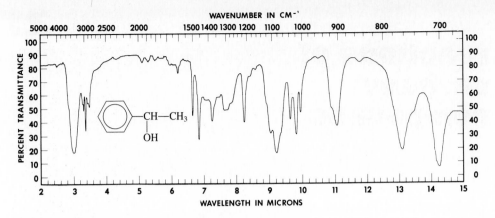

Figure 20.11 IR spectrum of 1-phenylethanol.

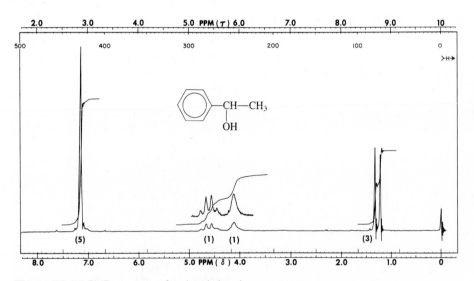

Figure 20.12 PMR spectrum of 1-phenylethanol.

chapter **21**

reactions of carbonyl compounds

21.1 Reactions Involving Nucleophilic Addition to the Carbonyl Group

The addition of a nucleophilic reagent (Nu—E) to a carbonyl compound may be written in the general form

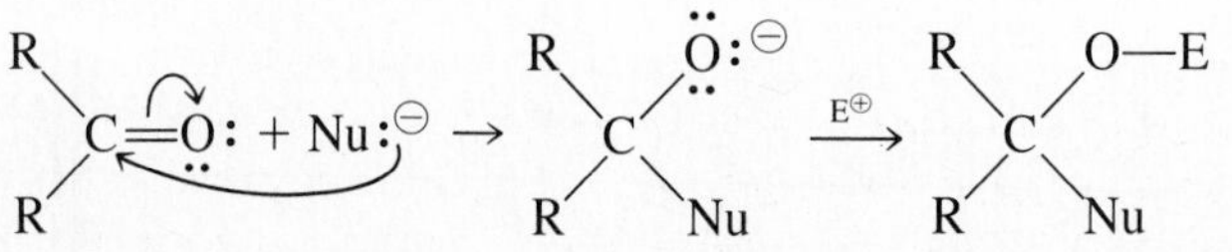

$$
\begin{array}{c}
R \\
\diagdown \\
C{=}\ddot{O}: + Nu:^{\ominus} \longrightarrow \\
\diagup \\
R
\end{array}
\quad
\begin{array}{c}
R \quad \ddot{O}:^{\ominus} \\
\diagdown \diagup \\
C \\
\diagup \diagdown \\
R \quad Nu
\end{array}
\quad \xrightarrow{E^{\oplus}} \quad
\begin{array}{c}
R \quad O{-}E \\
\diagdown \diagup \\
C \\
\diagup \diagdown \\
R \quad Nu
\end{array}
\qquad \textbf{(1)}
$$

where Nu is a nucleophile and E is an electrophile. One of the most important examples of this reaction is the addition of **Grignard reagents** in which Nu is an alkyl group and E is MgX. In Section 22.2 experiments are described in which phenylmagnesium bromide adds to carbon dioxide and methyl benzoate; the reactions of Grignard reagents with aldehydes, ketones, and various other carbonyl compounds also provide extremely valuable synthetic procedures.

(a) Addition of Derivatives of Ammonia

Nitrogen compounds that may be considered to be derivatives of ammonia constitute other important types of nucleophilic reagents that add to carbonyl compounds (equation 2).

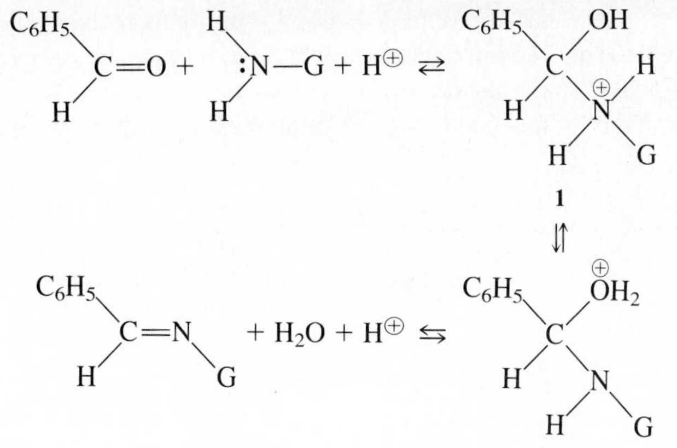

$$
\begin{array}{c}
\underset{H}{\overset{C_6H_5}{\diagdown}} C{=}O \;+\; \underset{H}{\overset{H}{\diagdown}} :N{-}G \;+\; H^{\oplus} \;\rightleftarrows\;
\underset{\underset{\overset{|}{N}}{\overset{|}{\underset{H\diagdown\diagup G}{\oplus}}}}{\overset{OH}{\underset{H}{\overset{C_6H_5\diagdown\diagup}{C}}}} H
\end{array}
$$

1

$$\Updownarrow \tag{2}$$

$$
\underset{H}{\overset{C_6H_5}{\diagdown}} C{=}\underset{G}{N} \;+\; H_2O \;+\; H^{\oplus} \;\rightleftharpoons\;
\overset{\overset{\oplus}{OH_2}}{\underset{\underset{H\diagup\diagdown G}{N}}{\overset{C_6H_5\diagdown\diagup}{\underset{H}{C}}}}
$$

3   **2**

In the case of *primary amines* (R = alkyl or aryl) the addition product, **2**, readily loses a molecule of water to produce an *imine*, **3**. Although aliphatic amines give imines that are unstable and polymerize, aromatic amines give stable imines, also called Schiff bases, that can be isolated. These imines are readily hydrolyzed to regenerate the carbonyl compound and amine; that is, the reaction of equation 2 is reversible.

Imines may be reduced either by catalytic hydrogenation or by the use of a metal hydride reagent such as $LiAlH_4$ or $NaBH_4$ (equation 3).

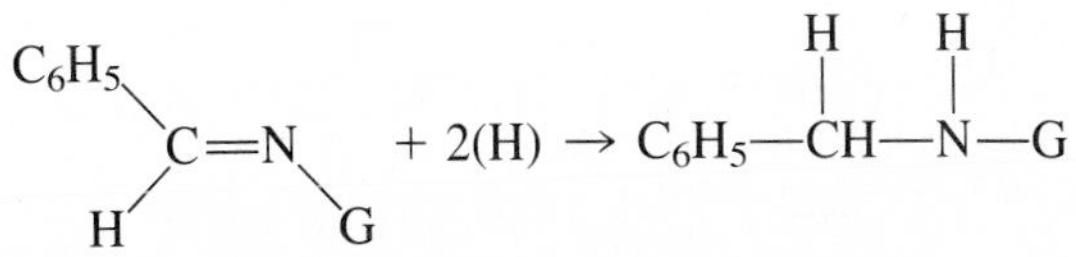

$$
\underset{H}{\overset{C_6H_5}{\diagdown}} C{=}\underset{G}{N} \;+\; 2(H) \;\rightarrow\; C_6H_5{-}\overset{\overset{\textstyle H}{|}}{C}H{-}\overset{\overset{\textstyle H}{|}}{N}{-}G \tag{3}
$$

3   **4**

An example of such a reduction is given in Section 20.3.

Other useful derivatives of ammonia that react with carbonyl compounds according to equation 2 are hydroxylamine (G = OH), which produces oximes, semicarbazide (G = NHCONH$_2$), which produces semicarbazones, and various arylhydrazines (G = NHAr), which produces arylhydrazones. The reaction of semicarbazide with cyclohexanone and 2-furaldehyde is presented in Chapter 16 to illustrate the principle of kinetic and equilibrium control. However, the main importance of the reaction of carbonyl compounds with these reagents is that the products are almost invariably crystalline solids, whereas the carbonyl compounds from which they are derived are often liquids. General procedures for preparing such derivatives, which can be useful in the identification of the parent carbonyl compounds, are given in Chapter 27.

(b) Addition of Wittig Reagents

In 1953 G. Wittig discovered that certain organophosphorus compounds, called *ylides*, **7**, add to carbonyl compounds to form unstable intermediates called *betaines*, **8**, which decompose to produce an alkene and a phosphine oxide, **9.** The highly reactive ylide is generated in the presence of the carbonyl compound by the action of a strong base, often an organolithium compound, on a phosphonium halide, **6.** The latter compound is produced most commonly by reaction of a primary or secondary organic halide with triphenylphosphine (**5,** equation 4; R may be alkyl, aryl, CN, or $CO_2C_2H_5$, among others).

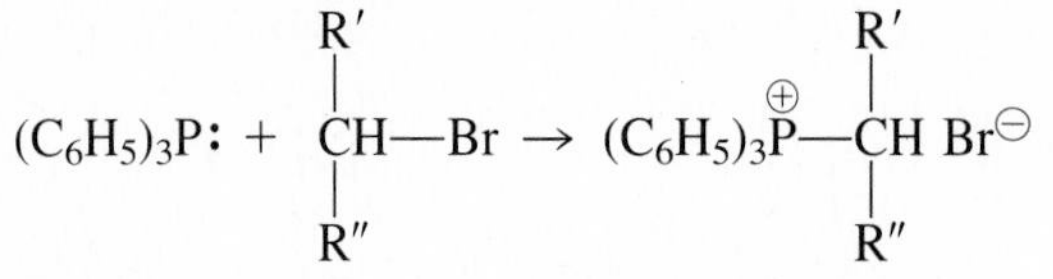

$$(C_6H_5)_3P: + \quad \overset{R'}{\underset{R''}{\overset{|}{\underset{|}{CH}}}}\!\!-\!\!Br \rightarrow (C_6H_5)_3\overset{\oplus}{P}\!\!-\!\!\overset{R'}{\underset{R''}{\overset{|}{\underset{|}{CH}}}} Br^{\ominus} \tag{4}$$

5   **6**

Triphenylphosphine   Phosphonium halide

$$\mathbf{6} \xrightarrow{\text{RLi}} \left[(C_6H_5)_3P\!\!=\!\!\overset{R'}{\underset{R''}{C}} \leftrightarrow (C_6H_5)_3\overset{\oplus}{P}\!\!-\!\!\overset{\ominus}{\underset{R''}{C}}\overset{R'}{} \right] + RH + LiBr \tag{5}$$

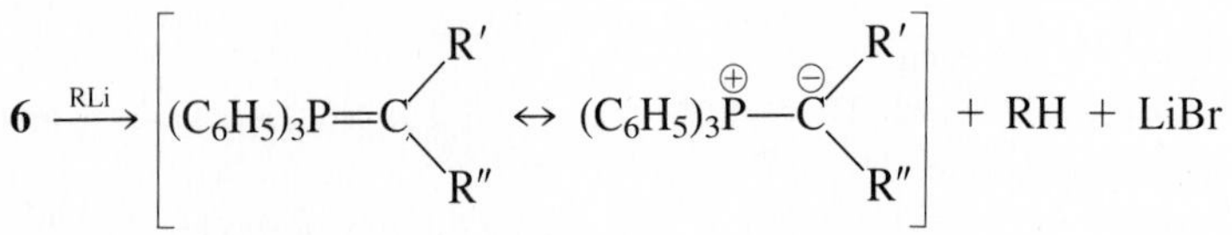

7

Ylide

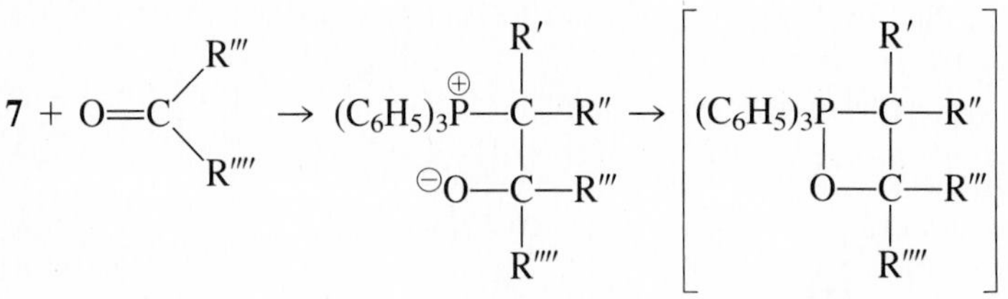

$$\mathbf{7} + O\!\!=\!\!C\overset{R'''}{\underset{R''''}{}} \rightarrow (C_6H_5)_3\overset{\oplus}{P}\!\!-\!\!\overset{R'}{\underset{\underset{R''''}{\overset{\ominus}{O}-\overset{|}{C}-R'''}}{\overset{|}{\underset{|}{C}}-R''}} \rightarrow \left[(C_6H_5)_3P\!\!-\!\!\overset{R'}{\underset{\underset{R''''}{\overset{|}{O}-\overset{|}{C}-R'''}}{\overset{|}{\underset{|}{C}}-R''}} \right] \tag{6}$$

8   **8a**

Betaine   Oxyphosphetane

$$(C_6H_5)_3\underset{O}{\overset{\|}{P}} \quad + \quad \overset{R'}{\underset{R''''}{\overset{\diagdown}{C}}}\overset{R''}{\underset{R'''}{\overset{\diagup}{\underset{\diagdown}{C}}}}$$

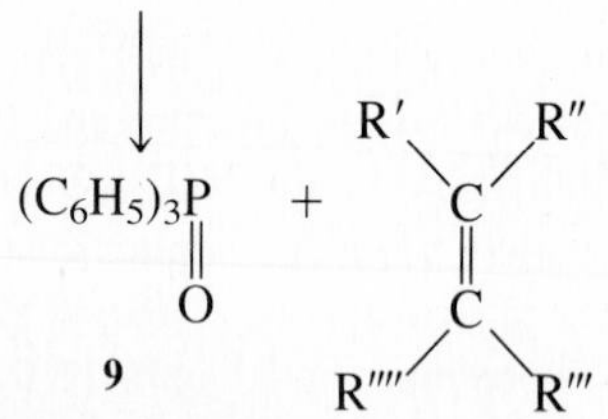

9

Triphenylphosphine
oxide

The electron-rich carbon atom in **7** is nucleophilic, as can be seen from the polarized resonance structure of the ylide, and it adds to a carbonyl group in the expected way to give the betaine. The decomposition of the betaine is thought to involve formation of an oxyphosphetane, **8a,** which then fragments to **9** and the alkene.

The overall effect of reactions 4–6 is conversion of a carbon-oxygen double bond to a carbon-carbon double bond, that is, $C\!\!=\!\!O \rightarrow C\!\!=\!\!C$. The Wittig synthesis thus represents a very general method of preparation of alkenes, one that has two important advantages over older procedures: (1) the carbonyl group is replaced *specifically* by a carbon-carbon double bond, without the formation of isomeric alkenes; and (2) the reactions are carried out under very mild conditions.

Various modifications of the original Wittig synthesis have been made. One modification that is convenient for an introductory experiment because it does not require an inert atmosphere and the use of the somewhat dangerous organolithium reagents involves the use of triethyl phosphite (**10**) in place of triphenylphosphine. Activated organic halides, such as benzyl and allyl chlorides and α-bromo esters, will react satisfactorily in an S_N2 manner with esters of phosphorous acid (equation 7). The phosphonium salt, **11**, produced is unstable toward heat; for example, the product from benzyl chloride evolves ethyl chloride and yields the phosphonate ester, **12**. The reaction between an alkyl halide and a phosphite to give a phosphonate is called the Arbuzov reaction. Treatment of 12 with sodium methoxide produces a nucleophile analogous to an ylide, and in the presence of benzaldehyde the reactions of equations 9 and 10 take place to yield an alkene (**14**, stilbene) and sodium diethylphosphate (**15**). The latter is water-soluble, as is the DMF used as solvent, so that the alkene product can readily be isolated by washing with aqueous methanol.

$$(C_2H_5O)_3P + C_6H_5CH_2\!\!-\!\!Cl \rightarrow (C_2H_5O)_3\overset{\oplus}{P}\!\!-\!\!CH_2C_6H_5 \ Cl^{\ominus} \qquad \textbf{(7)}$$

10		**11**
Triethyl phosphite	Benzyl chloride	Phosphonium salt

$$(C_2H_5O)_2\underset{\oplus}{P}\!\!-\!\!CH_2C_6H_5 \ Cl^{\ominus} \xrightarrow[\text{heat}]{} (C_2H_5O)_2\overset{O}{\overset{\|}{P}}\!\!-\!\!CH_2C_6H_5 + C_2H_5Cl \uparrow \qquad \textbf{(8)}$$

(with $O\!\!-\!\!CH_2CH_3$ group shown)

11	**12**	
	Phosphonate ester	Chloroethane

$$12 + CH_3O^{\ominus}Na^{\oplus} \rightarrow (C_2H_5O)_2\overset{O}{\overset{\|}{P}}\!\!-\!\!\overset{\ominus}{C}HC_6H_5 \ Na^{\oplus} + CH_3OH \qquad \textbf{(9)}$$

	Sodium methoxide	**13**

$$13 + O\!\!=\!\!CHC_6H_5 \rightarrow (C_2H_5O)_2\overset{O}{\overset{\|}{P}}\!\!-\!\!CHC_6H_5 \ Na^{\oplus} \qquad \textbf{(10)}$$

Benzaldehyde

$$^{\ominus}O\!\!-\!\!CHC_6H_5$$

$$\downarrow$$

$$(C_2H_5O)_2\overset{O}{\overset{\|}{P}}O^{\ominus}Na^{\oplus} + \overset{CHC_6H_5}{\overset{\|}{CHC_6H_5}}$$

15	**14**
Sodium diethylphosphate	Stilbene

There are two geometric isomers of stilbene: the *trans* isomer is a solid at room temperature, mp 126–127°C, whereas the *cis* isomer is a liquid, mp 6°C. The phosphonate ester modification of the Wittig reaction is reported to give the pure *trans* isomer, even though the original Wittig procedure (using triphenylphosphine as starting material) is reported to give a mixture of 30% *cis*-, 70% *trans*-stilbene. The IR, PMR, and UV spectra of the two isomers are distinctly different and may be used in both qualitative and quantitative analysis of mixtures.

Optional or Additional Experiments   Other aldehydes may be used in place of benzaldehyde in the phosphonate modification of the Wittig reaction. With *trans*-cinnamaldehyde (**16**), the product is *trans,trans*-1,4-diphenyl-1,3-butadiene (**17**).

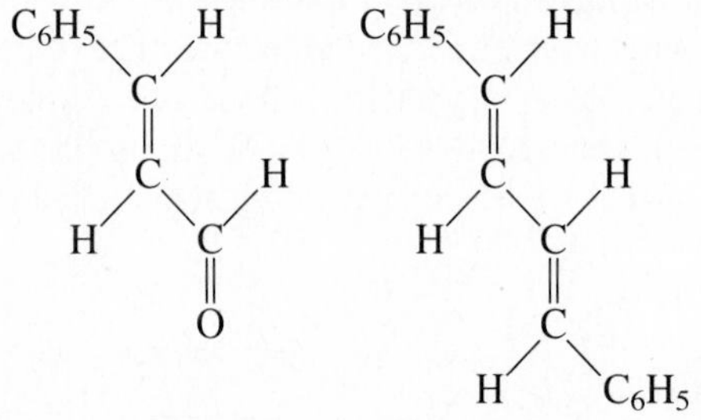

16
trans-Cinnamaldehyde

17
trans,trans-1,4-
Diphenyl-1,3-butadiene

If some students prepare this diene, it will be interesting to compare its UV absorption spectrum with that of stilbene to see the effect of the additional conjugated double bond. The UV spectra of *cis*- and *trans*-stilbene are provided in Figure 21.7; if it is not practical to record the UV spectrum of **17** in your laboratory, consult reference 3 at the end of this section.

When 2-furaldehyde (**18**) is substituted for benzaldehyde, the product is *trans*-2-styryl-furan (**19**).

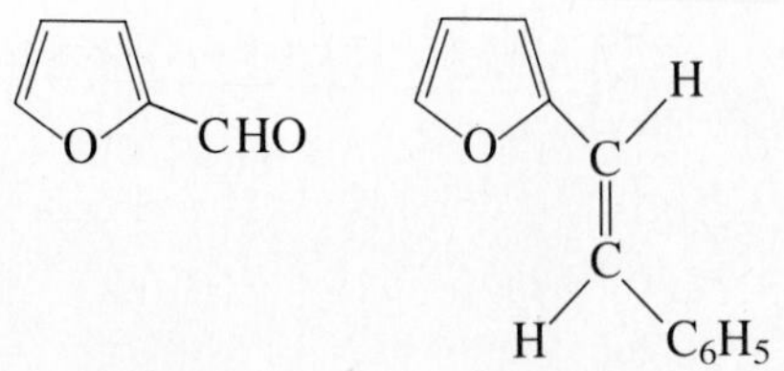

18
2-Furaldehyde

19
trans-2-Styrylfuran

Pre-lab exercises for Section 21.1, Reactions Involving Nucleophilic Addition to the Carbonyl Group, can be found on page PL. 85.

EXPERIMENTAL PROCEDURE

DO IT SAFELY

1 Organophosphorus compounds are toxic, and benzyl chloride is irritating and lachrymatory. Avoid contact of these substances with your skin and inhalation of their vapors. Experiments using these reagents should be performed in the hood if possible, and rubber gloves should be worn, particularly when measuring out the reagents. Should these substances come into contact with your skin, wash the area thoroughly with soap and warm water, and rinse with large quantities of water.

2 Carbon disulfide is highly volatile and extremely flammable. Avoid its use near open flames or hot plates.

A. *trans*-Stilbene

Weigh 0.045 mol of triethyl phosphite into a 25- or 50-mL round-bottomed flask, and add 0.045 mol of benzyl chloride and several small boiling stones. Connect a water-cooled condenser to the flask, and heat the mixture at gentle reflux for 1 hr. While this reaction mixture is cooling to room temperature, place 0.045 mol of sodium methoxide in a 125-mL Erlenmeyer flask, and immediately add 20 mL of *N,N*-dimethylformamide (DMF). Pour the *cool* phosphonate reaction mixture into the Erlenmeyer flask, and rinse the round-bottomed flask with 20 mL of DMF, adding the rinse solution to the rest of the reaction mixture. Cool the Erlenmeyer flask and its contents in an ice-water bath, and stir until the temperature of the solution is below 20°C. While continuing to stir the cooled solution, slowly add 0.045 mol of benzaldehyde, cooling as necessary so that the temperature of the mixture does not rise above 35°C. Remove the flask from the cooling bath, and allow it to stand at room temperature for about 10 min. Add 35 mL of water with stirring, collect the crystals on a Büchner funnel, and wash them with cold 1:1 methanol-water. Determine the weight and the melting point of the crystalline product. The reported melting point of *trans*-stilbene is 124°C. A mixture of ethanol and ethyl acetate may be used to recrystallize impure stilbene. Prepare a sample of the product in carbon disulfide or carbon tetrachloride solution or in a KBr pellet for infrared analysis, and compare the spectrum with those of Figures 21.3 and 21.4.

B. 1,4-Diphenyl-1,3-butadiene

This diene may be prepared by following the procedure above up to the point of addition of benzaldehyde. Use 0.045 mol of cinnamaldehyde in place of benzaldehyde, and continue to follow the procedure up to the point of collecting the crystals. Wash the crystals first with water and then with methanol until the filtrate is colorless. Weigh the product and determine its melting point. The melting point of *trans,trans*-1,4-diphenyl-1,3-butadiene is reported to be 150–151°C.

C. *trans*-2-Styrylfuran

Follow the procedure for *trans*-stilbene exactly except substitute an equimolar amount of freshly distilled 2-furaldehyde for benzaldehyde. The melting point of *trans*-2-styrylfuran is reported to be 54–55°C. This product may be recrystallized from methanol.

EXERCISES

1. Compare the mechanism of aldol addition to that of the Wittig synthesis, pointing out similarities and points of divergence.

2. Why are ammonium salts of the type $(C_6H_5)_3\overset{+}{N}CHR_2\ Br^-$ much less acidic than phosphonium salts such as **6**?

3. Write equations for the preparation of the following alkenes by the original Wittig synthesis from triphenylphosphine and the necessary organic halides and carbonyl compounds:

   a. $C_6H_5CH{=}C(CH_3)_2$

   b. 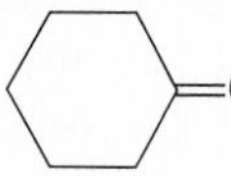 $={=}C(CH_3)C_6H_5$

   c. $CH_2{=}C(CH_3)C_6H_5$

4. Write equations for Wittig syntheses of the compounds of Exercise 3 using alternative organic halides and carbonyl compounds. Tell which pair of reactants would give the desired product in better yield in each case.

5. Write equations for the preparation of the following alkenes by the phosphonate ester modification (starting with triethyl phosphite):
   a. $C_6H_5C(CH_3){=}C(CH_3)C_6H_5$
   b. $CH_2{=}CH{-}CH{=}CH{-}C_6H_5$
   c. $(CH_3)_2C{=}CH{-}CO_2C_2H_5$

6. Would you expect an ylide, **7**, in which $R_1 = CN$, to be more or less stable than one in which $R_1 = $ alkyl?

7. Why should the sodium methoxide not be exposed to the atmosphere for more than a few minutes?

8. Why should the aldehydes used as starting materials in Wittig syntheses be free of carboxylic acids?

9. What peaks in the IR spectra of the stilbenes (Figures 21.3 and 21.4) would be most useful for quantitative analysis of a mixture of *cis* and *trans* isomers?

10. *N,N*-Dimethylformamide (DMF) is highly water-soluble. Why?

11. DMF is known to be an extremely effective solvator of cations, which is one reason that $NaOCH_3$ readily dissolves in it. How might you account for this property of the solvent?

12. a. Refer to Figure 21.7 and calculate log ϵ for the three maxima in the UV spectrum of *trans*-stilbene and for the two maxima in the UV spectrum of *cis*-stilbene.

 *b. Rationalize the differences in the UV spectra exhibited by these two geometric isomers.

SPECTRA OF STARTING MATERIALS AND PRODUCTS

The IR spectrum and NMR data for benzaldehyde are given in Figures 8.22a and 19.4, respectively. The IR spectrum and NMR data for cinnamaldehyde are given in Figures 20.5 and 20.6, respectively. The IR spectrum, NMR data, and UV spectrum of 2-furaldehyde are given in Figures 16.3, 16.4, and 16.5, respectively.

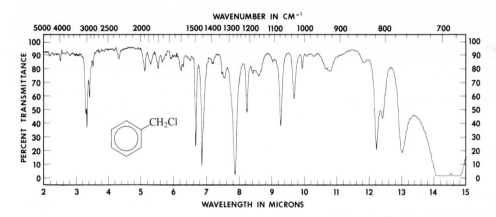

Figure 21.1 IR spectrum of benzyl chloride.

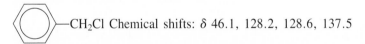

—CH$_2$Cl Chemical shifts: δ 46.1, 128.2, 128.6, 137.5

Figure 21.2 CMR data for benzyl chloride.

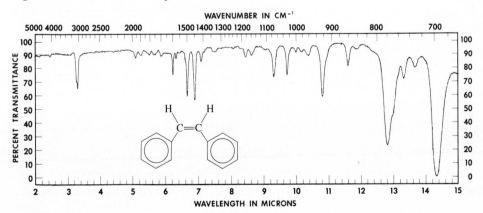

Figure 21.3 IR spectrum of *cis*-stilbene.

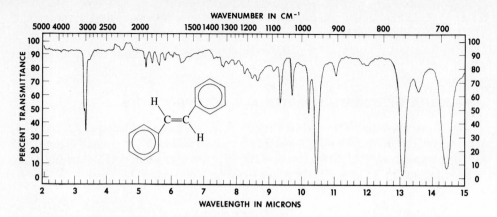

Figure 21.4 IR spectrum of *trans*-stilbene (KBr pellet).

(a) PMR spectrum.

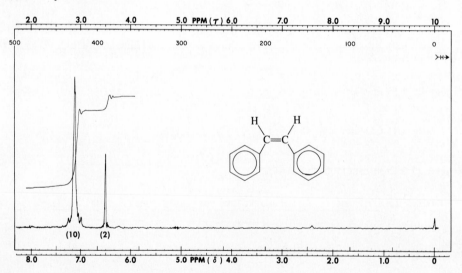

(b) CMR data. Chemical shifts: δ 127.1, 128.2, 128.9, 130.2, 137.4.

Figure 21.5 NMR data for *cis*-stilbene.

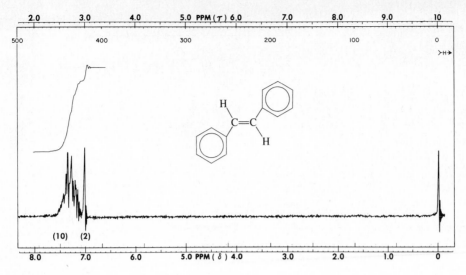

Figure 21.6 PMR spectrum of *trans*-stilbene.

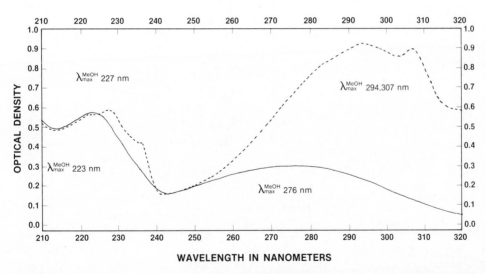

Figure 21.7 UV spectra of *cis*- (solid line) and *trans*- (dashed line) stilbene. Concentration 0.0500 g/L. Pathlength 0.1 cm.

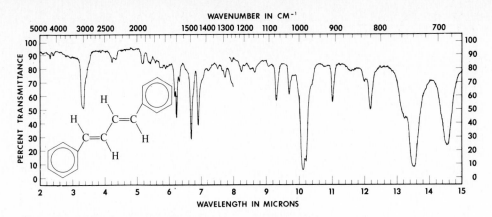

Figure 21.8 IR spectrum of *trans,trans*-1,4-diphenyl-1,3-butadiene (KBr pellet).

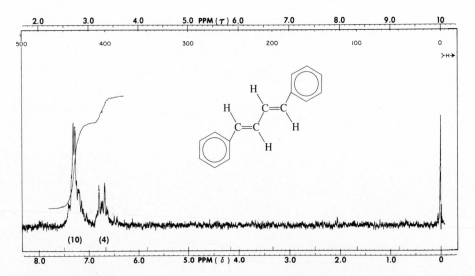

Figure 21.9 PMR spectrum of *trans,trans*-1,4-diphenyl-1,3-butadiene.

REFERENCES

1. Seus, E. J.; Wilson, C. V. *Journal of Organic Chemistry,* **1961,** *26,* 5243.

2. Wittig G.; Schollkopf, U. *Chemische Berichte,* **1954,** *87,* 1318.

3. Pinckard, J. H.; Willie, B.; Zechmeister, L. *Journal of the American Chemical Society,* **1948,** *70,* 1939.

4. Maercker, A., in *Organic Reactions,* A. C. Cope, editor, John Wiley & Sons, New York, 1965, Vol. 14, Chapter 3. A good general review.

21.2 **Reactions of Stabilized Carbanions Derived from Carbonyl Compounds**

The reactions studied in Chapter 19 and in the preceding section of this chapter—that is, oxidation to a carboxyl group, reduction to an alcohol group, or addition of various reactants across the carbon-oxygen π-bond—all take place at the carbonyl group. The nature of the carbonyl group also affects its neighboring α-carbon atoms. The partial positive charge on the carbonyl carbon tends to make the hydrogens on the α-carbons (the "α-hydrogens") easily removed as protons, and the resulting carbanion is stabilized by resonance with the carbonyl π-electron system to produce what is termed an *enolate anion* (**20**, equation 11).

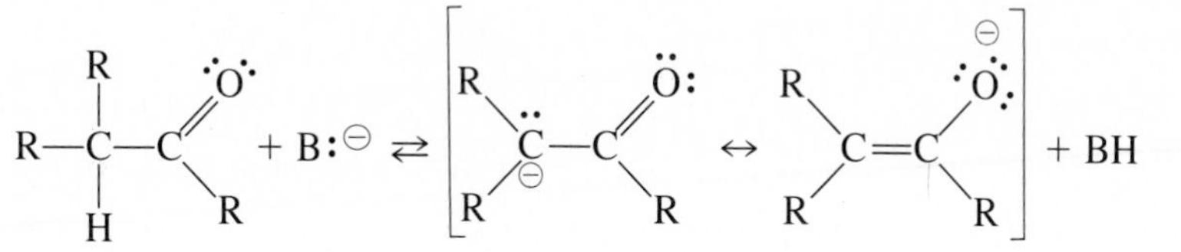

$$\text{(11)}$$

(a) Aldol Additions and Condensations

An important and general reaction of enolate anions is addition to the carbonyl group of the aldehyde or ketone from which the enolate anion was derived. In this way a dimeric anion, **21**, is produced (equation 12), which stabilizes itself by abstracting a proton from the solvent (water or alcohol), as shown in equation 13.

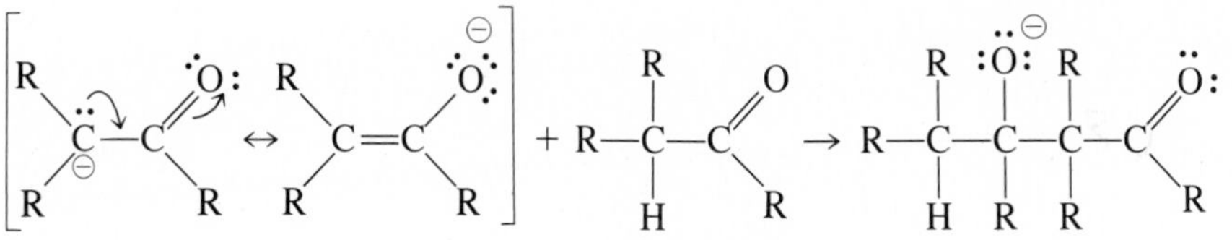

$$\text{(12)}$$

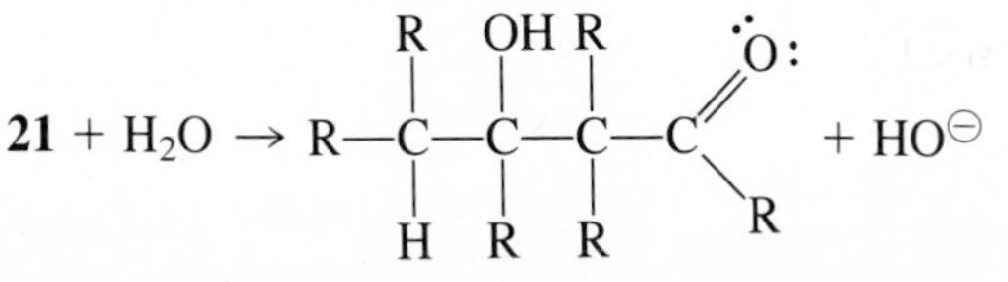

$$\text{(13)}$$

The β-hydroxy carbonyl compound of type **22** is called an "aldol," since it is both an *ald*ehyde and an alco*hol*. The term "aldol *addition*" is also applied generally to the self-addition of ketones as well as of aldehydes. The overall change for the reaction of acetaldehyde is given in equation 14. Most β-hydroxy aldehydes and ketones undergo dehydration readily to α,β-unsaturated aldehydes and ketones (equation 15).

$$2\ CH_3CHO \xrightarrow{\ HO^{\ominus}\ } \underset{\text{3-Hydroxybutanal}}{CH_3\overset{\displaystyle OH}{\overset{|}{CH}}-CH_2CHO} \qquad \textbf{(14)}$$

$$\underset{\text{2-Butenal}}{CH_3\overset{\displaystyle OH}{\overset{|}{CH}}-CH_2CHO \rightarrow CH_3CH{=}CHCHO + H_2O} \qquad \textbf{(15)}$$

In this case the overall reaction may be referred to as an "aldol *condensation*," since a molecule of water is eliminated from the adduct.

In general, ketones do not undergo self-addition as easily as aldehydes, and special conditions must usually be employed to obtain good yields in such reactions. (Refer to a textbook for a discussion of the reaction of acetone, for example). "Mixed (or crossed) aldol condensations" of two different aldehydes or of an aldehyde with a ketone are possible. Such mixed condensations are practical for synthesis in the case of (1) two aldehydes *if* one of the aldehydes has no α-hydrogens (and hence only serves as a carbonyl acceptor of the enolate anion derived from the other aldehyde) and (2) a ketone and aldehyde under conditions such that the ketone undergoes no appreciable self-condensation. A good example of the latter case is the reaction of anisaldehyde (*p*-methoxybenzaldehyde, **24**) with acetophenone in the presence of dilute alcoholic sodium hydroxide solution (equations 16–19). The product, **27**, is called anisalacetophenone. Under the conditions of the experiment, the dehydration of the "aldol," **26**, is spontaneous in the presence of base, owing to the stabilization gained by conjugation of the new carbon-carbon double bond with the aromatic ring (equation 19).

$$\underset{\text{Acetophenone}}{C_6H_5\overset{\displaystyle O}{\overset{\|}{C}}-CH_3} + HO^{\ominus} \xrightarrow{\ -H_2O\ } \left[\underset{\textbf{23}}{C_6H_5\overset{\displaystyle \cdot\overset{\cdot\cdot}{O}\cdot}{\overset{\|}{C}}-\overset{\ominus}{C}H_2 \leftrightarrow C_6H_5\overset{\displaystyle :\overset{\cdot\cdot}{O}:^{\ominus}}{\overset{|}{C}}{=}CH_2}\right] \qquad \textbf{(16)}$$

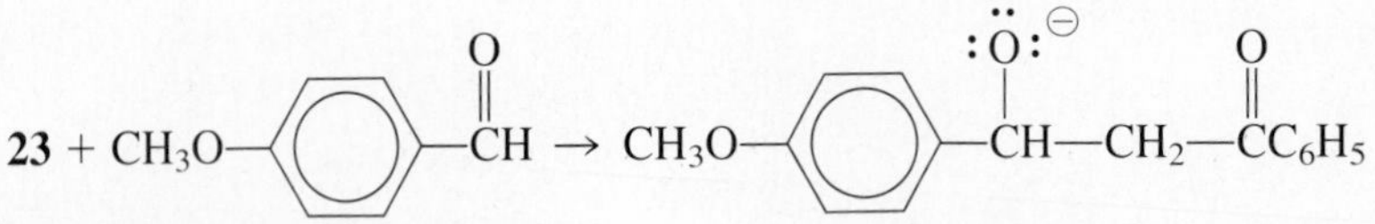

$$\textbf{23} + CH_3O-\!\!\left\langle\bigcirc\right\rangle\!\!-\overset{\displaystyle O}{\overset{\|}{C}}H \rightarrow \underset{\textbf{25}}{CH_3O-\!\!\left\langle\bigcirc\right\rangle\!\!-\overset{\displaystyle :\overset{\cdot\cdot}{O}:^{\ominus}}{\overset{|}{C}}H-CH_2-\overset{\displaystyle O}{\overset{\|}{C}}C_6H_5} \qquad \textbf{(17)}$$

24
p-Methoxybenzaldehyde
(anisaldehyde)

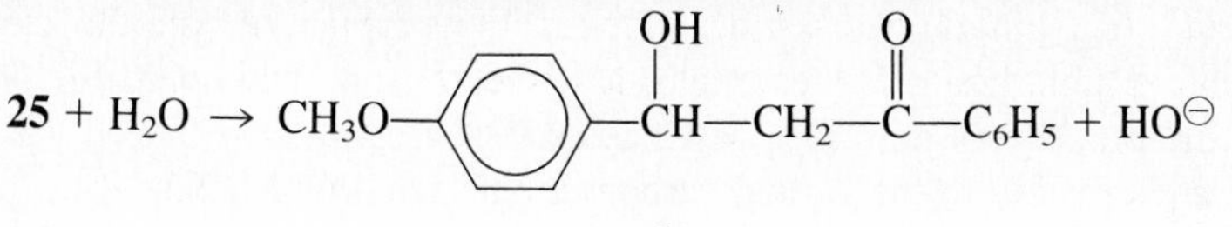

$$25 + H_2O \rightarrow CH_3O-\!\!\!\bigcirc\!\!\!-\underset{\displaystyle \overset{OH}{|}}{CH}-CH_2-\overset{\displaystyle O}{\overset{\|}{C}}-C_6H_5 + HO^{\ominus} \qquad \textbf{(18)}$$

26

Aldol

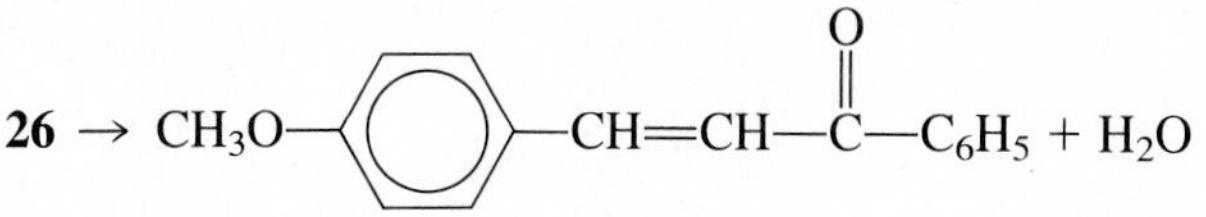

$$26 \rightarrow CH_3O-\!\!\!\bigcirc\!\!\!-CH=\!CH-\overset{\displaystyle O}{\overset{\|}{C}}-C_6H_5 + H_2O \qquad \textbf{(19)}$$

27

Anisalacetophenone

A second example of an aldol condensation is that of 2-furaldehyde **(28)** with *m*-nitroacetophenone **(29)** to yield 3-(2-furyl)-1-(3-nitrophenyl)propenone **(30)**. The overall reaction is given in equation 20.

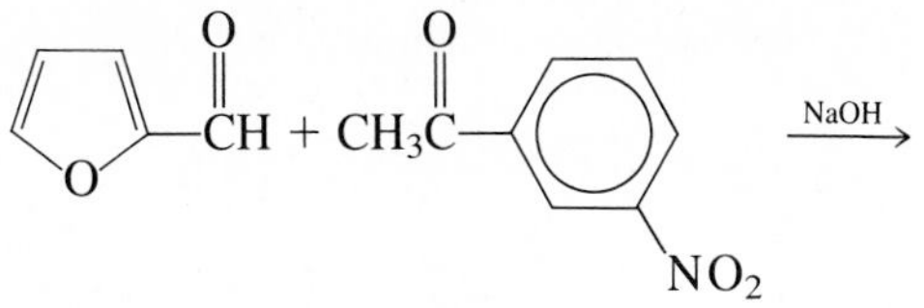

$$\qquad \textbf{(20)}$$

28

2-Furaldehyde

29

m-Nitroacetophenone

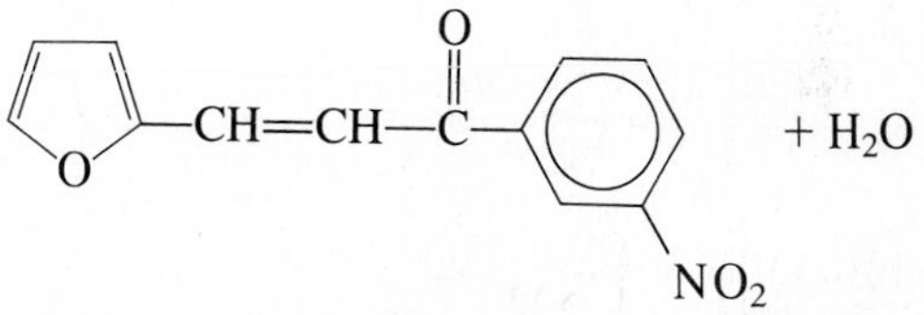

30

3-(2-Furyl)-1-(3-nitrophenyl)propenone

(b) Reactions of α,β-Unsaturated Ketones

The aldol condensation products **27** and **30** belong to a group of α,β-unsaturated ketones called *chalcones*. The parent member of the group is benzalacetophenone, **31,** but as this compound has been found to cause skin irritation to some individuals, the substituted chalcones **27** and **30** have been chosen for our experiments.

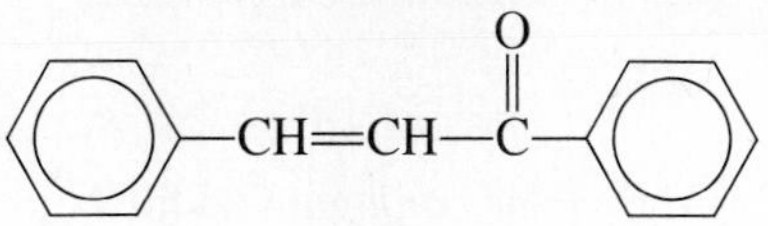

31

Benzalacetophenone

As expected, α,β-unsaturated ketones give reactions typical of both the ketone function and the alkenic double bond. For example, a semicarbazone and an oxime of **27** may be obtained by the procedures described in Chapter 27.

Bromine adds readily to the carbon-carbon double bond of **27**. Since two chiral carbon atoms are produced in this reaction, there should be four possible stereoisomers of the dibromide, two enantiomeric pairs:

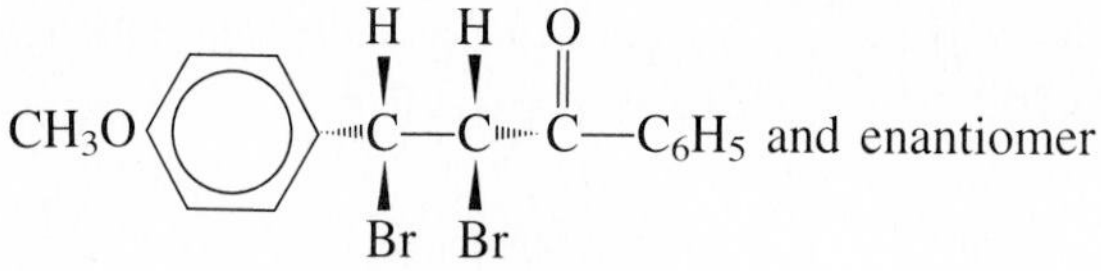

Erythro

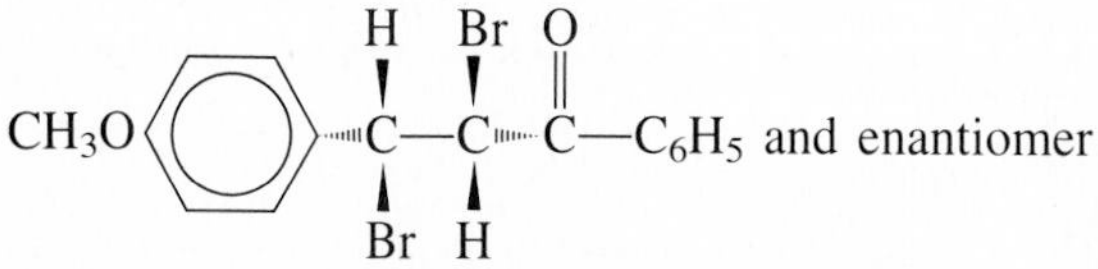

Threo

The product reported in the literature (mp 140°C) is assumed to have the *erythro* structure. This is the product expected by *anti* addition of bromine to *trans*-**27**, which is the more stable form of **27** that is normally obtained in the mixed aldol condensation. (The *cis* isomer of **27** can be obtained by ultraviolet irradiation of the *trans* isomer; its melting point is reported to be 33–33.5°C.)

A third type of reaction of α,β-unsaturated ketones is indicative of the interaction between the carbonyl group and the carbon-carbon double bond. Although nucleophiles do not normally attack isolated carbon-carbon double bonds, α,β-unsaturated carbonyl compounds will often undergo 1,4-addition. For example, ethylmagnesium bromide reacts with **27** to give 3-(*p*-methoxyphenyl)pentenophenone **(32)**.

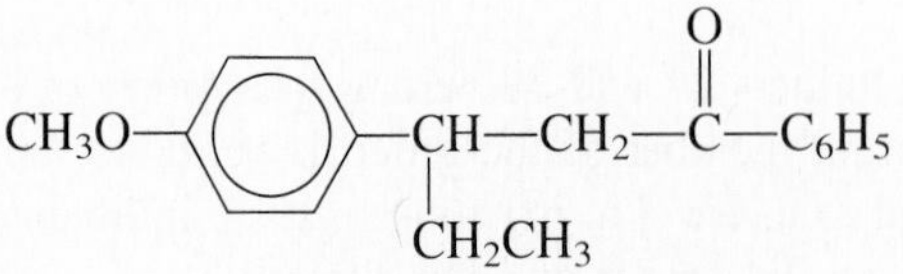

32

3-(*p*-Methoxyphenyl)pentenophenone

Another example of 1,4-addition, or *conjugate* addition, is given in Section 21.3, the acid-catalyzed reaction of 2-methylpropanal (isobutyraldehyde) with 3-buten-2-one (methyl vinyl ketone), and an explanation is given there for this type of reaction.

EXPERIMENTAL PROCEDURE

DO IT SAFELY

1 The Br_2/CCl_4 solution should be handled with care, because bromine, even in solution, is a dangerous chemical. Use the reagent in the hood if possible. Do not breathe the vapors of the solution or spill it on your skin. In case of accident, see the Do It Safely section in Section 12.2.

2 The concentrated solution of sodium hydroxide prepared from a few pellets of NaOH is highly caustic. Take care not to get it on your skin; if you do, wash the area with copious amounts of water.

3 If you warm the solutions in 95% ethanol to hasten solution, or if you recrystallize your products from this solvent, do not use a flame, and take care that there are no flames in the vicinity.

Pre-lab exercises for Section 21.2, Part A, Aldol Additions and Condensations, can be found on page PL. 87.

A. The Aldol Condensation

1. Preparation of Anisalacetophenone (27) In a 100-mL beaker place 0.05 mol of anisaldehyde and an equimolar amount of acetophenone. Add 20 mL of 95% ethanol and swirl to dissolve the reactants. In another small beaker place 2 or 3 pellets of NaOH and dissolve them in about 1 or 2 mL of water; mashing and stirring with a glass rod will expedite the dissolution. Pour the NaOH solution into the ethanolic solution of the reactants, swirl it for a minute or two until the solution is homogeneous, and allow it to stand at room temperature for 10 min.★ Cool the reaction mixture, in which crystals may have already begun to form, in ice water, then collect the crystals by vacuum filtration. Wash the light yellow crystals on the filter with a little cold 95% ethanol, dry them in air,★ and determine their melting point. The melting point of pure *trans*-anisalacetophenone is 77–78°C. It may be recrystallized from 95% ethanol.

2. Preparation of 3-(2-Furyl)-1-(3-nitrophenyl)propenone (30) This compound may be prepared by the same procedure as that for anisalacetophenone **(27)**, except that the reactants, 2-furaldehyde **(28)** and *m*-nitroacetophenone **(29)**, should be dissolved in 50 mL of 95% ethanol instead of 20 mL. The 2-furaldehyde should be freshly distilled before being used in this preparation. The melting point of the pure chalcone is 100–101°C. The crystals should be light yellow when washed free of the dark by-products from the 2-furaldehyde. They may be recrystallized from 95% ethanol.

Pre-lab exercises for Section 21.2, Part B, Reactions of α,β-Unsaturated Ketones, can be found on page PL. 89.

B. Reactions of α,β-Unsaturated Ketones

1. Preparation of a Semicarbazone and an Oxime of Anisalacetophenone See Section 27.5a, part 2, for directions for preparing these derivatives.

2. Preparation of Anisalacetophenone Dibromide Dissolve 2 g of anisalacetophenone in 5 mL of chloroform and add 10 mL of 1 M bromine in CCl_4 dropwise. Allow the mixture to stand at room temperature for 30 to 40 min★ and collect the crystals by vacuum filtration. Wash the white crystals on the filter paper with a little cold CCl_4, allow them to dry,★ and determine their melting point. The reported melting point of the dibromide (assumed to be *erythro*) is 139–140°C.

EXERCISES

1. Explain why the main reaction between acetophenone and anisaldehyde is the mixed aldol reaction rather than (a) self-condensation of acetophenone or (b) Cannizzaro reaction of anisaldehyde.

2. No more than 2 mL of water should be used to dissolve the 2 or 3 pellets of NaOH. Why is not a larger volume of water used?

3. Anisalacetophenone dibromide is much more soluble in $CHCl_3$ than in CCl_4. Suggest a reason for this difference.

4. Crystals of pure *erythro*-anisalacetophenone dibromide are *white*, whereas crystals of pure anisalacetophenone are light *yellow*. Suggest a reason for this difference in color.

5. Could *trans*-anisalacetophenone be isomerized to the *cis* isomer by irradiating it with UV light? Why or why not?

6. Write out a mechanism that rationalizes the stereochemistry of the formation of *erythro*-anisalacetophenone dibromide from bromine and *trans*-anisalacetophenone.

7. Write equations to show the major products from the reactions of the following sets of starting materials:

   a. propionaldehyde, dilute NaOH
   b. formaldehyde, concentrated KOH
   c. trimethylacetaldehyde, acetophenone, dilute NaOH
   d. *p*-methoxybenzaldehyde (2 mol), acetone, dilute KOH
   e. benzaldehyde, methyl propionate, sodium methoxide in methanol

8. Refer to Figures 21.14 and 21.17 and calculate log ϵ for the two maxima in the UV spectrum of *trans*-anisalacetophenone and for the single maxima in its dibromide.

*9. Why does *trans*-anisalacetophenone exhibit a maximum in its UV spectrum at substantially longer wavelength than does its dibromide? What electronic excitation is responsible for the absorption at longer wavelength?

SPECTRA OF STARTING MATERIALS AND PRODUCTS

The IR, NMR, and UV spectra of 2-furaldehyde are provided as Figures 16.3, 16.4, and 16.5. The IR spectrum and NMR data for acetophenone are given in Figures 8.23a and 20.10, respectively.

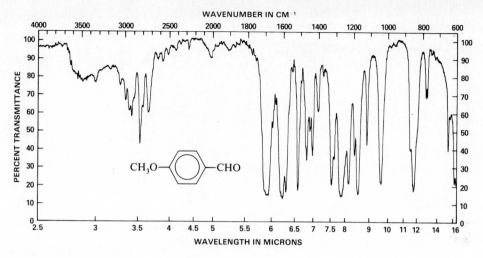

WAVENUMBER IN CM⁻¹

PERCENT TRANSMITTANCE

CH_3O—⟨benzene ring⟩—CHO

WAVELENGTH IN MICRONS

Figure 21.10 IR spectrum of anisaldehyde.

(a) PMR spectrum.

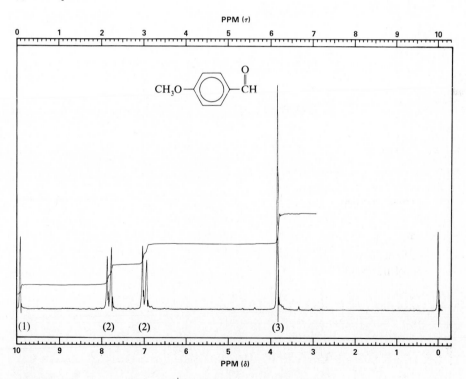

PPM (τ)

CH_3O—⟨benzene ring⟩—$\overset{\overset{\displaystyle O}{\|}}{C}H$

(1)   (2)   (2)   (3)

PPM (δ)

(b) CMR data. Chemical shifts: δ 55.5, 114.5, 130.2, 131.9, 164.6, 190.5.

Figure 21.11 NMR data for anisaldehyde.

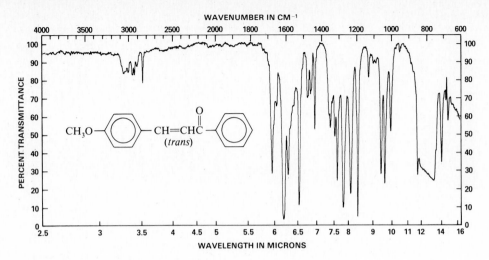

Figure 21.12 IR spectrum of *trans*-anisalacetophenone (in CCl₄ solution).

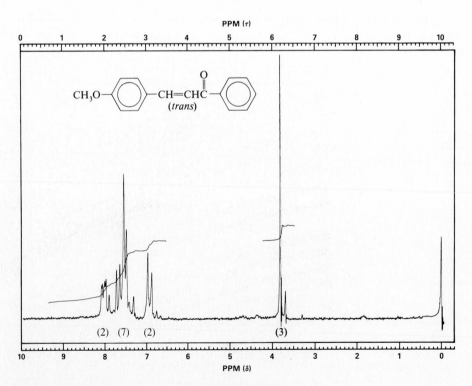

Figure 21.13 PMR spectrum of *trans*-anisalacetophenone.

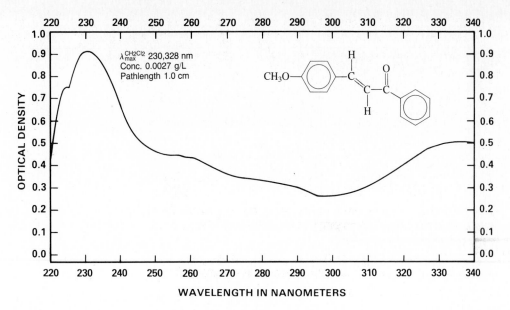

Figure 21.14 UV spectrum of *trans*-anisalacetophenone.

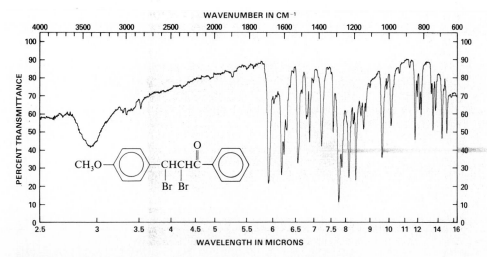

Figure 21.15 IR spectrum of anisalacetophenone dibromide (KBr pellet).

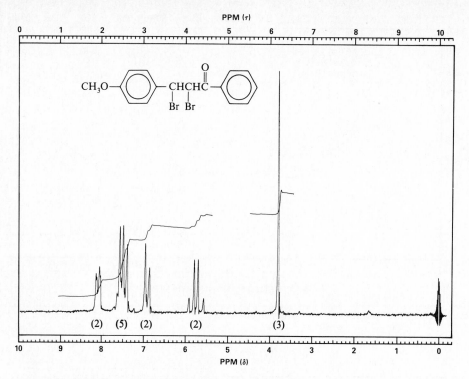

Figure 21.16 PMR spectrum of anisalacetophenone dibromide.

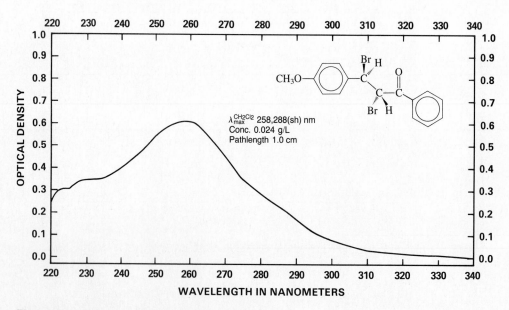

Figure 21.17 UV spectrum of anisalacetophenone dibromide.

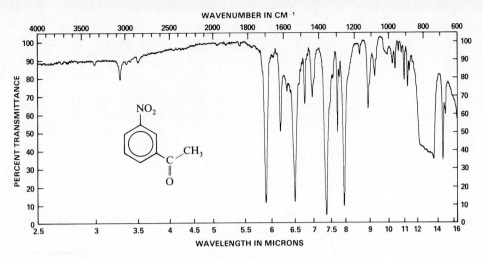

Figure 21.18 IR spectrum of *m*-nitroacetophenone (in CCl₄ solution).

(a) PMR spectrum.

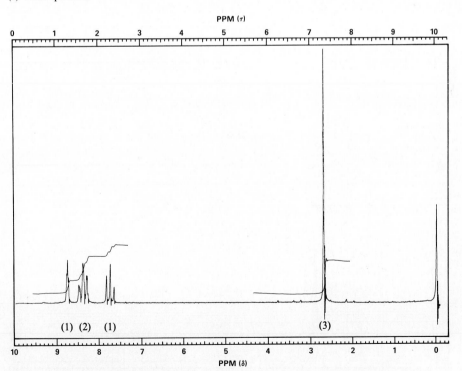

(b) CMR data. Chemical shifts: δ 26.6, 122.9, 127.2, 130.1, 133.9, 138.5, 148.6, 195.6.

Figure 21.19 NMR data for *m*-nitroacetophenone.

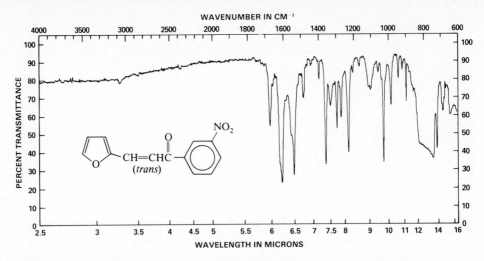

Figure 21.20 IR spectrum of 3-(2-furyl)-1-(3-nitrophenyl)propenone (in CCl₄ solution).

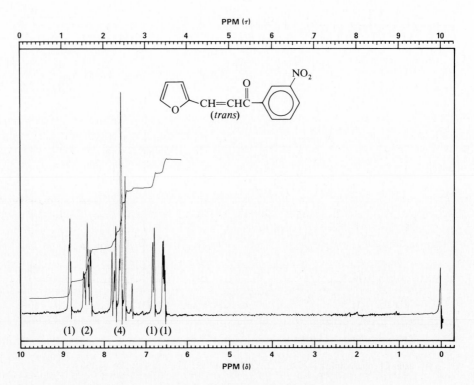

Figure 21.21 PMR spectrum of 3-(2-furyl)-1-(3-nitrophenyl)propenone.

(c) Alkylation of Enolate Ions

The aldol addition (equations 12 and 13) is of great value because it represents a method of producing a new carbon-carbon bond, either intra- or intermolecularly, and because the products contain functional groups on which a variety of further useful synthetic transformations can be performed. In a general way the crucial carbon-carbon bond-forming step (equation 12) in the process can be characterized as reaction of a nucleophile, the enolate ion, with an electrophile, the carbonyl carbon atom. It should not be surprising that organic chemists have studied the reaction of enolate ions with other potential carbon electrophiles in order to broaden the scope and utility of enolate chemistry; use of one such class of electrophiles is discussed in the following paragraphs.

That alkyl halides serve as electrophiles is evidenced by their tendency to undergo S_N1 and S_N2 reactions (Chapter 18). They do not function well for alkylation of enolate ions generated under the usual reaction conditions for the aldol addition (dilute aqueous or alcoholic base) for a variety of reasons, among which are two side reactions: (1) the competing self-addition of the carbonyl compound (the aldol addition) and (2) reaction of the base with the alkyl halide. Even if the enolate is first generated *irreversibly* and the alkyl halide is subsequently added, problems such as polyalkylation (equation 21) lower the yield of the reaction and detract from its utility.

From a mechanistic standpoint, the difficulty that causes polyalkylation is mainly that the alkylation step itself is a relatively sluggish process compared with proton transfer. Consequently, the desired monoalkylation product, **34,** formed initially, undergoes deprotonation with **33** serving as the base, and the new enolate ion(s) that result can then be alkylated. A mixture of mono-, di-, and further alkylated products along with the unalkylated starting material can thus result, as shown in equation 21.

Most of the experimental variations that have been developed to aid in overcoming the problems of self-addition and polyalkylation in the alkylation of ketones are too sophisticated to be used in an introductory course. However, the alkylation procedure in this section circumvents the difficulties mentioned by employing an enolate precursor that is less prone to self-addition than is a normal ketone or aldehyde and by using a specialized alkylating agent with which polyalkylation is a disfavored process.

The source of the enolate in our procedure is dimethyl malonate (**35**), a carbonyl-containing compound that is expected to be reasonably acidic owing to the presence of *two* carbonyl groups that can assist in the delocalization of the negative charge in the conjugate base (**36,** equation 22). This expectation is realized in that **35** has a pK$_a$ of about 13 (the pK$_a$ of acetone is about 20 and that of a *mono*ester such as methyl acetate is about 25). Consequently, if the diester **35** is treated with an appropriate base, ionization to the enolate ion **36** will occur.

$$CH_3O-\overset{\overset{\displaystyle O}{\|}}{C}-\overset{\overset{\displaystyle H}{|}}{\underset{\underset{\displaystyle H}{|}}{C}}-\overset{\overset{\displaystyle O}{\|}}{C}-OCH_3 + B{:}^{\ominus}M^{\oplus} \rightleftarrows B-H + CH_3O-\overset{\overset{\displaystyle O^{\cdots}M^{\oplus}O}{}}{C\underset{\underset{\underset{\displaystyle H}{|}}{\overset{\displaystyle \ominus}{C}}}{}C}-OCH_3 \quad \textbf{(22)}$$

35   **36**

The addition of this ion to the starting diester **35,** a process leading to self-condensation products, is less important than the similar reaction with aldehydes and ketones because of the lesser tendency of the carbonyl group of esters to undergo nucleophilic attack. This is due to electron delocalization in the ester, a type of stabilization that is lost upon addition of a nucleophile (equation 23).

$$\left[R-\overset{\overset{\displaystyle \ddot{O}\cdot}{\|}}{C}-\ddot{O}-R' \leftrightarrow R-\overset{\overset{\displaystyle :\ddot{O}:^{\ominus}}{|}}{C}=\overset{\oplus}{\ddot{O}}-R' \right] \xrightarrow{Nu:^{\ominus}} R-\overset{\overset{\displaystyle :\ddot{O}:^{\ominus}}{|}}{\underset{\underset{\displaystyle Nu}{|}}{C}}-\ddot{O}-R' \quad \textbf{(23)}$$

The alkylating agent used in the procedure is 1,3-dibromopropane **(37)**, which is chosen for two reasons. First, it circumvents the problems associated with polyalkylation of dimethyl malonate: Difficultly separable mixtures of mono- and dialkylated products normally result. Second, it leads to formation of a derivative of cyclobutane, illustrating the utility of enolate alkylation in producing cyclic compounds that might otherwise be difficult to synthesize. Thus, the dibromide **37** can be considered as a *dialkylating* agent in its reaction with **35** to produce dimethyl cyclobutane-1,1-dicarboxylate (**38**, equation 24).

$$CH_2(CO_2CH_3)_2 + Br-CH_2CH_2CH_2-Br \xrightarrow[\text{CH}_3\text{OH}]{\text{2NaOCH}_3} \overset{\displaystyle CO_2CH_3}{\underset{\displaystyle CO_2CH_3}{\bigbowtie}} \quad \textbf{(24)}$$

        35   **37**   **38**
Dimethyl   1,3-Dibromopropane   Dimethyl
malonate   cyclobutane-1,1-
                                                 dicarboxylate

The *intramolecular* reaction that produces **38** by way of the enolate **(40)** of the monoalkylation product **39** (equation 25) is favored by entropy over the *intermolecular* process that would convert **26** to **39** (equation 41). In any event, **38, 39,** and **41** are each separable by distillation because of the differences in their boiling points.

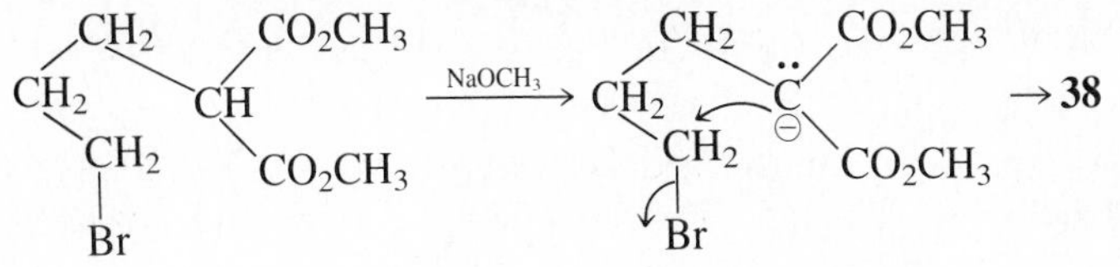

$$\begin{array}{c}\text{structure } \mathbf{39} \xrightarrow{\text{NaOCH}_3} \text{structure } \mathbf{40} \rightarrow \mathbf{38}\end{array} \tag{25}$$

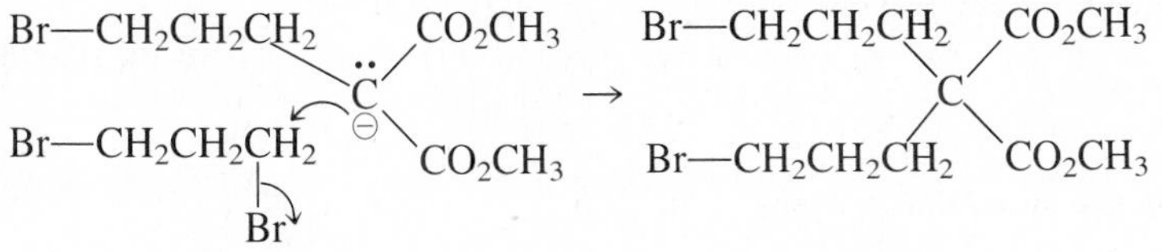

$$\text{structure } \mathbf{40} \rightarrow \text{structure } \mathbf{41} \tag{26}$$

The choice of the specific base to be used in the reaction is important since there is more than one site in the malonate **35** that is susceptible to attack by base. Thus instead of deprotonating **35** (the desired reaction, equation 22), a base could *add* to the carbonyl carbon of an ester function, with subsequent elimination of methoxide ion (equation 27).

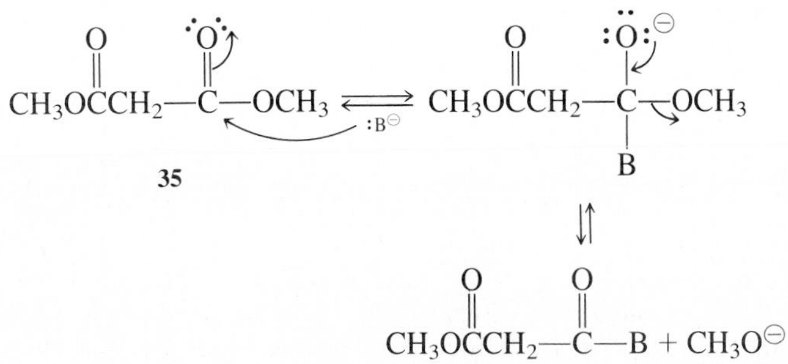

$$\underset{\mathbf{35}}{\text{CH}_3\text{OCCH}_2-\overset{\overset{\cdot\cdot}{\text{O}}}{\overset{\|}{\text{C}}}-\text{OCH}_3} \; \underset{:B^{\ominus}}{\rightleftharpoons} \; \text{CH}_3\text{OCCH}_2-\overset{:\overset{\cdot\cdot}{\text{O}}:^{\ominus}}{\underset{B}{\overset{|}{\text{C}}}}-\text{OCH}_3$$

$$\updownarrow$$

$$\underset{}{\text{CH}_3\text{OCCH}_2}-\overset{\text{O}}{\overset{\|}{\text{C}}}-\text{B} + \text{CH}_3\text{O}^{\ominus} \tag{27}$$

This potential reaction excludes the use of bases such as hydroxide and ethoxide since their reaction with **35** according to equation 27 would lead to hydrolysis and transesterification, respectively, of the ester functions of **35**. Transesterification would be of less concern with *t*-butoxide ion, owing to its steric bulk, but this base is too expensive for use in our procedure. Selection of methoxide ion for the alkylation is an obvious choice not only because this base is sufficiently strong to deprotonate the diester (the pK_a of methanol is about 16) but also because addition of it to the carbonyl group is harmless (equation 27, B = CH$_3$O).

The undesired effect that the use of hydroxide ion would have on the reaction (equation 27, B = HO) should give a clue regarding an important experimental precaution: namely, all reagents, solvents, and apparatus employed in the alkylation should be *scrupulously dry*. Water would, of course, react with the methoxide to generate the unwanted hydroxide, which would in turn promote hydrolysis of the ester.

Intermolecular alkylation of **39** (equation 26) represents one side reaction that is possible, but others must also be considered. Two examples are the reactions of methoxide ion with

1,3-dibromopropane (37) and with the monoalkylated intermediate 39 to produce ethers (equation 28).

$$CH_3O^\ominus + Br\overset{\frown}{}CH_2CH_2CH_2-R \rightarrow CH_3O-CH_2CH_2CH_2-R \qquad \text{(28)}$$

Fortunately, this type of S_N2 process is not competitive with the desired deprotonation reactions of 35 and of 39. The type of reaction represented by equation 28 may be further suppressed by *dropwise* addition of methoxide to the reaction mixture; this keeps the concentration of methoxide at a level throughout the reaction that is lower than it would be if the base were added to the reaction mixture in a single portion.

The malonate ion 36 can, of course, promote the same type of side reaction on the intermediate 39 as does methoxide (compare equations 28 and 29).

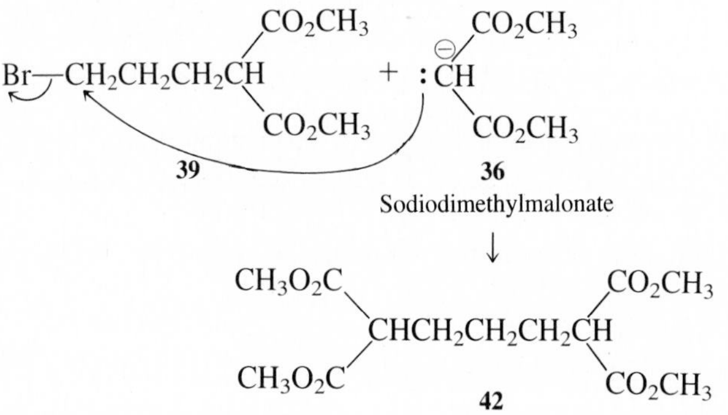

Sodiodimethylmalonate

(29)

Since the undesired reaction in equation 29 is enhanced by high concentrations of malonate 35 whereas the desired reaction in equation 25 is not, both the dimethyl malonate 35 and the solution of sodium methoxide in methanol should be added simultaneously in a dropwise fashion to the 1,3-dibromopropane. Ideally, then, the reaction would be effected by use of a three-necked flask equipped with *two* addition funnels for the simultaneous dropwise addition of each solution. Such apparatus is often unavailable in the beginning organic laboratory, so the procedure given utilizes a *single* addition funnel. Thus a methanolic solution of sodiodimethylmalonate (36) is prepared from 35 and sodium methoxide, and this is added in a dropwise manner to the 1,3-dibromopropane 37. This procedure, which is only slightly less efficient than that using more complex apparatus, takes advantage of the reluctance of the malonate 35 to undergo self-condensation in the presence of sodium methoxide.

A final potential side reaction is worth noting. The anion 36 of dimethyl malonate is an *ambident* ion; that is, because of charge delocalization, 36 has two sites that may function as nucleophilic centers in reacting with electrophiles. Although only reactions involving the nucleophilic carbon atom of 36 have been discussed thus far, attack of an oxygen on 37 would produce 43, a so-called O-alkylated product (equation 30). However, the reaction conditions chosen in our procedure suppress production of 43, largely because the protic molecules of the solvent (methanol) and the counter ion (sodium) tend to cluster about the oxygen atoms of 36. This is because there is greater charge density at oxygen than at carbon, owing to the difference in their electronegativities. Thus the molecules of solvent and the counter ion tend to inhibit addition of an electrophile at oxygen by steric and electronic effects.

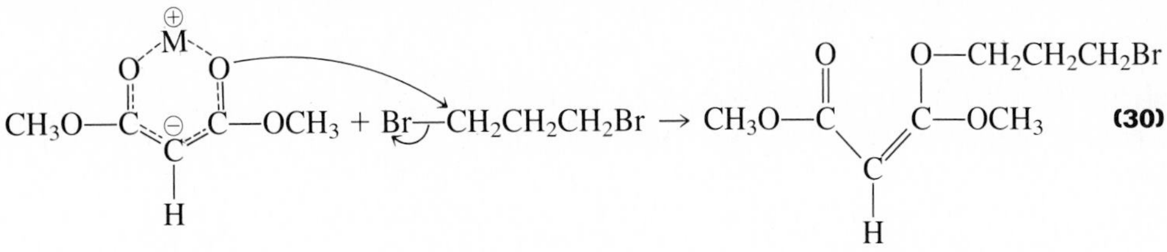

$$CH_3O-C \overset{\ominus}{\underset{\underset{H}{C}}{C}} - OCH_3 + Br-CH_2CH_2CH_2Br \rightarrow CH_3O-C \overset{O \quad O-CH_2CH_2CH_2Br}{\underset{\underset{H}{C}}{C}} - OCH_3 \qquad \textbf{(30)}$$

$$\quad\quad\quad \textbf{36} \quad\quad\quad\quad\quad\quad \textbf{37} \quad\quad\quad\quad\quad\quad\quad\quad\quad \textbf{43}$$

The substituted malonic ester **38**, produced by the alkylation reaction discussed in the preceding paragraphs, can readily be converted to the corresponding cyclobutane-1,1-dicarboxylic acid **44** by hydrolysis under basic conditions, followed by careful acidification of the dibasic salt that results (equation 31). A procedure for isolation of cyclobutane-1,1-dicarboxylic acid is given in the Experimental Procedure.

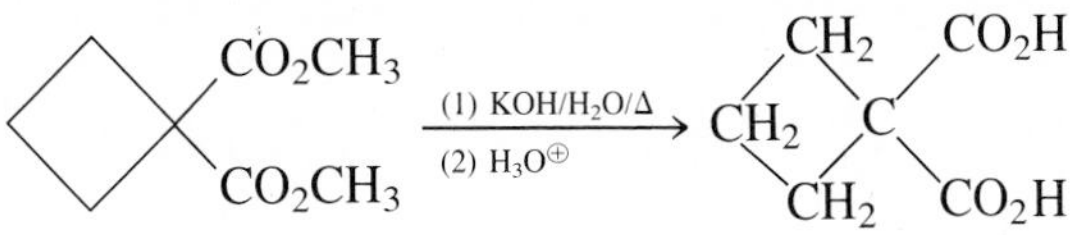

$$\overset{CO_2CH_3}{\underset{CO_2CH_3}{\diamondsuit}} \xrightarrow[\text{(2) } H_3O^\oplus]{\text{(1) KOH/H}_2\text{O/}\Delta} \overset{CH_2 \quad CO_2H}{\underset{CH_2 \quad CO_2H}{CH_2 \quad C}} \qquad \textbf{(31)}$$

$$\quad\quad \textbf{38} \quad\quad\quad\quad\quad\quad\quad\quad\quad\quad \textbf{44}$$

Dimethyl
cyclobutane-1,1-
dicarboxylate

Cyclobutane-1,1-dicarboxylic
acid

Alternatively, hydrolysis of **38** under acidic conditions provides **44**, which when further heated to about 180°C suffers *decarboxylation* to produce cyclobutanecarboxylic acid (**45**, equation 32). Synthesis of cyclobutanecarboxylic acid by the sequence of reactions shown in equations 24 and 32 represents an example of the "malonic ester synthesis" of substituted acetic acids. The synthesis of substituted acetic acids by the *direct* alkylation of the esters of acetic acid is much less efficient and more difficult than by the malonic ester synthesis. This is because the α-hydrogens of monoesters are much less acidic than those of malonic esters, so much stronger, more difficultly handled bases must be used.

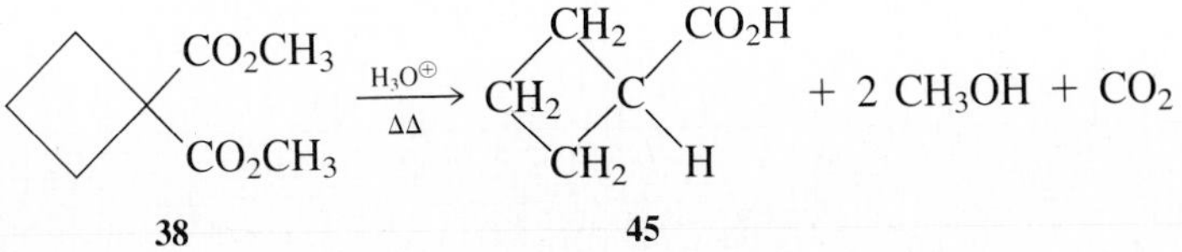

$$\overset{CO_2CH_3}{\underset{CO_2CH_3}{\diamondsuit}} \xrightarrow[\Delta\Delta]{H_3O^\oplus} \overset{CH_2 \quad CO_2H}{\underset{CH_2 \quad H}{CH_2 \quad C}} + 2\ CH_3OH + CO_2 \qquad \textbf{(32)}$$

$$\quad\quad \textbf{38} \quad\quad\quad\quad\quad\quad\quad\quad\quad\quad \textbf{45}$$

Cyclobutanecarboxylic
acid

The decarboxylation shown in equation 32 is an example of a general reaction wherein carboxylic acids having a carbonyl group at the β-position lose carbon dioxide on heating, presumably by way of a six-centered transition state (equation 33). The enediol intermediate **46** rapidly tautomerizes to the final product **45**.

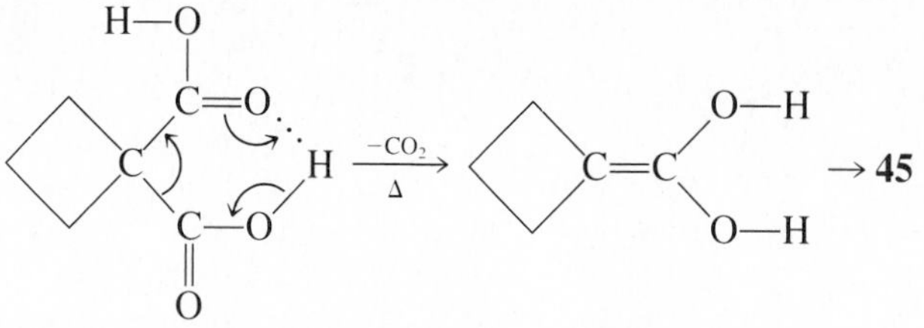

$$\qquad \qquad 44 \qquad \qquad \qquad \qquad \qquad 46 \qquad \qquad\qquad (33)$$

Pre-lab exercises for Section 21.2c, Alkylation of Enolate Ions, can be found on page PL. 91.

EXPERIMENTAL PROCEDURE

DO IT SAFELY

1 Methanolic sodium methoxide and ethanolic potassium hydroxide are strongly caustic solutions. If they should come into contact with your skin, wash the affected area immediately with large quantities of cold water.

2 Strong heating of mixtures containing undissolved solids can cause superheating that results in severe bumping. Heat such mixtures carefully, and if possible, stir them to minimize any bumping.

3 When performing a vacuum distillation, take care not to use any glassware that is cracked or otherwise damaged. Damaged glassware may implode under reduced pressure, especially when heated, and may result in spillage of hot chemicals and danger of cuts from broken glass.

A. Preparation of Dimethyl Cyclobutane-1,1-dicarboxylate

The methanol used in this experiment must be anhydrous and should be dried over 3-Å molecular sieves for 24 hr before use. It is best to distil dimethyl malonate just prior to its use; however, if this is not done, drying of the malonate over 3-Å molecular sieves for 24 hr will normally suffice. Do not unnecessarily expose the solid sodium methoxide to the atmosphere. Prepare a solution of methanolic sodium methoxide by adding 0.15 mol of solid sodium methoxide *in 5 or 6 portions* to 50 mL of ice-cold anhydrous methanol contained in a dry Erlenmeyer flask fitted with a stopper. The heat of solution is quite high, so that the mixture should be chilled with swirling in an ice-water bath as the portions of methoxide are added. Once all the methoxide has been added and dissolved, further add to the *chilled* solution 0.080 mol of dimethyl malonate to provide a methanolic solution of sodiodimethyl malonate (**36**, M = Na). Note that this solution also contains the additional equivalent of sodium methoxide needed for the final cyclization to product (equation 25). Keep the flask stoppered as much as possible throughout the preparation of this solution to avoid exposure to atmospheric moisture.

Equip a 250-mL round-bottomed flask with a magnetic stirring bar, if this is available, and a Claisen connecting tube with its two openings fitted with a condenser and an addition funnel. Protect both the condenser and the addition funnel with drying tubes filled with anhydrous calcium chloride. This glassware assembly must be thoroughly dried by heating with the flame of a burner before continuing (see Section 2.24). Once the glassware has dried and cooled to near room temperature, add to the flask through the addition funnel a solution of 0.075 mol of 1,3-dibromopropane in 15 mL of anhydrous methanol. Close the stopcock of the addition funnel, and add to the funnel the previously prepared methanolic solution of sodiodimethyl malonate. Commence stirring (if you are using magnetic stirring equipment), and heat the reaction mixture to reflux. Discontinue heating once the 1,3-dibromopropane solution is boiling, and begin dropwise addition of the sodiodimethyl malonate solution at such a rate that the total addition will be completed in about 30 min. If you are not using magnetic stirring, the mixture must be manually swirled frequently during the addition. The reaction is mildly exothermic, and reflux *may* occur without external heating. If it does not, use either a steam bath or a heating mantle to maintain gentle reflux of the reaction mixture. On completion of the addition, heat the mixture at reflux until addition of 3 or 4 drops of the reaction mixture to about 0.5 mL of water in a test tube gives a solution that is no longer alkaline to pH indicator paper.★ If neutrality has not been attained after 1.5 hr at reflux (as evidenced by this test), add sufficient glacial acetic acid to make the reaction mixture weakly acidic.★

Allow the mixture to cool slightly, and arrange the flask for simple distillation. Distil the methanol as rapidly as bumping caused by precipitated sodium bromide will allow; if possible, continue stirring during the distillation to minimize the force of bumping. Collect the methanol in a graduated cylinder. After 60–65 mL have been distilled, allow the residue to cool to room temperature.★ Add 75 mL of water, and swirl the mixture until all salts are dissolved. Transfer the two-phase mixture to a separatory funnel, and add 25 mL of technical diethyl ether to dissolve the organic products. Separate the layers, and extract the aqueous phase three additional times with 25-mL portions of diethyl ether. Wash the combined ethereal extracts once with about 20 mL of brine, and dry the solution over anhydrous magnesium sulfate.★ Remove the ether and any residual methanol by simple distillation, terminating the distillation when the head temperature reaches 80°C or so.★ Place the residue in an appropriately sized round-bottomed flask, and *vacuum*-distil the product with the aid of either a water aspirator or the house vacuum line. The desired product has bp 116–117°C (20 torr) or 174–176°C (165 torr) and should be colorless. Weigh the product and calculate the yield. Normally, yields of 40–60% may be anticipated. The distillation residue is largely the undesired tetraester **41,** bp 185–189°C (2 torr).

B. Cyclobutanecarboxylic Acid

Combine 3 mL of concentrated hydrochloric acid and 1.5 mL of water with each 0.006 mol of dimethyl cyclobutane-1,1-dicarboxylate that is to be hydrolyzed, and place this mixture in an appropriately sized flask that has previously been fitted with a reflux condenser and magnetic stirring bar, if available. Heat and stir or occasionally swirl the mixture at reflux until the solution becomes homogeneous,★ and then continue heating for an additional hour.★ Arrange the flask for simple distillation; once the methanol, water, and hydrochloric acid have been removed, continue heating the distillation residue until *gas* evolution ceases.★ If an oil bath has been used for heating, it must ultimately be heated to about 180°C to complete the decarboxylation. Crude product can be isolated by simple distillation of the residue at atmospheric pressure, bp 195–196°C. Redistillation under vacuum, bp 105–108C° (25 torr), gives pure acid in an overall yield of 60–70%.

C. Cyclobutane-1,1-dicarboxylic Acid

Combine 6.1 g of potassium hydroxide and 5 mL of 95% ethanol with each 0.025 mol of dimethyl cyclobutane-1,1-dicarboxylate that is to be hydrolyzed, and place this mixture in an appropriately sized flask equipped with a reflux condenser. Heat the mixture at reflux for 2 hr,★ and then arrange the apparatus for simple distillation. Remove nearly all of the methanol and ethanol by distillation,★ and then evaporate the residue to dryness on a steam bath.★ Dissolve the solid residue in a *minimum* amount of water, and then make this solution strongly acidic by *cautious* addition of concentrated hydrochloric acid.★ Extract the aqueous mixture four times with 15-mL portions of technical diethyl ether, dry the combined ethereal extracts over magnesium sulfate,★ filter, and remove most of the ether by simple distillation (*no flames*). Completion of the evaporation to dryness may conveniently be accomplished by fitting the flask with a stoppered vacuum adapter and attaching this apparatus to a water aspirator or house vacuum line; gentle warming of the flask on a steam bath will hasten the removal of the ether. Alternatively, the flask may be allowed to sit unstoppered until the next laboratory period. The resulting beautifully crystalline colorless solid is the desired product, mp 157–158°C. It can be recrystallized from ethyl acetate if desired (no flames). The yield should be 80–85%.

EXERCISES

1. What is the expected consequence of using methanolic sodium *ethoxide* rather than sodium *methoxide* in the alkylation reaction?

2. Write reactions showing the by-products anticipated if the alkylation is conducted with reagents and/or solvents contaminated with water.

3. Why might the reaction mixture in the alkylation not become neutral after the 1.5-hr period of reflux?

4. What consequences might be expected if the reaction mixture in the alkylation were still basic during the work-up procedure? If acid is added to achieve neutrality, why is it important that the reaction mixture not be made *strongly* acidic?

5. Explain why dimethyl malonate (35) is expected to be somewhat more acidic than the monoalkylation product 39.

6. Calculate the equilibrium concentration of sodiodimethyl malonate (36) present if 0.18 mol of dimethyl malonate (35) is added to 100 mL of 3 M sodium methoxide in methanol. The pK_a of dimethyl malonate is about 13 and that of methanol is about 16.

7. Draw the structure of the self-condensation product expected to arise from methoxide-promoted reaction of two molecules of dimethyl malonate (35). Predict whether this substance would be more or less acidic than 35. Why?

8. In the alkylation reaction, why is the bulk of the methanol removed by distillation *prior to* addition of water and subsequent extraction with diethyl ether? In other words, why does the work-up procedure not simply call for addition of water to the neutral reaction mixture followed by extraction of this solution with diethyl ether?

9. Why do you think the base-promoted rather than the acid-catalyzed hydrolysis is the preferred method for preparing cyclobutane-1,1-dicarboxylic acid?

10. Dipotassium cyclobutane-1,1-dicarboxylate does *not* decarboxylate when heated above 150°C, whereas the corresponding diacid does. Provide an explanation for this difference in reactivity.

11. Why might calcium chloride *not* be a particularly good choice as a drying agent for the ethereal solution of cyclobutane-1,1-dicarboxylic acid obtained from the procedure for base-catalyzed hydrolysis of the diester?

*12. Why does sodiomalonate react with alkyl bromide rather than with dimethyl malonate?

SPECTRA OF STARTING MATERIALS AND PRODUCTS

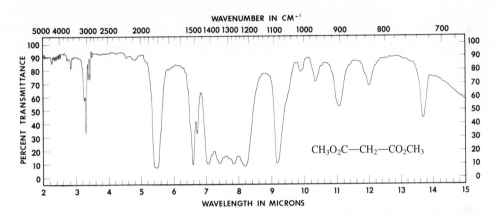

CH₃O₂C—CH₂—CO₂CH₃

Figure 21.22 IR spectrum of dimethyl malonate.

(a) PMR spectrum.

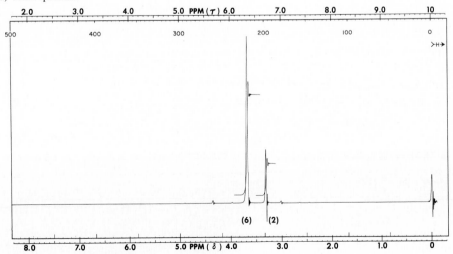

(6)   (2)

(b) CMR data. Chemical shifts: δ 41.2, 52.4, 167.3.

Figure 21.23 NMR data for dimethyl malonate.

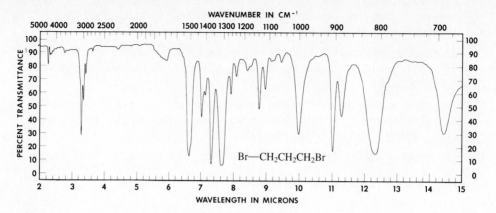

Figure 21.24 IR spectrum of 1,3-dibromopropane.

(a) PMR spectrum.

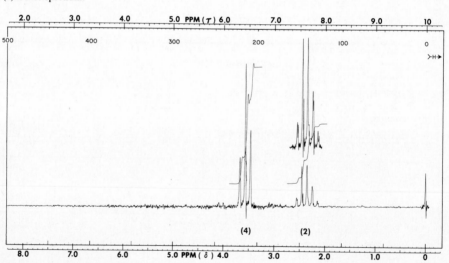

(b) CMR data. Chemical shifts: δ 31.1, 35.1.

Figure 21.25 NMR data for 1,3-dibromopropane.

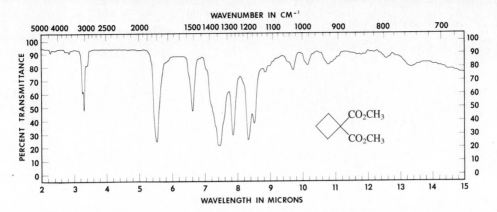

Figure 21.26 IR spectrum of dimethyl cyclobutane-1,1-dicarboxylate.

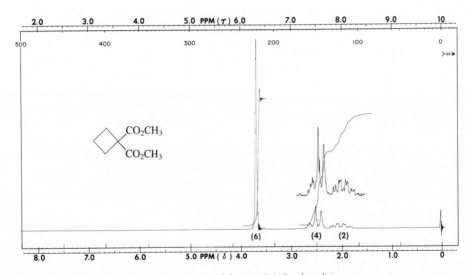

Figure 21.27 PMR spectrum of dimethyl cyclobutane-1,1-dicarboxylate.

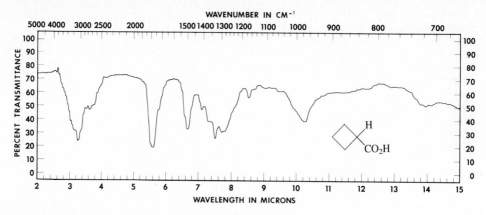

Figure 21.28 IR spectrum of cyclobutanecarboxylic acid.

Chemical shifts: δ 18.6, 25.4, 38.4, 181.9

Figure 21.29 CMR data for cyclobutanecarboxylic acid.

21.3 Conjugate Addition to an α,β-Unsaturated Ketone

In Section 21.2 we described the addition of bromine to anisalacetophenone (**27**, equation 19), a reaction analogous to the addition of bromine to an isolated carbon-carbon double bond in a simple alkene (Chapter 13). However, in anisalacetophenone, the carbon-carbon double bond is conjugated with the carbon-oxygen double bond. This makes the carbon-carbon double bond electron-deficient and prone to reaction with electron-rich reagents that normally do not attack double bonds. One example mentioned previously in Section 21.2 was the addition of a Grignard reagent. Other nucleophilic reagents such as cyanide ion, CN^-, and enolate anions also add to α,β-unsaturated carbonyl compounds, usually under basic conditions. (Consult Section 21.2 for a discussion of the stabilization of enolate anions by carbonyl groups.) An interesting example of an *acid*-catalyzed addition is that of 2-methylpropanal (isobutyraldehyde) to 3-buten-2-one (methyl vinyl ketone) (**47**). The initial product, which may be isolated if desired, is 2,2-dimethyl-5-oxohexanal (**48**, equation 34); however, it is convenient to convert this product into 4,4-dimethyl-2-cyclohexen-1-one (**49**) by an aldol-type cyclization followed by dehydration (equation 35).

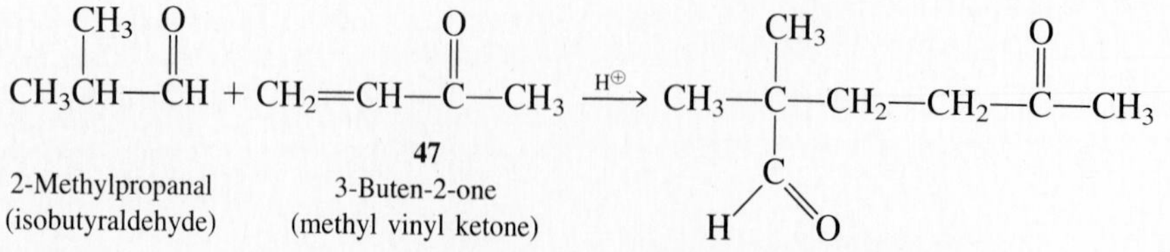

$$\underset{\substack{\text{2-Methylpropanal}\\\text{(isobutyraldehyde)}}}{\underset{\overset{|}{\text{CH}_3}}{\text{CH}_3\text{CH}}-\underset{\overset{\parallel}{\text{O}}}{\text{CH}}} + \underset{\substack{\textbf{47}\\\text{3-Buten-2-one}\\\text{(methyl vinyl ketone)}}}{\text{CH}_2{=}\text{CH}-\underset{\overset{\parallel}{\text{O}}}{\text{C}}-\text{CH}_3} \xrightarrow{\text{H}^{\oplus}} \underset{\substack{\textbf{48}\\\text{2,2-Dimethyl-5-}\\\text{oxohexanal}}}{\text{CH}_3-\underset{\underset{\text{H}\ \ \text{O}}{\overset{|}{\text{C}}}}{\overset{\overset{\text{CH}_3}{|}}{\text{C}}}-\text{CH}_2-\text{CH}_2-\underset{\overset{\parallel}{\text{O}}}{\text{C}}-\text{CH}_3} \qquad \textbf{(34)}$$

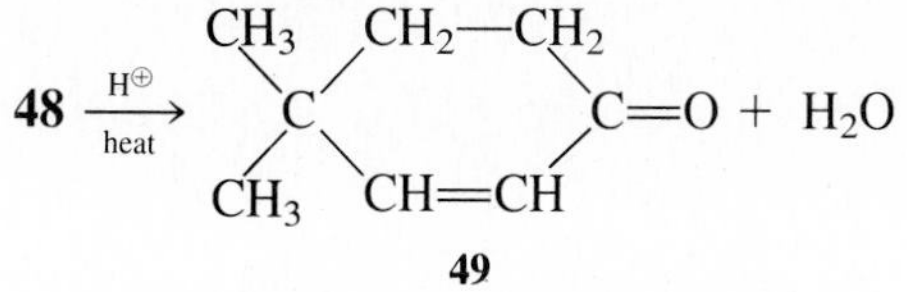

$$48 \xrightarrow[\text{heat}]{H^\oplus} \quad \text{(structure 49)} \quad + H_2O \qquad \textbf{(35)}$$

49

4,4-Dimethyl-2-cyclohexen-1-one

Plausible mechanisms for these reactions are shown in Figures 21.30 and 21.31. The first step is the acid-catalyzed tautomerization of 2-methylpropanal to its enol form, **50,** which makes the "α"-carbon nucleophilic, although less so than in an enolate. Protonation of 3-buten-2-one converts it into the resonance-stabilized cation, **51,** which is more subject to attack by the relatively weak nucleophile, the enol **50.** Addition of **50** to **51** produces **52,** and one can see at this stage that the aldehyde may be said to have undergone conjugate addition (or "1,4-addition") to 3-buten-2-one. The enol form **52** is unstable with respect to its keto form; tautomerization followed by loss of a proton produces the initial product, 2,2-dimethyl-5-oxohexanal, **48.**

Under the conditions of our experiment, **48** is not isolated, but an acid-catalyzed cyclic aldol-type addition occurs. Another enol form of **48** is possible **(53)** and protonation of this enol gives **54,** which is capable of an intramolecular addition analogous to the step in which

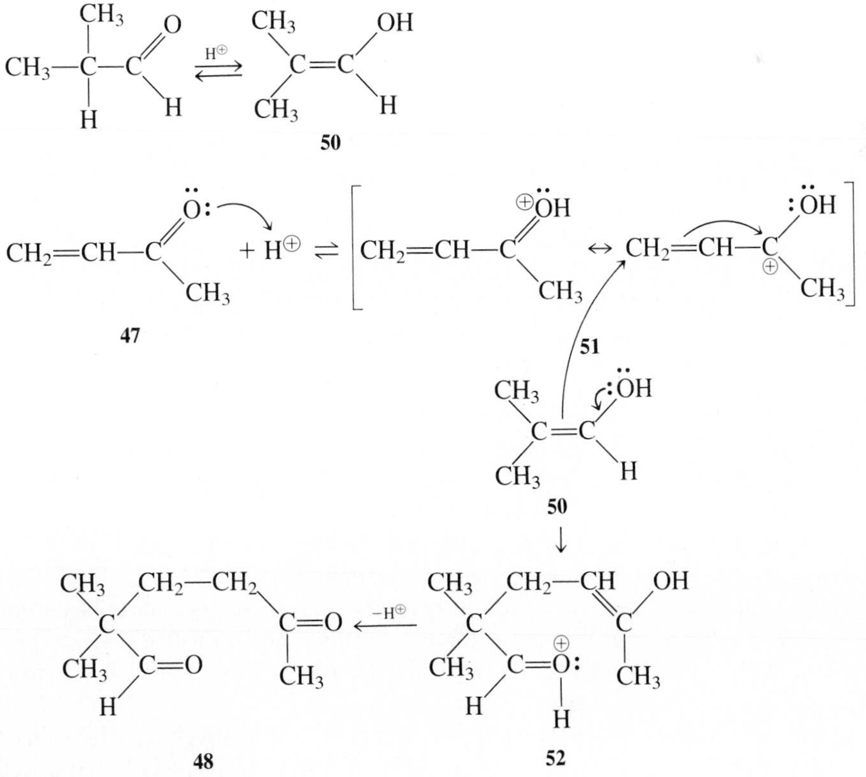

Figure 21.30 Mechanism for conjugate addition of 2-methyl propanal (isobutyraldehyde) to 3-buten-2-one (methyl vinyl ketone).

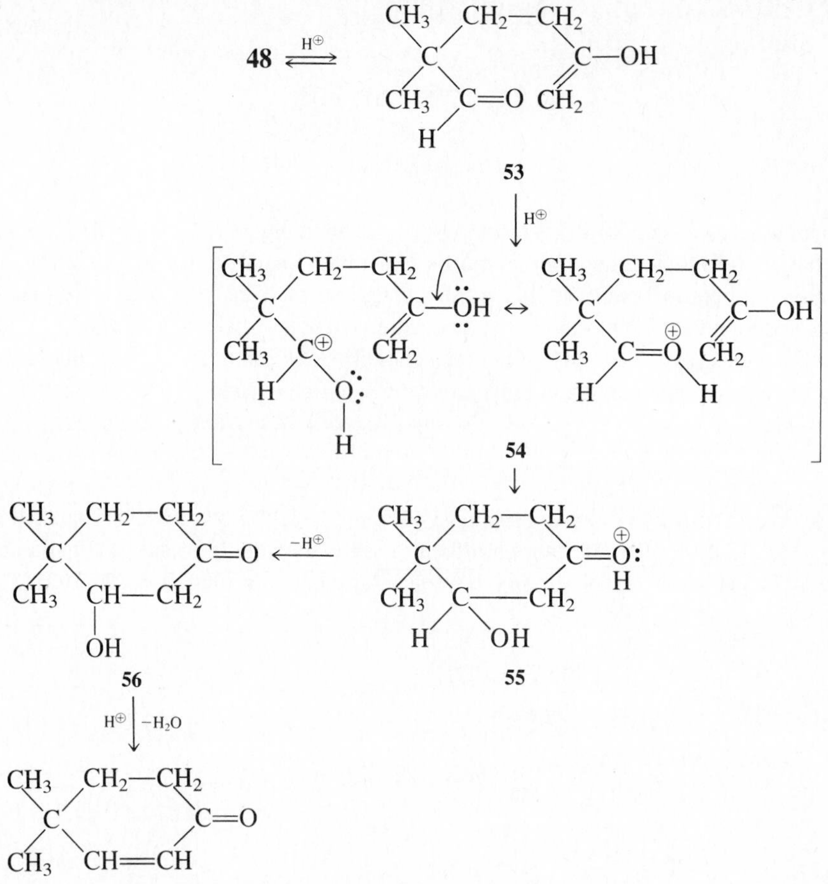

$$48 \quad \underset{}{\overset{H^{\oplus}}{\rightleftharpoons}}$$

53

↓ H⊕

54

↓

56 ←—$-H^{\oplus}$— **55**

H⊕ | $-H_2O$

49

Figure 21.31 Mechanism of aldol cycloaddition of 2,2-dimethyl-5-oxohexanal **(48)** to form 4,4-dimethyl-2-cyclohexen-1-one **(49).**

---

50 added to **51.** In this way **55** is produced; loss of a proton yields **56,** which may be recognized as an aldol (actually a ketol). The fact that **48** undergoes efficient cyclization rather than intermolecular condensations with the enolic form of 2-methylpropanal or other potential nucleophiles is mainly attributable to entropic considerations that tend to favor *intra*molecular processes over *inter*molecular ones.

Finally, dehydration of **56** is accomplished by azeotropic distillation of the water with toluene, which also serves as a solvent for the reactants. The apparatus is assembled in such a way as to form a simple trap to prevent the water from returning to the reaction mixture (Figure 21.32). A more elaborate device for accomplishing removal of water from a reaction mixture is called a Dean-Stark trap (Figure 21.33).

It should be noted that the choice of an aldehyde such as 2-methylpropanal, having only one α-hydrogen, is important to the success of this mixed aldol condensation. 2-Methylpropanal can undergo facile self-dimerization to **57,** but this process is reversible

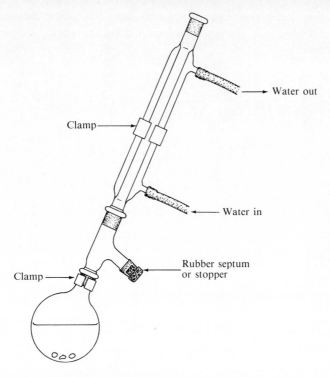

Figure 21.32 Apparatus for reaction of 2-methylpropanal (isobutyr-aldehyde) with 3-buten-2-one (methyl vinyl ketone).

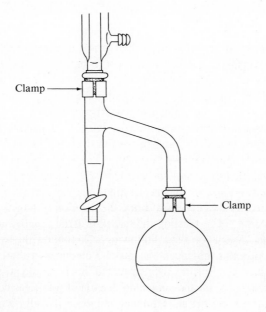

Figure 21.33 Dean-Stark trap for removal of water by distillation.

under the reaction conditions (equation 36). Because **57** *cannot* dehydrate to a stable α,β-unsaturated product, thus disrupting the equilibrium, self-addition of 2-methylpropanal is of no importance.

$$
2 \quad
\begin{array}{c}
CH_3 \\
\diagup \\
CH-C \\
\diagdown \\
CH_3
\end{array}
\begin{array}{c}
O \\
\parallel \\
\\
\diagdown \\
H
\end{array}
\quad \underset{}{\overset{H^{\oplus}}{\rightleftharpoons}} \quad
\begin{array}{c}
CH_3 \\
\diagup \\
CH- \\
\diagdown \\
CH_3
\end{array}
\begin{array}{c}
OH \\
\mid \\
C- \\
\mid \\
H
\end{array}
\begin{array}{c}
CH_3 \\
\mid \\
C- \\
\mid \\
CH_3
\end{array}
\begin{array}{c}
O \\
\parallel \\
C \\
\diagdown \\
H
\end{array}
\qquad \textbf{(36)}
$$

57

Pre-lab exercises for Section 21.3, Conjugate Addition to an α,β-Unsaturated Ketone, can be found on page PL. 93.

EXPERIMENTAL PROCEDURE

Conjugate Addition of 2-Methypropanal to 3-Buten-2-one

DO IT SAFELY

Volatile, flammable solvents are used in this experiment. Avoid the use of flames if possible. If not, be certain that all joints in your apparatus are firmly mated in order to prevent ignition of vapors escaping from it.

Equip a distillation head with a septum or rubber stopper firmly seated in the male joint, which is normally attached to a condenser, and attach it to a 250-mL round-bottom flask containing 0.2 g of 2-naphthalenesulfonic acid. (To be certain that the septum or rubber stopper will not come out during the reflux period, it may be wise to wire it on securely; have your instructor check the apparatus before beginning the heating.) The flask should be tilted at an angle of about 30° from the vertical so that when a condenser is attached, condensate will drip into the small sidearm of the distillation head, as illustrated in Figure 21.32. Add about 30 mL of toluene to the flask and then swirl the resulting slurry to effect at least partial solution. To this mixture add a solution of 0.067 mol of 3-buten-2-one (methyl vinyl ketone) and 0.086 mol of 2-methylpropanal (isobutyraldehyde). Rinse the flask that contained this solution with a 20-mL portion of toluene, and add the rinse to the reaction flask. Fit the distillation head with a reflux condenser, bring the solution to a brisk reflux, and continue reflux for at least 2.5 hr.★ The reaction mixture will darken, and water will collect in the plugged sidearm of the distillation head. Following reflux, allow the reaction mixture to cool and then carefully dismount the apparatus, making certain that the collected water is not spilled. Pour the water and solvent present in the sidearm into a graduated cylinder to determine what percentage of the theoretical production of water has been achieved.★ Transfer the reaction mixture to a separatory funnel, and wash it with 30 mL of 0.6 *M* aqueous sodium bicarbonate solution. Dry the organic layer over anhydrous magnesium sulfate for at least 0.5 hr* and then filter it into a 100-mL round-bottomed flask.* Equip the flask for simple distillation and carefully (*caution:* foaming) remove unchanged starting materials and solvent, collecting them as a single fraction boiling up to a temperature of about 130°C.*

If you are to isolate the 4,4-dimethyl-2-cyclohexen-1-one by distillation at atmospheric pressure, insulate the stillhead with glasswool or some equivalent material (if you have not done so already), stop the flow of water through the condenser, and increase the rate of heating of the stillpot. The ketone boils at about 200°C at 760 torr, but you may collect it as a fraction boiling at 130°C and higher; as the head temperature may never reach the stated boiling point, owing to the high boiling temperature and the small amount of sample being distilled. The distillate obtained will be light to dark yellow in color and should amount to about 3 g. Terminate heating when the very dark pot residue becomes viscous and starts to evolve fumes.

Alternatively, and preferably, the final stage of the distillation can be done under vacuum. If you are to do this, allow the dark liquid in the stillpot to cool somewhat, add a wooden stick for purposes of ebullition (see Section 2.9), and attach the vacuum adapter to a water aspirator. After a vacuum has been established in the apparatus, resume heating. The boiling point of the product will depend on the efficiency of the water aspirator. For example, it is reported to boil at 73–76°C (14 torr) and can be estimated to boil at approximately 100°C (40 torr) and 125°C (100 torr). Your aspirator should be capable of achieving a vacuum within the range of 40–100 torr if all joints in your apparatus are properly lubricated and tightly mated. The distillate should be colorless to slightly yellow and should amount to about 5 g. Terminate heating when the dark pot residue becomes viscous, and only a few bubbles are being evolved from it.

The pleasant-smelling liquid that is obtained will be contaminated with residual toluene (see Figures 21.37 and 21.38), the amount of which will be dependent upon the efficiency with which the first stage of the distillation has been performed. However, it should constitute no more than 30% of the distillate. The exact amount can be determined by GC or PMR analysis of the isolated product (see Exercise 4).

4,4-Dimethyl-2-cyclohexen-1-one forms a 2,4-DNP derivative, mp 140–142°C; confirm the identity of your product by converting a 200–500-mg portion of it to this derivative (Section 27.5a). The orange solid can be recrystallized from a minimum amount of 95% ethanol.

REFERENCE

Flaugh, M. E.; Crowell, T. A.; Farlow, D. S. *Journal of Organic Chemistry,* **1980,** *45,* 5399.

EXERCISES

1. In the literature reference given in this section, the solvent used in the reaction was benzene rather than toluene. For purposes of safety, toluene was chosen here (benzene is believed to be carcinogenic, whereas toluene is not). The reaction appears to go faster in toluene than in benzene, as judged by the rate at which water is collected in the side-arm of the distillation head. How can you account for this rate acceleration?

2. Why is it important that water not be returned to the reaction flask?

3. Product **58** potentially could be produced by aldol condensation of **48,** but it is not.

Provide a mechanism for formation of **58** under conditions of acid catalysis, and explain why this alternate product is not observed.

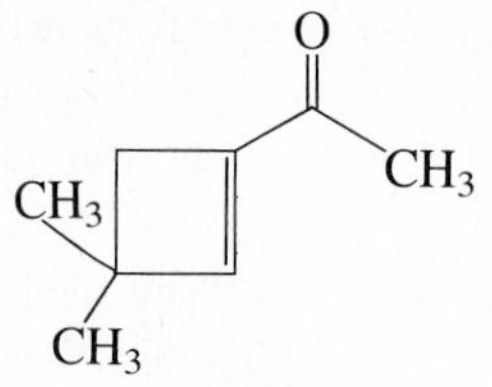

58

4. a. Figure 21.40 (page 544) is the GC trace of a sample of product obtained in this experiment. Calculate the ratio of toluene to **49** present, assuming the response factor of the detector is equivalent for the two compounds. Note that the GC column used in this analysis separates substances on the basis of their relative boiling points.
   b. Figure 21.38 (page 543) is the PMR spectrum of this same sample. Calculate the ratio of toluene to **49** present according to this method of analysis.
   c. How do you account for the discrepancy between the two ratios obtained?
   d. The distillate from which the samples for the above analyses were drawn weighed a total of 5.64 g. Based on the PMR analysis of the mixture, calculate how many grams of the desired **49** were present in the distillate, and calculate the yield of the reaction, assuming it was run on the same scale as described in the Experimental Procedure of this section.

5. Attempted acid-catalyzed reaction of propanal (propionaldehyde) with 3-buten-2-one fails to give a good yield of the desired product **(59).** Why might this be?

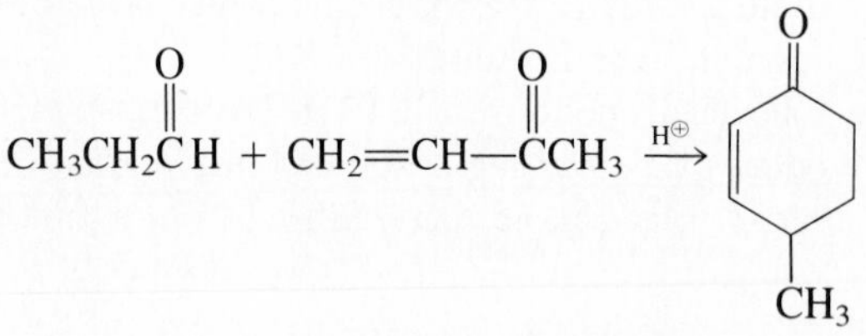

47   **59**

6. a. Refer to Figure 21.39 and calculate log ϵ for the two maxima in the UV spectrum of 4,4-dimethyl-2-cyclohexen-1-one.
  *b. What electronic excitation is responsible for each of these maxima?

SPECTRA OF STARTING MATERIALS AND PRODUCT

The IR spectrum and NMR data for 2-methylpropanal (isobutyraldehyde) are given in Figures 8.22b and 19.2, respectively.

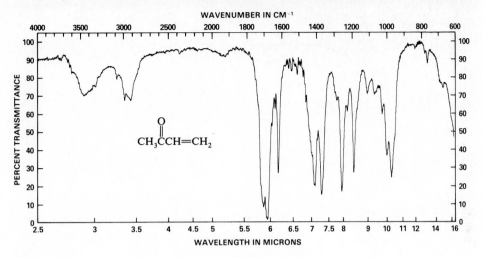

Figure 21.34 IR spectrum of 3-buten-2-one (methyl vinyl ketone).

(a) PMR spectrum.

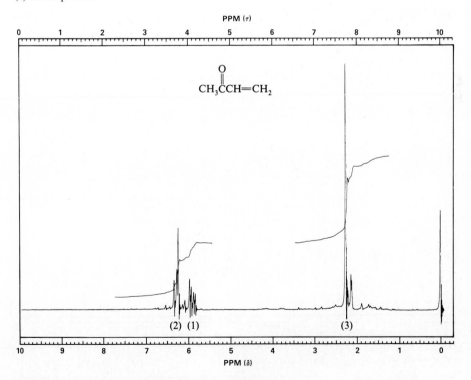

(b) CMR data. Chemical shifts: δ 127.4, 136.3, 196.9.

Figure 21.35 NMR data for 3-buten-2-one (methyl vinyl ketone).

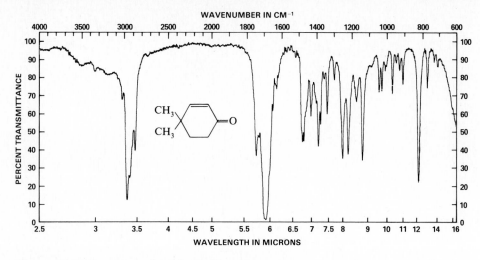

Figure 21.36 IR spectrum of 4,4-dimethyl-2-cyclohexen-1-one.

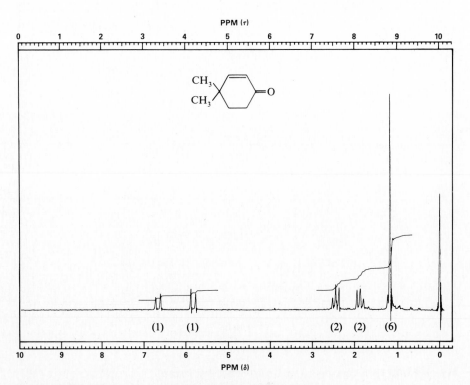

Figure 21.37 PMR spectrum of 4,4-dimethyl-2-cyclohexen-1-one.

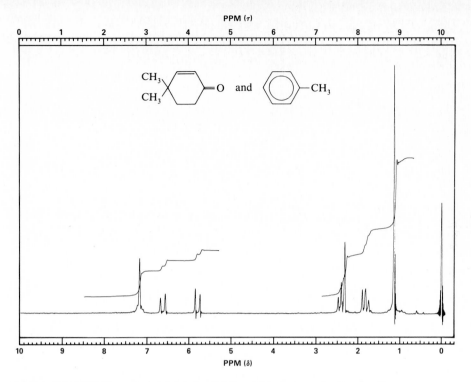

Figure 21.38 PMR spectrum of 4,4-dimethyl-2-cyclohexen-1-one for Exercise 4b.

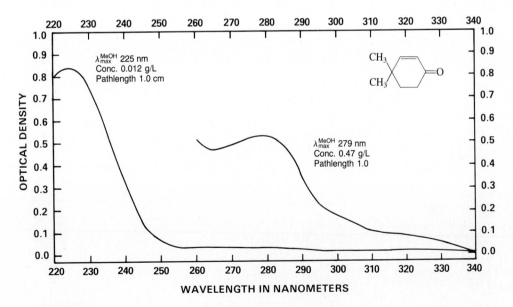

Figure 21.39 UV spectrum of 4,4-dimethyl-2-cyclohexen-1-one.

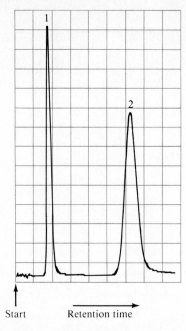

Start   Retention time

Figure 21.40 GC trace of reaction mixture for Exercise 4a. Peak 1 is toluene, and peak 2 is 4,4-dimethyl-2-cyclohexen-1-one.

organometallic chemistry

22.1 Introduction

During the past 20 years the applications of organometallic reagents in organic synthesis have expanded dramatically beyond the use of the organomagnesium compounds (**1,** M = Mg) first prepared by Victor Grignard in the early part of the twentieth century. The most important characteristic of these reagents rests in the polarization of the carbon-metal bond, which is *reversed* relative to that of the bonds of carbon to more electronegative elements such as oxygen, nitrogen, and halogen, as shown by comparison of **1** (M = Li, Mg, Hg, *etc.*) with **2** (X = halogen) and **3.** Significantly, the metal-bound carbon in **1** is electron-rich and *nucleophilic,* whereas the designated carbon atoms in **2** and **3** are electron-deficient and *electrophilic.* This complementary reactivity results in reactions of the types shown in equations 1 and 2, which are of particular value in organic synthesis because they lead to the formation of new *carbon-carbon* bonds.

$$\overset{\delta^{\ominus}}{(R}-\overset{\delta^{\oplus}}{CH_2)_n}M \qquad \overset{\delta^{\oplus}}{R}-\overset{\delta^{\ominus}}{CH_2}-X \qquad \underset{R}{\overset{R}{\diagdown}}\overset{\delta^{\oplus}}{C}=\overset{\delta^{\ominus}}{O}$$

$$\textbf{1} \qquad\qquad \textbf{2} \qquad\qquad \textbf{3}$$

$$R-CH_2-M + R'-CH_2-X \rightarrow R-CH_2-CH_2-R' + MX \qquad\qquad \textbf{(1)}$$

$$R-CH_2-M + \underset{R'}{\overset{R'}{\diagdown}}C=O \rightarrow R-CH_2-\underset{R'}{\overset{R'}{\underset{|}{\overset{|}{C}}}}-O^{\ominus}M^{\oplus} \tag{2}$$

The discussions in the following sections concern the use of organometallics involving magnesium, copper, and lithium as the metals, and in the cases of magnesium and copper, experiments are included to show the applicability of these reagents to organic synthesis.

22.2 The Grignard Reagent: Its Preparation and Reactions

(a) Preparation and Side Reactions

The preparation of the Grignard reagent, RMgX, is given in equation 3, where RX represents an organic halide. The *exact* nature of the Grignard reagent in solution is still unknown. The reagent is believed to be a mixture of a variety of complexes, but some of the principal species present have been shown to be *di*alkylmagnesium, R_2Mg, and magnesium dihalide, MgX_2, in equilibrium with two equivalents of the *mono*alkylmagnesium halide, RMgX. All of the species are solvated by the ethereal solvent and are also complexed with one another. Although it is customary to represent the Grignard reagent by the formula RMgX when writing chemical equations, it should be remembered that the actual species in solution have much more complex structures.

$$RX + Mg \xrightarrow[\text{(solvent)}]{\text{dry ether}} RMgX \tag{3}$$

The ethereal solvent is an essential ingredient for successful formation of the Grignard reagent. The oxygen atom of the solvent complexes with the electropositive magnesium atom of RMgX, thereby stabilizing the organometallic species. Cases are known where Grignard reagents have been prepared in the absence of ethers, but the yields were poor. The ethereal solvent most commonly used in the reaction is diethyl ether, $(C_2H_5)_2O$, owing to its low cost and ease of removal; its boiling point is 36°C. Other ethers such as di-*n*-butyl ether, bp 142°C, have been used successfully when higher-boiling solvents are required. The *cyclic* ether tetrahydrofuran, $(CH_2)_4O$, has been used with excellent success in the preparation of a number of Grignard reagents.

The organic halide in general may have either an aliphatic or aromatic moiety as the organic component, and the halide may be bromide, chloride, or iodide. Even fluoride can be used if highly activated magnesium metal has been prepared (see below). Alkyl iodides, bromides, and chlorides, and aryl iodides and bromides can be converted to the corresponding Grignard reagents using diethyl ether as solvent. Aryl chlorides generally do not react with magnesium in diethyl ether but readily do so when tetrahydrofuran is the solvent.

The alkyl or aryl halide must *not* contain any functionality that will react with the Grignard reagent once it is formed. Thus, hydroxy (OH), amino (NH_2), and carbonyl groups must not be present, since they will react with the nucleophilic, and basic, carbon atom bound to the metal to destroy the reagent. As an example, reaction of 5-bromo-1-pentanol with magnesium would fail to give the expected Grignard reagent because it would be converted to the corresponding bromomagnesium alkoxide by transfer of a proton from the *acidic* hydroxyl group to the *basic* carbon atom of the Grignard reagent (equation 4).

$$H—OCH_2(CH_2)_3CH_2—Br \xrightarrow[\text{(C}_2\text{H}_5)_2\text{O}]{\text{Mg}^0} [H—OCH_2(CH_2)_3CH_2—MgBr] \qquad \textbf{(4)}$$

5-Bromo-1-pentanol

$$\downarrow$$

$$BrMg—OCH_2(CH_2)_3CH_2—H$$

The magnesium metal used for the reaction is normally in the form of "turnings," which are thin shavings having a relatively high surface area as compared with chunks of the metal. If the turnings have been exposed to oxygen, their surface will be dull due to a coating of oxide, and their reactivity will be lessened as a result. Nevertheless, this type of magnesium is suitable for preparation of most Grignard reagents. A highly activated form of magnesium can be prepared in cases for which the turnings are ineffective. This alternate material is generated by reduction of magnesium chloride with potassium metal (equation 5) and results in formation of particles of magnesium that have an extremely high surface area and are free of oxide.

$$MgCl_2 + 2\ K^0 \xrightarrow{\text{(C}_2\text{H}_5)_2\text{O}} Mg^0 + 2\ KCl \qquad \textbf{(5)}$$

Initiation of the reaction between the halide and magnesium is sometimes difficult, with iodides being more reactive than bromides, which are in turn more reactive than chlorides. Aryl halides are less reactive than their alkyl counterparts, and it has been found that the aryl bromides and the alkyl chlorides react about equally well. In cases where one is dealing with a particularly unreactive halide, the initiation of the reaction may occasionally be accomplished by the addition of a small crystal of iodine (I_2) to the reaction mixture. This is believed to cause conversion of a small amount of the organic halide to the iodide, which is more reactive, and/or to activate the metal toward reaction by removing some of its oxide coating. Alternatively, a small amount of 1,2-dibromoethane can be added to the reaction mixture. This compound is very reactive toward magnesium metal and readily forms a Grignard reagent (equation 6), a process that serves to free the magnesium of its oxide coating. As shown in equation 6, no by-products attend the formation of this Grignard reagent as it decomposes with the evolution of ethylene.

$$BrCH_2CH_2Br \xrightarrow[\text{ether}]{\text{Mg}^0} [Br—CH_2—CH_2—MgBr] \rightarrow MgBr_2 + CH_2{=}CH_2 \qquad \textbf{(6)}$$

1,2-Dibromoethane $\qquad\qquad\qquad\qquad\qquad\qquad\qquad\qquad$ Ethene
(ethylene)

Preparation of the Grignard reagent *must* be carried out under *anhydrous* conditions and, if possible, in the absence of oxygen. It is *very* important to maintain completely dry conditions throughout, since the presence of water inhibits the initiation of the reaction by destroying the reagent as it is formed.

Formation of the Grignard reagent is a typical example of a highly exothermic process, and it is important to understand the techniques needed to control the exothermicity. In all cases, the key to performing such reactions safely is careful control of the rate at which the reaction is **allowed** to proceed. A typical way to do this is to add one of the reagents necessary in the rate-determining step of the reaction by slow dropwise addition in order to maintain that reagent at a low concentration. The rate of the reaction, and thus the rate of heat evolution, can thus be controlled by the rate of addition. Commonly the reagent being added in a dropwise fashion is in solution, the concentration of which can also be varied to aid in controlling the exothermicity of the process.

The heat generated in exothermic reactions is usually removed by reflux of a solvent, such as the diethyl ether in Grignard reaction, and transfer of the heat contained in the vapors to the water in the reflux condenser. Consequently, the rate of the reaction must be maintained within the capacity of the condenser to condense the refluxing solvent. Should it appear that the reaction is getting out of control, as evidenced by failure of all vapors to be condensed, the reaction mixture should be immediately cooled with an ice-water bath. It is prudent always to have such a bath prepared when performing an exothermic reaction!

The reaction that occurs when the Grignard reagent comes in contact with water is shown in equation 7 and is the *intermolecular* analog of the process given in equation 4. The reagent is a strong base, with the carbon atom bound to the metal bearing considerable negative charge ($R^- Mg^{2+} X^-$); its great basicity can be assessed by noting that, to a first approximation, a Grignard reagent is the conjugate base of the exceedingly weak acid, RH. It is no surprise, then, that the reagent can readily remove a proton from water, which is a weak acid but not nearly so weak as a hydrocarbon. The overall result is the destruction of the Grignard reagent, with the formation of a hydrocarbon, RH, and a basic magnesium salt. The hydrocarbon is inert, so that the net effect is the production of hydroxide ion in solution (equation 7). A practical use can be made of the neutralization of the Grignard reagent by protonation as it is possible to standardize a solution of such a reagent by adding a known volume of it to an excess, known volume of standardized mineral acid and back-titrating the excess acid with standardized base.

$$RM_gX + H_2O \rightarrow RH + HOM_gX \qquad (7)$$

The susceptibility of the Grignard reagent to reaction with water (equation 7) dictates that all reagents, solvents, and apparatus used for the preparation of the reagent be as dry as possible. Consequently, when diethyl ether is the solvent, as in the Experimental Procedures of this chapter, the *anhydrous* version rather than the *technical* grade must be used. The water content of anhydrous diethyl ether is typically less than 0.01%, whereas it is about 3% in the technical grade.

Anhydrous diethyl ether rapidly absorbs atmospheric moisture, so that if its container has been opened a number of times over a period of several days, the ether in it will no longer be suitable for use in a Grignard reaction. Only freshly opened cans of this solvent should be used if at all possible, and the cans should be tightly sealed immediately after the necessary volume of ether has been removed. Another property of anhydrous diethyl ether is that it readily forms peroxides, by a free-radical chain mechanism, upon exposure to air (equation 8). Because these peroxides are explosive, large volumes of anhydrous ether suspected to contain peroxides should not be evaporated to dryness, and emptied containers of this solvent should be thoroughly rinsed with water before being discarded.

$$\underset{\underset{H}{|}}{CH_3CHOC_2H_5} \xrightarrow{O_2} \underset{\underset{OOH}{|}}{CH_3CHOC_2H_5} \qquad \text{(8)}$$

A hydroperoxide

Formation of peroxides is also a problem when other *anhydrous* ethers, such as tetrahydrofuran, are used as solvents. However, since these other ethers are typically dried by distillation from alkali metals (sodium or potassium) or lithium aluminum hydride, $LiAlH_4$, and since these drying agents destroy the peroxides, ethers *freshly* distilled from these materials should be free of peroxides.

In addition to the reaction with water (equation 7), there are other side reactions that may occur during formation of the Grignard reagent, as shown in equations 9–11.

Reaction with oxygen:
$$2\ RMgX + O_2 \rightarrow 2\ ROMgX \qquad \text{(9)}$$

Reaction with carbon dioxide:
$$RMgX + CO_2 \rightarrow RCO_2MgX \qquad \text{(10)}$$

Coupling:
$$RMgX + RX \rightarrow R\text{—}R + MgX_2 \qquad \text{(11)}$$

These side reactions can be minimized by taking certain precautions when doing the experimental work. Reaction of the Grignard reagent with oxygen and carbon dioxide (equations 9 and10) may be avoided by performing the reaction under an inert atmosphere, such as nitrogen (N_2) or helium. In research work this is done routinely; however, when diethyl ether is used as the solvent, it excludes much of the air from the reaction vessel as a result of its very high vapor pressure. The coupling reaction (equation 11) is an example of a Wurtz-type reaction and sometimes is a desired process if synthesis of a *symmetrical* hydrocarbon, R—R or Ar—Ar, is the goal. It is not possible to eliminate this reaction completely, but it may be minimized by using dilute solutions so as to avoid localized high concentrations of halide. This can be done by very efficient stirring and by slowly adding the halide *to* the magnesium in the ethereal solvent. For example, when diethyl ether is used as the solvent, the rate of addition of the ethereal solution of the halide is normally adjusted so that it is about equal to the rate of reflux of the reaction mixture. Alkyl iodides are more prone to coupling than are the bromides and chlorides, so that the latter, although less reactive, are preferable for preparation of Grignard reagents.

The experiments appearing in this chapter involve the preparation of two Grignard reagents from organic bromides and three typical reactions of these reagents to produce a tertiary alcohol, a carboxylic acid, and a secondary alcohol.

(b) Preparation of Phenylmagnesium Bromide and Its Reaction with Methyl Benzoate

The preparation of phenylmagnesium bromide (**4**) theoretically requires equivalent amounts of bromobenzene and magnesium (equation 12), although a slight excess of the metal is normally used. The most important side reaction involves coupling (equation 13), but it is not a significant problem in this case; it is unnecessary to separate the small amount of the by-product, biphenyl **(5)**, from the Grignard reagent before using it.

$$C_6H_5Br + Mg \xrightarrow{(C_2H_5)_2O} C_6H_5MgBr \tag{12}$$
$$\qquad\qquad\qquad\qquad\qquad\qquad\quad \textbf{4}$$

Bromobenzene   Phenylmagnesium bromide

$$C_6H_5MgBr + C_6H_5Br \rightarrow C_6H_5{-}C_6H_5 + MgBr_2 \tag{13}$$
$$\qquad\qquad\qquad\qquad\qquad\quad \textbf{5}$$
$$\qquad\qquad\qquad\qquad\quad \text{Biphenyl}$$

Reaction of **4** with methyl or ethyl benzoate (**6**, R = CH_3 or CH_2CH_3) gives triphenyl-methanol **(7)** as the major product (equation 14).

$$2\ C_6H_5MgBr + C_6H_5CO_2R \xrightarrow[\text{(2) H}_3O^\oplus]{\text{(1) Combine}} (C_6H_5)_3COH \tag{14}$$
$$\quad\textbf{4}\qquad\qquad\quad\textbf{6}\qquad\qquad\qquad\qquad\qquad\textbf{7}$$

      Alkyl benzoate   Triphenylmethanol

The formation of this compound can be envisioned as arising from the sequence shown in Scheme 22.1. Attack of phenylmagnesium bromide **(4)** on the ester **6** gives benzophenone **(8)** as an intermediate; reaction of **8** with an additional mole of **4** then produces an alkoxide salt, which can be converted to triphenylmethanol by hydrolysis. In order to avoid precipitation of the basic magnesium salt, HOMgX, the hydrolysis is performed with an acidic solution, such as aqueous sulfuric acid, which keeps the magnesium ion in solution in the form of its water-soluble sulfate.

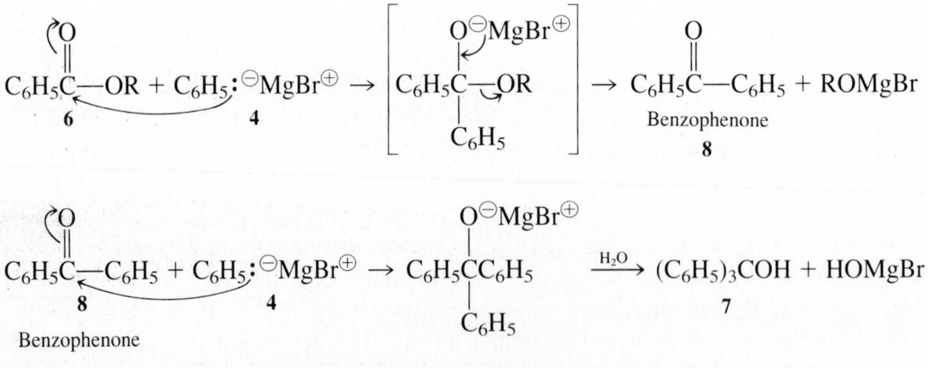

Scheme 22.1 Reaction of a Grignard reagent with an ester.

The synthesis of triphenylmethanol is accomplished experimentally by first preparing the Grignard reagent and then adding the ester to it with stirring. Cooling is required during the addition because the reaction is highly exothermic. Acidic hydrolysis of the reaction mixture affords an organic layer (ethereal solution) containing **5–8** and benzene, the latter coming from reaction of water with **4.** The principal organic products present are the desired triphenylmethanol (**7**) and the unwanted biphenyl (**5**). It is possible to separate these two, owing to their relative solubilities in nonpolar hydrocarbon solvents such as petroleum ether and hexane. Biphenyl is considerably more soluble than the polar alcohol in these solvents, so that recrystallization of the crude product mixture from them gives pure triphenylmethanol as the isolated solid.

(c) Esterification: Preparation of Methyl Benzoate

It may be necessary to synthesize the benzoate ester, **6,** that is required for the preparation of triphenylmethanol, so a brief discussion of a synthetic approach to esters follows. Most commonly, the esters of carboxylic acids are obtained by allowing the acid and an alcohol to react in the presence of a mineral acid (equation 15). The acid catalyst promotes attack of the nucleophilic oxygen of the alcohol upon the carbonyl carbon atom of the carboxylic acid by protonation of its carbonyl oxygen to give **9.**

The overall process of esterification is one involving equilibria among a variety of compounds, and for the reaction to give a high yield, the equilibria must be shifted toward the products, the desired ester and water. This can be accomplished either by removing one or more of the products from the reaction mixture as they are formed or by using a large excess of one of the starting reagents.

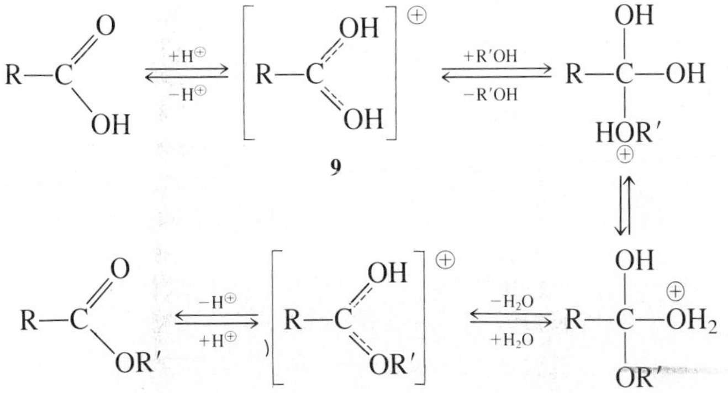

(15)

The effect of the latter approach is obvious from consideration of the mass law relating starting materials and products (equation 16). Increasing the amount of either the alcohol or the carboxylic acid will result in an increase in the amount of products formed since the equilibrium constant, K, for the reaction must remain constant at a given temperature, no matter what quantity of either reagent is used.

$$RCOOH + R'OH \underset{}{\overset{K}{\rightleftharpoons}} RCOOR' + H_2O$$ (16)

$$K = \frac{[RCOOR'][H_2O]}{[RCOOH][R'OH]}$$

Other methods must be employed when it is not economically feasible to use a large excess of either reagent. One method involves the removal of water by azeotropic distillation with the esterifying alcohol. Alternatively, a large excess of the strong acid catalyst is used; water is converted to its conjugate acid, H_3O^+, which is non-nucleophilic and cannot promote the reversion of the desired ester to reactants. A final ploy is the use of a solvent such as 1,2-dichloroethane ($ClCH_2CH_2Cl$) in which the carboxylic acid, alcohol, and ester are soluble but water is not. As it is formed, water is removed by formation of a separate layer in the reaction mixture. The procedure adopted in the esterification of benzoic acid with methanol (equation 16, $R = C_6H_5$, $R' = CH_3$), a reaction described in the Experimental Procedures, is to employ a large excess of the alcohol, which is inexpensive.

(d) Reaction of Phenylmagnesium Bromide with Carbon Dioxide

The transformation that occurs when phenylmagnesium bromide (**1**) reacts with carbon dioxide and the resulting magnesium salt is hydrolyzed is shown in equation 17; the major product is benzoic acid. The principal side reactions encountered in this process result in the formation of benzophenone (**8**) or triphenylmethanol (**7**), as shown by the steps in equation 18. Formation of these by-products can be minimized in several ways. First, the bromomagnesium salt of the carboxylic acid, because it is only slightly soluble in diethyl ether, precipitates from solution and is thus effectively prevented from undergoing further reaction. Second, Dry Ice is used as the source of carbon dioxide so that the temperature of the reaction mixture is maintained at $-78°C$ until all of the Grignard reagent has reacted. Finally, a large excess of carbon dioxide is always used, thus enhancing the likelihood that **1** will react with carbon dioxide rather than with the magnesium salt of the carboxylic acid to produce **8** or with the ketone **8** to give the magnesium salt of **7**.

$$C_6H_5\!:^{\ominus}\!MgBr^{\oplus} + O{=}C{=}O \rightarrow C_6H_5-\overset{\overset{\displaystyle O}{\|}}{C}-O^{\ominus}MgBr^{\oplus} \overset{H_2O}{\longrightarrow}$$
$$\mathbf{4}$$

$$C_6H_5\overset{\overset{\displaystyle O}{\|}}{C}-OH + HOMgBr \quad \textbf{(17)}$$
$$\text{Benzoic acid}$$

$$C_6H_5CO_2MgBr + C_6H_5MgBr \rightarrow (C_6H_5)_2CO + MgBr_2 + MgO \qquad \textbf{(18)}$$
$$\downarrow {\scriptstyle C_6H_5MgBr}$$
$$(C_6H_5)_3COMgBr \overset{H_2O}{\longrightarrow} (C_6H_5)_3COH + HOMgBr$$
$$\text{Triphenylmethanol}$$

Experimentally, this reaction is performed by first preparing the Grignard reagent and then pouring it slowly and with swirling onto an excess of finely powdered Dry Ice. The excess Dry Ice is then allowed to evaporate, the carboxylic acid is liberated from its salt by treatment of the reaction mixture with dilute mineral acid, and the ethereal solution containing all the organic products is separated from the aqueous layer. The desired benzoic acid is readily separated from the crude mixture, which comprises this acid, benzophenone (**8**), triphenylmethanol (**7**), benzene, and biphenyl (**5**), by extraction of the solution with dilute aqueous sodium hydroxide; the acid is converted to its water-soluble sodium salt, whereas the by-products remain in the organic layer. Cautious acidification of the aqueous layer regenerates the benzoic acid, which can be purified further if necessary.

(e) Preparation of *n*-Butylmagnesium Bromide and Reaction with 2-Methylpropanal

n-Butylmagnesium bromide can be prepared from 1-bromobutane by a procedure entirely analogous to the preparation of phenylmagnesium bromide (equation 12). Reaction of this Grignard reagent with 2-methylpropanal gives the magnesium salt of 2-methyl-3-heptanol, as shown in equation 19, and the salt gives the alcohol upon hydrolysis with dilute sulfuric acid. The same types of side reactions as discussed in the formation and reactions of phenyl-magnesium bromide are possible in this case as well.

$$
\text{CH}_3\text{CH}_2\text{CH}_2\text{CH}_2\text{MgBr} + \underset{\substack{\overset{\displaystyle |}{\text{CH}_3}}}{\overset{\displaystyle \overset{\text{O}}{\parallel}}{\text{CH}_3\text{CHCH}}} \longrightarrow \underset{\substack{\overset{\displaystyle |}{\text{CH}_3}}}{\text{CH}_3\text{CH}_2\text{CH}_2\text{CH}_2\overset{\overset{\displaystyle \text{O}^{\ominus}\text{MgBr}^{\oplus}}{\displaystyle |}}{\text{CH}}\text{CHCH}_3} \qquad \textbf{(19)}
$$

n-Butylmagnesium bromide   2-Methylpropanal

$$\text{H}_2\text{O} \Big| \text{H}_2\text{SO}_4 \downarrow$$

$$\underset{\substack{\overset{\displaystyle |}{\text{CH}_3}}}{\text{CH}_3\text{CH}_2\text{CH}_2\text{CH}_2\overset{\overset{\displaystyle \text{OH}}{\displaystyle |}}{\text{CH}}\text{CHCH}_3}$$

2-Methyl-3-heptanol

Note that the experiments described here involve two different techniques for bringing organometallic compounds into reaction with substrates. In the preparations of triphenyl-methanol and 2-methyl-3-heptanol, the carbonyl-containing substrates—methyl benzoate and 2-methylpropanal, respectively—are added *to* the Grignard reagent; this mode of addition is designated **normal** addition. On the other hand, in the preparation of benzoic acid the Grignard reagent is added to the substrate, carbon dioxide; this procedure is called **inverse** addition. The mode of addition to be used is dictated by the reactions involved. For example, if carbon dioxide gas were bubbled into the Grignard reagent (normal addition) in the preparation of benzoic acid, the reactions given in equation 18 would be the predominant ones; there would be an *excess* of the Grignard reagent **4** present relative to carbon dioxide, so that further reaction between **4** and the bromomagnesium salt of benzoic acid would be favored. On the other hand, in the preparation of triphenylmethanol the combination of the Grignard reagent with methyl benzoate could equally well be done in either the normal or inverse fashion. The latter, however, would involve the experimentally inconvenient transfer of the Grignard reagent to a dropping funnel prior to its addition to a reaction flask containing an ether solution of the ester.

Pre-lab exercises for Section 22.2, Part A, Preparation of Methyl Benzoate, can be found on page PL. 95.

EXPERIMENTAL PROCEDURE

A. The Preparation of Methyl Benzoate

DO IT SAFELY

1 Methanol is flammable. Be certain that all ground-glass joints of your apparatus are well lubricated and tightly mated to prevent escape of methanol during the period of reflux.

2 Handle containers of concentrated sulfuric acid with care. Should any of this strong acid come in contact with your skin, flood the affected area immediately with cold water and rinse it with dilute sodium bicarbonate solution.

3 When working with diethyl ether, be certain that there are no open flames in the **entire** laboratory. Ether is **extremely** flammable and volatile, and its vapors can easily travel several feet along the bench top or the floor and then be ignited.

4 Open containers of ether must *not* be kept at your laboratory bench. Estimate the total volume of ether you need, measure it out in the ventilation hood, and return it in a tightly stoppered container to your work area. Be certain to reclose the container from which you obtained the ether in order to prevent evaporation, absorption of moisture, and accidental fires.

5 Be certain to vent the separatory funnel frequently during the workup procedure as pressure develops, owing to the volatility of ether and the evolution of carbon dioxide.

Place 0.082 mol of benzoic acid and 0.62 mol of methanol in a 100-mL round-bottomed flask, and carefully pour 3 mL of concentrated sulfuric acid down the side of the flask. Swirl the flask to mix the reagents, attach a reflux condenser, and gently heat the mixture at reflux for 1 hr.★ Cool the solution and transfer it to a separatory funnel containing 50 mL of water. Rinse the flask with 40 mL of technical diethyl ether, and add the rinsings to the funnel. Shake the funnel thoroughly to facilitate extraction of methyl benzoate into the ether layer; *vent the funnel occasionally.* Remove the aqueous layer and wash the organic layer with a second 25-mL portion of water. Separate the layers, add 25 mL of 0.6 *M* aqueous sodium bicarbonate to the funnel *(Caution: foaming may occur!)*, and shake the mixture, *frequently* venting the funnel. Remove the aqueous washes and test to see that they are basic to litmus paper. If they are not, repeat the washing of the organic layer with additional aqueous bicarbonate until basic washes are obtained. After a final wash with saturated sodium chloride solution, dry the ether solution over *anhydrous* magnesium sulfate.★

Filter the solution and remove the ether by simple distillation using a steam or hot-water bath *(no flames)*. Then decant the methyl benzoate into a 100-mL round-bottomed flask, and attach the flask to an apparatus for simple distillation. Distil the ester using an air-cooled rather than a water-cooled condenser (which might crack because of the high boiling point, 199°C, of the ester). Collect the material boiling above 190°C in a tared Erlenmeyer flask. Determine the yield of the methyl benzoate obtained.

Should you be instructed to do so, recover unchanged benzoic acid by carefully acidifying the basic washes with concentrated hydrochloric acid (*Caution: heat* is evolved and *foaming* occurs!) and collecting by vacuum filtration any precipitate that results. Dry and weigh the precipitate. Ascertain that this solid is benzoic acid by determining its melting point, and recalculate the yield of methyl benzoate, making allowance for the amount of starting acid that was recovered.

Pre-lab exercises for Section 22.2, Part B, Preparation of the Grignard Reagent, can be found on page PL. 97.

B. Preparation of the Grignard Reagent

DO IT SAFELY

1 When flame-drying the experimental apparatus, avoid excessive heating in the vicinity of ring seals in the condenser and of the stopcock of the addition funnel, particularly if it is made of Teflon.

2 When working with diethyl ether, be certain that there are no open flames in the **entire** laboratory. Ether is **extremely** flammable and volatile, and its vapors can easily travel several feet along the bench top or the floor and then be ignited.

3 Open containers of ether must **not** be kept at your laboratory bench. Estimate the total volume of ether you need, measure it out in the ventilation hood, and return it to your work area in a stoppered container. Be certain to reclose the container from which you obtained the ether in order to prevent evaporation, absorption of moisture, and accidental fires.

4 Do not stopper flasks containing ether too tightly. Excessive pressure build-up may result because of temperature changes and create a hazard.

5 Do not store flasks containing ether in your laboratory drawer, because vapors may build up and create a fire hazard. Rather, place the flasks, properly labeled, in the ventilation hood.

6 Lubricate all ground-glass joints in the apparatus carefully and mate them tightly to prevent escape of ether during the reaction.

7 The formation of the Grignard reagent is **exothermic.** Have an ice-water bath prepared in the event that moderation of the rate of reaction becomes necessary, as measured by an excessively rapid rate of reflux.

Equip a 250-mL round-bottomed flask with a magnetic stirring bar, if this method of stirring is available, and with a Claisen adapter to which are attached a reflux condenser and a dropping funnel. Position the dropping funnel directly above the flask and the condenser on the side arm of the adapter (Figure 2.31b). Attach calcium chloride drying tubes to the top of the condenser and to the dropping funnel.

Dry the assembled apparatus using a flame (see Section 2.24). *Allow the apparatus to cool to room temperature and be sure that there are NO flames in the laboratory before continuing!*

Working quickly, remove the dropping funnel from the Claisen adapter and add 0.10 mol of magnesium turnings through the opening. Replace the funnel and add 10 mL of *anhydrous* diethyl ether through it. Prepare a solution of 0.12 mol of whichever halide has been assigned in 25 mL of *anhydrous* ether, swirl the solution to achieve homogeneity, and add it to the dropping funnel (stopcock *closed!*). Also prepare an ice-water bath in case it is needed. If you are using magnetic stirring, turn on the stirrer motor. Be sure that water is running through the condenser.

Add a 2- to 3-mL portion of the halide-ether solution from the dropping funnel onto the magnesium turnings. If you are not using magnetic stirring, loosen the clamp to the flask, and manually swirl the flask in order to mix the contents. A change in the appearance of the reaction mixture, as evidenced by the presence of a slightly cloudy (chalky) solution and by the formation of small bubbles at the surface of the magnesium turnings, indicates that the reaction has started. Once the reaction has started, the ether will be observed to reflux, and the flask will become slightly warm. If the reaction has started, disregard the instructions in the next paragraph.

If the reaction does not start on its own, obtain two or three additional turnings of magnesium, and crush them thoroughly with a heavy spatula or the end of a clamp. Remove the dropping funnel just long enough to add these broken pieces of magnesium to the flask, and then replace the funnel. The clean, unoxidized surfaces of magnesium that are exposed should aid in the initiation of the reaction. If the reaction still has not started after an additional 3 to 5 min, consult your instructor. The best remedy at this point is to warm the flask and add 2 or 3 drops of 1,2-dibromoethane to the mixture (equation 6).

Once the reaction has started, either on its own or as a result of using the remedies discussed above, and the ether is observed to be refluxing smoothly, add an extra 20-mL portion of anhydrous diethyl ether to the reaction mixture through the condenser. This serves to dilute the reaction mixture and to minimize the coupling reaction. The rest of the halide-ether solution should now be added *dropwise* to the reaction mixture at a rate that is just fast enough to maintain a gentle reflux. If it is added too fast the reaction may get out of control, and the yield will also be reduced, owing to increased coupling. If the reaction becomes too vigorous, cool the flask a bit with the ice-water bath, and reduce the rate of addition. This total addition should take about 15–30 min. If the spontaneous boiling of the mixture becomes too slow, increase the rate of addition slightly. If this does not serve to accelerate the rate of reflux, use a steam or hot-water bath or heating mantle to gently heat the mixture as necessary during the remainder of addition.

At the end of the reaction the solution will normally have a tan to brown, chalky appearance, and most of the magnesium will have disappeared, although residual bits of metal usually remain.

Use the Grignard reagent as soon as possible after its preparation, following one of the procedures given in part C. Either C1 or C2 applies if phenylmagnesium bromide was prepared; C3 applies if *n*-butylmagnesium bromide was prepared.

Pre-lab exercises for Section 22.2, Part C, Reactions of the Grignard Reagent, can be found on pages PL. 99–103.

C. Reactions of the Grignard Reagent

1. Preparation of Triphenylmethanol Once the reaction mixture for the preparation of phenylmagnesium bromide (part B) has cooled to room temperature, dissolve 0.045 mol of methyl (or ethyl) benzoate in about 20 mL of anhydrous diethyl ether, and place this solution in the dropping funnel (stopcock *closed*). Cool the reaction flask with an ice-water bath, and then begin slow, dropwise addition of the solution of methyl benzoate to the phenylmagnesium bromide solution. This reaction is exothermic; control the rate of reaction by adjustment of the addition rate and by occasional cooling of the reaction flask with the ice-water bath as necessary. If you are not using magnetic stirring, swirl the flask from time to time during the addition. Frequently a white solid forms during the reaction and is a sign that the reaction is proceeding normally. After the addition is complete and the exothermic reaction has subsided, *either* of the following may be done in order to complete the reaction. (1) Heat the reaction mixture at reflux for 30 min, using either a heating mantle or a steam bath. (2) Stopper the flask after cooling to room temperature, and allow the mixture to stand in the hood until the next laboratory period (no reflux required).★

In a 250-mL Erlenmeyer flask place about 50 mL of 6 M sulfuric acid and about 30 g of ice, and pour the entire reaction mixture into this flask with swirling. Continue swirling until the heterogeneous mixture is completely free of undissolved solids. It may be necessary to add more diethyl ether (technical) to dissolve all the organic material. Transfer the entire mixture to a 250-mL separatory funnel, shake it vigorously but carefully, venting the funnel often to relieve pressure, and remove the aqueous layer.★ Wash the organic layer with 3 M sulfuric acid and then with saturated sodium chloride solution. Repeat the washings with salt solution until the resulting aqueous layer is no longer acidic. Dry the organic layer with anhydrous sodium sulfate, and filter the solution into a round-bottom flask of suitable size. Remove the ether by simple distillation *(no flames)*. After the crude solid residue has dried, determine its melting range.★ It should weigh 11–12 g and may melt over a wide range.

The crude solid may be purified by recrystallization from a 2:1 mixture of cyclohexane:absolute ethanol. Use a minimum amount of the boiling solvent, carrying out the operation in the hood, or use a funnel inverted over the flask, which is attached to a vacuum source. Once all the material is in solution, evaporate the solvent slowly until small crystals of triphenylmethanol start to form. Remove the heat and allow crystallization to continue at room temperature; complete the process by cooling the solution to 0°C until no more crystals appear to form. Isolate the product by filtration; it should be colorless. Determine the melting point and yield of the product. The yield should be 5–6 g. The reported melting point of triphenylmethanol is 164°C.

An alternative method of recovering the triphenylmethanol from the crude mixture containing it and biphenyl is as follows. Add about 150 mL of petroleum ether and about 50 mL of technical-grade diethyl ether to the crude product. Heat the mixture on a steam bath (preferably in the hood) until the crystals dissolve; add more ether if needed. Boil off the solvent until the first crystals appear, and then stop heating. Allow the flask to cool, and collect the crystals that form. Proceed to determine the melting point of the crystals, and recrystallize if needed.

2. Preparation of Benzoic Acid In a 1-L Erlenmeyer flask place 25–30 g of solid carbon dioxide (Dry Ice) that has been coarsely crushed and protected from moisture as much as possible. Add the phenylmagnesium bromide solution to the Dry Ice slowly with gentle swirling; the mixture normally becomes somewhat viscous. After the addition is complete, allow the excess carbon dioxide to evaporate by letting the flask stand in the hood, properly labeled with your name, until the next laboratory period.★

If desired, the process of removing the excess CO_2 can be expedited in one of several ways: (1) shake or swirl the flask while warming it *very slightly* in a warm-water bath, (2) stir the viscous mixture that forms during the reaction, or (3) add small amounts of warm water to the reaction mixture. All these methods may cause a sudden loss of CO_2 gas, and great care should be taken in hastening this process. At no time should the flask be stoppered. Any warming that is done also causes the ether to evaporate, and care should be taken to replace the ether before going to the next step.

After the excess Dry Ice is gone, add 100 mL of technical ether and treat the mixture with about 50 mL of 3 *M* sulfuric acid that has been mixed with 25–30 g of ice. The addition of the acid to the crude reaction mixture should be done with care in order to avoid foaming. If the ether has evaporated appreciably, more technical diethyl ether may be added so that the total volume of ether is about 100 mL. Transfer the entire mixture to a 500-mL separatory funnel after mixing it thoroughly by swirling. Rinse the flask with a small portion of technical diethyl ether, and add the rinse to the separatory funnel. Shake the funnel cautiously, *with venting,* and separate the layers. Extract the aqueous layer two additional times with 15-mL portions of diethyl ether, and *combine all* the ether extracts.

Extract the ethereal solution with three 20-mL portions of a 1 *M* solution of sodium hydroxide, venting the funnel frequently during the extractions. Treat the combined alkaline extracts with decolorizing carbon and filter by gravity. Add 6 *M* hydrochloric acid until precipitation is complete and the aqueous mixture is acidic to litmus paper. Isolate the solid by vacuum filtration.★ Benzoic acid may be recrystallized from water if needed (see Section 3.1). After drying the product, determine the melting point and yield; the yield should be about 6 g.

3. Preparation of 2-Methyl-3-heptanol After the reaction involving the preparation of *n*-butylmagnesium bromide (part B) has subsided and the reaction mixture has cooled to room temperature, prepare 2-methyl-3-heptanol as follows. Dissolve 0.090 mol of freshly distilled 2-methylpropanal in 15 mL of *anhydrous* diethyl ether, and place this solution in the dropping funnel (stopcock *closed!*) of the apparatus in which the Grignard reagent was prepared. Begin the dropwise addition of the solution of 2-methylpropanal *to* that of the *n*-butylmagnesium bromide. Control the resulting highly exothermic reaction by adjusting the rate of addition so that the highest point at which condensation can be discerned is no more than one-third to one-half way up the reflux condenser. Cool the reaction flask with a previously prepared ice-water bath if necessary. Either stir or swirl the reaction mixture occasionally during the course of the addition, which may require 15–20 min. After completion of the addition, allow the reaction mixture to stand about 15 min.★ If the mixture is to be stored until the next laboratory period, reread comments 4 and 5 in the Do It Safely section of part B.

Place about 150 mL of crushed ice in a 500-mL beaker and add 9 mL of concentrated sulfuric acid. Pour the reaction mixture slowly and with stirring into the ice-acid mixture. After the addition is complete, transfer the cold mixture, which may contain some precipitate, to a 250- or 500-mL separatory funnel and shake it gently. The precipitate should dissolve. Separate the layers, extract the aqueous layer with two 25-mL portions of technical diethyl ether, and add these extracts to the main ether layer. Wash the combined ethereal solution contained in the separatory funnel with 30 mL of saturated sodium bisulfite solution, venting the funnel frequently to relieve pressure, then with two 30-mL portions of 1.2 *M* aqueous sodium bicarbonate, and finally with 30 mL of saturated sodium chloride solution. Dry the ethereal solution over anhydrous magnesium sulfate.★ If this solution is to be stored until the next laboratory period, reread comments 4 and 5 in the Do It Safely section of part B.

Filter the solution by gravity from the drying agent into a 250-mL round-bottomed flask equipped for simple distillation, and remove most of the ether by distillation (*Caution:* no flames! Use a steam or hot-water bath or a heating mantle). Transfer the residue, which contains the product along with some ether, to a 50-mL round-bottomed flask equipped for fractional distillation. Insulate the top of the distilling flask and the stillhead with glasswool or aluminum foil to ensure steady distillation of the rather high-boiling product. Collect separately any forerun boiling below 165°C and the 2-methyl-3-heptanol, bp 165–168°C. Determine the yield of product.

EXERCISES

General Questions

1. Arrange the following compounds in order of increasing reactivity toward attack of the Grignard reagent at the carbonyl carbon atom: methyl benzoate, benzoic acid, benzaldehyde, acetophenone, benzoyl chloride. Explain the basis for your decision, making use of mechanisms where needed.

2. What is (are) the product(s) of reaction of each of the above carbonyl-containing compounds with *excess* Grignard reagent, RMgBr?

3. How might primary, secondary, and tertiary alcohols be prepared from a Grignard reagent, RMgBr, and a suitable carbonyl-containing compound? Write chemical reactions for these preparations and indicate stoichiometry where important.

Questions for Part A

4. Assuming that the equilibrium constant for esterification of benzoic acid with methanol is 3, calculate the theoretical yield of methyl benzoate expected using the molar amounts employed in this experiment.

5. Concentrated sulfuric acid is added as a catalyst in the esterification procedure even though another acid, benzoic acid, is one of the organic reagents used. Why is the sulfuric acid necessary?

6. Would the use of an increased amount of sulfuric acid be expected to increase the yield of the desired ester? Explain.

7. Suppose that the washes with aqueous base were omitted in the workup of methyl benzoate. What problems could result?

8. Why would the choice of solid potassium hydroxide as a drying agent for the ester be a poor one?

Questions for Part B

9. Why were you cautioned not to heat excessively in the vicinity of ring seals when drying the apparatus for this experiment?

10. Ethanol is often present in the technical grade of diethyl ether. If this grade rather than anhydrous were used, what effect, if any, would the ethanol have on the formation of the Grignard reagent? Explain.

11. Give a plausible, three-dimensional structure for the complex $RMgBr \cdot 2(C_2H_5)_2O$. How do you think the ether molecules are bound to the Grignard reagent?

12. Suppose bromocyclohexane were used to assist in the initiation of the preparation of the Grignard reagents phenylmagnesium bromide and n-butylmagnesium bromide. Why would it lead to potential experimental difficulties that are avoided if 1,2-dibromoethane is used instead?

13. Why is it unwise to allow the solution of the Grignard reagent to remain exposed to air for an unnecessary period of time even if it is protected from moisture by drying tubes?

Questions for Part C.1

14. Why is it unwise to begin addition of the solution of methyl (or ethyl) benzoate to the Grignard reagent before the latter has cooled to room temperature?

15. Why should anhydrous rather than technical diethyl ether be used to prepare the solution of methyl (or ethyl) benzoate that is added to the Grignard reagent?

16. What is the solid that forms during the addition of the ester to the Grignard reagent?

17. Why does pressure develop when the separatory funnel containing aqueous sulfuric acid and the ethereal solution of organic products is shaken?

18. Comment on the use of steam distillation as a possible alternative procedure for the purification of the crude triphenylmethanol. Consider the possible unchanged starting materials and the products that are formed, and indicate which of these should steam-distil and which should not. Explain in detail how this method of purification would yield pure triphenylmethanol, if it would at all.

Questions for Part C.2

19. Why does pressure develop when the separatory funnel containing aqueous sulfuric acid and the ethereal solution of organic products is shaken?

20. The yield of benzoic acid obtained by students who add only enough acid to the aqueous solution of sodium benzoate to bring its pH to 7 is consistently lower than that obtained if the pH is brought below 5. Why is this?

21. What function does extracting the ethereal solution of organic products with aqueous base have in the purification of benzoic acid?

Questions for Part C.3

22. Given that commercially available aldehydes usually are produced by oxidative methods, what contaminant(s) might be present in the 2-methylpropanal to cause a diminution in yield?

23. Why is it unwise to begin addition of the solution of 2-methylpropanal to the Grignard reagent before the latter has cooled to room temperature?

24. What is the solid that forms upon reaction of *n*-butylmagnesium bromide with 2-methylpropanal?

25. Why does pressure develop when the separatory funnel containing aqueous sulfuric acid and the ethereal solution of organic products is shaken?

26. The work-up in this reaction calls for successive washes of an ethereal solution of the product with aqueous sodium bisulfite, sodium bicarbonate, and sodium chloride. What is the purpose of each of these steps?

27. The IR spectrum of 2-methyl-3-heptanol prepared by the procedure of Part C.3 sometimes shows an absorption at 1720 cm^{-1}. Give a possible source for this absorption.

*28. Explain why there are eight peaks in the CMR spectrum of 2-methyl-3-heptanol (Figure 22.6b).

SPECTRA OF STARTING MATERIALS AND PRODUCTS

The IR, PMR, and CMR spectra of benzoic acid are given in Figures 8.24a and 19.5, respectively. The IR and CMR spectra of methyl benzoate are presented in Figures 8.25a and 9.26, respectively. The IR, PMR, and CMR spectra of 2-methylpropanal are contained in Figures 8.22b and 19.2, respectively.

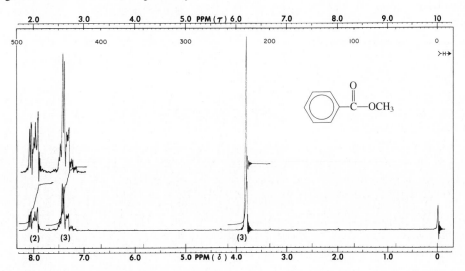

Figure 22.1 PMR spectrum of methyl benzoate.

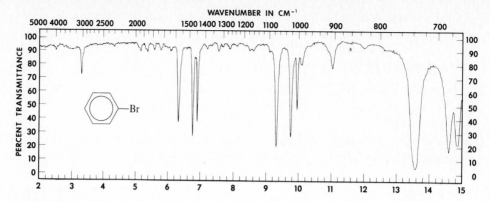

Figure 22.2 IR spectrum of bromobenzene.

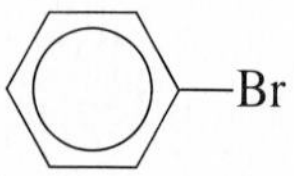

Br   Chemical shifts: δ 122.5, 126.7, 129.8, 131.4.

Figure 22.3 CMR data for bromobenzene.

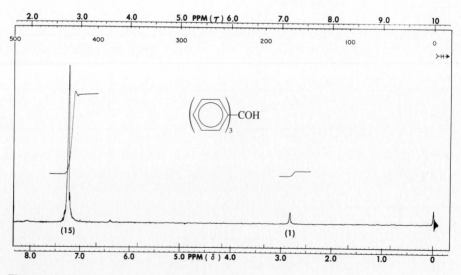

Figure 22.4 PMR spectrum of triphenylmethanol.

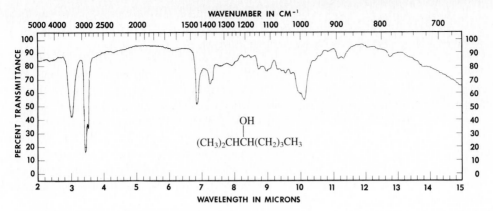

Figure 22.5 IR spectrum for 2-methyl-3-heptanol.

(a) PMR spectrum.

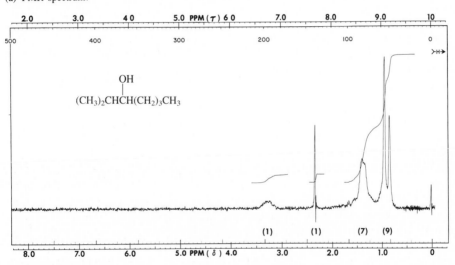

(b) CMR data. Chemical shifts: δ 14.2, 17.5, 19.3, 23.1, 28.6, 33.8, 34.1, 76.6.

Figure 22.6 NMR data for 2-methyl-3-heptanol.

22.3 The Organolithium Reagent

Alkyl and aryl lithium compounds play a role in organic synthesis second only to that of the Grignard reagents discussed in Section 22.2. Because of their importance we feel it is desirable to include a brief discussion of their preparation and reactivity; however, owing to possible difficulties in working with the reactive lithium metal, we have opted not to include any experiments involving the preparation of organolithium reagents.

The preparation of the lithium reagent, RLi, is indicated in equation 20, where RX represents an alkyl or aryl halide.

$$RX + 2\,Li \xrightarrow[\substack{\text{or} \\ \text{hydrocarbon solvent}}]{\text{dry ether solvent}} RLi + LiX \tag{20}$$

The solvent does not appear to play as important a role in the preparation of organolithium reagents as it does in the preparation of Grignard reagents. This has been shown by the fact that it is possible to prepare organolithium reagents in both hydrocarbon solvents, such as pentane and hexane, and ether solvents. However, there is a difference in reactivity of the reagents toward the solvent. Ethereal solutions of the Grignard reagent may be stored under anhydrous conditions for some time, but organolithium reagents may be kept only when a hydrocarbon solvent is used. The latter cleave ethers, even though the reagent can be prepared in ether solution. When ether is used, the reagent must be used immediately after its preparation and may not be stored.

It should be noted that 2 moles of lithium are required for each mole of organic halide, in contrast to the one-to-one mole ratio of magnesium to halide. Pure lithium metal can be used effectively in preparing the lithium reagent, although the use of lithium containing 0.8% by weight of sodium metal often increases the ease with which the reaction starts.

The reactivity of the alkyl and aryl halides toward lithium metal is the same as that for the Grignard reagent. The major exception is the necessity of using tetrahydrofuran as the solvent for the preparation of the Grignard reagent from aryl chlorides. Most alkyl halides form lithium reagents by reaction with lithium metal either in diethyl ether or in a hydrocarbon solvent.

The side reactions arising from the preparation of the lithium reagent are the same as those found for the Grignard reagents (see Section 22.2, equations 9–11). The same conditions of extreme dryness *must* be utilized; the side reactions may be minimized by use of an inert atmosphere and dilute solutions.

The lithium reagents usually react in a manner similar to Grignard reagents. In some instances lithium reagents are preferable because they are generally more reactive than their halomagnesium counterparts. Furthermore, once the lithium reagent has reacted with a substrate, the resulting salt that forms is generally more soluble in the solvent than the corresponding salt resulting from the Grignard reagent.

An advantage of the greater reactivity of the lithium reagent is realized in *metalation reactions,* in which the lithium reagent, RLi, abstracts a proton from a hydrocarbon, R'H, as shown in equation 21.

$$RLi + R'H \rightarrow RH + R'Li \tag{21}$$

This reaction has been used to determine the acidity of hydrocarbons and can also be used to introduce substituents into various aromatic compounds. *n*-Butyllithium (equation 22) has been used extensively for this purpose, and some typical examples are shown in equations 23–25.

$$n\text{-}C_4H_9Br + 2\,Li \xrightarrow[-10°C]{\text{diethyl ether}} n\text{-}C_4H_9Li + LiBr \tag{22}$$

n-Butyl bromide   *n*-Butyllithium
85% yield

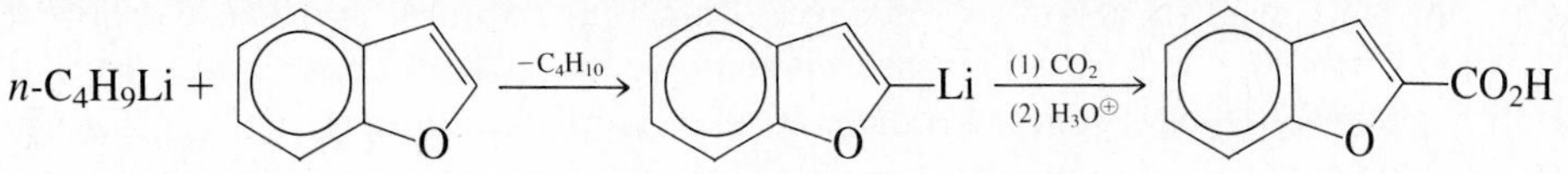

$$n\text{-}C_4H_9Li + \quad \xrightarrow{-C_4H_{10}} \quad \text{—Li} \quad \xrightarrow[\text{(2) } H_3O^{\oplus}]{\text{(1) } CO_2} \quad \text{—CO}_2H \quad \textbf{(23)}$$

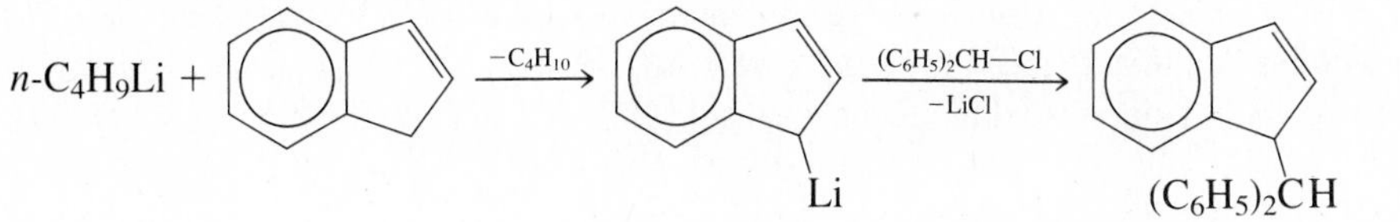

$$n\text{-}C_4H_9Li + \quad \xrightarrow{-C_4H_{10}} \quad \text{Li} \quad \xrightarrow[\text{—LiCl}]{(C_6H_5)_2CH\text{—Cl}} \quad (C_6H_5)_2CH \quad \textbf{(24)}$$

$$+ n\text{-}C_4H_9Li \xrightarrow{-C_4H_{10}} \underset{\text{Li}}{CH_2} \xrightarrow[\text{—LiBr}]{Br\text{—Br}} \underset{\text{Br}}{CH_2} \quad \textbf{(25)}$$

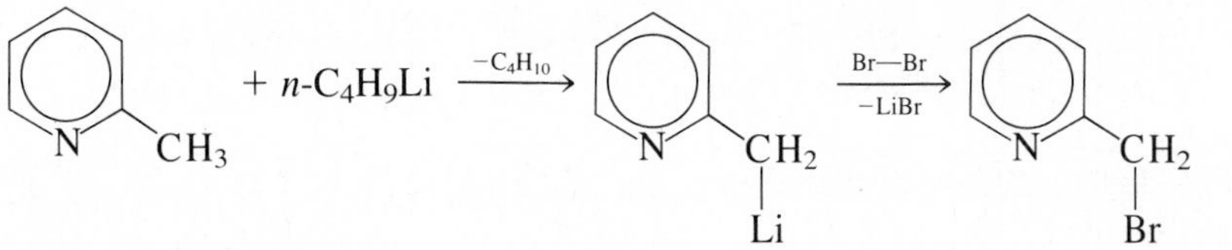

22.4 Organocopper Reagents

Organocopper reagents cannot be prepared by reaction of metallic copper with an alkyl or aryl halide, but rather are generated indirectly through reaction of either cuprous bromide or cuprous iodide with Grignard reagents or, more frequently, organolithium reagents. For example, reaction of methyllithium with one molar equivalent of cuprous iodide yields the polymeric methylcopper, an *organocopper* (**10,** equation 26), and addition of a second molar equivalent of methyllithium provides lithium dimethylcuprate, an *organocuprate* (**11,** equation 27). The summation of these two reactions is shown in equation 28.

$$(x) \; CH_3Li + (x) \; CuI \rightarrow (CH_3\text{—Cu})_x + (x) \; LiI \quad \textbf{(26)}$$

Methyllithium   **10**
                                  Methylcopper

$$(CH_3\text{—Cu})_x + (x) \; CH_3Li \rightarrow (x) \; (CH_3)_2Cu^{\ominus}Li^{\oplus} \quad \textbf{(27)}$$

11

$$2 \; CH_3Li + CuI \rightarrow (CH_3)_2Cu^{\ominus}Li^{\oplus} + LiI \quad \textbf{(28)}$$

Lithium dimethylcuprate

The synthetic utility of these two types of copper reagents relative to the organometallic reagents described in the preceding two sections is exemplified in the following discussion.

When alkylcuprates react with primary alkyl halides, coupling products arising from S_N2-type reactions may be isolated in good yield, as shown in equation 29, for example. The yields of coupling products are much higher than those obtained with Grignard reagents or alkyllithium compounds, so that this is the chosen procedure for preparing hydrocarbons by this type of approach. Alkylcopper reagents react similarly, although normally akylcuprates are used.

$$2 \; (CH_3)_2CH\text{—Li} \xrightarrow{CuI} [(CH_3)_2CH]_2Cu^{\ominus}Li^{\oplus} \xrightarrow{2 \; CH_3CH_2Br} 2 \; (CH_3)_2CH\text{—CH}_2CH_3 \quad \textbf{(29)}$$

Isopropyllithium   Lithium diisopropylcopper   2-Methylbutane

Organocopper reagents also react effectively with acid chlorides in the preparation of ketones (equation 30). Such reactions provide good yields because organocopper reagents react only sluggishly with ketones. Reaction of acid chlorides with Grignard reagents and alkyllithium compounds does not often provide acceptable yields of ketonic products because of the tendency of ketones to undergo *further* reaction with the organometallic reagent to produce alcohols. In this application, copper provides the same advantage as does cadmium in the similar use of organocadmium reagents.

$$2 \ \underset{\substack{\| \\ O}}{R-C-Cl} + (CH_3)_2Cu^{\ominus}Li^{\oplus} \rightarrow 2 \ \underset{\substack{\| \\ O}}{R-C-CH_3} + CuCl + LiCl \qquad \textbf{(30)}$$

Lithium dimethylcopper

Usually Grignard reagents react with α,β-unsaturated ketones to produce both **12** and **13**, products of 1,2- (normal) and 1,4- (conjugate) addition, respectively (equation 31).

$$R-CH=CH-\underset{\substack{\| \\ O}}{C}-R' \xrightarrow[\text{(2) } H_3O^{\oplus}]{\text{(1) } CH_3MgX} R-CH=CH-\underset{\substack{| \\ CH_3}}{\overset{OH}{C}}-R' + R-\underset{\substack{| \\ CH_3}}{CH}-CH_2-\underset{\substack{\| \\ O}}{C}-R' \quad \textbf{(31)}$$

$$\text{12} \qquad\qquad\qquad\qquad \text{13}$$

The reaction is very sensitive to steric effects, and examples range from those in which the product of conjugate addition is the major isomer formed to those in which it is either the minor product or is not produced at all. In contrast, alkyllithium reagents do not give conjugate addition, so that only products of addition to the carbonyl group (equation 32) are observed.

$$RCH=CH-\underset{\substack{\| \\ O}}{C}-R' \xrightarrow{CH_3Li} RCH=CH-\underset{\substack{| \\ CH_3}}{\overset{O^{\ominus}Li^{\oplus}}{C}}-R' \xrightarrow{H_3O^{\oplus}} RCH=CH-\underset{\substack{| \\ CH_3}}{\overset{OH}{C}}-R' \quad \textbf{(32)}$$

If the product of conjugate addition is desired, alkyllithium reagents cannot be used, and the unpredictability of the Grignard addition in giving this type of product is a nuisance, However, both alkylcopper and alkylcuprate reagents react with α,β-unsaturated ketones to provide either totally or predominantly the conjugate addition product. Thus, reaction of 3,5,5-trimethylcyclohex-2-enone (**14**, isophorone) with lithium dimethylcuprate or methyl copper provides 3,3,5,5-tetramethylcyclohexanone in excellent yield (equation 33).

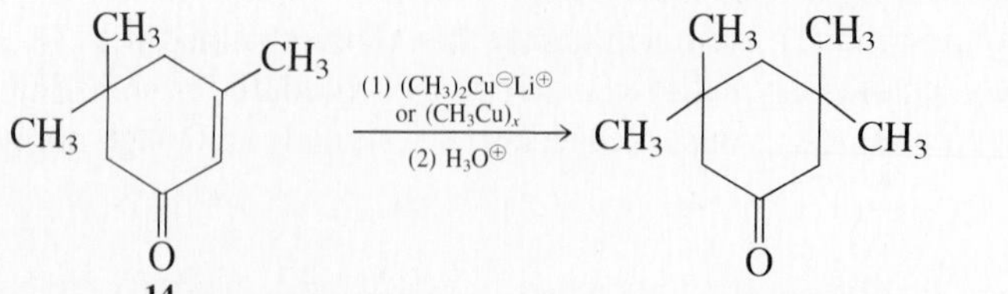

14
3,5,5-Trimethylcyclohex-2-enone

3,3,5,5-Tetramethylcyclohexanone

$$\textbf{(33)}$$

The corresponding reaction of **14** with methylmagnesium iodide, in contrast, produces the tertiary alcohol **15** in 90% yield (equation 34).

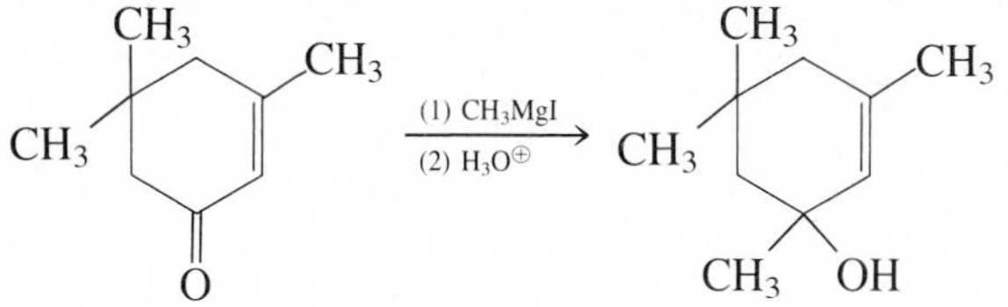

$$\text{(34)}$$

In the following experiment, 3,5,5-trimethylcyclohex-2-enone (**14**) will be allowed to react with a solution of methylmagnesium iodide to which has been added a *catalytic* amount of cuprous bromide. Methylcopper is produced *in situ* (equation 35), and then reacts with **14** to produce the enolate ion **16** (equation 36), regenerating the cuprous ion for further reaction with additional methylmagnesium iodide (equation 35).

$$(x)\ CH_3MgI + (x)\ CuBr \rightarrow (CH_3\!-\!Cu)_x + (x)\ MgIBr \qquad \text{(35)}$$

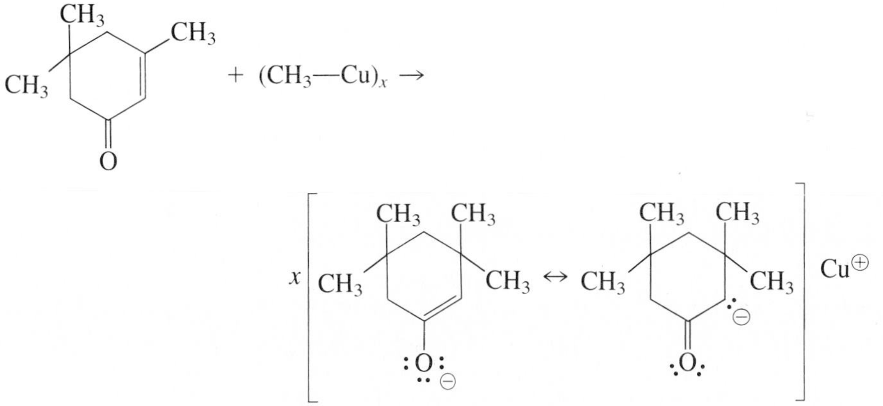

$$\text{(36)}$$

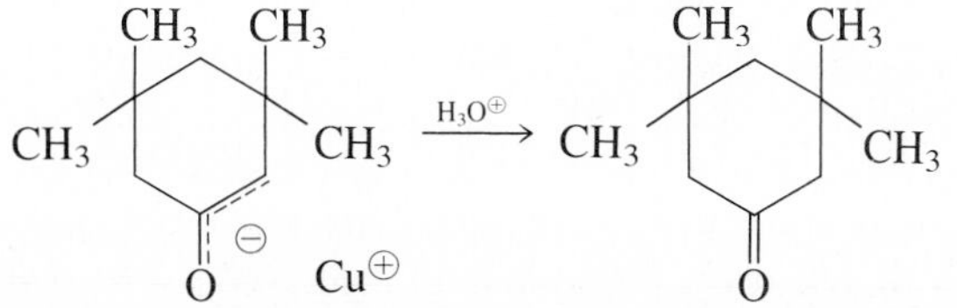

$$\text{(37)}$$

Following neutralization (equation 37), 3,3,5,5-tetramethylcyclohexanone may ultimately be isolated in up to 80% yield. Methylcopper is apparently somewhat more reactive toward **14** than methylmagnesium iodide, since the reaction shown in equation 34 does not compete effectively. *Small* amounts of 1,3,5,5-tetramethyl-1,3-cyclohexadiene (**17**) may be isolated during distillation of the final product; the diene presumably arises by dehydration of the tertiary alcohol **15** (equation 38).

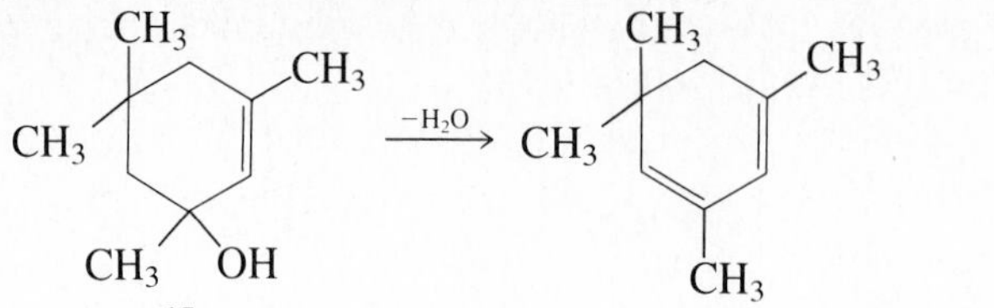

(38)

1,3,5,5-Tetramethyl-1,3-cyclohexadiene

Pre-lab exercises for Section 22.4, Organocopper Reagents, can be found on page PL. 105.

EXPERIMENTAL PROCEDURE

DO IT SAFELY

1 Before preparing methylmagnesium iodide, read the Do It Safely directions in Section 22.2A.

2 Iodomethane is an alkylating agent. As with all alkylating agents of moderate or greater reactivity, care should be taken to avoid breathing its vapors or allowing contact with the skin. Although iodomethane has **not** been implicated, several other alkylating agents are either suspected or proven carcinogens.

According to the procedures given in Parts A and B of the Experimental Procedure in Section 22.2, prepare a solution of methylmagnesium iodide. Use 0.10 mol of magnesium turnings and 0.12 mol of iodomethane. If small amounts of magnesium metal remain at the end of the preparation, add a few more drops of iodomethane, and allow the reaction to proceed a little longer. When the magnesium has completely dissolved, add 0.2 g of cuprous bromide to the solution of methylmagnesium iodide, and cool the flask in an ice-water bath to 5°C.

Prepare a solution of 0.08 mol of 3,5,5-trimethylcyclohex-2-enone (**14**) in 15 mL of anhydrous diethyl ether, and place this solution in the dropping funnel (stopcock *closed*) of the apparatus used to make the Grignard reagent. While continuing to cool the flask, begin dropwise addition of the solution of **14** to the reaction mixture; the dropping rate should be adjusted to require about 20 min to complete the addition. After the addition is complete, remove the cooling bath, and allow the mixture to stand or stir at room temperature for about 15 min. Then heat the mixture at reflux for a period of about 30 min, using either a steam bath or a heating mantle. Cool the mixture to room temperature with the ice-water bath.

Place 60 g of ice and 6 g of glacial acetic acid in a 500-mL beaker, and with swirling slowly pour the cooled reaction mixture over the ice. When most of the ice has melted, decant the liquid into a 250-mL separatory funnel, and separate the ether layer. Extract the aqueous layer with an additional 25 mL of technical diethyl ether, and combine this extract with the original ether layer.★ Extract the ethereal solution twice with 15-mL portions of 0.5 *M* sodium carbonate solution, followed by two washings with 15-mL portions of water and finally with 15 mL of saturated aqueous sodium chloride. Dry the organic solution over anhydrous sodium sulfate for about 15 min with occasional swirling, and filter.★

Remove the ether by simple distillation, using a steam or hot-water bath or a heating mantle, and then transfer the crude product to a smaller flask.★ Owing to the high-boiling nature of the product, the final simple distillation will need to be carried out with either a heating mantle or a burner. If heating mantles are not available so that a burner must be used, first connect the flask to either a house vacuum line or an aspirator, and evacuate the flask with swirling for a few minutes. It will help to warm the flask. This will ensure *complete* removal of flammable ether before using the burner.

Distil the residual oil, collecting 3,3,5,5-tetramethylcyclohexanone in the boiling range of 193–197°C; use no water in the condenser to avoid cracking it, owing to thermal stresses on the glass resulting from the hot vapors of product contacting the condenser. If care is taken, a small forerun of 1,3,5,5-tetramethyl-1,3-cyclohexadiene (**17**) may be collected in the range 151–155°C. Calculate the yield of product. Obtain an IR spectrum of the product, and compare it with that of 3,5,5-trimethylcyclohex-2-enone (Figure 22.7).

EXERCISES

1. Why were you cautioned not to heat excessively in the vicinity of ring seals when drying the apparatus for this experiment?

2. Ethanol is often present in the technical grade of diethyl ether. If this grade rather than anhydrous were used, what effect, if any, would the ethanol have on the formation of the Grignard reagent? Explain.

3. Give a plausible, three-dimensional structure for the complex $CH_3MgI \cdot 2(C_2H_5)_2O$. How do you think the ether molecules are bound to the Grignard reagent?

4. Why is it unwise to allow the solution of methylmagnesium iodide to remain exposed to air for an unnecessary period of time even if it is protected from moisture by drying tubes?

5. Why should anhydrous rather than technical diethyl ether be used to prepare the solution of 3,5,5-trimethylcyclohex-2-enone that is added to the mixture of the Grignard reagent and cuprous bromide?

6. Explain the role of the acetic acid used in the procedure.

7. At what point in the work-up procedure is the crude product isolated from most of the copper and magnesium salts present in the reaction mixture?

8. Explain how 0.08 mol of 3,5,5-trimethylcyclohex-2-enone can react with methylcopper when only about 0.02 mol of cuprous ion was added to the methylmagnesium iodide.

SPECTRA OF STARTING MATERIAL AND PRODUCT

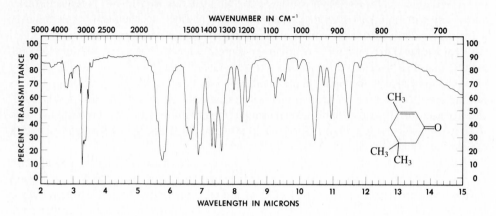

Figure 22.7 IR spectrum of 3,5,5-trimethylcyclohex-2-enone.

(a) PMR spectrum.

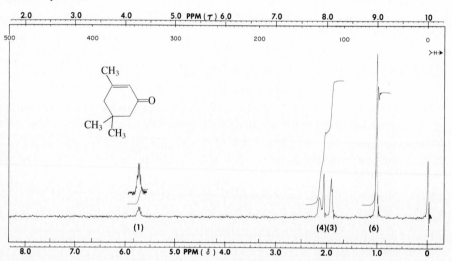

(b) CMR data. Chemical shifts: δ 24.2, 28.2, 33.3, 45.1, 50.7, 125.4, 159.4, 198.5.

Figure 22.8 NMR data for 3,5,5-trimethylcyclohex-2-enone.

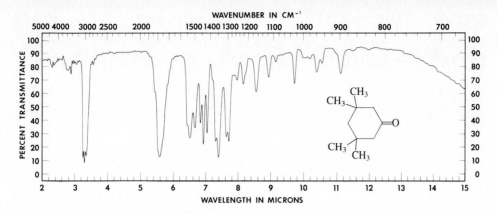

Figure 22.9 IR spectrum of 3,3,5,5-tetramethylcyclohexanone.

(a) PMR spectrum.

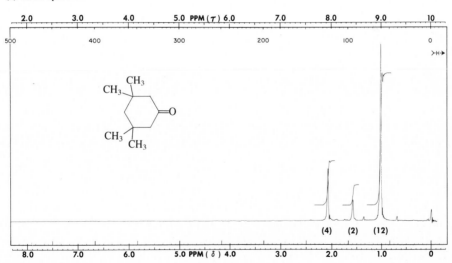

(b) CMR data. Chemical shifts: δ 31.3, 36.0, 51.5, 53.8, 212.0.

Figure 22.10 NMR data for 3,3,5,5-tetramethylcyclohexanone.

chapter **23**
multistep organic syntheses

23.1 Introduction

The synthesis of complex molecules from simpler starting materials is one of the most important aspects of organic chemistry. Some syntheses require as many as 20 or 30 sequential reactions or "steps." Multistep syntheses of such lengths are usually economically feasible only for end products that are biologically active but cannot be obtained readily from natural sources, as in the case of the valuable anti-inflammatory drug cortisone. This is because of the cumulative effect on *overall* yield of less-than-quantitative yields of individual steps when there are many such steps. For example, even a five-step synthesis in which each step occurs in 80% yield gives an overall yield of only 33% $[(0.8)^5 \times 100 = 32.8]$. To offset this cumulative effect, the synthetic chemist searches for reaction sequences of minimal length that are capable of high yields and seeks to develop optimal experimental techniques to minimize losses in each individual step. A remarkably successful example of one such effort is the synthesis of the enzyme ribonuclease by R. B. Merrifield. This enzyme is a protein having 124 amino acid units (see Chapter 26). Its synthesis required putting the amino acids together in the proper sequence by 369 chemical reactions, in which nearly 12,000 experimental operations such as additions of reagents, filtrations, and washings were involved. To accomplish this efficiently, Merrifield and his coworkers developed a "synthesis machine" that could be programmed to carry out the operations automatically. Although the *overall* yield of the biologically active enzyme was only 17%, because of the enormous number of steps, this represented an *average yield* of more than 99% for each *individual* step!

    Perhaps the most complicated natural product ever synthesized is vitamin B_{12} (Figure 23.1), whose total synthesis was announced in 1972 by R. B. Woodward and A. Eschenmoser as a result of collaborative work at Harvard and Zurich by 100 chemists from 19

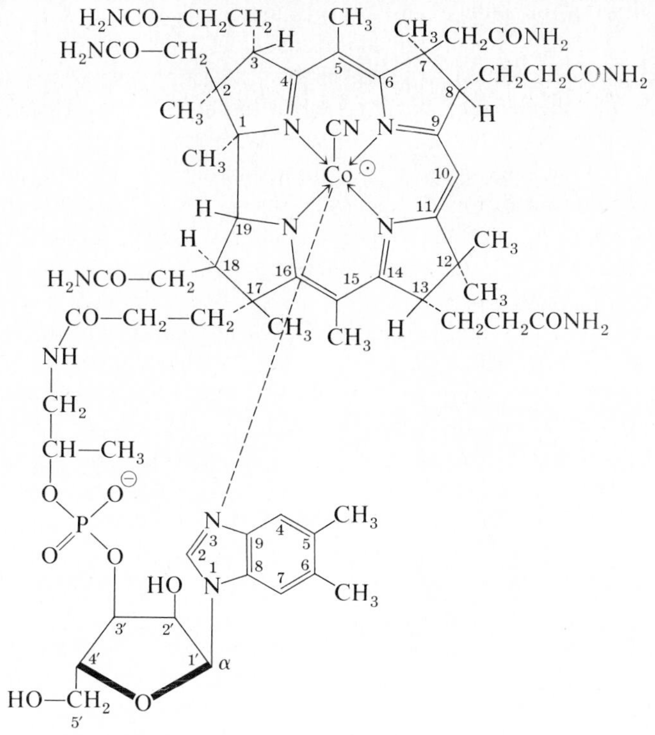

Figure 23.1 Structure of vitamin B_{12}.

different countries over a period of 11 years. Although the synthesis will never be used as a practical source of the vitamin, it was a landmark in organic synthesis and was of significant value because new reactions, techniques, and theories were developed in the course of the work, including the Woodward-Hoffmann principle of conservation of orbital symmetry (see references 7 and 8 in Chapter 15).

The synthesis with the smallest number of steps is not invariably the best route to a target molecule. As a simple example, 2-chloropropane can be produced from propane in one step by direct chlorination, but it must be separated from the 1-chloropropane formed concurrently. If the mixture of 1- and 2-chloropropane is treated with a base, both isomers undergo dehydrochlorination to propene, and subsequent addition of hydrogen chloride produces 2-chloropropane as the *sole* product. In this case the three-step process is the preferred way to obtain pure 2-chloropropane. It should be clear that an important consideration in the design of a synthesis is the ease of separation and purification of the desired product from by-products. Other considerations include the availability and cost of starting materials, simplicity of equipment and instrumentation, energy costs (for example, low-temperature reactions are desirable), activity of catalysts, selectivity of reactions in polyfunctional molecules, and stereochemical control.

Since so many variables must be considered in planning a synthesis for a complex molecule, it is not surprising that organic chemists are investigating the possibility of using computers to design and analyze multistep syntheses. The memory capacity and data-retrieval capabilities of computers can be put to use in handling an enormous amount of information, and it is hoped that optimal sequences of reaction steps can some day be predicted by computer programs. In considering this fascinating prospect, however, it should be apparent that it is the information provided to the computer by humans, specifically chemists, that will be vital to the success of such a project.

Some insight into the kind of information needed in the planning of syntheses for complex molecules can be gained from doing a synthesis consisting of a fairly small number of steps. This chapter provides such an introduction by means of three examples, the syntheses of sulfanilamide (Section 23.2), of 4-*m*-chloroanilino-7-chloroquinoline (Section 23.3), and of lidocaine (Section 23.4). There is also the possibility of devising additional synthetic sequences by combining some of the separate experiments described in other chapters in this book. Suggestions for two such sequences are given in Section 23.5.

These syntheses provide experience in using the product of one reaction in subsequent steps, an experience that emphasizes the importance of good experimental technique and gives some insight into the excitement of producing a complex organic substance from simpler starting materials. In some of the procedures there are alternatives either of isolating intermediate products in pure form or of using unpurified ones in the next step. This is the sort of choice the synthetic organic chemist constantly faces. In general, purification of each intermediate in a sequence is avoided so long as the impurities can eventually be removed and their presence does not interfere with the course of the desired reactions. However, even if it is decided not to purify the entire amount of a particular intermediate compound, it is still good scientific practice to purify a small sample of it for purposes of characterization by spectroscopic (IR, UV, NMR, MS) and physical (bp, mp, *etc.*) methods.

23.2 The Synthesis of Sulfanilamide

(a) General Discussion

Some of the most effective antibacterial drugs currently available belong to a general class called "sulfa" drugs, all members of which share the structural feature of being derivatives of the parent compound, *p*-aminobenzenesulfonamide **(1)**, commonly called sulfanilamide. Note that this compound contains two different —NH_2 moieties, one as the amino group and the other as the sulfonamido (—SO_2NH_2) group; it is the introduction of different substituents on the sulfonamido nitrogen atom that gives rise to the family of sulfa drugs. An example of one such substituent is seen in **2,** which is sulfathiazole.

1
p-Aminobenzenesulfonamide
(sulfanilamide)

2
Sulfathiazole

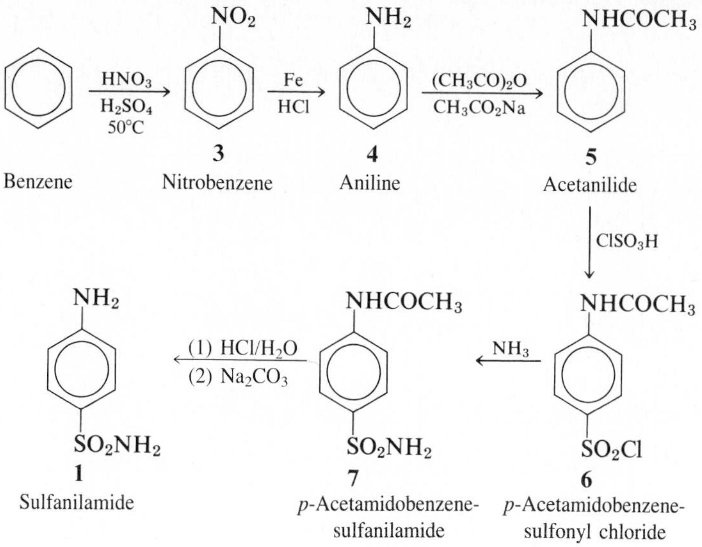

Figure 23.2 Laboratory synthesis of sulfanilamide **(1).**

Sulfanilamide itself is an antibacterial drug, and its total synthesis using benzene as the source of the aromatic ring is outlined in Figure 23.2. Owing to the toxicity of benzene, the experimental sequence provided here will commence with the second step shown, using the considerably less toxic nitrobenzene as the starting material. The preparations of several interesting intermediate compounds are included, and each of them can be isolated, if desired, or used in subsequent steps without extensive purification. A discussion of the reactions involved in each step follows.

(b) Aniline; Reduction of Aromatic Nitro Compounds

A variety of methods are available for the reduction of an *aromatic* nitro compound, which in turn is readily prepared by electrophilic aromatic nitration (Section 17.3) of the corresponding hydrocarbon. The analogous reduction of *aliphatic* nitro compounds to an amine is seldom encountered, mainly because of the difficulty of introducing this functional group into aliphatic molecules. The general methods of reduction include catalytic hydrogenation and electrolytic and chemical reduction. The most commonly used method in the laboratory is chemical reduction, in which various metals are used in acidic solution, but the most important commercial method is probably catalytic hydrogenation.

The mechanisms of these reductions are not well understood. The isolation of various stable intermediate compounds, some of which are shown in equation 1, suggests that a stepwise reaction is occurring. Differences in rates of reduction of the various intermediates may be important, but the nature of the reducing agent dictates the final product. For example, the reduction of nitrobenzene **(3)** with tin metal and hydrochloric acid gives only aniline **(4)**, whereas the use of zinc metal and ammonium chloride gives only *N*-phenyl-hydroxylamine **(11)**. Note that in equation 1, the symbol [H] means reduction by any of the possible reducing agents.

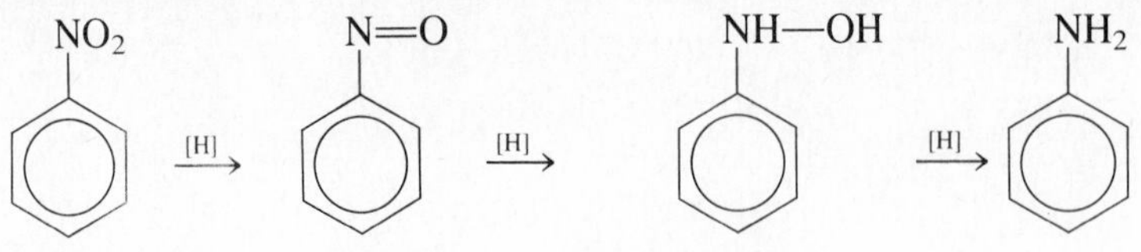

3	**10**	**11**	**4**
Nitrobenzene	Nitrosobenzene	*N*-Phenylhydroxylamine	Aniline

(1)

In this experiment iron powder with a small amount of hydrochloric acid will be used to reduce nitrobenzene to aniline. The main reaction is shown in equation 2, in unbalanced form. This is a typical oxidation-reduction equation that can be easily balanced by considering the two half-reactions given in equations 2a and 2b (see Exercise 9).

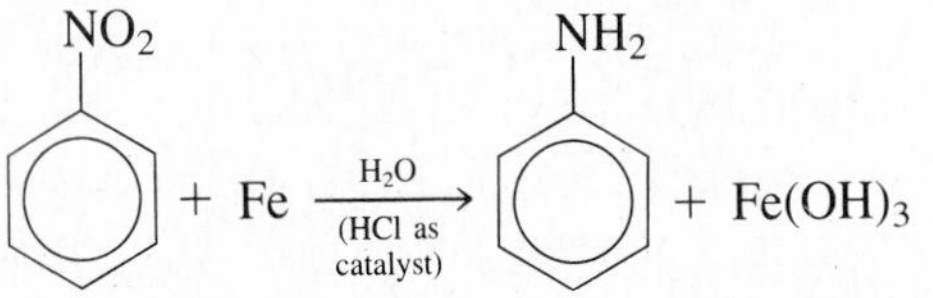

$$\text{NO}_2\text{-benzene} + \text{Fe} \xrightarrow[\substack{\text{(HCl as} \\ \text{catalyst)}}]{\text{H}_2\text{O}} \text{NH}_2\text{-benzene} + \text{Fe(OH)}_3 \qquad \textbf{(2)}$$

$$6\ e^{\ominus} + 6\ \text{H}^{\oplus} + \text{NO}_2\text{-benzene} \rightarrow \text{NH}_2\text{-benzene} + 2\ \text{H}_2\text{O} \qquad \textbf{(2a)}$$

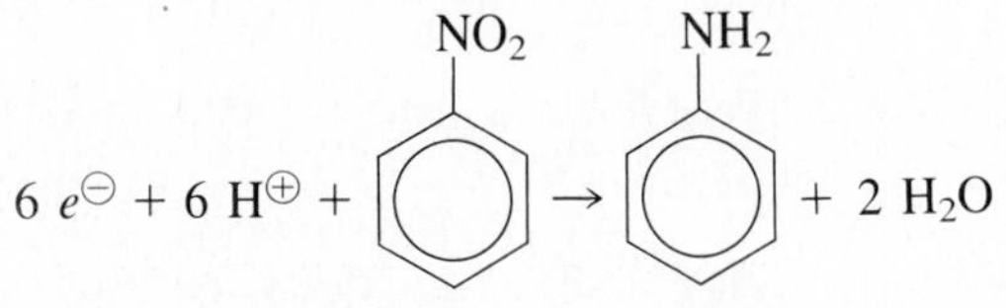

$$3\ \text{H}_2\text{O} + \text{Fe} \rightarrow \text{Fe(OH)}_3 + 3\ \text{H}^{\oplus} + 3\ e^{\ominus} \qquad \textbf{(2b)}$$

Several possible side reactions occur but only to a minor extent under the reaction conditions used. These side reactions (equation 3) occur because of the presence of some of the intermediate products (equation 1), which then react with one another. Nitrosobenzene (**10**) can react with aniline (4) to give azobenzene (**12**), which can be reduced to hydrazobenzene (**13**) by iron. Hydrazobenzene can also undergo rearrangement in the presence of the acid catalyst to yield benzidine (**14**); this rearrangement is often referred to as the *benzidine rearrangement*. *N*-Phenylhydroxylamine (**11**) can undergo acid-catalyzed rearrangement to *p*-aminophenol (**15**).

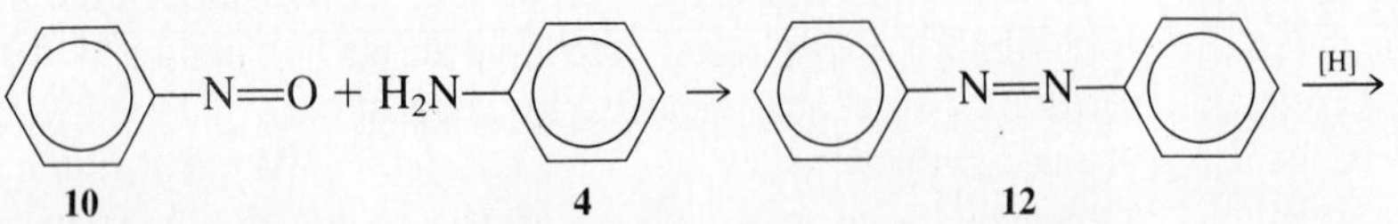

10	**4**	**12**	
		Azobenzene	

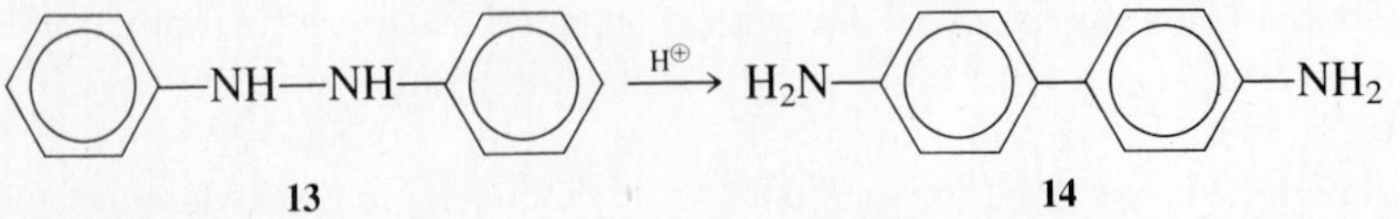

(3a)

13	**14**
Hydrazobenzene	Benzidine

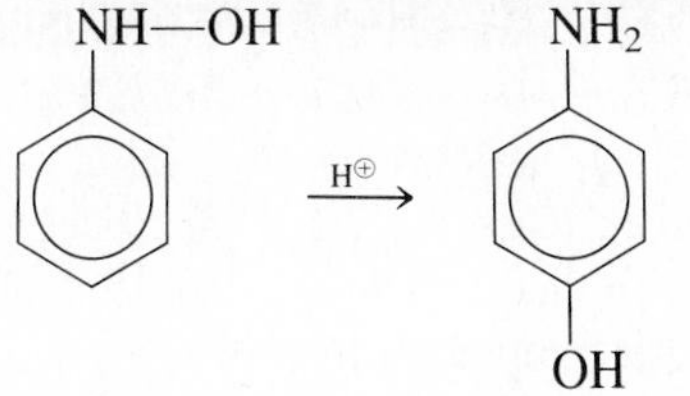

NH—OH

11

$\xrightarrow{\text{H}^{\oplus}}$

NH$_2$

OH

15

p-Aminophenol

(3b)

The isolation and purification of the aniline represent an interesting example of how the dependence on pH of the water solubility of organic compounds and the technique of steam distillation (see Section 2.10) can be combined to effect the isolation of a single pure product. In this case, it is necessary to remove aniline from its principal impurities, which are unchanged nitrobenzene **(3)**, benzidine **(14)**, and *p*-aminophenol **(15)**. Some of the aniline will be present as its hydrochloric acid salt upon completion of the reduction step. The reaction mixture is made *basic* and then steam-distilled. The phenol **(15)**, which will have been converted to its water-soluble, nonvolatile salt, will remain in the aqueous residue along with the higher molecular weight benzidine. The aniline and nitrobenzene will steam-distil as a mixture. To separate these two, it is necessary to convert the amine to its soluble hydrochloric acid salt and steam-distil this solution. The nitrobenzene will now steam-distil, whereas the anilinium salt will remain in the aqueous residue. The aniline may then be recovered by making this acidic residue *basic* and extracting the amine into an organic solvent such as dichloromethane.

Alternatively, the aniline could be isolated from the unchanged nitrobenzene by treating the mixture of the two, as obtained by steam distillation, with dilute acid and extracting the neutral nitrobenzene with an organic solvent. Aniline will have been converted to the salt and will remain in the aqueous layer. Making the aqueous layer *basic,* followed by extraction with dichloromethane, will liberate the aniline, thus allowing its transfer into the second dichloromethane layer.

(c) Acetanilide

Aniline **(4)** can be converted to acetanilide **(5)** by acetylation in aqueous media, using a mixture of acetic anhydride and sodium acetate (equation 4).

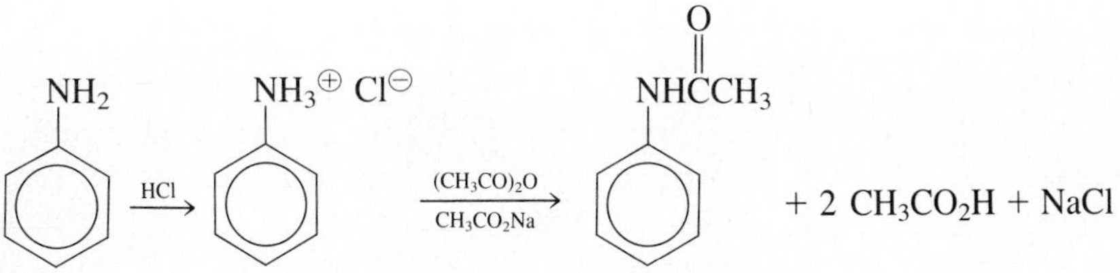

NH$_2$

4

$\xrightarrow{\text{HCl}}$

NH$_3^{\oplus}$ Cl$^{\ominus}$

$\xrightarrow[\text{CH}_3\text{CO}_2\text{Na}]{\text{(CH}_3\text{CO)}_2\text{O}}$

$$\overset{\overset{\displaystyle O}{\parallel}}{\text{NHCCH}_3}$$

Acetanilide

5

$+ \ 2 \ \text{CH}_3\text{CO}_2\text{H} + \text{NaCl}$ **(4)**

The water-insoluble aniline is first converted to its water-soluble hydrochloric acid salt, and acetic anhydride is added to the aqueous solution of the salt that results. Addition of sodium acetate to this solution reconverts the anilinium salt to aniline, which is rapidly acetylated by the acetic anhydride. This provides an easy method of acetylation, and the yield is high. Under the conditions used, acetanilide (5) is produced as a solid, which can be isolated by filtration and purified by recrystallization from water.

A side reaction that might occur is diacetylation (equation 5). It appears that this is minimized by the procedure described, whereas acetylation of aniline in pure rather than aqueous acetic anhydride frequently leads to the production of the diacetyl compound.

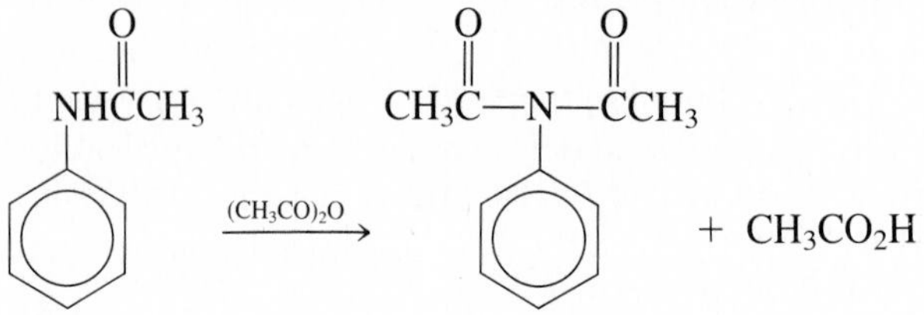

$$ \text{(5)} $$

(d) *p*-Acetamidobenzenesulfonyl Chloride; Chlorosulfonation Reactions

A simple, one-step reaction can be used to introduce the sulfonyl group, —SO_2Cl, onto an aromatic ring. This reaction, known as chlorosulfonation, involves the use of chlorosulfonic acid, $ClSO_3H$, and an aromatic compound. As applied to acetanilide (5), the major product is *p*-acetamidobenzenesulfonyl chloride (6, equation 6). The reaction is known to proceed through the sulfonic acid, which is converted to the sulfonyl chloride on further reaction with chlorosulfonic acid (equation 7). At least two equivalents of the latter acid per equivalent of acetanilide (5) are needed as a result.

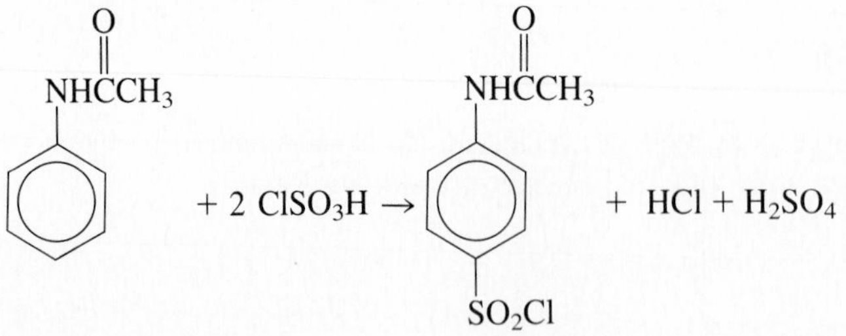

$$ + 2\ ClSO_3H \rightarrow \qquad + HCl + H_2SO_4 \qquad \text{(6)} $$

p-Acetamido-
benzene-
sulfonyl chloride

The sulfonation of acetanilide is a typical electrophilic aromatic substitution reaction with SO_3 probably being the electrophile. The SO_3 may arise from chlorosulfonic acid according

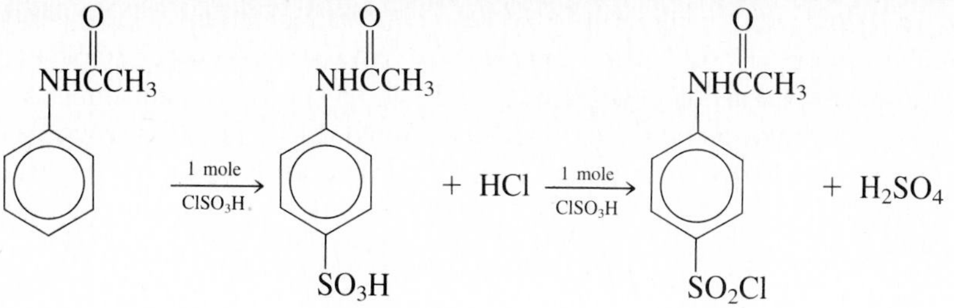

$$\text{5} \xrightarrow[\text{ClSO}_3\text{H}]{\text{1 mole}} \textit{p-Acetamido-benzene-sulfonic acid (SO}_3\text{H)} + \text{HCl} \xrightarrow[\text{ClSO}_3\text{H}]{\text{1 mole}} \textbf{6} + \text{H}_2\text{SO}_4 \qquad \textbf{(7)}$$

to the equilibrium shown in equation 8. The acetamido group directs an incoming group predominantly to the *para* position, and indeed virtually none of the *ortho* isomer is observed in the present case (equation 6). This effect presumably is due to the steric bulk of the acetamido group.

$$\text{ClSO}_3\text{H} \rightleftharpoons \text{SO}_3 + \text{HCl} \qquad \textbf{(8)}$$

In order to isolate the product, the reaction mixture is poured into ice water, and the product is obtained as a precipitate. The water serves to hydrolyze the excess chlorosulfonic acid (equation 9). In general, sulfonyl chlorides are much less reactive toward water than are carboxylic acid chlorides, but it is unwise to leave them moist for any length of time, since they do hydrolyze slowly to give the corresponding sulfonic acids (equation 10). Drying or purifying the *p*-acetamidobenzenesulfonyl chloride (**6**) nonetheless is unnecessary for the preparation of sulfanilamide (**1**), since this reaction step involves treatment of **6** with *aqueous* ammonia immediately after isolation of the crude acid chloride.

$$\text{ClSO}_3\text{H} + \text{H}_2\text{O} \rightarrow \text{HCl} + \text{H}_2\text{SO}_4 \qquad \textbf{(9)}$$

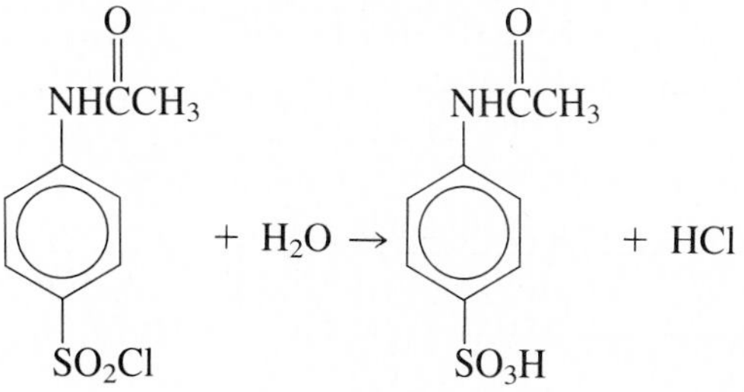

$$\textbf{6 (SO}_2\text{Cl)} + \text{H}_2\text{O} \rightarrow \text{(SO}_3\text{H)} + \text{HCl} \qquad \textbf{(10)}$$

(e) Sulfanilamide (*p*-Aminobenzenesulfonamide)

The preparation of sulfanilamide (**1**) involves treatment of *p*-acetamidobenzenesulfonyl chloride (**6**) with an excess of aqueous ammonia, followed by removal of the acetyl group. This latter process can be carried out in acidic or basic solution without affecting the sulfonamido group, which hydrolyzes much more slowly than the acetamido group. Acid hydrolysis will be used in this experiment, and the resulting amine consequently forms a soluble hydrochloric acid salt under the reaction conditions. In order to obtain free sulfanilamide, the acidic solution must be neutralized with a base such as sodium carbonate. This sequence of reactions is given in equation 11. The sulfanilamide is purified further by recrystallization from water.

(11)

A possible side reaction is the hydrolysis of the sulfonyl chloride during the first step shown in equation 11. This problem can be avoided by treatment of the sulfonyl chloride **6** with ammonia *immediately* after its isolation. Consequently, the conversion of acetanilide (**5**) to **6** *and* the ammonolysis of **6** to **7** *must be completed within a single laboratory period* or the yield in this part of the overall sequence will be unacceptable. Careful planning and budgeting of time is therefore required if success is to be achieved!

An important synthetic strategy has been introduced in the preparation of sulfanilamide **(1),** namely, the use of a **protecting group,** in the form of the acetyl moiety, which is subsequently removed. The importance of this strategy can be seen from the following discussion.

It would appear that one reasonable method for preparing sulfanilamide might be formation of *p*-aminobenzenesulfonyl chloride (**16**) followed by its reaction with ammonia. However, a chlorosulfonyl group *cannot* be successfully generated in the presence of an amino group contained in the same molecule; this is because the amino group of one molecule would react with the chlorosulfonyl group of another molecule to give a *polymeric* material containing sulfonamide linkages (equation 12). Thus, to prepare the sulfanilamides, it is necessary first to ''protect'' the free amino group in aniline (**4**) *before* introducing the chlorosulfonyl group. Such protection is a general requirement when a molecule contains two or more functional groups that are reactive toward the same reagent. In the present synthesis this is most easily done by acetylating the amine. The free amine can be regenerated by removing the acetyl group after the chlorosulfonyl moiety has been introduced and converted to the sulfonamide group.

If the technique of group protection is to be successful, care must be taken to ensure that the protected functionality can be deprotected by a process that does not affect the other functional groups in the molecule. In the present case the acetyl group can be easily removed by controlled hydrolysis without concomitant hydrolysis of the sulfonamido group because the electrophilic sulfur atom of the latter is *tetrahedrally* substituted. This makes the sulfur atom more sterically shielded from attack by the nucleophilic hydroxide ion than is the *trigonally* substituted carbonyl carbon atom of the acetamido group.

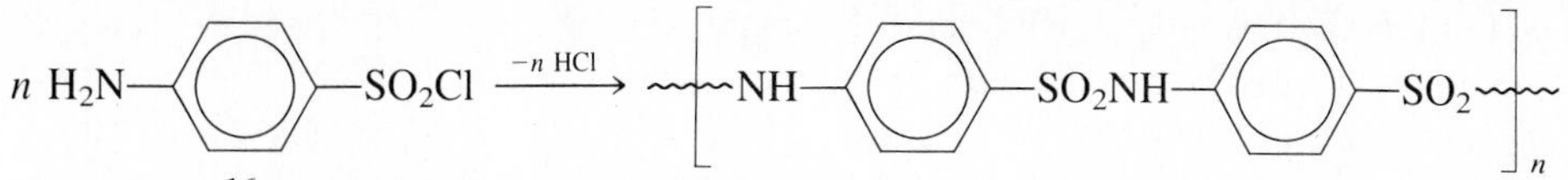

$$n \ H_2N - \langle \bigcirc \rangle - SO_2Cl \xrightarrow{-n \ HCl} \left[\sim\!\!\sim\!\!NH - \langle \bigcirc \rangle - SO_2NH - \langle \bigcirc \rangle - SO_2\sim\!\!\sim \right]_n \quad (12)$$

16
p-Aminobenzenesulfonyl chloride

Pre-lab exercises for Section 23.2, The Synthesis of Sulfanilamide, can be found on pages PL. 107–113.

EXPERIMENTAL PROCEDURE

The experimental procedures start with nitrobenzene and carry through to sulfanilamide. The instructor will indicate at which point to start and how far along the synthesis to proceed. Quantities of reagents should be adjusted according to the amount of starting material available. Beginning with nitrobenzene, for example, some of the reactions may give better yields than those indicated, whereas others may give poorer yields. This must be taken into account as you go to the next step; however, do *not* change the reaction *times* unless told to do so. It may be more convenient to use flasks of sizes different from those specified in the procedures.

A. Aniline

> **DO IT SAFELY**
>
> **1** In all parts of this experiment, examine the glassware carefully for "star cracks" or other weaknesses.
>
> **2** Nitrobenzene is toxic, and care should be taken to avoid breathing its vapors or allowing it to come in contact with your skin. It is good practice to wear rubber gloves when transferring this chemical.
>
> **3** Concentrated hydrochloric acid can cause burns if it comes in contact with the skin. Should this occur, flood the affected area with water and rinse it thoroughly with dilute aqueous sodium bicarbonate solution. Wear rubber gloves when transferring this acid.

Place 0.54 mol of iron powder[1] in a 500-mL round-bottomed flask; add to it 0.24 mol of nitrobenzene and 100 mL of water. Attach a reflux condenser and add about 0.5 mL of

---

[1]Iron powder should be finely divided and should be free of an oxide coating. The reduction reaction is a heterogeneous surface process, so that a large area of metal surface is desirable. Iron metal obtained by reduction of the oxide with hydrogen is termed "reduced with hydrogen" and is most suitable for the present reduction. Tin powder may be used in place of iron.

concentrated hydrochloric acid through the top of the condenser. Swirl and agitate the reaction mixture vigorously. If the reaction does not start soon thereafter, as evidenced by the production of heat, heat the flask gently, but be prepared to *cool* the flask in a pan of water if it appears that the reaction is getting out of control, as evidenced by vigorous evolution of gas. (What is the gas?) After the reaction has started, add another 0.5 mL of concentrated hydrochloric acid and shake the flask vigorously. When the reaction has subsided, bring the mixture to gentle reflux. After 15 min add 1 mL of concentrated hydrochloric acid and then continue heating under reflux for an additional 45 min. If the reaction has required heating for its initiation, heat the mixture under reflux for at least 90 additional minutes.

When the reflux period is complete, add 5 mL of 6 *M* sodium hydroxide solution directly to the reaction mixture and equip the flask for steam distillation (see Section 2.10). Steam-distil the mixture until the distillate emerging from the condenser no longer contains any visible amount of organic product. The distillate may still be slightly cloudy, but if there is no visible amount of oil in it, it may be assumed that the distillation is complete.★ Add 20 mL of concentrated hydrochloric acid to the distillate, and steam-distil this mixture until the *residue* in the distilling flask is clear and free from oily material.★

Make the acidic distillation *residue* basic with a minimal volume of 12 *M* sodium hydroxide solution. Use care in this step because heat is evolved. Saturate the basic solution with sodium chloride (roughly 25 g of salt per 100 mL of solution), cool the mixture, and transfer it to a separatory funnel. Extract the organic product with two 50-mL portions of diethyl ether, using the first portion to rinse the flask in which the neutralization was done. Separate the aqueous layer from the organic layer as thoroughly as possible; transfer the combined organic extracts to a small Erlenmeyer flask, and swirl the solution with several sodium hydroxide pellets until it is clear.★ Transfer the solution by decantation into a distilling flask of appropriate size and perform a simple distillation. Collect three fractions having boiling ranges of 35–90°C, 90–180°C, and 180–185°C, respectively. Discard the first fraction, and if the second is of significant volume, redistil it to obtain more of the high-boiling product.★ Pure aniline is colorless but may darken immediately following distillation owing to air oxidation. Aniline is often redistilled just before use to remove colored oxidation products. The yield of aniline should be 85–90%.

B. Acetanilide

All reagents needed for this procedure should be weighed out and the solution of sodium acetate prepared *before* the reaction is begun.

Dissolve 0.22 mol of aniline in 500 mL of water to which 18 mL of concentrated hydrochloric acid has been added; use a 1-L Erlenmeyer flask. Swirl the mixture to aid in dissolving the aniline. If the solution is dark-colored, add about 1 g of decolorizing carbon to it, swirl the flask, and filter the solution. Prepare a solution containing 0.24 mol of sodium acetate trihydrate dissolved in 100 mL of water, and measure out 24 mL of acetic anhydride. Warm the solution containing the dissolved aniline to 50°C and add the acetic anhydride. Swirl the flask to effect dissolution and then add the solution of sodium acetate *immediately and in one portion*. Cool the reaction mixture to 5°C in an ice-water bath and stir it (do *not* use a thermometer as a stirring rod!) while the product crystallizes. Collect the acetanilide by vacuum filtration, wash it with a small portion of cold water, and dry the product.★ Determine the melting point of this product and its yield, which should be 65–75%. If impure or slightly colored acetanilide is obtained, it may be recrystallized from a minimal volume of hot water, using decolorizing carbon if necessary to give a colorless product. See Section 3.1 (Part B2) for a discussion of the recrystallization of acetanilide from water.

C. *p*-Acetamidobenzenesulfonyl Chloride

This experiment *as well as the next one,* up to the first stopping point, must be completed in a *single* laboratory period unless the product sulfonyl chloride is purified. No matter how much acetanilide you have available from previous reaction steps, use *no more* than 0.1 mol of it in this procedure.

If possible, either do the following reaction in the hood or use the gas-removal apparatus described to prevent escape of hydrogen chloride or oxides of sulfur (SO_2, SO_3) into the laboratory.

DO IT SAFELY

1 Be **extremely** careful in handling and transferring the highly corrosive chlorosulfonic acid. Rubber gloves should be worn for all such operations. Chlorosulfonic acid reacts **vigorously** with water; use *dry* glassware and avoid contact of the acid with your skin. Should this occur, flood the affected area *immediately* with cold water, and then rinse it with 0.6 *M* sodium bicarbonate solution.

2 Because open containers of chlorosulfonic acid will fume from reaction with atmospheric moisture to give HCl and SO_3, both of which are noxious gases, the acid should be measured and transferred *only in the ventilation hood.* It is recommended that several "community" graduated cylinders be kept in the hood for all students to use for obtaining the proper volume of chlorosulfonic acid.

3 To destroy residual chlorosulfonic acid in graduated cylinders and other glassware that has contained it, add cracked ice to the glassware *in the ventilation hood*, and let the glassware remain there until the ice has melted. Then rinse the apparatus with copious amounts of cold water.

4 Lubricate the ground-glass joints of the apparatus carefully and be certain that the joints are intimately mated. Otherwise, noxious gases will escape.

Equip a *dry* 250-mL round-bottomed flask with a Claisen connecting tube. Place a vacuum adapter that either is filled with 4–6 mesh calcium chloride or is fitted with a drying tube containing this drying agent on the side arm of the Claisen tube, and connect the vacuum adapter to a vacuum trap (see Figure 2.32). Grease all joints carefully, because an airtight seal is required at all connections. Place 0.1 mol of *dry* acetanilide in the flask. *In the hood* measure 0.53 mol of chlorosulfonic acid into a 125-mL separatory funnel; be certain that the stopcock of the funnel is firmly seated *prior* to this transfer. *Use care in handling chlorosulfonic acid. Be careful to avoid contact with the skin and with moisture. Chlorosulfonic acid reacts vigorously with water; care should be used in washing any equipment that has contained the acid.* See items 1–3 in the Do It Safely section that precedes this experiment. Stopper the funnel and place it on the straight arm of the Claisen tube so that the chlorosulfonic acid will drop directly onto the acetanilide contained in the reaction flask.

Cool the flask to 10–15°C in a water bath containing a little ice, but do not allow the temperature to go below 10°C. Turn on the vacuum so that a *gentle* flow of air occurs through the vacuum adapter. It is not necessary to have the maximal vacuum to achieve the desired result; however, if fumes are evolved through the open end of the gas trap, adjust the vacuum until such emissions no longer occur. Fully open the stopcock of the funnel so that the chlorosulfonic acid is added as rapidly as possible to the flask; it may be necessary to lift the stopper on the funnel to equalize the pressure in the system if the flow of acid becomes erratic or slow. The reaction mixture becomes black as the addition proceeds. After completion of the addition, *gently* swirl the apparatus from time to time to speed the rate of dissolution of the acetanilide; maintain the temperature of the mixture below 20°C. After most of the solid has dissolved, allow the reaction mixture to warm to room temperature, and then heat it, with the aid of a heating mantle or a hot-water bath, until moderate agitation of the apparatus produces no increase in the rate of gas evolution; 10–20 min of heating will be required.

Cool the mixture to room temperature or slightly below using an ice-water bath. *Working at the hood,* place 600 g of cracked ice and 100 mL of water in a 1-L beaker and pour the reaction mixture slowly and with stirring onto the ice. *Use care in this step; do not add the mixture too quickly, and avoid splattering of the chlorosulfonic acid.* After the addition is complete, rinse the flask with a little ice water and transfer this to the beaker. The precipitate that forms in the beaker is crude *p*-acetamidobenzenesulfonyl chloride, which may be white to pink in color. It may soon become a hard mass, and any lumps that form should be broken up with a stirring rod. Take care not to break the bottom of the flask in this operation! Collect the crude material by vacuum filtration. Wash the solid with a small amount of cold water, and press it as dry as possible with a cork.

The crude sulfonyl chloride should be used *immediately* for the preparation of sulfanilamide. However, a small sample of the product can be purified in the following way, if desired.[2] Dissolve the sample of the sulfonyl chloride in a *minimal* amount of boiling dichloromethane or chloroform contained in a test tube. Using a pipet, remove the upper, aqueous layer as quickly as possible; be very careful in separating the layers. On cooling of the organic layer, colorless crystals of pure product should be obtained, and these can be isolated by vacuum filtration and air-dried. The reported melting point of pure *p*-acetamidobenzenesulfonyl chloride is 149–150°C.

---

[2]Purification of the entire sample of crude sulfonyl chloride can be done as follows. Working at the hood, dissolve the product in a *minimal* amount of boiling dichloromethane or chloroform contained in a 250-mL flask; no more than 150 mL of solvent should be required. Transfer the boiling solution to a separatory funnel that has been warmed on a steam cone or in an oven. If dichloromethane has been used as the solvent, bear in mind that its boiling point is 35°C, so the funnel should be no hotter than this! Remove the organic layer from the aqueous layer as quickly as possible; be very careful in separating the layers. Rinse the funnel with an additional 40-mL portion of hot solvent and combine this rinse with the main organic layer. Allow the solution to cool and isolate the purified product by vacuum filtration.

D. Sulfanilamide

Transfer the crude p-acetamidobenzenesulfonyl chloride to a 500-mL Erlenmeyer flask, and add 75 mL of concentrated (28%) ammonium hydroxide. A very rapid exothermic reaction should occur, presumably the result of the presence of acidic contaminants in the chloride. Purified sulfonyl chloride does not produce similar evolution of heat when combined with ammonium hydroxide. With a stirring rod, break up any lumps of solid that may remain; the reaction mixture should be thick but homogeneous. Heat the mixture on a steam or hot-water bath for about 30 min. Because some vapors of ammonia will be evolved, heating preferably should be done in a hood; alternatively, invert a funnel over a flask and attach the funnel to a vacuum source.★ Cool the flask in an ice-water bath, and add 6 M sulfuric acid to the cool reaction mixture until it is acidic to Congo red paper. Cool the reaction mixture in an ice-water bath and collect the product by vacuum filtration. Wash the crystals with cold water and air-dry them.★ The melting point of the pure product is 218–220°C. Do not use an oil bath to determine the melting point of your product; a hot-stage apparatus (see Section 2.3d) should be used instead.

Weigh the p-acetamidobenzenesulfonamide and transfer it to a 250-mL round-bottomed flask. Prepare a solution of dilute hydrochloric acid by mixing equal volumes of concentrated hydrochloric acid and water. Add to the reaction flask an amount of this dilute solution that is *twice* the weight of the amide. Equip the flask for reflux, and heat the reaction mixture to a gentle reflux for 30 min. At the beginning, boiling should be particularly gentle and accompanied by swirling so as to aid in the dissolution of the solid. To the homogeneous reaction mixture, add an equal quantity of water, and transfer the resulting mixture to a 600-mL beaker. Neutralize the excess acid that is present by adding small quantities of *solid* sodium carbonate; continue addition of the base until the solution is slightly alkaline to litmus paper. (*Caution:* Add the sodium carbonate in small quantities because foaming will occur!) A precipitate should form during neutralization. Cool the mixture in an ice-water bath to complete the precipitation of the product. Collect the crystals by vacuum filtration, wash them with a small amount of cold water, and allow them to air-dry.★ Purify the crude product by recrystallization from hot water; 12–15 mL of hot water per gram of compound will be required. Decolorize the hot solution, if necessary, and filter it using a funnel that has been preheated so that the product will not crystallize in the funnel. Cool the aqueous solution of sulfanilamide in an ice-water bath. The product should separate as long, white needles that can be isolated by vacuum filtration. Determine the melting point, reported to be 164.5– 166.5°C, and the yield, which should be about 0.7 g per gram of starting p-acetamidobenzenesulfonamide.

Test the solubility of sulfanilamide in 1.5 M hydrochloric acid solution and in 1.5 M sodium hydroxide solution.

EXERCISES

General Questions

1. Calculate the overall yield of cortisone, an important drug, produced in a 33-step synthesis, assuming an average yield of 90% in each step.

2. Outline a possible synthesis for the compound shown here, using benzene as the only source of an aromatic ring. Use any needed aliphatic or inorganic reagents.

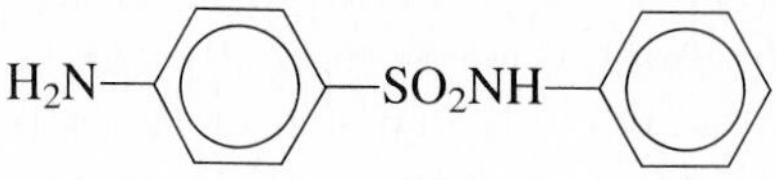

$$H_2N-\langle\bigcirc\rangle-SO_2NH-\langle\bigcirc\rangle$$

3. Suppose that the two hypothetical sequences shown below for conversion of A to E give the desired product in the same overall yield.
   a. What is the overall yield?
   b. As the production manager responsible for selecting the more *economical* of the two routes, what factor(s) other than yield might you consider in reaching your decision? Assume that no capital investment for new equipment would be required for either sequence.

$$A \xrightarrow{30\%} B \xrightarrow{49\%} C \xrightarrow{62\%} D \xrightarrow{57\%} E$$

$$A \xrightarrow{62\%} F \xrightarrow{57\%} G \xrightarrow{49\%} H \xrightarrow{30\%} E$$

Questions for Part A

4. Outline in a flow diagram the procedure for the purification of aniline. Indicate the importance of each step in the procedure, and give reasons for doing the steam distillation first with a basic solution and then with an acidic solution. Write equation(s) for reactions that occur when base and then acid are added.

5. In the purification of aniline, sodium hydroxide pellets are used as the drying agent. Why is this compound used rather than magnesium sulfate, calcium chloride, or other common drying agents? Would potassium hydroxide be an acceptable substitute for sodium hydroxide?

6. Write the balanced equation for reduction of nitrobenzene to aniline with iron powder and aqueous hydrochloric acid.

Questions for Part B

7. Why should the acetic anhydride *not* be allowed to stay in contact with the aqueous solution of the hydrochloric acid salt of aniline (anilinium hydrochloride) for an extended period of time before the solution of sodium acetate is added?

8. Why is aqueous sodium acetate preferred to aqueous sodium hydroxide for the conversion of anilinium hydrochloride to aniline?

9. Why is aniline soluble in aqueous hydrochloric acid whereas acetanilide is not?

10. Give a stepwise reaction mechanism for the reaction of aniline with acetic anhydride.

Questions for Part C

11. Explain why *p*-acetamidobenzenesulfonyl chloride is much less susceptible to hydrolysis of the acid chloride function than is *p*-acetamidobenzoyl chloride.

12. Why is calcium chloride present in the gas-trap apparatus used in this experiment?

13. What materials, organic or inorganic, may contaminate the crude sulfonyl chloride prepared in this reaction? Which, if any, of them are likely to react with the ammonia used in the next reaction step of the sequence?

Questions for Part D

14. In the preparation of sulfanilamide from *p*-acetamidobenzenesulfonamide, only the acetamido group is hydrolyzed. Give an explanation for this difference in reactivity of the acetamido and sulfonamido groups toward aqueous acid.

15. Following hydrolysis of *p*-acetamidobenzenesulfonamide with aqueous acid, the reaction mixture is homogeneous, although mixtures of the sulfonamide in aqueous acid are heterogeneous. How can you explain the change in solubility that occurs as a result of the hydrolysis?

16. Explain the results obtained when the solubility of sulfanilamide was determined in 1.5 *M* hydrochloric acid and in 1.5 *M* sodium hydroxide. Write equations for any reaction(s) that occur.

17. What would be observed if *p*-acetamidobenzenesulfonamide were subjected to vigorous hydrolysis conditions, such as *concentrated* hydrochloric acid and heat, for a long period of time? Write an equation for the reaction that would occur.

18. What acids might be present in the crude sulfonyl chloride to cause the exothermic reaction with ammonia?

19. Why would you expect the yield of sulfanilamide to be lowered if the crude sulfonyl chloride were not combined with ammonia until the laboratory period following its preparation?

20. Calculate the *overall* yield of sulfanilamide obtained in the sequence of reactions that you performed.

SPECTRA OF STARTING MATERIALS AND PRODUCTS

The IR spectra of nitrobenzene and aniline are given in Figures 8.30 and 8.33. The PMR spectrum of acetanilide is presented in Figure 9.24.

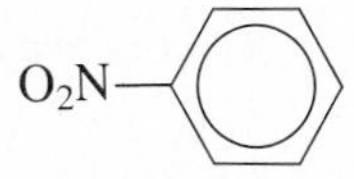

O_2N— Chemical shifts: δ 123.5, 129.5, 134.8, 148.3.

Figure 23.3 CMR data for nitrobenzene.

(a) PMR spectrum.

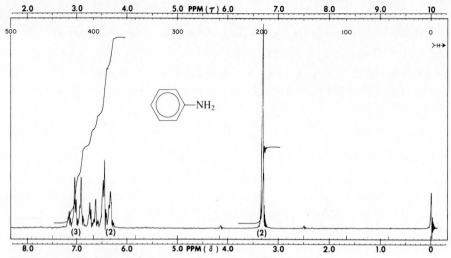

(b) CMR data. Chemical shifts: δ 115.1, 118.2, 129.2, 146.7.

Figure 23.4 NMR data for aniline.

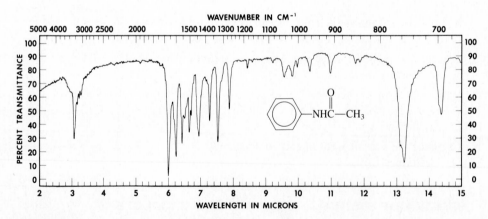

Figure 23.5 IR spectrum of acetanilide (KBr pellet).

Chemical shifts: δ 24.0, 119.7, 123.3, 128.5, 139.2, 168.7

Figure 23.6 CMR data for acetanilide.

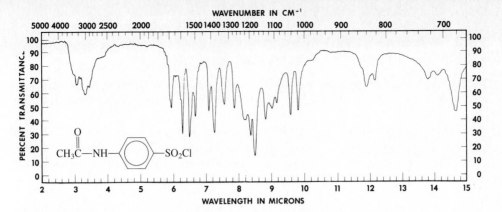

Figure 23.7 IR spectrum of *p*-acetamidobenzenesulfonyl chloride (in CCl$_4$ solution).

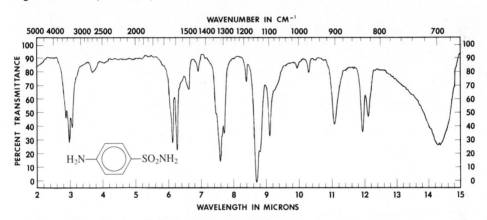

Figure 23.8 IR spectrum of sulfanilamide (KBr pellet).

(a) PMR spectrum.

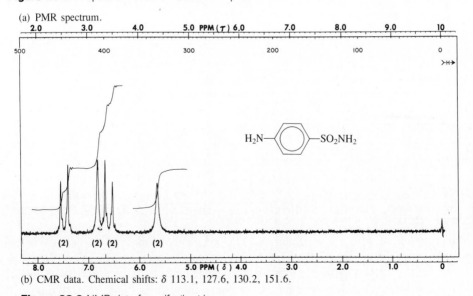

(b) CMR data. Chemical shifts: δ 113.1, 127.6, 130.2, 151.6.

Figure 23.9 NMR data for sulfanilamide.

23.3 4-Aminoquinolines

The *heterocyclic* natural product quinine (**17**) is obtained from the bark of the cinchona tree and has long been used as an antimalarial drug. One of the first successful *synthetic* antimalarial drugs was quinacrine (**18**), also called Atebrin, first produced in Germany. During and after World War II, thousands of heterocyclic compounds structurally related to quinine and quinacrine were synthesized by organic chemists and tested for their effectiveness as antimalarials. Many of these were derivatives of 4-aminoquinoline. One of the most successful was chloroquine (**28a**), which has structural similarities to both quinine and quinacrine.

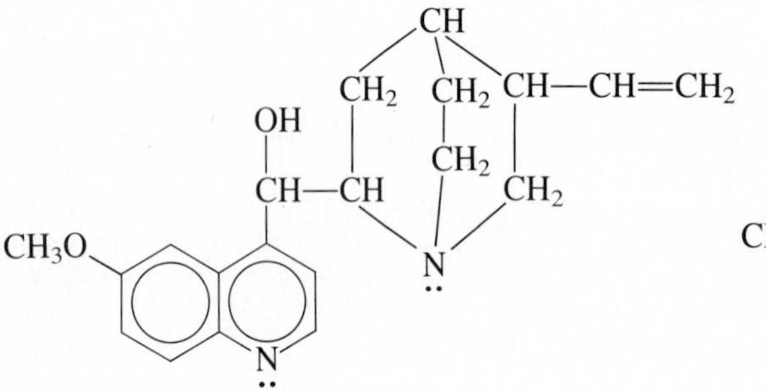

17
Quinine

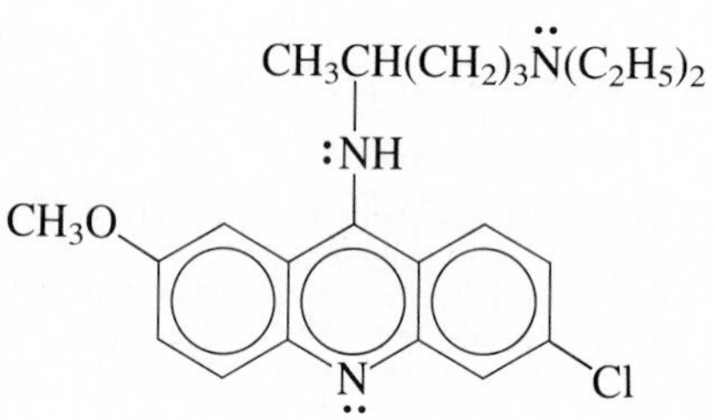

18
Quinacrine

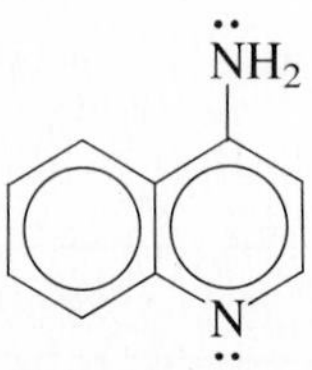

19
4-Aminoquinoline

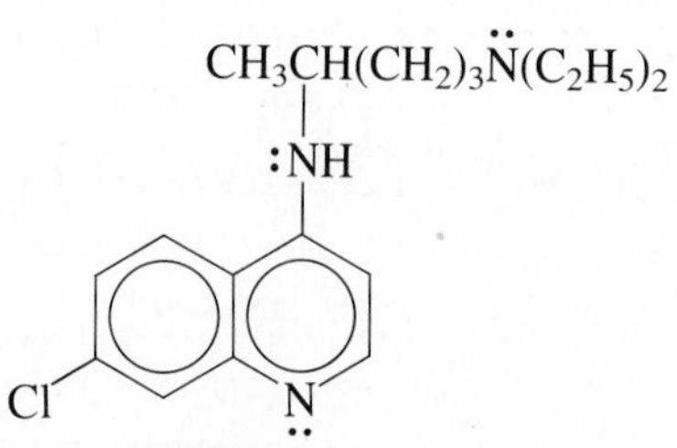

28a
Chloroquine

A practical scheme for the synthesis of chloroquine and of other 4-aminoquinoline derivatives is outlined in Figure 23.10. This synthesis is typical of the majority of quinoline syntheses in that a cyclization is carried out to produce the *pyridine* ring rather than the *benzene* ring. Besides *m*-chloroaniline (**20**), many other substituted anilines may be used to produce quinolines having other substituents in the benzene ring, and the amine side chain may be varied by the reaction of different primary amines with 4,7-dichloroquinoline (**26**) in the final step.

In our experiment the primary amine used in the last step of the synthesis is *m*-chloroaniline (**20**), and it was selected for several reasons. (1) It is relatively inexpensive. (2) It is the same compound needed for the *first* step of the synthesis. (3) It does not have the extremely disagreeable odor of the diamine required to produce chloroquine (**28a**). (4) The final product, 4-*m*-chloroanilino-7-chloroquinoline (**28b**), is a crystalline compound that is easily puri-

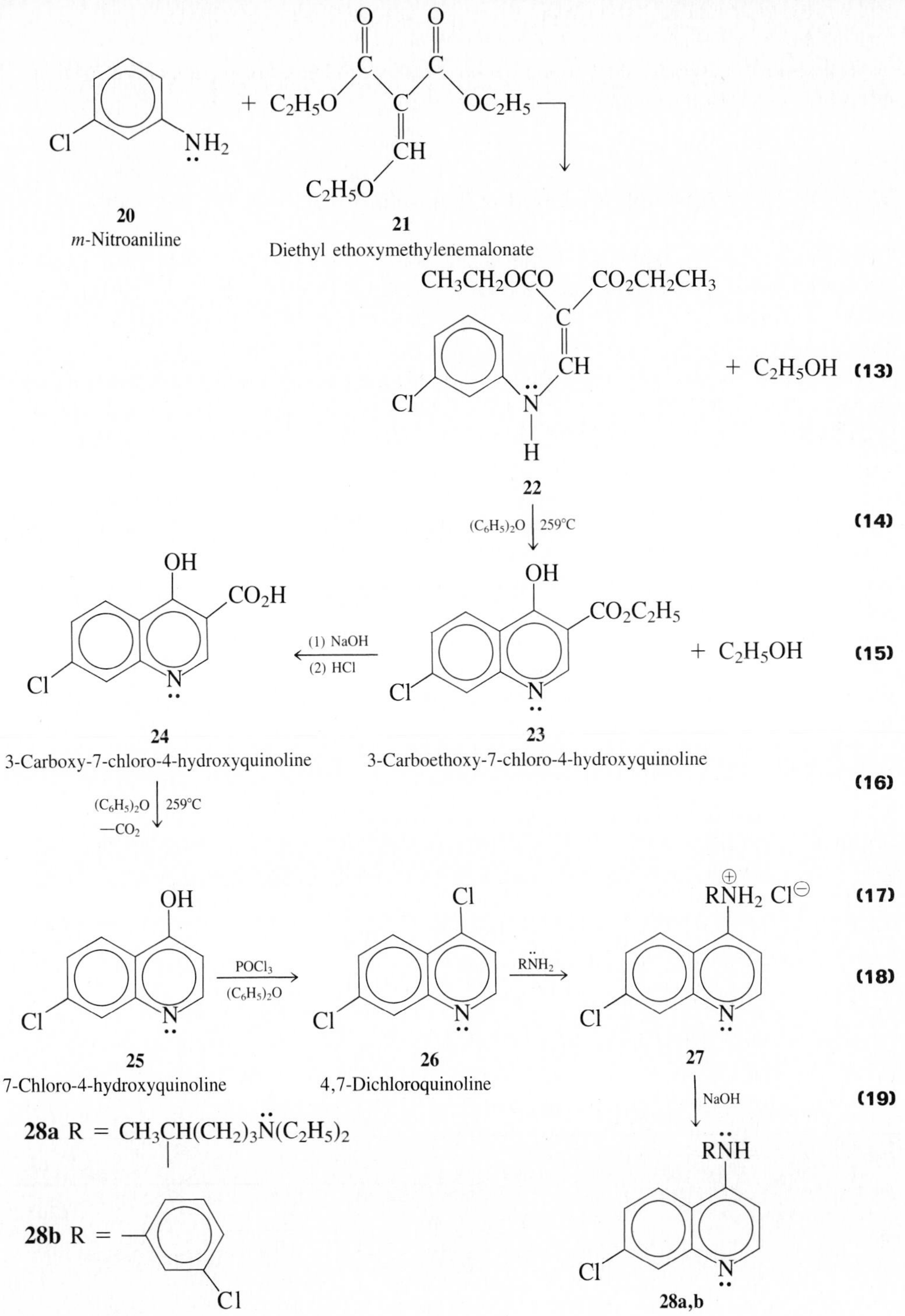

Figure 23.10 Synthesis of 4-amino-7-chloroquinolines.

(13)

(14)

(15)

(16)

(17)

(18)

(19)

20
m-Nitroaniline

21
Diethyl ethoxymethylenemalonate

22

24
3-Carboxy-7-chloro-4-hydroxyquinoline

23
3-Carboethoxy-7-chloro-4-hydroxyquinoline

25
7-Chloro-4-hydroxyquinoline

26
4,7-Dichloroquinoline

27

28a,b

28a R = $CH_3CH(CH_2)_3\ddot{N}(C_2H_5)_2$

28b R =

fied by recrystallization. Although preparation of **28b** is a readily obtained synthetic target, this substance is *not* an effective antimalarial drug.

The various individual steps in the overall synthetic scheme for preparation of **28b** are discussed in the following paragraphs.

(a) 3-Carboethoxy-7-chloro-4-hydroxyquinoline

The first step in the synthesis is the conversion of *m*-chloroaniline (**20**) and diethyl ethoxymethylenemalonate (**21**) to 3-carboethoxy-7-chloro-4-hydroxyquinoline (**23**), and this occurs in two stages. The first involves a condensation reaction between **20** and **21** to produce **22** and ethanol (equation 13). Mechanistically, this process results from a *nucleophilic* attack of **20** upon the alkene **21**, which is made susceptible to such an attack by the presence of the ester groups (see Sections 13.3 and 21.3 for further discussion of nucleophilic attack on alkenes). In principle, the ethanol could add to **22** and cause regeneration of starting materials, but establishment of such an equilibrium is prevented by performing the condensation under vacuum so that the ethanol is removed from the reaction mixture as soon as it is formed; the desired reaction is thus driven to completion. The second stage of the reaction involves cyclization of **22** and loss of a second molecule of ethanol (equation 14) and requires higher temperatures than does the first. Consequently, diphenyl ether, a high-boiling polar solvent (bp 259°C), is added to the crude **22,** which need not be isolated, and the resulting solution is brought to reflux. Cyclization and loss of ethanol then proceed, as evidenced by the separation of crystalline **23** from the boiling mixture. Isolation of the product is quite simple, requiring only filtration and then washing with petroleum ether, to remove residual diphenyl ether.

The cyclization step may be rationalized as involving an orbital symmetry–allowed electrocyclic reaction, as shown in equation 20 (see Chapter 15 and references 7 and 8 at the end of that chapter for further information on reactions involving orbital symmetry). Enolization of **22** gives **22a**, which can be considered as analogous to a 1,3,5-hexatriene. Such trienes are known to cyclize thermally to 1,3-cyclohexadienes (equation 21), and a corresponding reaction with **22a** would give **29**. Loss of ethanol from **29** (equation 22) completes the conversion of **22** to **23**. Although cyclizations of the type shown in equation 21 normally occur at temperatures below 150°C, the substantially higher temperature required to produce **23** presumably reflects the loss of *aromaticity* associated with the formation of **29** from **22a**.

22   22a   29

(20)

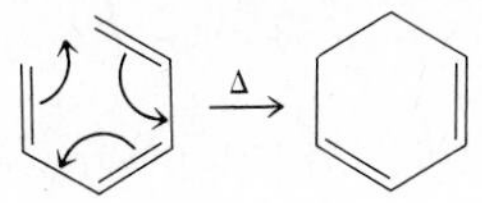

(21)

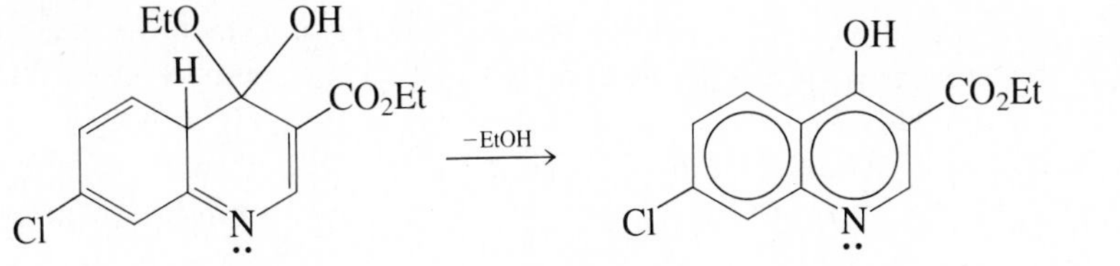

(22)

29

23
3-Carboethoxy-7-chloro-4-hydroxyquinoline

The second carboethoxy group present in **22** may appear to have no function in the cyclization reaction, and it is in fact *removed* in subsequent steps. However, it apparently does serve a useful purpose in influencing the *direction* of the cyclization. If ring closure of **22a** occurred *ortho* to the chlorine atom, the 5-chloroquinoline isomer **30** would result (equation 23). The isomer of chloroquine (**28a**) having chlorine at the 5-position is known *not* to be an effective antimalarial drug, so any cyclization in this direction represents an undesirable side reaction. Fortunately, little of this occurs, and the 5-chloroquinoline isomer has been detected only when the synthesis has been conducted on a very large scale. It is interesting to speculate on the explanation for the selectivity of the ring closure and on the possible role the aforementioned second carboethoxy group plays in it (see Exercise 1c).

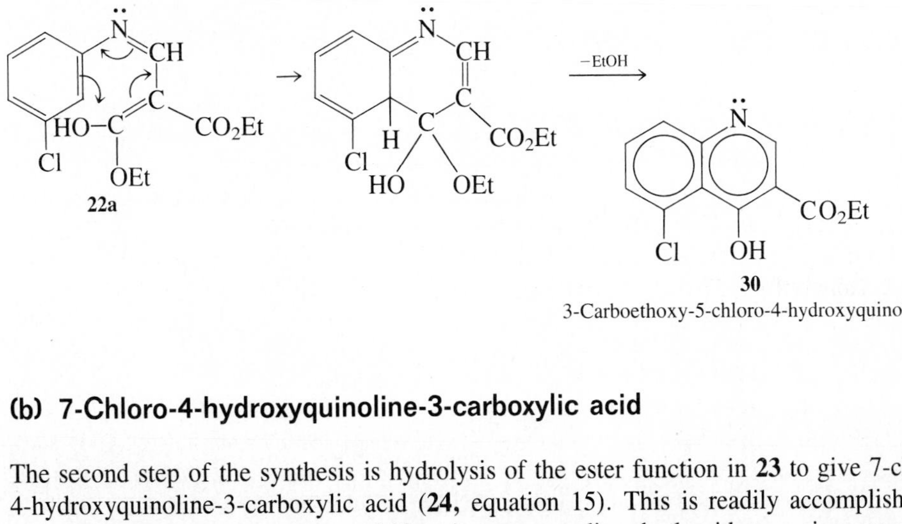

(23)

22a

30
3-Carboethoxy-5-chloro-4-hydroxyquinoline

(b) 7-Chloro-4-hydroxyquinoline-3-carboxylic acid

The second step of the synthesis is hydrolysis of the ester function in **23** to give 7-chloro-4-hydroxyquinoline-3-carboxylic acid (**24**, equation 15). This is readily accomplished by heating a heterogeneous mixture of **23** and aqueous sodium hydroxide to a vigorous reflux. As hydrolysis occurs, the carboxylic acid **24** produced goes into solution as its disodium salt, whereas all nonacidic organic by-products of this and the previous reaction step do *not* dissolve. The salt of **24** can then be separated from such by-products by filtration, and the acid itself is recovered by acidification of the filtrate.

(c) 4,7-Dichloroquinoline

The next step in the sequence of Figure 23.10 is the conversion of the carboxylic acid **24** to 4,7-dichloroquinoline (**26**). Just as with step a of the sequence, this transformation involves two discrete stages. The first is decarboxylation of **24** to **25** (equation 16), which can be achieved successfully with the crude acid by heating it in diphenyl ether, the same high-boiling solvent used in step a. A clue to the ease of this decarboxylation may be obtained by consideration of the tautomeric form **24a** that results from enolization of the carboxylic acid **24** (equation 24). This tautomer is a β-ketoacid, and these types of acids are well known to be quite sensitive to decarboxylation, owing to the availability of a transition state that allows for simultaneous intramolecular proton transfer to the ketonic oxygen along with cleavage of the carbon-carbon bond that involves the carboxy group (equation 24).

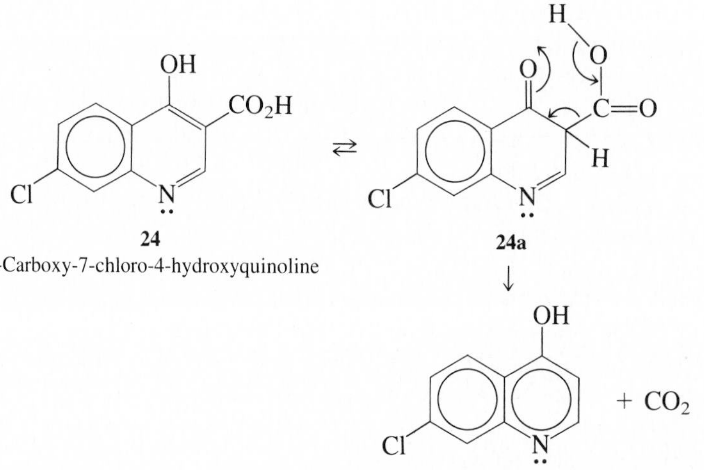

24
3-Carboxy-7-chloro-4-hydroxyquinoline

24a

$+ CO_2$ (24)

25
7-Chloro-4-hydroxyquinoline

The next stage in the overall reaction is one in which reaction of crude **25** with phosphorus oxychloride replaces the phenolic hydroxyl group of **25** with chlorine to produce 4,7-dichloroquinoline (**26**, equation 17). A possible mechanism for this transformation involves initial attack of the hydroxyl group of **25** on the phosphorus oxychloride to give **31** and hydrogen chloride, which combine to give the salt **32** (equation 25). This intermediate then undergoes nucleophilic aromatic substitution by chloride ion, a process that is facilitated by the presence of the positively charged nitrogen atom, which becomes neutralized upon addition of chloride to the ring, as shown in **33,** the immediate precursor of **26.** Note that heating **25** with a source of chloride ion, for example, sodium chloride, would *not* result in formation of **26.** This is because the reaction requires activation toward nucleophilic attack and because the leaving group for nucleophilic aromatic substitution in this case, hydroxide, is a *strong* base, and therefore a much poorer leaving group than the *weak* base, $Cl_2PO_2^-$, that is produced under our reaction conditions (equation 25).

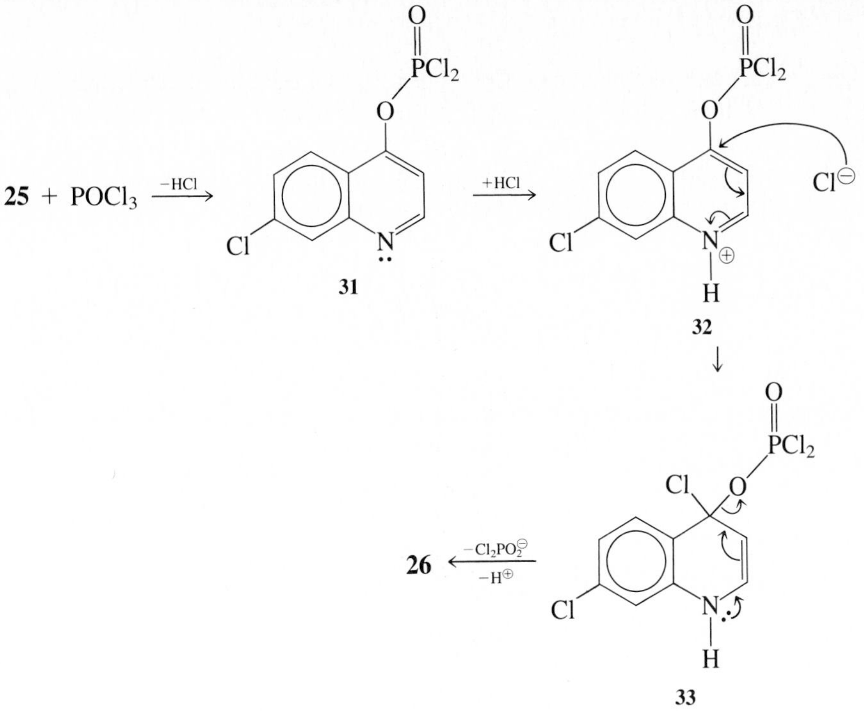

$$25 + POCl_3 \xrightarrow{-HCl} \quad 31 \quad \xrightarrow{+HCl} \quad 32 \quad \downarrow \quad 33 \quad \xrightarrow[-H^\oplus]{-Cl_2PO_2^\ominus} 26 \tag{25}$$

The desired dichloroquinoline **26** is soluble in diphenyl ether, and the solvent cannot readily be removed by distillation, owing to its very high boiling point. However, the separation of **26** from the solvent is easily accomplished by taking advantage of the basic properties of quinolines and extracting **26** into aqueous hydrochloric acid, in which the diphenyl ether and all nonbasic organic by-products of the reaction are insoluble. Making the aqueous acidic solution basic liberates the 4,7-dichloroquinoline, which precipitates from solution.

(d) 4-*m*-Chloroanilino-7-chloroquinoline

The final step in the synthesis involves another nucleophilic aromatic substitution, with *m*-chloroaniline (**20**) serving as the nucleophile (equations 18 and 19). Both of the chlorine-bearing carbon atoms of 4,7-dichloroquinoline (**26**) are potentially susceptible to nucleophilic attack, but only substitution at the 4-position results in the desired product, **28b**. Because the reaction is under kinetic control, and the rate-determining step is attack of the amine upon **26**, the high selectivity that is actually observed for reaction at the 4-position must depend on a wide difference in the susceptibility toward nucleophilic substitution of the carbon atoms at positions 4 and 7 in **26**.

A basis for the necessary difference can be seen by consideration of the relative stabilities of the charged intermediates **34** and **35** derived from attack of an amine at C-4 and C-7 (equations 26 and 27, respectively). The nitrogen atom is able to stabilize the negative charge in **34** *without* complete disruption of aromaticity, as illustrated in resonance structure **34a,** whereas the corresponding stabilization in **35,** as represented in resonance structure **35a,** does *not* allow for retention of aromaticity. Consequently, **34** and the transition state leading to it (see Chapter 16) are *lower* in energy than **35** and its corresponding transition state. The result, then, is that the 7-position has the low reactivity toward nucleophilic aromatic substitution characteristic of chlorobenzene, whereas the 4-position is quite reactive toward such a reaction.

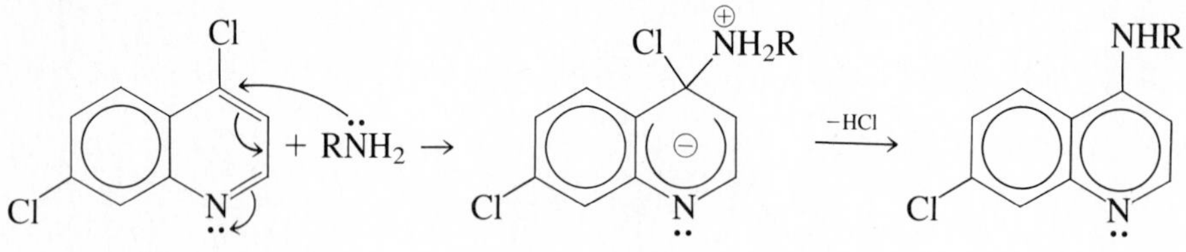

26
4,7-Dichloroquinoline

34

28

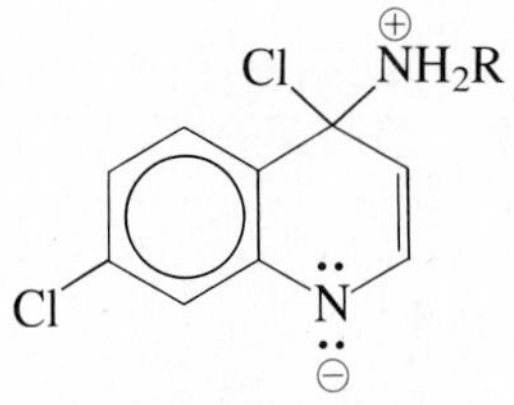

(26)

34a

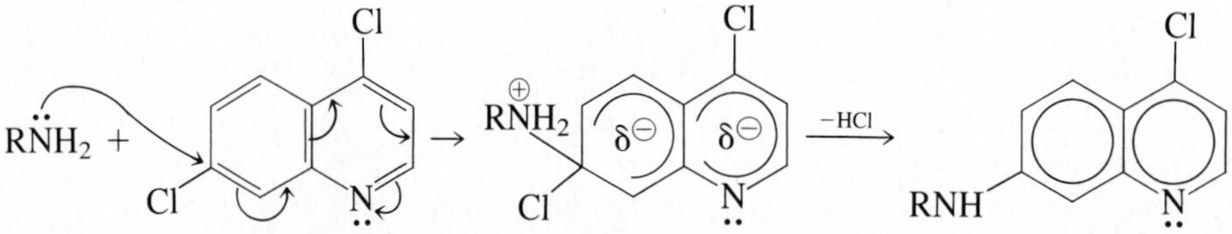

26

35

Not formed

(27)

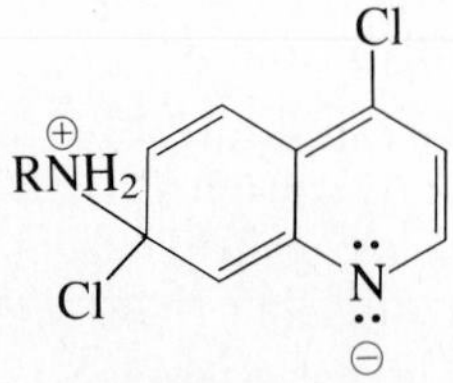

35a

This substitution reaction is best carried out in the presence of hydrochloric acid, which serves as a catalyst (equation 28). Protonation of the basic nitrogen of the ring makes the system more electrophilic, further enhancing it toward nucleophilic aromatic substitution, as discussed under step c. The crystalline hydrochloric acid salt, **27b,** precipitates from the aqueous mixture in which the reaction is run and can be isolated by filtration. Stirring **27b** with aqueous base liberates the final product, **28b** (equation 19), which also precipitates from the aqueous mixture.

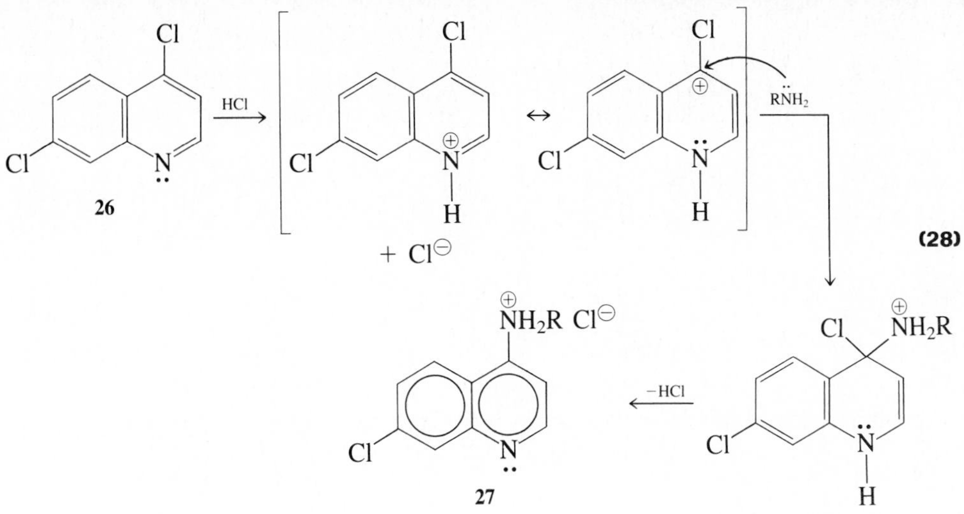

(28)

Steps a and c of the above discussion each involve two-stage transformations in which the stable intermediates **22** and **25,** respectively, are produced but are not isolated and purified as a prelude to the subsequent reaction step. This type of "one-pot" multiple reaction strategy is a common one in a multistep synthesis and has great value, since mechanical losses of the intermediate products and the investment of time associated with workup of the reaction mixture are decreased as a result. Of course, this strategy must be used cautiously, since by-products are also carried along in the process. If these cannot be separated from the ultimate product of the one-pot reaction sequence, but can be separated from one of the intermediates, then isolation and purification of the intermediate is required.

Pre-lab exercises for Section 23.3, 4-Aminoquinolines, can be found on pages PL. 115–121.

EXPERIMENTAL PROCEDURE

A. 3-Carboethoxy-7-chloro-4-hydroxyquinoline (**23**)

DO IT SAFELY

1 During a portion of this experiment it is necessary to use a burner to attain the required reaction temperature. Therefore, in assembling the apparatus pictured in Figure 23.11, be sure that all ground-glass joints are thoroughly lubricated and tightly mated to avoid escape of flammable vapors.

2 Be certain that there are **no** open flames in the vicinity when you are working with petroleum ether.

3 Some of the compounds in this and following experiments have melting points above 200°C. Do **not** attempt to take the melting point of any compound having such a high melting point with a liquid-bath apparatus; use a metal block apparatus instead (see Figure 2.5 in Section 2.3).

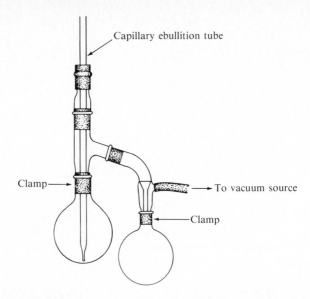

Figure 23.11 Apparatus for preparation of 3-carboethoxy-7-chloro-4-hydroxyquinoline **(23).**

Equip a 250-mL round-bottomed flask with a stillhead to which a capillary ebullition tube is attached by means of a neoprene thermometer adapter. Connect a vacuum adapter to the stillhead, and attach a 50- or 100-mL round-bottomed receiving flask to the vacuum adapter, as shown in Figure 23.11. In the 250-mL flask place 0.050 mol of *m*-chloroaniline and 0.051 mol of diethyl ethoxymethylenemalonate. Connect the vacuum adapter to a vacuum source and heat the reaction mixture under vacuum on a steam or hot-water bath for 1 hr. During this time the ethanol produced according to equation 13 will evaporate but will not condense, owing to the vacuum in the system. Release the vacuum and then remove the stillhead from the reaction flask.★

If desired, a small sample of the condensation product **22** may be removed by pipet. Upon cooling to 0°C, it should crystallize. The crude **22** can be purified by recrystallization from petroleum ether (bp 60–80°C). The melting point of pure **22** is 55–56°C.

Add 50 mL of diphenyl ether to the reaction mixture, attach a reflux condenser *without water hoses connected,* and heat the mixture to reflux, using a flame diffused by a wire gauze. Continue heating for 15 min after reflux temperature has been attained. During this period, crystals of 3-carboethoxy-7-chloro-4-hydroxyquinoline **(23)** should separate from the boiling solution. Remove and extinquish the flame and allow the reaction mixture to cool to room temperature. Collect the crystalline ester **23** by vacuum filtration, and press the filter cake with a clean cork or stopper to remove as much solvent as possible. Transfer the crystals to a 250-mL beaker, and stir them well with about 75 mL of petroleum ether (bp 60–80°C). Collect the crystals by vacuum filtration and wash them with a little more petroleum ether. Resuspend the crystals once again in about 50 mL of petroleum ether, stir, and reisolate them by vacuum filtration.★

A sample of 3-carboethoxy-7-chloro-4-hydroxyquinoline **(23)** may be purified by recrystallization from pyridine, if desired. This recrystallization should be performed in a hood because of the toxicity and unpleasant odor of this solvent. The reported melting point of **23** is 295–297°C. (*Caution:* See item 3 of the Do It Safely comments.)

B. 7-Chloro-4-hydroxyquinoline-3-carboxylic Acid (**24**)

Place the crystals of crude 3-carboethoxy-7-chloro-4-hydroxyquinoline (**23**) from Part A in a 250-mL round-bottomed flask, and add 50 mL of 3 *M* aqueous sodium hydroxide and a few boiling chips. Attach a water-cooled reflux condenser and heat the mixture to vigorous reflux for 1 hr.★ Discontinue heating and allow the reaction mixture to cool for a few minutes. Disconnect the condenser and add about 0.5 g of decolorizing carbon. Replace the condenser and heat the mixture to reflux again for 5 min; swirl the flask if bumping occurs.★

Allow the reaction mixture to cool almost to room temperature and filter it using a fluted filter. Wash the filter with 20 mL of hot water, adding the wash to the main filtrate. Acidify the filtrate to pH 4 with 3 *M* hydrochloric acid. Collect the 7-chloro-4-hydroxyquinoline-3-carboxylic acid (**24**) by vacuum filtration and wash it on the filter with a little water.★ If instructed to do so, recrystallize a small sample of the acid from ethanol; a large volume of solvent will be required. The reported melting point of pure **24** is 273–274°C. (*Caution:* See item 3 of the Do It Safely comments preceding Part A.)

C. 4,7-Dichloroquinoline (**26**)

DO IT SAFELY

1   Phosphorus oxychloride is a lachrymator and produces hydrogen chloride when it comes in contact with atmospheric water and moist mucous membranes. Handle and transfer this chemical in the ventilation hood and use rubber gloves when doing so. Avoid inhalation of its vapors. Should this chemical come in contact with your skin, flood the affected area with water and rinse it thoroughly with dilute aqueous sodium bicarbonate.

2   4,7-Dichloroquinoline (**26**) is also a lachrymator and irritant to the skin. Avoid inhalation of its vapors and contact with your skin. It would be wise to wear rubber gloves for this experiment.

Place the crude 7-chloro-4-hydroxyquinoline-3-carboxylic acid (**24**) from part B in a 250-mL round-bottomed flask and add 50 mL of diphenyl ether and a few boiling chips. Attach a reflux condenser, *without* water cooling, and heat the mixture to reflux for 1 hr using a burner. The acid should dissolve in the hot solvent as the decarboxylation proceeds. Allow the reaction mixture to cool to room temperature.★

If isolation of a sample of pure 7-chloro-4-hydroxyquinoline (**25**) is desired, pour out 10 mL of the reaction mixture into a flask containing 10 mL of petroleum ether (bp 60–80°C). Stir the mixture, collect the crystals by vacuum filtration, and wash them on the filter with more petroleum ether. This crude product can be recrystallized from water, using some decolorizing carbon if necessary; a large volume of solvent will be required. The melting point of pure **25** is reported to be 270–272°C, sintering from 260°C. (*Caution:* See item 3 of the Do It Safely comments preceding Part A.)

Add 5 mL of phosphorus oxychloride and some fresh boiling chips to the reaction mixture. Attach a water-cooled reflux condenser to the flask, and suspend a 360°C thermometer from the top of the condenser by means of a copper wire so that the bulb of the thermometer is in the liquid reaction mixture. Heat the mixture slowly to 135–140°C, and swirl the flask from time to time during 1 hr of heating.★

Allow the reaction mixture to cool to room temperature, transfer it to a separatory funnel with the aid of about an equal volume of diethyl ether (No flames!) and extract it with four 50-mL portions of 3 *M* hydrochloric acid. Take care to vent the funnel immediately after each shaking operation because some heat may be produced in the extraction. Add a little crushed ice to the combined acid extracts, and neutralize them with 3 *M* aqueous sodium hydroxide. Collect the precipitated 4,7-dichloroquinoline (**26**) by vacuum filtration. Resuspend the crystals in water, stir well, and recollect them by vacuum filtration. The product may be air-dried in a ventilation hood or, better, in a vacuum desiccator. Determine the melting point and yield of this product, basing the calculation on the amount of diethyl ethoxymethylenemalonate originally used. The melting point of pure **26** is 83.5–84.5°C.

D. 4-*m*-Chloroanilino-7-chloroquinoline (28b)

Place 2.0 g of the 4,7-dichloroquinoline (**26**) from Part C and 1.3 g of *m*-chloroaniline in 100 mL of water contained in a round-bottomed flask, and add a few drops of concentrated hydrochloric acid. Equip the flask for reflux, and heat the mixture to a gentle boil. Within a few minutes the hydrochloric acid salt **27b** should begin to precipitate. Heat for an additional 10 min, allow the mixture to cool to room temperature, and collect the product by vacuum filtration. Resuspend the crystalline hydrochloride in about 20 mL of 3 *M* aqueous sodium hydroxide contained in an Erlenmeyer flask, and heat the mixture to boiling for about 5 min while stirring. Collect the solid product by vacuum filtration and recrystallize it from ethanol. Shiny white plates of 4-*m*-chloroanilino-7-chloroquinoline (**28b**) should be obtained. Determine the yield and melting point of this final product. The reported melting point of pure **28b** is 223–225°C. (*Caution:* See item 3 under the Do It Safely comments in Part A.)

EXERCISES

General Questions

1. 7-Chloro-4-hydroxyquinoline (**25**) has also been synthesized starting with *m*-chloroaniline and diethyl oxaloacetate, which condense to give compound **36** (equation 29). The sequence by which **36** is converted into 7-chloro-4-hydroxyquinoline is parallel to that in the present synthesis. However, the decarboxylation of the intermediate 7-chloro-4-hydroxy-2-carboxylic acid (**37**) is more difficult than that of its isomer **24** produced in the sequence of Figure 23.10. Moreover, a significant amount of 5-chloro-4-hydroxy-quinoline (**38**) is found along with the desired product, **25**.

    a. Outline all the steps in the synthesis of **25** from **36,** giving stepwise reaction mechanisms for each of these steps as well as for that of the reaction by which **36** is formed (equation 29).

b. Explain why decarboxylation of **37** is more difficult than that of **24**.

*c.** Explain why substantially more **38** is produced in this sequence, as compared with that of Figure 23.10. (*Hint:* Consider the selectivity of the cyclization steps involving **22** and **36**, respectively.)

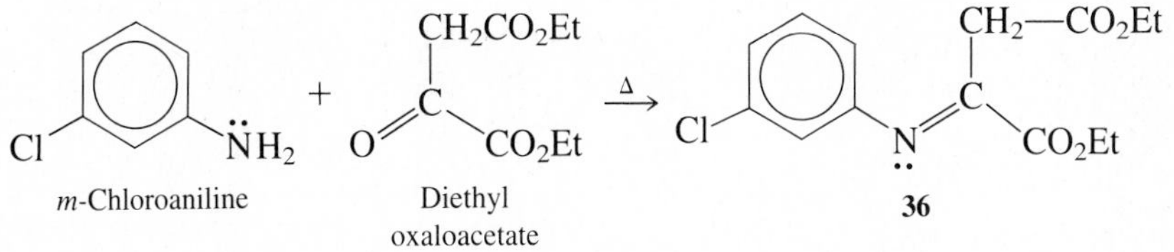

m-Chloroaniline

Diethyl oxaloacetate

36

(29)

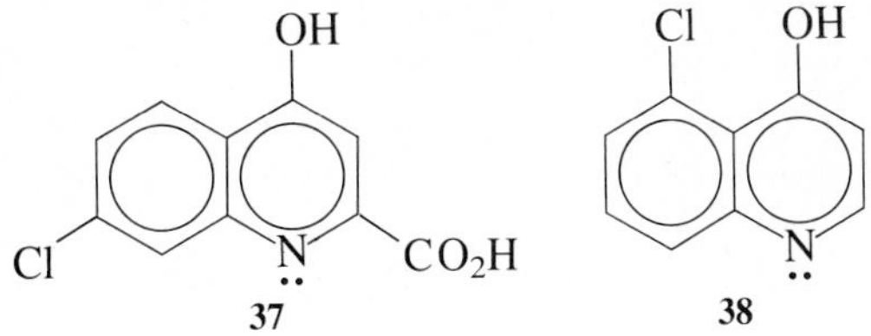

37

38

5-Chloro-4-hydroxyquinoline

2. It was asserted that an important advantage of doing multi-step conversions in a "one-pot" fashion was a decrease in "mechanical" losses of products. Describe the meaning of this term with regard to isolation of organic products.

Questions for Part A

3. Why would it be unwise to use a mineral-oil bath to bring diphenyl ether to reflux?

4. Why would diethyl ether be a poor choice as the solvent in which to effect the cyclization of **22**?

5. Compound **23** is amphoteric and potentially could exist as the zwitterion **39**. It does not, however.

   a. Define *amphoteric*.

   *b.** (1) Assume that the pKa's of the phenolic group of **23** and of the quinolinium moiety of **39** are the same as those of the parent species, phenol and quinolinium ion (the conjugate acid of quinoline). Given that the pKa's of phenol and quinolinium ion are about 10 and 5, respectively, calculate the equilibrium constant between **23** and **39**. Is the zwitterionic form, **39**, calculated to be more or less stable than the nonionic form, **23**?

   (2) Explain why the assumption made above may be invalid and indicate how your conclusion about the relative stabilities of **23** and **39** might be changed as a result.

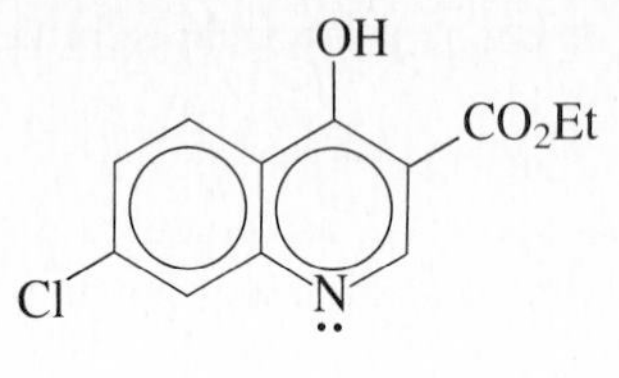

23 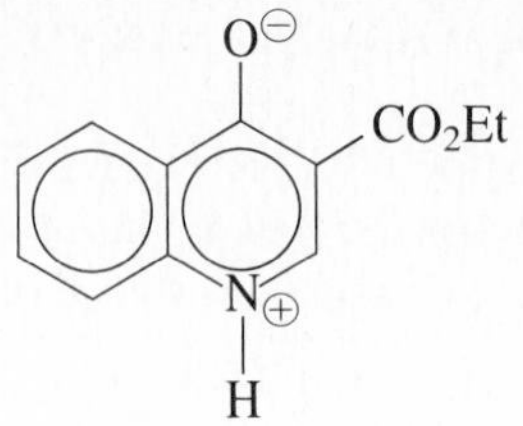   39 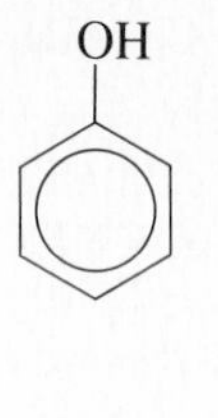   Phenol 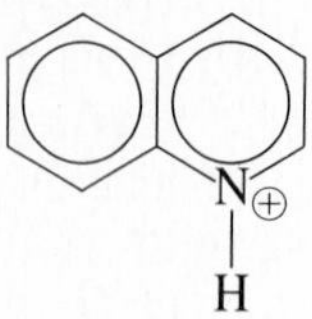   Quinolinium ion

*6. Compound **29** (equation 20) could potentially lose water instead of ethanol to give **40** instead of **23**.

   a. Explain why elimination of water occurs preferentially, given the following assumptions:
      (1) The elimination reaction requires an *anti*-coplanar relationship between the proton that is lost and the group, either ethoxy or hydroxy, that leaves.
      (2) The nonbenzenoid carbon-carbon double bond of **22a** has the *E* stereochemistry.
      (3) The cyclization of **22a** to **29** is controlled by orbital symmetry.

   b. Give structures, including stereochemistry, for all intermediates involved in the transformation of **22a** to **23**.

   c. What explanation might account for the preference of **22a** to adopt the *E* rather than the *Z* geometry?

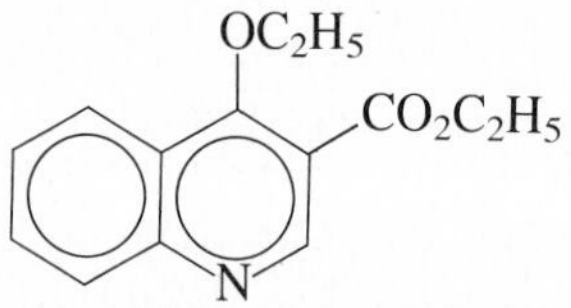

40

7. a. Assign the three major PMR absorptions in the spectrum of diethyl ethoxymethyl-enemalonate (Figure 23.12a) to the hydrogen atoms responsible for them.

   b. Rationalize the very low field position of the absorption at δ 7.5.

   c. What does the PMR spectrum tell you about the equivalence or nonequivalence of the three ethyl moieties in this molecule?

Questions for Part B

8. When **23** is added to aqueous sodium hydroxide the monosodium salt is produced. This salt apparently is not very soluble in the medium, since the mixture is heterogeneous. Yet, as hydrolysis proceeds, the reaction mixture tends to become homogeneous.

   a. Write the structure of the monosodium salt of **23**.

   b. Explain why hydrolysis of the ester function of **23** promotes homogeneity of the reaction mixture.

   c. Would you expect the rate of hydrolysis of the monosodium salt of **23** to be greater or less than that of the analogous compound shown below? Explain.

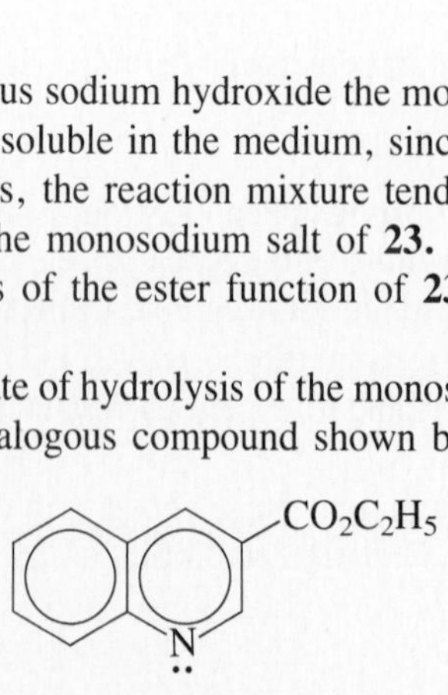

9. Given the workup used in this procedure, would you expect **24** to be contaminated with either neutral or basic organic by-products? Explain.

10. Suppose the alkaline reaction mixture were brought only to pH 7 by the addition of acid. What would be the consequence? Explain.

11. Suppose the hydrolysis of **23** were performed in acidic rather than basic medium. What experimental problems might result?

12. In taking the melting point of 7-chloro-4-hydroxyquinoline-3-carboxylic acid **(24)**, a student noticed that the sample began to froth and bubble as it melted in the capillary tube. Explain.

Questions for Part C

13. Why would it be unwise to use a mineral-oil bath to bring diphenyl ether to reflux?

14. How is separation of **26** from diphenyl ether achieved in the workup of the reaction mixture?

Questions for Part D

15. Calculate the overall yield of **28b** obtained by the sequence of reactions that you performed. Assume that had all of the **26** that you made been used in Part D it would have been converted to **28b** in the same percentage yield as was the amount that you actually did use.

16. It is known that the ring nitrogen of **28b** is more basic than is that of **26**. Rationalize this.

SPECTRA OF REACTANT

(a) PMR spectrum.

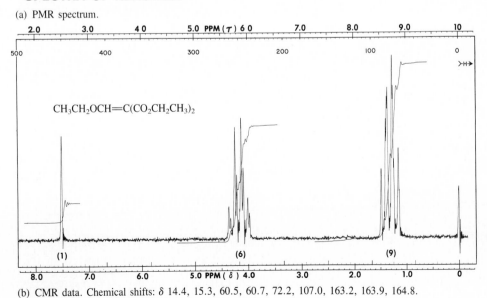

(b) CMR data. Chemical shifts: δ 14.4, 15.3, 60.5, 60.7, 72.2, 107.0, 163.2, 163.9, 164.8.

Figure 23.12 NMR data for diethyl ethoxymethylenemalonate.

23.4 The Synthesis of Lidocaine

Lidocaine **(41)** is the *generic* (common) name of an important member of the category of drugs widely used as local anesthetics. *Trade* names for this substance include Xylocaine, Isocaine, and Anestacon, and its systematic chemical name is 2-(diethylamino)-*N*-(2',6'-dimethylphenyl)acetamide. Two other members of this same family are procaine **(42),** known more commonly by the trade name Novocain, and isocaine **(43).** Isocaine, in addition to having applications as a topical anesthetic, is commonly found in "sunscreens" that are applied to the skin to prevent sunburn because it absorbs the ultraviolet rays that are responsible for the burning of skin. Similarly, **41** and **42** are used in ways other than for anesthesia. For example, both are effective in the treatment of arrhythmia, a condition involving erratic beating of the heart, although **42** must be converted to the amide **44** in order to maximize its effectiveness in this application; this simple chemical transformation increases the drug's half-life under biological conditions and suppresses the rate at which it enters the central nervous system, an undesired property. Interestingly, the antiarrhythmic properties of these compounds were discovered quite accidentally by cardiologists during the course of surgical procedures for which their anesthetic applications were needed.

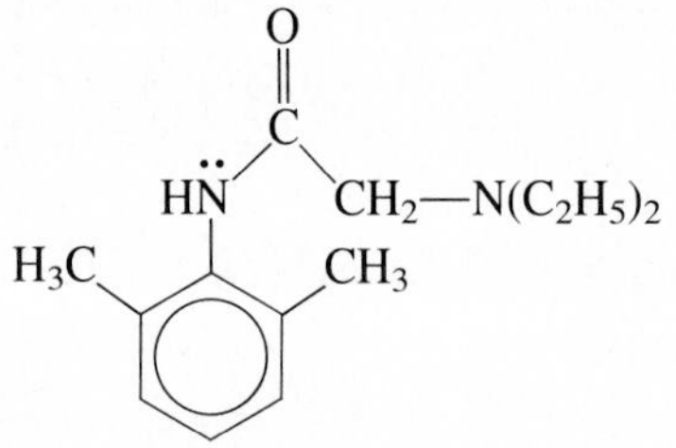

41
Lidocaine

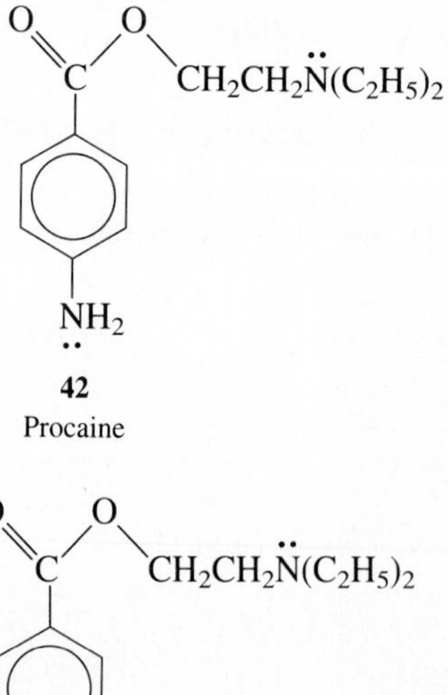

42
Procaine

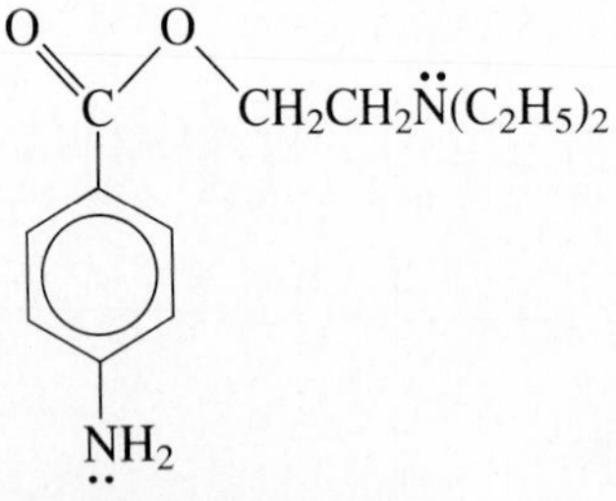

43
Isocaine

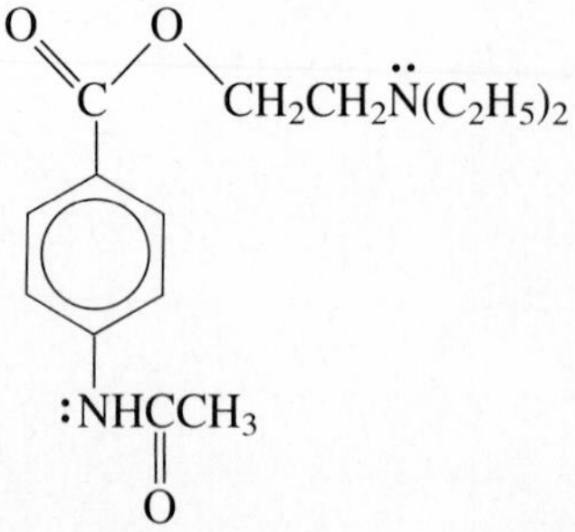

44

Compounds **41–43** are but three examples of the many hundreds of substances that have been synthesized and found to have uses as anesthetics. A substantial number of these share the common structural characteristic of having a trisubstituted nitrogen atom separated by one to four carbon atoms from an anilide function, as in lidocaine, or a benzoate ester group, as in procaine. The dialkylamino moiety, R_2N—, presumably is essential for binding of the compound to enzymes and is present in both naturally and non-naturally occurring drugs having a broad spectrum of medicinal applications. Examples include the *tranquilizers* reserpine **(45)** and perazine **(46)**, the *antidepressant* imipramine **(47)**, the *antihistamine* chlorpheniramine **(48)**, the *psychomimetic* psilocybin **(49)**, and the *antimalarials* quinine and chloroquine **(17 and 28a**, respectively, in Section 23.3). Compounds **17, 45,** and **49** are all natural products, but the others are synthetic, "manmade" substances.

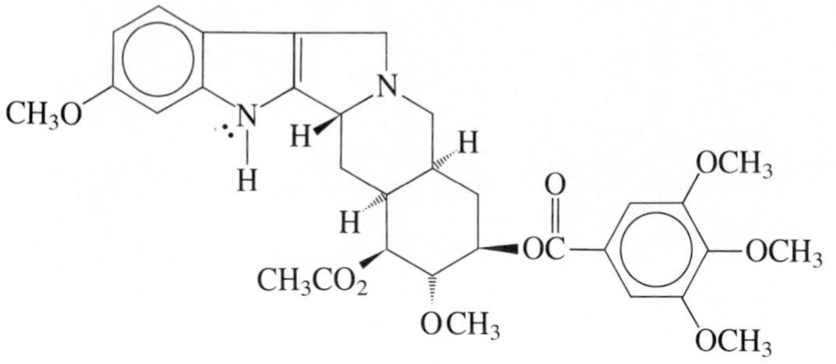

42
Reserpine

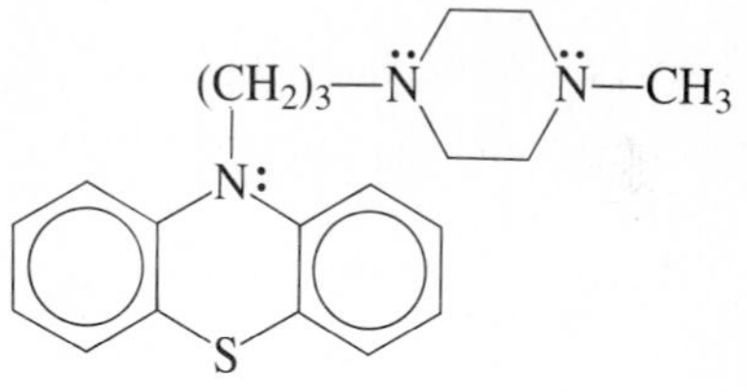

43
Perazine

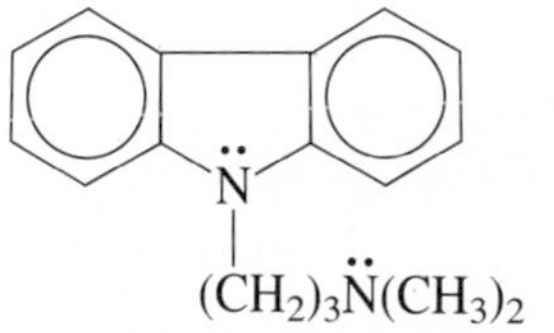

47
Imipramine

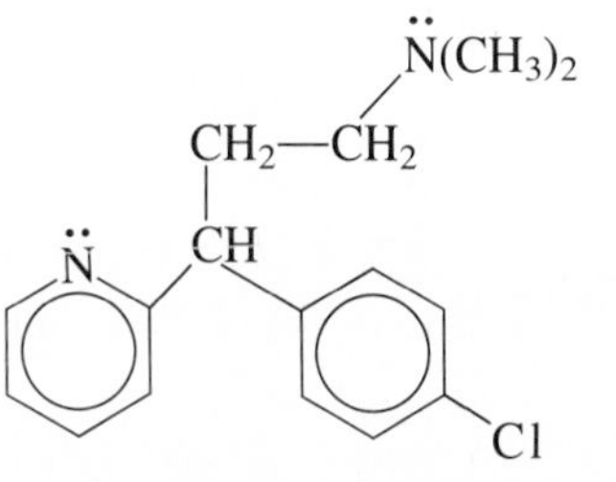

48
Chlorpheniramine

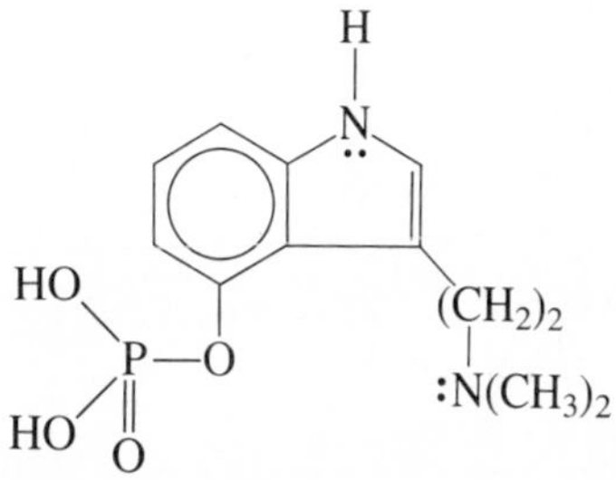

49
Psilocybin

It should not be too surprising to learn that many biologically active compounds that are available today as a result of the synthetic skills of chemists but are not found in nature have molecular features that mimic those found in natural products. In fact, alert scientists have developed new drugs on the basis of folk medicine, noting the medicinal uses that native populations make of various natural sources, plants for example, and then isolating the component responsible for the pharmacological medicinal activity from these sources. Compounds **41–43** can be considered to be structural mimics of cocaine **(50)**, a heterocyclic natural product found in the coca plant that is native to South America. Unfortunately, cocaine has addictive as well as anesthetic properties, so one of the more compelling reasons to develop synthetic analogs of it was to eliminate this undesired property of the natural product. The separation of desired and undesired effects due to a natural or unnatural substance destined for use as a drug is a major goal and challenge to the medicinal chemist and is often achieved by the preparation of a variety of compounds, each of which represents a structural modification of the parent substance of interest.

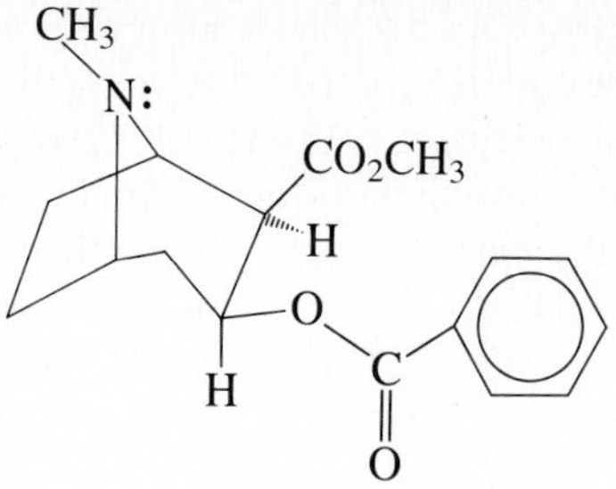

50
Cocaine

The present synthesis of lidocaine from 2,6-dimethylnitrobenzene (**51**) is given in Figure 23.13. Some of the details of each step in the sequence are discussed in the following paragraphs.

(a) 2,6-Dimethylaniline (**52**)

The preparation of lidocaine starts with the conversion of **51** to the dimethylaniline **52** by use of stannous chloride as a reducing agent (equation 30). This equation can be balanced by using the technique of half-reactions as discussed in Section 23.2 in connection with the reduction of nitrobenzene to aniline (equation 2). The side reactions associated with this particular type of reduction are also discussed in that section.

$$51 + SnCl_2 + HCl \rightarrow 52 + SnCl_4 + H_2O \tag{30}$$

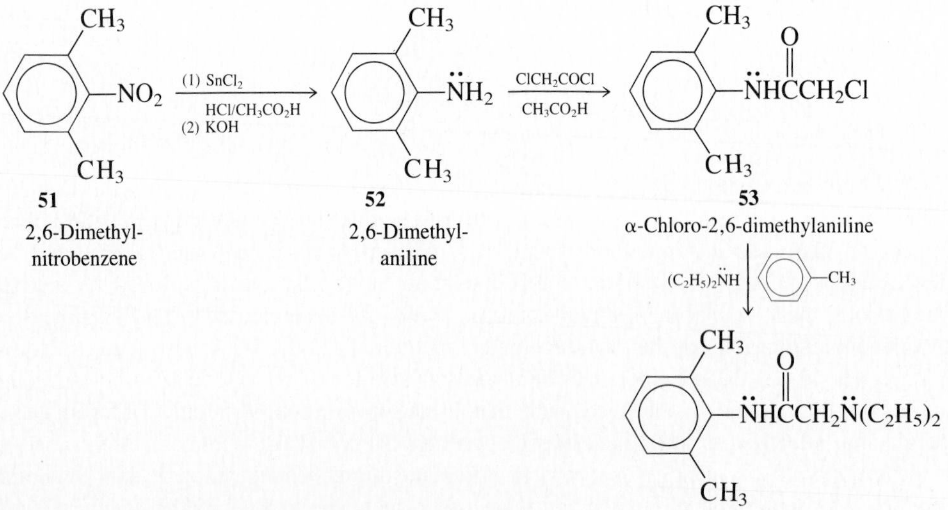

Figure 23.13 Preparation of lidocaine (**41**) from 2,6-dimethylnitrobenzene.

Purification of **52** is uncomplicated. The hydrochloric acid salt of **52** is formed upon reduction of the nitro group and precipitates from the reaction mixture (equation 31). Isolation of the salt by filtration frees it from contaminants such as unchanged **51** and all by-products that are soluble in the reaction medium. Liberation of the salt by treatment with aqueous base gives the aniline **52** (equation 31) in a form sufficiently pure for use in the next step of the reaction. It is instructive to compare the steps required for the isolation of **52** with those needed in the case of aniline (**4**) itself (see Section 23.2, Part A).

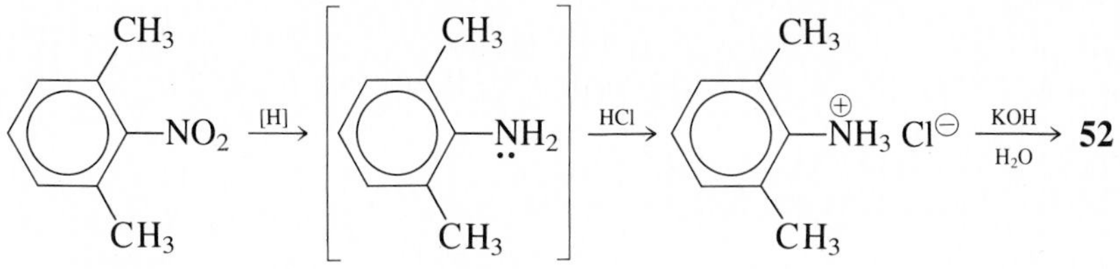

51	52
2,6-Dimethyl-nitrobenzene	2,6-Dimethyl-aniline

(b) α-Chloro-2,6-dimethylacetanilide (**53**)

The substituted aniline **52** is next converted to **53**, the immediate precursor of lidocaine, by treatment with the difunctional reagent α-chloroacetyl chloride, ClCH$_2$COCl (equation 32). Selective substitution at the acyl carbon atom in this step is a reflection of the substantially greater reactivity of nucleophiles with acid chlorides relative to alkyl chlorides because of the difference in electrophilicities and steric environments of the two possible sites for nucleophilic attack. Therefore, reaction at the α-carbon atom to give **54** is at best a minor competing reaction.

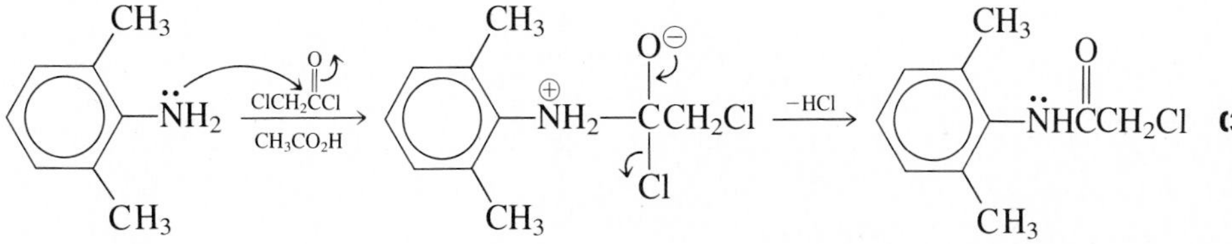

52

53
α-Chloro-2,6-dimethylacetanilide

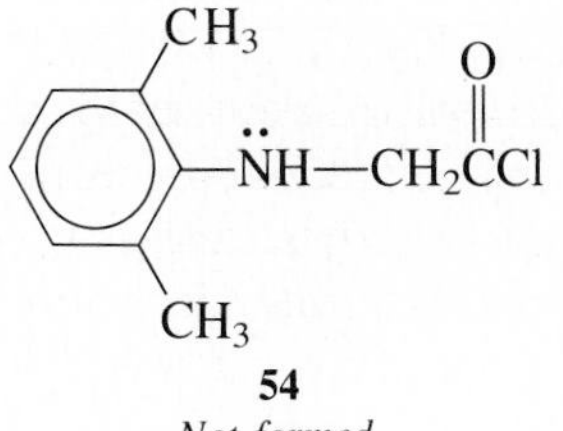

54
Not formed

This reaction is performed in glacial (anhydrous) acetic acid, which means that **52** is in equilibrium with the salt **55** (equation 33). As the reaction proceeds, hydrochloric acid is liberated so that **52** is also partially converted to its salt with this acid (equation 31). Should any of this salt remain at the end of the reaction period, it would contaminate the precipitated **53** because both the salt and **53** are insoluble in cold acetic acid. To avoid coprecipitation of **53** and the hydrochloric acid salt of **52**, aqueous sodium acetate is added to the warm reaction mixture to consume the hydrochloric acid and reestablish the process of equation 33 as the only significant acid-base equilibrium involving **52**. The acetate salt **55** *is* soluble in cold aqueous acetic acid, so that filtration allows isolation of crystalline **53**, with **52** and **55** appearing in the filtrate. Possible by-products like **56** and **57,** which arises by reaction of **54** with water, are also soluble in cold, aqueous acetic acid and therefore are removed from the crystalline **53** by filtration.

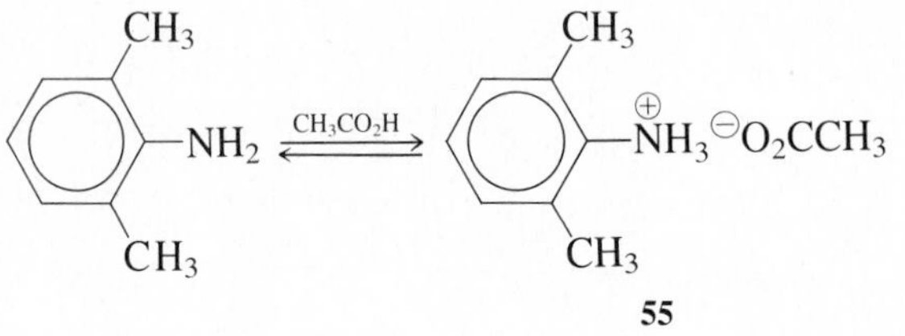

$$\text{equation with structures} \tag{33}$$

55

56

57

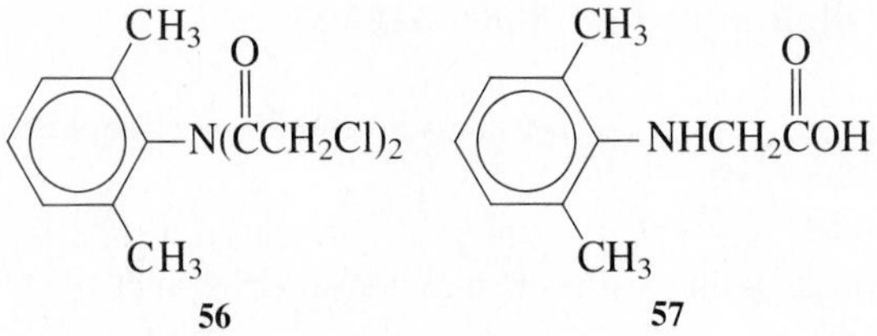

(c) Lidocaine (**41**)

The reaction of the chloride **53** with diethylamine completes the synthetic sequence and is another example of a selective reaction. In this case, nucleophilic attack at the carbonyl function of the amido group is disfavored relative to reaction at the α-carbon atom, a result that is anticipated in view of the disruption of amide resonance that would accompany attack at the carbonyl group. The diethylamine serves the dual roles of acting as a nucleophile and as a base in this final step, not only displacing the chloride ion from **53** but also reacting with the hydrogen chloride formed in the reaction (equation 34). The latter reaction makes the use of an excess of diethylamine a necessity if high yields of lidocaine (**41**) are to be obtained.

Isolation of **41** in pure form involves filtration of the reaction mixture to remove the hydrochloric acid salt of diethylamine, followed by extraction of the basic **41** into aqueous hydrochloric acid. All nonbasic contaminants, such as unchanged **53**, remain in the toluene solution. Liberation of lidocaine by treatment of its hydrochloric acid salt with base, extraction of it into petroleum ether, removal of solvent, and crystallization complete the synthesis.

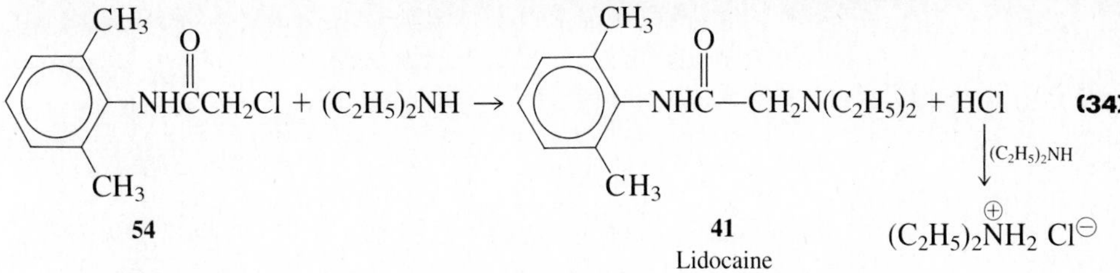

$$
\underset{\textbf{54}}{\text{(2,6-dimethylphenyl)NHCCH}_2\text{Cl}} + (\text{C}_2\text{H}_5)_2\text{NH} \rightarrow \underset{\substack{\textbf{41}\\ \text{Lidocaine}}}{\text{(2,6-dimethylphenyl)NHC—CH}_2\text{N(C}_2\text{H}_5)_2} + \text{HCl} \quad \textbf{(34)}
$$

$$\Big\downarrow {}_{(C_2H_5)_2NH}$$

$$(C_2H_5)_2\overset{\oplus}{N}H_2\ Cl^{\ominus}$$

The option of converting lidocaine to the salt **(58)** with sulfuric acid (equation 35) is also available in the Experimental Procedures. Many drugs are sold in the form of salts because the salts often are more stable and more compatible with the biological media—that is, the stomach, bloodstream, *etc.*—into which they are delivered. As is the case with **41**, drugs containing a basic site are converted to their salts with hydrochloric or sulfuric acid, whereas those containing an acidic site—aspirin **(59),** for example—are commonly transformed to the corresponding sodium salts by reaction with sodium carbonate or sodium hydroxide.

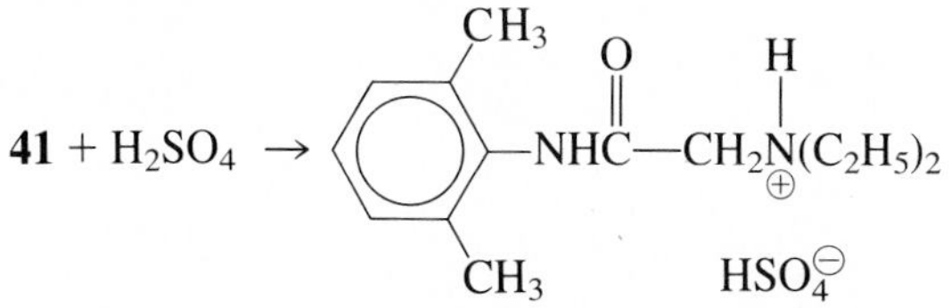

$$
\textbf{41} + \text{H}_2\text{SO}_4 \rightarrow \underset{\substack{\textbf{58}\\ \text{Lidocaine hydrogen sulfate}}}{\text{(2,6-dimethylphenyl)NHC—CH}_2\overset{\oplus}{\underset{H}{N}}(\text{C}_2\text{H}_5)_2 \quad \text{HSO}_4^{\ominus}} \quad \textbf{(35)}
$$

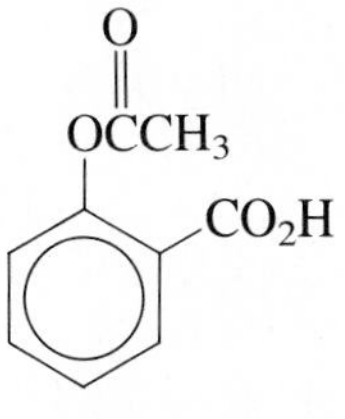

59
Aspirin

Pre-lab exercises for Section 23.4, The Synthesis of Lidocaine, can be found on pages PL. 123–127.

EXPERIMENTAL PROCEDURES

> **DO IT SAFELY**
>
> **1** Use care in handling and transferring concentrated hydrochloric and glacial acetic acids. Should these acids come in contact with the skin, flood the affected area with cold water and thoroughly rinse it with dilute aqueous sodium bicarbonate solution.
>
> **2** α-Chloroacetyl chloride is irritating to the mucous membranes and to the skin. Rubber gloves should be worn when handling and transferring containers of this material, and such work should be done at the ventilation hood. Should this chemical come in contact with the skin, flood the affected area with cold water and thoroughly rinse it with dilute aqueous sodium bicarbonate solution.
>
> **3** Diethylamine is an unpleasant-smelling liquid. Measure it out in the ventilation hood. Should any of this chemical come in contact with the skin, flood the affected area with cold water.
>
> **4** Be certain that there are no open flames in the vicinity when you are working with diethyl and petroleum ethers.
>
> **5** Be certain that all ground-glass joints are well lubricated and tightly mated before heating solutions to reflux or performing distillations.

A. 2,6-Dimethylaniline (52)

Dissolve 0.10 mol of $SnCl_2 \cdot 2\ H_2O$ in 40 mL of *concentrated* hydrochloric acid, using an appropriately sized Erlenmeyer flask; heating on a steam bath may be required for complete dissolution. Add this solution in one portion to a solution of 0.033 mol of 2,6-dimethyl-nitrobenzene in 50 mL of glacial acetic acid contained in a 250-mL Erlenmeyer flask. Swirl the resulting mixture briefly and let the resulting warm solution stand for 15 min.★ Cool the reaction mixture and collect the precipitate that has formed by vacuum filtration.★ Place the damp product in another flask, add 25 mL of water, and make the resulting solution strongly basic by careful addition of 40–50 mL of 8 *M* aqueous KOH solution.★ Upon cooling of the warm basic solution to room temperature, extract the aqueous mixture first with a 25-mL portion and then with an additional 10-mL portion of diethyl ether. Combine the ether extracts, wash this solution twice with two 10-mL portions of water, and dry it over anhydrous K_2CO_3.★ After gravity filtration, remove the ether by simple distillation (Use no flames!), transfer the residue to a pre-weighed test tube, and determine the weight of the oily 2,6-dimethylaniline.

B. α-Chloro-2,6-dimethylacetanilide (53)

Combine the dimethylaniline (52) from Part A with 25 mL of glacial acetic acid and 0.033 mol of α-chloroacetyl chloride, in that order, in an appropriately sized Erlenmeyer flask. With the aid of a steam or hot-water bath, warm the solution to 40–50°C, remove the flask from the bath, and add a solution of 5 g of sodium acetate trihydrate dissolved in 100 mL of water. Cool the resulting mixture in an ice-water bath and collect the product by vacuum filtration. Rinse the filter cake with water until the odor of acetic acid can no longer be detected, and dry it as completely as possible by pressing on it with a clean cork while the vacuum source is still attached. Transfer the solid to a fresh sheet of filter paper and allow it to air-dry for at least 24 hr. Determine the percent yield and melting point of the product (reported mp, 145–146°C).

C. Lidocaine (**41**)

Note: All reagents and equipment used in the first paragraph of the procedure must be dry. Place the α-chloro-2,6-dimethylacetanilide from Part B in an appropriately sized one-necked round-bottomed flask and add 45 mL of toluene, followed by *three* moles of diethylamine per mole of the acetanilide. Equip the flask with a reflux condenser and bring the reaction mixture to a vigorous reflux. After 90 min at reflux, allow the mixture to cool, and isolate the crystalline solid that forms by vacuum filtration. Rinse the filter cake with a little cold petroleum ether (30–60°C), and air-dry and weigh this product (see Exercise 11).

Transfer the filtrate to a separatory funnel and extract it with two 25-mL portions of 3 *M* HCl. Shake vigorously, with venting. Combine the acidic aqueous extracts in a 250-mL Erlenmeyer flask and add 50 mL of 8 *M* KOH solution to make the mixture strongly basic to pHydrion paper. If the mixture is not strongly basic, add additional small portions of 8 *M* KOH until it becomes so. You should anticipate seeing a thin, dark-yellow oil layer in the flask. Cool the alkaline mixture *thoroughly* by *immersion* into an ice-water bath; the use of an ice-salt water bath may shorten this process. Once chilled, agitation by vigorous swirling or stirring should initiate crystallization of crude lidocaine; scratching with a stirring rod may also help. If no crystals form, consult the directions in the next paragraph. Collect the crude lidocaine by vacuum filtration, and wash the filter cake with a little cold water.★ Allow the crystals to dry. Save a small amount of the crude lidocaine for possible use later as seed crystals. Recrystallize lidocaine by dissolving the crude product in boiling petroleum ether (bp 30–60°C), using about 10 mL per g of solid. Allow the solution to cool, add decolorizing carbon, and reheat to boiling. Following hot gravity filtration, allow the solution to cool to room temperature and then chill in an ice-water bath. Beautiful long, white needles of lidocaine should form. Determine the yield and melting point (reported 68–69°C) of the product.

If no crystals form with agitation of the chilled alkaline solution, transfer it to a separatory funnel and extract with two 50-mL portions of petroleum ether (bp 30–60°C). The extractions should be carried out with vigorous shaking and frequent venting. Wash the combined petroleum ether extracts with 25 mL of water and dry them with anhydrous K_2CO_3. Add decolorizing carbon to the filtrate, heat the solution to boiling, and then filter using hot gravity filtration. Concentrate the filtrate to a volume of about 40 mL by simple distillation and allow to cool to room temperature, followed by chilling in an ice-water bath. Crystalline white needles of lidocaine should form. Determine the yield and melting point (reported 68–69°C) of the product.

Alternatively, and preferably, the crude, lidocaine-containing oil can be reconverted to the crystalline salt, lidocaine bisulfate, by first dissolving it in diethyl ether (10 mL of solvent per g of solute; *no flames!*) and then adding a solution of 2 mL of 2.2 *M* sulfuric acid in ethanol per g of solute. Mix the solutions thoroughly and scratch at the air-liquid interface to induce crystallization. Dilute the mixture with an equal volume of reagent-grade acetone to facilitate filtration, and isolate the precipitated salt by vacuum filtration. Rinse the filter cake with a few milliliters of cold reagent-grade acetone and then air-dry and weigh the product. The salt may be recrystallized by dissolving it in an equal weight of hot water and adding 20 times this volume of reagent-grade acetone *in one portion*. Swirl to effect mixing, then allow the solution to stand until crystallization is complete. Isolate the lidocaine bisulfate and determine its melting point and percent yield (reported mp, 210–212°C). (*Caution:* Do *not* use a liquid-bath apparatus for determination of the melting point; use a metal block apparatus instead. See Figure 2.5 in Section 2.3.)

EXERCISES

General Questions

1. Propose a synthesis of **51** using benzene and any inorganic compounds that are needed. The synthesis should avoid reactions that would lead to formation of isomers of **51**.

2. By reference to chemical catalogs, determine the costs of **51** and of the reagents and solvents necessary to convert it to **41** by way of the sequence of Figure 23.13. Calculate the cost of the chemicals and solvents needed to produce one mole of **41**, given the yields that you obtained in the laboratory.

Questions for Part A

3. What is the precipitate that is originally collected in the reduction of 2,6-dimethylnitrobenzene (**51**) by stannous chloride?

4. What would be the consequence of failing to make the aqueous solution basic prior to extracting the 2,6-dimethylaniline (**52**) into ether?

5. Write structures for at least two organic by-products expected to be produced by the reduction of 2,6-dimethylnitrobenzene with stannous chloride.

Questions for Part B

6. Why would ethanol be a poor choice of a solvent for the reaction between 2,6-dimethylaniline and α-chloroacetyl chloride?

7. Why is the anilide **53** much less basic than the dimethylaniline **52**?

8. What organic by-product(s) might be formed if **52** were allowed to react with two moles of α-chloroacetyl chloride?

Questions for Part C

9. What would be the expected effect on yield if only one mole of diethylamine per mole of the anilide **53** were used in the final step of the preparation of lidocaine? Explain.

10. What side reaction(s) would be expected if the anilide **53** were wet when the reaction with diethylamine was performed?

11. What is the solid that is isolated initially in the reaction between **53** and diethylamine?

12. Why does sulfuric acid protonate the nitrogen atom of the diethylamino group preferentially to that of the amido group, as is shown by formation of **58**?

SPECTRA OF STARTING MATERIALS AND PRODUCTS

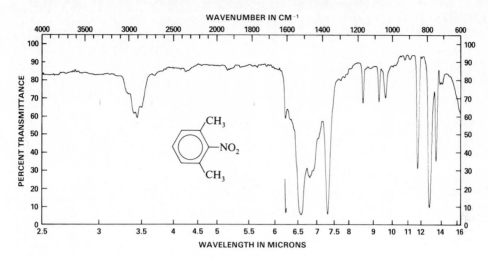

Figure 23.14 IR spectrum of 2,6-dimethylnitrobenzene.

(a) PMR spectrum.

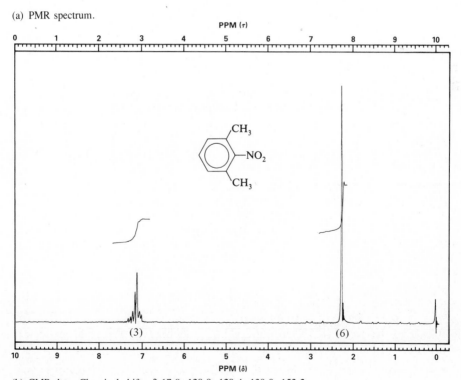

(b) CMR data. Chemical shifts: δ 17.0, 128.9, 129.4, 130.0, 152.2.

Figure 23.15 NMR data for 2,6-dimethylnitrobenzene.

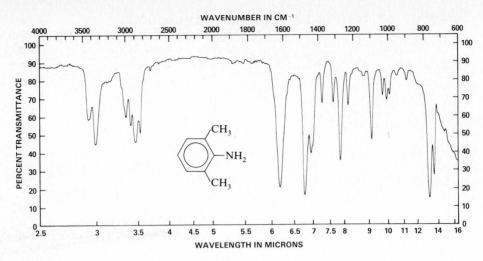

Figure 23.16 IR spectrum of 2,6-dimethylaniline.

(a) PMR spectrum.

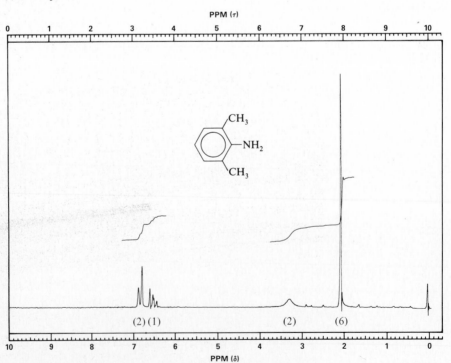

(b) CMR data. Chemical shifts: δ 17.3, 117.8, 121.4, 128.2, 142.9.

Figure 23.17 NMR data for 2,6-dimethylaniline.

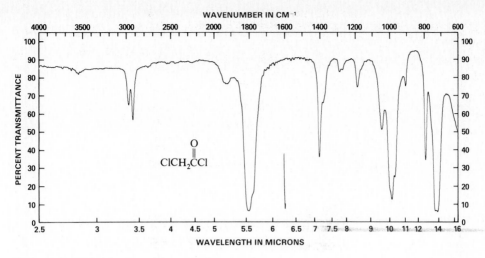

Figure 23.18 IR spectrum of α-chloroacetyl chloride.

$$ClCH_2\overset{\displaystyle O}{\overset{\|}{C}}Cl \qquad \text{Chemical shifts: } \delta\ 49.0,\ 168.1$$

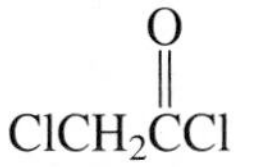

Figure 23.19 CMR data for α-chloroacetyl chloride.

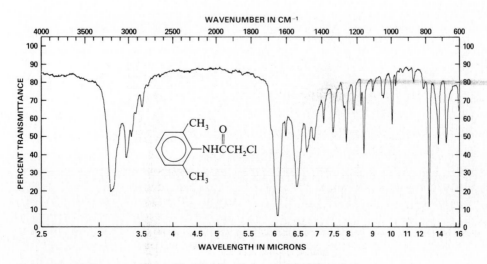

Figure 23.20 IR spectrum of α-chloro-2,6-dimethylacetanilide (KBr pellet).

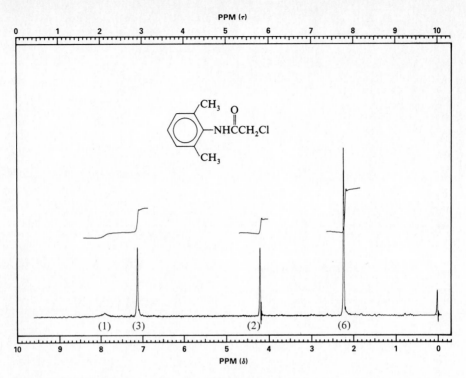

Figure 23.21 PMR spectrum of α-chloro-2,6-dimethylacetanilide.

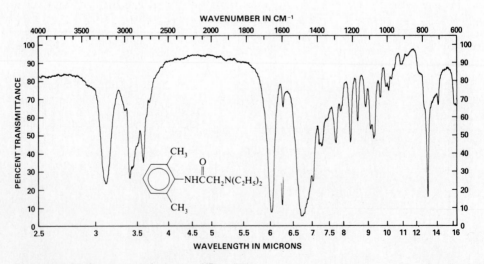

Figure 23.22 IR spectrum of lidocaine (KBr pellet).

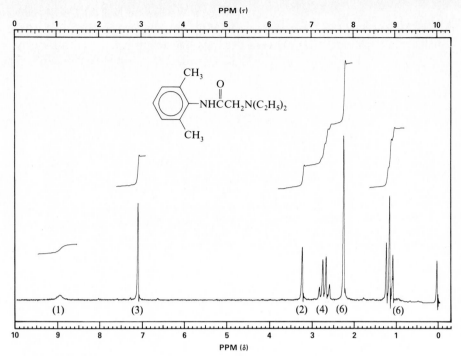

Figure 23.23 PMR spectrum of lidocaine.

23.5 Additional Multistep Synthetic Sequences

In Chapter 15 the preparation of 4-cyclohexene-*cis*-1,2-dicarboxylic anhydride **(60)** from maleic anhydride and 1,3-butadiene is described, as well as hydrolysis of **(60)** to the corresponding acid **(61)**. In Chapter 20 directions are given for hydrogenation of **61** to cyclohexane-*cis*-1,2-dicarboxylic acid **(62)**. These preparations may be combined as shown in Figure 23.24 to comprise a three-step sequence.

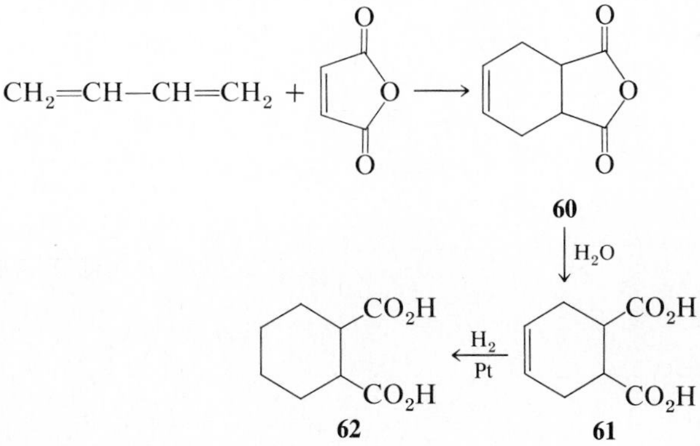

Figure 23.24 Synthesis of cyclohexane-*cis*-1,2-dicarboxylic acid.

The preparation of 1-bromobutane **(63)** is described in Section 18.2, and the oxidation of 2-methyl-1-propanol **(65)** to 2-methylpropanal **(65)** is given in Section 19.2, Part A. 1-Bromobutane may be converted to the Grignard reagent **(66)** and its reaction with 2-methylpropanal carried out as described in Section 22.2e to produce 2-methyl-3-heptanol **(67).** These preparations may be combined as shown in Figure 23.25.

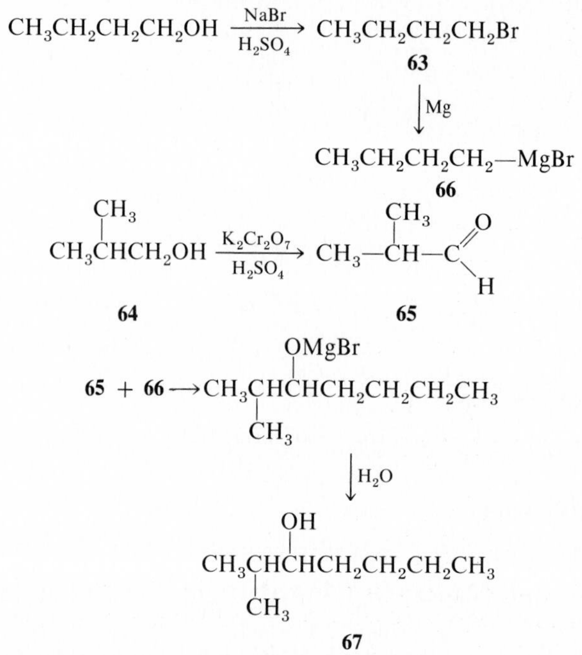

$$CH_3CH_2CH_2CH_2OH \xrightarrow[H_2SO_4]{NaBr} CH_3CH_2CH_2CH_2Br$$
$$\textbf{63}$$

$$\Big\downarrow Mg$$

$$CH_3CH_2CH_2CH_2—MgBr$$
$$\textbf{66}$$

$$\underset{\textbf{64}}{\overset{\overset{\displaystyle CH_3}{|}}{CH_3CHCH_2OH}} \xrightarrow[H_2SO_4]{K_2Cr_2O_7} \underset{\textbf{65}}{\overset{\overset{\displaystyle CH_3}{|}}{CH_3—CH—C}}\overset{\displaystyle O}{\underset{\displaystyle H}{\diagup}}$$

$$\textbf{65} + \textbf{66} \longrightarrow \underset{\underset{\displaystyle CH_3}{|}}{\overset{\overset{\displaystyle OMgBr}{|}}{CH_3CHCHCH_2CH_2CH_2CH_3}}$$

$$\Big\downarrow H_2O$$

$$\underset{\underset{\displaystyle CH_3}{|}}{\overset{\overset{\displaystyle OH}{|}}{CH_3CHCHCH_2CH_2CH_2CH_3}}$$
$$\textbf{67}$$

Figure 23.25 Synthesis of 2-methyl-3-heptanol.

chapter 24

polymers

Polymer is the name given to a class of molecules characterized by molecular weights in the range of thousands to hundreds of thousands and by recurring structural units that are related to simpler molecules (monomers) from which the polymer may be considered to be derived. Because of the size of their molecules, polymers are also often referred to as *macromolecules*. Important examples of polymers are found in nature, for example, proteins, polysaccharides, rubber, and nucleic acids. Polymers may also be produced synthetically, as is attested by the myriad of synthetic plastics, elastomers, and fibers that have become commonplace in our contemporary society. Additional mention of some of the natural polymers may be found in Chapters 25 and 26; in this chapter we are concerned with typical examples of synthetic polymers.

24.1 Chain-Reaction Polymerization

Two major types of polymerization methods are used to convert small molecules (monomers) into synthetic polymers. These methods were originally referred to as *addition* and *condensation* polymerization. Because of certain ambiguities in these terms, the preferred names now are *chain-reaction* polymerization and *step-growth* polymerization. The major distinctions between these two methods result from the differences in the kinetics of the polymerization reactions, as will be seen from the following discussions.

The term *addition polymerization* has been used because the polymers are produced by the self-addition of a large number of monomers. For example, as the name implies, polyethylene is produced by the catalyzed self-addition of thousands of ethylene molecules (equation 1).

$$(n) \; CH_2{=}CH_2 \xrightarrow[\text{and/or heat}]{\text{catalyst}} \; {-}(CH_2{-}CH_2)_n{-} \qquad \textbf{(1)}$$

Ethene $n = 10{,}000{-}30{,}000$

(ethylene)   Polyethene

(polyethylene)

It may be noted that the molecular formula of the polymer is essentially the same as that of the monomer, since the material at the ends of the polymer chain is insignificant in comparison with the rest of the polymer molecule, because n is such a large number.

Polystyrene (**2**) is a polymer of ethenylbenzene, or styrene (**1**, equation 2).

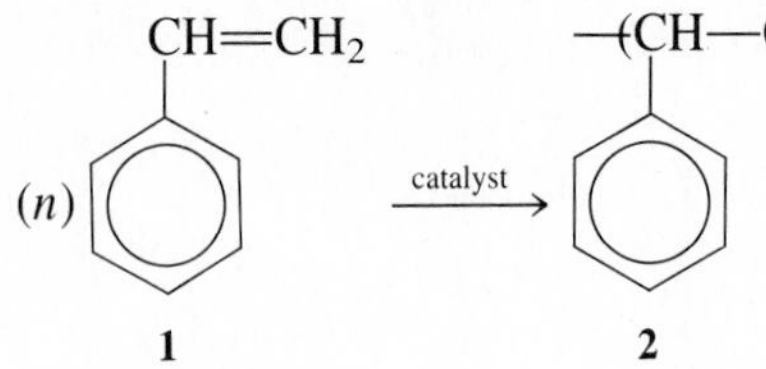

$$(n) \qquad \xrightarrow{\text{catalyst}} \qquad \qquad \textbf{(2)}$$

1   **2**

Ethenylbenzene   Polyethenylbenzene

(styrene)   (polystyrene)

Some other common addition polymers have trade names that do not indicate their structure; for example, Teflon is a polymer of tetrafluoroethylene (**3**), and Plexiglas and Lucite are polymers of methyl 2-methylpropenoate (**4**, methyl methacrylate).

$$CF_2{=}CF_2 \qquad CH_2{=}\underset{\underset{\textstyle CH_3}{|}}{C}{-}CO_2CH_3$$

3   **4**

Tetrafluoroethene   Methyl 2-methylpropenoate

(tetrafluoroethylene)   (methyl methacrylate)

Copolymers may be produced by the polymerization of a mixture of monomers. For example, a useful plastic film (such as Saran Wrap) is made by polymerizing a mixture of chloroethene (vinyl chloride, **5**) and 1,1-dichloroethene (vinylidene chloride, **6**, equation 3).

$$(n) \; CH_2{=}\underset{\underset{\textstyle Cl}{|}}{CH} + (m) \; CH_2{=}CCl_2 \rightarrow \left({-}CH_2{-}\underset{\underset{\textstyle Cl}{|}}{CH}{-}\right)_n\left({-}CH_2{-}\underset{\underset{\textstyle Cl}{|}}{\overset{\overset{\textstyle Cl}{|}}{C}}{-}\right)_m \qquad \textbf{(3)}$$

5   **6**   **7**

Chloroethene   1,1-Dichloroethene

(vinyl chloride)   (vinylidene chloride)

The abbreviated formula of **7** is not meant to imply that all of the two monomer units are bunched together in two blocks, although such "block copolymers" can be produced by special techniques. In the more common copolymers the different monomer units are distributed randomly in the chain.

Chain-reaction polymerization is, as the name implies, a *chain* reaction in which the initiator may be a cation, anion, or free radical. Examples of cationic polymerization include that of 2-methylpropene (isobutylene) by protic and Lewis acid catalysts to give low molecular weight polymers (oligomers) and polymers. The polymerization of propenenitrile (acrylonitrile) by the very strong base sodium amide is an anion-catalyzed process. The enormously valuable stereoregulated polymerizations with Ziegler-Natta catalysts take place by a mechanism that involves the carbanion character of the C—Al bond in an aluminum alkyl (AlR_3) and the ability of a transition metal such as titanium (in $TiCl_3$, for example) to coordinate with the π-bonds of the alkene.

However, free-radical polymerization is by far the most widely used. The mechanism of this type of polymerization as described below should be compared with that of the chlorination of hydrocarbons discussed in Section 12.1. The reaction is initiated by thermal decomposition of the catalyst, which in our experiment is *tert*-butyl peroxybenzoate **(8)**; this compound produces the free radicals **9** and **10** when heated (equation 4).

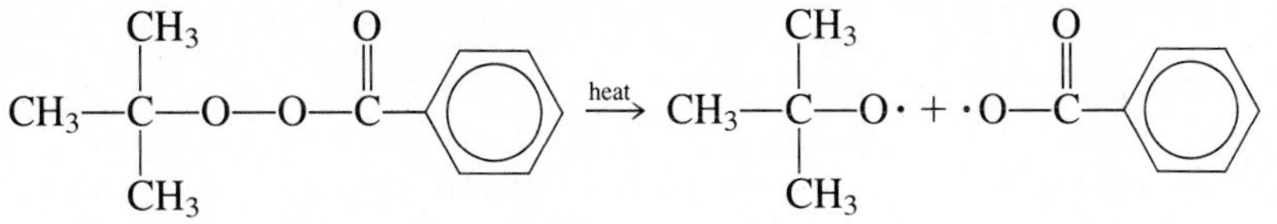

(4)

8 *tert*-Butyl peroxybenzoate **9** **10**

If In· stands for either or both of these free radicals, the course of the polymerization may be indicated as shown in equations 5–8. Equation 5 indicates the function of the free radicals in *initiating* the polymerization.

$$\text{In}\cdot + CH_2{=}\underset{C_6H_5}{\underset{|}{CH}} \rightarrow \text{In}{-}CH_2{-}\underset{C_6H_5}{\underset{|}{CH}}\cdot \qquad (5)$$

$$\text{In}{-}CH_2{-}\underset{C_6H_5}{\underset{|}{CH}}\cdot + n\ CH_2{=}\underset{C_6H_5}{\underset{|}{CH}} \rightarrow \underset{\mathbf{11}}{\text{In}{-}CH_2{-}\underset{C_6H_5}{\underset{|}{CH}}{-}(CH_2{-}\underset{C_6H_5}{\underset{|}{CH}})_n\cdot} \qquad (6)$$

$$2\ \mathbf{11} \rightarrow \underset{\mathbf{12}}{\text{In}{-}CH_2{-}\underset{C_6H_5}{\underset{|}{CH}}{-}(CH_2{-}\underset{C_6H_5}{\underset{|}{CH}})_{n-1}CH_2{-}\underset{C_6H_5}{\underset{|}{CH_2}}} \qquad (7)$$

$$+\ \underset{\mathbf{13}}{\text{In}{-}CH_2{-}\underset{C_6H_5}{\underset{|}{CH}}{-}(CH_2{-}\underset{C_6H_5}{\underset{|}{CH}})_{n-1}CH{=}\underset{C_6H_5}{\underset{|}{CH}}}$$

$$\mathbf{11} + R\cdot \rightarrow \text{In}{-}CH_2{-}\underset{C_6H_5}{\underset{|}{CH}}{-}(CH_2{-}\underset{C_6H_5}{\underset{|}{CH}})_n{-}R \qquad (8)$$

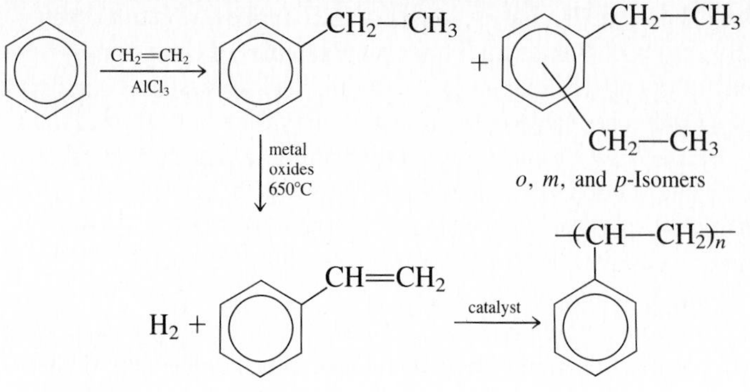

Figure 24.1 Industrial synthesis of polystyrene.

Equation 6 represents the *propagation* of the growing polymer chain. Equations 7 and 8 show possible *termination* processes. In equation 7 the free-radical end of one growing polymer chain abstracts a hydrogen atom with its electron from the carbon atom next to the end of another polymer radical to produce one polymer molecule that is saturated at the end **(12)** and one polymer molecule that is unsaturated at the end **(13),** a process termed *disproportionation*. In equation 8 R· may be one of the initiating radicals, In·, or another growing polymer chain.

An industrial process for the manufacture of styrene is outlined in Figure 24.1. The production of styrene in the United States in 1983 was over 6 billion pounds, about 85% of it being monopolymerized to polystyrene and about 10% copolymerized with 1,3-butadiene to give a synthetic rubber.

The commercially available styrene that is used in our polymerization experiments contains a stabilizer, *tert*-butylcatechol **(14),** a compound that acts as a "scavenger" for free radicals, converting them into substances that are unreactive as initiators.

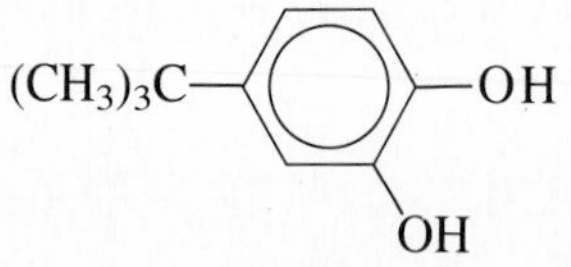

14
tert-Butyl catechol

The presence of the stabilizer is necessary to prevent premature polymerization of styrene during storage or shipment, because it is so readily polymerized by traces of catalytic substances such as even the oxygen in the air. In the following procedure, directions are given for producing the polymer in the form of an amorphous solid, a film, and a clear "glass."

Pre-lab exercises for Section 24.1, Chain-Reaction Polymerization, can be found on page PL. 129.

EXPERIMENTAL PROCEDURE

> **DO IT SAFELY**
>
> The free-radical catalyst (initiator), *tert*-butyl peroxybenzoate, is a very safe material to use in this experiment, since it decomposes at a moderate rate when heated. A catalyst that is used industrially, **benzoyl peroxide, should not be used,** because it is sensitive to vibration and gentle heat and may explode unexpectedly.

A. Removal of the Inhibitor from Commercial Styrene

Place about 20 mL of commercial styrene in a small separatory funnel and add 8 mL of 3 M sodium hydroxide and 30 mL of water. Shake the mixture thoroughly, allow the layers to separate, and withdraw and discard the lower aqueous layer. Wash the upper layer (styrene) with two 15-mL portions of water and then separate the water layers carefully and discard them. Dry the styrene by pouring it into a small Erlenmeyer flask containing a little anhydrous calcium chloride and then swirling the mixture, and then allowing it to stand for 5 or 10 min. Decant the styrene from the calcium chloride and use it in the following experiments.

B. Polymerization of Pure Styrene

Place about 5 mL of pure, dry styrene in an 18 × 150-mm soft-glass test tube, and add 4 or 5 drops of *tert*-butyl peroxybenzoate. Clamp the test tube in a vertical position over a wire gauze, insert a 360°C thermometer so that its bulb is in the liquid, and heat the styrene and catalyst with a *small* burner flame. When the temperature reaches 140°C, remove the flame temporarily. If boiling stops, replace the flame to maintain gentle boiling. Since the polymerization is an exothermic reaction and since free radicals are produced by thermal decomposition of the catalyst, polymerization begins to occur rapidly. *Caution:* Be on the lookout for a rapid increase in the rate of boiling, and be prepared to remove the flame if the refluxing liquid rises to the top of the test tube.

After the onset of polymerization the temperature may be seen to rise to 180°C or 190°C, much above the boiling point of styrene, 145°C. The viscosity of the liquid may also be observed to increase rapidly during this time. As soon as the temperature begins to decrease, remove the thermometer and pour the polystyrene onto a watch glass. (*Caution:* Do *not* touch the thermometer *before* the temperature decreases because movement of the thermometer in the boiling liquid might cause a sudden "bump," which would throw hot liquid out of the tube.) Note the formation of fibers as the thermometer is pulled out of the polymer.

The rate of solidification of the polystyrene will depend on the amount of catalyst used, the temperature, and the length of time the mixture was heated. It may be instructive for some students to use only 1 or 2 drops of catalyst per 5 g of styrene for comparison purposes. The properties of the polystyrene will also vary with the conditions of the polymerization process. (See Exercise 3.)

C. Solution Polymerization of Styrene

Place about 10 mL of pure, dry styrene and 30 mL of xylene (commercial mixture of isomers) in a 100-mL round-bottomed flask and add 10 drops of *tert*-butyl peroxybenzoate.
Connect a reflux condenser and heat the mixture to reflux with a small burner for 30 min.
Cool the solution to room temperature, and then pour about half of it into 150 mL of methanol. Collect the white precipitate of polystyrene by decantation or by vacuum filtration, if
decantation is not practical. Resuspend the polystyrene in fresh methanol and stir it vigorously; collect the polystyrene on a filter and allow it to dry.

Pour the remaining half of the polystyrene solution on a watch glass or the bottom of a
large inverted beaker, and allow the solvent to evaporate. A clear film of polystyrene should
result.

EXERCISES

1. The use of phenols such as *tert*-butylcatechol (**14**) as free-radical "scavengers" is based
   on the fact that phenolic hydrogens are readily abstracted by radicals, producing relatively
   stable phenoxyl radicals, which interrupt chain processes of oxidation and polymerization. Write an equation to illustrate the function of *tert*-butylcatechol in stabilizing styrene
   toward radical-catalyzed polymerization.

2. Write an equation for the reaction involved in the removal of *tert*-butylcatechol from
   styrene by extraction with sodium hydroxide.

3. Why is the polymerization of styrene an exothermic reaction? Explain in terms of a
   calculation based on the equation given below.

$$PhCH-H + Ph-CH=CH_2 \rightarrow Ph-CH_2-CH_2-CH-Ph$$
$$\;\;\;|\qquad\qquad\qquad\qquad\qquad\qquad\qquad\qquad |$$
$$\;CH_3\qquad\qquad\qquad\qquad\qquad\qquad\qquad\; CH_3$$

4. Explain why polystyrene is soluble in xylene but not in methanol.

5. What effect on the average molecular weight of polystyrene would you expect to be
   produced by using a smaller proportion of catalyst to styrene?

REFERENCE

Wilen, S. H.; Kremer, C. B.; Waltcher, I. *Journal of Chemical Education,* **1961,** *38,* 304.

SPECTRA OF STARTING MATERIALS AND PRODUCTS

The IR spectrum of polystyrene is provided as Figure 8.12.

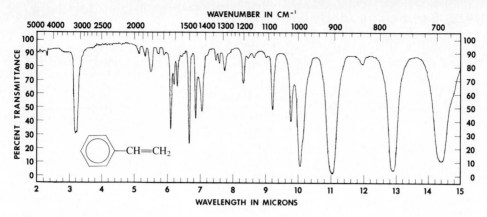

Figure 24.2 IR spectrum of styrene.

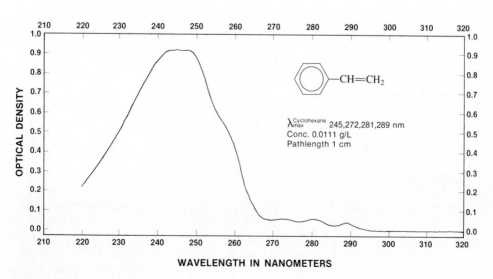

Chemical shifts: δ 113.5, 126.2, 127.8, 128.5, 137.0, 137.7.

Figure 24.3 CMR data for styrene.

Figure 24.4 UV spectrum of styrene.

24.2 Step-Growth Polymerization

This type of polymerization typically employs two different monomers that are capable of undergoing ordinary organic reactions. For example, a diacid (terephthalic acid, **15**) can react with a diol (ethylene glycol, **16**) in the presence of an acid catalyst to produce a polyester, as shown in equation 9.

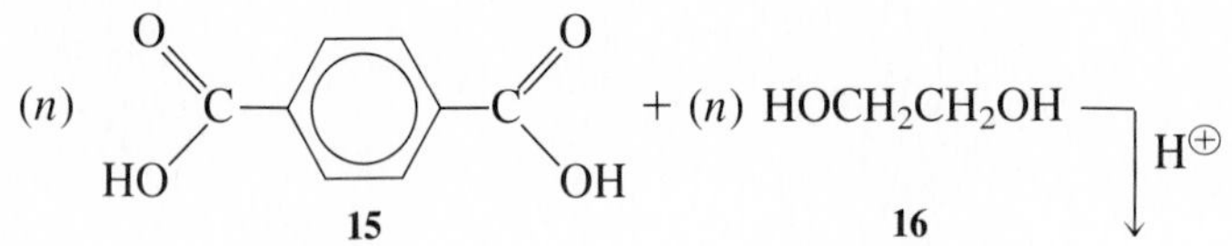

Chain growth is initiated by the reaction of one of the diacid's carboxyl groups with one of the diol's hydroxyl groups. The free carboxyl group or hydroxyl group of the resulting dimer can then react with an appropriate functional group in another monomer or dimer, and the process is repeated in such *steps* until all of the monomer molecules are converted into dimers, trimers, tetramers, and, eventually, polymers. The steps in the polymer growth involve intermolecular elimination of water molecules, which has led to this type of polymerization being called ''condensation polymerization.'' However, as may be seen in Section 24.3, not all step-growth polymerizations involve condensation reactions. What is more generally characteristic of step-growth processes, in contrast to chain-reaction processes, is a much slower rate of polymer growth, a higher activation energy for the reactions so that heating is often required for satisfactory rates of polymerization, and a lower final average molecular weight of the polymers.

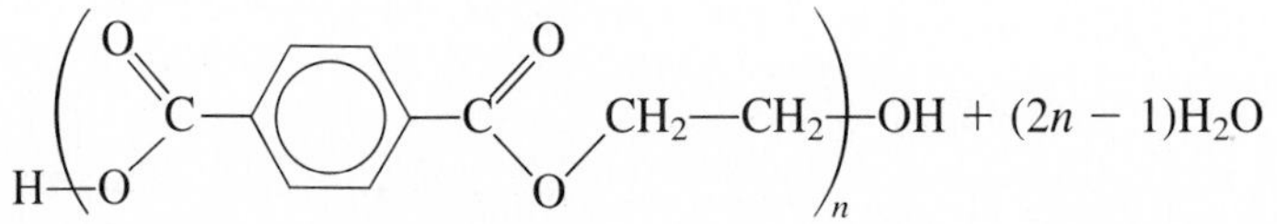

In the production of polyesters for use in textile fibers, the polymerization may be carried out as a transesterification rather than an esterification reaction. For example, dimethyl terephthalate and ethylene glycol are used as monomers, and, in the condensation reaction, the molecules eliminated in the monomer-linking steps are methanol rather than water. Methanol is removed more easily than water from the growing polymer because it is more volatile, and, hence, higher molecular weight polymer is obtained. Dacron, Kodel, and Terylene are trade names of commercial polyesters used as textile fibers, and Mylar is a polyester used as film for audio and video tape, among other uses.

Polyamides are another type of useful polymer produced by a step-growth process. A whole spectrum of such polymers has been produced from various diacids and diamines. Nylon-6,6 was the first commercially successful polyamide. The numbers in the name derive from the six carbon atoms in each of the monomer molecules. Nylon-6,6 is made from hexanedioic acid (adipic acid, **17**) and 1,6-hexanediamine (hexamethylenediamine, **18**), as shown in equation 10.

$$(n)\ \text{HO}-\overset{\overset{\text{O}}{\|}}{\text{C}}(\text{CH}_2)_4\overset{\overset{\text{O}}{\|}}{\text{C}}-\text{OH} + (n)\ \text{H}_2\text{N}(\text{CH}_2)_6\text{NH}_2 \rightarrow \tag{10}$$

17
Hexanedioic acid
(adipic acid)

18
1,6-Hexanediamine
(hexamethylenediamine)

$$\text{HO}-\left(\overset{\overset{\text{O}}{\|}}{\text{C}}(\text{CH}_2)_4\overset{\overset{\text{O}}{\|}}{\text{C}}-\text{NH}(\text{CH}_2)_6\text{NH}\right)_{\!n}\!\!-\text{H} + (2n - 1)\ \text{H}_2\text{O}$$

Nylon-6,6

In the industrial process equimolar amounts of the diacid and diamine are mixed to give the salt, which is then heated to high temperature under vacuum to eliminate the water. The polymer so produced has a molecular weight of about 10,000 and a melting point of about 250°C. Fibers can be spun from melted polymer, and if the fibers are stretched to several times their original length, they become very strong. This "cold drawing" serves to orient the polymer molecules parallel to one another so that hydrogen bonds form between C=O and N—H groups on adjacent polymer chains, greatly increasing the strength of the fibers. The strength of the fibers of silk, a well-known natural polymer of the protein type, is attributed to the same factor: the hydrogen bonds between the natural polyamide molecules. An interesting aspect of the Du Pont Company's tremendous success with nylon is that it stems from their patent on the cold drawing process rather than from a patent on the molecular composition of the polymer!

Two other industrially important polyamides are Qiana and Nomex. Qiana is produced from the monomers dodecanedioic acid, which is synthesized from 1,3-butadiene, and the diamine **19,** which is synthesized from aniline and formaldehyde. Qiana is said to be a superior textile fiber material. Nomex has achieved considerable publicity as the insulator between the ceramic tiles and the aluminum surface of the space shuttles. Its strength and especially its high melting point make it ideal for this purpose. Nomex belongs to a family of "aramides," which are produced from aromatic diacyl chlorides and diamines; the Nomex monomers are **20** and **21.**

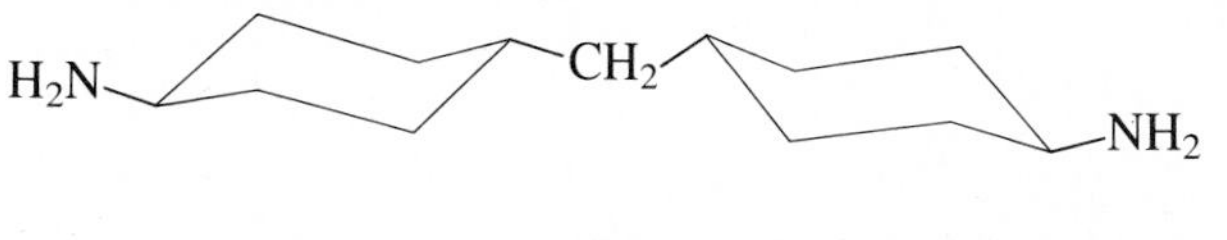

19
(trans, trans-)

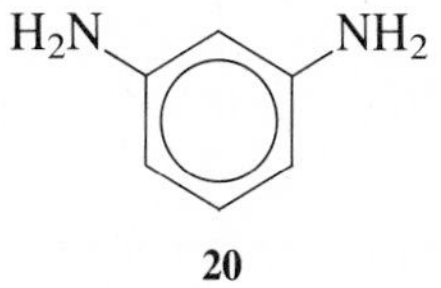

20

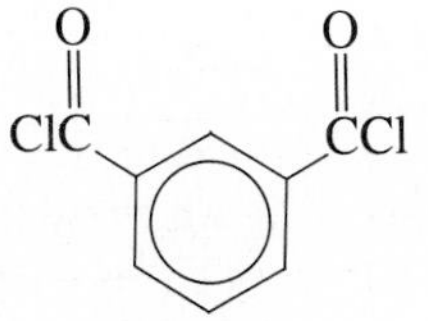

21

The polyamide produced from decanedioic acid (sebacic acid, **22**) and 1,6-hexanediamine **(18)** is called nylon-6,10 (equation 11).

$$(n)\ HO-\overset{\overset{O}{\|}}{C}(CH_2)_8\overset{\overset{O}{\|}}{C}-OH + (n)\ H_2N(CH_2)_6NH_2 \rightarrow$$

22   **18**

Decanedioic acid   1,6-Hexanediamine

(sebacic acid)

(11)

$$HO-\left(\overset{\overset{O}{\|}}{C}(CH_2)_8\overset{\overset{O}{\|}}{C}-NH(CH_2)_6NH\right)_n H + (2n-1)\ H_2O$$

Nylon-6,10

For our experiment, the preparation of nylon-6,10 rather than nylon-6,6 is chosen. To produce a polyamide under simple laboratory conditions, the diacyl chloride derivative of the diacid is used because it is more reactive, and decanedioyl chloride is more stable toward hydrolysis than the corresponding six-carbon compound in contact with moist air or the aqueous phase of the reaction mixture. When the diacyl chloride is employed, the small molecule eliminated in the polymerization process is hydrogen chloride rather than water (equation 12).

$$-CH_2-\overset{\overset{O}{\|}}{C}-Cl + H_2N-CH_2- \rightarrow -CH_2\overset{\overset{O}{\|}}{C}-NH-CH_2- + HCl \qquad \textbf{(12)}$$

Sodium carbonate is added to neutralize the acid formed by the reaction to avoid using an excess of the expensive diamine.

Using the reactive diacyl chloride makes it possible to carry out a polymerization under very mild conditions. When a solution of the diacyl chloride in a water-immiscible solvent is brought into contact with an aqueous solution of the aliphatic diamine at room temperature, a film of high molecular weight polymer forms at once where the two solutions meet. The film is thin but strongly coherent, and it can be pulled out of the interface between the two solutions, where it is immediately and continuously replaced. In this way a long cord or "rope" of polyamide can be produced in much the same way as a magician pulls a string of silk handkerchiefs out of a top hat. When this experiment was first described by two Du Pont chemists, they characterized it as the "Nylon Rope Trick."[1] It does seem to be almost magic that a polymer formed in fractions of a second can attain an average molecular weight in the range 5,000 to 20,000!

To do justice to this experiment, equipment should be arranged to allow the polymer rope to be pulled out of the reaction zone as rapidly as it is formed. A convenient way to do this is illustrated in Figure 24.5. A can such as that in which coffee, fruit juice, or motor oil is packaged, preferably with a diameter of 10 cm or more, makes a good drum on which to wind the polymer. If the can has been emptied by punctures at the edges, it can then be

[1]The present experiment is adapted from the original article by Morgan, P. W.; Kwolek, S. L. *Journal of Chemical Education,* **1959,** *36,* 182.

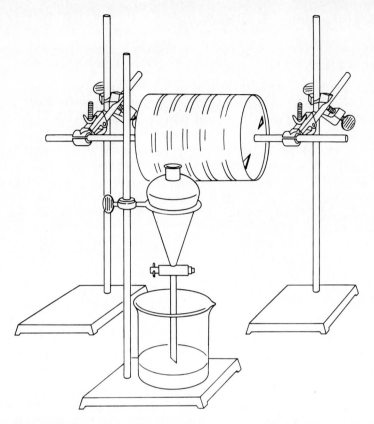

Figure 24.5 Apparatus for the "Nylon Rope Trick."

punctured in the center of each end, and a wooden or metal rod can be passed through the center holes to make an axle for the drum. The rod can be supported horizontally by clamps attached to ring stands in the usual way. To be able to estimate the length of the nylon rope produced, a piece of string should be passed around the drum to measure its circumference; when the rope is being wound on the drum, the revolutions of the drum may be counted and the total length of the rope calculated. A length of 12 m or more can usually be obtained with the procedure described here.

Pre-lab exercises for Section 24.2, Step-Growth Polymerization, can be found on page PL. 131.

EXPERIMENTAL PROCEDURE

DO IT SAFELY

1 If pipets are used instead of syringes to measure the reactants, use a rubber bulb to draw up the liquid; **never use a pipet with your mouth.**

2 Do not handle the polymer rope with your hands any more than is necessary until it has been washed free of solvent and reagents. Use rubber gloves, tongs, or forceps to manipulate it. If you touch the crude polymer, wash your hands with soap and warm water immediately thereafter.

3 After all the rope has been drawn from the reaction mixture, stir the remaining mixture thoroughly until no more polymer forms. Any additional polymer so formed should then be separated from the liquid, washed well with water, and disposed of in the solid waste receptacle.

4 If formic acid is used to form a film, take care not to get it on your skin. It will cause deep skin burns that are not immediately apparent.

Measure into a 250-mL beaker 2 mL (0.0093 mol) of decanedioyl chloride by means of a syringe or pipet. (*The size of the beaker is important.* In smaller beakers the polymer tends to stick to the walls, whereas in larger beakers poor "ropes" are obtained unless larger amounts of reagents are used.) Dissolve the decanedioyl chloride in 100 mL of dichloromethane. Place 1.1 g (0.0095 mol) of crystalline 1,6-hexanediamine (or 1.3 mL of a commercially available 80–95% aqueous solution) in a 125- or 250-mL separatory funnel, add 2.0 g of sodium carbonate, and dissolve both substances by adding 50 mL of water and shaking gently. Arrange the drum on which the polymer is to be wound at a height such that the beaker containing the decanedioyl chloride solution can be placed on the lab bench about 40 cm beneath and slightly in front of the drum.

Support the separatory funnel containing the other reagents in such a way that the lower tip of the funnel is centered no more than a centimeter above the surface of the dichloromethane solution of the decanedioyl chloride. Open the stopcock of the separatory funnel slightly so that the aqueous solution runs *slowly and gently* onto the surface of the organic solution. A film of polymer will form immediately at the interface of the two solutions. Use a long, forceps or tongs to grasp the *center* of the polymer film and pull the rope that forms up to the front of the drum, loop it over the drum, and rotate the drum away from you so as to wind the rope onto the drum. For the first turn or two it may be necessary for you to use your fingers to secure the rope to the drum. If it is, rinse your hands as soon as possible thereafter. Continue to rotate the drum and wind the nylon rope onto the drum at a rapid rate until the reactants are used up, remembering to count the revolutions of the drum as you wind; counting may be facilitated by previously marking a spot on the drum with an adhesive label or a marking pen.

Replace the beaker with a large dish or pan containing about 200 mL of 50% aqueous ethanol, and unwind the nylon rope into the wash solution. After stirring the mixture gently, decant the wash solution, and transfer the polymer to a filter on a Büchner funnel. Press the polymer as dry as possible, and then place it in your desk to dry until the next laboratory period. Dispose of the residual reaction mixture as described in the Do It Safely section.

Examination of Dry Nylon-6,10   You will probably encounter two surprises in this experiment. The first will be the apparently enormous amount of nylon rope obtained from about 3 g of starting material. The second surprise will be the decrease in bulk of the polymer on drying. You will learn that the "rope" was really a delicate *tube* that appeared much larger when it was swollen with solvents. When the nylon is thoroughly dry, weigh it and calculate the yield.

Film Formation   The dry polymer may be dissolved in about 10 times its weight of 90–100% formic acid (*Caution:* see Do It Safely, item 4) by stirring at room temperature. (Heating to dissolve will degrade the polymer.) If the viscous solution is spread on a glass plate, a film of nylon-6,10 will be left by evaporation of the formic acid. The plate should be left *in a hood* from one laboratory period to the following one.

Fiber Formation   The dry polymer obtained in this experiment does not appear to have the properties expected of nylon; it is fragile and of low density. However, if the product is carefully melted by *gentle* heating in a metal spoon or spatula over a very small burner flame or an electric hot plate, fibers may be drawn from the melt with a small glass rod. It will probably be necessary for several students to combine their yields to provide enough polymer to be melted and drawn successfully. The polymer should not be heated much above the melting temperature, or it will become discolored and charred. When the molten polymer cools, it will be found to be much more dense, and its appearance will be more characteristic of a typical polyamide.

EXERCISES

1. Write an equation for the formation of a salt that might be produced from one molecule of hexanedioic acid and two molecules of 1,6-hexanediamine.

2. Explain the purpose of the sodium carbonate in the reaction.

3. Using full structural formulas, draw a typical portion of a nylon-6,6 molecule; that is, expand a portion of the formula given in equation 10. Show at least two hexanedioic acid units and two 1,6-hexanediamine units.

4. Draw formulas that illustrate the hydrogen bonding that may exist between two polyamide molecules after fibers have been "cold drawn."

5. Nylon-6 is produced from caprolactam by adding a small amount of aqueous base and then heating to about 270°C.
   a. Draw a representative portion of the polyamide molecule.
   b. Suggest a mechanism for the polymerization and decide whether it is of the chain-reaction or step-growth type.

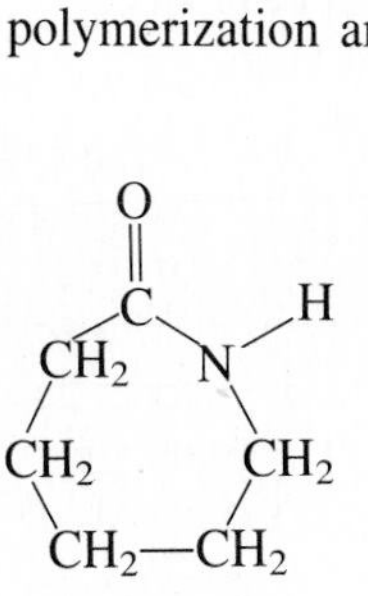

Caprolactam

24.3 Polyurethanes

A simple urethane is the product of addition of an alcohol to an isocyanate, as shown in equation 13.

$$\text{ROH} + \text{R}'\text{N}=\text{C}=\text{O} \rightarrow \underset{\substack{| \\ \text{OR}}}{\text{R}'\!-\!\text{NH}\!-\!\overset{}{\text{C}}=\text{O}} \tag{13}$$

An isocyanate

A urethane

Polyurethanes are produced by a step-growth polymerization mechanism, but it is one involving an *addition* reaction between two different monomers. A typical polyurethane-forming reaction is the addition reaction between an aliphatic *diol* or *triol* and an aromatic *di*isocyanate, as in equation 14.

$$(n)\ \text{HO}\!-\!\boxed{\text{R}}\!-\!\text{OH} + (n)\ \text{O}=\text{C}=\text{N}\!-\!\boxed{\text{Ar}}\!-\!\text{N}=\text{C}=\text{O} \rightarrow \tag{14}$$

A diol   A diisocyanate

$$\text{HO}\!\!\left(\!\boxed{\text{R}}\!-\!\text{O}\!-\!\overset{\displaystyle O}{\overset{\|}{\text{C}}}\!-\!\text{NH}\!-\!\boxed{\text{Ar}}\!\right)_{\!\!n}\!\!-\!\text{N}=\text{C}=\text{O}$$

A polyurethane

The reaction involves nucleophilic addition of an alcohol function to an electrophilic carbon, in this case the isocyanate carbon. Since no small molecules are eliminated in the polymer-forming step, the reaction cannot be considered a condensation.

An aromatic diisocyanate used industrially is toluene diisocyanate **(23).** In 1983 about 650 million pounds of this monomer were produced in the United States.

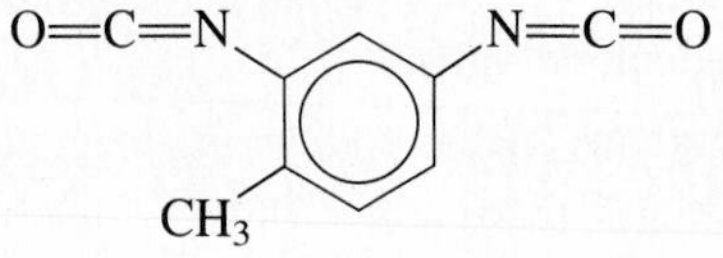

$\text{O}=\text{C}=\text{N}$ $\text{N}=\text{C}=\text{O}$

CH_3

23

Toluene diisocyanate

If a small amount of water is added to an alcohol-isocyanate starting mixture, the polyurethane may be produced in the form of a plastic *foam*. The water reacts with some of the isocyanate groups to form unstable carbamic acid groups that spontaneously undergo decarboxylation, yielding carbon dioxide and amine groups (equation 15).

$$\text{R}\!-\!\text{N}=\text{C}=\text{O} + \text{H}_2\text{O} \rightarrow \left[\underset{\substack{| \\ \text{H}}}{\text{R}\!-\!\text{N}\!-\!\overset{\displaystyle O}{\overset{\|}{\text{C}}}\!-\!\text{OH}}\right] \rightarrow \text{RNH}_2 + \text{CO}_2 \tag{15}$$

A carbamic acid

The carbon dioxide forms bubbles in the developing polymer, thus producing a foam; the newly created amine groups react with residual isocyanate groups so as to extend the polymer chains through urea linkages (equation 16) as well as urethane linkages.

$$
R—N{=}C{=}O + RNH_2 \rightarrow R—N—\overset{\displaystyle O}{\overset{\displaystyle \|}{C}}—N—R \tag{16}
$$

$\underset{H}{|}$ $\underset{H}{|}$

A disubstituted urea

The polyurethane foams may be made *flexible* or *rigid,* depending on the specific nature of the starting materials. Flexible foams are extensively used as cushioning materials for furniture, mattresses, and automobile seats. Rigid foams have more recently found wide application as strong, lightweight building and insulating materials.

Rigid Polyurethane Foams

The key to making a rigid polyurethane foam is to produce a polymer with a highly branched, cross-linked molecular structure. A cross-linked polymer is one in which individual chains are connected to one another by covalent bonds at various points along the chains. This can be done as illustrated in Figure 24.6, in which a symbolized *triol* and *triisocyanate* are the starting materials. After the initial reaction between two molecules, subsequent reactions indicated by dashed arrows will quickly lead to a highly cross-linked polymer. Although some foaming may be produced by carbon dioxide generated by reaction of water as de-

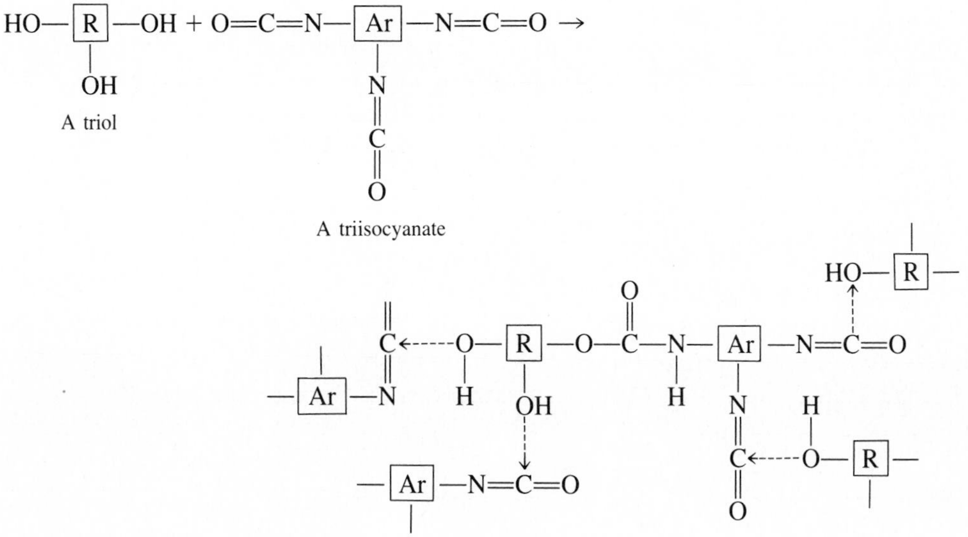

Figure 24.6 Possible reactions between a triol and a triisocyanate.

scribed, an additional foaming agent is usually added. A halocarbon such as $CFCl_3$ is often used because it is just volatile enough to be vaporized by the heat generated by the exothermic polymerization reaction. Other desirable properties of the foaming agent are a low thermal conductivity, which increases the insulating value of the foam, and a slow rate of diffusion through the polymer cell walls.

A great deal of chemistry and technology is involved in polyurethane foam production. For example, a *surfactant* (detergent) is usually added to the mixture of reactants and foaming agent to aid in the nucleation and stabilization of the bubbles in the foam because uniform small bubbles are critical to the structural and insulating properties of the foam. The rates of polymerization and bubble production must be precisely coordinated so that the bubbles are trapped by the hardening polymer at the optimal time. This is done by using a catalyst to control the rate of the polymerization reaction between the alcohol and the isocyanate functional groups.

In this experiment for the production of a rigid polyurethane foam, the alcohol and isocyanate components are not as simple as the triol and triisocyanate shown in Figure 24.6 but are actually a polyol and a polyisocyanate. The polyol is derived from *sucrose* (structure **11,** Section 25.3) by reaction of propylene oxide **(24)** with the hydroxyl groups of sucrose, as illustrated in equation 17.

$$R\text{—OH} + (n)\ \overset{\displaystyle O}{\overset{\displaystyle \triangle}{CH_2\text{—}CHCH_3}} \rightarrow R\text{—O}\!\left(\!CH_2\text{—}\underset{\underset{\displaystyle CH_3}{|}}{CHO}\!\right)_{\!n}\!\!H \qquad (17)$$

(Sucrose)   **24**   **25**
1,2-Epoxypropane
(propylene oxide)

Although each sucrose molecule has eight hydroxyl groups that might react according to equation 17, the reaction is controlled so that an *average* of four or five "propoxylated" groups are introduced per sucrose molecule to give a suitable polyol **(25).** The polyol thus has the potential of reacting with isocyanate groups with its primary hydroxyl groups on the ends of the propoxyl chains, as well as the hydroxyl groups remaining on the sucrose moiety.

The polyisocyanate can be represented by a formula such as **26.**

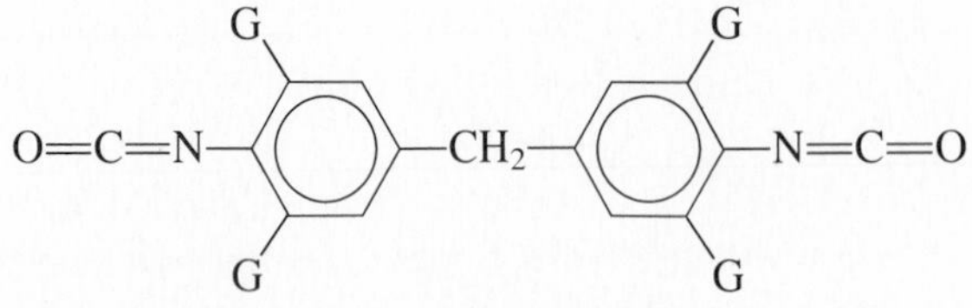

26

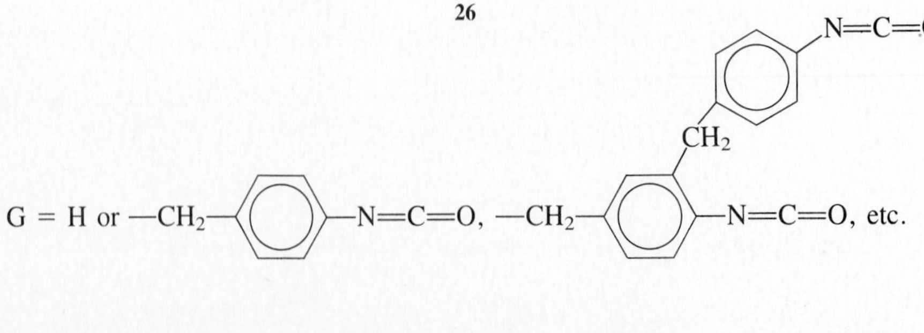

$$G = H\ \text{or}\ \text{—CH}_2\text{—}\!\!\!\bigcirc\!\!\!\text{—N=C=O},\ \text{—CH}_2\text{—}\!\!\!\bigcirc\!\!\!\text{—N=C=O, etc.}$$

The reagent used in this experiment is a complex one that consists mainly of molecules having three isocyanate groups per molecule (that is, in formula **26,** three Gs may be Hs and one G may be —CH$_2$—⟨◯⟩—N=C=O) but that also contains molecules having two, four, or more isocyanate groups. The polyfunctionality of both the alcohol and the isocyanate components of the reaction mixture leads to extensive cross-linking in the polyurethane product.

Pre-lab exercises for Section 24.3, Polyurethanes, can be found on page PL. 133.

EXPERIMENTAL PROCEDURE

DO IT SAFELY

Although simple aryl isocyanates are toxic, the polyisocyanate used in this experiment is quite safe to handle.

Note: This experiment may be presented as an impressive demonstration by two instructors working as a team. It may also be done by students, but if it is, it had also better be done by pairs of students working as teams, for the following reasons: First, the amounts of the major starting materials are rather large, and although the scale could be reduced, the amounts of the minor components are small and must be measured accurately and quickly, which would be more difficult on a smaller scale. Second, it will be helpful for the partners to cooperate in the weighings, observations, and timings involved. For the most efficient operation, the partners should decide in advance what the duties of each will be.

After the reactants have been mixed at room temperature, four stages of the polymer foam formation are interesting to observe and time with a stopwatch or any watch with a second hand. In the industrial vernacular these are (1) *cream time,* the time of the first change in appearance of the complete reaction mixture to a light-brown creamy consistency as foaming begins; (2) *gel time,* the time when the rising foam changes from a bubble-filled liquid to an expanding gel and becomes sticky on the rising surface; (3) *tack-free time,* the time when the surface of the gel is no longer sticky or ''tacky'' to the touch; (4) *rise time,* the time when the foam has reached its maximum height.

The five components of the reaction mixture should be weighed, in the exact sequence given, into a 400-mL beaker, using a top-loading balance that is accurate to 0.1 g. First record the tare weight of the beaker, then add 73.2 g of the polyol,[2] 1.0 g of the silicone surfactant, 1.0 g of the catalyst, and 28.0 g of the fluorocarbon blowing agent. Record the total weight of the beaker and the first four components. Using a wooden tongue depressor, thoroughly mix the components, as in whipping cream, for 3–4 min until a creamy, emulsified mixture is obtained. Return the beaker to the balance, and add more fluorocarbon to

---

[2]The starting materials for this experiment are all industrial products. Instructors should refer to the Instructor's Manual for satisfactory sources and ordering instructions.

bring the mixture back to the original weight, since some of the volatile fluorocarbon will have been lost during the mixing; stir lightly. Now add 96.8 g of the polyisocyanate. *Note the time,* thoroughly stir the mixture for 10–15 sec, and then pour it into a 1-gal cardboard container such as an ice cream carton. Watch the surface of the liquid, and record the **cream time.** The foam will begin to rise almost immediately after this time; as the rate of rising decreases after about 70–80 sec, touch the surface of the rising foam with a clean tongue depressor and then draw it away. At **gel time** strings of foam will attach to the tongue depressor and stretch as it is pulled away. Repeat the touching of the surface until the foam fails to adhere to the tongue depressor; this is the **tack-free time.** Continue to observe the foam as its rate of rise decreases, and record the time that maximum height is finally attained; this is the **rise time.**

The formation of the rigid foam should be complete in an additional 2–5 min. After this time the firm foam can be cut with a razor blade to allow observation of the size, shape, and uniformity of the gas-filled cells.

EXERCISES

1. Write an equation for the reaction of 1,2,3-propanetriol (glycerol) with toluene diisocyanate **(23)** showing how a cross-linked polymer might be formed.

2. Toluene diisocyanate is prepared from toluene, phosgene ($COCl_2$), and inorganic reagents. Indicate the steps in its synthesis.

3. A possible structure for a polyol made from sucrose and propylene oxide would be one produced by reaction of 4 moles of propylene oxide with each of the three primary alcohol groups in sucrose and with one of the secondary alcohol groups. Write a formula for this hypothetical polyol. The formula of sucrose is given in Section 25.3 (structure **11**).

4. A polyisocyanate such as the one used in this experiment is prepared from aniline, formaldehyde, phosgene ($COCl_2$), and inorganic reagents. Suggest possible steps in the synthesis.

5. What effect would the presence of aqueous acid or base have on the production of a polyurethane foam? Show with equations.

chapter **25**

carbohydrates

25.1 Introduction

Carbohydrates, also referred to as "sugars" or saccharides (Sanskrit, *sárkarā*, grit, gravel, sugar), are an extremely important class of naturally occurring polyhydroxy aldehydes (aldoses) and ketones (ketoses), or substances that yield aldoses or ketoses on hydrolysis. Many of the simplest carbohydrates, *monosaccharides,* have the general formula $C_nH_{2n}O_n$. Alternatively, their composition has been expressed as $C_n(H_2O)_n$, a historical and misleading representation of the general formula, because it is hardly true that these substances are hydrates of carbon. This is because some monosaccharides, for example, the deoxy-sugars such as deoxyribose, and compounds containing heteroatoms such as nitrogen, sulfur, or phosphorus do not adhere to this formula.

More complex carbohydrates yield monosaccharides upon hydrolysis. Thus the disaccharide sucrose (table sugar) hydrolyzes to provide one molecule of D-glucose and one of D-fructose.[1] Polysaccharides, depending on their constitution, may be degraded to a mixture of monosaccharides or to only a single product; for example, starch is a mixture of the polymers amylose and amylopectin, each of which is made up *only* of D-glucose units. The experiments in this chapter necessarily utilize the very simplest of carbohydrates, but the impression should be avoided that these may be the most abundant, or even the most significant, of the carbohydrates. In fact, polysaccharides are by far the most ubiquitous bioorganic compounds on earth.

[1] The D is a symbol used to designate the configuration of these sugars relative to that of D-glyceraldehyde, the standard of configuration for carbohydrates. See any modern organic textbook for further explanation.

Carbohydrates provide the ultimate energy source in the food chain. D-Glucose is synthesized in green leaves from carbon dioxide and water by the process of photosynthesis and the action of chlorophyll. This thermodynamically unfavorable process is made possible by the energy of sunlight. D-Glucose is combined in the plant to provide starch and cellulose. After ingestion starch, and in some animals cellulose, are broken down again into D-glucose. One function of the liver is to repolymerize D-glucose from the blood stream to form glycogen, which serves to store energy within the body. Glycogen as needed is reconverted into D-glucose, which is metabolized ultimately to carbon dioxide and water (equation 1), providing the animal with the energy originally stored during photosynthesis. The other monosaccharides that are found, usually in combined form, in living systems are thought to be produced from D-glucose by the actions of various enzymes.

$$C_6H_{12}O_6 + 6\ O_2 \xrightarrow{\text{enzymes}} 6\ CO_2 + 6\ H_2O + \text{energy} \qquad (1)$$
D-Glucose

Carbohydrates are utilized within a living organism in many ways in addition to those of storage and transference of energy. In plants, for example, the polysaccharide cellulose is an important structural component providing rigidity and form. In animals, polysaccharides in combination with protein are important constituents of connective and other tissues. For example, the chondroitin sulfates are found in mammalian cartilages, tendons, heart valves, and cornea. A segment of the structure of chondroitin A, one of the three chondroitin sulfates that have been isolated, is shown here. Protein linkages are found at certain points along the chain of the polysaccharide.

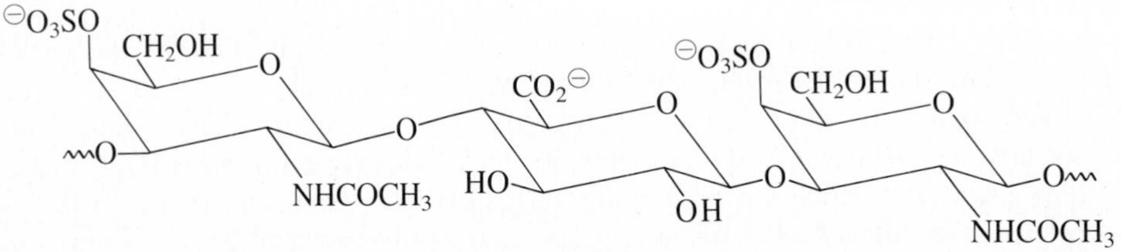

Segment of chondroitin A

Moreover, carbohydrates serve as precursors in the biochemical formation of several other important bioorganic compounds. For example, they ultimately become involved in the biosynthesis of certain α-amino acids, the accepted pathway for which is shown in Figure 25.1. The pyridine derivative **3** is pyridoxine, called vitamin B_6. The Schiff bases **4** and **5**, formed by condensation of **3** with an α-ketoacid, **1** or **2**, interconvert through enzyme-catalyzed tautomerization. Hydrolysis of either of these Schiff bases produces either the glycolysis product (**1** or **2**) or the nonessential amino acids alanine (**6**) or aspartic acid (**7**). Although this equation is vastly oversimplified because it omits the crucial role of enzymatic catalysis in all steps, it does serve to exemplify the significance of carbohydrates as precursors to a living system. Many other examples of the use of carbohydrates as biochemical "building blocks" may be cited; these include the occurrence of ribose and deoxyribose as structural constituents of nucleosides and deoxynucleosides necessary in the formation of RNA and DNA and the incorporation of ribitol in the biosynthesis of riboflavin, one of the B vitamins.

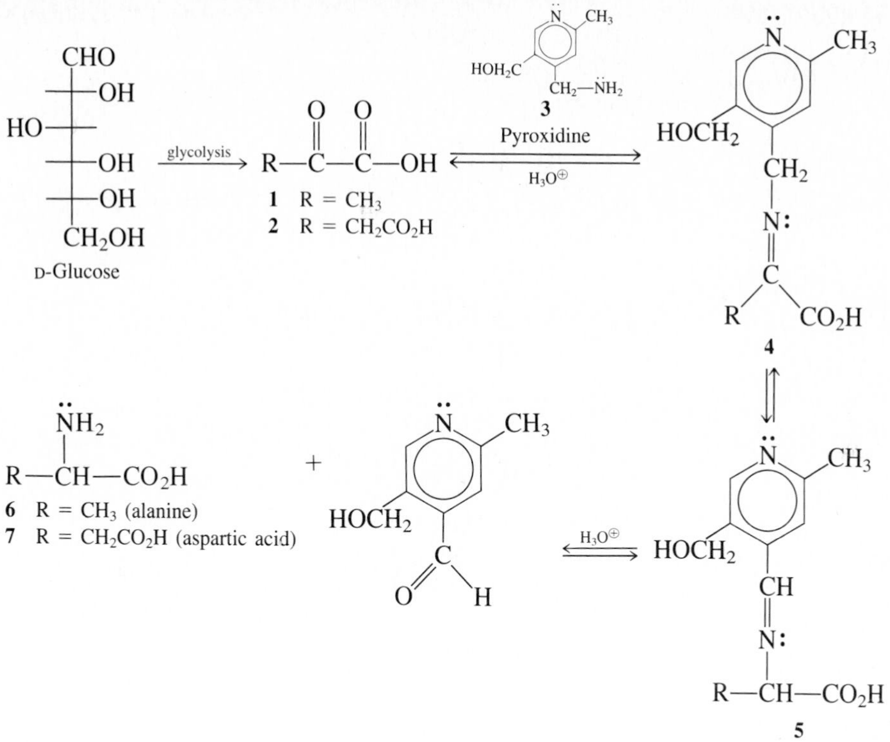

Figure 25.1 The chemical role of vitamin B_6 **(3)** in metabolism of D-glucose.

Although the foregoing discussion has been cursory, it provides an appreciation for the multiple functionality of carbohydrates in biochemistry and suggests that the proper and balanced "operation" of an organism as a chemical system is dependent on this class of compounds in many ways. In the following sections some of the chemical and physical properties of carbohydrates are investigated, and certain techniques used in the isolation and proof of structure of this interesting group of substances are described.

25.2 Mutarotation of D-Glucose

Figure 25.2 shows structures in Fischer projection form of several of the more common monosaccharides; only the D isomers are shown. The sign of optical rotation, shown as (+) or (−), is *independent* of the absolute configuration (D or L). Note that although these sugars are written in their *open-chain* form, those containing a chain of at least four carbon atoms exist predominately in *cyclic* hemiacetal or hemiketal form (see below). Only a few of the monosaccharides shown are known to be naturally occurring, but the remainder have been synthesized. By examination of the structures, it can be seen that monosaccharides generally

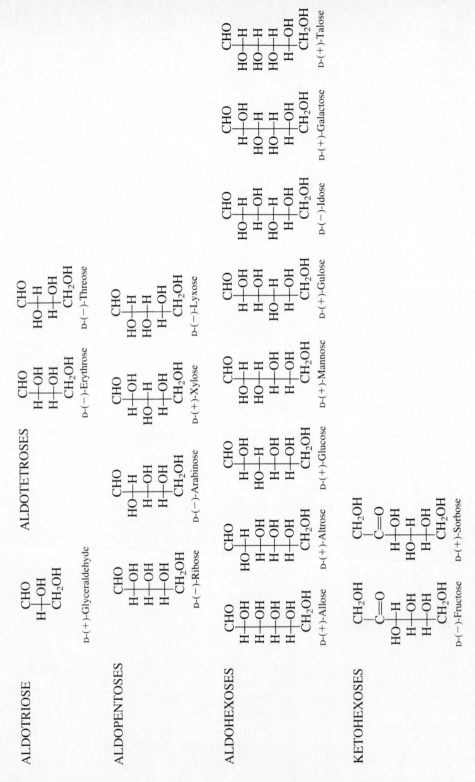

Figure 25.2 Structures of several monosaccharides.

contain more than one asymmetric carbon atom and are therefore subject to extensive stereo-isomerism. An aldohexose, for example, has four chiral carbon atoms and consequently may exist in 16 different stereoisomeric forms ($2^4 = 16$). Within this group of 16 isomers there are 8 enantiomeric pairs.

As is shown in Figure 25.2, D-glucose is one specific stereoisomer of the 16 aldohexoses. Consideration of the functional groups present in D-glucose and recollection that aldehydes react with alcohols to produce hemiacetals and acetals suggest that the open-chain form of this sugar might undergo an intramolecular reaction to produce a hemiacetal. Note that the conversion of an open-chain structure with its four chiral carbon atoms to a cyclic hemiacetal by, for example, reaction between the aldehyde group at C-1 and the C-5 hydroxyl group creates an additional chiral center at C-1. Because there are two possible configurations about this new center, there must be possible two isomeric hemiacetals: **8,** the α-form, and **9,** the β-form (see Figure 25.3). Saccharides such as **8** and **9** that differ only in configuration at the carbon atom involved in the cyclization, C-1 for aldoses and C-2 for ketoses, are called *anomers*.

D-Glucose does in fact exist in the two *diastereomeric* cyclic forms, **8** and **9,** the two isomers being in equilibrium with one another in aqueous solution by way of the inter-mediacy of the open-chain structure **10** (Figure 25.3). The cyclic forms are the major compo-nents of the equilibrium mixture of **8, 9,** and **10.** Because compounds **8** and **9** are *diastereo-mers* rather than *enantiomers,* they have different physical properties and can be separated from one another by rather specific techniques of crystallization, as is done in the experi-ment.

The α-form has a specific rotation of $[\alpha]_D^{25°} + 112°$. When it is placed in water solution the specific rotation gradually changes until it reaches a constant $[\alpha]_D^{25°} +52.7°$. The β-form, in water solution, undergoes a change of specific rotation from $[\alpha]_D^{25°} + 19°$ to $[\alpha]_D^{25°} + 52.7°$. The value $+52.7°$ represents the specific rotation of an *equilibrium mixture* of **8** and **9.** The equilibration of diastereomeric isomers that gives rise to an equilibrium rotation is called *mutarotation.* It is the fact that the equilibrium mixture of the anomers of D-glucose rotates plane-polarized light to the right, in a dextrorotatory fashion, that accounts for this sugar also being called dextrose.

By following the rate of change of optical rotation of α- or β-D-glucose, information concerning the *rate* of mutarotation can be obtained. For a reaction of the type shown in

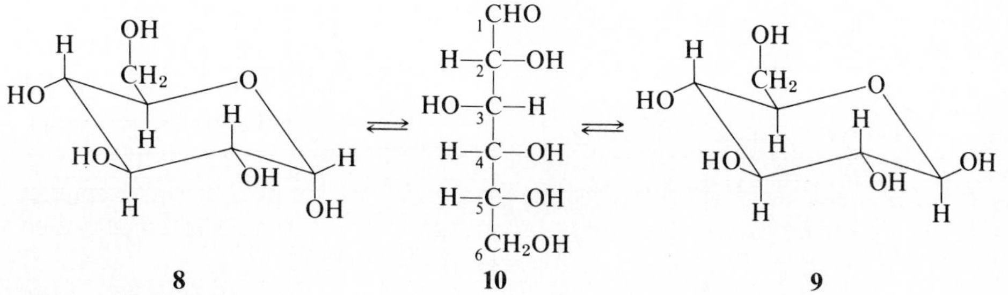

8

α-D-Glucose

10

D-Glucose

9

β-D-Glucose

Figure 25.3 Solution equilibria of D-glucose

equation 1, where A may be α-D-glucose (**8**) and B may be β-D-glucose (**9**), the expression of equation 2 can be written.[2]

$$A \underset{k_2}{\overset{k_1}{\rightleftharpoons}} B \qquad (1)$$

$$2.303 \log \frac{A_e - A_0}{A_e - A_t} = (k_1 + k_2)t \qquad (2)$$

The symbol A_e is the concentration of A at equilibrium, A_0 is the concentration of A at time $t = 0$, and A_t is the concentration at time t during the equilibration process. Note that the term $(A_e - A_0)$ represents the total extent of reaction from the beginning to the final establishment of equilibrium, that is, the total change in concentration of A, whereas the term $(A_e - A_t)$ represents the extent of reaction remaining at time t. It is possible to substitute for these terms any other terms that also represent the ratio of the total extent of reaction and the extent remaining at time t. Any terms utilized must vary linearly during the course of reaction, as does concentration. Discussion in Chapter 7 indicates that optical rotation is linearly related to concentration; thus we may substitute the optical rotation of the solution measured at the beginning, end, and at time t for the concentrations in equation 2 to give equation 3.

$$2.303 \log \frac{\alpha_e - \alpha_0}{\alpha_e - \alpha_t} = (k_1 + k_2)t \qquad (3)$$

Thus we see that measurements of the optical rotation of the initial solution of α-D-glucose, of the rotation of the same solution at equilibrium, and of the rotations at a series of times in between give the data needed for calculation of $(k_1 + k_2)$. This is accomplished by plotting the left-hand term versus t, which should give a straight line whose slope is $(k_1 + k_2)$. It is not possible to obtain k_1 and k_2 individually unless the value of the equilibrium constant is also known (see Experimental Procedure).

Pre-lab exercises for Section 25.2, Mutarotation of D-Glucose, can be found on page PL. 135.

EXPERIMENTAL PROCEDURE

A. Preparation of α-D-Glucose

DO IT SAFELY

1 The vapors of *glacial* acetic acid are extremely irritating to the nasal passages and the eyes. This material should be measured in the hood and brought to your desk only when you are ready for it.

2 Glacial acetic acid will cause blistering of the skin. If this acid comes in contact with the skin, flood the affected area immediately with cold water and then apply a paste of sodium bicarbonate for a few minutes.

---

[2]The interested student can find the derivation of this expression in Frost, A. A.; Pearson, R. G. *Kinetics and Mechanism*, 2d ed., John Wiley & Sons, New York, 1961, p. 186.

α-D-Glucose is the form obtained when D-glucose is crystallized slowly at room temperature from a mixture of acetic acid and water. Dissolve 50 g of anhydrous D-glucose (dextrose) in 25 mL of distilled water contained in a 500-mL Erlenmeyer flask. Heat the mixture on a steam bath to effect complete solution. The solution will be quite viscous and should be continuously stirred during this step. When solution is complete and *no* crystals remain, remove the syrup from the steam bath, add 100 mL of *glacial* acetic acid that has been previously cooled in an ice-water bath to 17°C, and swirl the flask until the mixture becomes homogeneous. Cooling below 17°C may result in crystallization of acetic acid, whose freezing point is 16°C. Stopper the flask with a cork, and allow it to remain undisturbed in the desk until the next laboratory period.

Collect the α-D-glucose by vacuum filtration, and wash the crystals thoroughly with 50–60 mL of 95% ethanol, followed by 50–60 mL of absolute ethanol. Either air-dry the crystals or dry them in an oven at approximately 80°C for 1–2 hr.

B. Rate of Mutarotation

Before coming to class, prepare a format in your laboratory notebook for recording the following information during the performance of this experiment: temperature, concentration, path length of sample tube, blank optical rotation, and the time of mixing of the solution. Also prepare a table for collecting a series of measurements of time and optical rotation. Consult Section 7.4 for detailed information about the use of a polarimeter.

Carefully fill the sample tube for the polarimeter with distilled water; be sure that no air bubbles remain trapped within. Place the sample tube in the polarimeter, and determine the blank reading of optical rotation for solvent. Record the blank rotation and the length of the sample tube. Empty and carefully dry the sample tube.

Accurately weigh 10–15 g of α-D-glucose to the nearest 0.05 g, and transfer the sample *completely* to a dry 100-mL volumetric flask. Fill the flask with distilled water to within a few milliliters of the volumetric mark, tightly stopper the flask, and shake the contents to effect complete solution. The fine solid should dissolve completely within about 30 sec. When the solid is approximately one-half dissolved, note the time to the nearest minute and record it. As soon as the solid is completely dissolved, carefully fill the flask to the mark with distilled water, using a dropper. Again stopper and shake the flask until the solution is homogeneous (approximately 10–15 sec). Working carefully but rapidly, fill the polarimeter sample tube with this solution and leave no air bubbles. Place a thermometer in the volumetric flask, and as soon as there is time, read and record the temperature. Insert the sample tube in the polarimeter, and determine the optical rotation of the solution. Record the rotation and the time of measurement to the nearest minute. During the next 6 or 7 min take a similar reading each minute, recording the data obtained. Then take additional readings every 3 or 4 min for the next 40–50 min.

Approximately 4.5–5 hr will be required for the rotation to drop to within experimental error of the equilibrium value. It will be most convenient to store the remainder of the glucose solution in the tightly stoppered volumetric flask to avoid evaporation until the next laboratory period. At this time and using the same sample tube and polarimeter if possible, refill the sample tube with the equilibrated solution and take a final reading, which will be α_e.

Treatment of Data

1 Define the first recorded time at which a measurement was taken as time $t = 0$. The observed rotation at that time is then α_0. For each reading following the defined zero point, determine and record the *total* elapsed time t from $t = 0$ in minutes.

2 Perform the calculations needed for the graphical plot in the next operation.

3 Plot $2.303 \log (\alpha_e - \alpha_0)/(\alpha_e - \alpha_t)$ (vertically) *versus* total elapsed time t (horizontally) on graph paper. Determine the slope of the best straight line that can be drawn through these points;[3] the value of the slope is equivalent to the value $(k_1 + k_2)$ (in min^{-1}).

4 From equation 4,[4] calculate the value of $(k_1 + k_2)$ to have been expected at the temperature at which the kinetic determination was made. Compare this with the value obtained experimentally. The experimental and calculated values of $(k_1 + k_2)$ should not differ by more than a factor of two.

$$\log (k_1 + k_2) = 11.0198 - 3873/T(\text{K}) \tag{4}$$

5 Plot α_t *(vertically) versus* t (horizontally) on a second sheet of graph paper. A curve should be obtained. Draw the best curved line through these points, and extrapolate the curve from time $t = 0$ back to the time of initial mixing. Using the extrapolated value of α obtained at that time, calculate the initial specific rotation, using equation 6 of Section 7.4, of the α-D-glucose used in the experiment. Recall that pure α-D-glucose has a specific rotation of $[\alpha]_D^{25°} +112°$.

6 Using the value of the equilibrium constant, K_e, for the equilibrium between α- and β-D-glucose given below and the value of $(k_1 + k_2)$ obtained in this experiment, calculate the individual values of k_1 and k_2.

$$K_e = 1.762 = \frac{k_1}{k_2} \tag{5}$$

C. Specific Rotation of Saccharides

Prepare aqueous solutions of accurately known concentration (approximately 0.1 g/mL) of one or more pure sugars, as assigned by the instructor. Following the directions for the measurement of optical rotation given in Part B, determine the specific rotations of each of these sugars.

EXERCISES

*1. Why are the crystals of α-D-glucose washed sequentially with *both* 95% ethanol and absolute ethanol? Why is 95% ethanol used at all?

2. Would any different experimental results have been expected if one had started with β-D-glucose rather than the α-anomer? Explain.

---

[3]See footnote 1 in Section 18.4b.

[4]Hudson, C. S.; Dale, J. K. *Journal of the American Chemical Society*, **1917**, *39*, 320.

3. From the specific rotation values of $+112°$ for α-D-glucose, $+19°$ for β-D-glucose, and $+52.7°$ for the equilibrium mixture, show by calculation that the value for the equilibrium constant given in equation 5 is correct.

4. Write a stepwise reaction mechanism for the isomerization of α- to β-D-glucose.

5. Suppose a solution is found to have a rotation of $+60°$?
   a. How could you determine that the rotation was in fact $+60°$ rather than $-300°$?
   b. Assuming that the rotation is actually in the positive direction, how could you demonstrate that it was not really $+420°$ or $+780°$?
   c. Assuming that the rotation is actually in the negative direction, how could you demonstrate that it was not really $-300°$ or $-660°$?

6. Suppose a solution gave an observed rotation of $0°$. This result *does not* require that the solution contain a racemic mixture or an achiral compound. Why is this so, and how could you show that the solution was not solely constituted of such a mixture or compound?

25.3 The Hydrolysis of Sucrose

Sucrose, a familiar foodstuff, is a carbohydrate having the structure **11**. As shown in the structure, the sucrose molecule is formed by a linkage between the hemiacetal form of α-D-glucose and one of the hemiketal forms of D-fructose (**12**); consequently, an **acetal** moiety is present.

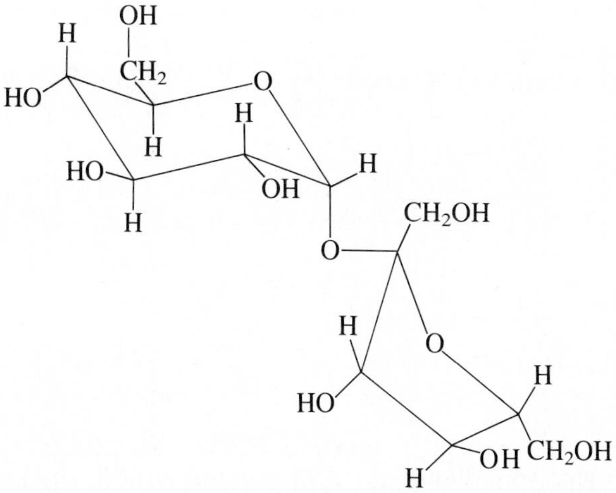

11
Sucrose

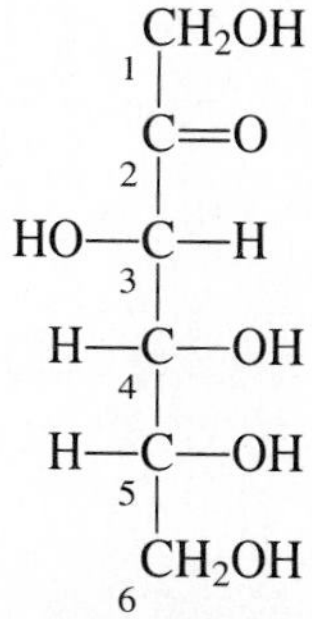

12
D-Fructose

(6)

The observation that sucrose does *not* undergo mutarotation is evidence that the linkage is through the glycosidic hydroxyl groups of each sugar. Note that saccharides like sucrose that do not contain a hemiacetal or hemiketal group are called **nonreducing** sugars (see Section 25.5, Part A).

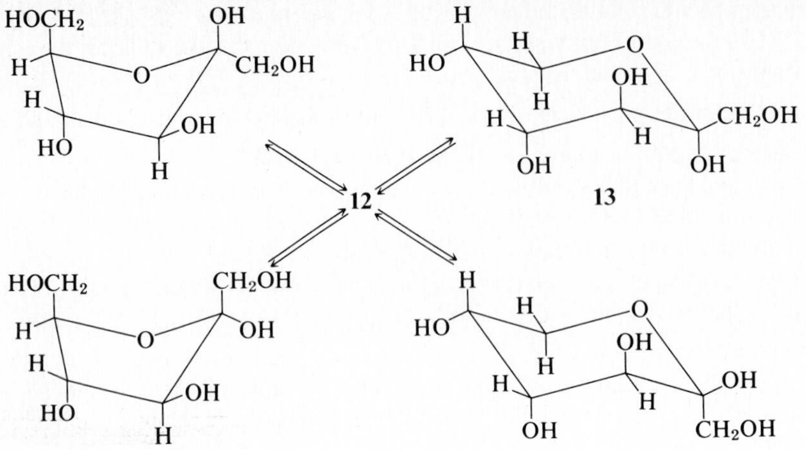

Figure 25.4 Solution equilibria of D-fructose.

Sucrose, $[\alpha]_D^{25°}$ +66.5°, undergoes acid-catalyzed hydrolysis to give a mixture of D-glucose and D-fructose (equation 6).

$$\text{Sucrose} \xrightarrow{\text{H}_3\text{O}^{\oplus}} \text{D-Glucose} + \text{D-Fructose}$$

Under the conditions of the hydrolysis the glucose undergoes rapid mutarotation to give the equilibrium mixture of α- and β-forms, $[\alpha]_D^{25°}$ +52.7°, whereas the fructose is formed as an equilibrium mixture of the isomers shown in Figure 25.4 and has $[\alpha]_D^{25°}$ −92°. Isomer **13**, a six-membered ring hemiketal, predominates in the equilibrium mixture. The observation that the sign of optical rotation changes on hydrolysis of sucrose has led to the name *invert sugar* for the product mixture. The enzyme *invertase* accomplishes the same chemical result as does the acid-catalyzed hydrolysis of sucrose.

Pre-lab exercises for Section 25.3, The Hydrolysis of Sucrose, can be found on page PL. 137.

EXPERIMENTAL PROCEDURE

Place 15 g of pure sucrose, whose weight has been accurately determined, in a 250-mL round-bottomed flask. Add about 80 mL of water, swirl the contents of the flask to effect solution, and then add about 0.5 mL of concentrated hydrochloric acid. Heat the solution at reflux for about 2 hr. During this time determine the specific rotation of sucrose following the general directions of Parts B and C of the Experimental Procedure in Section 25.2 and the more specific directions of Section 7.4 and your instructor regarding the use of a polarimeter.

At the end of the period of reflux, cool the reaction mixture to room temperature, and carefully transfer *all* the solution to a 100-mL volumetric flask. Use small amounts of water to rinse the round-bottomed flask, and add the rinses to the rest of the solution. Dilute the mixture to a volume of 100 mL. Using a polarimeter, determine the specific rotation of the product mixture from the hydrolysis of sucrose. Compare this value with the specific rotation of sucrose that you determined earlier.

EXERCISES

1. Explain the change in sign of the optical rotation that occurs as sucrose undergoes hydrolysis.

2. Calculate the specific rotation of *invert* sugar from the known rotations of the equilibrium mixtures of the anomers of D-glucose and D-fructose. How does this number compare with that determined experimentally?

3. In what way is the specific rotation of invert sugar analogous to the specific rotation of a racemic mixture?

4. Write a stepwise reaction mechanism for the acid-catalyzed hydrolysis of sucrose to D-glucose and D-fructose.

5. Determine which of the structures in Figure 25.4 are α-anomers and which are β-anomers.

6. Write a stepwise reaction mechanism for acid-catalyzed isomerization of **13** to **12**, followed by cyclization to produce a five-membered ring.

25.4 Isolation of α,α-Trehalose

Trehalose is the name given to the D-glucosyl-D-glucosides. It derives from the isolation of the disaccharide α,α-trehalose (**14**) from the *trehala manna,* an oval shell built by certain insects, which has been shown to consist of 25–30% α,α-trehalose. The anomeric forms α,β- and β,β-trehalose have not been found in nature.

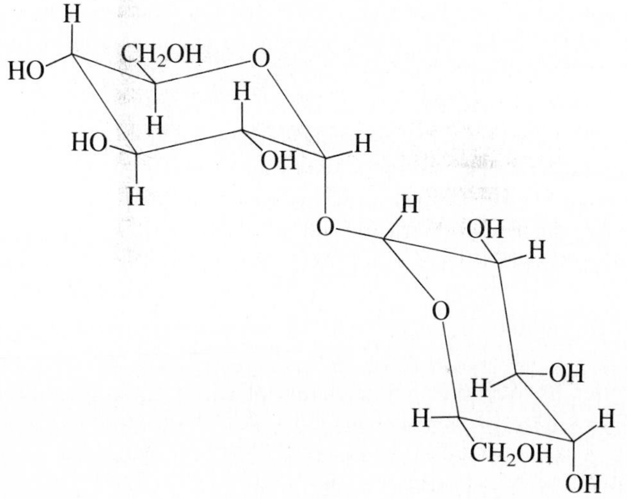

14
Trehalose

This disaccharide was probably first isolated from the ergot of rye, a fungus, in 1832. It has since been shown to occur in other fungi, bacteria, the blood of insects, certain algae and lichens, some of the higher-order plants, such as the resurrection plant, and yeast, as well as the trehala manna. α,α-Trehalose also occurs in combined form in human tubercle bacilli. The lipids from these bacilli may be separated into free fatty acids and natural fats, the fats containing no glycerol and being esters of fatty acids with α,α-trehalose. The esters of sugars with fatty acids are termed *microsides*.

This experiment involves the isolation of α,α-trehalose from dried baker's yeast, in which its content may reach 10–15%. α,α-Trehalose is formed biosynthetically by yeast enzymes from D-glucose and stored within the cell, as is glycogen. These cells also contain an enzyme, *trehalase,* capable of enzymatically hydrolyzing the glycosidic linkage. Proliferating and active yeast cells utilize available D-glucose almost exclusively, producing carbon dioxide and water, and metabolize stored α,α-trehalose only after D-glucose has been consumed. Older yeast cells, on the other hand, ferment α,α-trehalose at least as fast as, and perhaps faster than, D-glucose. These observations probably offer at least a partial explanation for the observation that the α,α-trehalose content of baker's yeast decreases on storage.

Interestingly, baker's yeast has been shown to ferment α,α-trehalose, which has been added to the yeast, while leaving unaffected the α,α-trehalose stored within the yeast. Apparently there is a spatial separation in the cell between *trehelase* and its stored trehalose. There is some evidence that the enzyme may be at the cell surface.

The procedure for the isolation of trehalose from yeast offers interesting insight into the requirements for separating cellular components. Most of the materials extracted from the yeast are systematically removed before the trehalose is finally precipitated. Insoluble materials, such as polysaccharides and fibrous protein and various aromatic compounds, such as aromatic amino acids and heterocyclic compounds, are separated from the extract by filtration and treatment with activated charcoal. Globular proteins, including the enzyme *trehalase,* are removed by heating to cause coagulation (denaturation) followed by precipitation as their insoluble zinc salts. Phosphorylated sugars are separated from the extract through the addition of barium hydroxide and filtration of their insoluble barium salts. Thus α,α-trehalose is essentially the only ethanol-insoluble component in the extract at the final precipitation.

α,α-Trehalose, like sucrose (Section 25.3) has its monosaccharidic components, the two units of D-glucose, joined by way of an acetal linkage involving the two anomeric carbon atoms. As a consequence, this disaccharide does not undergo mutarotation and is classified as a nonreducing sugar. The latter term means that α,α-trehalose is *not* oxidized by mild oxidizing agents like the copper(II) ion present in Benedict's reagent (see Section 25.5, Part A). Thus, this sugar should give no reaction when treated with such oxidants, an outcome that is easily determined visually, because the color change normally undergone by the oxidant as it is reduced does not occur.

Pre-lab exercises for Section 25.4, Isolation of α,α-Trehalose, can be found on page PL. 139.

EXPERIMENTAL PROCEDURE

A. Isolation of α,α-Trehalose

Prepare a paste from 32 g of dried baker's yeast and 68 mL of water. Add 250 mL of 95% ethanol and, with occasional stirring, allow the mixture to stand for about 30 min. Filter the mixture by vacuum filtration, and wash the filter cake with three 30-mL portions of 70% ethanol. Combine the washings with the main solution. To the filtrate add 20 mL of 1.2 M aqueous zinc sulfate, 1 mL of 1% phenolphthalein solution, and a sufficient quantity of saturated barium hydroxide solution to make the solution basic; about 50 mL will be required. Add 2 g of activated charcoal, and heat the mixture to 70°C on a steam or hot-water bath. Filter the hot solution through a Büchner funnel previously layered with filter-aid (see Section 2.16). Adjust the filtrate to about pH 7 with 0.1 M hydrochloric acid, and concentrate the solution to approximately 10 mL by *gentle* heating under vacuum. Slowly stir 80 mL of 95% ethanol into the resulting syrup, stopper the flask, and leave it undisturbed. Crystals will normally form within a day or two; however, a week or more may occasionally be required. The crystals are frequently quite large, owing to their slow growth. If desired, crystallization may be hastened by the addition of a little more ethanol or by cooling the solution in an ice-water bath.

Verify that you have successfully isolated α,α-trehalose by determination of its decomposition point (203°C; dihydrate, 97°C), by subjecting it to Benedict's and Fehling's tests (see Section 25.5, Part A) and by measurement of its specific rotation (see Section 7.4 and Parts B and C of the Experimental Procedure in Section 25.2); the $[\alpha]_D^{20°}$ of α,α-trehalose is reported to be $+178.3°$ (H$_2$O). Hydrolyze a portion of the trehalose by preparing a 0.5% solution in 1 M HCl and boiling it for 20 min. That D-glucose is the sole monosaccharidic constituent of trehalose may be demonstrated by thin-layer chromatographic analysis (Part B).

B. Thin-Layer Chromatography of Monosaccharides[5]

Obtain a 12-cm strip of cellulose chromatogram sheet without fluorescent indicator. Spot the strip about 1 cm from the bottom with the dilute solution of an unknown sugar or sugar mixture and also with solutions of any desired known sugars for comparison purposes (use glucose for the trehalose hydrolysate). Spots should be separated by about 1 cm. Develop the plate, using as developing solvent a mixture of pyridine–ethyl acetate–acetic acid–water in the respective ratios 5:5:1:3. Development may require nearly 1.5 hr. Make the spots visible either by spraying with p-anisidine phthalate reagent or by leaving the plate in contact with iodine vapor. The reagent spray, for reducing sugars, can be prepared by dissolving 1.23 g of p-anisidine and 1.66 g of phthalic acid in 100 mL of 95% ethanol. With the spray reagent, hexoses yield green spots and pentoses give red-violet spots after the plate is heated

---

[5]For an excellent compilation of techniques and procedures for the thin-layer chromatographic analysis of saccharides and their derivatives, consult Lewis, B. A.; Smith, F. in *Thin-Layer Chromatography*, 2d ed., E. Stahl, editor, Springer-Verlag, New York, 1969, Chapter 10.

at 100°C for 10 min. Record the results by drawing a picture of the developed plate in your notebook.

EXERCISES

1. In the step involving concentration of an aqueous solution of α,α-trehalose followed by crystallization of the desired product, what would be the consequence if the aqueous solution had inadvertently been made acidic rather than neutral?

2. Why does the α,α-trehalose content of baker's yeast decrease on storage?

3. Write a stepwise reaction mechanism for acid-catalyzed hydrolysis of α,α-trehalose to D-glucose.

4. Would α,α-trehalose be expected to undergo mutarotation? Explain.

5. Why does α,α-trehalose give negative Fehling's and Benedict's tests?

25.5 Carbohydrates: Their Characterization and Identification

The determination of the complete structure of an unknown monosaccharide was a formidable problem to the early organic chemists owing to the stereochemical questions that had to be answered. For example, it was insufficient to know that, in its open-chain form, (+)-glucose was a 2,3,4,5,6-pentahydroxyhexanal; rather, the configuration at each of the four chiral carbon atoms had to be defined. Furthermore, once it was realized that intramolecular hemiacetal or hemiketal formation occurred, it became necessary to determine the size of the ring that was formed; that is, did the sugar exist in the form of a five-membered furanose or six-membered pyranose ring? Nevertheless, despite these difficulties, the complete structures of most monosaccharides containing as many as seven carbon atoms have now been determined.

If the question of structure elucidation is extended to polysaccharides, the complexity of the problem is greatly increased. Such questions as how many and which monosaccharides constitute the unknown, which carbon atoms are bonded to the oxygen atom that serves as the intermolecular linkage between the monosaccharide units, and the stereochemistry of linkages at the anomeric carbon atoms require answers. The size of the ring formed by each monosaccharide contained in the polysaccharide must also be determined, since the ring size adopted by a particular monosaccharide unit in a polysaccharide may not be the same as the size found for the monosaccharide itself!

Organic chemists have developed a variety of techniques for eliciting information about unknown carbohydrates that will aid in their systematic characterization. The following paragraphs discuss some of the various possible tests that may be used for the classification of carbohydrates according to structural types. Note that the judicious interpretation of the results of these tests and other more specific ones can provide much, if not all, of the information required to prove the structure of an unknown carbohydrate.

A. Reducing Sugars

All monosaccharides and *many* disaccharides are **reducing sugars.** This property is specifically based on the presence of an aldehyde or α-hydroxyketo group that may participate in an oxidation-reduction reaction with various oxidizing agents. Although all aldoses of four or more carbon atoms exist predominantly in cyclic hemiacetal form, their reversible equilibrium with the aldehydo form allows the reaction to proceed, with a shift in the equilibrium (see Figure 25.3). Disaccharides in which one of the rings is a hemiacetal are reducing sugars for the same reason (equation 6).

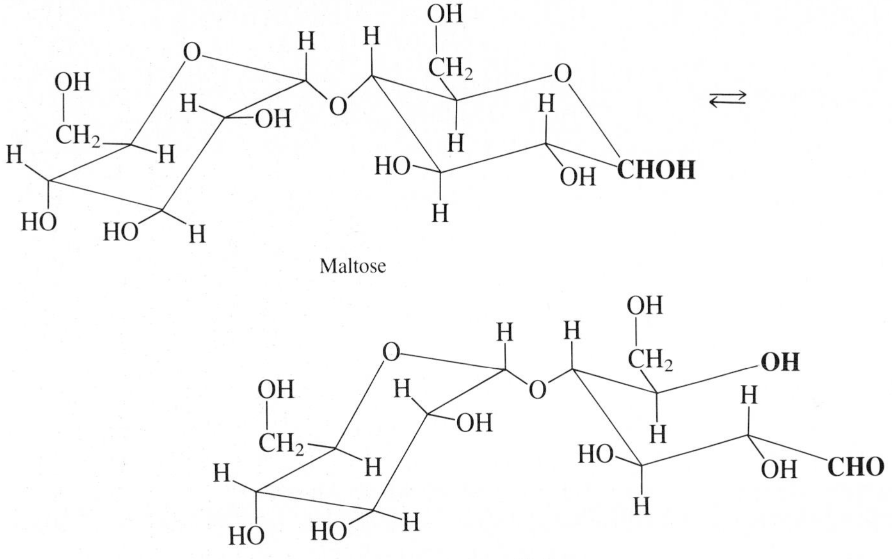

Maltose

(6)

Maltose

Disaccharides in which both rings are in acetal or ketal form are *not* reducing sugars because they cannot be in equilibrium with aldehydic or keto forms. Note, for example, that in sucrose **(11)** the potential aldehyde and α-hydroxyketo functions are both tied up in the glycosidic, intermonosaccharidic linkage.

Ketoses yield positive tests for reducing sugars because of the base-catalyzed tautomerization of the α-hydroxyketo functionality. Thus, for example, the base-catalyzed enolization of fructose, a nonreducing sugar, provides an enediol that may reketonize to provide a mixture of glucose and mannose, both of which are reducing sugars. Equation 7 shows with partial structures the processes that occur.

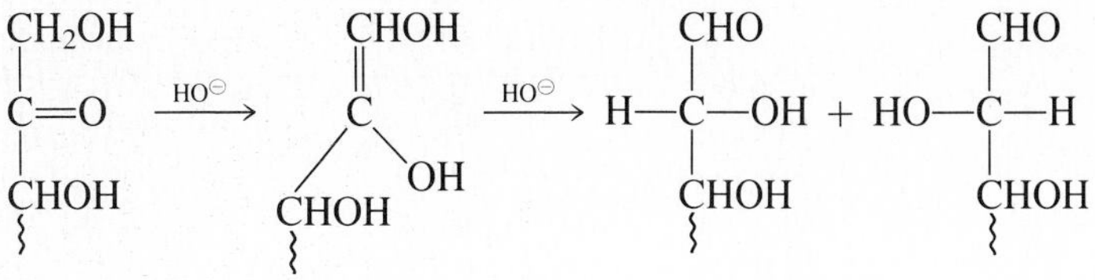

A ketose   An enediol

(7)

Note that each of the tests for reducing sugars is carried out under basic conditions that allow the equilibria shown in equation 7 to be established.

Tollens' Test   This procedure is included within the classification tests for aldehydes in Section 27.5, Part A2.

Benedict's Test   The reagent that is used in this test is prepared as a solution of cupric sulfate, sodium citrate, and sodium carbonate. The citrate is added so that a cupricitrate complex ion is formed that prevents $Cu(OH)_2$ from precipitating from the basic solution. Cupric ion serves as an oxidizing agent of aliphatic aldehydes, including α-hydroxyaldehydes as in aldoses, but not of aromatic aldehydes.[6] A positive test for these types of aldehydes is evidenced by the formation of a yellow to red precipitate (equation 8) of cuprous oxide, Cu_2O. The yellow precipitate occasionally observed has apparently not been characterized; its formation seems to depend on the amount of oxidizing agent present.

$$\text{RCHO} + 2\ \text{Cu}^{2+} + 5\ \text{HO}^{\ominus} \xrightarrow{\text{citrate}} \text{RCO}_2^{\ominus} + \underset{\substack{\text{brick-}\\\text{red}}}{\text{Cu}_2\text{O}} + 3\ \text{H}_2\text{O} \tag{8}$$

EXPERIMENTAL PROCEDURE

Benedict's Test   A stock solution of Benedict's reagent can be prepared by dissolving 86.5 g of hydrated sodium citrate and 50 g of anhydrous sodium carbonate in 400 mL of water. To this solution is added with stirring a solution of 8.65 g of cupric sulfate in 50 mL of water. This solution is diluted to 500 mL and filtered if necessary.

Place about 0.2 g of the compound in a test tube and add *ca.* 5 mL of water. Stir to dissolve and then add 5 mL of Benedict's reagent. Heat the solution to boiling. The formation of a yellow (green when viewed in the blue solution of the reagent) to red precipitate is a positive test for aliphatic aldehydes and α-hydroxyaldehydes. For comparison, perform the test simultaneously on an unknown, on glucose, and on sucrose.

Barfoed's Test for Monosaccharides

As is true of the tests just described, this test likewise depends on the reducing properties of the saccharides being tested. The conditions for it, however, allow a significant selectivity between monosaccharides and disaccharides. The test reagent consists of an aqueous solution of cupric acetate and acetic acid. Thus, in contrast to the previous tests, the reaction is carried out under *acidic* conditions. A positive test for monosaccharides is constituted by the formation of the brick-red precipitate of Cu_2O within *two* or *three* minutes. Disaccharides require a longer time, providing the precipitate only after about ten minutes or more. Nonreducing sugars, *e.g.*, sucrose, apparently undergo slow hydrolysis under the test conditions, as they also give a precipitate after *extended* time. It is not fully clear why reducing disaccharides oxidize more slowly than monosaccharides; however, the test is based on this fact.

[6]Morrison, J. D. *Journal of Chemical Education,* **1965,** *42,* 554.

EXPERIMENTAL PROCEDURE

Barfoed's Test   A stock solution of Barfoed's reagent can be prepared by dissolving 66 g of cupric acetate and 10 mL of glacial acetic acid in water and diluting this solution to 1 L.

Place 3 mL of Barfoed's reagent and 1 mL of a 1% sugar solution in a test tube; place the test tube in a beaker of boiling water for *5 min (no longer)*. Remove the test tube and cool it under running water. A red precipitate of cuprous oxide is a positive test. It may be necessary to view the tube against a dark background in good light. For comparison, run tests simultaneously on the unknown, on glucose, and on lactose. Record the results.

B. Formation of Osazones

Most carbohydrates, particularly when impure, crystallize from solution only with difficulty, tending instead to form syrups; this frequently makes their direct characterization difficult. Fortunately, many carbohydrates react with phenylhydrazine to form bright yellow crystalline derivatives called *osazones*. These derivatives may usually be identified readily by both their melting points or temperatures of decomposition and their crystalline forms.

Not all sugars form osazones, for only those that have an aldehydo or keto carbonyl group, either free or in equilibrium with a hemiacetal or hemiketal, will react with phenylhydrazine. For example, α- and β-D-glucose (**8** and **9**) in water solution are in dynamic equilibrium with the ring-opened form **10,** which, because of its aldehydo group, will react with phenylhydrazine (see Section 25.5a).

Equation (9) shows the formation of glucosazone **(15),** the osazone derivative of glucose. Note that three equivalents of phenylhydrazine are required and that two of these reagent molecules are incorporated into the osazone structure. Additional products include aniline and ammonia.

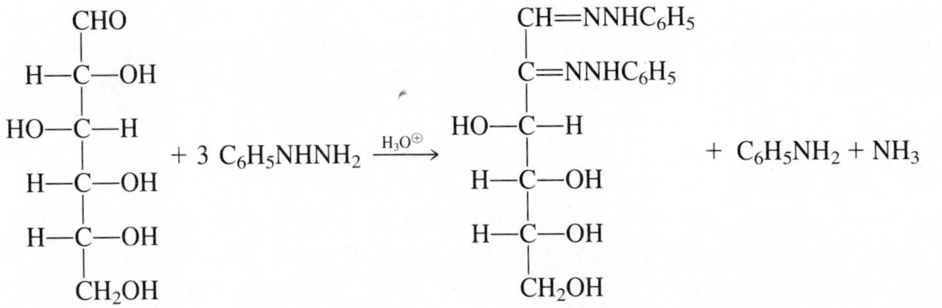

10	15
D-Glucose	Osazone of D-glucose

The accepted mechanism for the reaction is presented in Figure 25.5. Following the formation of a phenylhydrazone, **16,** an internal oxidation-reduction reaction is accomplished by tautomeric migration of two hydrogens from C-2 to the hydrazone moiety to give **17.** The newly formed carbonyl group condenses with a second equivalent of phenylhydrazine to give **18,** which undergoes subsequent tautomerization to **19.** Following a 1,4-elimina-

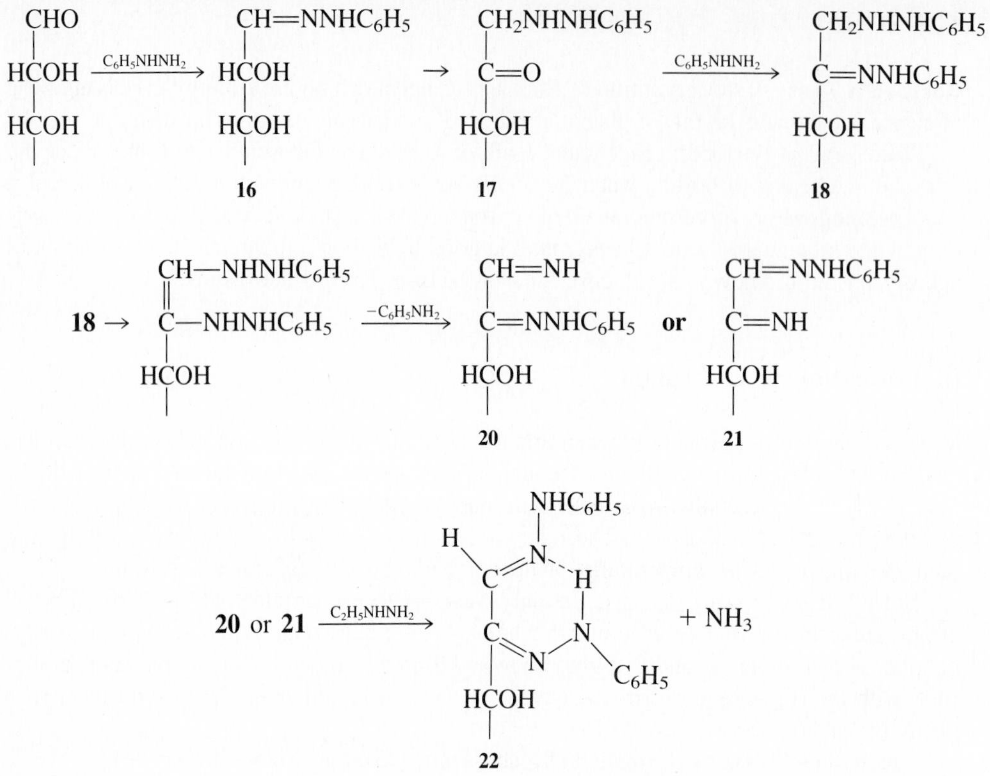

$$\underset{16}{\overset{\text{CHO}}{\underset{\big|}{\text{HCOH}}}\underset{\big|}{\text{HCOH}}} \xrightarrow{\text{C}_6\text{H}_5\text{NHNH}_2} \underset{\big|}{\overset{\text{CH}=\text{NNHC}_6\text{H}_5}{\underset{\big|}{\text{HCOH}}}\underset{}{\text{HCOH}}} \rightarrow \underset{17}{\overset{\text{CH}_2\text{NHNHC}_6\text{H}_5}{\underset{\big|}{\text{C}=\text{O}}}\underset{\big|}{\text{HCOH}}} \xrightarrow{\text{C}_6\text{H}_5\text{NHNH}_2} \underset{18}{\overset{\text{CH}_2\text{NHNHC}_6\text{H}_5}{\underset{\big|}{\text{C}=\text{NNHC}_6\text{H}_5}}\underset{}{\text{HCOH}}}$$

$$\mathbf{18} \rightarrow \underset{\big|}{\overset{\text{CH}-\text{NHNHC}_6\text{H}_5}{\text{C}-\text{NHNHC}_6\text{H}_5}}\text{HCOH} \xrightarrow{-\text{C}_6\text{H}_5\text{NH}_2} \underset{20}{\overset{\text{CH}=\text{NH}}{\underset{\big|}{\text{C}=\text{NNHC}_6\text{H}_5}}\underset{}{\text{HCOH}}} \quad \textbf{or} \quad \underset{21}{\overset{\text{CH}=\text{NNHC}_6\text{H}_5}{\underset{\big|}{\text{C}=\text{NH}}}\underset{}{\text{HCOH}}}$$

$$\mathbf{20 \text{ or } 21} \xrightarrow{\text{C}_2\text{H}_5\text{NHNH}_2} \quad \mathbf{22} \quad + \text{ NH}_3$$

Figure 25.5 Mechanism for osazone formation.

tion of aniline, which produces either **20** or **21,** a third equivalent of phenylhydrazine condenses with the imine group to give the osazone, **22,** and ammonia. Although it may appear that **22** should undergo further reaction by intramolecular oxidation-reduction between the secondary alcohol group at C-3 and the hydrazone group at C-2, the reaction stops at this point so that only two phenylhydrazine units are introduced. The formation of the intramolecular hydrogen bond shown in **22** has been established as the cause of limitation of the reaction to the first two carbons of the chain.[7]

As indicated above, osazones may frequently be identified by their crystalline forms. Osazones from different saccharides tend to form highly distinctive crystalline clusters that may most conveniently be observed under a microscope. Strong supporting evidence for the identification of an unknown saccharide may be collected by comparing the osazone of an unknown with those of several known saccharides. Further, under closely controlled conditions, the time required for precipitation of many osazones may be significant in tentative identification of the saccharide.

[7]When 1-methylphenylhydrazine [$H_2NN(CH_3)C_6H_5$] is used in place of phenylhydrazine, the reaction proceeds down the chain readily at least as far as C-5. Note that in this reagent, the N—H proton involved in the hydrogen bonding in **22** has been replaced with a methyl group. Chapman, O. L.; Welstead, W. J., Jr.; Murphy, T. J.; King, R. W. *Journal of the American Chemical Society,* **1964,** *86,* 732, 4968.

EXPERIMENTAL PROCEDURE

Using the procedure described below, attempt preparation of osazone derivatives of D-glucose, D-fructose, and sucrose. It will be most efficient to carry out these reactions simultaneously.

Heat a large beaker of water to boiling. In each of three test tubes, separately dissolve 0.2 g of one of the above sugars in 4 mL of water and 0.5 mL of saturated sodium bisulfite solution (this prevents oxidation of the phenylhydrazine during the reaction and the contamination of the osazone by the resulting tarry products), then add to this solution *either* 0.6 mL of glacial acetic acid, 0.6 g of sodium acetate, and 0.4 g of phenylhydrazine *or* 0.6 g of sodium acetate and 0.6 g of phenylhydrazine hydrochloride. (*Caution:* Phenylhydrazine is toxic, and care should be exercised in its handling.) Stir the solutions thoroughly, place the test tubes in the beaker of boiling water, and discontinue heating. Allow the test tubes to remain in the water bath for 30 min.

After *ca.* 30 min, remove the test tubes, cool the contents to room temperature, and collect any precipitates by vacuum filtration. Recrystallize each of the osazones obtained from ethanol-water and determine their melting points. Determine the melting point of a *mixture* of the osazone derivatives of glucose and fructose. Insofar as the melting points of osazones may depend on the rate of heating, it is advisable to carry out these melting-point determinations simultaneously.

Prepare an osazone derivative of the product of hydrolysis of sucrose. Substitute 4 mL of the solution that was diluted for rotation measurements (Section 25.3) for the 0.2 g of sugar and 4 mL of water used in the above procedure. Determine the melting points of this osazone *and* of a mixture of this osazone with the osazone of fructose.

EXERCISES

1. What conclusions can be drawn from the results of the mixture melting-point determinations on the osazones of glucose and fructose?
2. Answer Exercise 1 for the osazones from fructose and the hydrolysis products of sucrose.

amino acids and peptides

26.1 Introduction

The amino acids constitute a highly important class of naturally occurring organic compounds. They are the monomeric units that are joined through amide linkages, called *peptide bonds,* to produce the important biopolymers, the proteins **(1),** on which every living system depends.

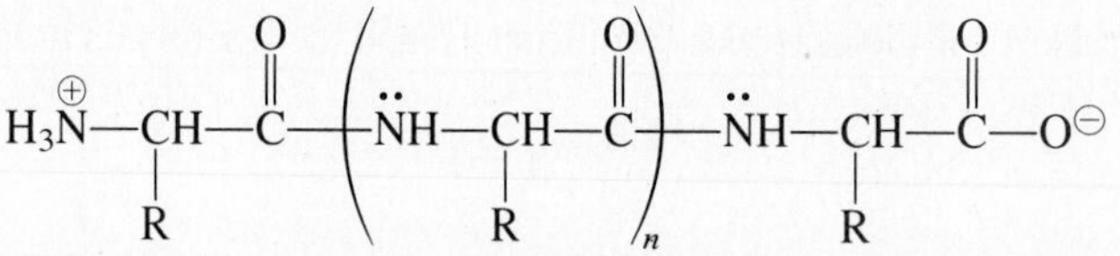

$$
\overset{\oplus}{H_3N}-CH-\overset{\displaystyle\overset{O}{\|}}{C}\!\!\left(-\ddot{N}H-CH-\overset{\displaystyle\overset{O}{\|}}{C}\right)_{\!\!n}\!\!-\ddot{N}H-CH-\overset{\displaystyle\overset{O}{\|}}{C}-O^{\ominus}
$$

$$
\quad\;\; R \qquad\qquad\quad R \qquad\qquad\quad R
$$

1

Proteins are polyamides in the molecular weight range above 5000; those polyamides of molecular weight below 5000 are more usually referred to as polypeptides. These types of compounds serve a variety of biological functions. Some, the *fibrous proteins,* compose such tissues as hair, skin, and muscle fiber. They possess quite appreciable mechanical strength, are insoluble in water, and chemically are relatively inert. Fibrous proteins generally possess very high, somewhat indefinite molecular weights and are sometimes polymerlike. The *globular proteins* and smaller natural *peptides* have much smaller molecular weights and exist as discrete chemical entities, often obtainable in crystalline form. They are water-soluble and have characteristic reactivity. The globular proteins serve a variety of roles ranging from catalytic functions **(enzymes)** and overall regulatory functions **(hormones)** through immunological defense functions (**antibodies**). To a limited extent, differences in function are reflected in differences in molecular weight, as with fibrous and globular proteins, for example. However, particularly among the globular proteins, differences in biological properties are more completely determined by the exact sequence of different amino acids in the peptide chain: the *primary structure*.

The number of such possible arrangements is vast. For example, in a pentapeptide (**1**, $n = 3$) composed of five *different* amino acids, the number of different sequential arrangements of amino acids is 120; that is, 120 different primary structures are possible. Each of these, in principle, would possess different biochemical reactivities. Moreover, it is not possible to predict the biological properties of a peptide or protein solely on the basis of its primary structure, as these properties are made more complex as a result of the three-dimensional structure of the polyamide chain.

Although each individual amide linkage is coplanar, owing to conjugation as shown in **2**, conformational differences in structure may arise through rotation about the remaining single bonds, namely, those to C_α in **2.** These rotations allow the chain to coil and to achieve

2

stabilization through hydrogen-bonding between amido hydrogens and carbonyls on separated peptide units. Such coiling constitutes the *secondary structure* of peptides. Furthermore, owing to convolutions and gross foldings of the coiled chain, amino acid residues (peptide units) in widely separated positions of the chain may be brought into close proximity so as to act in concert and provide the peptide with its characteristic reactivity. This folding, which may be the result of a variety of structural influences, constitutes the *tertiary structure* of the peptide. Finally, the spatial relationship of one polypeptide chain to another results in the *quaternary structure* of the overall protein structure.[1] For example, the tobacco mosaic virus, with an overall molecular weight of 41,000,000, is composed of many identical polypeptide subunits, each with a molecular weight of 17,500, held together by noncovalent interactions.

A complete understanding of the biochemical behavior of a peptide from a molecular and mechanistic point of view must depend on the determination of its total structure. Of the various levels of structural complexity, determination of the primary structure, called *sequencing,* must be deemed the most important. This is a logical consequence of the realization that the higher degrees of structural complexity are in the first instance dependent on the primary structure. It is this sequence that is genetically coded in DNA. Determination of the primary structure requires initial knowledge of the numbers of each kind of amino acid involved in the chain. This may be accomplished by the total hydrolysis of the peptide to provide a mixture of the amino acids constituting the structure (equation 1).

---

[1]See any modern organic textbook for further information about the three-dimensional structural properties of peptides.

$$\overset{\oplus}{N}H_3-CH-\overset{\overset{O}{\parallel}}{C}-\overset{..}{N}H-CH-\overset{\overset{O}{\parallel}}{C}-\overset{..}{N}H-CH-\overset{\overset{O}{\parallel}}{C}-O^{\ominus} \xrightarrow{H_3O^{\oplus}} \overset{\oplus}{N}H_3-CH-CO_2H$$

$$\underset{R}{\qquad} \underset{R'}{\qquad} \underset{R''}{\qquad} \underset{R}{\qquad} \tag{1}$$

$$+ \overset{\oplus}{N}H_3-CH-CO_2H + \overset{\oplus}{N}H_3-CH-CO_2H$$
$$\underset{R'}{\qquad} \underset{R''}{\qquad}$$

Qualitative and quantitative analyses of this mixture can be used to determine the identities and relative numbers of each amino acid present. If the molecular weight is known, then the exact numbers of each amino acid in the chain may be determined. For example, if the hydrolysis of a peptide of unknown structure provided a mixture of amino acids analyzed to contain only alanine (ala) and glycine (gly, see Table 26.1) in a ratio of 2:1, respectively, the peptide could be a tripeptide (ala$_2$, gly), a hexapeptide (ala$_4$, gly$_2$), and so on. The abbreviated formulae in parentheses show this ratio; the comma between the specific amino acids present, by convention, indicates that the *sequence* of these units in the peptide is *unknown*. If the unknown peptide was found to have a molecular weight of approximately 200, a general structure of unknown sequence would be established. Note that the molecular weight of (ala$_2$, gly) is 203 (2 ala + gly − 2 H$_2$O), whereas the molecular weight of (ala$_4$, gly$_2$) is 388 (4 ala + 2 gly − 5 H$_2$O). Experimental approaches to the determination of sequence will be discussed in Section 26.3.

Peptide bonds may be hydrolyzed under either acid- or base-catalyzed conditions. Although both procedures have some disadvantages, the acid-catalyzed hydrolysis is preferable, primarily because alkaline conditions result in extensive racemization of the chiral center at the α-position as well as in degradation of some amino acid residues, for example, arginine (arg) and threonine (thr, see Table 26.1). Although certain amino acid residues are sensitive to acid and undergo partial destruction during acid-catalyzed hydrolysis, as with serine and tryptophan, for example, these effects are well understood, and quantitative corrections may be applied during careful work in the research laboratory. The mechanism of acid hydrolysis of the peptide linkage, which is simply an amide bond, is qualitatively similar to that for the hydrolysis of an ester.

Total hydrolysis of a peptide is normally accomplished by treatment with 6 M HCl at 100–110°C over a period of 16–20 hr. The hydrolysis is effected in a sealed tube in order to avoid evaporation and the charring that would result. An experiment involving peptide hydrolysis in the determination of structure of an unknown dipeptide is included in the experimental part of Section 26.3.

26.2 Analysis of Amino Acids

Although the number of conceivable structures containing both amino and carboxylic acid functional groups on the same carbon atom is vast indeed, fortunately only 20 or so are actually found in polypeptides from living sources. Nearly all these are α-amino acids of the type shown in **3**. The occasional exception contains an α-amino function as part of a ring as in proline (pro, see Table 26.1). All have an α-hydrogen, so that the α-carbon atom is chiral,

TABLE 26.1 The Common Amino Acids

Name	Abbreviation	Formula	Isoelectric Point	Color from Pyridine-Isatin Reagent	Numerical Key to Figures 26.3 and 26.4
Alanine	ala	$CH_3CH(NH_2)CO_2H$	6.0	Pink-red	10
Arginine	arg	$\underset{H_2N}{\overset{HN}{}}C{-}NH(CH_2)_3CH(NH_2)CO_2H$	11.2	Deep pink	4
Asparagine	asn	$NH_2COCH_2CH(NH_2)CO_2H$	5.4	...	...
Aspartic acid	asp	$HO_2CCH_2CH(NH_2)CO_2H$	2.8	Bright red	6
Cysteine	cys	$HSCH_2CH(NH_2)CO_2H$	5.1	Yellow-brown	1
Glutamic acid	glu	$HO_2C(CH_2)_2CH(NH_2)CO_2H$	3.2	Bright red	9
Glutamine	gln	$NH_2CO(CH_2)_2CH(NH_2)CO_2H$	5.7	...	...
Glycine	gly	$H_2NCH_2CO_2H$	6.0	Orange-red	7
Histidine	his	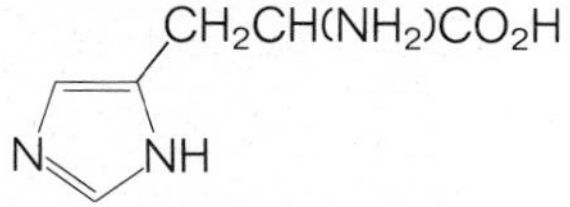	7.5	Orange-red	3
Isoleucine	ile	$CH_3CH_2CH(CH_3)CH(NH_2)CO_2H$	6.0	Bright red	16
Leucine	leu	$(CH_3)_2CHCH_2CH(NH_2)CO_2H$	6.0	Orange-red	18
Lysine	lys	$NH_2(CH_2)_4CH(NH_2)CO_2H$	9.6	Red	2
Methionine	met	$CH_3S(CH_2)_2CH(NH_2)CO_2H$	5.7	Pink	14
Phenylalanine	phe	$C_6H_5CH_2CH(NH_2)CO_2H$	5.5	Red-brown	17
Proline	pro	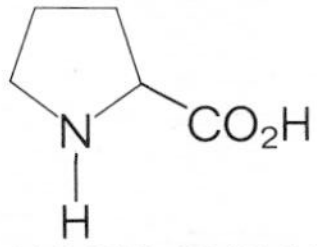	6.3	Intense blue	11
Serine	ser	$HOCH_2CH(NH_2)CO_2H$	5.7	Pink	5
Threonine	thr	$CH_3CH(OH)CH(NH_2)CO_2H$	5.6	Pink	8
Tryptophan	try	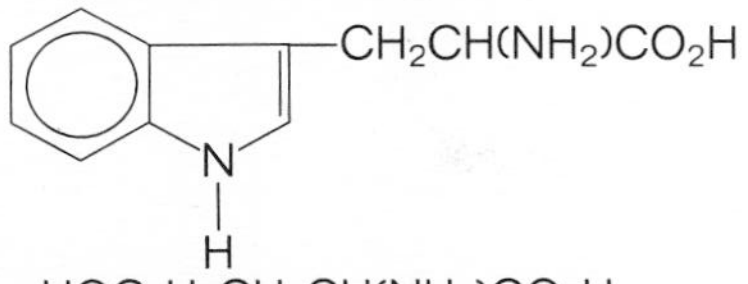	5.9	Red-brown	15
Tyrosine	tyr	$p\text{-}HOC_6H_4CH_2CH(NH_2)CO_2H$	5.7	Light brown	12
Valine	val	$(CH_3)_2CHCH(NH_2)CO_2H$	6.0	Red	13

except in the case of glycine, the simplest α-amino acid, in which there are two α-hydrogen atoms. With the exception of glycine and of a few D-amino acids derived from microorganisms, all the important amino acids found in polypeptides from living sources are of the L-configuration **(4)**. Table 26.1 includes many of the common, naturally occurring amino acids.

$$\overset{\oplus}{H_3N}\text{—}\overset{*}{CH}\text{—}CO_2^{\ominus} \qquad \overset{\overset{\oplus}{NH_3}}{\underset{R}{\overset{H}{\diagdown}}}C\text{—}CO_2^{\ominus}$$

$$\underset{R}{|}$$

$$\mathbf{3} \qquad\qquad\qquad \mathbf{4}$$

(a) Amino Acids as Acids and Bases

Note that the amino acids have both basic and acidic functional groups. As a result, in their crystalline forms they exist as the *zwitterions,* **3,** which are internal salts. Consequently, they are high-melting solids that are generally insoluble in organic solvents but soluble in water.

Because of the acidic and basic character of an amino acid, there are established in aqueous solution pH-dependent equilibria among the forms shown in equation 2.

$$\overset{\oplus}{NH_3}\text{—}\underset{\underset{R}{|}}{CH}\text{—}CO_2H \rightleftarrows \overset{\cdot\cdot}{NH_2}\text{—}\underset{\underset{R}{|}}{CH}\text{—}CO_2H \rightleftarrows$$

$$\mathbf{5} \qquad\qquad\qquad \mathbf{6}$$

$$\overset{\oplus}{NH_3}\text{—}\underset{\underset{R}{|}}{CH}\text{—}CO_2^{\ominus} \rightleftarrows \overset{\cdot\cdot}{NH_2}\text{—}\underset{\underset{R}{|}}{CH}\text{—}CO_2^{\ominus} \qquad \textbf{(2)}$$

$$\mathbf{7} \qquad\qquad\qquad \mathbf{8}$$

The equilibria are displaced toward **8** in more alkaline solutions, whereas in more acidic solutions **5** becomes more predominant. The equilibrium between **6** and **7** results in no change in hydrogen ion concentration, so that the ratio of **7** to **6** in solution is pH-*independent*. Thus, the pH-*dependent* component of the equilibria in equation 2 is the relative concentrations of the species **5** and **8.** If electrodes are placed in a solution of an amino acid, there will be a net migration of the solute toward either the cathode or the anode, depending on whether **5** or **8,** respectively, is predominant.

At a certain pH, specific for each amino acid, the concentrations of **5** and **8** will be equal, and there will be *no* net migration of the solute toward the electrodes. This pH value is called the *isoelectric point*. Table 26.1 lists the isoelectric points for the common amino acids. Note that the values fall into three ranges: 2–3 for those amino acids containing additional acid groups as part of the side-chain R (the acidic amino acids), 5.5–6.5 for those amino acids containing neutral side chains, and 9–11 for those amino acids containing an additional basic site in the side chain (the basic amino acids). Amino acids have their *minimum* solubility in water at the isoelectric point.

Chromatographic techniques are utilized nearly universally for the analysis of mixtures of amino acids. In order of importance, as gauged from work carried out in research laboratories, these procedures involve ion-exchange chromatography, paper chromatography, and thin-layer chromatography. Because amino acids are colorless, each of these techniques necessarily requires methods of detecting the separated amino acids. The most important detecting agent in use is ninhydrin (**9**).

(b) The Ninhydrin Color-Forming Reaction of Amino Acids

Ninhydrin (**9**) reacts with amino acids of type **3** to produce characteristic blue-violet colors. The sensitivity and reliability of the test are such that 0.1 μmol of amino acid gives a color intensity that is reproducible to a few percent, so long as a reducing agent such as stannous chloride is present to prevent oxidation of the colored salt by dissolved oxygen. Consequently, this color-forming reaction can be used for quantitative as well as qualitative purposes. Although not all amino acids give the same color (for example, proline gives a pale-yellow color), most do, indicating that the colored product formed is the same in most cases, irrespective of the structure of the original amino acid. The sequence of steps involved in the color-forming reaction is shown in Figure 26.1.

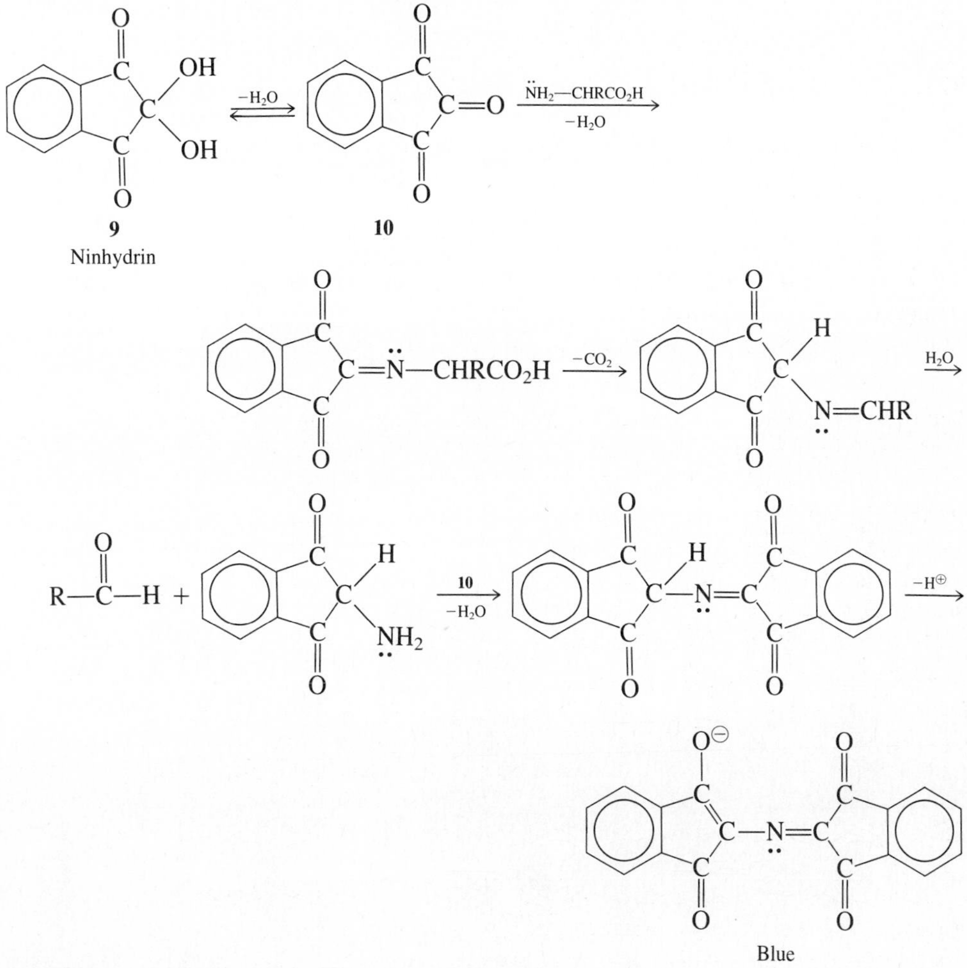

Figure 26.1 Reactions involved in the ninhydrin color test.

(c) The Isatin Color-Forming Reaction of Amino Acids

Isatin (**11**) also reacts with α-amino acids to produce compounds having a range of colors. As can be seen by reference to Table 26.1, the color that is generated is characteristic of the particular amino acid that is present, and such information can be useful in making tentative assignments of structure, particularly in cases of amino acids having similar R_f values but different colors with the isatin reagent. The fact that a variety of colors is observed suggests that a number of different types of products are produced in this color-forming reaction, in contrast to the corresponding reaction with ninhydrin (**9**); this makes visualization of amino acids with isatin unsuitable if quantitative measurements are desired. The structure of one such product, that derived from proline and isatin, and an abbreviated mechanism for its formation are shown in equation 3.

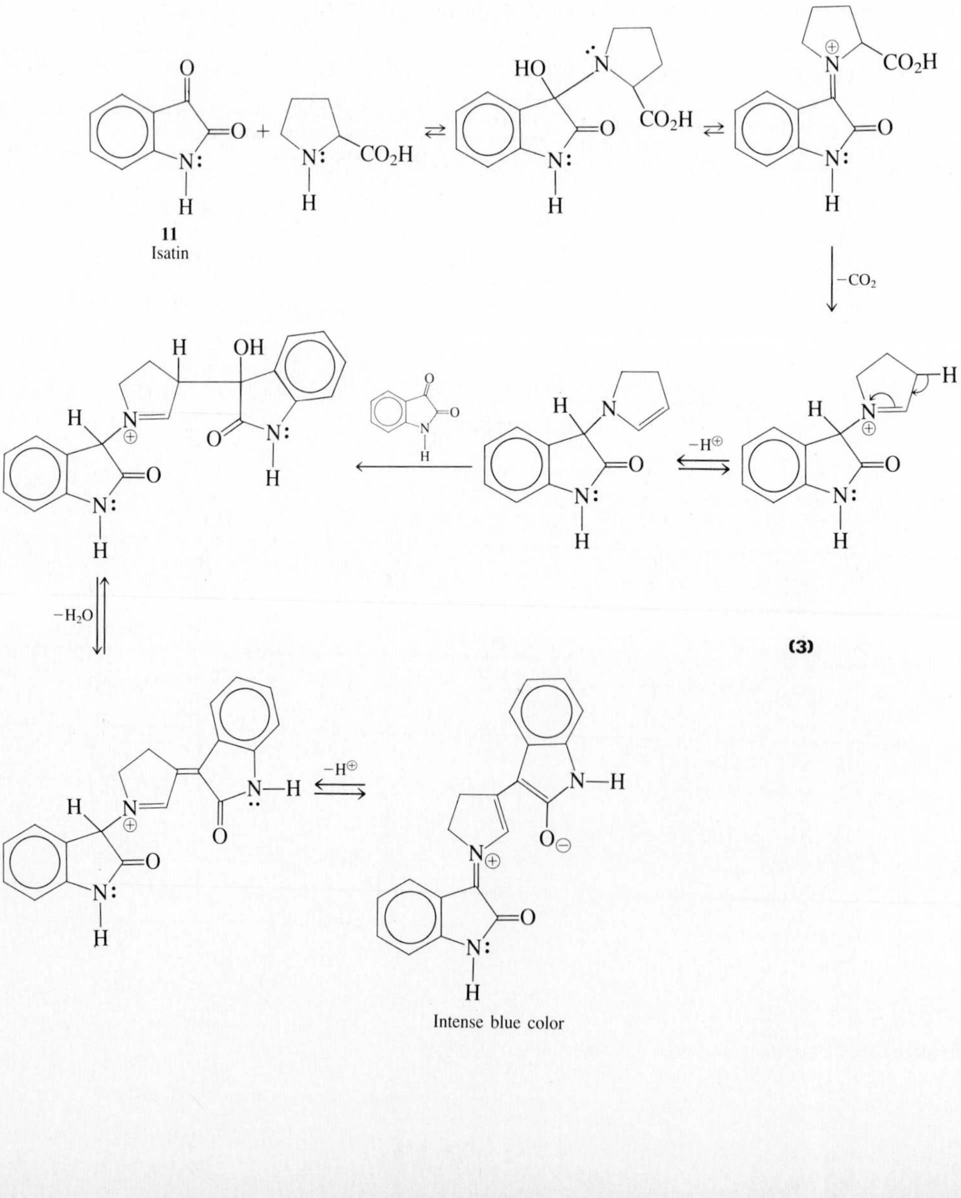

11
Isatin

(3)

Intense blue color

(d) The Automatic Amino Acid Analyzer

In the past, *quantitative* amino acid analyses were highly time-consuming and extremely tedious and required considerable amounts of peptide (about 25 g of protein for a full amino acid analysis). More recently, however, the use of ion-exchange chromatography in conjunction with the *automatic amino acid analyzer* has revolutionized the practice of protein analysis. A complete analysis on automated, microprocessor-controlled equipment may be completed in 4–5 hr on no more than 0.1 μmole of protein! The amino acids are separated by elution ion-exchange chromatography in which the eluent is passed through the column at a constant rate. Because the amino acids elute at different rates, the flow of eluent at the base of the column contains different amino acids at different elapsed-time intervals. As it leaves the column, the effluent is admixed with a solution of ninhydrin and then passed through a Teflon tube immersed in a boiling-water bath to speed up the color-forming reaction (Figure 26.1). The eluent stream then continues through a photoelectric colorimeter that continuously measures the color intensity; an electronic recorder is used to record this intensity as a function of time. Because the color intensity in the effluent stream as a function of time is directly related to the elution times for the various amino acids, the recorder produces a chromatogram that is qualitatively similar to a gas chromatogram (see Figure 6.9). When correction factors are applied that relate the sensitivity of each amino acid to the ninhydrin color-forming reaction, the areas under the peaks are proportional to the molar ratios of the amino acids in the original mixture. Thus, this procedure allows both qualitative and quantitative analyses of mixtures of amino acids.

The ion-exchange resin consists of a sulfonated cross-linked polymer produced by copolymerization of styrene and *para*-divinylbenzene (Figure 26.2).

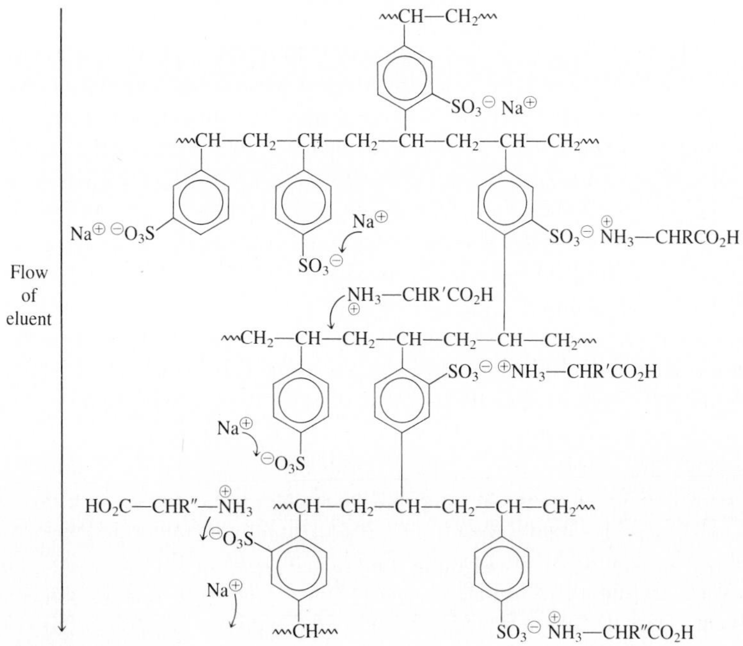

Figure 26.2 Cationic exchange chromatography of amino acids. The arrows represent displacement of one ion by another as the eluent passes down the column.

The sample is applied as a solution buffered at pH 2 to a column containing this resin. At this pH, which is below the isoelectric point of all amino acids, the amino acids are present in the conjugate acid form, **5.** Acting as cations, they displace sodium ions and are held at the head of the column by the sulfonate groups of the resin. An empirical pattern of elution involving aqueous buffered solutions of sodium ion of increasing pH has been worked out that allows separation of all the common amino acids as a result of their individual abilities to be displaced from the resin by sodium ion. Although the factors that control resolution of the amino acids are complex, to a first approximation the differential elution pattern of two amino acids is controlled by the relative extent to which, at a given pH, they are present in the cationic form **5.** The more basic amino acids would therefore elute at higher pH values than the neutral or acidic amino acids.

(e) Paper Chromatography of Amino Acids

Paper chromatography is an especially valuable tool for the relatively rapid qualitative analysis of mixtures of amino acids. As noted in Chapter 6, paper chromatography is a type of partition chromatography in which the substrate is partitioned between the water, which is tightly bound within the cellulose fibers of paper (the stationary phase) and an organic solvent (the mobile phase), which is allowed to migrate upward along the paper by capillary action. Those acids having the highest solubility in the organic solvent relative to their solubility in water will have the greatest mobility and will migrate upward on the paper most rapidly. Because the various amino acids will migrate at different rates, they will separate at different vertical displacements on the paper. Although the R_f values (see Figure 6.3) that may be used to identify the amino acid components present in the mixture are fairly reproducible, there is sufficient variation in the quality and type of paper used, in the temperature from run to run, and in the purity and exact composition of solvent as to make sole reliance on the values published by other workers somewhat risky. Consequently, it is standard practice to run samples of known amino acids simultaneously with the unknown mixture so that the R_f values under identical conditions can be compared.

A very large variety of organic solvent systems and types of paper have been investigated for the purpose of separating mixtures of amino acids. The solvent and paper used are generally defined by the nature of the particular determination to be performed; that is, the experimental conditions are dependent on which amino acids are present in the mixture.

Because the separability of different amino acids depends greatly on the solvent system used, it is frequently found that some amino acids will separate into individual spots while others will remain unresolved in overlapping spots (see Figure 26.3a). Two-dimensional paper chromatography is used to overcome the problem of overlapping. In this procedure the mixture is spotted on the paper in the lower left-hand corner, and the paper is developed with one solvent. The paper is then removed from the developing chamber and allowed to dry. It is then rotated 90° relative to the original direction of solvent flow and developed with a different solvent, chosen for its ability to separate the amino acids that did not separate with the first solvent. Thus the spots are displaced in two directions from the original spot rather than in one.

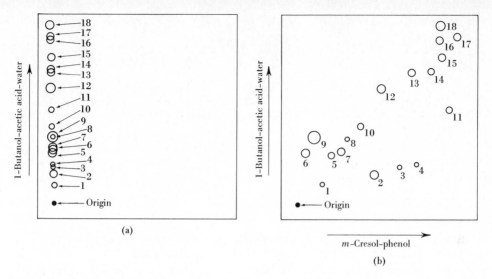

Figure 26.3 Paper chromatography of a mixture of 18 amino acids. (a) In one dimension. (b) In two dimensions using different solvents. The number corresponding to each of the amino acids is keyed in Table 26.1.

Figure 26.3 shows idealized reproductions of actual paper chromatograms. In Figure 26.3a a mixture of 18 amino acids has been developed in one dimension using a mixture of 1-butanol–acetic acid–water as solvent in a ratio of 4 : 1 : 5 (by volume). Note that although there is significant separation of amino acids, several of the spots remain unresolved and overlapped [serine (5), aspartic acid (6), and glycine (7), for example]. In Figure 26.3b the same chromatogram is rotated 90° and developed with m-cresol–phenol in a 1 : 1 ratio (by weight) buffered with a borate solution of pH 9.3. Observe that all the spots have now been resolved by the two-dimensional chromatography. The numbers that identify the amino acid responsible for each spot are keyed in Table 26.1.

Although two-dimensional paper chromatography provides greater resolution of the amino acids in a mixture, it has the disadvantage of allowing only one sample to be run at a time. Because it is desirable to run samples of known amino acids simultaneously with the unknown to facilitate identification, the following experiment involves only one-dimensional chromatography. The solvent system 1-butanol–acetic acid–water (4 : 1 : 5) is quite good for this purpose and is the one used.

Pre-lab exercises for Section 26.2, Analysis of Amino Acids, can be found on page PL. 141.

EXPERIMENTAL PROCEDURE[2]

> **DO IT SAFELY**
>
> **1** The solvent mixture used in the development of the paper chromatogram in this experiment may be somewhat irritating to nose and eyes. Avoid excessive exposure to these vapors and work at a hood if possible.
>
> **2** When spraying the chromatogram with either the ninhydrin or the isatin reagent, be careful neither to breathe the aerosol mist nor to allow contact with your skin. Spraying should be done in a hood behind an appropriate shield. It is advisable to wear rubber gloves. If these reagents do contact the skin, wash the affected area at once with soap and warm water.

Pour approximately 20 mL of solvent consisting of a 4:1:5 (by volume) mixture of 1-butanol, glacial acetic acid, and water, respectively, into an 800-mL or 1-L beaker; avoid splashing the liquid onto the sides of the beaker. The depth of solvent in the beaker should be about 6 mm. Cover the beaker with a watch glass. If a piece of cotton is available, use it to close the opening at the lip of the beaker. Alternatively, cover the top of the beaker with aluminum foil. Allow *at least 10 min* after preparing the chamber for the atmosphere inside to become saturated with the solvent vapor.

Standard aqueous amino acid solutions (approximately 0.07 *M*) will be provided in the laboratory. Among those amino acids that may be included are DL-aspartic acid, L-isoleucine, L-arginine, glycine, DL-serine, L-proline, DL-phenylalanine, DL-methionine, DL-threonine, L-lysine, L-cysteine, L-histidine, DL-alanine, L-leucine, DL-tryptophan, L-tyrosine, L-glutamic acid, and DL-valine. Two unknown amino acid mixtures will be provided, each of which will contain two or more amino acids chosen from those listed here. The instructor will indicate whether any additional amino acids, other than those listed, have been included.

Obtain a piece of chromatographic paper measuring 13 × 15 cm. The size may vary in length, depending on the number of samples to be run. (*Caution:* Fingerprints contain significant amounts of amino acids, often enough to be easily detected by the methods used in this experiment. This is particularly true after the solutions of amino acids have been handled. Avoid touching the surface of the chromatographic paper. Handle it as little as possible, and then only along a thin strip on the edge *opposite* to that along which the samples are to be spotted.) Place a series of light pencil dots about 1–1.5 cm apart along a line parallel to a long edge of the paper and about 2 cm from this edge. The outermost dots should be about 1 cm from the short edges of the paper. Label each of the spots so that the solution spotted at each point may readily be identified later from information recorded in your notebook. Avoid losing track of the positions at which given solutions are spotted. Such errors eventually lead to confusion and probable misidentification of the components of the mixture.

Now apply spots to the paper with a capillary pipet. If pipets are unavailable, the instructor will demonstrate how to draw them from capillary melting-point tubes. In spotting, place

---

[2]The experimental directions provided are adapted in part from those given by Slaten, L.; Willard, M.; Green, A. A. *Journal of Chemical Education,* **1956,** *33,* 140, which may be consulted for additional experimental details.

a small drop of the solution at the mark indicated and allow it to dry before development. If the drop is too small, the amount of sample may be insufficient for eventual detection of some of the components; if it is too large, it may lead to poor resolution during development. The ideal size is 2–3 mm in diameter. Spots will increase in size by a factor of three or four times in the direction of mobility during development; they should not spread laterally very much. Practice spotting samples on scrap filter paper, and when you feel comfortable with your technique, spot the chromatographic paper with unknowns and any samples of knowns to be run. Be sure to use a *new* capillary pipet for each spot to avoid contamination of the samples being spotted.

After spotting, allow the spots to dry, and then coil the paper into a cylinder, fasten it with staples or paper clips, and insert it, sample edge down, into the developing chamber. The paper should not touch the sides of the beaker and the samples must not be below the surface of the eluting solvent. When the solvent front has migrated to within about 2 cm from the upper edge of the paper, remove the paper and lightly mark the position of the solvent front. Stand the cylinder on a watch glass in the hood for a few minutes to dry. The drying should then be completed by spreading the paper out under a heat lamp, by placing it in an oven at about 105°C, or by hanging it in a gentle stream of air in the hood.

When the paper is dry, spray it lightly and evenly with ninhydrin solution, which has been prepared as a 0.1% solution in 95% ethanol, and redry it either under a heat lamp or in the oven; heat is necessary to the color-forming reaction. The colors should be readily visible after 20–30 min.

It may be desirable to run additional chromatograms in order to obtain R_f values for additional standard samples, or to develop the chromatogram using different detecting agents such as isatin. The colors produced by isatin spray for the common amino acids are provided in Table 26.1. These colors develop after the chromatogram is heated to 80–85°C for 30 min. A stock solution of the isatin spray reagent can be prepared by dissolving 1 g of isatin and 1.5 g of zinc acetate in 100 mL of 2-propanol and 1 mL of pyridine. The solution is effected by warming it on a water bath at 80°C, after which the solution must be kept cool.

Calculate R_f values for each of the standard samples and for each of the spots resolved in the unknowns, consulting, if necessary, the legend to Figure 6.3 for the procedure of calculation. Systematically tabulate in your notebook the colors observed for all spots and the calculated R_f values. Identify the constituents of the unknowns, giving the justification for your conclusions. Fasten the chromatograms in your notebook as a permanent record.

EXERCISES

1. Why should the initial spots of amino acids not be too large?

2. Why is the central carbonyl group of **10,** Figure 26.1, more susceptible to nucleophilic attack than are the adjacent carbonyl groups?

3. Why is the carbonyl group of isatin (**11**) that is adjacent to the aromatic ring more susceptible to nucleophilic attack than is the carbonyl group adjacent to nitrogen?

4. Why is there a large difference in the isoelectric points of lysine (9.6) and glutamic acid (3.2)?

5. Proline reacts with ninhydrin to produce a yellow color; the other amino acids in Table 26.1 produce a blue color. What structural feature of proline do you expect is

responsible for this distinction in behavior? (*Hint:* Consult the mechanism of the color-forming reaction with ninhydrin as given in Figure 26.1.)

6. Why is the solubility of an α-amino acid at a minimum in a solution having a pH corresponding to the isoelectric point of the acid?

26.3   Determination of Primary Structure of a Polypeptide

The most difficult aspect of the determination of primary structure of a polypeptide is the establishment of the *sequence* of amino acid residues in the chain. The total hydrolysis and amino acid analyses discussed in the preceding section allow determination of the number and types of amino acids present, but all information regarding sequence is lost at the hydrolysis step. The standardized approach for establishing sequence is discussed in the following paragraphs.

(a) Terminal Residue Analyses

Note that the amino acid residues at the termini of the polypeptide chain differ from the remainder of the residues: One, the *N-terminal residue,* is the only residue that contains a free *alpha* amino group; the other, the *C-terminal residue,* is the only residue that contains a free carboxyl group *alpha* to a peptide linkage (see **1,** for example). The special significance of these residues is that it is relatively simple to determine their identity. The importance of identifying these residues may be illustrated in the following example. There are 720 possible sequential arrangements for a hexapeptide containing six *different* amino acid residues. If *either* of the terminal residues is determined, the remaining number of possible sequences is only 120; if both are known, this number is reduced to 24. Thus, in this example, terminal residue analyses would result in a reduction of the number of structures to be considered by a factor of 30! The results of terminal residue analysis, together with information gained by partial hydrolysis, will usually allow the sequence of polypeptides to be determined.

A very successful method of identifying the *N*-terminal residue utilizes 2,4-dinitrofluorobenzene (DNFB). DNFB reacts by nucleophilic aromatic substitution in weakly alkaline aqueous solutions with free amino (*N*-terminal and lysyl), phenol (tyrosyl), and imidazole (histidyl) groups to provide dinitrophenyl (DNP) derivatized peptides (**12,** equation 4). Excess DNFB may easily be removed from the alkaline reaction mixture by extraction with diethyl ether. The DNP-peptide remains water-soluble at alkaline pH, as does any 2,4-dinitrophenol that has been formed by hydrolysis of DNFB during the reaction. The DNP-peptide and, unfortunately, 2,4-dinitrophenol are separated from the aqueous solution by adjustment of the reaction mixture to pH 1 and extraction with diethyl ether. The DNP-peptide no longer contains basic amino groups and consequently is not soluble in the now-acidic aqueous medium. Following removal of the ether, the DNP-peptide is totally hydrolyzed with 6 *M* HCl at 100°C for 24 hr to provide a mixture of both derivatized and underivatized amino acids. Table 26.2 portrays the chemical results at this stage, with especial attention drawn to lysine, histidine, and tyrosine, which are the amino acids containing side-chain functional groups reactive to DNFB.

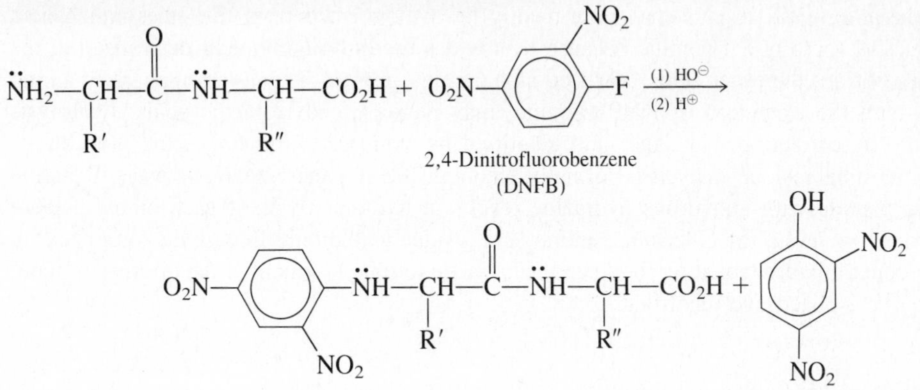

$$\ddot{N}H_2{-}CH{-}\overset{\displaystyle O}{\overset{\|}{C}}{-}\ddot{N}H{-}CH{-}CO_2H \ + \ O_2N{-}\overset{\displaystyle NO_2}{\underset{}{\bigcirc}}{-}F \xrightarrow[\text{(2) } H^{\oplus}]{\text{(1) } HO^{\ominus}}$$

$$O_2N{-}\bigcirc{-}\ddot{N}H{-}CH{-}\overset{\displaystyle O}{\overset{\|}{C}}{-}\ddot{N}H{-}CH{-}CO_2H \ + \ \bigcirc + \qquad \textbf{(4)}$$

2,4-Dinitrofluorobenzene
(DNFB)

12

2,4-Dinitrophenol

TABLE 26.2   Chemical Results of 2,4-Dinitrophenylation and Hydrolysis of the Resulting DNP-Peptide

Amino Acid Residue	As N-Terminal Residue	Other
Lysyl	DNP—N̈H—CH—CO₂H (CH₂)₄ DNP—NH	⁺NH₃—CH—CO₂H (CH₂)₄ DNP—NH
Histidyl	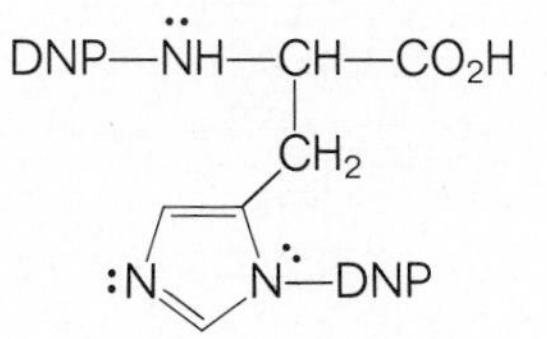 DNP—N̈H—CH—CO₂H CH₂ (imidazole):N̈—N̈—DNP	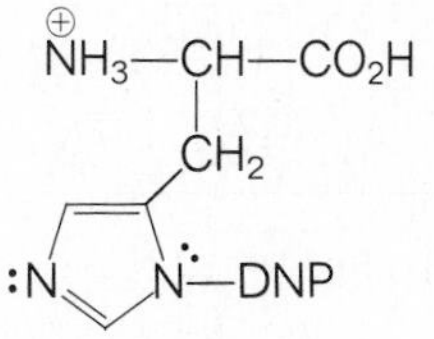 ⁺NH₃—CH—CO₂H CH₂ (imidazole):N—N̈—DNP
Tyrosyl	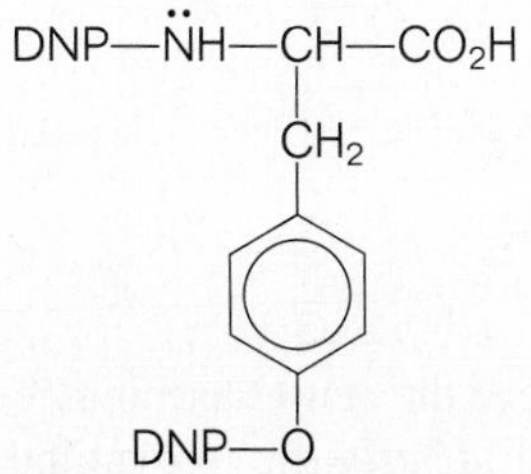 DNP—N̈H—CH—CO₂H CH₂ (phenyl ring) DNP—O	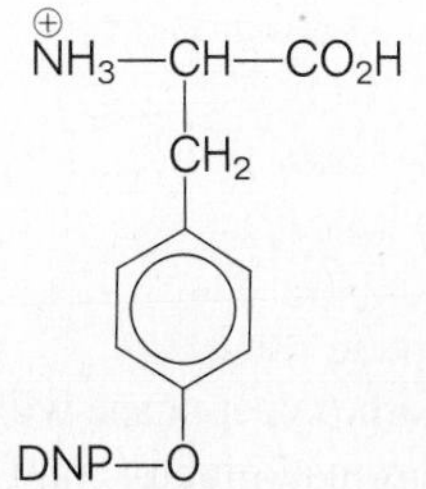 ⁺NH₃—CH—CO₂H CH₂ (phenyl ring) DNP—O
All other amino acids	DNP—N̈H—CH—CO₂H R	⁺NH₃—CH—CO₂H R

The *N*-terminal residue may now readily be distinguished from the other amino acid residues of the original peptide because it provides the only amino acid derivatized at the alpha position; the other acids bear free alpha amino groups. The *N*-terminal DNP–amino acid, with the exception of DNP-arginine, may be separated from the acidic hydrolysis mixture by extraction with ether and identified by standard chromatographic procedures.

C-terminal residue analysis is normally accomplished in either one of two ways. Reaction of the peptide with anhydrous hydrazine results in hydrazinolysis of each of the peptide linkages, providing the C-terminal amino acid residue as the only free *alpha* amino acid in the product mixture (equation 5). As the only water-soluble fragment of the original peptide, it may be isolated and identified.

$$
\ddot{N}H_2\text{—}CH\text{—}\overset{\overset{\displaystyle O}{\|}}{C}\text{—}\ddot{N}H\text{—}CH\text{—}CO_2H \xrightarrow{\ H_2NNH_2\ }
$$
$$
\underset{R}{|} \qquad\qquad \underset{R'}{|}
$$

$$
\ddot{N}H_2\text{—}CH\text{—}\overset{\overset{\displaystyle O}{\|}}{C}\text{—}\ddot{N}HNH_2 + \overset{\oplus}{N}H_3\text{—}CH\text{—}CO_2{}^{\ominus} \qquad \textbf{(5)}
$$
$$
\underset{R}{|} \qquad\qquad\qquad\qquad \underset{R'}{|}
$$

A second method of C-terminal residue determination makes use of the enzyme *carboxypeptidase,* a pancreatic enzyme whose characteristic reactivity is the hydrolysis of peptide bonds adjacent to free *alpha*-carboxyl groups. Because the peptide bond to the C-terminal residue is the only such bond in a peptide, carboxypeptidase selectively removes this residue as a free alpha amino acid, producing a shortened peptide chain (equation 6).

$$
\overset{\oplus}{N}H_3\text{—}CH\text{—}\overset{\overset{\displaystyle O}{\|}}{C}\text{—}\ddot{N}H\text{—}CH\text{—}\overset{\overset{\displaystyle O}{\|}}{C}\text{—}\ddot{N}H\text{—}CH\text{—}CO_2{}^{\ominus} \xrightarrow{\ carboxypeptidase\ }
$$
$$
\underset{R}{|} \qquad\qquad \underset{R'}{|} \qquad\qquad \underset{R''}{|}
$$

$$
\overset{\oplus}{N}H_3\text{—}CH\text{—}\overset{\overset{\displaystyle O}{\|}}{C}\text{—}\ddot{N}H\text{—}CH\text{—}CO_2{}^{\ominus} + \overset{\oplus}{N}H_3\text{—}CH\text{—}CO_2{}^{\ominus} \qquad \textbf{(6)}
$$
$$
\underset{R}{|} \qquad\qquad \underset{R'}{|} \qquad\qquad\qquad \underset{R''}{|}
$$

Carboxypeptidase will then remove the *new* C-terminal residue, and so on. Analysis by paper chromatography may be used to identify the free amino acids present in the mixture. The amino acid corresponding to the original C-terminal residue is the first residue removed and thus will develop maximum concentration, as judged from spot color intensity on the chromatogram, at a shorter elapsed reaction time than acids from positions successively farther from that end of the peptide chain.

(b) Partial Hydrolysis and Sequence Determination

In principle, the sequence of residues in a polypeptide might be determined by devising a procedure for selectively removing a terminal residue, identifying it, and then removing and identifying the next, until all have been identified (a terminal sequence determination). A variety of chemical procedures are available that may be used in just this way. These procedures, however, are generally feasible only for relatively short peptides; for example, the primary structures of peptides containing up to 60 amino acid residues have been determined in this fashion with the aid of automated apparatus. Consequently, the sequence of larger polypeptides and proteins is determined by effecting only their *partial* hydrolysis to produce a mixture of smaller peptides (dipeptides and tripeptides, for example), which are separated and whose sequences are determined by terminal sequence analysis. When the sequences of a sufficient number of fragments are known, the primary structure of the original polypeptide may be logically deduced. The interested student may consult any modern organic or biochemistry textbook for additional information, details, and examples of these and other procedures and their applications.

The following experiment involves the determination of structure of an unknown dipeptide. It is apparent that the extent of accumulated information necessary to deduce the structure of a dipeptide is somewhat less than would be required for larger peptides, yet the procedures in the experiment demonstrate many of the techniques commonly used by protein chemists. The structural determination is accomplished with milligram quantities of the unknown peptide, so this technique provides experience in the microtechniques of handling materials and solutions.

EXPERIMENTAL PROCEDURE

A. Hydrolysis of an Unknown Dipeptide

Obtain a commercially available ampoule of soft glass having a length of about 10 cm and an internal diameter of 1.0 to 1.5 mm for use as a hydrolysis tube. If this is not available, make an ampoule from a piece of soft glass tubing of these dimensions by sealing one end by drawing it out, using a microburner as a heat source. (*Caution:* The seal must be complete.) Place about 1 mg of an unknown dipeptide either on a porcelain spot plate or in a small test tube. Using either a syringe or a 0.1-mL pipet, add $30\mu L$ (0.03 mL) of 6 M hydrochloric acid to the sample. Mix the solution well and, using a disposable pipet, transfer it to the hydrolysis tube. Seal the tube by drawing it out, affix an identification label, and heat the tube for 10–12 hr in an oven set at 110°C.

After allowing the hydrolysis tube to cool, carefully open it, and with a disposable pipet transfer the solution to either a small watch glass or a spot plate. Evaporate the sample to dryness with a heat lamp, add 20 μL of water, and reevaporate to remove the last traces of hydrogen chloride. In each of these evaporative steps, be careful not to char the sample. Add 50 μL of water, and use 5–10 μL of this solution per spot in analyzing for the component amino acids of the dipeptide by paper chromatography, using the procedure provided in Section 26.2.

B. *N*-Terminal Residue Analysis

> **DO IT SAFELY**
>
> Dinitrofluorobenzene (DNFB) is a vesicant; that is, it will cause blistering and burns when it comes in contact with the skin. Operations involving the transfer of DNFB or its solutions should be carried out in the hood if possible, and rubber gloves should be worn. The DNFB or its solutions should be handled *only* by means of a small pipet or a syringe. **Do not pipet DNFB solutions by mouth.** If DNFB comes in contact with your skin, wash the affected area immediately with soap and water, and then rinse it with 0.6 *M* sodium bicarbonate solution.

To a 12-mL conical centrifuge tube add 2 mg of an unknown dipeptide, 0.2 mL of water, 0.05 mL of 0.5 *M* aqueous sodium bicarbonate solution, and 0.4 mL of a stock solution of 2,4-dinitrofluorobenzene. This solution can be prepared by dissolving 0.25 g of DNFB in 4.8 mL of absolute ethanol. Stopper the tube and shake the mixture frequently during 1 hr. The pH should be maintained at 8–9 by adding more 0.5 *M* sodium bicarbonate solution as necessary. Large amounts of precipitates indicate that the pH is too low.

After the 1-hr reaction period, add 1 mL of water and 0.05 mL of 0.5 *M* aqueous sodium bicarbonate solution to the reaction mixture in the centrifuge tube. Extract this solution three times with equal volumes of peroxide-free diethyl ether to remove unchanged DNFB. Carry out these extractions directly in the conical centrifuge tube by adding the portion of ether, stirring vigorously with a stirring rod, and removing the ether layer with a pipet. If necessary, centrifuge the solution to hasten the separation of layers.

Using pH paper as a guide, adjust the pH of the aqueous layer to about pH 1 by adding approximately 0.1 mL of 6 *M* hydrochloric acid, and extract three times with 2-mL portions of diethyl ether. Combine the extracts in a test tube and evaporate the ether. The evaporation may be accomplished conveniently by placing the test tube in a beaker of warm water and blowing a gentle stream of air into the test tube.

All traces of ether must be removed from the DNP-peptide before its hydrolysis. To accomplish the removal of traces of ether simultaneously with the transfer of the DNP-peptide to a hydrolysis tube, which may be prepared by sealing one end of a 10-cm length of 5-mm glass tubing, add 0.2 mL of acetone to the dried DNP-peptide, and transfer the resulting solution to the tube. Evaporate the solution in the hydrolysis tube to dryness as before, using a disposable pipet to channel a gentle air stream into the tube. Add 0.5 mL of 6 *M* hydrochloric acid, seal and label the tube, and heat it at 100°C in an oven for 10–12 hr.

Open the hydrolysis tube, and transfer the solution to a small test tube. After adding 1 mL of water, extract the hydrolysis solution three times with 2-mL portions of diethyl ether. Combine the ether extracts and evaporate to dryness as before. Dissolve the DNP–amino acid in 0.5 mL of acetone, and use this solution for chromatographic analysis.

If it is desired to confirm the identity of the C-terminal residue, evaporate the aqueous phase of the hydrolysate to dryness with a heat lamp. After adding 0.1 mL of water and redrying, dissolve the residue, which contains the C-terminal amino acid, in 50 μL of water and identify it by paper chromatography, following the procedure of Section 26.2. Note that if the C-terminal residue is lysine, histidine, or tyrosine (see Table 26.2), the procedure of Section 26.2 is not applicable.

C. Identification of N-Terminal DNP–Amino Acids by Thin-Layer Chromatography[3]

DO IT SAFELY

If the dinitrophenyl derivatives of the two amino acid residues of your unknown have similar R_f values with formic acid–water as a developing solvent, it will be necessary to use the benzene–acetic acid solvent system. Insofar as benzene has been implicated as a leukemia-causing carcinogenic substance, do both the chromatographic development and the subsequent drying of the plate *in the hood*, if you must use this solvent system. Do not remove any of the solvent from the hood. You should wear rubber gloves when handling the wet plate and should transfer the solution with a pipet. These precautions should be followed in order to avoid release of benzene vapors into the room and to prevent adsorption of benzene through the skin.

Obtain a 4×10-cm strip of polyamide chromatogram sheet for qualitative analysis of the N-terminal DNP–amino acid. On a line about 1.5 cm from a narrower side of the plate and with 1-cm spacings from each other and from the longer sides, spot 5 μL each of the unknown DNP–amino acid–acetone solution and the two appropriate standard DNP–amino acid solutions provided in the laboratory. Note that the appropriate solutions may be identified from the results of the amino acid analysis performed. Also spot the paper with a solution of 2,4-dinitrophenol.

One of two solvents may be used to develop the chromatogram, according to the requirements of the analysis. Consult Figure 26.4 for information to aid in this decision. Use either toluene–glacial acetic acid (80:20, by volume) or 90% formic acid–water (50:50, by volume). The first solvent will require about 1.5 hr for development and the second about 1 hr. Because DNP–amino acids are light-sensitive, the chromatographic development should be performed in the dark by placing the chamber in the desk drawer or by covering the chamber with a cardboard box. The DNP–amino acids produce yellow spots, as does 2,4-dinitrophenol. The phenol, in contrast to the DNP–amino acids, is colorless below pH 4. If you are uncertain which spot from the unknown is 2,4-dinitrophenol, add a drop of dilute hydrochloric acid to each spot. If the DNP spots are hard to find because of low concentration, examination under ultraviolet light may be helpful. The colors of the spots will fade with time, so they should be outlined with a pencil after development. A drawing of the plate should be recorded in your notebook.

Using the information collected in these procedures, assign a structure for the unknown peptide.

[3]For additional information, consult Wang, K.; Wang, I. S. Y. *Journal of Chromatography and Data*, **1967,** 27, 318.

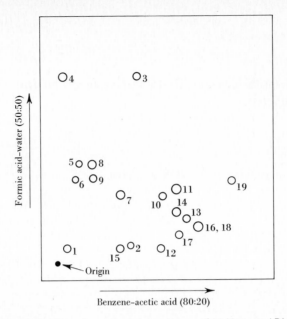

Figure 26.4 Two-dimensional chromatogram of 18 N-terminal DNP–amino acids (see second column of Table 26.2) and 2,4-dinitrophenol (circle 19). The number corresponding to each of the derivatized amino acids is keyed in Table 26.1.

EXERCISES

Questions for Part A

1. In the color-forming reaction with α-amino acids, why is the central carbonyl group of **10**, Figure 26.1, more susceptible to nucleophilic attack than are the adjacent carbonyl groups?

2. Why is the carbonyl group of isatin (**11**) that is adjacent to the aromatic ring more susceptible to nucleophilic attack than is the carbonyl group adjacent to nitrogen?

3. The rate of hydrolysis of gelatin, a protein, increases linearly with acid concentration over the range of 3.0–10.4 M hydrochloric acid. With reference to the mechanism of amide hydrolysis, explain why this should be the case.

4. Account for the observation that glycylvaline is hydrolyzed under acidic conditions much more rapidly than valylglycine.

5. When asparagine is present in peptides, an acid-catalyzed rearrangement is sometimes observed during hydrolysis to a mixture of α- (**13**) and β-aspartyl (**14**) peptides. Account mechanistically for these transformations.

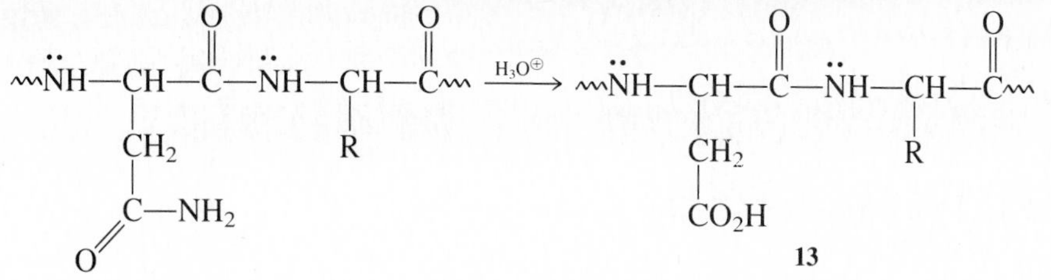

Questions for Part B

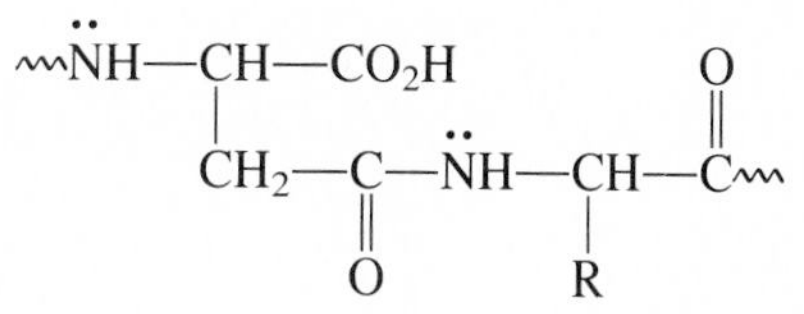

6. Explain why maintaining the pH at 8–9 during the reaction of a peptide with 2,4-dinitrofluorobenzene prevents formation of a precipitate. What would the precipitate be?

7. Provide a mechanism for the reaction of 2,4-dinitrofluorobenzene with a peptide to produce an *N*-terminal DNP-peptide (**12**, equation 4). Would you expect 2,4-dinitrochlorobenzene to require more or less vigorous conditions if used in place of DNFB? Explain.

8. Why is the reaction with DNFB performed in aqueous base rather than in aqueous acid?

9. In the workup of the hydrolysate, you were instructed to adjust the pH to about 1 prior to extracting with ether. What would be the consequence of failing to do this and instead bringing the pH to only about 7 or so?

10. Provide a flow diagram that outlines the various separations that are effected on the basis of solubilities during the preparation of the *N*-terminal DNP–amino acid from a dipeptide.

REFERENCES

1. Dickerson, R. E.; Geis, I. *The Structure and Actions of Proteins*, Harper & Row, New York, 1969.

2. Blackburn, S. *Protein Sequence Determination: Methods and Techniques*, Marcel Dekker, New York, 1970.

3. Bailey, J. L. *Techniques in Protein Chemistry*, Elsevier Publishing Company, New York, 1962.

chapter **27**

identification of organic compounds

One of the greatest challenges to the chemist is identifying the materials that are obtained from chemical reactions or isolated from natural sources. However, structure elucidation is often difficult and time-consuming, and it can be done today by using various spectroscopic techniques in combination with chemical methods that involve performing laboratory work. The approaches that can be used to accomplish this goal are described in this chapter.

Systematic procedures for the identification of organic compounds were developed much later than those for inorganic compounds, ions, and elements. The first successful scheme of qualitative organic analysis was developed by Professor Oliver Kamm and culminated in the publication of his textbook in 1922. This scheme, and the modifications and modernizations that have been made to it, is one on which most textbooks are still based (see references 1 and 2 at the end of Section 27.4) and will be referred to as *classical qualitative organic analysis*.

In recent years, the development of chromatographic methods of separation (Chapter 6) and analysis by spectroscopic techniques (Chapters 8–11) has revolutionized the laboratory practice of organic chemistry. Nevertheless, interest in classical qualitative organic analysis remains high because it is recognized by teachers and students alike as an effective and interesting means of teaching fundamental organic chemistry. For this reason, this chapter contains an outline of the classical scheme, and experimental procedures associated with it, that is adequate for obtaining a good understanding of the approach. Although used extensively in the past, the classical scheme is seldom employed by practicing organic chemists today because a wealth of information is readily available from spectroscopic techniques. Consequently, an introduction to identification and structure determination based on data from modern spectroscopic methods is included in Section 27.4. The advantages of the combined use of the classical and instrumental methods are also described in that section.

27.1 **Overview of Organic Analysis**

A discussion starting with an overview of a systematic procedure that can be used to identify an unknown *pure* organic compound will serve to introduce the classical organic analysis scheme. The first step is to ensure that the compound is in fact pure, and this can be done in several ways. In the case of a liquid, gas chromatography (Section 6.5) can be used to demonstrate that only one component is likely to be present; a pure compound should produce only a single peak in a gas chromatogram, assuming no decomposition under conditions of analysis. If enough liquid is available, it can be distilled (Chapter 5), and the observation of a narrow boiling-point range of 1–2°C implies that it is pure; however, gas chromatography provides a better proof of purity. The purity of a solid substance can be ascertained from its melting point (Chapter 3), the observation of a narrow melting-point range of 1–2°C usually indicating purity except in the rare instance of a eutectic mixture. Impure liquids can be purified by simple or fractional distillation or by gas chromatography, while solids can be purified by recrystallization or by column or thin-layer chromatography. The purification and identification of gases is not included in this text because of the difficulties in handling them.

Once purity is established, various physical properties are determined. The melting point of a solid or boiling point of a liquid is considered essential. Optionally, the density and/or refractive index of a liquid may be useful, and for certain compounds, either liquid or solid, the specific rotation can be determined if a substance is thought to be optically active.

Knowing what elements other than carbon and hydrogen are present may provide an important clue in the ultimate identification of the compound. This so-called *elemental analysis* is described later. Other observations, such as molecular weight, as determined by cryoscopic techniques or mass spectrometry (Chapter 11), or percentage by weight composition of the elements present, can provide data useful to the process of identification. The solubility of the unknown compound in water, in dilute acids and bases, or in various organic solvents can be determined, and the results provide information regarding various functional groups that may or may not be present.

The next and perhaps most important step in the identification of an unknown substance is the determination of the functional group(s) present. Infrared spectroscopy may also be used for this purpose, since the correct analysis of an infrared spectrum yields much information about the presence or absence of certain functional groups. Before the development of infrared spectroscopy as a routine experimental technique, functional group determination involved performing many qualitative chemical tests, which had to be done for each possible group. Although infrared analysis may not provide an unequivocal answer about the presence of certain groups, it at least permits one to narrow the possibilities to a small number, and then one or two chemical tests can be performed to complete the determination of the groups present.

Final assignment of a structure to the unknown can be achieved by one of several procedures. The classical method involves the conversion of the substance into a solid, called a *derivative*. The success of this technique depends on the availability of information about the unknown and its various derivatives. Of prime importance is knowledge of the melting or boiling points of possible candidates for the unknown as well as the melting points of solid derivatives. Many tabulations of organic compounds, arranged by functional group and in order by melting point or boiling point and giving melting points of their derivatives, are available for this purpose. Often, however, the identification of the compound can be completed by determining and analyzing the NMR spectra of the compound. Spectroscopic proof based *only* upon NMR spectra is usually not possible unless they are being used to distinguish among a limited number of possible compounds or unless the molecular formula of the compound is known. On the other hand, comparison of IR and NMR spectra, particularly the former, of an unknown and a known will suffice.

The following sections describe the stepwise procedures that may be used to identify an unknown compound using classical methods alone or a combination of classical and spectroscopic methods. However, it is unwise to try any of these procedures unless the compound is pure. Since it is possible that you may be given a mixture of unknown compounds to identify, a procedure is provided in Section 27.3 for the separation of a mixture into its individual components so that each one can be identified.

27.2 Classical Qualitative Analysis Procedure for Identification of a Pure Compound

The classical system of qualitative organic analysis consists of six steps. The first four, which can be carried out in any order, should be completed before performing the qualitative tests for functional groups. The final step must always be the preparation of one or more solid derivatives.

a **Preliminary examination** of physical and chemical characteristics

b Determination of **physical constants**

c **Elemental analysis** to determine the presence of elements other than carbon, hydrogen, and oxygen

d **Solubility tests** in water, dilute acid, and dilute base

e **Functional group analysis** using classification tests

f **Preparation of derivatives**

It is a tribute to the effectiveness of the system that one can identify an unknown organic compound with certainty, in view of the fact that it may be one of several million known compounds. However, with the exception of a few general guidelines, there are no rigid directions to be followed. You, the investigator, must rely on good judgment and initiative in selecting a course of attack on the unknown, and it is particularly important to observe and consider each experimental result. Negative results can be as useful as positive ones in the quest to reveal the identity of an unknown.

(a) The Preliminary Examination

The preliminary examination may provide more information with less effort than any other part of the identification procedure, if it is carried out intelligently. The simple observation that the unknown is a *crystalline* solid, for example, eliminates from consideration a major fraction of all organic compounds, since most of them are liquids at room temperature. The *color* is also informative; most pure organic compounds are white or colorless. A brown color is most often characteristic of small amounts of impurities; for example, aromatic amines and phenols quickly become discolored by the formation of trace amounts of highly colored air-oxidation products. Color in a pure organic compound is usually attributable to conjugated double bonds.

The *odor* of many organic compounds is highly distinctive, particularly among those of lower molecular weight. A conscious effort should be made to learn and recognize the odors that are characteristic of several classes of compounds such as the alcohols, esters, ketones, and aliphatic and aromatic hydrocarbons. The odors of certain compounds demand respect, even when they are encountered in small amounts and at considerable distance; for example, the unpleasant odors of thiols (mercaptans), isonitriles, and higher carboxylic acids and diamines cannot be described definitively, but they are recognizable once encountered. *Be cautious* in smelling unknowns, since some compounds are not only disagreeable but also irritating to the mucous membranes. Large amounts of organic vapors should never be inhaled because many compounds are toxic.

The *ignition test* is a highly informative procedure. Heat 1 drop of a liquid or about 50 mg of a solid gently on a small spatula or crucible cover, at first above or to the side of a microburner flame. Make a note as to whether a solid melts at low temperature or only upon heating more strongly. Observe the flammability and the nature of any flame. A yellow, sooty flame is indicative of an aromatic or a highly unsaturated alphatic compound; a yellow but nonsooty flame is characteristic of aliphatic hydrocarbons. Oxygen content in a substance makes its flame more colorless or blue; extensive oxygen content lowers or prevents flammability, as does halogen content. The unmistakable odor of sulfur dioxide indicates the presence of sulfur in the compound.

If a white, nonvolatile residue is left after ignition, add a drop of water and test the solution with litmus or pH paper; a sodium (or other metal) salt is indicated by an alkaline test.

(b) Physical Constants

If the unknown is a solid, determine its melting point by the capillary tube method (Section 2.3). If the melting range is more than 2–3°C, recrystallize the sample (Section 3.1).

If the unknown is a liquid, determine its boiling point by the micro boiling-point technique (Section 2.4). If the boiling point is indefinite or nonreproducible or if the unknown sample is discolored or inhomogeneous, distil it (Section 2.7); the boiling point will be obtained as a result of distillation.

Other physical constants that *may* be of use for liquids are the *refractive index* and the *density*. Consult your instructor about the advisability of making these measurements; full directions are given in references 1 and 2 at the end of Section 27.4.

(c) Elemental Analysis

The technique of elemental analysis involves the determination of *which* elements may be present in a compound. The halogens, sulfur, oxygen, phosphorus, and nitrogen are the elements other than carbon and hydrogen that are most commonly found in organic molecules. Although there is no simple way to test for the presence of oxygen, it is fairly easy to determine the presence of the other hetero atoms, and appropriate procedures for doing this are provided for the halogens, sulfur, and nitrogen.

The basis of the procedures is as follows. Because the bonding found in organic compounds is principally covalent, there seldom are direct methods analogous to those applicable to ionic inorganic compounds for determining the presence of the aforementioned elements. The covalent bonds to these hetero atoms may, however, be broken by heating an organic compound with sodium metal. This process, called *sodium fusion,* results in the formation of inorganic ions involving these elements if they were present in the original compound: halide ions, X^-, from halogens, sulfide ion, S^{2-}, from sulfur, and cyanide ion, CN^-, from nitrogen. After the organic compound has been heated with sodium metal, the residue is hydrolyzed with *distilled water* to destroy the excess sodium and dissolve the inorganic ions that were formed. If suitable experimental procedures are selected, the aqueous solution can then be analyzed for the presence of halide ion, sulfide ion, and/or cyanide ion.

A portion of the solution is acidified with dilute nitric acid and boiled *in the hood* to remove any sulfide or cyanide ions that are expelled as hydrogen sulfide or hydrogen cyanide, respectively. Sulfide and cyanide must be removed because they interfere with the test for halogens. Silver nitrate solution is added, and the presence of halide is detected by the formation of a precipitate of silver halide (equation 1).

$$Ag^{\oplus} + X^{\ominus} \rightarrow \underline{AgX} \text{ (solid)} \tag{1}$$

The color of the precipitate provides a tentative indication of the halogen present; AgCl is *white,* AgBr is *light yellow,* and AgI is *dark yellow.* Positive identification must be made by standard inorganic qualitative analysis procedures or by means of thin-layer chromatography.

Sulfur may be detected by taking a second portion of the sodium fusion solution and carefully acidifying it. Any sulfide ion present is converted to H_2S gas, which forms a dark precipitate of PbS when allowed to come in contact with a strip of paper that has been saturated with lead acetate solution (equation 2).

$$Pb(OAc)_2 + H_2S \rightarrow \underline{PbS} \text{ (solid)} + 2 \text{ HOAc} \tag{2}$$
$$\textit{Black}$$

The following qualitative test, if positive, indicates the presence of nitrogen in the unknown. A third portion of the aqueous solution is carefully acidified, followed by addition of ferrous ion, Fe^{2+}, and ferric ion, Fe^{3+}. this converts the cyanide ion into potassium ferric ferrocyanide, which precipitates as an intense blue solid called *Prussian blue* (equation 3).

$$4 \text{ CN}^{\ominus} + Fe^{(2+)} \rightarrow Fe(CN)_6^{(4-)} \xrightarrow{K^{\oplus} \text{ and } Fe^{(3+)}} \underline{KFeFe(CN)_6} \tag{3}$$
$$\textit{Prussian Blue}$$

EXPERIMENTAL PROCEDURE

DO IT SAFELY

1 Sodium fusion involves heating sodium metal or a sodium-lead alloy to a high temperature. Be very careful when doing this, and make sure the open end of the test tube is pointed away from you and those working around you. Never let your eyes come near the test tube.

2 It is preferable to perform the sodium fusion in the hood.

3 Check the test tube for cracks or other imperfections before doing the sodium fusion.

4 Be very careful when adding the unknown solid or liquid into the test tube containing hot metal. The organic material may burst into flame when it contacts the hot metal.

5 If sodium metal is used, the residue must be hydrolyzed **very** carefully as directed below, because any excess metal reacts **vigorously** with alcohol or water.

6 Be careful when handling the test tube after the fusion is complete, and remember that it may still be hot.

Sodium fusion may be carried out using either sodium metal or sodium-lead alloy, which is called "dri-Na" and contains nine parts lead and one part sodium. For safety reasons, we recommend using the sodium-lead alloy method because it is easier to handle and poses less potential danger during hydrolysis.

Sodium-Lead Alloy Method   Support a small Pyrex test tube in a vertical position using a clamp with either an asbestos liner or no liner *(no rubber)*. Weigh a 0.5 g sample of sodium-lead alloy ("dri-Na") and place it into the test tube. Heat the alloy with a flame until it melts and fumes of sodium are seen 1–2 cm up the walls of the test tube. *Do not* heat the test tube to redness. Add to the hot alloy 2–3 drops of a liquid sample or about 10 mg of a solid sample, being careful during the addition not to allow any of the sample to touch the sides of the hot test tube. If there is no visible reaction, heat the fusion mixture gently to initiate the reaction, then discontinue the heating and allow the reaction to subside. Next, heat the test tube to redness for a minute or two, and then let it cool. Add 3 mL of *distilled water,* and heat gently for a few minutes to decompose the excess sodium with water. Filter the solution, washing the filter paper with about 2 mL of water. (If the filtration is not done, dilute the decanted solution with about 2 mL of water.) Discard in an appropriate container the lump of metallic lead that remains. Use this fusion solution in the tests for sulfur, nitrogen, and the halogens given below.

Sodium Metal Method *Caution: Safety glasses must be worn throughout this procedure, and the face should be kept away from the mouth of the test tube at all times. Avoid pointing the test tube in the direction of anybody else.*

Support a small *Pyrex* test tube in a vertical position using a clamp with either no liner or an asbestos liner. *The clamp must not have a rubber liner.* Place a clean cube of sodium metal about 3 to 4 mm on an edge in the tube and heat it gently with a microburner flame until the sodium melts and the vapors rise about 1 cm. Then remove the flame and add to the hot test tube 2 or 3 drops of a liquid or about 10 mg of a solid sample, being careful not to allow any of the sample to come in contact with the side of the hot test tube during the addition. A brief flash of fire is normally observed. Heat the bottom of the tube again, then remove the flame and add a second equivalently sized portion of the organic compound. Now heat the bottom of the tube until it is a dull red color. Next, remove the flame and allow the tube to cool. Add about 1 mL of 95% ethanol to decompose the excess sodium; stir the contents of the tube with a stirring rod. After a few minutes, add another 1 mL of ethanol and stir again. After the reaction has subsided, apply gentle heat to boil the ethanol and place an inverted funnel connected to a vacuum line or aspirator over the mouth of the test tube to keep the ethanol vapors from the room. When the ethanol has been removed, allow the tube to cool and add about 10 mL of *distilled water* to it; stir the mixture and pour it into a small beaker. Rinse the tube with an additional 5 mL of distilled water, and combine the rinse with the main solution. The total amount of water should be 15 to 20 mL. Boil the aqueous mixture, filter, and use the filtrate in the tests for sulfur, nitrogen, and the halogens, as described below.

B. Qualitative Analysis for Sulfur, Nitrogen, and Halogens

Sulfur Acidify a 1- to 2-mL sample of the solution with *acetic acid,* and add a few drops of 0.15 M lead acetate solution. A black precipitate of PbS indicates the presence of sulfur in the original organic compound.

Nitrogen Check the pH of a 1-mL sample of the fusion solution with Hydrion E indicator paper. The pH should be about 13. If the pH is definitely above 13, add a *small* drop of 3 M sulfuric acid to bring the pH down to about 13. If the pH of the fusion solution is definitely below 13, add a *small* drop of 6 M NaOH to bring the pH up to about 13. Add 2 drops each of a saturated solution of ferrous ammonium sulfate and of 5 M potassium fluoride. Boil the mixture gently for about 30 sec, cool it, and add 2 drops of 5% ferric chloride solution. Then carefully add 3 M sulfuric acid to the mixture, 1 drop at a time until the precipitate of iron hydroxide *just* dissolves. Avoid an excess of acid. At this point the appearance of the deep-blue color of potassium ferric ferrocyanide (Prussian blue) indicates the presence of nitrogen in the original organic compound. If the solution is green or blue-green, filter it; a blue color remaining on the filter paper is a weak but positive test for nitrogen.

Halogens Acidify about 2 mL of the fusion solution by dropwise addition of 6 M nitric acid. Follow the acidification with blue litmus paper. Boil the solution gently for 2–3 min to expel any hydrogen sulfide or cyanide that may be present. (Sulfide and cyanide, if present, will interfere with the test for the halogens.) Cool the solution and add several drops of 0.3 M

aqueous silver nitrate solution. A *heavy precipitate* of silver halide indicates the presence of chlorine, bromine, or iodine in the original organic compound. A faint turbidity should not be interpreted as a positive test. *Tentative* identification of the particular halogen may be made on the basis of color: silver chloride is white, silver bromide is pale yellow, and silver iodide is yellow. *Positive* identification must be made by standard inorganic qualitative procedures,[1] or by means of thin-layer chromatography (Section 6.2), the procedure for which follows.

Obtain from the instructor a 2.5×7.5-cm strip of fluorescent silica gel chromatogram sheet. About 1 cm from one end, place four equivalently spaced spots as follows. At the left, using a capillary to provide the sample, spot the original test solution. Because this solution is likely to be relatively dilute in halide ion, it may need to be respotted several times; allow the spot to dry following each application. This may be hastened by blowing on the plate. Keep the spot as small as possible. Next, in order, spot samples of 1 *M* potassium chloride, 1 *M* potassium bromide, and 1 *M* potassium iodide. Develop the plate in a solvent mixture of 2-propanone, 1-butanol, concentrated ammonium hydroxide, and water in the volume ratio of $13:4:2:1$ (see Section 6.2 for details). Following development, allow the plate to air-dry, and in a hood spray the plate lightly with an indicator spray prepared by dissolving 1 g of silver nitrate in 2 mL of water and adding this solution to 100 mL of methanol containing 0.1 g of fluorescein and 1 mL of concentrated ammonium hydroxide. Allow the yellow strip to dry, and then irradiate it for several minutes with a long-wavelength ultraviolet lamp (366 nm). Compare the spots formed from the test solution with those formed from the solutions of known halides. (*Note:* Iodide gives two spots.)

(d) Solubility Tests

The solubility of an organic compound in water, dilute acid, or dilute base can provide useful, but not definitive, information about the presence or absence of certain functional groups. Note that the assignment of an unknown to a formal solubility class is rather arbitrary because of the large number of compounds that exhibit *borderline* behavior. It is recommended that the solubility tests be done in the order presented below.

Water Test the solubility of the unknown in water. For the present purposes, a compound is said to be soluble if it dissolves to the extent of 3 g in 100 mL of water, or more practically, 100 mg in 3 mL of water.

Several structural features of the unknown can be deduced if it is water-soluble. It must be of low molecular weight and will usually contain no more than four to five carbon atoms, unless it is polyfunctional. It must contain a polar group that will hydrogen-bond with water, as for example, an OH group (alcohol), a CO_2H group (carboxylic acid), an amino group, or a carbonyl group (aldehyde or ketone). Esters, amides, and nitriles dissolve to a lesser extent, and acid chlorides or anhydrides will react with water. On the other hand, alkanes, alkenes, alkynes, and alkyl halides are water-insoluble.

If the unknown is water-soluble to the extent of 100 mg/3 mL of water, test its aqueous solution with litmus paper or pH paper. If the solution is acidic, the unknown is likely to be

---

[1]See references 1, 2, and 3 at the end of Section 27.4.

a low molecular weight carboxylic acid such as acetic acid. If it is basic, a low molecular weight organic base such as diethylamine is possible. A neutral solution suggests the presence of a neutral polar compound such as ethanol or acetone. In general, very few organic compounds exhibit appreciable water solubility as defined here.

The borderline for water solubility of *monofunctional* organic compounds is most commonly at or near the member of the homologous series containing five carbon atoms. Thus, butanoic acid is soluble, pentanoic acid is borderline, and hexanoic acid is insoluble; 1-butanol is soluble, 1-pentanol is borderline, and 1-hexanol is insoluble. The relation between molecular surface area and solubility is demonstrated, however, by the observation that 2-methyl-2-butanol (*tert*-pentyl alcohol) is water-soluble to the extent of 12.5 g/100 mL of water even though it contains five carbon atoms.

If an unknown is insoluble in water, test its solubility first in sodium hydroxide, then in sodium bicarbonate, and finally in hydrochloric acid. Solubility in one or more of these acids and bases is defined in terms of the compound being *more soluble in acid or base than in water* and reflects the presence of an acidic or basic functional group in the water-insoluble unknown compound.

In each of the following solubility tests, *the unknown should be shaken with the test reagent at room temperature*. If it does not dissolve, warm the mixture for several minutes in a hot-water bath and continue shaking. If the substance still does not appear to dissolve, decant or filter the liquid from the undissolved sample, and carefully *neutralize* the filtrate. The formation of a precipitate or turbidity is indicative of greater solubility in the reagent than in water. It is important that the filtrate only be *neutralized* because an unknown may show enhanced solubility in both acid and base solutions if it contains *both* basic and acidic functional groups. Hence *all* the solubility tests should be performed.

Sodium Hydroxide If the compound is water-insoluble, test its solubility in dilute (1.5 M) NaOH solution. Carboxylic acids and sulfonic acids (strong acids) and phenols (weak acids) dissolve in sodium hydroxide because they are converted into their water-soluble sodium salts (equation 4).

$$\text{Carboxylic acids: } RCO_2H \xrightarrow{\text{NaOH}} RCO_2^{\ominus}Na^{\oplus} \tag{4a}$$
Water-insoluble   Water-soluble

$$\text{Phenols: } ArOH \xrightarrow{\text{NaOH}} ArO^{\ominus}Na^{\oplus} \tag{4b}$$
Water-insoluble   Water-soluble

An unknown that is more soluble in NaOH than in water can be either a phenol or a carboxylic acid, and it should then be tested for solubility in the weaker base, 0.6 M NaHCO$_3$, which may permit distinction between these two functional groups. If it does not exhibit solubility in NaOH, its solubility in NaHCO$_3$ need not be tried; it should then be tested for solubility in 1.5 M HCl.

Sodium Bicarbonate The solubility of the unknown in 0.6 M NaHCO$_3$ should be determined, and if it is soluble, the presence of a carboxylic acid group may be tentatively concluded, owing to the formation of the water-soluble sodium salt (equation 5).

$$\text{Carboxylic acids: RCO}_2\text{H} \xrightarrow{\text{NaHCO}_3} \text{RCO}_2^{\ominus}\text{Na}^{\oplus} \tag{5a}$$

Water-
insoluble   Water-
soluble

$$\text{Phenols: ArOH} \xrightarrow{\text{NaHCO}_3} \text{no reaction} \tag{5b}$$

Water-
insoluble

Carboxylic acids are usually soluble in $NaHCO_3$ and in $NaOH$, whereas phenols usually dissolve only in $NaOH$. Caution must be exercised in making definitive conclusions about the presence of a carboxylic acid or phenol based upon solubility in $NaHCO_3$. For example, a phenol containing one or more strong electron-withdrawing groups, such as nitro, can be as acidic as a carboxylic acid, or more so.

Hydrochloric Acid   The solubility of the unknown in 1.5 M hydrochloric acid should be determined, and if it is soluble, the presence of an amino group in the compound is indicated. This is because amines are organic bases and react with dilute acids to form ammonium salts, which are usually water-soluble (equation 6). However, this solubility test does not permit the distinction between weak and strong organic bases.

$$\text{RNH}_2 \xrightarrow{\text{HCl}} \text{RNH}_3^{\oplus}\text{Cl}^{\ominus} \tag{6}$$

Water-
insoluble   Water-
soluble

Concentrated Sulfuric Acid   Many compounds that are too weakly basic or acidic to dissolve in dilute aqueous acid or base will dissolve in or react with concentrated H_2SO_4. Such solubility is often accompanied by the observation of a dark solution or the formation of a precipitate; any detectable reaction such as evolution of a gas or formation of precipitate is considered "solubility" in concentrated H_2SO_4. This behavior can usually be attributed to the presence of carbon-carbon double or triple bonds, oxygen, nitrogen, or sulfur in the unknown. The solubility is usually due to some type of reaction of one of these functional groups with the concentrated acid, which results in the formation of a salt that is soluble in the reagent. For example, an alkene adds the elements of sulfuric acid to form an alkyl hydrogen sulfate (equation 7) that is soluble in the acid, and an oxygen-containing compound becomes protonated in concentrated acid to form an oxonium salt (equation 8) that is likewise soluble.

$$\text{R}_2\text{C}{=}\text{CR}_2 + \text{H}_2\text{SO}_4 \rightarrow \underset{\overset{|}{\text{H}}\quad\overset{|}{\text{OSO}_3\text{H}}}{\text{R}_2\text{C}{-}\text{CR}_2} \tag{7}$$

$$\text{R}_2\text{C}{=}\text{O or ROR} + \text{H}_2\text{SO}_4 \rightarrow \text{R}_2\text{C}{=}\overset{\oplus}{\text{O}}\text{H HSO}_4^{\ominus} \text{ or } \underset{\overset{|}{\text{H}}}{\text{R}{-}\overset{\oplus}{\text{O}}{-}\text{R}} \text{ HSO}_4^{\ominus} \tag{8}$$

Substances that exhibit this solubility behavior are termed "neutral" compounds. Obviously, there is no point in testing compounds that are soluble in dilute HCl or neutral water-insoluble compounds containing N or S in concentrated H_2SO_4, since they will invariably dissolve in or react with it. Note that solubility or insolubility of an unknown in concentrated H_2SO_4 does *not* yield a great deal of evidence for the presence or absence of any *specific* group, whereas the solubility of a compound in dilute HCl or NaOH or $NaHCO_3$ provides a strong indication of the type of functional group present.

Assuming that a compound contains only *one* functional group, the following scheme classifies compounds according to their solubility in acid or base. This picture can be changed dramatically if a compound contains several similar polar functional groups that cause its solubility properties to be different than expected or if an acidic or basic compound contains one or more strong electron-withdrawing groups.

> *Acidic Compounds Soluble in NaOH and NaHCO₃*
> Carboxylic acids
> Phenols
> *Acidic Compounds Soluble in NaOH but not in NaHCO₃*
> Phenols
> *Basic Compounds Soluble in HCl:*
> Amines
> *Neutral Compounds Soluble in Concentrated H₂SO₄*
> Carbonyl compounds (aldehydes and ketones)
> Unsaturated compounds (alkenes and alkynes)
> Alcohols
> Esters
> Amides
> Nitriles
> Nitro compounds
> Ethers
> *Neutral Compounds Insoluble in Concentrated H₂SO₄*
> Alkyl halides
> Aryl halides
> Aromatic hydrocarbons

After determining the physical constants, elemental analysis, and solubility properties, students may be requested to present a *preliminary report* of these findings to their instructor, who should advise them of any errors to prevent unnecessary loss of time.

(e) Classification Tests; Functional Group Identification

The next step in the identification of a compound is to determine which functional groups are present. The classical scheme involves performing a number of chemical tests on a substance, each of which is specific for a type of functional group. These tests can normally be done fairly quickly and are designed so that the *observation* of a color change or the formation of a precipitate indicates the presence of a particular type of functional group. The results of these tests usually allow the assignment of the unknown to a structural class such as alkene, aldehyde, ketone, or ester, for example. The following factors should be considered when performing qualitative classification tests for functional groups:

1. A compound may contain more than one functional group, so the complete series of tests should be performed unless you have been told that the compound is monofunctional.

2. Very careful *observation* is required when the functional group tests are performed. Note and record all observations, for example, the formation of a solid and its color.
3. Some of the color-forming tests occur for several different functional groups, and although the expected colors are given in the experimental procedures, a color may be affected by the presence of other functional groups.
4. It is of utmost importance to perform a qualitative test on *both the unknown and a known compound that contains the group being tested.* Some functional groups may appear to give only a slightly positive test, and it may be helpful to determine how a compound known to contain a given functional group reacts under the conditions of the test being performed. It is most efficient and reliable to do the tests on standards *at the same time* as on the unknown. In this manner, inconclusive positive tests may be interpreted correctly. Because aliphatic compounds are sometimes more reactive than aromatic ones, it is good practice to perform a test on both of these types of standards along with the unknown.
5. The results obtained from the elemental analysis and the solubility tests can be used to advantage in deciding which functional group tests should be performed initially or which should not be done at all. The following examples illustrate the use of the preliminary work in making these decisions:
   a. If a compound is found to be soluble in dilute hydrochloric acid and to contain nitrogen, a classification test for an amine should be applied first.
   b. The test for a phenol should be performed on an unknown that is soluble in dilute sodium hydroxide but insoluble in dilute sodium bicarbonate.
   c. If the elemental analysis indicates the absence of halogen, the tests for alkyl or aryl halides can be omitted. The absence of nitrogen means that tests for amines, amides, nitriles, and nitro compounds need not be performed.
6. It is suggested that a logical approach be adopted in deciding which tests need be performed. The result obtained from one test, whether positive or negative, often has a bearing on which additional tests should be done. A random, hit-or-miss approach is wasteful of time and may even lead to erroneous results. A common error made in qualitative organic analysis is to ''short cut'' the tests for functional groups and go directly to the preparation of a derivative, with no success. For example, attempting to make a derivative of a ketone is sure to fail if the unknown is actually an alcohol.

Qualitative tests for most of the common functional groups are presented in Section 27.5, which includes the following structural classes along with references:

Neutral Compounds
    Alcohols, Section 27.5e
    Aldehydes, Section 27.5a
    Alkenes, Section 27.5b
    Alkyl halides (Haloalkanes), Section 27.5c
    Alkynes, Section 27.5b
    Amides, Section 27.5l
    Aromatic hydrocarbons, Section 27.5d
    Aryl halides, Section 27.5d
    Esters, Section 27.5j
    Nitriles, Section 27.5k
    Nitro compounds, Section 27.5i
Acidic Compounds
    Carboxylic acids, Section 27.5g
    Phenols, Section 27.5f
Basic Compounds
    Amines, Section 27.5h

(f) Preparation of Derivatives

As mentioned earlier, in the classical approach it is usually necessary to convert a suspected unknown liquid or solid into a second compound that is a solid, the latter being called a *derivative* of the first compound. Although not a requirement, it is most convenient to prepare a derivative that is a solid rather than a liquid because solids can be obtained in pure form by recrystallization and because the melting point of a solid can be determined on a small quantity of material, whereas a larger amount of a liquid derivative would be required in order to determine its boiling point.

It is preferable to prepare two solid derivatives of an unknown compound in order to double-check its identity. The melting points of the derivatives, along with the melting point or boiling point of the unknown compound, usually serve to identify the unknown completely. However, the success of this type of identification depends upon the availability of tables listing the melting points or boiling points of known compounds and the melting points of suitable derivatives. Tables containing this information for selected compounds are provided in Section 27.6, and more extensive listings of compounds may be found in references 1, 2, and 5 at the end of Section 27.4. These tabulations are by no means comprehensive, and many other compounds that have been identified on the basis of derivatives appear in the scientific literature (see Chapter 28).

Each of the classes of compounds listed in part e above are considered in the following manner in Section 27.5: *Classification tests* for the functional group characteristic of the type of compound will be given, and these will be followed by experimental procedures for preparing several different solid *derivatives*. References to the tables in Section 27.6, which list various specific compounds and their derivatives, are also made for each functional group.

One warning should be emphasized. Do not go directly to the preparation of a derivative on the basis of a hunch about the class of compound to which your unknown belongs. Rather, make certain of the type of functional group present by obtaining one *or more* positive classification tests before attempting the preparation of any derivative. For example, it is frustrating to try to make a derivative of an alcohol when the unknown is actually an ester or a ketone!

The following example illustrates how preparation of derivatives can be used to identify an unknown compound, for which the following experimental data were obtained.

> *Elemental analysis:* No X, S, or N
> *Boiling point:* 119–120°C
> *Solubility tests:* Slightly soluble in water; no increased solubility in dilute HCl, NaOH, or NaHCO$_3$; soluble in concentrated H$_2$SO$_4$
> *Functional group analysis:* Positive test for ketone, positive test for methyl ketone, negative tests for all other functional groups

Because of experimental errors inherent in the determination of melting or boiling points, it is good practice to consider compounds melting or boiling within ±5°C of the observed melting or boiling point. In the case of the unknown liquid being discussed in this example, one must examine Table 27.15a for liquid ketones, from which the following list of possible substances for the unknown (assuming that the unknown is selected *only* from compounds appearing in the tables presented in this text) is taken:

1. 1-methoxy-2-propanone (bp 115°C), $CH_3OCH_2\overset{\displaystyle O}{\overset{\|}{C}}CH_3$

2. 4-methyl-2-pentanone (bp 117°C), $(CH_3)_2CHCH_2\overset{\displaystyle O}{\overset{\|}{C}}CH_3$

3. 3-methyl-2-pentanone (bp 118°C), $CH_3CH_2\underset{\underset{\displaystyle CH_3}{|}}{CH}\overset{\displaystyle O}{\overset{\|}{C}}CH_3$

4. chloroacetone (bp 119°C), $ClCH_2\overset{\displaystyle O}{\overset{\|}{C}}CH_3$

5. 3-penten-2-one (bp 122°C), $CH_3CH{=}CH\overset{\displaystyle O}{\overset{\|}{C}}CH_3$

6. 2,4-dimethyl-3-pentanone (bp 124°C), $(CH_3)_2CH\overset{\displaystyle O}{\overset{\|}{C}}CH(CH_3)_2$

From the experimental facts that were given, some of these compounds can be eliminated as possibilities for the following reasons: compound 4 because it contains a halogen, compound 5 because it contains a carbon-carbon double bond, and compound 6 because it is not a methyl ketone. Now suppose that two derivatives, the semicarbazone (Section 27.5b) and the 2,4-dinitrophenylhydrazone (Section 27.5b), were prepared from the unknown and found to melt at 93–95°C and 69–71°C, respectively. Further examination of Table 27.15a indicates that the derivatives of only one of the liquid ketones under consideration, namely 3-methyl-2-pentanone, melt at these temperatures. Hence, the identity of the unknown can be deduced. Although it is possible that other ketones with similar boiling points *may* exist and may *not* be listed in Table 27.15a, it is unlikely that any of these will give two derivatives with the same melting points as those of the derivatives obtained from 3-methyl-2-pentanone. This emphasizes the desirability of preparing *two* derivatives.

The preceding analysis of the data obtained for an unknown indicates how positive *and* negative information can be utilized in determining the identity of the substance. Note the importance of the functional group classification tests, which must be done carefully and thoroughly in order to exclude the possible presence of all groups other than the keto group in the unknown.

27.3 Separation of Mixtures of Organic Compounds

The preceding section described the identification of a *pure* organic compound. However, when a chemist is faced with the problem of identifying an organic compound, it is seldom pure but is often mixed with by-products or starting materials if it has been synthesized in the laboratory. Modern methods of separation, particularly chromatographic techniques, make the isolation of a pure compound easier than it used to be, but one must not lose sight of the importance of classical techniques of separation, which are treated in detail in Chapters 3–5.

The common basis of the procedures most often used to separate mixtures of organic compounds is the difference in *polarity* that exists or may be induced in the components of the mixture. This difference in polarity is exploited in nearly all the separation techniques, including distillation, recrystallization, extraction, and chromatography. The greatest differences in polarity, which make for the simplest separations, are those that exist between salts and nonpolar organic compounds. Whenever one or more of the components of a mixture can be converted to a salt, it can be separated easily and efficiently from the nonpolar components by extraction (Chapter 5) or distillation (Chapter 4).

In your introduction to qualitative organic analysis, you might be given a *mixture* of unknown compounds, each of which is to be identified. Before this can be done, each component of the mixture must be obtained in pure form. The following general approach illustrates how this may be done, utilizing the principles of separation based on differences in polarity.

General Scheme for Separating Simple Mixtures of Water-Insoluble Compounds

A procedure that is adequate for separating mixtures of liquid or solid carboxylic acids, phenols, amines, and neutral compounds is outlined schematically in Figure 27.1. For this scheme to be applicable, each component of the mixture must have a low solubility in water and must not undergo appreciable hydrolysis by reaction with dilute acids or bases at room temperature. The procedure is based primarily on the partition of compounds of significantly different polarities between diethyl ether and water and the separation of these liquid layers in a separatory funnel. The underlying concept of this process involves extraction, the theory of which is discussed in Section 5.4.

Assuming that a mixture contains a carboxylic acid, a phenol, an amine, and a neutral compound, the separation is initiated by dissolving the mixture in a suitable organic solvent such as diethyl ether. The ether solution of the mixture is extracted first with sodium bicarbonate solution, which removes the carboxylic acid by converting it to its water-soluble sodium salt. Second, the ether solution is extracted with sodium hydroxide solution, which removes the water-soluble sodium salt of the phenol. Third, the ether solution of the mixture is treated with hydrochloric acid, which reacts with the amine and converts it into a water-soluble ammonium salt. After removal of the aqueous solution, the remaining ethereal solution contains the neutral compound.

It should be pointed out that each extraction is performed on the *same* original solution of the mixture, and the *sequence* of extraction (with $NaHCO_3$, then $NaOH$, and finally HCl) is highly important. The bases and acid each remove one type of organic compound from the mixture and leave the neutral compound in the ether layer when the extractions have been completed. Each of the basic and acidic extracts is subsequently treated with acid or base to liberate the carboxylic acid, phenol, and amine from its salt, and then each of these compounds is removed from the respective aqueous solutions by extraction with ether. In the experimental procedures that follow, it should be noted that many different layers and solu-

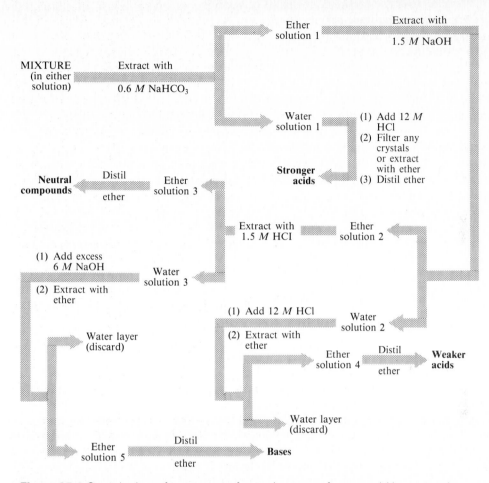

Figure 27.1 General scheme for separation of a simple mixture of water-insoluble compounds.

tions are obtained. For the successful completion of the separations it is suggested that the flasks containing each liquid be labeled as to identity of their contents, and that the flasks containing all layers and solutions be retained until it is certain that they are no longer needed.

EXPERIMENTAL PROCEDURE

DO IT SAFELY

Diethyl ether is removed from solutes after several extraction steps in this procedure. The ether should be dried over sodium or calcium sulfate, and the usual precautions against fire should be exercised, whether the ether is distilled using a safe heating source or is removed by using a water aspirator at room temperature.

The following experimental procedure, based on the scheme of Figure 27.1, may be used to separate the components of a mixture consisting of three or four compounds, 5 to 6 g of each, of the types described.

The mixture (15 to 25 g) is stirred with 50 mL of technical diethyl ether at room temperature. Any solid material that does not dissolve is collected on a filter and discarded. The ether filtrate is extracted with 20-mL portions of 0.6 M NaHCO$_3$ solution until the aqueous extract remains slightly basic. *Caution:* The separatory funnel should be vented quickly after the first mixing because of possible buildup of pressure from carbon dioxide. This solution, **water solution 1** of Figure 27.1, should contain the sodium salt of any carboxylic acid present in the mixture. The free organic acid is regenerated by careful acidification with 12 M hydrochloric acid. If a solid acid separates, collect it by vacuum filtration; otherwise, the aqueous acid solution is extracted with several 20-mL portions of diethyl ether, and the ether is distilled or evaporated.

The ether solution (**ether solution 1**) that was extracted with bicarbonate solution is extracted with two 20-mL portions of 1.5 M NaOH solution to remove a phenol or any other weak acid from the ether solution. The combined aqueous extracts (**water solution 2**) are acidified with 12 M hydrochloric acid, and the regenerated weak acid is extracted with several 20-mL portions of diethyl ether to yield **ether solution 4.** If amphoteric compounds were included in the unknown mixture, they would be carried through to the water layer separated from **ether solution 4.** The ether is removed from this solution as already described, leaving the weak acid as a residue.

Ether solution 2 is extracted with one or more 20-mL portions of 1.5 M hydrochloric acid until the aqueous extract (**water solution 3**) remains acidic. The ether layer, referred to in Figure 27.1 as **ether solution 3,** should contain any neutral organic compounds, which are recovered by removing the ether as described.

To **water solution 3** is added 5 M NaOH solution until the solution is strongly basic, and then it is extracted with several 20-mL portions of diethyl ether. The combined ether extracts (**ether solution 5**) should contain any organic base present in the original unknown mixture; the base may be isolated by removal of the ether.

It should be understood that this very generalized procedure may not give complete separation of all compounds of even the limited types for which it is intended. The products separated should be tested for purity by the usual methods, that is, by melting- or boiling-point determinations and, if possible, by gas or thin-layer chromatography. Before attempting identification of the individual compounds by any of the classical or modern instrumental methods described later, the samples should be purified by recrystallization, distillation, or chromatography.

27.4 Applications of Spectroscopic Methods to Qualitative Organic Analysis

One of the major limitations inherent to the classical system of qualitative organic analysis is that only *known compounds* can be identified. The research chemist is constantly faced with the task of identifying *new* compounds. Although information about the *type* of compound can be derived from the classical system, the complete identification of a new organic compound previously required a combination of degradation and synthesis in order to achieve correlation with a known substance and was usually a lengthy and laborious task. However, the advent of spectroscopic techniques has changed this picture dramatically; new and unknown compounds may be identified quickly and with certainty by using a combination of spectroscopic methods such as those described in Chapters 8–11. A number of examples might be cited of structural elucidations that were completed in a matter of days or weeks of molecules of greater complexity than those of compounds that defied the lifework of several of the great nineteenth- and twentieth-century organic chemists. One of the significant early examples of this was the application of IR spectroscopy and X-ray diffraction to the determination of the structure of the penicillin-G molecule during World War II.

Penicillin-G

The contemporary procedures that are used to identify organic compounds usually involve a combination of spectroscopic and classical methods. Ideally, a student should be introduced to application of spectroscopy in organic chemistry through the use of the instruments that produce the spectra, but this is not feasible in many instances because some of these instruments are very expensive. The next-best alternative is for students to have access to the spectra of typical known compounds and then to be provided with the spectra of ''unknowns'' to identify. This text contains more than 200 IR, PMR, and UV spectra of starting materials and of the products that are obtained from various preparative experiments. A careful study of these spectra, aided by the material presented in Chapters 8–11 and by additional discussion on the part of the instructor, should enable a student to make meaningful use of IR, PMR, and UV spectra in the identification of unknown organic compounds. Infrared and nuclear magnetic resonance spectroscopy are perhaps the most useful in this regard. Chapter 8 presents a collection of the IR spectra of compounds containing some of the more commonly encountered functional groups in organic chemistry, and the PMR spectra of the same compounds are given in Chapter 9. These spectra should be useful in helping deduce which functional groups may be present in an unknown substance.

The spectral data may serve to complement or supplement the ''wet'' qualitative classification tests and in many instances may substitute for these tests. For example, a strong IR absorption in the 1690–1760 cm^{-1} region is as indicative of the presence of a carbonyl group as formation of a 2,4-dinitrophenylhydrazone, and absorptions in the δ 6.0–8.5 region of a PMR spectrum are a more reliable indication of the presence of an aromatic ring than a color test with $CHCl_3$ and $AlCl_3$.

It must be emphasized that spectroscopic analysis requires *careful interpretation of the data*. Students should be cautioned not to go overboard in their enthusiasm with modern spectroscopy. Although some problems can be solved quickly and uniquely by modern spectroscopy, others require the intelligent application of both the modern methods and the classical methods. The remainder of this section illustrates the use of spectroscopy in structure determination.

Since the primary use of infrared spectroscopy is the identification of functional groups, the observation of certain IR absorptions provides information about the presence of particular functional groups in a compound. Conversely, the absence of certain peaks is useful in excluding the possibility of other groups. For example, the presence of a strong IR band in the $1650-1760$ cm^{-1} region suggests that a substance contains a carbon-oxygen double bond of an aldehyde, a ketone, a carboxylic acid, or a derivative of a carboxylic acid. On the other hand, the absence of such an absorption suggests that the compound contains none of these functional groups. Furthermore, a strong band in the $3500-3650$ cm^{-1} region points to a compound being an alcohol, a phenol, a carboxylic acid, an amine, or an amide, whereas the complete absence of such an absorption can be used to exclude these functional groups from further consideration. Because certain structural features in a molecule cause small shifts in the expected absorptions of some functional groups, infrared spectroscopy does not always provide a unique answer about the presence of a particular group. However, it is very useful in limiting the possibilities to a very few groups, and the uncertainty can then be resolved by performing just a few qualitative tests in the laboratory.

Although PMR spectra do not permit the direct observation of certain functional groups, they do provide indirect evidence regarding the presence or absence of some groups. The three major features of PMR spectra that are useful in compound identification are the chemical shift, the splitting pattern, and the relative abundance of each type of hydrogen, as determined by the peak areas. A proposed structure must be consistent with *each* of these features. When a PMR spectrum is obtained, the usual practice is first to determine the peak areas before trying to interpret the chemical shifts and splitting patterns.

It may be desirable to obtain a UV spectrum of the compound, but this is normally done *only* after it has been determined that the substance contains structural features that are amenable to UV studies. UV spectroscopy (Section 11.1) is useful for studying conjugated systems such as α,β-unsaturated carbonyl compounds and aromatic compounds, and the IR and PMR spectra of the compound often provide clues about the presence of these structural features, as is illustrated by Example 2 below.

The research chemist will usually obtain an IR and a PMR spectrum of a *pure* substance even before obtaining an elemental analysis or performing solubility tests. These types of spectra can be determined easily and quickly on a small amount of sample, and only after analyzing them does the researcher undertake other experimental work. This process can save many hours of unnecessary laboratory work. Other useful information about a compound can be obtained from mass spectrometry (Chapter 11), which can provide the molecular weight of the compound in question, and from elemental analysis, which gives the percentage composition by weight of the elements present and thus the empirical formula for the substance. If both the empirical formula and the molecular weight are known, the molecular formula of the compound can be determined.

In the following paragraphs, two examples are given to illustrate the value of the complementary application of classical qualitative analysis and spectral analysis for the determination of the structures of unknown compounds.

Example 1

A student was given a liquid of unknown structure, and the following information was obtained:

Boiling range: 143–145°C
Elemental analysis: No halogen, sulfur, or nitrogen
Solubility tests: Water-soluble to give a neutral solution
Spectra: IR and PMR spectra are shown in Figure 27.2.

Analysis Owing to its water-solubility, the compound most probably contains oxygen-bearing polar functional groups; it also probably contains a relatively small number of carbon atoms because compounds of more than five to six carbon atoms are usually water-insoluble. It is immediately evident from the IR spectrum shown in Figure 27.2 that the liquid contains

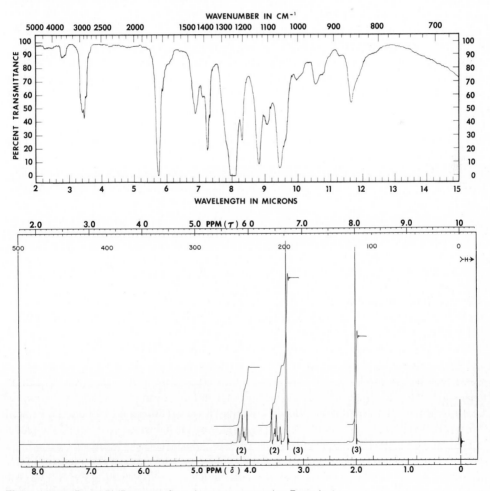

Figure 27.2 IR and PMR spectra for unknown compound in Example 1.

an ester group because of the strong absorption peaks at about 1750 and 1230 cm^{-1}; the former peak is due to C=O stretching and the latter is caused by the C—O—C bond. Although the 1750 cm^{-1} band could be indicative of a five-membered cyclic ketone, this and other possibilities are negated by the presence of numerous peaks in the PMR spectrum. The student confirmed the presence of an ester by obtaining a positive qualitative test for this group (Section 27.5j).

The PMR spectrum shows four regions of absorption: a multiplet centered at δ 4.1, another multiplet centered at δ 3.5, a singlet at δ 3.3, and another singlet at δ 2.0. Measurement of the height (in mm) of the integration peaks above each absorption gives values of 16:16:24:24, respectively, which corresponds to a relative hydrogen abundance of 2:2:3:3 for these four peaks. The two singlets at δ 2.0 and 3.3 each integrate for three protons and must represent two methyl groups with no neighboring hydrogens. The higher-field peak at δ 2.0 is characteristic of a methyl ketone or acetate, the latter being consistent with the deduction provided by the IR spectrum and the qualitative test that the unknown is an ester. The second methyl peak at δ 3.3 is shifted downfield, as would be expected if the methyl group were bonded to oxygen. This peak is too far upfield for a methyl ester, since methyl esters normally show the methyl absorption at about δ 3.7–4.1. However, it is within the δ 3.3–4.0 range frequently observed for aliphatic α-hydrogens of an alcohol or ether. Because neither spectrum shows any evidence for an —OH group, the compound probably contains a CH$_3$O— group, and the presence of an aliphatic ether is consistent with the C—O—C absorption observed at 1050 cm^{-1} in the IR spectrum. The multiplets centered at δ 4.1 and 3.5 each integrate for two hydrogens and show mutual spin-spin coupling. This pattern is diagnostic of chemically nonequivalent, adjacent methylene groups such as X—CH$_2$CH$_2$—Y. The downfield absorption for each of the CH$_2$ groups indicates that each is bonded to oxygen.

By reference to Table 27.13a for liquid esters, only a single structure is seemingly consistent with all of the spectral observations: *2-methoxyethyl acetate* (**1**, bp 145°C).

$$CH_3O—CH_2CH_2—O—\overset{\overset{\textstyle O}{\|}}{C}—CH_3$$

1

2-Methoxyethyl acetate

The student confirmed this structural assignment by hydrolyzing the ester and converting the resulting alcohol into a α-naphthyl urethane derivative (Section 27.5e), which exhibited the correct melting point (Table 27.3a).

Now suppose the student had also been provided with the elemental percentage composition by weight for the unknown: 50.83% C, 8.53% H, and 40.63% O. The empirical formula could then be determined as follows. Assuming a 100-g sample of the liquid, it therefore contains 50.83 g of oxygen, 8.53 g of hydrogen and 40.63 g of oxygen. First, the weights of each element are converted into moles by dividing the weight of each by its atomic weight. Then the number of moles of each element is divided by the number of moles of that element present in the smallest amount to obtain a molar ratio, which is often a whole-number ratio; if not, the ratio must be multiplied by some number before whole numbers are obtained. In this example, oxygen is present in the smallest molar amount, so the number of moles of each other element is divided by 2.54. These calculations are summarized on the next page.

Element	Weight (g)	Conversion to Moles	Conversion to Molar Ratio
C	50.83	$\dfrac{50.83 \text{ g}}{12.01 \text{ g/mole}} = 4.23$ moles	$\dfrac{4.23}{2.54} = 1.67$
H	8.53	$\dfrac{8.53 \text{ g}}{1.01 \text{ g/mole}} = 8.54$ moles	$\dfrac{8.54}{2.54} = 3.33$
O	40.63	$\dfrac{40.63 \text{ g}}{16.00 \text{ g/mole}} = 2.54$ moles	$\dfrac{2.54}{2.54} = 1.00$

Example 2

The following information was obtained from laboratory work that was performed on an unknown compound:

Boiling range: 227–230°C
Elemental analysis: No halogen, sulfur, or nitrogen
Solubility tests: Soluble only in concentrated sulfuric acid
Spectra: IR and PMR spectra are shown in Figure 27.3; UV spectrum of compound in methanol had two maxima at 235 nm ($\epsilon = 8630$) and at 304 nm ($\epsilon = 579$).

Analysis In order to identify potential functional groups, the spectra were analyzed as follows. The PMR spectrum shows four regions of absorption at δ 6.75, 4.78, 2.1–3.0, and 1.75, with relative areas (in mm) of 14:28:75:88, respectively. This pattern of integration is consistent with a relative hydrogen abundance of 1:2:5:6. The large singlet at δ 1.75 represents six protons and is probably caused by two methyl groups in closely similar chemical environments. They are probably attached to sp^2-hybridized carbon atoms because of the value of their chemical shifts and the presence in the IR spectrum of absorptions at 3000–3100, 1645, and 895 cm^{-1}, which are all characteristic of alkenic absorptions. The band at 895 cm^{-1} and the two-proton PMR absorption at δ 4.78 are at the positions expected for a terminal double bond (C=CH$_2$). The strong IR band at 1680 cm^{-1} indicates that the compound probably contains either an α,β-unsaturated ketone or aldehyde group, the latter being effectively eliminated from consideration owing to the absence of any additional evidence for —CHO in either of the spectra. The conclusion that an α,β-unsaturated ketone is present is supported by the UV spectrum; the maximum occurring at the shorter wavelength is attributed to the $\pi \rightarrow \pi^*$ excitation, whereas the weaker absorption at longer wavelength is assigned to the $n \rightarrow \pi^*$ transition (see Section 11.1).

Although the PMR peak at δ 6.75, representing one proton, is shifted downfield considerably, its position is in keeping with the possible presence of a conjugated ketone. Olefinic hydrogens beta to the carbonyl in these functions are strongly deshielded because of charge separation in contributing resonance structures:

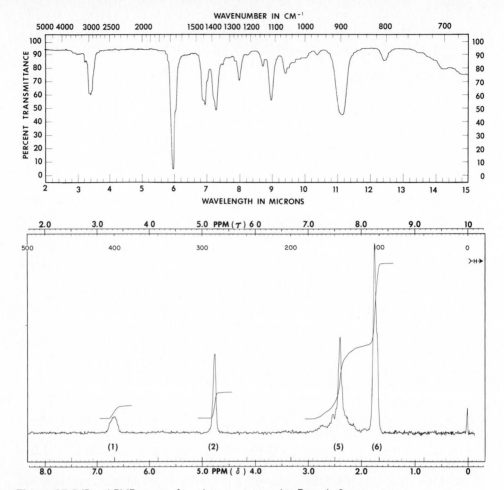

Figure 27.3 IR and PMR spectra for unknown compound in Example 2.

The four-proton multiplet at δ 2.1–3.0 represents aliphatic hydrogens deshielded by adjacent groups such as vinyl or carbonyl and, on the basis of the chemical shift and general appearance of the multiplet, are probably methylene and/or methine hydrogens.

The compound was chemically shown to contain *at least* one carbon-carbon double bond through the observation that it decolorized both bromine in CCl_4 and aqueous potassium permanganate solutions. That the compound was a ketone was verified through the formation of a 2,4-dinitrophenylhydrazone derivative (mp 190–191°C). Reference to Table 27.15a led to the identification of the unknown as carvone **(2)**, an essential oil that can be isolated from caraway seed or spearmint oil, as described in Section 8.5.

As further confirmation of this structure, a quantitative elemental analysis was obtained with the following percentage by weight composition: C 79.97%, H 9.39%, and O 10.65%. It is left as an exercise to determine the empirical formula of this compound from the data given.

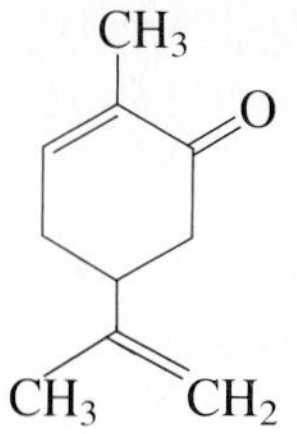

2
Carvone

The preceding examples should foster an appreciation of the complementary aspects of spectral analysis and the ''wet'' classification scheme, including the use of physical properties. Although the spectra were not in themselves specifically definitive in the assignment of the structures of the unknowns, they were used to implicate the possible presence of certain functional groups and suggested appropriate classification tests. Thus, the spectral analyses helped avoid the effort and time that might conceivably have been spent in performing numerous classification tests, many of which would have been negative. The final decisions regarding the structures of the unknowns came from derivative preparation. On the other hand, it is often possible to find the IR and PMR spectra of the suspected compounds in one of the catalogues of spectra given in the references at the end of Chapters 8 and 9, and if the reported spectra are identical to the ones obtained for the unknowns, the identity of the unknowns is confirmed.

REFERENCES

1. Shriner, R. L.; Fuson, R. C.; Curtin, D. Y.; Morrill, T. C. *The Systematic Identification of Organic Compounds,* 6th ed., John Wiley & Sons, New York, 1980.

2. Cheronis, N. D.; Entrikin, J. B.; Hodnett, E. M. *Semimicro Qualitative Organic Analysis,* 3d ed., Interscience Publishers, New York, 1965.

3. Hogness, T. R.; Johnson, W. C.; Armstrong, A. R. *Qualitative Analysis and Chemical Equilibrium,* 5th ed., Holt, Rinehart and Winston, New York, 1966.

4. Pasto, D. J.; Johnson, C. R. *Organic Structure Determination,* Prentice-Hall, Englewood Cliffs, N.J., 1969.

5. *Handbook of Tables for Identification of Organic Compounds,* 3d ed., Zvi Rapoport, editor, Chemical Rubber Company, Cleveland, 1967. Gives physical properties and derivatives for more than 4000 compounds, arranged according to functional groups.

27.5 Qualitative Classification Tests and Preparation of Derivatives

This section contains discussions about and experimental procedures for the qualitative classification tests and preparation of derivatives for compounds having most of the commonly encountered functional groups. References are also provided for the tables of compounds in Section 27.6, which list boiling points and melting points of many compounds and the melting points of suitable derivatives.

(a) Aldehydes and Ketones (Tables 27.4 and 27.15)

1. CLASSIFICATION TESTS

A. 2,4-Dinitrophenylhydrazine Aryl hydrazines, $ArNHNH_2$, are commonly used to make crystalline derivatives of carbonyl compounds (Section 21.1), and they include phenylhydrazine, *p*-nitrophenylhydrazine, and 2,4-dinitrophenylhydrazine. The formation of a 2,4-dinitrophenylhydrazone (**3**) is represented by equation 9.

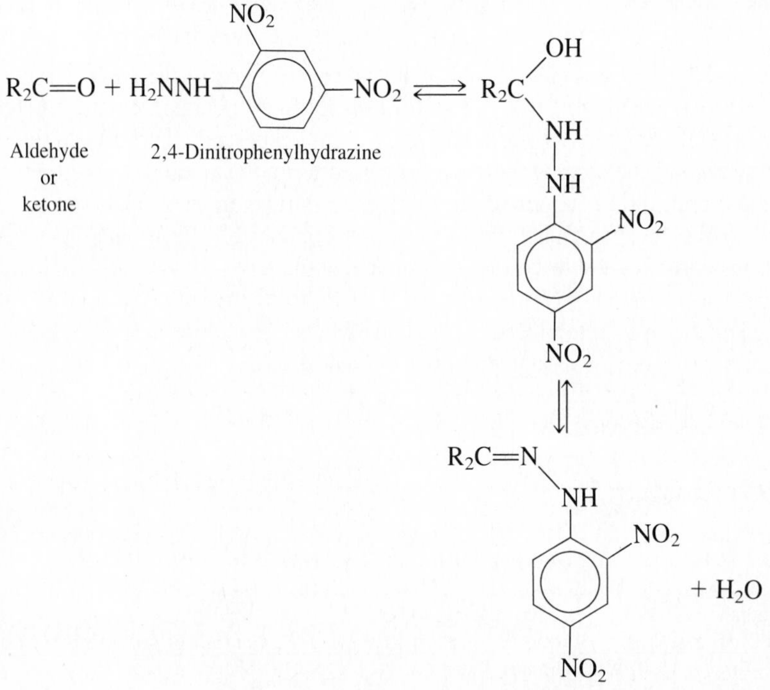

A 2,4-dinitrophenylhydrazone

2,4-Dinitrophenylhydrazine will give a positive test for *either* an aldehyde or a ketone. Since the products of the tests may be used as derivatives of the aldehyde or ketone, 2,4-dinitrophenylhydrazine is a valuable reagent for both classification and derivative formation. The Tollens' test given below provides a method for distinguishing between these two types of compounds.

Experimental Procedure

If the reagent is not supplied, it is prepared by dissolving 1 g of 2,4-dinitrophenylhydrazine in 5 mL of concentrated sulfuric acid. This solution is then added, with stirring, to 7 mL of water and 25 mL of 95% ethanol. After being stirred vigorously, the solution is filtered from any undissolved solid.

Dissolve 1 or 2 drops of a liquid (or about 100 mg of a solid) in 2 mL of 95% ethanol, and add this solution to 2 mL of the 2,4-dinitrophenylhydrazine reagent. Shake the mixture vigorously; if a precipitate does not form immediately, let the solution stand for 15 min.

If more crystals are desired for a melting-point determination, dissolve 200–500 mg of the carbonyl compound in 20 mL of 95% ethanol, and add this solution to 15 mL of the reagent. The product may be recrystallized from aqueous ethanol.

B. Tollens' Test   A method for distinguishing between aldehydes and ketones is the Tollens' test. A positive test indicates the presence of an aldehyde function, whereas no reaction occurs with ketones. Tollens' reagent consists of silver ammonia complex, $Ag(NH_3)_2^{\oplus}$, in an ammonia solution. This reagent is reduced on reaction with an aliphatic or aromatic aldehyde, whereas the aldehyde is oxidized to the corresponding carboxylic acid; the silver is reduced from the +1 state to elemental silver and frequently is deposited as a silver mirror on the glass wall of the test tube. Thus, the formation of the silver mirror is considered a positive test. Equation 10 shows the reaction that occurs.

$$RCHO + 2\ Ag(NH_3)_2^{\oplus} + 2\ HO^{\ominus} \rightarrow 2\ \underline{Ag} + RCO_2^{\ominus}NH_4^{\oplus} + H_2O + 3\ NH_3 \quad \textbf{(10)}$$

Similar tests for aldehydes make use of Fehling's and Benedict's reagents, which contain complex salts (tartrate and citrate, respectively) of cupric ion as the oxidizing agents. With these reagents a positive test is the formation of a brick-red precipitate of cuprous oxide (Cu_2O), which forms when Cu^{2+} is reduced to Cu^+ by the aldehyde. These two tests are more useful in distinguishing between aliphatic and aromatic aldehydes, since the aliphatic compounds give a fast reaction. They have been used widely to detect reducing sugars (Chapter 25), whereas the Tollens' test is used to distinguish between aldehydes (aliphatic or aromatic) and ketones.

Experimental Procedure

Prepare Tollens' reagent by mixing solution A (2.5 g of silver nitrate in 43 mL of distilled water) and solution B (3 g of potassium hydroxide in 42 mL of distilled water) according to the following directions. Obtain 3 mL of both solutions A and B. To 3 mL of solution A, add concentrated ammonium hydroxide solution dropwise until the initial brown precipitate begins to clear. The solution should be grayish and almost clear. Add 3 mL of solution B, and again add concentrated ammonium hydroxide dropwise until the solution is almost clear.

To carry out the test, add 0.5 mL of the reagent to 3 drops or 50–100 mg of the unknown compound; the formation of a silver mirror or black precipitate constitutes a positive test. The silver will deposit so as to form a mirror only on a *clean* glass surface. A black precipitate, although not so aesthetically pleasing, still constitutes a positive test. If no reaction occurs at room temperature, warm the solution slightly in a beaker of warm water.

At the end of the laboratory period discard any unused Tollens' reagent; *do not store the solution because it decomposes on standing and yields an explosive precipitate.*

C. Chromic Acid Test   Another method for distinguishing between aldehydes and ketones is the chromic acid test, which utilizes chromic acid, prepared by dissolving chromic anhydride in sulfuric acid (see Section 19.2 for discussion) and used in acetone solution. This reagent oxidizes primary and secondary alcohols and all aldehydes with a distinctive color change, but it gives no visible reaction with tertiary alcohols and ketones under the conditions of the test. The reactions are shown in equation 11.

$$RCH_2OH \xrightarrow{H_2CrO_4} RCHO \xrightarrow{H_2CrO_4} RCO_2H \qquad \textbf{(11a)}$$

$$R_2CHOH \xrightarrow{H_2CrO_4} R_2CO \xrightarrow{H_2CrO_4} \text{no visible reaction} \qquad \textbf{(11b)}$$

$$R_3COH \xrightarrow{H_2CrO_4} \text{no visible reaction} \qquad \textbf{(11c)}$$

Thus the chromic acid reagent gives a clear-cut distinction between primary and secondary alcohols and aldehydes on the one hand and tertiary alcohols and ketones on the other. Aldehydes may be distinguished from primary and secondary alcohols by means of Tollens', Benedict's, or Fehling's test, and lower molecular weight primary and secondary alcohols may be differentiated on the basis of their rates of reaction with concentrated hydrochloric acid containing zinc chloride—the Lucas reagent (Section 27.5e, part 1B).

Experimental Procedure

The chromic acid reagent, if not available, is prepared as follows. Add 25 g of chromic anhydride (CrO_3) to 25 mL of concentrated sulfuric acid and stir until a smooth paste is obtained. Dilute the paste *cautiously* with 75 mL of distilled water, and stir until a clear orange solution is obtained.

Place in a test tube 1 mL of *reagent-grade* acetone (or solvent-grade that has been distilled from potassium permanganate), and dissolve in it 1 drop of a liquid or about 10 mg of a solid alcohol or carbonyl compound. Add 1 drop of the acidic chromic anhydride reagent to the acetone solution, and shake the tube to mix the contents. A positive oxidation reaction is indicated by disappearance of the orange color of the reagent and the formation of a green or blue-green precipitate or emulsion.

Primary and secondary alcohols and aliphatic aldehydes give a positive test within 5 sec. Aromatic aldehydes require 30 to 45 sec. Color changes occurring after about 1 min should not be interpreted as positive tests; other functional groups such as ethers and esters may slowly hydrolyze under the conditions of the test, releasing alcohols that in turn provide "false-positive" tests. Tertiary alcohols and ketones produce no visible change in several minutes. Phenols and aromatic amines give dark precipitates, as do aromatic aldehydes having hydroxyl or amino groups on the aromatic ring.

D. Iodoform Test When an aldehyde or ketone that has α-hydrogens is treated with a halogen in basic medium, halogenation occurs on the α-carbon (equation 12).

$$R_2C-\overset{\displaystyle O}{\overset{\|}{C}}\diagdown_R \xrightarrow{\text{base, } X_2} R_2C-\overset{\displaystyle O}{\overset{\|}{C}}\diagdown_R \qquad \textbf{(12)}$$

Aldehyde or ketone

This reaction involves the formation of an α-stabilized carbanion (equation 11, Section 21.1), which subsequently reacts with halogen to produce the substitution product. In the case of acetaldehyde or a methyl ketone, all three of the α-hydrogens of the methyl group are replaced by halogen to give **4** (equation 13), which then reacts with base to give products **5** and **6** (equation 14).

$$\underset{\text{R = H, alkyl, or aryl}}{H-\overset{\overset{\displaystyle H}{|}}{\underset{\underset{\displaystyle H}{|}}{C}}-\overset{\displaystyle O}{\underset{\displaystyle R}{C}}} \xrightarrow{X_2,\ NaOH} \underset{\textbf{4}}{X-\overset{\overset{\displaystyle X}{|}}{\underset{\underset{\displaystyle X}{|}}{C}}-\overset{\displaystyle O}{\underset{\displaystyle R}{C}}} \qquad \textbf{(13)}$$

$$\underset{\textbf{4}}{X_3C-\overset{\displaystyle O}{\underset{\displaystyle R}{C}}} \xrightarrow{NaOH} \underset{\substack{\textbf{5}\\ \text{Haloform}}}{CHX_3} + \underset{\substack{\textbf{6}\\ \text{Sodium salt}\\ \text{of a carboxylic acid}}}{RCO_2^{\ominus}Na^{\oplus}} \qquad \textbf{(14)}$$

Although chlorine, bromine and iodine react in this manner, the qualitative test for a methyl ketone utilizes iodine because it is safer to use and because the product is *iodoform*, HCI_3, a highly insoluble crystalline *yellow* solid that is readily observed and identified on the basis of its characteristic odor and melting point. It should be pointed out that a positive iodoform test involves two experimental observations: (1) the characteristic red-brown color of iodine must be discharged as the test reagent is added to the compound being tested, *and* (2) a yellow precipitate of iodoform must form. The reason is that all aldehydes and ketones containing α-hydrogens react with iodine in base and thus decolorize the test reagent. Only the methyl ketones, after being trisubstituted with iodine on the methyl group, react with base to produce iodoform and the salt of a carboxylic acid.

This qualitative test also occurs for compounds having a hydroxymethyl ($-CHOHCH_3$) functionality, and it is observed for two specific compounds, ethanol and ethanal (acetaldehyde). The reason that a hydroxymethyl compound gives a positive iodoform test is that this group is first oxidized to a methyl ketone by the reagent, and the methyl ketone then reacts with iodine in the presence of base, as discussed above. Thus, it is always necessary to consider this alternative before making a final decision. This ambiguity in interpretation of a positive iodoform test is solved by determining whether a keto group is present or absent by using another qualitative test such as 2,4-dinitrophenylhydrazine.

The iodoform test is sometimes called the *hypoiodite test* because sodium hypoiodite, NaOI, is formed by the reaction between the base, NaOH, and iodine: $I_2 + 2\ NaOH \rightarrow NaI + NaOI + H_2O$. In more general terms, the reactions described above can be called the *haloform test* or the *sodium hypohalite test* when the specific halogen is not specified.

Experimental Procedure

If not already prepared, the iodine reagent is prepared by dissolving 25 g of iodine in a solution of 50 g of potassium iodide in 200 mL of water. The potassium iodide is added to increase the solubility of iodine in water, owing to the formation of potassium triiodide, KI_3, by the reaction: $I_2 + KI \rightarrow KI_3$.

If the substance is water-soluble, dissolve 2 to 3 drops of a liquid or an estimated 50 mg of a solid in 2 mL of water in a small test tube, add 2 mL of 3 M sodium hydroxide, and then slowly add 3 mL of iodine solution. In a positive test the brown color disappears, and yellow iodoform separates. If the substance tested is insoluble in water, dissolve it in 2 mL of dioxane, proceed as above, and at the end dilute with 10 mL of water.

Iodoform can be recognized by its odor and yellow color and, more definitely, by its melting point, 119°C. The substance can be isolated by vacuum filtration of the test mixture or by adding 2 mL of chloroform, shaking the stoppered test tube to extract the iodoform into the small lower layer, withdrawing the clear part of this layer with a capillary dropping tube, and evaporating it in a small tube on the steam bath. The crude solid is recrystallized from methanol-water.

2. DERIVATIVES

Two of the most useful solid derivatives of aldehydes and ketones are the *2,4-dinitrophenylhydrazones* and the *semicarbazones*. Oximes are also sometimes useful, but they often form as oils rather than solids.

A. 2,4-Dinitrophenylhydrazones   The qualitative test for aldehydes and ketones is described in part 1A above, and the solid that forms may be isolated and purified as indicated in that section.

B. Semicarbazones   Semicarbazide reacts with aldehydes and ketones to produce derivatives that are called *semicarbazones* (equation 15).

$$R_2C{=}O + H_2N{-}NHCNH_2 \rightarrow R_2C{=}N{-}NHCNH_2 + H_2O \qquad \textbf{(15)}$$

Aldehyde   Semicarbazide   A semicarbazone
or
ketone

Because semicarbazide is unstable as the free base, it is usually stored in the form of the hydrochloric acid salt, or "hydrochloride." In the procedure that follows, it is liberated from the hydrochloride by addition of sodium acetate.

Experimental Procedure

Dissolve 0.5 g of semicarbazide hydrochloride and 0.8 g of sodium acetate in 5 mL of water in a test tube, and then add about 0.5 mL of the carbonyl compound. Stopper and shake the tube vigorously, remove the stopper, and place the test tube in a beaker of boiling water. Discontinue heating the water, and allow the test tube to cool to room temperature in the beaker of water. Remove the test tube to an ice-water bath, and scratch the side of the tube with a glass rod at the interface between the liquid and air. The semicarbazone may be recrystallized from water or aqueous ethanol.

If the carbonyl compound is insoluble in water, dissolve it in 5 mL of ethanol. Add water until the solution becomes turbid, then add a little ethanol until the turbidity disappears. Add the semicarbazide hydrochloride and sodium acetate, and continue as above from this point.

C. Oximes   Hydroxylamine reacts with aldehydes or ketones to yield derivatives that are called *oximes* (equation 16), and these have the limitation that they are frequently not crystalline solids. Hydroxylamine is usually stored as the hydrochloric acid salt or "hydrochloride" because it is not stable as the free base; it is liberated from the hydrochloride by addition of base, as indicated in the Experimental Procedure.

$$R_2C{=}O \; + \; H_2N{-}OH \; \rightarrow \; R_2C{=}N{-}OH + H_2O \qquad \text{(16)}$$

| Aldehyde | Hydroxylamine | Oxime |
| or ketone | | |

Experimental Procedure

Dissolve 0.5 g of hydroxylamine hydrochloride in 5 mL of water and 3 mL of 3 M sodium hydroxide solution, and then add 0.5 g of the aldehyde or ketone. If the carbonyl compound is insoluble in water, add just enough ethanol to give a clear solution. Warm the mixture on a steam bath or boiling-water bath for 10 min, and then cool it in an ice-water bath. If crystals do not form immediately, scratch with a glass rod the side of the container at and below the liquid level. The oxime may be recrystallized from water or aqueous ethanol.

In some cases the use of 3 mL of pyridine and 3 mL of absolute ethanol in place of the 3 mL of 3 M solution hydroxide solution and 5 mL of water will be found to be more effective. A longer heating period is often necessary. After the heating is finished, pour the mixture into an evaporating dish, and remove the solvent with a current of air in a hood. Grind the solid residue with 3–4 mL of cold water and filter the mixture. Recrystallize the oxime from water or aqueous ethanol.

(b) Alkenes and Alkynes (Table 27.6)

1. CLASSIFICATION TESTS

Two common types of unsaturated compounds are alkenes and alkynes, characterized by the carbon-carbon double and triple bond, respectively, as the functional group. There are no simple direct ways to prepare solid derivatives of unsaturated aliphatic compounds having no other functional groups, but it is often useful to detect the presence of these two functional groups. The two common qualitative tests for unsaturation are the reaction of the compounds with *bromine in carbon tetrachloride* and with *potassium permanganate*. In both cases a positive test is denoted by decoloration of the reagent.

A. Bromine in Carbon Tetrachloride   Bromine will add to the carbon-carbon double bond of alkenes to produce dibromoalkanes (equation 17) and with alkynes to produce tetrabromoalkanes (equation 18). When this reaction occurs, molecular bromine is consumed, and its characteristic dark red-brown color disappears if bromine is not added in excess. The *rapid* disappearance of the bromine color is a positive test for unsaturation.

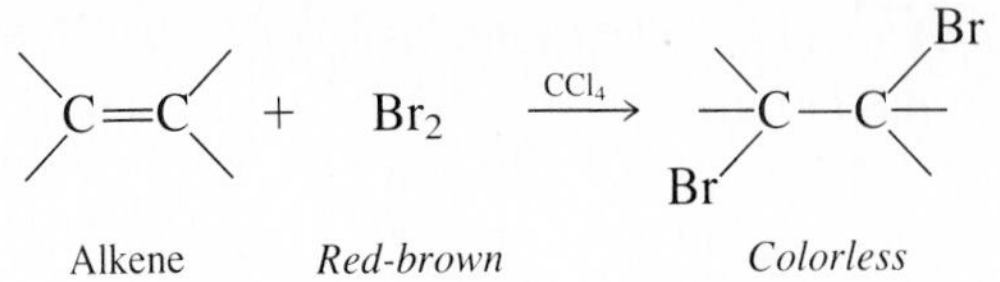

$$\begin{array}{c}\diagup\kern-0.5em \diagdown \\ C=C \\ \diagup\kern-0.5em \diagdown \end{array} + Br_2 \xrightarrow{CCl_4} \begin{array}{c} Br \\ | \\ -C-C- \\ | \\ Br \end{array} \quad (17)$$

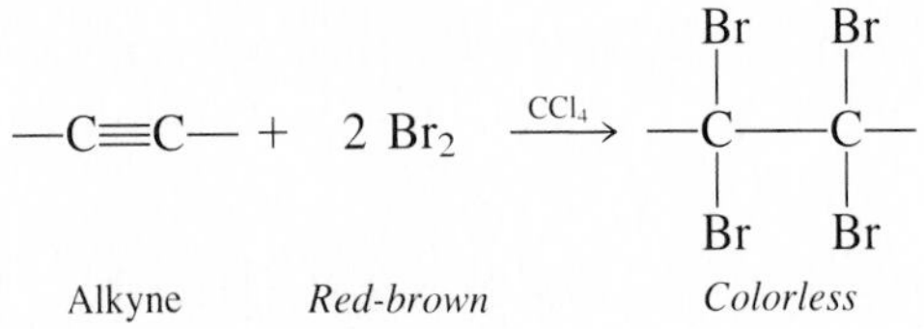

$$\underset{\text{Alkene}}{} \quad \underset{\text{Red-brown}}{} \quad \underset{\text{Colorless}}{}$$

$$-C\equiv C- + 2\ Br_2 \xrightarrow{CCl_4} \begin{array}{cc} Br & Br \\ | & | \\ -C-C- \\ | & | \\ Br & Br \end{array} \quad (18)$$

$$\underset{\text{Alkyne}}{} \quad \underset{\text{Red-brown}}{} \quad \underset{\text{Colorless}}{}$$

The test is not unequivocal, however, because some alkenes do not react with bromine, and some react very slowly. In the case of a negative test, therefore, the potassium permanganate test should be performed.

More details about the reaction of bromine in carbon tetrachloride with alkenes can be found in Section 13.3.

Experimental Procedure

To 1 or 2 mL of 0.1 *M* bromine in carbon tetrachloride solution, add 1 or 2 drops of the unknown. *Rapid* disappearance of the bromine color to give a colorless solution is a positive test for unsaturation.

B. Potassium Permanganate (the Baeyer Test) A second qualitative test for unsaturation, the Baeyer test, depends on the ability of potassium permanganate to oxidize the carbon-carbon double bond to give alkanediols (equation 19) or the carbon-carbon triple bond to give carboxylic acids (equation 20).

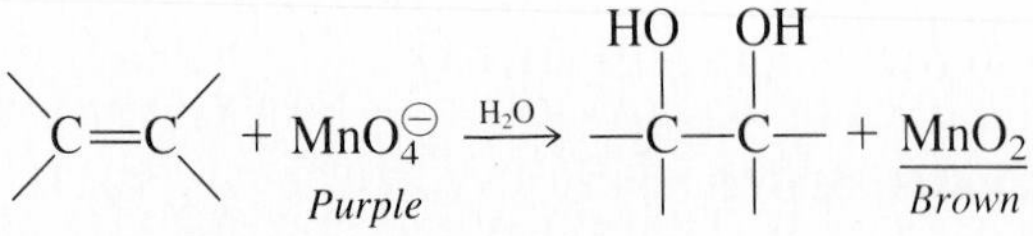

$$\begin{array}{c}\diagup\kern-0.5em \diagdown \\ C=C \\ \diagup\kern-0.5em \diagdown \end{array} + \underset{Purple}{MnO_4^\ominus} \xrightarrow{H_2O} \begin{array}{cc} HO & OH \\ | & | \\ -C-C- \\ | & | \end{array} + \underset{Brown}{MnO_2} \quad (19)$$

$$R-C\equiv C-R' + \underset{Purple}{MnO_4^\ominus} \xrightarrow{H_2O} R-CO_2H + HO_2C-R' + \underset{Brown}{MnO_2} \quad (20)$$

The permanganate is destroyed in the reaction, and a brown precipitate of MnO_2 is produced. The disappearance of the characteristic color of the permanganate ion is a positive test for unsaturation. However, care must be taken, since compounds containing certain other types of functional groups (for example, aldehydes, containing the $-CH{=}O$ group) also decolorize permanganate ion.

Experimental Procedure

Dissolve 1 or 2 drops of the unknown in 2 mL of 95% ethanol, then add 0.1 M $KMnO_4$ solution dropwise, observing the results. Count the number of drops added before the permanganate color persists. For a *blank determination,* count the number of drops that may be added to 2 mL of 95% ethanol before the color persists. A significant difference in the number of drops required in the two cases is a positive test for unsaturation.

(c) Alkyl Halides (Haloalkanes) (Table 27.7)

1. CLASSIFICATION TESTS

Qualitative tests for alkyl halides are useful in deciding whether the compound in question is a primary, secondary, or tertiary halide. In general it is quite difficult to prepare solid derivatives of alkyl halides, so we limit this discussion to the two qualitative tests: (a) the reaction with *alcoholic silver nitrate* solution and (b) the reaction with *sodium iodide in acetone.*

A. Alcoholic Silver Nitrate If a compound is known to contain a halogen (bromine, chlorine, or iodine), information concerning its environment may be obtained from observation of its reaction with alcoholic silver nitrate. The overall reaction is shown in equation 21.

$$RX + AgNO_3 \xrightarrow{\text{ethanol}} \underline{AgX} + RONO_2 \qquad \textbf{(21)}$$

Such a reaction will be of the S_N1 type. As asserted in Section 18.1, tertiary halides are more reactive in an S_N1 reaction than secondary halides, which are in turn more reactive than primary halides. Differing rates of silver halide precipitation would be expected from halogen in each of these environments, namely, primary $<$ secondary $<$ tertiary. These differences are best determined by testing in separate test tubes authentic samples of primary, secondary, and tertiary halides with silver nitrate and observing the results.

Alkyl bromides and iodides react more rapidly than chlorides, and the latter may require warming to produce a reaction in a reasonable period.

Aryl halides are unreactive toward the test reagent, as are any vinyl or alkynyl halides generally. Allylic and benzylic halides, even when primary, show reactivities as great as or greater than tertiary halides because of resonance stabilization of the resulting allyl or benzyl carbocations.

Experimental Procedure

Add 1 drop of the alkyl halide to 2 mL of a 0.1 M solution of silver nitrate in 95% ethanol. If no reaction is observed within 5 min at room temperature, warm the mixture in a beaker of boiling water and observe any change. Note the color of any precipitates; silver chloride is white, silver bromide is pale yellow, and silver iodide is yellow. If there is any precipitate, add several drops of 1 M nitric acid solution to it, and note any changes; the silver halides are insoluble in acid. To determine expected reactivities, test known primary, secondary, and tertiary halides in this manner. If possible, use alkyl iodides, bromides, and chlorides so that differences in halogen reactivity can also be observed.

B. Sodium Iodide in Acetone   Another method for distinguishing between primary, secondary, and tertiary halides makes use of sodium iodide dissolved in acetone. This test complements the alcoholic silver nitrate test, and when these two tests are used together, it is possible to determine fairly accurately the gross structure of the attached alkyl group.

The test depends on the fact that both sodium chloride and sodium bromide are not very soluble in acetone, whereas sodium iodide is. The reactions that occur (equations 22 and 23) are S_N2 substitutions in which iodide ion is the nucleophile; the order of reactivity is primary > secondary > tertiary.

$$RCl + NaI \xrightarrow{\text{acetone}} RI + \underline{NaCl} \tag{22}$$

$$RBr + NaI \xrightarrow{\text{acetone}} RI + \underline{NaBr} \tag{23}$$

With the reagent, primary bromides give a precipitate of sodium bromide in about 3 min at room temperature, whereas the primary and secondary chlorides must be heated to about 50°C before reaction occurs. Secondary and tertiary bromides react at 50°C, but the tertiary chlorides fail to react in a reasonable time. It should be noted that this test is necessarily limited to bromides and chlorides.

Experimental Procedure

Place 1 mL of the sodium iodide–acetone test solution in a test tube, and add 2 drops of the chloro or bromo compound. If the compound is a solid, dissolve about 50 mg of it in a minimum volume of acetone, and add this solution to the reagent. Shake the test tube, and allow it to stand for 3 min at room temperature. Note whether a precipitate forms; if no change occurs after 3 min, warm the mixture in a beaker of water at 50°C. After 6 min of heating, cool to room temperature, and note whether any precipitate forms. Occasionally a precipitate forms immediately after combination of the reagents; this represents a positive test only if the precipitate remains after shaking the mixture and allowing it to stand for 3 min.

Carry out this reaction with a series of primary, secondary, and tertiary halides, both chlorides and bromides. Note in all cases the differences in reactivity as evidenced by the rate of formation of sodium bromide or chloride.

(d) Aromatic Hydrocarbons and Aryl Halides (Tables 27.10 and 27.11)

1. CLASSIFICATION TEST

This test for the presence of an aromatic ring should be performed only on compounds that have been shown to be insoluble in concentrated sulfuric acid (see Solubility Tests). The test involves the reaction between an aromatic compound and chloroform in the presence of anhydrous aluminum chloride catalyst. The colors produced in this type of reaction are often quite characteristic for certain aromatic compounds, whereas aliphatic compounds give little or no color with this test. Some typical examples are tabulated below. Often these colors change with time and ultimately yield brown-colored solutions. Carbon tetrachloride may be used in place of chloroform; it yields similar colors.

Type of Compound	Color
Benzene and homologs	Orange to red
Aryl halides	Orange to red
Naphthalene	Blue
Biphenyl	Purple

The test is based upon a series of Friedel-Crafts alkylation reactions; for benzene, the ultimate product is triphenylmethane (equation 24).

$$3 \ C_6H_6 + CHCl_3 \xrightarrow{\text{AlCl}_3} (C_6H_5)_3CH + 3 \ HCl \qquad \textbf{(24)}$$

The colors arise owing to formation of species such as triphenylmethyl cations $(C_6H_5)_3C^+$, which remain in the solution as $AlCl_4^-$ salts; ions of this sort are highly colored owing to the extensive delocalization of charge that is possible throughout the three aromatic rings.

The test is significant if positive, but a negative test does not rule out an aromatic structure; some compounds are so unreactive that they do not readily undergo Friedel-Crafts reactions.

Positive tests for aryl halides are difficult to obtain directly, and some of the best evidence for their presence involves indirect methods. Elemental analysis will indicate the presence of halogen. If *both* the silver nitrate and sodium iodide–acetone tests are negative, then the compound is most likely a vinyl or an aromatic halide, both of which are very unreactive toward silver nitrate and sodium iodide. Distinction between a vinyl and an aromatic halide can be made by means of the aluminum chloride–chloroform test.

Experimental Procedure

Heat about 100 mg of *anhydrous* aluminum chloride in a Pyrex test tube held almost horizontally until the material has sublimed to 3 or 4 cm above the bottom of the tube. Allow the tube to cool until it is almost comfortable to touch, and then add down the side of the tube about 20 mg of a solid, or 1 drop of a liquid, unknown, followed by 2 or 3 drops of chloroform. The appearance of a bright color ranging from red to blue where the sample and chloroform come in contact with the aluminum chloride is a positive indication of an aromatic ring.

2. DERIVATIVES

Two types of derivatives can be used to characterize aromatic hydrocarbons and aryl halides. These are prepared by (a) nitration and (b) side-chain oxidation. The second method involves oxidation of a side chain to a carboxylic acid group. Since carboxylic acids are often solids, they themselves serve as suitable derivatives.

Experimental Procedure

A. Nitration Some of the best solid derivatives of aryl halides are mono- and dinitration products. Two general procedures can be used for nitration. Whenever nitrating a compound, whether it is known or unknown, use care, since many of these compounds react vigorously under typical nitration conditions.

Method A This method yields *m*-dinitrobenzene from benzene or nitrobenzene and the *p*-nitro derivative from chloro- or bromobenzene, benzyl chloride, or toluene. Dinitro derivatives are obtained from phenol, acetanilide, naphthalene, and biphenyl. Add about 1 g of the compound to 4 mL of concentrated sulfuric acid. Add dropwise to this mixture 4 mL of concentrated nitric acid; shake after each addition. Carry out the reaction in a large test tube or small Erlenmeyer flask, and heat at 45°C for about 5 min, using a beaker of water to supply the heat. Then pour the reaction mixture onto 25 g of ice, and collect the precipitate on a filter. The solid may be recrystallized from aqueous ethanol if needed.

Method B This method is the best one to use for nitrating halogenated benzenes, since dinitration occurs to give compounds that have higher melting points and are easier to purify than are mononitration products obtained from method A. The xylenes, mesitylene, and pseudocumene yield trinitro compounds. Follow the procedure used for method A, except that 4 mL of *fuming* nitric acid should be used in place of the *concentrated* nitric acid and the mixture should be warmed using a steam bath for 10 min. If little or no nitration occurs, substitute *fuming* sulfuric acid for the *concentrated* sulfuric acid. *Carry out this reaction in a hood.*

B. Side-Chain Oxidation
(1) Permanganate Method Add 1 g of the compound to a solution prepared from 80 mL of water and 4 g of potassium permanganate. Add 1 mL of 3 *M* sodium hydroxide solution, and heat the mixture at reflux until the purple color characteristic of the permanganate has disappeared; this will normally take from 30 min to 3 hr. At the end of the reflux period, cool the mixture and carefully acidify it with 3 *M* sulfuric acid. Now heat the mixture for an additional 30 min and cool it again; remove excess brown manganese dioxide (if any) by addition of sodium bisulfite solution. The bisulfite serves to reduce the manganese dioxide to manganous ion, which is water-soluble. Collect the solid acid that remains by vacuum filtration. Recrystallize the acid from toluene or aqueous ethanol. If little or no solid acid is formed, this may be due to the fact that the acid is somewhat water-soluble. In this case, extract the aqueous layer with chloroform, diethyl ether, or dichloromethane, dry the organic extracts, and remove the organic solvent by means of a steam bath in the hood. Recrystallize the acid that remains. In this particular method the presence of base during the oxidation often means that some silicic acid will form on acidification; thus, purification before determining the melting point is necessary.

(2) Chromic Acid Method Dissolve 7 g of sodium dichromate in 15 mL of water, and add 2 to 3 g of the compound to be oxidized. Add 10 mL of concentrated sulfuric acid to the mixture with mixing and cooling. Attach a reflux condenser to the flask, and heat gently until a reaction ensues; as soon as the reaction begins, remove the flame and cool the mixture if

necessary. After spontaneous boiling subsides, heat the mixture at reflux for 2 hr. Pour the reaction mixture into 25 mL of water, and collect the precipitate by filtration. Transfer the solid to a flask, add 20 mL of 2 M sulfuric acid, and then warm the flask on a steam cone with stirring. Cool the mixture, collect the precipitate, and wash it with about 20 mL of cold water. Dissolve the residue in 20 mL of 1.5 M sodium hydroxide solution and filter the solution. Add the filtrate, with stirring, to 25 mL of 2 M sulfuric acid. Collect the new precipitate, wash with cold water, and recrystallize from either toluene or aqueous ethanol.

Lists of possible aromatic hydrocarbons and aryl halides appear in Tables 27.10 and 27.11 along with derivatives.

(e) Alcohols (Table 27.3)

1. CLASSIFICATION TESTS

The tests for the presence of a hydroxy group not only detect the presence of the group but may also indicate whether it occupies a primary, secondary, or tertiary position.

A. Chromic Acid in Acetone This test may be used to detect the presence of a hydroxy group, provided that it has been shown previously that the molecule does not contain an aldehyde function. The reactions and experimental procedures for this test are given in Section 27.5a, part C. It has been pointed out that chromic acid does not distinguish between primary and secondary alcohols, since primary and secondary alcohols *both* give a positive test, whereas tertiary alcohols do not.

B. The Lucas Test This test is used to distinguish among primary, secondary, and tertiary alcohols. The reagent used is a mixture of concentrated hydrochloric acid and zinc chloride, which on reaction with alcohols converts them to the corresponding alkyl chlorides. With this reagent primary alcohols give no appreciable reaction, secondary alcohols react more rapidly, and tertiary alcohols react very rapidly. A positive test depends on the fact that the alcohol is soluble in the reagent, whereas the alkyl chloride is not; thus the formation of a second layer or an emulsion constitutes a positive test. The *solubility of the alcohol in the reagent places limitations on the utility of the test,* and in general only monofunctional alcohols with six or fewer carbon atoms, as well as polyfunctional alcohols, can be used.

Primary: $\qquad$ $RCH_2OH + HCl \xrightarrow{ZnCl_2}$ No reaction $\qquad$ **(25)**

Secondary: $\qquad$ $R_2CHOH + HCl \xrightarrow{ZnCl_2} R_2CHCl + H_2O$ $\qquad$ **(26)**

Tertiary: $\qquad$ $R_3COH + HCl \xrightarrow{ZnCl_2} R_3CCl + H_2O$ $\qquad$ **(27)**

Note the similarity of this reaction with the nucleophilic displacement reactions between alcohols and hydrohalic acids that are discussed in Section 18.1. In the Lucas test the presence of zinc chloride, which is a Lewis acid, greatly increases the reactivity of alcohols toward hydrochloric acid.

Experimental Procedure

Add 10 mL of the hydrochloric acid–zinc chloride reagent (Lucas reagent) to about 1 mL of the compound in a test tube. Stopper the tube and shake; allow the mixture to stand at room temperature. Try this test with known primary, secondary, and tertiary alcohols, and note the *time* required for the formation of an alkyl chloride, which will appear either as a second layer or as an emulsion. Repeat the test with an unknown, and compare the result with the results from the knowns.

C. The Ceric Nitrate Test This reagent can also be used as a qualitative test for alcohols. Although it has been used primarily for phenols, it does give a positive test with alcohols. Discussion about and experimental procedures for this test are given under Phenols in this chapter.

2. DERIVATIVES

Two common derivatives of alcohols are the urethanes and the benzoate esters; the former are best for primary and secondary alcohols, whereas the latter are useful for all types of alcohols.

A. Urethanes When an alcohol is allowed to react with an aryl substituted isocyanate, $ArN{=}C{=}O$, addition of the alcohol occurs to give a urethane (equation 28).

$$ArN{=}C{=}O + R'OH \rightarrow ArNH{-}\overset{\displaystyle O}{\overset{\displaystyle \|}{C}}{-}OR' \qquad \textbf{(28)}$$

Aryl isocyanate   A urethane

Some commonly used isocyanates are α-naphthyl, *p*-nitrophenyl, and phenyl isocyanate. A major side reaction is that of water with the isocyanate; water hydrolyzes the isocyanate to an amine, and the amine reacts with more isocyanate to give a disubstituted urea (equations 29 and 30).

$$ArN{=}C{=}O + H_2O \rightarrow (ArNH{-}CO_2H) \rightarrow ArNH_2 + CO_2 \qquad \textbf{(29)}$$

A carbamic acid
(unstable)

$$ArN{=}C{=}O + ArNH_2 \rightarrow ArNH{-}\overset{\displaystyle O}{\overset{\displaystyle \|}{C}}{-}NHAr \qquad \textbf{(30)}$$

A disubstituted urea

Since the ureas are high-melting amides owing to their symmetry, their presence makes purification of the desired urethane quite difficult. In using this procedure, take precautions to ensure that the alcohol is *anhydrous*. The procedure works best for water-insoluble alcohols, which can therefore be obtained easily in anhydrous form.

This type of derivative can also be useful for phenols; the procedure given here has been generalized so that it can be used for alcohols and phenols. Other derivatives of phenols are given later.

Experimental Procedure

> **DO IT SAFELY**
>
> Aryl isocyanates such as those used in the preparation of these derivatives are toxic. Take normal precautions in handling them. Quickly washing your hands with soap and warm water will remove the isocyanates from your skin.

Place 1 g of the *anhydrous* alcohol or phenol in a small round-bottomed flask, and add 0.5 mL of phenyl isocyanate or α-naphthyl isocyanate (recap the bottle of isocyanate tightly). If you are preparing the derivative of a phenol, also add 2 or 3 drops of dry pyridine as a catalyst. Affix a calcium chloride drying tube to the flask. Warm the reaction mixture with a heating mantle or a steam bath for 5 min. Cool the mixture in an ice-water bath, and scratch the mixture with a stirring rod to induce crystallization. Recrystallize the crude derivative from either carbon tetrachloride or petroleum ether. *Note:* 1,3-Di(α-naphthyl)urea has mp 293°C, and 1,3-diphenylurea (carbanilide) has mp 237°C; if your product shows one of these melting points, repeat the preparation, taking greater care to maintain anhydrous conditions.

B. 3,5-Dinitrobenzoates The reaction between 3,5-dinitrobenzoyl chloride and an alcohol gives the corresponding ester (equation 31). This method is useful for primary, secondary, and tertiary alcohols, especially those that are water-soluble and that are likely to contain traces of water.

$$\text{ROH} + \text{Cl}-\underset{\underset{\text{3,5-Dinitrobenzoyl chloride}}{}}{\overset{\text{O}}{\underset{\|}{\text{C}}}}\underbrace{}_{\substack{\text{NO}_2\\\text{NO}_2}} \xrightarrow{\text{pyridine}} \text{RO}-\underset{\underset{\text{A 3,5-dinitrobenzoate}}{}}{\overset{\text{O}}{\underset{\|}{\text{C}}}}\underbrace{}_{\substack{\text{NO}_2\\\text{NO}_2}} + \text{HCl} \qquad \textbf{(31)}$$

Experimental Procedure

Method A (*Note:* 3,5-dinitrobenzoyl chloride is reactive toward water; it should be used immediately after weighing. Take care to minimize its exposure to air and to keep the bottle tightly closed.) In a small flask bearing a reflux condenser mix 2 mL of the alcohol with about 0.5 g of 3,5-dinitrobenzoyl chloride and 0.5 mL of pyridine. Boil the mixture gently for 30 min (15 min is sufficient for a primary alcohol). Cool the solution and add about 10 mL of 0.6 *M* aqueous sodium bicarbonate solution. Cool this solution in an ice-water bath, and collect the crude crystalline product. Recrystallize the product from aqueous ethanol. A minimum volume of solvent should be used; adjust the composition of the solvent by adding just enough water to the alcohol so that the product dissolves in the hot solution but yields crystals when cooled.

Method B Carry out this reaction in the hood, if possible, or use a gas trap such as that shown in Figure 5.1. To a 50-mL flask fitted with a reflux condenser (bearing a gas trap if a hood is not available) add 1 g of 3,5-dinitrobenzoic acid, 3 mL of thionyl chloride, and 1 drop of pyridine. Heat the mixture at reflux until the acid has dissolved and then for an additional 10 min. The total reflux time should be about 30 min. Equip the flask for simple distillation. Cool the receiving flask with an ice-salt bath, and attach the vacuum adapter to an aspirator by means of a safety trap such as that shown in Figure 2.32. Evacuate the system and distil the excess thionyl chloride by heating with a steam bath. When the thionyl chloride has been removed, cautiously release the vacuum. Discard the excess thionyl chloride by pouring slowly down a drain *in the hood* or by putting it in a container for recovered thionyl chloride. To the residue in the stillpot, which is 3,5-dinitrobenzoyl chloride, add in the same flask 2 mL of the alcohol and 0.5 mL of pyridine. Fit the flask with a reflux condenser bearing a drying tube. Proceed with the period of reflux following the procedure of method A.

A list of alcohols and their derivatives is given in Table 27.3.

(f) Phenols (Table 27.18)

1. CLASSIFICATION TESTS

Several tests can be used to detect the presence of a phenolic hydroxy group: (a) bromine water, (b) ceric nitrate reagent, and (c) ferric chloride solution. In addition to these, solubility tests give a preliminary indication of a phenol, since phenols are soluble in 1.5 M sodium hydroxide solution but generally insoluble in 0.6 M sodium bicarbonate solution. Care must be exercised here, however, because phenols containing highly electronegative groups are stronger acids and may be soluble in 0.6 M sodium bicarbonate. Examples are 2,4,6-tribromophenol and 2,4-dinitrophenol.

A. Bromine Water Phenols are generally highly reactive toward electrophilic substitution and consequently are brominated readily by bromine water (see equation 32, for example).

$$\text{(32)}$$

The rate of bromination is much greater in water than in carbon tetrachloride solution. The water, being more polar than carbon tetrachloride, increases the ionization of bromine and thus enhances the ionic bromination mechanism. Although hydrogen bromide is liberated, it is not observed when water is used as the solvent. Phenols are so reactive that all unsubstituted positions *ortho* and *para* to the hydroxy group are brominated. The brominated compounds so formed are often solids and can also be used as derivatives. Aniline and substituted anilines are also very reactive toward bromine and react analogously; however, solubility tests can be used to distinguish between anilines and phenols.

Experimental Procedure

Prepare a 1% aqueous solution of the unknown. If necessary, dilute sodium hydroxide solution may be added dropwise to bring the phenol into solution. Add dropwise to this solution a saturated solution of bromine in water; continue addition until the bromine color remains. Note how much bromine water was used, and try this experiment on phenol and aniline for purposes of comparison.

B. Ceric Nitrate Reagent Alcohols and phenols are capable of replacing nitrate ions in complex cerate anions, resulting in a change from a yellow to a red solution (equation 33).

$$(NH_4^{\oplus})_2Ce(NO_3)_6^{\ominus} + ROH \rightarrow (NH_4^{\oplus})_2Ce(OR)(NO_3)_5^{\ominus} + HNO_3 \qquad \textbf{(33)}$$
Yellow   Red

Alcohols and phenols having no more than ten carbon atoms give a positive test. Alcohols give a red solution. Phenols give a brown to greenish-brown precipitate in aqueous solution; in dioxane a red-to-brown solution is produced. Aromatic amines may be oxidized by the reagent and give a color indicating a positive test.

Experimental Procedure

Dissolve about 20 mg of a solid or 1 drop of a liquid unknown in 1–2 mL of water, and add 0.5 mL of the ceric ammonium nitrate reagent; shake and note the color. If the unknown is insoluble in water, dissolve it in 1 mL of dioxane, and add 0.5 mL of the reagent.

C. Ferric Chloride Test Most phenols and enols react with ferric chloride to give colored complexes. The colors vary, depending not only on the nature of the phenol or enol but also on the solvent, concentration, and time of observation. Some phenols that do not give coloration in aqueous or alcoholic solution do so in chloroform solution, especially after addition of a drop of pyridine. The nature of the colored complexes is still uncertain; they may be ferric phenoxide salts that absorb visible light to give an excited state in which electrons are delocalized over both the iron atoms and the conjugated organic system. The production of a color is typical of phenols and enols; however, many of them do *not* give colors, so a negative ferric chloride test must not be taken as significant without supporting information (for example, the ceric nitrate and bromine water tests).

Experimental Procedure

Dissolve 30 to 50 mg of the unknown compound in 1–2 mL of water (or a mixture of water and 95% ethanol if the compound is not water-soluble), and add several drops of a 0.2 M aqueous solution of ferric chloride. Most phenols produce red, blue, purple, or green coloration; enols give red, violet, or tan coloration.

2. DERIVATIVES

Two useful solid derivatives of phenols are α-naphthyl urethane and bromo derivatives. The preparation of urethanes has already been discussed in Section 27.5e, part 2. Although either substituted urethane could be prepared, the majority of the derivatives reported are the α-naphthyl urethanes, and they are suggested as the urethanes of choice.

Experimental Procedure

The solution for bromination in this preparation should be supplied for you; it is prepared by dissolving 10 g of potassium bromide in 60 mL of water and adding 6 g of bromine. Dissolve 1 g of the phenolic compound in water or 95% ethanol, and add the brominating solution to it *dropwise*. Continue the addition until the reaction mixture begins to develop a yellow color, indicating excess bromine. Let the mixture stand for about 5 min; if the yellow coloration begins to fade, add another drop or two of the brominating solution. Add 50 mL of water to the mixture and then a few drops of 0.5 M sodium bisulfite solution to destroy the excess bromine. Shake the mixture vigorously, and remove the solid derivative by vacuum filtration. It may prove necessary to neutralize the solution with concentrated hydrochloric acid to achieve precipitation of the solid derivative. This derivative may be purified by recrystallization from 95% ethanol or aqueous ethanol.

A listing of phenols and their derivatives is given in Table 27.18.

(g) Carboxylic Acids (Table 27.12)

1. CLASSIFICATION TEST

One of the best qualitative tests for the carboxylic acid group is solubility in basic solutions. Carboxylic acids are soluble both in 1.5 M sodium hydroxide solution and in 0.6 M sodium bicarbonate solution, from which they can be regenerated by addition of acid. Solubility properties were discussed earlier in this section under Solubility Tests.

The relatively high acidity of carboxylic acids enables ready determination of the **equivalent weight** or **neutralization equivalent** of the acid by titration with standard base. The equivalent weight of an acid is that weight, in grams, of acid that reacts with one equivalent of base. As an example, suppose that 0.1000 g of an unknown acid required 16.90 mL of 0.1000 N sodium hydroxide solution to be titrated to a phenolphthalein end point. This means that 0.1000 g of the acid corresponds to (16.90 mL)(0.1000 equivalent/1000 mL) or 0.0016901 equivalent of the acid, or that one equivalent of the acid weighs 0.1000/0.00169 or 59.201 g. Thus the following expression applies:

$$\text{equivalent weight} = \frac{\text{grams of acid}}{(\text{volume of base consumed in liters})(N)}$$

where N is the normality of the standard base.

Because each carboxylic acid function in a molecule will be titrated with base, the equivalent weight corresponds to the molecular weight of the acid divided by n, where n is the number of acid functions present in the molecule. Thus, for the example given, the molecular weight may be 59.20 if a single acid function is present, 118.4 if two are present, and 177.6 if three are present. If the molecular weight of an unknown compound is known, then the number of acid groups in the molecule can be calculated by dividing the molecular weight by the equivalent weight. Hence, if the molecular weight of the unknown compound is 118 and its equivalent weight is 59.201, the unknown must have *two* titratable acid functions.

Experimental Procedure

Dissolve an accurately weighed sample (about 0.2 g) of the acid in 50 to 100 mL of water or 95% ethanol or a mixture of the two. It may be necessary to warm the mixture to dissolve the compound completely. Titrate the solution with a *standardized* sodium hydroxide solution having a concentration of about 0.1 M. Use phenolphthalein as the indicator, and from these data calculate the equivalent weight.

2. DERIVATIVES

Three good solid derivatives of carboxylic acids are (a) amides, (b) anilides, and (c) *p*-toluidides. These derivatives are prepared from the corresponding acid chloride by treatment of the latter with either ammonia, aniline, or *p*-toluidine. The amides are generally less satisfactory than the other two because they tend to be more soluble in water and as a result are harder to isolate. The acid chlorides are most conveniently prepared from the acid, or its salt, and thionyl chloride.

$$\text{RCO}_2\text{H (or RCO}_2^{\ominus}\text{ Na}^{\oplus}) + \text{SOCl}_2 \rightarrow \text{RCOCl} + \text{SO}_2 + \text{HCl (or NaCl)} \qquad \textbf{(34)}$$

Amides: $\quad \text{RCOCl} + 2\ \text{NH}_3 \xrightarrow{\text{cold}} \text{RCONH}_2 + \text{NH}_4^{\oplus}\ \text{Cl}^{\ominus}$ $\qquad\qquad\qquad$ **(35)**

Anilides: $\quad \text{RCOCl} + 2\ \text{C}_6\text{H}_5\text{NH}_2 \rightarrow \text{RCONHC}_6\text{H}_5 + \text{C}_6\text{H}_5\text{NH}_3^{\oplus}\ \text{Cl}^{\ominus}$ $\qquad$ **(36)**

Toluidides: $\quad \text{RCOCl} + 2\ p\text{-CH}_3\text{C}_6\text{H}_4\text{NH}_2 \rightarrow \text{RCONHC}_6\text{H}_4\text{CH}_3\text{-}p$ $\qquad\qquad$ **(37)**
$$+\ p\text{-CH}_3\text{C}_6\text{H}_4\text{NH}_3^{\oplus}\ \text{Cl}^{\ominus}$$

Experimental Procedure

To prepare the acid chloride from either a carboxylic acid or its salt, place 1 g of the acid or its sodium salt in a small round-bottomed flask with 1 mL of thionyl chloride and 5 drops of dimethyl formamide (DMF). Attach a calcium chloride drying tube *directly* to the flask, and clamp the flask into a 55–65°C water bath *in the hood*. Bubbling or fuming usually begins shortly after the addition of the DMF. The reaction is sufficiently complete when the bubbling has slowed down greatly, and the mixture containing the acid chloride can be used to make the amide, anilides, or *p*-toluidide as described below.[2]

A. Amides *In the hood* pour the mixture containing the acid chloride and unchanged thionyl chloride into 15 mL of *ice-cold*, concentrated ammonium hydroxide solution. Be very careful on this addition; the reaction is quite vigorous. Collect the precipitated amide derivative of the carboxylic acid on a filter, and recrystallize from water or aqueous ethanol.

B. Anilides and *p*-Toluidides Prepare the acid chloride of the carboxylic acid from 1 g of the acid and 2 mL of thionyl chloride. After the 30-min period of reflux, cool the mixture. Dissolve 1 g of either aniline or *p*-toluidine in 30 mL of cyclohexane; slight warming may be

[2]This procedure using dimethyl formamide is discussed in an article by Long, K. P. *Journal of Chemical Education*, **1979**, *56*, 420.

necessary to effect complete solution. Pour the cooled acid chloride into the cyclohexane solution of the amine, and heat the resulting mixture on a steam bath for 2–3 min. A heavy white precipitate of the amine hydrochloride will form; this precipitate should be removed by vacuum filtration and *set aside* (do not discard). In a separatory funnel wash the filtrate with 5 mL of water, followed by 5 mL of 1.5 M HCl, 5 mL of 1.5 M NaOH, and finally with 5 mL of water. In some cases precipitation may occur in the organic layer during one or more of these washings. If this occurs, warm the solution gently with a warm-water bath to redissolve the precipitate. Following the washings remove the cyclohexane from the organic layer either by distillation or by evaporation on a steam bath in the hood. Recrystallize the derivative of the carboxylic acid from aqueous ethanol. *Note:* If little residue remains following evaporation of the cyclohexane, dissolve the precipitate that was removed earlier in about 10 mL of water. Stir and remove any undissolved solid by filtration, combining this solid with the residue obtained from the cyclohexane.

(h) Amines (Table 27.9)

1. CLASSIFICATION TESTS

Two common qualitative tests for amines are the Hinsberg test and the nitrous acid test. We have not included the nitrous acid test because recent research indicates that the *N*-nitroso derivatives of secondary amines constitute one of the four or five major classifications of carcinogenic substances. The risk of producing an as-yet-unrecognized carcinogenic material in this test outweighs any possible benefit of a test that can be misleading and difficult to interpret. The modified sodium nitroprusside test has been included as an alternative.

A. The Hinsberg Test The reaction between primary or secondary amines and benzenesulfonyl chloride (equations 38 and 40, respectively) yields the corresponding substituted benzenesulfonamide. The reaction is carried out in excess base; if the amine is primary, the sulfonamide, which has an acidic amido hydrogen, is converted by base (equation 39) to the normally soluble potassium salt. Thus with few exceptions, which are discussed in the next paragraph, primary amines react with benzenesulfonyl chloride to provide homogeneous reaction mixtures. Acidification of this solution regenerates the insoluble primary benzenesulfonamide. On the other hand, the benzenesulfonamides of secondary amines bear no acidic amido hydrogens: they typically are insoluble in both acid and base. Therefore, secondary amines react to yield heterogeneous reaction mixtures, with production of either an oily organic layer or a solid precipitate.

Primary:

Benzenesulfonyl chloride   Insoluble in water

(38)

Soluble in water

(39)

Secondary:

$$R_2\ddot{N}H + \langle\bigcirc\rangle\!-\!SO_2Cl \xrightarrow{\text{KOH}} \langle\bigcirc\rangle\!-\!SO_2\ddot{N}R_2 + KCl + H_2O \qquad \textbf{(40)}$$

$$\underset{\text{Insoluble in water}}{}$$

$$\downarrow \overset{\text{excess}}{\underset{\text{KOH}}{}}$$

No reaction

The distinction between primary and secondary amines then depends on the different solubility properties of their benzenesulfonamide derivatives. However, the potassium salts of *certain* primary sulfonamides are not completely soluble in basic solution. Examples are generally found among those primary amines of higher molecular weight and those having cyclic alkyl groups.[3] To avoid confusion and possible misassignment of a primary amine as secondary, the basic solution is separated from the oil or solid and acidified. The formation of an oil or a precipitate indicates that the derivative is partially soluble and that the amine is primary. It is important not to overacidify the solution because this may precipitate certain side products that may form, resulting in an ambiguous test. The original oil or solid should be tested for solubility in water and acid to substantiate the test for a primary or a secondary amine.

Tertiary amines behave somewhat differently.[4] Typically, under the conditions of the Hinsberg test the processes shown in equation 41 provide for the conversion of benzenesulfonyl chloride to potassium benzenesulfonate with recovery of the tertiary amine. Because tertiary amines are nearly always insoluble in the aqueous potassium hydroxide solution, the test mixture remains heterogeneous. It is worthwhile to note relative densities of the oil layer and of the test solution. Benzenesulfonamides are generally more dense than the solution, whereas the amines are less dense. The oil is separated and tested for solubility in aqueous acid; solubility usually indicates a tertiary amine.

Tertiary:

$$R_3\ddot{N} + \langle\bigcirc\rangle\!-\!SO_2Cl \rightarrow \langle\bigcirc\rangle\!-\!SO_2\!-\!\overset{\oplus}{N}R_3\ Cl^{\ominus}$$

$$\downarrow \text{KOH}$$

$$\langle\bigcirc\rangle\!-\!SO_3^{\ominus}K^{\oplus} + \ddot{N}R_3 + H_2O \qquad \textbf{(41)}$$

$$\xrightarrow{NR_3} \langle\bigcirc\rangle\!-\!SO_2NR_2 + NR_4^{\oplus}Cl^{\ominus} \qquad \textbf{(42)}$$

[3]Fanta, P. E.; Wang, C. S. *Journal of Chemical Education,* **1964,** *41,* 280.

[4]Historically and almost invariably, sources of information have asserted that tertiary amines do not react with benzenesulfonyl chloride. For an interesting refutation of this widely accepted myth, see Gambill, C. R.; Roberts, T. D.; Shechter, H. *Journal of Chemical Education,* **1972,** *49,* 287.

$$\text{Ar—SO}_2\text{Cl} \xrightarrow{\text{KOH}} \text{Ar—SO}_3^{\ominus}\text{K}^{\oplus} + \text{KCl} + \text{H}_2\text{O} \tag{43}$$

$$\xrightarrow[\text{Ar}\ddot{\text{N}}\text{R}_2]{} \begin{array}{l}\text{Complex}\\ \text{mixture}\\ \text{including:}\end{array} \quad \text{Ar—SO}_2\text{—}\ddot{\text{N}}\text{RAr} \tag{44}$$

The test procedure should be followed as closely as possible. It is designed to minimize complications that may arise because of side reactions of tertiary amines with benzenesulfonyl chloride. As shown in equation 42, the initial adduct is subject to further reaction with another molecule of amine to produce the benzenesulfonamide of a secondary amine. The relative competing rates of reaction of the adduct with hydroxide ion and with amine do not favor equation 42, particularly when excess amine is avoided, yet the formation of *small* amounts of an insoluble product may, through confusion, cause an amine to be incorrectly designated as secondary. Adduct formation such as shown in equation 41 is generally less of a problem with tertiary arylamines because they are normally much less soluble in the test solution and are less nucleophilic than trialkylamines. The competing hydrolysis of benzene-sulfonyl chloride by hydroxide ion (equation 43) allows the recovery of most of the amine. Moreover, tertiary arylamines are often subject to other side reactions producing a complex mixture of mainly insoluble products (equation 44). Because benzenesulfonyl chloride reacts more slowly with tertiary arylamines than with hydroxide ion, it is also possible to minimize the attendant ambiguity caused by these reactions of the tertiary amine by keeping the reaction time short and the temperature low.

To summarize the discussion, tertiary amines may produce small amounts of insoluble products if the concentration of the amine in the test solution is too high and if the reaction time is too long. If the directions of the procedure are followed and *care is taken not to interpret small amounts of insoluble product as a positive test for secondary amines,* the Hinsberg test may be used with confidence to designate an amine as primary, secondary, or tertiary. Tertiary amines often contain quantities of secondary amines as impurities. If it was not possible to obtain a reliable boiling point and the amine was not carefully distilled, small quantities of precipitate may form for this reason also, obscuring the test results.

Experimental Procedure

Mix 10 mL of 2 M aqueous potassium hydroxide, 0.2 mL or 0.2 g of the amine, and 0.7 mL of benzenesulfonyl chloride (*Caution:* It is a lachrymator) in a test tube. Stopper the tube and shake the mixture *vigorously,* with cooling if necessary, until the odor of benzenesulfonyl chloride is gone. In even the slowest case this should take no more than about 5 min. Test the solution to see that it is still basic; if it is not, add sufficient 2 M potassium hydroxide solution until it is.

If the mixture has formed two layers or a precipitate, note the relative densities, and separate the oil or solid by decantation or filtration. Test an oil for solubility in 0.6 M hydrochloric acid. The sulfonamide of a secondary amine is insoluble, whereas an amine is at least partially soluble. If this solubility test indicates an amine, it may be either a tertiary amine or one of certain secondary amines that react with benzenesulfonyl chloride only very slowly because of steric bulk. Test a solid for solubility in water and in dilute acid. The potassium salt of a sulfonamide that is insoluble in base solution is usually soluble in water; the sulfonamide that forms from the potassium salt when placed in acid is insoluble in that medium. A solid sulfonamide of a secondary amine is insoluble in both water and acid. Acidify the solution from the original reaction mixture to pH 4 using pH indicator paper or a few drops of Congo red indicator solution; the formation of a precipitate or oil indicates a primary amine.

If the original mixture has not formed two layers, the test is indicative of a primary amine. Acidify the solution to pH 4; a sulfonamide of a primary amine will either separate as an oil or precipitate as a solid.

B. The Sodium Nitroprusside Tests

Two color tests to distinguish primary and secondary *aliphatic* amines have been available for many years,[5] although they have not been widely used. Recently, through a change in the solvent system and the introduction of a $ZnCl_2$ catalyst, these tests have been extended to primary, secondary, and tertiary *aromatic* amines.[6] Both the original and the modified tests are inconclusive for tertiary aliphatic amines. No attempt is made here to explain the complex color-forming reactions that occur. However, they most likely involve the reaction of the amine with either acetone (the Ramini test) or acetaldehyde (the Simon test) and the interaction of the products of these reactions with sodium nitroprusside to form colored complexes.

To apply these tests on an unknown amine, the conventional Ramini and/or the conventional Simon tests should first be performed. These will give positive results in the cases of primary and secondary aliphatic amines. If these tests are negative and an aromatic amine is suspected, then the modified versions of these tests may be performed. Figure 27.4 will be useful in interpreting the results of these tests.

	1° Aliphatic	2° Aliphatic	1° Aromatic	2° Aromatic	3° Aromatic
Ramini	Deep red	Deep red			
Simon	Pale yellow to red-brown	Deep blue			
Modified Ramini			Orange-red to red-brown	Orange-red to red-brown	Green
Modified Simon			Orange-red to red-brown	Purple	Usually green

Figure 27.4 Colors formed in the Ramini and the Simon tests.

[5]Simon, L. *Comptes Rendus,* **1897,** *125,* 534; Ramini, E. *Chemisches Zentralblatt,* **1898,** *11,* 132.
[6]Baumgarten, R. L.; Dougherty, C. M.; Nercessian, O. *Journal of Chemical Education,* **1977,** *54,* 189.

Experimental Procedure

The sodium nitroprusside reagent, for use in both the *conventional* Ramini and Simon tests, is prepared by dissolving 3.9 g of sodium nitroprusside ($Na_2[Fe(NO)(CN)_5] \cdot 2H_2O$) in 100 mL of 50% aqueous methanol.

A. Ramini Test To 1 mL of the sodium nitroprusside reagent add 1 mL of water, 0.2 mL of acetone, and then about 30 mg of an amine. In most cases the characteristic colors given in Figure 27.4 appear in a few seconds, although in some instances up to about 2 min may be necessary.

B. Simon Test To 1 mL of the sodium nitroprusside reagent add 1 mL of water, 0.2 mL of 2.5 M aqueous acetaldehyde solution, and then about 30 mg of an amine. As in the Ramini test, color formation will normally occur in a few seconds, although occasionally up to 2 min may be necessary.

C. Modified Ramini Test To 1 mL of the *modified* sodium nitroprusside reagent[7] add in the following order: 1 mL of saturated aqueous zinc chloride solution, 0.2 mL of acetone, and then about 30 mg of an amine. Primary and secondary aromatic amines provide orange-red to red-brown colors within a period of a few seconds to 5 min. Tertiary aromatic amines give a color that changes from orange-red to green over a period of about 5 min.

D. Modified Simon Test To 1 mL of the *modified* sodium nitroprusside reagent[7] add in the following order: 1 mL of saturated aqueous zinc chloride solution, 0.2 mL of 2.5 M aqueous acetaldehyde solution, and then about 30 mg of an amine. Primary aromatic amines give an orange-red to red-brown color within 5 min; secondary aromatic amines give a color changing from red to green within 5 min; tertiary aromatic amines give a color that changes from orange-red to green over a period of 5 min.

2. DERIVATIVES

Suitable derivatives of primary and secondary amines are the benzamides and benzenesulfonamides (equations 45 and 46, respectively).

$$RNH_2 \text{ (or } R_2NH) + C_6H_5COCl \xrightarrow{\text{pyridine}} C_6H_5\overset{\text{O}}{\overset{\|}{C}}-NHR \text{ (or } C_6H_5\overset{\text{O}}{\overset{\|}{C}}-NR_2) \quad \textbf{(45)}$$

$$RNH_2 \text{ (or } R_2NH) + C_6H_5SO_2Cl \longrightarrow C_6H_5SO_2-NHR \text{ (or } C_6H_5SO_2NR_2) \quad \textbf{(46)}$$

Although one or both of the above methods are satisfactory with most primary and secondary amines, tertiary amines do not undergo the same reactions. In general, one must take advantage of the fact that tertiary amines do form salts. Two useful crystalline salts are

---

[7]This reagent, for use in both the *modified* Ramini and Simon tests, is prepared by dissolving 3.9 g of sodium nitroprusside in a solution containing 80 mL of dimethylsulfoxide and 20 mL of water. To avoid decomposition of the reagent, it should be stored in the refrigerator and dispensed in small dropper bottles as needed.

the ones from methyl iodide (methiodides) and picric acid (picrates) (equations 47 and 48, respectively).

$$R_3N\colon + CH_3I \rightarrow R_3\overset{\oplus}{N}CH_3\ I^{\ominus} \qquad (47)$$

$$R_3N\colon + HO\!-\!\underset{NO_2}{\overset{NO_2}{\bigcirc}}\!-\!NO_2 \rightarrow R_3\overset{\oplus}{N}H\ \ ^{\ominus}O\!-\!\underset{NO_2}{\overset{NO_2}{\bigcirc}}\!-\!NO_2 \qquad (48)$$

Picric acid   Picrate salt

Experimental Procedure

A. Benzenesulfonamides The method of preparing the benzenesulfonamides has been discussed under the Hinsberg Test. The derivatives can be prepared using that method, but sufficient amounts of material should be used so that the final product can be purified by recrystallization from 95% ethanol. If the derivative is obtained as an oil, it *may* crystallize by scratching in the presence of the mother liquor with a stirring rod. If the oil cannot be made to crystallize, separate it and dissolve it in a minimum quantity of hot ethanol and allow to cool. Note that some amines do not give *solid* benzenesulfonamide derivatives.

B. Benzamides In a 50-mL round-bottom flask dissolve 0.5 g of the amine in 5 mL of dry pyridine. *Slowly* add 0.5 mL of benzoyl chloride to this solution. Affix a drying tube to the flask and heat the reaction mixture to 60–70°C for 30 min, using a water bath. Following the heating period, pour the mixture into 50 mL of water. If the solid derivative precipitates at this time, collect it on a filter, and when it is nearly dry, dissolve it into 20 mL of diethyl ether. If no precipitate forms, extract the aqueous mixture twice with 15-mL portions of diethyl ether. Combine the ether extracts. Wash the ether solution of the derivative in sequence with equal volumes of water, 1.5 M HCl, and 0.6 M sodium bicarbonate solution. Dry the ether layer over anhydrous magnesium sulfate, filter, and remove the ether by simple distillation. The solid residue, which constitutes the derivative, may be recrystallized from one of the following solvents: cyclohexane-hexane mixtures, cyclohexane-ethyl acetate mixtures, 95% ethanol, or aqueous ethanol.

C. Methiodides To prepare the methyl iodide derivative, mix 0.5 g of the amine with 0.5 mL of methyl iodide, and warm the test tube with a water bath for several minutes. Cool the test tube in an ice-water bath; the tube may be scratched with a rod to help induce crystallization. Purify the product by recrystallization from absolute ethanol or methanol or from ethyl acetate.

D. Picrates The picric acid derivative may be prepared by mixing 0.3 to 0.5 g of the compound with 10 mL of 95% ethanol. If the solution is not complete, remove the excess solid by filtration. Add to the mixture 10 ml of a saturated solution of picric acid in 95%

ethanol, and heat the mixture to boiling. Cool the solution slowly, and remove the yellow crystals of the picrate salt by filtration. Recrystallize the salt from 95% ethanol.

A list of amines and their derivatives can be found in Table 27.9.

(i) Nitro Compounds (Table 27.17)

1. CLASSIFICATION TEST

Ferrous Hydroxide Test   Organic compounds that are oxidizing agents will oxidize ferrous hydroxide (blue) to ferric hydroxide (brown). The most common organic compounds that function in this way are the *nitro compounds,* both aliphatic and aromatic, which are in turn reduced to amines in the reaction (equation 49).

$$\text{RNO}_2 + 6\ \underset{Blue}{\underline{\text{Fe(OH)}_2}} + 4\ \text{H}_2\text{O} \rightarrow \text{RNH}_2 + 6\ \underset{Brown}{\underline{\text{Fe(OH)}_3}} \qquad \textbf{(49)}$$

Other less common types of compounds that give the same test are nitroso compounds, hydroxylamines, alkyl nitrates, alkyl nitrites, and quinones.

Experimental Procedure

In a 10×75-mm or smaller test tube, mix about 20 mg of a solid or 1 drop of a liquid unknown with 1.5 mL of freshly prepared 5% ferrous ammonium sulfate solution. Add 1 drop of 3 M sulfuric acid and 1 mL of 2 M potassium hydroxide in methanol. Stopper the tube immediately and shake it. A positive test is indicated by the blue precipitate turning rust-brown within 1 min. A slight darkening or greenish coloration of the blue precipitate should not be considered a positive test.

2. DERIVATIVES

Two different types of derivatives of nitro compounds can be prepared. Aromatic nitro compounds can be di- and trinitrated with nitric acid and sulfuric acid. Discussion of and procedures for nitration have been given under Aryl Halides. Refer to Section 27.5d for additional information.

The other method for preparation of a derivative can be utilized for both aliphatic and aromatic nitro compounds. This involves the reduction of the nitro compound to the corresponding primary amine (equation 50), followed by conversion of the amine to a benzamide or benzenesulfonamide, as described under Amines (Section 27.5h). The reduction is most often carried out with tin and hydrochloric acid.

$$\text{RNO}_2 \text{ or ArNO}_2 \xrightarrow[\text{(2) NaOH}]{\text{(1) Sn, HCl}} \text{RNH}_2 \text{ or ArNH}_2 \qquad \textbf{(50)}$$

Experimental Procedure

Carry out the reduction of the nitro compound by combining 1 g of the compound and 2 g of granulated tin in a small flask. Attach a reflux condenser, and add, in small portions, 20 mL of 3 M hydrochloric acid. Shake after each addition. After addition is complete, warm the mixture for 10 min, using a steam bath. If the nitro compound is insoluble, add 5 mL of 95% ethanol to increase its solubility. Decant the warm, homogeneous solution into 10 mL of water, and add enough 12 M sodium hydroxide solution so that the tin hydroxide completely dissolves. Extract the basic solution with several 10-mL portions of diethyl ether. Dry the ether solution over potassium hydroxide pellets, and remove the ether by distillation.

The residue contains the primary amine. Convert it to one of the derivatives described under Amines (Section 27.5h).

A list of some nitro compounds and their derivatives appears in Table 27.17.

(j) Esters (Table 27.13)

1. CLASSIFICATION TEST

A test for the presence of the ester group involves the use of hydroxylamine and ferric chloride. The former converts the ester to a hydroxamic acid, which then complexes with Fe(III) to give a colored species (equations 51 and 52).

$$\underset{\text{Hydroxylamine}}{R-\overset{\overset{\textstyle O}{\|}}{C}-OR' + H_2NOH} \rightarrow \underset{\substack{\text{A hydroxamic} \\ \text{acid}}}{R-\overset{\overset{\textstyle O}{\|}}{C}-NHOH + R'OH} \qquad \textbf{(51)}$$

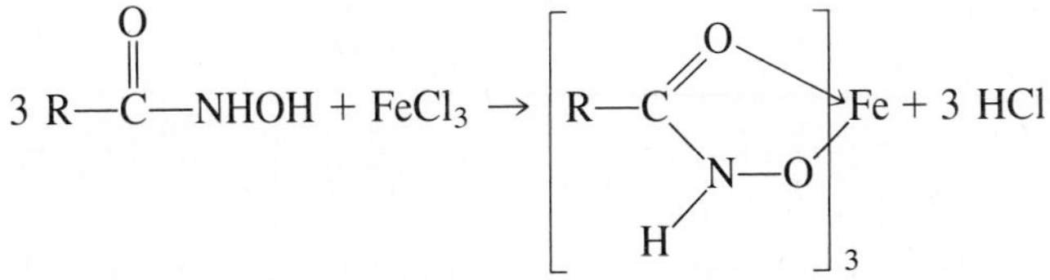

$$3\ R-\overset{\overset{\textstyle O}{\|}}{C}-NHOH + FeCl_3 \rightarrow \left[\underset{\substack{\\ H}}{R-C\underset{N-O}{\overset{O}{\diagdown}}}\!\!\!\diagup Fe\right]_3 + 3\ HCl \qquad \textbf{(52)}$$

Colored

All carboxylic acid esters (including polyesters and lactones) give magenta colors that vary in intensity depending on structural features in the molecule. Acid chlorides and anhydrides also give positive tests. Formic acid produces a red color, but other free acids give negative tests. Primary or secondary aliphatic nitro compounds give a positive test because ferric chloride reacts with the *aci* form (equivalent to the enol form of a ketone) that is present in basic solution. Most imides give positive tests. Some amides, but not all, give light magenta coloration, whereas most nitriles give a negative test. A modification of the following procedure, which will yield a positive test for amides and nitriles, is given later.

Experimental Procedure

Before the final test is performed, it is necessary to run a preliminary (or "blank") test, since some compounds will give a positive test even though they do not contain an ester linkage.

A. Preliminary Test   Mix 1 mL of 95% ethanol and 50–100 mg of the compound to be tested, and add 1 mL of 1 M hydrochloric acid. Note the color that is produced when 1 drop of 0.6 M aqueous ferric chloride solution is added. If the color is orange, red, blue, or violet, the following test for the ester group does not apply and cannot be used.

B. Final Test Mix 40–50 mg of the unknown, 1 mL of 0.5 M hydroxylamine hydrochloride in 95% ethanol, and 0.2 mL of 6 M sodium hydroxide. Heat the mixture to boiling, and after cooling it slightly, add 2 mL of 1 M hydrochloric acid. If the solution is cloudy, add more (about 2 mL) 95% ethanol. Add 1 drop of 0.6 M ferric chloride, and observe the color. Add more ferric chloride solution if the color does not persist, and continue to add it until it does. Compare the color obtained here with that from the preliminary test. If the color is burgundy or magenta, as compared to the yellow color in the preliminary experiment, the presence of an ester group is indicated.

2. SAPONIFICATION EQUIVALENT

It is possible to carry out the hydrolysis of an ester with alkali in a quantitative manner so that a value, termed the **saponification equivalent,** can be derived. This value is analogous to the equivalent weight of an acid in that it is the molecular weight of the ester divided by the number of ester functions in the molecule. Therefore the saponification equivalent is the number of grams of ester required to react with one gram-equivalent of alkali. This equivalent is determined by hydrolyzing a weighed amount of the ester with an excess of standardized alkali and then titrating the excess alkali to a phenolphthalein end-point with standardized hydrochloric acid. The saponification equivalent is then given as follows:

$$\text{Saponification equivalent} = \frac{\text{grams of ester}}{\text{equivalents of alkali consumed}}$$

$$= \frac{\text{grams of ester}}{(\text{volume of alkali in liters})(N) - (\text{volume of acid in liters})(N')}$$

where N is the normality of the standard base and N' is the normality of the standard acid.

Experimental Procedure

Dissolve approximately 3 g of potassium hydroxide in 60 mL of 95% ethanol. Allow the small amount of insoluble material to settle to the bottom, and fill a 50-mL buret with the clear solution by decantation. Measure exactly 25.0 mL of the alcoholic solution into each of two flasks. Weigh *accurately* into one of the flasks a 0.3- to 0.4-g sample of pure, dry ester; the other basic solution will be used as a blank. Fit each flask with a reflux condenser.

Heat the solutions in both flasks at a gentle reflux for 1 hr. When the flasks have cooled, rinse each condenser with about 10 mL of distilled water, catching the rinse water in the flask. Add phenolphthalein, and separately titrate the solutions in each flask with *standardized* hydrochloric acid that is approximately 0.5 M.

The difference in the volumes of hydrochloric acid required to neutralize the base in the flask containing the sample and in the flask containing the blank corresponds to the amount of potassium hydroxide that reacted with the ester. The volume difference (in milliliters) multiplied by the molarity of the hydrochloric acid equals the number of *milli*moles of potassium hydroxide consumed. Using the titration data, calculate the saponification equivalent of the unknown ester.

If the ester does not completely saponify in the allotted time as evidenced by a *nonho-mogeneous* solution, heat under reflux for longer periods (2–4 hr). In some cases higher temperatures may be required; if so, diethylene glycol must be used as a solvent *in place of* the original 60 mL of 95% ethanol.

3. DERIVATIVES

To characterize an ester completely, it is necessary to prepare solid derivatives of both the acid and the alcohol components. The problem here is to isolate both of these components in pure form so that suitable derivatives can be prepared. One such way is to carry out the ester hydrolysis in base (equation 53) in a high-boiling solvent. If the alcohol is low-boiling, it can be distilled from the reaction mixture and characterized. The acid that remains in the mixture can also be isolated. Derivatives of acids and alcohols have already been discussed.

$$RCO_2R' + HO^{\ominus} \rightarrow RCO_2^{\ominus} + R'OH \qquad \text{(53)}$$

Experimental Procedure

In a small reaction vessel mix 3 mL of diethylene glycol, 0.6 g (2 pellets) of potassium hydroxide, and 10 drops of water. Heat until the solution is homogeneous, and cool to room temperature. Add 1 mL of the ester and equip the apparatus with a condenser. Heat to boiling again, with swirling, and after the ester layer dissolves (3–5 min), recool the solution. Equip for a simple distillation, and heat the flask strongly so that the alcohol distils; *all but high-boiling alcohols can be removed by direct distillation*. The distillate, which should be fairly pure and dry, can be used for the preparation of a solid derivative.

The residue that remains after distillation contains the salt of the carboxylic acid. Add 10 mL of water to the residue and mix thoroughly. Acidify the solution with 6 *M* sulfuric acid. Allow the mixture to stand, and collect any crystals by filtration. If crystals do not form, extract the aqueous acidic solution with diethyl ether or dichloromethane, dry the organic solution, and evaporate the solvent. Use the residual acid to prepare a derivative.

A list of esters and their boiling or melting points is given in Table 27.13. Alcohols and carboxylic acids and their derivatives are given in Tables 27.3 and 27.12, respectively.

(k) Nitriles (Table 27.16)

1. CLASSIFICATION TEST

A qualitative test that may be used for nitriles is similar to that for esters. Common nitriles, as well as amides, give a colored solution on treatment with hydroxylamine and ferric chloride (equation 54).

$$R-C\equiv N + H_2NOH \rightarrow R-\overset{\overset{\displaystyle NH}{\|}}{C}-NHOH \xrightarrow{FeCl_3} \left[R-C\underset{\underset{H}{\overset{\displaystyle \diagdown}{N}}-O}{\overset{\displaystyle \diagup NH}{\diagdown}} \right]_3 Fe + 3\ HCl \qquad \text{(54)}$$

Experimental Procedure

Prepare a mixture consisting of 2 mL of 1 M hydroxylamine hydrochloride in propylene glycol, 30–50 mg of the compound that has been dissolved in a minimum amount of propylene glycol, and 1 mL of 1 M potassium hydroxide. Heat the mixture to boiling for 2 min, and cool to room temperature; add 0.5–1.0 mL of a 0.5 M *alcoholic* ferric chloride solution. A red-to-violet color is a positive test. Yellow colors are negative, and brown colors and precipitates are neither positive nor negative.

2. DERIVATIVES

On hydrolysis, in either acidic or basic solution, nitriles are ultimately converted to the corresponding carboxylic acids. Using methods given previously, it is then possible to prepare a derivative of the acid.

$$\text{Basic hydrolysis:} \quad \text{RCN} + \text{NaOH} \xrightarrow{\text{H}_2\text{O}} \text{RCO}_2^{\ominus}\text{Na}^{\oplus} + \text{NH}_3 \qquad \textbf{(55)}$$

$$\text{Acidic hydrolysis:} \quad \text{RCN} \xrightarrow[\text{H}_2\text{SO}_4]{\text{H}_2\text{O}} \text{RCONH}_2 \xrightarrow[\text{H}_2\text{SO}_4]{\text{H}_2\text{O}} \text{RCO}_2\text{H} + \text{NH}_4^{\oplus} \qquad \textbf{(56)}$$

Experimental Procedure

A. Basic Hydrolysis Mix 10 mL of 3 M sodium hydroxide solution and 1 g of the nitrile. Heat the mixture to boiling, and note the odor of ammonia, or hold a piece of moist red litmus paper over the container, and note the color change. After the mixture is homogeneous, cool it and make it acidic to litmus. If the acid is a solid, collect the crystals by filtration. If it is a liquid, extract the acidic solution with diethyl ether; after drying the ether solution, remove the ether by distillation. The residue that remains is the acid. Prepare a suitable derivative of the acid, using procedures given previously.

B. Acidic Hydrolysis Treat 1 g of the nitrile with 10 mL of concentrated sulfuric acid or concentrated hydrochloric acid, and warm the mixture to 50°C for about 30 min. Dilute the mixture with water (*Caution:* add the mixture slowly to water if sulfuric acid has been used), and heat the mixture at gentle reflux for 30 min to 2 hr. The organic layer will be the acid. Cool the mixture and either collect the crystals by filtration or extract the liquid with diethyl ether. Prepare derivatives of the acid.

A list of some nitriles is given in Table 27.16; carboxylic acids and their derivatives are given in Table 27.12.

(l) Amides (Table 27.8)

1. CLASSIFICATION TEST

A qualitative test for an amide group is the same as that given for a nitrile (equation 57). Follow exactly the procedure given for nitriles. The colors observed with amides are the same as those with nitriles.

$$R-\overset{\overset{\displaystyle O}{\|}}{C}-NH_2 + H_2NOH \rightarrow R-\overset{\overset{\displaystyle O}{\|}}{C}-NHOH \xrightarrow{FeCl_3} \left[R-C\underset{\underset{\displaystyle H}{N-O}}{\overset{\displaystyle O}{\diagdown}}\right]_3 Fe + 3\ HCl \qquad \textbf{(57)}$$

Colored

2. DERIVATIVES

Like nitriles, amides must be hydrolyzed (acidic or basic) to give an amine and a carboxylic acid (equation 58). In the case of unsubstituted amides, ammonia is liberated, but with substituted amides a substituted amine is obtained. In those cases it is necessary to classify the amine as being primary or secondary and to prepare derivatives of both the acid and the amine.

$$RCONR_2' \xrightarrow[\text{H}_2\text{O}]{\text{H}^\oplus \text{ or HO}^\ominus} RCO_2H + HNR_2' \qquad \textbf{(58)}$$

$$(R' = \text{alkyl, aryl, or H})$$

Experimental Procedure

A. Basic Hydrolysis Carry out the procedure described for the hydrolysis of nitriles. Distil the ammonia or volatile amine from the alkaline solution into a container of dilute hydrochloric acid. Neutralize this acidic solution, carry out the Hinsberg test, and prepare a derivative of the amine. If the amine is not volatile, it may be extracted from the aqueous layer with diethyl ether, the solution dried over potassium hydroxide pellets, and the ether removed to give the amine. After the amine has been obtained, by either distillation or extraction from the hydrolysis mixture, make the alkaline solution acidic, and isolate the acid (either by filtration if a solid or by extraction if a liquid). Characterize the acid by preparing a suitable solid derivative.

B. Acidic Hydrolysis Carry out the hydrolysis, using the method described for nitriles. In this case the free acid is liberated and can be removed by filtration or extraction with diethyl ether. Prepare a derivative of the acid. Make the acidic hydrolysis mixture alkaline to liberate the amine. Collect the amine by distillation or extraction, characterize it by the Hinsberg test, and make a derivative.

A list of amides is given in Table 27.8. Lists of amines and acids and their derivatives are given in Tables 27.9 and 27.12.

27.6 Tables of Compounds and Derivatives

The following tables of organic compounds and their derivatives are arranged according to classes of compounds. Tables are included for some functional groups for which derivatives are not described in Section 27.5. Some instructors may assign unknowns that are to be identified on the basis of physical properties and spectroscopic methods without requiring the preparation of derivatives, and because of this possibility, extensive tables of compounds for most common functional groups have been provided.

The melting points and boiling points of the compounds listed represent the highest point in the range actually observed. Some compounds and derivatives may decompose at their boiling or melting points, and these are indicated by *dec* by that temperature. The abbreviation *di* by some melting points indicates that the value is for the *di*substituted derivative.

Acid Anhydrides, Table 27.1
Acid Chlorides, Table 27.2
Alcohols, Table 27.3
Aldehydes, Table 27.4
Alkanes, Table 27.5
Alkenes and Alkynes, Table 27.6
Alkyl Halides (Haloalkanes), Table 27.7
Amides, Table 27.8
Amines, Table 27.9
Aromatic Hydrocarbons, Table 27.10
Aryl Halides, Table 27.11
Carboxylic Acids, Table 27.12
Esters, Table 27.13
Ethers, Table 27.14
Ketones, Table 27.15
Nitriles, Table 27.16
Nitro Compounds, Table 27.17
Phenols, Table 27.18

TABLE 27.1 Acid Anhydrides

Name of Compound	Boiling Point (°C)	Melting Point (°C)	Corresponding Carboxylic Acid	
			bp (°C)	mp (°C)
Acetic anhydride	140		118	
Propionic anhydride	167		141	
Isobutyric anhydride	182		155	
Butyric anhydride	198		164	
Dichloroacetic anhydride	216*dec*		194	
cis-1,2-Cyclohexanedicarboxylic anhydride		34		192
Benzoic anhydride		42		122
Chloroacetic anhydride		46	189	63
Maleic anhydride		54		130
Glutaric anhydride		56		98
4-Methylbenzoic anhydride		95		179
Succinic anhydride		120		189
Phthalic anhydride		132		210
trans-1,2-Cyclohexanedicarboxylic anhydride		147		230
3-Nitrophthalic anhydride		162		218
4-Nitrobenzoic anhydride		189		241
d,l-Camphoric anhydride		225		187
Tetrachlorophthalic anhydride		256		250*dec*
1,8-Naphthalic anhydride		274		270*dec*
Tetrabromophthalic anhydride		280		266

TABLE 27.2   Acid Chlorides

Name of Compound	Boiling Point (°C)	Melting Point (°C)	Corresponding Carboxylic Acid bp (°C)	mp (°C)
Acetyl chloride	52		118	
Oxalyl chloride	64			101
Methyl chloroformate	72			
Propionyl chloride	80		141	
Isobutyryl chloride	92		155	
Ethyl chloroformate	93			
Methacrylyl chloride	95		163	
Butyryl chloride	102		164	
Chloroacetyl chloride	108		189	63
Methoxyacetyl chloride	113		204	
Isovaleryl chloride	115		177	
Trichloroacetyl chloride	118		197	58
trans-Crotonyl chloride	126		189	72
Pentanoyl chloride	126		186	
Isobutyl chloroformate	129			
Hexanoyl chloride	153		205	
Fumaryl chloride	162			289(200)
Cyclohexanecarboxylic acid chloride	184			31
Succinyl chloride	190dec	20		189
Octanoyl chloride	196		239	16
Benzoyl chloride	197			122
Phenylacetyl chloride	210			76
Nonanoyl chloride	215		255	12
Glutaryl chloride	218			98
4-Chlorobenzoyl chloride	222			243
3-Chlorobenzoyl chloride	225			158
Phenoxyacetyl chloride	226			99
4-Methylbenzoyl chloride	226			179
2-Chlorobenzoyl chloride	238			142
Phthaloyl chloride	280	15		210
Adipoyl chloride	dec			154
Sebacoyl chloride	dec			134
trans-Cinnamoyl chloride	258	35		133
4-Bromobenzoyl chloride	247	42		251
Isophthaloyl chloride	276	44		348
2,4-Dinitrobenzoyl chloride		46		183
3,5-Dinitrobenzoyl chloride		69		207
4-Nitrobenzoyl chloride		75		241
Terephthaloyl chloride		84		300

TABLE 27.3 Alcohols (a) Liquids

Name of Compound	Boiling Point (°C)	Melting Points of Derivatives (°C)	
		3,5-Dinitrobenzoate	α-Naphthylurethane
Methanol	65	108	124
Ethanol	78	93	79
2-Propanol	82	123	106
2-Methyl-2-propanol (*tert*-butyl alcohol)	83	142	101
3-Buten-2-ol	95	54	
2-Propen-1-ol (allyl alcohol)	97	50	108
1-Propanol	97	74	80
2-Butanol	100	76	97
2-Methyl-2-butanol	102	116	72
2-Methyl-1-propanol (isobutyl alcohol)	108	87	104
3-Buten-1-ol	113	59	
3-Methyl-2-butanol	114	76	109
3-Pentanol	116	101	95
1-Butanol	117	64	71
2-Pentanol	120	62	75
3,3-Dimethyl-2-butanol	120	107	
2,3-Dimethyl-2-butanol	120	111	101
3-Methyl-3-pentanol	123	96	84
2-Methyl-2-pentanol	123	72	104
2-Methoxyethanol	125		113
1-Chloro-2-propanol	127	77	
2-Methyl-3-pentanol	128	85	
2-Methyl-1-butanol	129	70	82
2-Chloroethanol	130	95	101
4-Methyl-2-pentanol	132	65	88
3-Methyl-1-butanol	132	61	68
2-Ethoxyethanol	135	75	67
3-Hexanol	136	97	72
2,2-Dimethyl-1-butanol	137	51	81
1-Pentanol	138	46	68
2-Hexanol	139	39	61
2,4-Dimethyl-3-pentanol	140		99
Cyclopentanol	141	115	118
4-Methyl-1-pentanol	153	72	58
4-Heptanol	156	64	80
1-Hexanol	157	60	62
2-Heptanol	159	49	54
Cyclohexanol	161	112	129
3-Chloro-1-propanol	161	77	76
Furfuryl alcohol	172	81	130
1-Heptanol	177	47	62
2-Octanol	179	32	63
2-Ethyl-1-hexanol	185		60
1,2-Propanediol	187		
1-Octanol	195	62	67
1,2-Ethanediol	198	169	176

(Continued on next page)

Name of Compound	Boiling Point (°C)	Melting Points of Derivatives (°C)	
		3,5-Dinitrobenzoate	α-Naphthylurethane
2-Nonanol	198	43.	56
1-Linalool	199		53
1-Phenylethanol	202	95	106
Benzyl alcohol	206	113	134
1-Nonanol	213	52	66
1,3-Propanediol	215	178	164
2-Phenylethanol	220	108	119
Geraniol	230	63	48
1-Decanol	231	58	73
3-Phenyl-1-propanol	237	92	

TABLE 27.3   Alcohols   (b) Solids

Name of Compound	Melting Point (°C)	Melting Points of Derivatives (°C)	
		3,5-Dinitrobenzoate	α-Naphthylurethane
Cinnamyl alcohol	33	121	114
α-Terpineol	36	79	152
1-Tetradecanol	39	67	82
Menthol	44	153	126
1-Hexadecanol	50	66	82
2,2-Dimethyl-1-propanol	52		100
4-Methylbenzyl alcohol	60	118	
1-Octadecanol	60	77	
Benzhydrol	68	141	139
4-Nitrobenzyl alcohol	93	157	
Benzoin	137		140
Lanosterol	140	201	
Cholesterol	148	195	176
Triphenylmethanol	161		

TABLE 27.4 Aldehydes **(a) Liquids**

Name of Compound	Boiling Point (°C)	Melting Points of Derivatives (°C)		
		Semi-carbazone	2,4-Dinitro-phenylhydrazone	Oxime
Ethanal (acetaldehyde)	20	169	168	47
Propanal (propionaldehyde)	48	89	150	40
Glyoxal	50	270	328	178
2-Propenal (acrolein)	52	171	165	
2-Methylpropanal (isobutyraldehyde)	64	126	187	oil
2-Methyl-2-propenal	68	198	206	
Butanal (n-butyraldehyde)	75	96	123	oil
Trimethylacetaldehyde	75	190	210	41
Chloroacetaldehyde	86	148		oil
3-Methylbutanal	93	107	123	49
Pentanal	103		98	52
2-Butenal	104	199	190	119
2-Ethylbutanal	117	99	129	
4-Methylpentanal	121	127	99	oil
Paraldehyde (m.p. = 12°C)	125	169	168	47
Hexanal	131	106	104	51
5-Methylhexanal	144	117	117	
Heptanal	155	109	108	57
Furfural	162	202	230	92
Octanal	171	101	106	60
Benzaldehyde	179	222	237	35
Nonanal	185	100	100	64
Glutaraldehyde	189			178
Phenylethanal (phenylacetaldehyde)	194	153	121	100
Salicylaldehyde	197	231	252	63
3-Methylbenzaldehyde (m-tolualdehyde)	199	204	195	60
2-Methylbenzaldehyde (o-tolualdehyde)	200	209	194	49
4-Methylbenzaldehyde (p-tolualdehyde)	205	234	234	80
Decanal	207	102	104	69
2-Chlorobenzaldehyde	214	230	209	103
3-Chlorobenzaldehyde	214	228	256	70
3-Methoxybenzaldehyde	230	233		40
3-Bromobenzaldehyde	234	228		72
2-Ethoxybenzaldehyde	247	219		59
4-Methoxybenzaldehyde	248	210	254	64
Cinnamaldehyde	252	216	255	65

TABLE 27.4 Aldehydes   (b) Solids

Name of Compound	Melting Point (°C)	Melting Points of Derivatives (°C)		
		Semi-carbazone	2,4-Dinitro-phenylhydrazone	Oxime
1-Naphthaldehyde	34	221		98
Phenylethanal (phenylacetaldehyde)	34	163	121	100
Piperonal	37	234	266	146
2-Methoxybenzaldehyde	39	215	254	92
4-Diethylaminobenzaldehyde	41	241		93
3,4-Dichlorobenzaldehyde	44		301	120
2-Nitrobenzaldehyde	44	256	265	102(154)
3,4-Dimethoxybenzaldehyde	45	177	261	95
4-Chlorobenzaldehyde	48	233	254	110(146)
2,3-Dimethoxybenzaldehyde	54	231		99
4-Bromobenzaldehyde	57	229	128(257)	157(111)
3-Nitrobenzaldehyde	58	246	293	122
2-Naphthaldehyde	60	245	270	156
3,5-Dichlorobenzaldehyde	65			112
2,6-Dichlorobenzaldehyde	71			150
2,4-Dimethoxybenzaldehyde	71			106
4-Aminobenzaldehyde	72	153		124
2-Chloro-4-nitrobenzaldehyde	74	234	247*dec*	
2,4-Dichlorobenzaldehyde	74			137
4-Dimethylaminobenzaldehyde	74	222	325	185
3,4,5-Trimethoxybenzaldehyde	78	219		84
2-Chloro-5-nitrobenzaldehyde	79		277	176
4-Hydroxy-3-methoxybenzaldehyde	81	230	271*dec*	122
3,5-Dibromosalicylaldehyde	85			220
Isophthaldehyde	89			180
3-Hydroxybenzaldehyde	104	198	257*dec*	90
5-Bromosalicylaldehyde	106	297*dec*		126
4-Nitrobenzaldehyde	106	221	322	133
4-Hydroxybenzaldehyde	116	224	271	72(112)
Terphthalaldehyde	116			200
2,4,6-Trimethoxybenzaldehyde	118			203
5-Nitrosalicylaldehyde	126			218
2,4-Dihydroxybenzaldehyde	136	260*dec*	286	192
3,4-Dihydroxybenzaldehyde	154	230*dec*	275*dec*	157
3,5-Dihydroxybenzaldehyde	156	223		
Benzaldehyde-3-carboxylic acid	175	265		188*dec*

TABLE 27.5 Alkanes

Name of Compound	Boiling Point (°C)
Pentane	36
Cyclopentane	49
2,2-Dimethylbutane	50
2,3-Dimethylbutane	58
2-Methylpentane	60
3-Methylpentane	63
Hexane	69
Methylcyclopentane	72
2,2-Dimethylpentane	79
2,4-Dimethylpentane	80
Cyclohexane	81
3,3-Dimethylpentane	86
2,3-Dimethylpentane	89
2-Methylhexane	90
3-Methylhexane	92
3-Ethylpentane	93
Heptane	98
2,2,4-Trimethylpentane	99
Methylcyclohexane	101
2,5-Dimethylhexane	109
2-Methylheptane	118
Cycloheptane	119
trans-1,4-Dimethylcyclohexane	119
1,1-Dimethylcyclohexane	120
trans-1,2-Dimethylcyclohexane	123
Octane	126
2,2-Dimethylheptane	133
Cyclooctane	151
Nonane	151
Isopropylcyclohexane	154
1-Isopropyl-4-methycyclohexane	169
Decane	174
trans-Decahydronaphthalene (trans-decalin)	187
Undecane	196

TABLE 27.6   Alkenes and Alkynes

Name of Compound	Boiling Point (°C)	Melting Point (°C)
1-Pentene	30	
2-Methyl-1-buten-3-yne	32	
2-Methyl-1,3-butadiene	34	
2-Methyl-2-butene	38	
1-Pentyne	40	
1,3-Pentadiene	41	
Cyclopentene	44	
4-Methyl-1-pentene	54	
3-Methyl-1-pentene	54	
2-Pentyne	56	
1,5-Hexadiene	59	
1-Hexene	63	
2-Ethyl-1-butene	65	
2,3-Dimethyl-1,3-butadiene	69	
1-Hexyne	71	
1,4-Hexadiene	72	
2,3-Dimethyl-2-butene	73	
2,4-Hexadiene	82	
3-Hexyne	82	
Cyclohexene	83	
2-Hexyne	84	
1-Heptene	94	
1-Heptyne	100	
2,4,4-Trimethyl-1-pentene	101	
4-Methylcyclohexene	103	
2,4,4-Trimethyl-2-pentene	105	
Cycloheptene	114	
1,3,5-Cycloheptatriene	116	
1-Octene	121	
1-Octyne	126	
4-Vinyl-1-cyclohexene	129	
2,5-Dimethyl-2,4-hexadiene	134	
1,3,5,7-Cyclooctatetraene	141	
Phenylethyne (phenylacetylene)	142	
Phenylethene (styrene)	145	
Cyclooctene	145	
1-Nonene	147	
1-Nonyne	151	
3-Phenylpropene (allylbenzene)	157	
d,l-Camphene	160	50
β-Pinene	164	
2-Phenylpropene	165	
Myrcene	166	
1-Phenylpropene	170	
Dicyclopentadiene	170	
1-Decene	171	
1-Decyne	174	
Limonene	178	
Indene	182	
1-Tetradecene	251	13
1,2-Diphenylethyne	298	62
trans-1,2-Diphenylethene (stilbene)		124

TABLE 27.7 Alkyl Halides (Haloalkanes)

Name of Compound	Boiling Point (°C)
(a) Alkyl Chlorides	
2-Chloropropane	36
1-Chloro-1-propene	37
Dichloromethane	41
3-Chloro-1-propene	45
1-Chloropropane	47
trans-1,2-Dichloroethene	48
2-Chloro-2-methylpropane	51
2-Chloro-1,3-butadiene	59
Chloroform	61
2-Chlorobutane	68
1-Chloro-2-methylpropane (isobutyl chloride)	68
1,1,1-Trichloroethane	74
Carbon tetrachloride	77
1-Chlorobutane	78
1-Chloro-2,2-dimethylpropane	84
1,2-Dichloroethane	84
2-Chloro-2-methylbutane	86
1,2-Dichloropropane	96
2-Chloropentane	97
3-Chloropentane	98
1-Chloro-3-methylbutane	101
1-Chloropentane	108
Chlorocyclopentane	115
3-Chloro-3-methylpentane	116
1,1,2,2-Tetrachloroethene	121
3-Chlorohexane	123
2-Chlorohexane	125
1,3-Dichloropropane	125
1-Chlorohexane	134
Chlorocyclohexane	143
1-Chloro-3-bromopropane	143
1-Chloroheptane	159
Benzyl chloride	179
1-Chlorooctane	180
2-Chloro-1-phenylethane	198
1-Chlorononane	203
α,α-Dichlorotoluene	205
m-Chlorobenzyl chloride	216
o-Chlorobenzyl chloride	217
p-Chlorobenzyl chloride	222
1-Chlorodecane	223

(Continued on next page)

Name of Compound	Boiling Point (°C)
(b) Alkyl Bromides	
Bromoethane	38
2-Bromopropane	60
1-Bromopropane	71
3-Bromo-1-propene (allyl bromide)	71
2-Bromo-2-methylpropane (*tert*-butyl bromide)	73
2-Bromobutane	91
1-Bromo-2-methylpropane (isobutyl bromide)	93
Dibromomethane	99
1-Bromobutane	102
2-Bromopentane	117
3-Bromopentane	119
1-Bromo-3-methylbutane	120
3-Bromo-3-methylpentane	130
1-Bromopentane	130
1,2-Dibromoethane	132
Bromocyclopentane	137
1-Bromo-3-chloropropane	143
Bromoform	151
1-Bromohexane	155
Bromocyclohexane	165
1,3-Dibromopropane	168
1,4-Dibromobutane	198
1-Bromooctane	201
1-Bromononane	221
1-Bromodecane	241
(c) Alkyl Iodides	
Iodomethane	42
Iodoethane	72
2-Iodopropane	90
3-Iodo-1-propene (allyl iodide)	102
1-Iodopropane	103
2-Iodobutane	118
1-Iodo-2-methylpropane (isobutyl iodide)	120
1-Iodobutane	131
1-Iodopentane	157
Iodocyclopentane	167
Iodocyclohexane	179
1-Iodohexane	181
Diiodomethane	181
1-Iodoheptane	204
1-Iodooctane	225
Iodoform	(mp =119°C)

TABLE 27.8   Amides   (a) Liquids

Name of Compound	Boiling Point (°C)
N,N-Dimethylformamide	153
N,N-Dimethylacetamide	165
N,N-Diethylformamide	178
N-Methylformamide	185
N,N-Diethylacetamide	186
Formamide	193(195dec)
N-Ethylformamide	199
N-Methyl-2-pyrrolidinone	202
N-Ethylacetamide	205
N-Methylformanilide	244
2-Pyrrolidinone (γ-butyrolactam)	250

TABLE 27.8   Amides   (b) Solids

Name of Compound	Melting Point (°C)
N-Methylacetamide	31
δ-Valerolactam (2-piperidone)	39
Ethyl urethane	49
Formanilide	50
Methyl urethane	52
Phenyl urethane	53
N-Ethylacetanilide	54
Butyl urethane	54
Acetoacetamide	54
Propyl urethane	60
N-Benzylformamide	60
Pentananilide	63
Heptananilide	70
Decananilide	70
ε-Caprolactam	71
3-Butenamide	73
N,N-Diphenylformamide	73
Oleamide	76
N-Acetylacetamide	79
α-Chloropropionamide	80
Propionamide	81
N-Methylbenzamide	82
Acetamide	82
Acrylamide	85
Acetoacetanilide	86
3-Bromoacetanilide	87
2-Chloroacetanilide	88
N-Phenylmaleimide	91
Bromoactamide	91
2-Nitroacetanilide	92

(Continued on next page)

Name of Compound	Melting Point (°C)
Maleimide	93
Hexananilide	95
Iodoacetamide	95
Heptanamide	96
Butyranilide	96
m-Toluamide	97
Dichloroacetamide	98
Nonanamide	99
Hexanamide	100
N-Methylacetamide	102
Undecanamide	103
Isobutyranilide	105
Propionanilide	106
Hexadecanamide (palmitamide)	106
Pentanamide	106
Tetradecanamide (myristamide)	107
Decanamide	108
Heptadecanamide	108
Anthranilamide	109
Octadecanamide (stearamide)	109
Dodecanamide (lauramide)	110
Octanamide	110
β-Bromopropionamide	111
4-Aminobenzamide	114
Acetanilide	114
Butyramide	115
Methacrylamide	116
Chloroacetamide	118
Cyanoacetamide	120
Succinimide	126
Isobutyramide	129
2-Methoxybenzamide	129
Benzamide	130
2-Ethoxybenzamide	130
Urea	133
3-Chlorobenzamide	134
Phenacetin	134
3-Methylbutanamide	136
Salicylanilide	136
2-Chlorobenzamide	142
Salicylamide	142
o-Toluamide	143
3-Nitrobenzamide	143
Cinnamamide	148
Trimethylacetamide (pivalamide)	155
2,5-Dichlorobenzamide	155
3-Brombenzamide	155
Phenylacetamide	156
Succinic acid monoamide	157
N-(1-Naphthyl)acetamide	159
p-Toluamide	159

(Continued on next page)

(Continued)

Name of Compound	Melting Point (°C)
2-Bromobenzamide	161
4-Hydroxybenzamide	162
Benzanilide	163
3,4-Dimethoxybenzamide	164
4-Bromoacetanilide	167
4-Methoxybenzamide	167
4-Hydroxyacetanilide	169
3-Hydroxybenzamide	170
Malonamide (diamide)	170
N-Bromosuccinimide	173
2-Nitrobenzamide	176
4-Chloroacetanilide	179
4-Aminobenzamide	183
3,5-Dinitrobenzamide	183
2-Iodobenzamide	184
3-Iodobenzamide	186
4-Bromobenzamide	189
4-Nitrobenzamide	200
1-Naphthamide	202
2,6-Dichlorobenzamide	202
2,4-Dinitrobenzamide	203
4-Nitroacetanilide	215
4-Iodobenzamide	217
Hydantoin	218
Phthalamide (diamide)	220
2,4-Dihydroxybenzamide	222
Phthalimide	238
sym-Diphenylurea	240
Succinamide (diamide)	260*dec*

TABLE 27.9 Amines (a) Liquids

Name of Compound	Boiling Point (°C)	Melting Points of Derivatives (°C)			
		Benzamide	Benzene-sulfon-amide	Methyl iodide	Picric acid
Isopropylamine	33		26		
Ethylmethylamine	36				
tert-Butylamine	46	134			
n-Propylamine	49	84	36		
Diethylamine	56	42	42		
sec-Butylamine	63	76	70		
Isobutylamine	69	57	53		
n-Butylamine	77	42			
Diisopropylamine	84		94		
Pyrrolidine	89				
Triethylamine	89				173
2-Aminopentane	92				
Isopentylamine	96				
n-Pentylamine	104				
Piperidine	106	48	93		
Di-n-propylamine	110		51		
Ethylenediamine	116	244	168		
Pyridine	116			117	167
2-Methylpyridine (2-picoline)	129			230	169
Morpholine	130	75	118		
n-Hexylamine	132	40	96		
Cyclohexylamine	134	149	89		
2-Dimethylaminoethyl alcohol	135				96
1,3-Diaminopropane	136	148	96		
Diisobutylamine	139		55		
2,6-Dimethylpyridine (2,6-lutidine)	143			233	168
3-Methylpyridine (3-picoline)	143				150
4-Methylpyridine (4-picoline)	146				167
n-Heptylamine	156				
Tri-n-propylamine	157				116
Di-n-butylamine	159				
1,4-Diaminobutane	159	177			
2-Aminoethanol	171				
2,4,6-Trimethylpyridine (2,4,6-collidine)	172				155
1,5-Diaminopentane	178	135	119		
n-Octylamine	180				
Benzyldimethylamine	181			179	
Benzylmethylamine	181				
Aniline	184	160	112		
Benzylamine	185	105	88		
1-Amino-1-phenylethane	187	120			
N,N-Dimethylaniline	193			228dec	163
N-Methylaniline	196	63	79		
2-Amino-1-phenylethane	198	116	69		
2-Methylaniline (o-toluidine)	200	146	124		

(Continued on next page)

(Continued)

Name of Compound	Boiling Point (°C)	Melting Points of Derivatives (°C)			
		Benzamide	Benzene-sulfon-amide	Methyl iodide	Picric acid
n-Nonylamine	201	49			
3-Methylaniline (m-toluidine)	203	125	95		
N-Ethylaniline	205	60			
2-Chloroaniline	208	99	129		
4-Methylbenzylamine	208	137			
Tri-n-butylamine	211			186	105
2,6-Dimethylaniline	215	168			
2,5-Dimethylaniline	215	140	138		
2,4-Dimethylaniline	216	192	130		
N,N-Diethylaniline	218			102	142
3,5-Dimethylaniline	220	144			
2,3-Dimethylaniline	221	189			
2-Methoxyaniline (o-anisidine)	225	60	89		
4-Isopropylaniline	225	162			
2,4,6-Trimethylaniline	229	204	137		
3-Chloroaniline	230	119	121		
Quinoline	237			72,* 133[†]	203
2-Chloro-6-methoxyaniline	246	135			
4-Ethoxyaniline	248	173	143		
3-Bromoaniline	251	120			
3-Methoxyaniline (m-anisidine)	251				
Dicyclohexylamine	255	153			
Tri-n-pentylamine	257				
Dibenzylamine	300	112	68		

*Hydrated.
[†]Anhydrous.

TABLE 27.9 Amines   **(b) Solids**

Name of Compound	Melting Point (°C)	Melting Points of Derivatives (°C)			
		Benzamide	Benzene-sulfon-amide	Methyl iodide	Picric acid
2-Bromoaniline	32	116			
3-Iodoaniline	33	157			
N-Benzylaniline	37	107	119		
2,6-Dichloroaniline	39		157		
1,6-Diaminohexane	42	155	154		
4-Methylaniline (p-toluidine)	44	158	120		
3,4-Dimethylaniline	49	185	118		
2,5-Dichloroaniline	50	120			
3,5-Dichloroaniline	50	147			
Indole	52	68	254		

(Continued on next page)

		Melting Points of Derivatives (°C)			
Name of Compound	Melting Point (°C)	Benzamide	Benzene-sulfon-amide	Methyl iodide	Picric acid
Diphenylamine	53	180	124		
2-Aminopyridine	57	165			
4-Methoxyaniline (p-anisidine)	58	154	95		
4-Iodoaniline	62	222			
1,3-Diaminobenzene (m-phenylenediamine)	63	125 240	194		
2,4-Dichloroaniline	63	117	128		
4-Bromoaniline	66	204	134		
2-Nitroaniline	71	110	104		
4-Chloroaniline	72	192	122		
3,4-Dichloroaniline	72	144	130		
8-Hydroxyquinoline	75			143dec	204
4-Methyl-3-nitroaniline	78	172	160		
2,4,6-Trichloroaniline	78	174	154		
2,4-Dibromoaniline	79	134			
3,4-Diaminotoluene	89	264	179		
Tribenzylamine	91			184	190
2-Methyl-3-nitroaniline	92	168			
2-Methyl-6-nitroaniline	97	167			
2,4-Diaminotoluene	99	224	192		
1,2-Diaminobenzene (o-phenylenediamine)	102	301	185		
2-Bromo-4-nitroaniline	105	160			
4-Aminoacetophenone	106	205	128		
2-Chloro-4-nitroaniline	107	161			
2-Methyl-5-nitroaniline	107	186	172		
3-Nitroaniline	114	157	136		
4-Methyl-2-nitroaniline	115	148	102		
4-Chloro-2-nitroaniline	116	133			
2,4,6-Tribromoaniline	122	198			
Triphenylamine	127				
2-Nitro-4-methoxyaniline	129	140			
2-Methyl-4-nitroaniline	130		158		
2-Methoxy-4-nitroaniline	139	150	181		
1,4-Diaminobenzene (p-phenylenediamine)	142	300	247		
4-Nitroaniline	147	199	139		
2-Aminobenzoic acid (anthranilic acid)	147	182	214		
4-Nitro-N-methylaniline	152	112	121		
4-Aminopyridine	159	202			
2-Hydroxyaniline	174	167	141		
3-Aminobenzoic acid	174	113			
2,4-Dinitroaniline	180	202			
4-Aminophenol	184	216	125		
4-Aminobenzoic acid	188	278	212		
2,4,6-Trinitroaniline	190	196	211		

TABLE 27.10 Aromatic Hydrocarbons

Name of Compound	Melting Point (°C)	Boiling Point (°C)	Nitration Product Position	Nitration Product mp (°C)
Benzene	5	80	1,3	89
Toluene		111	2,4	70
Ethylbenzene		136	2,4,6	37
p-Xylene	13	138	2,3,5	139
m-Xylene		139	2,4,6	183
o-Xylene		142	4,5	118
Isopropylbenzene (cumene)		153	2,4,6	109
n-Propylbenzene		159	2,4	liquid
1,3,5-Trimethylbenzene		165	2,4	86
			2,4,6	235
tert-Butylbenzene		169	2,4	62
			2,4,6	124
1,2,4-Trimethylbenzene		169	3,5,6	185
1,2,3-Trimethylbenzene		176		
Indane		177	5	40
4-Isopropyltoluene		177	2,6	54
Indene		182		
1,2,3,5-Tetramethylbenzene		198	4,6	181(157)
1,3-Diisopropylbenzene		203	4,6	77
1,2,3,4-Tetramethylbenzene		205	5,6	176
1,3-Dimethyl-5-tert-butylbenzene		206	2,4,6	107(114)
1,2,3,4-Tetrahydronaphthalene (tetralin)		207	5,7	95
1,4-Diisopropylbenzene		210		
1-Phenylhexane		226		
Cyclohexylbenzene		236		
1-Methylnaphthalene		245	4	71
Diphenylmethane	27	264	2,2',4,4'	172
2-Methylnaphthalene	38	240	1	81
1,2-Diphenylethane	53	284	4,4'	180
			2,2',4,4'	169
Pentamethylbenzene	54	232	6	154
Biphenyl	69	254	4,4'	237
			2,2',4,4'	150
1,2,4,5-Tetramethylbenzene (durene)	80	198	3,6	205
Naphthalene				
Triphenylmethane	92	358	4,4',4"	206
Acenaphthene	96	278	5	101
Phenanthrene	101	340		
2,3-Dimethylnaphthalene	104	266		
2,6-Dimethylnaphthalene	111	262		
Fluorene	114	295	2	156
			2,7	199
trans-Stilbene	124			
1,4-Diphenyl-1,3-butadiene	152			
Anthracene	216			

TABLE 27.11 Aryl Halides

Name of Compound	Boiling Point (°C)	Melting Point (°C)	Nitration Product	
			Position	mp (°C)
Chlorobenzene	132		2,4	52
Bromobenzene	156		2,4	70
2-Chlorotoluene	159		3,5	63
3-Chlorotoluene	162		4,6	91
4-Chlorotoluene	162	7	2	38
1,3-Dichlorobenzene	173		4,6	103
1,2-Dichlorobenzene	180		4,5	110
2-Bromotoluene	182		3,5	82
3-Bromotoluene	184		4,6	103
4-Bromotoluene	184	28	2	47
Iodobenzene	188		4	171
2,6-Dichlorotoluene	199		3	50
2,4-Dichlorotoluene	200		3,5	104
3,4-Dichlorotoluene	201		2,6	91
3,5-Dichlorotoluene	201	26	2,6	99
3-Iodotoluene	204		4,6	108
1,2,4-Trichlorobenzene	213	17	5	56
1,2-Dibromobenzene	225	7	4,5	114
1-Chloronaphthalene	259		4,5	180
1-Bromonaphthalene	281	6	4	85
3,5-Dichlorotoluene	201	26	2,6	99
4-Bromotoluene	184	28	2	47
2,4,6-Trichlorotoluene		34	3	54
4-Iodotoluene	211	35		
1,2,3,4-Tetrachlorobenzene	275	46	5	64
			5,6	151
1,2,3,5-Tetrachlorobenzene	246	51	4	41
			4,6	162
1,4-Dichlorobenzene	173	53	2	54
			2,6	106
1,3,5-Trichlorobenzene	208	63	2	68
			2,4	131
1-Bromo-4-chlorobenzene	197	67	2	72
4-Chlorobiphenyl	293	77		
2,4,5-Trichlorotoluene		82	3	92
			3,6	227
1,4-Dibromobenzene	219	89	2,5	84
1,3,5-Tribromobenzene	271	120		
4,4'-Dichlorobiphenyl	315	149	2	102
			2,2'	138

TABLE 27.12 Carboxylic Acids   (a) Liquids

Name of Compound	Boiling Point (°C)	Melting Points of Derivatives (°C)		
		Anilide	p-Toluidide	Amide
Methanoic acid (formic acid)	101	50	53	
Ethanoic acid (acetic acid)	118	114	147	82
Propenoic acid	140	105	141	85
Propanoic acid (propionic acid)	141	106	126	81
2-Methylpropanoic acid (isobutyric acid)	155	105	107	129
2-Methylpropenoic acid	163			109
Butanoic acid	164	96	75	115
3-Butenoic acid	164	58		73
cis-2-Butenoic acid	169	102	132	102
3-Methylbutanoic acid (isovaleric acid)	177	110	107	137
3,3-Dimethylbutanoic acid	184	132	134	132
2-Chloropropanoic acid	186	92	124	80
Pentanoic acid (valeric acid)	186	63	74	106
Dichloroacetic acid	194	118	153	98
4-Methylpentanoic acid (isocaproic acid)	199	112	63	121
2-Bromopropanoic acid	205dec	99	125	123
Hexanoic acid (caproic acid)	205	95	75	100
2-Bromobutanoic acid	217dec	98	92	112
Heptanoic acid	223	70	81	96
2-Ethylhexanoic acid	228			102
Octanoic acid	239	57	70	110

TABLE 27.12 Carboxylic Acids   (b) Solids

Name of Compound	Melting Point (°C)	Boiling Point (°C)	Melting Points of Derivatives (°C)		
			Anilide	p-Toluidide	Amide
2-Bromopropanoic acid	26	205dec	99	125	123
Undecanoic acid	29	284	71	89	103
Cyclohexanecarboxylic acid	31	233	146		186
Decanoic acid	32	270	70	78	108
2-Oxobutanoic acid	32				117
4-Oxopentanoic acid	34	245	102	108	108dec
Pivalic acid (trimethylacetic acid)	36	164	130	120	157
Acetoacetic acid (3-oxobutanoic acid)	37	dec	86	95	54
3-Chloropropanoic acid	42	204			101
2-Phenylbutanoic acid	42	270			85
Dodecanoic acid	44	299	78	87	100
Hydrocinnamic acid (3-phenyl-propanoic acid)	48	280	98	135	105
Bromoacetic acid	50	208	131	91	91
4-Phenylbutanoic acid	52	290			84
Tetradecanoic acid	54		84	93	107
Trichloroacetic acid	58	198	97	113	141
5-Phenylpentanoic acid	60		90		109
3-Bromopropanoic acid	61				111
Hexadecanoic acid	62		91	98	106
Chloroacetic acid	63	189	137	162	118

(Continued on next page)

Name of Compound	Melting Point (°C)	Boiling Point (°C)	Melting Points of Derivatives (°C)		
			Anilide	p-Toluidide	Amide
cis-2-Methyl-2-butenoic acid	65	199	77	71	76
Cyanoacetic acid	66		198		120
Benzoylformic acid	66				91
Octadecanoic acid	70		96	102	109
trans-2-Butenoic acid (crotonic acid)	72	189	118	132	160
m-Methoxyphenylacetic acid	73				125
Phenylacetic acid	76		117	136	156
Glycolic acid (hydroxyacetic acid)	79		97	143	120
α-Methylcinnamic acid	81				128
Iodoacetic acid	83		144		95
p-Methoxyphenylacetic acid	87				189
o-Methylphenylacetic acid	90				161
3-Chloro-2-butenoic acid	94		124		101
o-Chlorophenylacetic acid	95		139	170	175
p-Methylphenylacetic acid	95				185
3,4-Dimethoxyphenylacetic acid	95				147
Glutaric acid (pentanedioic acid)	98		224	218	176
3-Phenoxypropionic acid	98				119
Phenoxyacetic acid	99		99		102
Citric acid hydrate	100		192	189	210dec
2-Methoxybenzoic acid (o-anisic acid)	101	200	131		129
Malic acid (2-hydroxybutanedioic acid)	101		197	207	157
Oxalic acid dihydrate	101		254	268	419dec
o-Toluic acid (2-methylbenzoic acid)	104		125	144	143
Heptanedioic acid	105			206	175
Nonanedioic acid	107		186	201	172
3-Methoxybenzoic acid (m-anisic acid)	110				136
Ethylmalonic acid	111		150		214
m-Toluic acid	112		126	118	95
2-Phenylbenzoic acid	113				177
2-Phenoxybenzoic acid	113				131
p-Bromophenylacetic acid	114				194
2-Acetylbenzoic acid	115				116
Methylsuccinic acid	115		200	164	225
β-Benzoylpropionic acid	116		150		146
2,6-Dimethylbenzoic acid	116				139
4-Isopropylbenzoic acid	117				133
Benzylmalonic acid	117dec		217		225
2-Phenyl-3-hydroxy-propanoic acid	117				169
Mandelic acid	118		152	172	132
m-Nitrophenylacetic acid	120				110
3-Furoic acid	121				169
Benzoic acid	122		160	158	130
Picric acid	123				
2,4-Dimethylbenzoic acid	127		141		180
2-Benzoylbenzoic acid	127		195		165
Dodecanedioic acid	128		191	165	185
Maleic acid (cis-butenedioic acid)	130		187	142	260

(Continued on next page)

(Continued)

Name of Compound	Melting Point (°C)	Boiling Point (°C)	Melting Points of Derivatives (°C)		
			Anilide	*p*-Toluidide	Amide
1-Naphthylacetic acid	132		155		180
2,5-Dimethylbenzoic acid	132		140		186
m-Chlorocinnamic acid	133		135	142	76
2-Furoic acid	133		124	108	143
trans-Cinnamic acid	133		153	168	148
Decanedioic acid (sebacic acid)	134		202	201	210
Malonic acid (propanedioic acid)	135		230	253	170
O-Acetylsalicylic acid	135		136		138
1,3-Acetonedicarboxylic acid	135*dec*		155		
Pyridine-2-carboxylic acid	137		76	104	107
Phenylpropynoic acid	137		126	142	100
Methylmalonic acid	138*dec*		182	228	217
5-Chloro-2-nitrobenzoic acid	139		164		154
3-Nitrobenzoic acid	140		154	162	143
meso-Tartaric acid	140				187
2-Chloro-4-nitrobenzoic acid	141		168		172
o-Nitrophenylacetic acid	141				161
2,4-Dichlorophenoxyacetic acid	141				130
4-Chloro-2-nitrobenzoic acid	142				172
2-Naphthylacetic acid	142				200
2-Chlorobenzoic acid	142		118	131	142
Octanedioic acid	144		186	218	217
2,4,5-Trimethoxybenzoic acid	144		155		185
2,6-Dichlorobenzoic acid	144				202
o-Chlorophenoxyacetic acid	146		121		150
2-Nitrobenzoic acid	146		155		176
2-Aminobenzoic acid	147		131	151	109
Diphenylacetic acid	148		180	173	168
p-Hydroxyphenylacetic acid	148				175
2-Bromobenzoic acid	150		141		155
Benzilic acid	150		175	190	154
Citric acid (anhydrous)	153		192	189	210*dec*
p-Nitrophenylacetic acid	153		198	210	198
2,5-Dichlorobenzoic acid	153				155
Hexanedioic acid (adipic acid)	154		241	239	220
3-Bromobenzoic acid	155		136		155
2,4,6-Trimethylbenzoic acid	155				188
p-Chlorophenoxyacetic acid	156		125		133
Hydroxymalonic acid	157*dec*				198
3-Chlorobenzoic acid	158		123		134
Salicylic acid (2-hydroxybenzoic acid)	158		136	156	142
1-Naphthoic acid	162		163		205
2-Iodobenzoic acid	162		141		110
4-Nitrophthalic acid	164		192		200*dec*
2,4-Dichlorobenzoic acid	164				194
3,4-Dinitrobenzoic acid	165		189		166
Propene-2,3-dicarboxylic acid	166				192*di*
5-Bromosalicylic acid	165		222		232
2-Chloro-5-nitrobenzoic acid	165				178

(Continued on next page)

(Continued)

Name of Compound	Melting Point (°C)	Boiling Point (°C)	Melting Points of Derivatives (°C)		
			Anilide	*p*-Toluidide	Amide
1,2,3-Propane-tricarboxylic acid	166		252		207*dec*
3,4-Dimethylbenzoic acid	166		104		130
3-Methylsalicylic acid	166		83		112
3,5-Dimethylbenzoic acid	166				133
d-Tartaric acid	170		264*dec*		196*dec*
3,4,5-Trimethoxybenzoic acid	171				177
5-Chlorosalicylic acid	172				227
3-Aminobenzoic acid	174		140		111
3,5-Dinitrosalicylic acid hydrate	174				181
Acetylenedicarboxylic acid	179				249*dec*
p-Toluic acid (4-methylbenzoic acid)	179	275	145	160	160
3,4-Dimethoxybenzoic acid	181		154		164
4-Chloro-3-nitrobenzoic acid	182		131		156
2,4-Dinitrobenzoic acid	183				203
4-Methoxybenzoic acid (*p*-anisic acid)	185		171	186	167
2-Naphthoic acid	185		171	192	192
3-Iodobenzoic acid	187				186
Coumarin-3-carboxylic acid	188		250		236
p-Nitrophenoxyacetic acid	188		170		158
4-Aminobenzoic acid	188				183
Succinic acid (butanedioic acid)	189		230	255	260*dec*
Hippuric acid	190		208		183
Dimethylmalonic acid	193				269*di*
4-Ethoxybenzoic acid	198		170		202
trans-m-Nitrocinnamic acid	199				196
3,4-Dihydroxybenzoic acid	200*dec*		166		212
Fumaric acid	200		314		266
3-Hydroxybenzoic acid	200		157	163	170
2,5-Dihydroxybenzoic acid	204				218
d,l-Tartaric acid	204		236		226
2,3-Dihydroxybenzoic acid	204				175
3,5-Dinitrobenzoic acid	207		234	147	183
3,4-Dichlorobenzoic acid	209				168
Phthalic acid	210		253	201	220
o-Chlorocinnamic acid	212		176		168
2,4-Dihydroxybenzoic acid	213		126		222
trans-o-Hydroxycinnamic acid	214*dec*				209*dec*
4-Hydroxybenzoic acid	215		197	204	162
3-Nitrophthalic acid	218		234	226	201*dec*
4-Cyanobenzoic acid	219		179		223
4-Phenylbenzoic acid	226				223
Piperonylic acid	229				169
5-Nitrosalicylic acid	230		224		225
3-Chloro-2-nitrobenzoic acid	235		186		
trans-o-Nitrocinnamic acid	240				185
4-Nitrobenzoic acid	241		211	204	201
4-Chlorobenzoic acid	243		194		179
4-Dimethylaminobenzoic acid	245		183		206
4-Bromobenzoic acid	251		197		190

TABLE 27.13 Esters   (a) Liquids

Name of Compound	Boiling Point (°C)
Methyl formate	31
Ethyl formate	54
Methyl acetate	57
Isopropyl formate	71
Vinyl acetate	73
Ethyl acetate	77
Methyl propionate	80
Methyl acrylate	80
Propyl formate	81
Isopropyl acetate	90
Methyl carbonate	91
Methyl isobutyrate	93
Isopropenyl acetate	94
tert-Butyl acetate	98
Methyl methacrylate	100
Ethyl propionate	100
Ethyl acrylate	101
Propyl acetate	102
Methyl butyrate	102
Allyl acetate	104
Ethyl isobutyrate	110
Isopropyl propionate	110
sec-Butyl acetate	112
Methyl isovalerate	117
Isobutyl acetate	117
Ethyl pivalate (ethyl trimethylacetate)	118
Methyl crotonate	119
Ethyl butyrate	122
Propyl propionate	123
Butyl acetate	126
Diethyl carbonate	127
Methyl valerate	127
Methyl methoxyacetate	130
Methyl chloroacetate	131
Ethyl isovalerate	135
Methyl pyruvate	137
Methyl α-hydroxyisobutyrate	137
Ethyl crotonate	138
3-Methylbutyl acetate (isoamyl acetate)	142
Methyl lactate	145
Ethyl chloroacetate	145
2-Methoxyethyl acetate	145
Ethyl valerate	146
Ethyl α-chloropropionate	146
Diisopropyl carbonate	147
Pentyl acetate	149
Methyl hexanoate	151
Cyclopentyl acetate	153
Ethyl lactate	154

(Continued on next page)

Name of Compound	Boiling Point (°C)
Ethyl pyruvate	155
Ethyl dichloroacetate	158
Ethyl α-bromopropionate	162
Butyl butyrate	167
Ethyl hexanoate	168
Ethyl trichloroacetate	168
Methyl acetoacetate	170
Hexyl acetate	172
Methyl heptanoate	172
Cyclohexyl acetate	175
Furfuryl acetate	176
Ethyl β-bromopropionate	179
Ethyl acetoacetate	181
Methyl furoate	181
Dimethyl malonate	182
Diethyl oxalate	185
Ethyl δ-chlorobutyrate	186
Ethyl heptanoate	187
Ethylene glycol diacetate	190
Heptyl acetate	192
Methyl octanoate	193
Dimethyl succinate	196
Ethyl cyclohexanecarboxylate	196
Dimethyl methylsuccinate	196
Phenyl acetate	197
Diethyl malonate	199
Methyl benzoate	199
γ-Butyrolactone	204
Dimethyl maleate	204
Ethyl levulinate	206
γ-Valerolactone	207
Ethyl octanoate	208
Octyl acetate	210
Ethyl benzoate	212
Dimethyl glutarate	215
Methyl nonanoate	215
Benzyl acetate	217
Diethyl succinate	217
Diethyl fumarate	218
Methyl phenylacetate	220
Diethyl maleate	223
Methyl salicylate	224
Methyl decanoate	225
Ethyl phenylacetate	228
Propyl benzoate	231
Diethyl glutarate	234
Ethyl salicylate	234
Methyl β-phenylpropionate	238
Propylene carbonate	240
Diethyl adipate	245
Methyl undecylenate	248

(Continued on next page)

(Continued)

Name of Compound	Boiling Point (°C)
Diethyl pimelate	255
Ethyl benzoylacetate	265
Dimethyl suberate	268
Ethyl cinnamate	271
Methyl 2-nitrobenzoate	275
Diethyl tartrate	280
Diethyl suberate	282
Dimethyl phthalate	284
Diethyl phthalate	290
Ethyl 3-aminobenzoate	294
Diethyl benzylmalonate	300
Methyl myristate (m.p. = 18°C)	323
Diisobutyl phthalate	327
Dibutyl phthalate	340

TABLE 27.13   Esters   (b) Solids

Name of Compound	Melting Point (°C)
Dimethyl succinate	18
Methyl myristate	18
Diethyl tartarate	18
Benzyl benzoate	21
Methyl anthranilate (methyl 2-aminobenzoate)	24
Dimethyl sebacate	27
Bornyl acetate	29
Methyl palmitate	30
Ethyl 2-nitrobenzoate	30
Methyl 4-toluate	33
Ethyl sterate	33
Ethyl 2-furoate	36
Methyl cinnamate	36
Ethylene carbonate	37
Ethyl mandelate	37
Dimethyl itaconate	39
Methyl stearate	39
Phenyl salicylate	42
Diethyl terephthalate	44
Ethyl 3-nitrobenzoate	47
Dimethyl tartrate	49
1-Naphthyl acetate	49
Methyl mandelate	53
Dimethyl oxalate	54
Ethyl 4-nitrobenzoate	56
Coumarin	67
Dimethyl isophthalate	68
Phenyl benzoate	69

(Continued on next page)

Name of Compound	Melting Point (°C)
Methyl 3-hydroxybenzoate	70
Diphenyl phthalate	74
Diphenyl carbonate	78
Methyl 3-nitrobenzoate	78
Methyl 4-bromobenzoate	83
Ethyl 4-aminobenzoate	90
Dimethyl *d,l*-tartarate	90
3-Carbethoxycoumarin	93
Methyl 4-nitrobenzoate	96
Propyl 4-hydroxybenzoate	96
Dimethyl fumarate	102
Cholesteryl acetate	114
Ethyl 4-hydroxybenzoate	116
Hydroquinone diacetate	124
Methyl 4-hydroxybenzoate	131
Ethyl 4-nitrocinnamate	137
Dimethyl terephthalate	141
Propyl gallate	150

TABLE 27.14 Ethers

Name of Compound	Boiling Point (°C)	Melting Point (°C)
Furan	31	
Ethyl ether	35	
Ethyl vinyl ether	36	
Methyl *n*-propyl ether	39	
Ethyl isopropyl ether	53	
tert-Butyl methyl ether	55	
Ethyl *n*-propyl ether	64	
Tetrahydrofuran	66	
Isopropyl ether	68	
2-Methyltetrahydrofuran	79	
1,2-Dimethoxyethane	85	
3,4-Dihydropyran	86	
Tetrahydropyran	88	
n-Propyl ether	90	
n-Butyl vinyl ether	94	
1,4-Dioxane	101	
β-Chloroethyl ethyl ether	107	
1,2-Epoxy-3-chloropropane	117	
Isobutyl ether	123	
n-Butyl ether	142	
Anisole	155	
Diethylene glycol dimethyl ether	162	
o-Methylanisole	171	
Ethoxybenzene	172	

(Continued on next page)

(Continued)

Name of Compound	Boiling Point (°C)	Melting Point (°C)
p-Methylanisole	174	
m-Methylanisole	176	
2,2'-Dichloroethyl ether	178	
n-Pentyl ether	188	
3-Chloroanisole	194	
2-Chloroanisole	195	
4-Chloroanisole	200	
1,2-Dimethoxybenzene	207	
Butyl phenyl ether	210	
1,3-Dimethoxybenzene	217	
n-Hexyl ether	229	
Safrole	233	
4-Propenylanisole	235	
2-Nitroanisole	277	
Benzyl ether	298	
Phenyl ether	258	28
2-Ethoxynaphthalene	282	36
3-Nitroanisole	258	39
1,2,3-Trimethoxybenzene	241	47
4-Iodoanisole	240	52
1,3,5-Trimethoxybenzene	255	53
4-Nitroanisole	274	54
1,4-Dimethoxybenzene	213	56

TABLE 27.15   Ketones   (a) Liquids

Name of Compound	Boiling Point (°C)	Melting Points of Derivatives (°C)		
		Semi-carbazone	2,4-Dinitro-phenylhydrazone	Oxime
Acetone	56	190	128	59
3-Buten-2-one (methyl vinyl ketone)	81	141		
2-Butanone	82	136	117	
3-Butyn-2-one	86		181	
3-Methyl-2-butanone	94	114	120	
3-Methyl-3-buten-2-one	98	173	181	
Cyclobutanone	100		146	
3-Pentanone	102	139	156	
2-Pentanone	102	112	144	
1-Penten-3-one	103		129	
3,3-Dimethyl-2-butanone (pinacolone)	106	158	125	79
1-Methoxy-2-propanone	115		163	
4-Methyl-2-pentanone	117	135	95	
3-Methyl-2-pentanone	118	95	71	
Chloroacetone	119	150	125	
3-Penten-2-one	122	142	155	

(Continued on next page)

Name of Compound	Boiling Point (°C)	Melting Points of Derivatives (°C)		
		Semi-carbazone	2,4-Dinitro-phenylhydrazone	Oxime
2,4-Dimethyl-3-pentanone	124	160	88	
3-Hexanone	125	113	130	
4,4-Dimethyl-2-pentanone	125		100	
2-Hexanone	128	125	106	49
5-Hexen-2-one	130	102	108	
4-Methyl-3-penten-2-one (mesityl oxide)	130	164	206	49
Cyclopentanone	131	210	146	56
5-Methyl-3-hexanone	136	152		
2-Methyl-3-hexanone	136	119		
2,4-Pentanedione (acetylacetone)	139	209	209	149
4-Heptanone	144	132	75	
1-Hydroxy-2-propanone	146	196	129	
3-Heptanone	148	103		
2-Heptanone	151	123	89	
Cyclohexanone	156	167	162	91
2,3-Hexanedione	158			175
3,5-Dimethyl-4-heptanone	162	84		
2-Methylcyclohexanone	165	191	137	43
2,6-Dimethyl-4-heptanone	168	126	66	
4-Octanone	170	96	41	
4-Methylcyclohexanone	171	199		39
2-Octanone	173	123	58	
2,2,6-Trimethylcyclohexanone	179	209	141	
Ethyl acetoacetate	181	133	93	
5-Nonanone	186	90		
3-Nonanone	187	112		
2,5-Hexanedione	194	224	257	137
2-Nonanone	195	119	56	
Acetophenone	202	199	240	60
Menthone	209	189	146	59
2-Methylacetophenone	214	205	159	61
1,5,5-Trimethylcyclohexen-3-one	215	200		79
1-Phenyl-2-propanone	216	200	156	70
Propiophenone	220	174	191	54
3-Methylacetophenone	220	198	207	55
Isobutyrophenone	222	181	163	94
1-Phenyl-2-butanone	226	135		
3-Chloroacetophenone	228	232		88
2,4-Dimethylacetophenone	228	187		63
2-Chloroacetophenone	229	160		113
n-Butyrophenone	230	188	190	50
d-Carvone	230	163	191	73
4-Chloroacetophenone	232	204	231	95
3,5-Dimethylacetophenone	237			114
2-Methoxyacetophenone	239	183		83
3-Methoxyacetophenone	240	196		
n-Valerophenone	248	160	166	52
2,5-Dichloroacetophenone	251			130

TABLE 27.15 Ketones (b) Solids

Name of Compound	Melting Point (°C)	Melting Points of Derivatives (°C)		
		Semi-carbazone	2,4-Dinitro-phenylhydrazone	Oxime
4-Methylacetophenone	28	205	260	88
2-Hydroxyacetophenone	28	210	212	118
2,6-Dimethyl-2,5-heptadien-4-one	28	221	118	48
2,4-Dichloroacetophenone	34	208		148
4-Chloropropiophenone	36	176	223	63
4-Methoxyacetophenone	38	198	220	87
2-Hydroxybenzophenone	39			143
2-Methoxybenzophenone	39		251	148
3-Bromopropiophenone	40	183		
4-Phenyl-3-buten-2-one (benzalacetone)	41	187	227	115
1-Indanone	42	233	258	146
4-Bromopropiophenone	46	171		91
Benzophenone	48	165	239	143
4-Bromoacetophenone	51	208	230	128
3,4-Dimethoxyacetophenone	51	218	207	140
Methyl 2-naphthyl ketone	54	235	262	149
4-Methyl benzophenone	57	121	200	154
Benzalacetophenone (chalcone)	58	170	245	140
α-Chloroacetophenone	59	156	214	89
Desoxybenzoin (benzyl phenyl ketone)	60	148	204	98
Benzoylacetone	61		151	
1,1-Diphenylacetone	61	170		165
4-Methoxybenzophenone	62		180	116
Cinnamalacetone	68	186	223	153
2,6-Dimethyl-1,4-benzoquinone	73			175
4-Chlorobenzophenone	78		185	105
1,4-Cyclohexanedione	79	231	240	188
3-Nitroacetophenone	80	257	228	132
4-Nitroacetophenone	81		258	174
4-Bromobenzophenone	82	350	230	116
9-Fluorenone	83		284	196
4,4'-Dimethylbenzophenone	95	143	219	163
Benzil	95	243	189	237
3-Hydroxyacetophenone	96	195	257	
1,3-Cyclohexanedione	104			156
4-Hydroxyacetophenone	109	199	261	145
3,4-Dihydroxyacetophenone	116			184dec
1,4-Benzoquinone	116	243	231di	240
1,4-Naphthoquinone	125	247	278	198
4-Hydroxybenzophenone	135	194	242	81
Benzoin	137	206dec	245	152
2,4-Dihydroxyacetophenone	147	218	208	200dec
4,4'-Dichlorobenzophenone	148		241	135
Camphor	178	248dec	177	119
4,4'-Dihydroxybenzophenone	210		192	

TABLE 27.16   Nitriles

Name of Compound	Boiling Point (°C)	Melting Point (°C)
Acrylonitrile	77	
Acetonitrile	81	
Propanenitrile	97	
2-Methylpropanenitrile	108	
Butanenitrile	117	
4-Methylbutanenitrile	130	
Pentanenitrile	141	
4-Methylpentanenitrile	155	
Hexanenitrile	165	
3-Chloropropanenitrile	178	
5-Methylhexanenitrile	180	
Heptanenitrile	183	
Benzonitrile	190	
4-Chlorobutanenitrile	196	
2-Methylbenzonitrile (2-tolunitrile)	205	
Octanenitrile	206	
Ethyl cyanoacetate	207	
3-Methylbenzonitrile (3-tolunitrile)	212	
Nonanenitrile	224	
Phenylacetonitrile (benzyl cyanide)	234	
Decanenitrile	245	
1,3-Dicyanopropane	286	
Cinnamonitrile	256	20
4-Methylbenzonitrile (4-tolunitrile)	217	27
Malononitrile	219	30
4-Chlorobenzyl cyanide	267	30
1-Cyanonaphthalene	299	34
3-Bromobenzonitrile	225	38
3-Chlorobenzonitrile		41
2-Chlorobenzonitrile		43
4-Cyanobutanoic acid		45
3-Cyanopropanoic acid		48
2-Aminobenzonitrile		51
2-Bromobenzonitrile		53
2,4,6-Trimethylbenzonitrile		55
Succinonitrile		57
4-Methoxybenzonitrile		62
3,5-Dichlorobenzonitrile		65
Cyanoacetic acid		67
3,4-Dichlorobenzonitrile		72
Diphenylacetonitrile		75
4-Cyanopyridine		78
3-Cyanobenzaldehyde		80
2-Chloro-6-methylbenzonitrile		82
4-Aminobenzonitrile		86
4-Chlorobenzonitrile		96
2-Cyanophenol		98

(Continued on next page)

(Continued)

Name of Compound	Boiling Point (°C)	Melting Point (°C)
2,4-Dinitrobenzonitrile		104
2-Nitrobenzonitrile		110
4-Bromobenzonitrile		112
4-Cyanophenol		113
p-Nitrophenylacetonitrile		116
3-Nitrobenzonitrile		118
2,5-Dichlorobenzonitrile		130
1,2-Dicyanobenzene		141
2,6-Dinitrobenzonitrile		145
4-Nitrobenzonitrile		147
1,3-Dicyanobenzene		162
2-Cyanobenzoic acid		187
3-Cyanobenzoic acid		217
4-Cyanobenzoic acid		219

TABLE 27.17 Nitro Compounds

Name of Compound	Boiling Point (°C)	Melting Point (°C)	Nitration Product Position	Nitration Product mp (°C)
Nitromethane	101			
Nitroethane	115			
2-Nitropropane	120			
1-Nitropropane	131			
2-Nitrobutane	140			
1-Nitrobutane	153			
1-Nitropentane	173			
Nitrobenzene	211		1,3	90
2-Nitrotoluene	222		2,4	71
1,3-Dimethyl-2-nitrobenzene	226	13	1,3,5	182
3-Nitrotoluene	233	16		
1,4-Dimethyl-2-nitrobenzene	241		1,2,4	139
1,3-Dimethyl-4-nitrobenzene	246	2	1,3,5	182
1,2-Dimethyl-3-nitrobenzene	248	15	1,2	82
1-Isopropyl-4-methyl-2-nitrobenzene	264		2,6	54
2-Nitroanisole	273	10	2,4,6	68
2-Methyl-2-nitropropane	127	26		
1,2-Dimethyl-4-nitrobenzene		31	1,2	82
2-Chloronitrobenzene	246	32	2,4	52
2,4-Dichloronitrobenzene	258	33		
2-Chloro-6-nitrotoluene	238	37	4,6	49
3-Nitroanisole		38	3,5	106
4-Chloro-2-nitrotoluene		38	2,6	77
2-Bromonitrobenzene		43	1,3	72
2,4,6-Trimethylnitrobenzene		44	1,3	86
3-Chloronitrobenzene		45		

(Continued on next page)

Name of Compound	Boiling Point (°C)	Melting Point (°C)	Nitration Product	
			Position	mp (°C)
4-Nitrotoluene	234	52	2,4	70
1-Chloro-2,4-dinitrobenzene		52	2,4,6	183
4-Nitroanisole		53	2,4	89
3-Nitrobromobenzene		56	3,4	59
1-Nitronaphthalene		57		
3,4-Dinitrotoluene		61		
2,6-Dinitrotoluene		66	2,4,6	82
2,4-Dinitrotoluene		70	2,4,6	82
3,5-Dimethylnitrobenzene		75		
4-Chloronitrobenzene	242	84	2,4	52
1,3-Dinitrobenzene		90		
1-Chloro-8-nitronaphthalene		94		
2,4-Dinitroanisole		95	2,4,6	68
4-Nitrobiphenyl		114	4,4'	240
1,2-Dinitrobenzene		118		
4-Nitrobromobenzene		126		
9-Nitroanthracene		146		
1,8-Dinitronaphthalene		170		
1,4-Dinitrobenzene		173		
1,5-Dinitronaphthalene		217		
4,4'-Dinitrobiphenyl		240		

TABLE 27.18   Phenols

Name of Compound	Melting Point (°C)	Boiling Point (°C)	Melting Points of Derivatives (°C)	
			α-Naphthyl-urethane	Bromo Derivative
2-Chlorophenol	7	176	120	76 *di*
Phenol	42	182	132	95 *tri*
2-Methylphenol (*o*-cresol)	31	191	142	56 *di*
2-Bromophenol	5	195	129	95 *tri*
Salicylaldehyde	2	197		
3-Methylphenol (*m*-cresol)	12	202	128	84 *tri*
4-Methylphenol (*p*-cresol)	35	202	146	198 *tetra*
2-Ethylphenol		207		
2,4-Dimethylphenol	27	212	135	
Methyl salicylate		224		
3-Methoxyphenol		243	129	104 *tri*
4-Allyl-2-methoxyphenol		255	122	118 *tetra*
2-Methoxy-4-propenylphenol		268	150	
2-Methoxyphenol	32	205	118	116 *tri*
3-Bromophenol	32	236	108	
3-Chlorophenol	33	214	158	
4-Methylphenol (*p*-cresol)	35	202	146	198 *tetra*

(Continued on next page)

(Continued)

Name of Compound	Melting Point (°C)	Boiling Point (°C)	Melting Points of Derivatives (°C)	
			α-Naphthyl-urethane	Bromo Derivative
2-Nitro-4-methylphenol	36			
2,4-Dibromophenol	36			95 *tri*
Phenol	42		132	95 *tri*
4-Chlorophenol	43		166	
2,4-Dichlorophenol	43			68 *mono*
2-Nitrophenol	45		113	117 *di*
4-Ethylphenol	47		128	
4-Chloro-2-methylphenol	49			
2,6-Dimethylphenol	49		176	79 *mono*
5-Methyl-2-isopropylphenol (thymol)	49		160	55 *mono*
4-Methoxyphenol	55			
2,5-Dichlorophenol	59			
3,4-Dimethylphenol	63		142	171 *tri*
4-Bromophenol	64		169	95 *tri*
4-Chloro-3-methylphenol	66		153	
2,6-Dichlorophenol	67			
3,5-Dimethylphenol	68			166 *tri*
3,5-Dichlorophenol	68			189 *tri*
3,4-Dichlorophenol	68			
2,4,5-Trichlorophenol	68			
2,4,6-Trichlorophenol	68			
2,4,6-Trimethylphenol	69			158 *di*
2,6-Di-*tert*-butyl-4-methylphenol	70			
2,5-Dimethylphenol	75		173	178 *tri*
8-Hydroxyquinoline	76			
4-Hydroxy-3-methoxybenz-aldehyde	81			
1-Naphthol	94		152	105 *di*
2,3,5-Trimethylphenol	96			
2-Methyl-4-nitrophenol	96			
3-Nitrophenol	97		167	91 *di*
4-*tert*-Butylphenol	100		110	50 *mono*
1,2-Dihydroxybenzene (catechol)	105		175	193 *tetra*
3,5-Dihydroxytoluene	106		160	104 *tri*
1,3-Dihydroxybenzene (resorcinol)	110			112 *tri*
2-Chloro-4-nitrophenol	111			
4-Nitrophenol	114		151	145 *di*
2,4-Dinitrophenol	114			118 *mono*
4-Hydroxybenzaldehyde	117			
1,3,5-Trihydroxybenzene	117			151 *tri*
2,3,5,6-Tetramethylphenol	118			118 *mono*
2,4,6-Trinitrophenol	122			
2-Naphthol	123		157	84 *mono*
3-Methyl-4-nitrophenol	129			
1,2,3-Trihydroxybenzene	133			158 *di*

(Continued on next page)

Name of Compound	Melting Point (°C)	Boiling Point (°C)	Melting Points of Derivatives (°C)	
			α-Naphthyl-urethane	Bromo Derivative
2,4-Dihydroxyacetophenone	147			
Salicylic acid	158			
2,3-Dihydroxynaphthalene	160			
1,4-Dihydroxybenzene (hydroquinone)	171			186 *di*
3,5-Dinitrosalicylic acid	173			
2-Aminophenol	174			
1,4-Dihydroxynaphthalene	176		220	
4-Aminophenol	184			
2,7-Dihydroxynaphthalene	190			
Pentachlorophenol	190			
3-Hydroxybenzoic acid	200			
4-Hydroxybenzoic acid	215			
1,3,5-Trihydroxybenzene (phloroglucinol anhydrous)	217			151 *tri*
1,5-Dihydroxynaphthalene	265			
4,4'-Biphenol	274			

chapter **28**

the literature of organic chemistry

The purpose of this chapter is to assist the interested student and teacher in obtaining additional information on experimental organic chemistry—information that will be useful in amplifying, modifying, and extending the introductory organic laboratory course for which this book is designed. The chapter is not intended to be a comprehensive guide to the literature of organic chemistry. An excellent series of articles that serves this purpose has been written by Professor J. E. H. Hancock, and an updated review of the subject may be found in an appendix to an advanced textbook by Professor J. March. Another recent brief survey may be found in H. J. E. Loewenthal's *Guide for the Perplexed Organic Experimentalist*. References to these sources are given at the end of this chapter.

28.1 Classification of the Literature of Organic Chemistry

Hancock divided the literature of organic chemistry into 18 classes; the following seven of these classes will be of the most value to the organic laboratory student (and probably also to the practicing organic chemist).

Class A: Primary research journals
Class B: Review journals
Class C: Encyclopedias and dictionaries
Class D: Abstract journals
Class E: Advanced textbooks
Class F: Reference works on synthetic procedures and techniques
Class G: Catalogs of physical data

In this section, selected examples from each of these classes are given, along with brief explanatory notes. In Section 28.2, two examples are given to illustrate how to use the literature to find information about a specific organic compound.

A. Primary Research Journals

These journals publish original research results, with theoretical discussion and experimental details.

1 *Journal of the American Chemical Society.* In recent years, the articles on organic chemistry have been limited to those that are especially timely (Communications to the Editor[1]) or of wide interest to all chemists.

2 *Journal of Organic Chemistry.* Articles, communications, and notes on organic chemistry.

3 *Tetrahedron.* Articles and reviews on organic chemistry, some in French or German as well as English.

4 *Tetrahedron Letters.* Brief communications in English, French, or German.

5 *Journal of the Chemical Society. Perkin Transactions.* Divided into two sections: I. Organic and bio-organic chemistry. II. Physical organic chemistry. Briefer articles are published in a separate journal entitled *Chemical Communications.*

6 *Angewandte Chemie. International Edition in English.* Concurrent translation of the German journal *Angewandte Chemie.* Some articles approach the length and scope of reviews.

7 *Journal of Organic Chemistry of the U.S.S.R.* The English translation of a Russian journal first appeared in 1965. The English edition appears about six months after the Russian original.

8 *Helvetica Chimica Acta.* Published in Switzerland. Articles in English, French, or German.

9 *Recueil des Travaux Chimiques des Pays-Bas.* Published in the Netherlands. Most articles in English.

10 *Bulletin of the Chemical Society of Japan.* Articles in English.

11 *Chemistry Letters.* Communications (brief articles), mostly in English. Published in Japan.

12 *Heterocycles.* Reviews, communications, and abstracts of heterocyclic chemistry. Published in Japan; mostly in English.

13 **and 14** *Chemische Berichte* and *Annalen der Chemie.* Articles only in German, published in West Germany.

---

[1]Communications are published with less delay than articles and notes. Since they usually contain few experimental details, they may be followed by a later article in which full descriptions of experiments are given.

B. Review Journals

Some review journals publish reviews in all areas of chemistry, while others cover only specific areas. All give references to primary journal articles.

1 *Chemical Reviews.* Published bimonthly since 1924 by the American Chemical Society. General in scope.

2 *Chemical Society Reviews.* Published quarterly since 1972; succeeded *Quarterly Reviews* of the Chemical Society (British). General in scope.

3 *Annual Reports on the Progress of Chemistry.* Published since 1904 by the Chemical Society. Since 1968, divided into sections; Section B covers organic chemistry. Supplemented by a series of *Specialist Periodical Reports,* which cover specific topics such as alkaloids, nuclear magnetic resonance, and organometallic compounds.

4 *Journal of Chemical Education.* Often contains reviews written by experts at a level that students and others unfamiliar with the subject may understand. New, tested experiments and modifications of old experiments suitable for organic laboratory courses are frequently published in the monthly issues.

5 *Accounts of Chemical Research.* Published monthly by the American Chemical Society since 1968. Gives concise reviews of active research areas.

6 *Angewandte Chemie. International Edition in English.* (See number 6 in Section 28.1A.)

7 *Synthesis.* Reviews and communications on organic synthetic methods, published monthly in English or German.

8 *Index of Reviews in Organic Chemistry.* Published by the Chemical Society. Second cumulative edition in 1976; supplements in 1979 and 1981. A bibliography of published reviews, organized by topic; useful for locating a review on a particular subject.

C. Encyclopedias and Dictionaries

1 *Beilstein's Handbuch der Organischen Chemie,* first published in 1881–1883 in two volumes, is perhaps the most complete reference work in any branch of science. The fourth edition (1918–1938) contains data on the 140,000 organic compounds known in 1909. This edition is known as the "Hauptwerk," or main work. It is continued by a series of supplements ("Ergänzungwerke"), the fourth of which is not yet completed. The first supplement (EI) covers the published literature of the years 1910–1919, the second supplement (EII), that of the years 1920–1929. Supplements three (EIII) and four (EIV) cover the years 1930–1959. However, in order to expedite the work, these supplements were combined beginning with Vol. 17. These, then, cover the following years:

EIII	Vols. 1–16	1930–1949
EIII/IV	Vols. 17–27	1930–1959
EIV	Vols. 1–16	1950–1959

The earlier part of EIV is still incomplete, having reached Vol. 11 in 1984.

Beilstein is organized by functional classes into a series of "systems," and a compound is always treated in the same "system," no matter what supplement it may appear in. This feature of the set facilitates location of information; once a compound's System Number has been obtained, it is easy to trace it through the whole set. There are molecular-formula and compound-name indexes for the main set and the first two supplements (Vols. 28 and 29). In addition, molecular-formula and compound-name indexes are now available for several of the volumes of EIV which have been completed. These are called "Gesamtregisters." The organic compound tables in the recent *CRC Handbooks of Chemistry and Physics* (see below) have included references to *Beilstein*, which further enhance access to the set. There are several recent guides to the use of *Beilstein*. The third article by Hancock, already cited, and the article by Reiner Luckenbach furnish a good introduction, as does the guide *How to Use Beilstein*. See the references at the end of this chapter.

2 *Dictionary of Organic Compounds,* 5th ed., J. Buckingham and editorial board, editors. Chapman and Hall, New York, 1982. This edition is in seven volumes. Volumes one–five contain the data for the compounds, while volume six is a name index with cross-references, and volume seven contains a molecular-formula index, a heteroatom index, and Chemical Abstracts Service registry number index. The set will be updated in annual supplements, the first of which appeared in 1983.

3 *CRC Handbook of Chemistry and Physics,* annual editions, CRC Press, Boca Raton, Florida. Gives physical properties and Beilstein references for about 15,000 organic compounds.

4 *Lange's Handbook of Chemistry,* 12th ed, J. A. Dean, editor. McGraw-Hill, New York, 1979. Gives physical properties for about 6500 organic compounds.

5 *CRC Atlas of Spectral Data and Physical Constants for Organic Compounds,* J. G. Grasselli and W. M. Ritchey, editors. Chemical Rubber Co., Cleveland, 1975. Six volumes. Gives data on some 21,000 organic compounds.

6 *Handbook of Tables for Identification of Organic Compounds,* 3d ed., Z. Rapoport, editor. Chemical Rubber Co., Cleveland, 1967. Gives physical properties and derivatives for more than 4000 compounds, arranged according to functional groups.

7 *Merck Index of Chemicals and Drugs,* 10th ed, Merck and Co., Rahway, New Jersey, 1983. Gives a concise summary of the physical and biological properties of more than 10,000 compounds, with some literature references. Organization is alphabetical by name; synonyms and trade names; cross-index and index of Chemical Abstracts Service registry numbers.

D. Abstract Journals

Abstract journals provide concise summaries of articles in Class A journals, listings of reviews (Class B), and announcements of new books (classes E and F), with reference to the original articles or books. Although abstracts are the quickest way to locate chemical information, they are always incomplete and sometimes even misleading, so they should not be relied on as the final source. A complete literature search should always include reference to the primary research publication.

1 *Chemical Abstracts.* This publication, "the key to the world's chemical literature," began in 1907 and now abstracts nearly one-half million items each year (more than

457,000 in 1982) from some 20,000 sources. *Chemical Abstracts* appears weekly; at the end of each six months, indexes by author, general subject, chemical substance, and molecular formula appear. These indexes are cumulated at five-year intervals.

Effective use of *Chemical Abstracts,* especially the chemical-substance indexes, requires an understanding of past and current nomenclature systems used. Current practice is summarized in the 1982 *Index Guide,* which is a listing of indexing terms and cross references; the *Guide* has also the indexing rules and procedures currently used. Supplementing the regular indexes are the *Parent Compound Handbook,* containing information on the various organic ring compounds, and the *Chemical Abstracts Service Registry Number Handbook,* which lists a computer-generated numbering system for unique unambiguous chemical substances, providing something like a Social Security number for each chemical substance.

In the 1970s Chemical Abstracts Service (CAS) and other abstracting agencies began to rely increasingly on sophisticated computer technology to manage the huge volume of information generated by scientific disciplines in general and by chemists in particular. Consequently, a new approach to locating information in scientific literature has been developed: the use of remote data bases for information retrieval.

Chemical Abstracts Service regularly provides machine-readable files of current issues of *Chemical Abstracts* to licensed Information Centers and "vendors" located throughout the United States and the rest of the world. One of these vendors is Lockheed Information Systems, which provides a data base known as DIALOG. These data bases may be reached ("accessed") and read, for a fee, by remote terminals at any location serviced by telephone. This service is also available at many colleges and universities.

Fundamentally, the approach to using remote data bases of *Chemical Abstracts* is similar to use of the conventional indexes. Key words involving subject topics, compound names, and/or authors are identified, and the index files are scanned by the vendor's computer to determine if relevant information is present. The data bases are constructed in such a way that complex search terms consisting of several "linked" key words (such as "naphthalene/oxidation/phthalate") may be employed. Thus, the scope of information obtained may be *inclusive* or *exclusive,* depending on the way in which the search terms are linked. Assistance by a trained search analyst in devising a proper linking strategy facilitates the quality and the cost-effectiveness of the search.

Chemical Abstracts data bases from 1967 to the present may be searched in minutes, and the information is available in the data base *before* it appears in the printed journal. There are other data bases available for searching specific and diverse topics relating to organic chemistry, such as patents, polymers, pharmacology, toxicology, agricultural research, and environmental science.

2 *Chemisches Zentralblatt* (1830–1970) is the second major abstract journal; it is published in German. It predates *Chemical Abstracts* by more than 70 years and prior to 1940 was more complete and more reliable than the latter. For this reason, in an exhaustive search for an organic compound in the years between 1929 (the last year covered by *Beilstein's* cumulative formula index) and 1940, it would be good to consult the collective formula indexes of *Chemisches Zentralblatt,* one for the years 1929–1934 and one for the years 1935–1939.

3 *Chemical Titles,* published biweekly by Chemical Abstracts Service since 1961, lists *titles* of articles in more than 700 chemical journals. The unique value of this publication derives from the fact that not only the title but also every significant word in the title is listed in alphabetical order. Although it is not an abstract journal, it serves a similar function, and a title appears much sooner than an abstract.

E. Advanced Textbooks

The advanced textbooks are subdivided according to subject and function.

1 General

   a *Rodd's Chemistry of Carbon Compounds*, 2d ed., S. Coffey, editor. Elsevier, Amsterdam; 1964– . A comprehensive survey of all classes of organic compounds, giving properties and syntheses for many individual compounds; consists of four volumes in 30 parts, as of 1983.

   b *Comprehensive Organic Chemistry*, D. Barton *et al.*, editors. Pergamon, Oxford, 1978. A six-volume treatise on the synthesis and reactions of organic compounds, written by more than 100 authors with over 20,000 references to the original literature. Intended to fill the gap between existing multivolume series, such as *Rodd*, and smaller books, such as the following.

   c March, J. *Advanced Organic Chemistry: Reactions, Mechanisms, and Structures*, 2d ed., McGraw-Hill, New York, 1977. Many references to the original literature are given.

   d Carey, F. A.; Sundberg, R. J. *Advanced Organic Chemistry; Part A: Structure and Mechanisms; Part B: Reactions and Synthesis*, 2 volumes. Plenum, New York, 1977.

   e Lowry, T. H.; Richardson, K. S. *Mechanism and Theory in Organic Chemistry*, 2nd ed. Harper and Row, New York, 1981.

   f House, H. O. *Modern Synthetic Reactions*, 2d ed. W. A. Benjamin, Menlo Park, California, 1972. Three general classes of reactions—those used for reduction, for oxidation, and for the formation of new carbon-carbon bonds—are surveyed in terms of scope, limitations, stereochemistry, and mechanisms.

2 Identification and analysis of organic compounds (see references at end of Section 27.4).

3 Instrumental techniques of analysis (see references at end of Sections 8.7, 9.10, 11.1, and 11.2).

F. Reference Works on Synthetic Procedures and Techniques

1 Buehler, C. A.; Pearson, D. E. *Survey of Organic Syntheses*. Wiley-Interscience, New York, Vol. 1, 1970; Vol. 2, 1972. This extensive two-volume work covers the principal methods of synthesizing the main types of organic compounds. The limitations of the reactions, the preferred reagents, the newer solvents, and experimental conditions are considered.

2 *Organic Syntheses*, H. Gilman, editor, Wiley, New York, 1932–present, 61 volumes through 1984. Every ten volumes have been collected, indexed, and published as *Collective Volumes*, through volume 49 as of 1983. Detailed directions for synthesis of more than 1000 compounds are given. Procedures have all been thoroughly checked by independent investigators before publication. Many of the general methods may be applied to the synthesis of related compounds other than those described. The collective volumes contain indexes of formulas, names, types of reaction, types of compounds, purification of solvents and reagents, and illustrations of special apparatus. A cumulative index to the five collective volumes was published in 1976.

3 *Organic Reactions*, by various contributors. Wiley, New York, 1942–present, 31 volumes through 1984. Each volume contains from 5 to 12 chapters, each of which deals with an organic reaction of wide applicability. Typical experimental procedures are given in detail, and extensive tables of examples with references are given. Each volume contains a cumulative author and chapter-title index.

4 Fieser, M.; Fieser, L. F. *Reagents for Organic Synthesis.* Wiley-Interscience, New York, 1967–present, ten volumes as of 1984. These volumes list some 8000 reagents and solvents, described in terms of methods of preparation or source, purification, and utilization in typical reactions. Ample references to original literature are given.

5 Harrison, I. T.; Hegedus, L. S.; Wade, L. G., Jr. *Compendium of Organic Synthetic Methods,* Wiley, New York, 1971–1984. Five volumes. These volumes present in outline form possible interconversions among the major functional groups of organic compounds. References to the primary literature sources are given.

6 *Synthetic Methods of Organic Chemistry,* W. Theilheimer and A. F. Finch, editors, Interscience, New York, Karger, Basel, 1948–present, 38 volumes through 1984. Emphasis is laid on functional group chemistry and ring reactions of a general type. The reactions are classified by symbols that are arranged systematically. There are cumulative indexes to recent volumes.

7 *Vogel's Elementary Practical Organic Chemistry,* 3rd ed. Vol. 1; B. V. Smith and N. M. Waldron, editors. Longman, London, 1980. Good coverage of basic experimental techniques, with a selection of contemporary laboratory preparations.

8 *Vogel's Textbook of Practical Organic Chemistry,* 4th ed., B. S. Furniss *et al.,* editors. Longman, London, 1978. For the more experienced researcher; furnishes additional tips on technique and a wider range of preparations to work with.

9 *Techniques of Organic Chemistry,* 3d ed., A. Weissberger, editor. Interscience, New York, 1959–present; some volumes have been revised. Title widened in 1970 to *Techniques of Chemistry.* Examples of useful volumes are *Elucidation of Organic Structures,* Vol. 4, parts 1–3; *Investigation of Rates and Mechanisms of Reactions,* Vol. 6, parts 1 and 2; *Separation and Purification,* Vol. 12; and *Thin Layer Chromatography,* Vol. 14.

G. Catalogs of Physical Data

1 ^{1}H NMR Spectra

a *High Resolution NMR Spectra Catalog,* compiled by the staff of Varian Associates, Palo Alto, Calif., 1962–1963. 2 vols. Hydrogen NMR spectra of 587 representative organic molecules are depicted, and the peaks are assigned to the hydrogen nuclei responsible for the absorptions.

b *Nuclear Magnetic Resonance Spectra,* Sadtler Research Laboratories, Philadelphia. Hydrogen NMR spectra of more than 40,000 compounds have been published as of 1984, with 1000 being added annually. Assignment of peaks are made as in the Varian spectra, and integration of the signals is shown on many of the spectra.

c *Aldrich Library of NMR Spectra,* 2d ed., C. J. Pouchert, editor. Aldrich Chemical Company, Milwaukee, Wisconsin, 1983. 2 vols. Contains about 37,000 spectra.

2 ^{13}C NMR Spectra

a *Carbon-13 NMR Nuclear Magnetic Resonance Spectra,* Sadtler Research Laboratories, Philadelphia, beginning in 1976. By 1984, 16,800 proton-decoupled spectra had been published.

   b Breitmaier, E.; Haas, G.; Voelter, W. *Atlas of Carbon-13 NMR Data.* 2 vols. plus index. Heyden, Philadelphia, 1979. Tabular data on 3017 compounds, with chemical shifts for ^{13}C given and ^{1}H–^{13}C multiplicities indicated.

3 IR Spectra
   a *Sadtler Standard Spectra, Midget Edition,* Sadtler Research Laboratories, Philadelphia. As of 1983, the prism series had 67,000 spectra and the grating series had 67,000 spectra.
   b *Aldrich Library of Infrared Spectra,* 2d ed., C. J. Pouchert, editor. Aldrich Chemical Company, Milwaukee, Wisconsin, 1975. A collection of about 10,000 spectra.

4 UV Spectra
   a *Sadtler Standard Spectra: Ultraviolet Spectra,* Sadtler Research Laboratories, Philadelphia. Through 1984, contains 75,000 spectra.

Some of the texts of Class E, such as those listed at the end of Chapters 8 and 9, contain numerous PMR and IR spectra with molecular assignments. Several "problem books" of spectroscopic analysis have been published more recently; these give various combinations of IR, PMR, UV, and mass spectra of "unknown" organic compounds, with answers provided.

28.2 Use of the Literature of Organic Chemistry in an Introductory Laboratory Course

The literature outline given in this chapter may be used in a variety of ways, according to the aims and needs of different courses and the library facilities available, ranging from no use at all to extensive application. Even in those cases where the pressure of time and/or lack of facilities preclude the use of literature beyond the pages of this textbook itself, we feel that this chapter may be valuable to the serious students who may decide to go further in the study of organic chemistry.

In many organic laboratory courses instructors are interested in making part of the experimentation open-ended—encouraging the students to plan and carry out experiments with some independence. Although this is highly desirable, it has an element of danger unless the plans are checked and the work is monitored carefully. In several chapters of this text, additional or alternative experiments are provided or suggested. The inclusion of the literature outline of this chapter now provides a wide source of information for additional experiments.

The most likely class of literature to yield appropriate synthetic experiments is Class F. The experiments from *Organic Syntheses* are particularly suitable; although they are usually on a large scale, they can easily be scaled down. Also deserving special mention are the experiments that appear from time to time in the *Journal of Chemical Education* (Class B). For more experience in identification of unknown organic compounds, the books of Class E.2 will be most useful. The catalogs of spectra listed in Class G represent a vast reservoir from which to draw for paper unknowns and problems. They should be used with discretion, however, because many molecules give IR and PMR spectra that are not easily interpreted by beginners. If you wish to learn how to make a comprehensive search for a specific compound in the literature, to learn its properties or a preferred method of synthesis, refer to the second and third articles by Hancock for a fuller introduction to the use of the Class C and Class D literature.

Example 1

The following example is given as an illustration of how one might proceed to solve problems such as those proposed in Exercise 11. ''Mustard gas'' is one of the names that has been applied to the compound $ClCH_2CH_2$—S—CH_2CH_2Cl. Find the answers to the following questions: (1) Has this compound been synthesized or isolated? (2) By whom? (3) When? (4) Where can the most recent information on this compound be found?

First, write the molecular formula as $C_4H_8Cl_2S$, and look in *Beilstein's General Formelregister, Zweites Ergänzungswerk*. On page 65 will be found the entry ''β,β'-Dichlor-diäthylsulfid, Senfgas **1**, 349, I 175, II 348, 940,'' and just below it the entry ''α,α'-Dichlor-diäthylsulfid **1** II 685.'' The first entry will be recognized as that pertaining to the subject compound. The references are to page 349 in Volume 1 of the main work, page 175 in Volume 1 of the first supplement, and to pages 348 and 940 in Volume 1 of the second supplement. Although the third supplement was published after the general index, by noting the ''System No.'' of the subject compound, 23, one may locate it on page 1382 of the third supplement using the index in the appropriate volume.

Referring to page 349 in Volume 1 of the main work, one finds ''β,β'-Dichlor-diäthylsulfid $C_4H_8Cl_2S = (CH_2Cl \cdot CH_2)_2S$.B. Aus Thiodiglykol S $(CH_2 \cdot CH_2 \cdot OH)_2$ und PCl_3 (V. Meyer, *B*. **19**, 3260).-.'' This translates: ''B. = Bildung, Formation. From thiodiglycol and PCl_3 (V. Meyer, *Berichte*, **19**, 3260).'' Looking up the reference in the *Berichte der Deutschen Chemischen Gesellschaft* (Vol. 19, p. 3260, published in 1886) one finds that Victor Meyer first prepared this compound in two steps as follows:

$$2\ \text{Cl—CH}_2\text{CH}_2\text{—OH} \xrightarrow{K_2S} \text{HO—CH}_2\text{CH}_2\text{—S—CH}_2\text{CH}_2\text{—OH}$$

$$\text{HO—CH}_2\text{CH}_2\text{—S—CH}_2\text{CH}_2\text{—OH} \xrightarrow{PCl_3} \text{Cl—CH}_2\text{CH}_2\text{—S—CH}_2\text{CH}_2\text{—Cl}$$

The entry in the main work of *Beilstein* (Vol. 1, p. 349) gives in a total of six lines the boiling point and solubility properties of the subject compound and one chemical reaction. The final two words are *Sehr giftig*, ''very poisonous''—a terse commentary on a material that was much feared as a lethal military weapon in World War II but was never used. It is interesting to note the statement by Meyer in his *Berichte* article that although his laboratory assistant developed skin eruptions and eye inflammation after preparing this compound, he himself suffered no ill effects even though he took no precautions in handling it!

In the first supplement (p. 175) 15 lines are devoted to ''β,β'-diäthylsulfid'' and in the second supplement (p. 348), five and one-half pages, indicating the increased interest in this compound during 1920–1929. Turning next to the *Chemical Abstracts Collective Formula Index* for 1920–1946, under $C_4H_8Cl_2S$ is found the entry ''(See also Sulfide, bis(chloroethyl).) Sulfide, 1-chloroethyl 2-chloroethyl, **25**: 2114[8].'' Since ''Sulfide, 1-chloroethyl 2-chloroethyl'' is not the compound of interest, we look in the *Chemical Abstracts Decennial Subject Index*, 1917–1926, for the entry ''Sulfide, bis (β-chloroethyl),'' which is followed by the names *mustard gas; yperite* and by six general references and two columns of more specific references beginning with "absorption by skin, mechanism of, **14:** 300[4]" and ending with "toxicity and skin-irritant effects of, **15:** 1943[6]."

The *Chemical Abstracts Decennial Subject Indexes* could presumably be used for more recent decades, but one must be on guard for changes in nomenclature. For this reason it is usually advantageous to use formula indexes first. For example, when we go to the January–June 1972 *Formula Index,* under $C_4H_8Cl_2S$ we find "Ethane, 1,1'-thiobis[2-chloro-" as the name for our compound, with seven references to abstracts. Under this name in the *Chemical Substance Index* for the same period, we find the same seven references, but with specific subject headings: for example, "DNA, cross-linking induced by, 95344w."

Example 2

A second example of a search will illustrate how *Beilstein* may be used to find compounds reported after 1929, the last year covered by the Formula Index of Volume 29. The preparation of the compound 3-(2-furyl)-1-(3-nitrophenyl) propenone **(1)** is described in Section 21.2. If one looks for this compound in the *Beilstein Formelregister* (Vol. 29), under

1

the molecular formula $C_{13}H_9NO_4$, no name that fits the structure is found, indicating that this compound was not reported through 1929. However, the analogous compound without the nitro group is listed under the formula $C_{13}H_{10}O_2$ in the *Formelregister,* with the notation *17* 353; II 377, which means that it is described in Vol. 17 of the *Hauptwerk* on p. 353 and in the *Zweites Ergänzungswerk* on p. 377. Looking up these entries, one finds that the "System Number" for this compound is 2467. Now, going to Vol. 17 of the combined *Drittes und Viertes Ergänzungswerk* (EIII/IV), one finds that System Number 2467 compounds are in Part 6 of Vol. 17. In the formula index of Part 6 under $C_{13}H_9NO_4$ one finds the name of the compound sought, followed by the page number 5262, where a description of the preparation and properties of this compound is given. Alternatively, the *Gesamtregister Formelregister* for Volumes 17 and 18 directs one to the same page in Volume 17.

As mentioned earlier, *Chemical Abstracts Collective Formula Indexes* may conveniently be used for locating compounds appearing in the literature after 1929. The 1920–1946 *Index* has no entry for the compound located in *Beilstein* as described above, but the 1947–1956 *Index* lists under the formula $C_{13}H_9NO_4$ the name "Acrylophenone,3-(2-furyl)-3'-nitro", which may (or may not!) be recognized as another name for our compound. Both *Beilstein* EIII/IV *17* and *Chemical Abstracts 43*, 3429g (1949) refer to the primary source, the *Journal of the American Chemical Society, 71*, 612 (1949), where D. L. Turner describes the first preparation of this compound.

This example illustrates the advantage of using formula indexes, either *Beilstein* and/or *Chemical Abstracts*, for locating an organic compound in the literature. Several different names may be used for the same compound, but the formula will be more distinctive, and one can be on the lookout for different ways of naming the compound.

EXERCISES

1. Find the melting points of the following crystalline derivatives (none of these are listed in the tables of Chapter 27): (a) 2,4-dinitrophenylhydrazone of trichloroacetaldehyde, (b) semicarbazone of 3-methylcyclohexanone, (c) 3,5-dinitrobenzoate of 1,3-dichloro-2-propanol, (d) amide of 2-methyl-3-phenylpropanoic acid, (e) benzamide of 4-fluoroaniline

2. Locate an article or a chapter on each of the following types of organic reactions: (a) the aldol condensation, (b) the Wittig reaction, (c) reactions of diazoacetic esters with unsaturated compounds, (d) hydration of alkenes and alkynes through hydroboration, (e) metalation with organolithium compounds

3. Give a reference for a practical synthetic procedure for each of the following compounds and state the yield that may be expected: (a) 1,2-dibromocyclohexane, (b) α-tetralone, (c) 3-chlorocyclopentene, (d) 2-carboethoxycyclopentanone, (e) norcarane, (f) tropylium fluoborate, (g) 1-methyl-2-tetralone, (h) adamantane

4. Locate descriptions of procedures for the preparation or purification of the following reagents and solvents used in organic syntheses: (a) Raney nickel catalysts, (b) sodium borohydride, (c) dimethyl sulfoxide, (d) sodium amide, (e) diazomethane

5. Find IR spectra for the following compounds: (a) N-cyclohexylbenzamide, (b) 4,5-dihydroxy-2-nitrobenzaldehyde, (c) benzyl acetate, (d) diisopropyl ether, (e) 3,6-diphenyl-2-cyclohexen-1-one, (f) 4-amino-1-butanol

6. Find PMR spectra of the following compounds: (a) benzyl acetate, (b) diisopropyl ether, (c) 4-amino-1-butanol, (d) 1-propanol, (e) indan

7. N-Mesityl-N'-phenylformamidine (**2**)

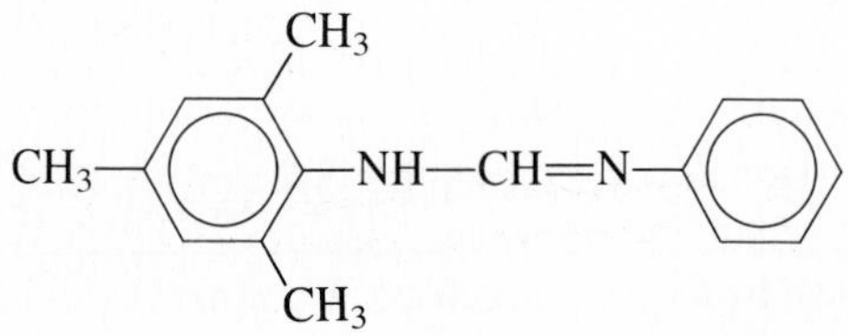

2

was first synthesized between 1950 and 1960. Find the primary research article in which this compound is described, and write an equation for the reaction used to prepare it.

8. N-Phenyl-N'-p-tolylformamidine (**3**) (the German name for this compound is the same as in English except that the final "e" is omitted)

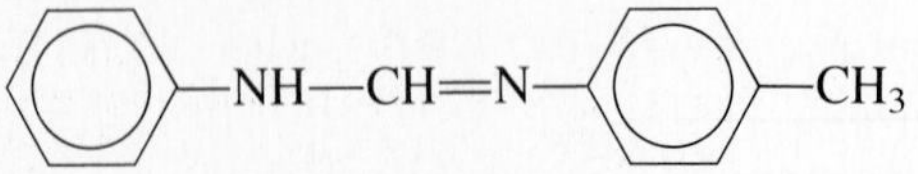

3

is reported in *Beilstein* to have a melting point of 86°C. If you check the first reference given in *Beilstein,* however, you will find the surprising fact that the same chemist who reported this pure compound to have a melting point of 86°C had described it as melting at 103.5–104.5°C two years previously. The discrepancy between these reports was not explained until the ambiguity was reexamined in the period 1947–1956.

Find the article that solved this mystery.

9. The benzoyl derivative of α-phenylethanamine (formula **3** in Chapter 7) was first described in the form of the optically active (−)-isomer in 1905. Using the formula index of *Beilstein* and given the German name "benzosäure-1-α-phenyläthylamid," find the first reference to this compound in a primary research journal. If you can read the German, give (a) the method of preparation by writing the equation, (b) the melting point of the pure compound and the recrystallization solvent, and (c) the $[\alpha]_D$. For a description in English, find an article published between 1910 and 1920.

10. The name used for the compound described in Exercise 9 in the formula indexes of *Chemical Abstracts* is "benzamide, N-methylbenzyl" or "benzamide, N-α-methylbenzyl." (a) Find a second reference to the (−)-isomer that was published between 1930 and 1940, and compare the physical constants given there with the earlier data. (b) Find a reference to a paper published in Czechoslovakia between 1950 and 1960 giving data on the racemic form of the compound. (c) Find a reference to data on the (+)-isomer of the compound in a paper published between 1960 and 1970.

11. Determine whether or not each of the following compounds has ever been synthesized, and if it has, give the reference to the first appearance of its synthesis in the literature.

(a) Vitamin A   (b) Strychnine

(c)

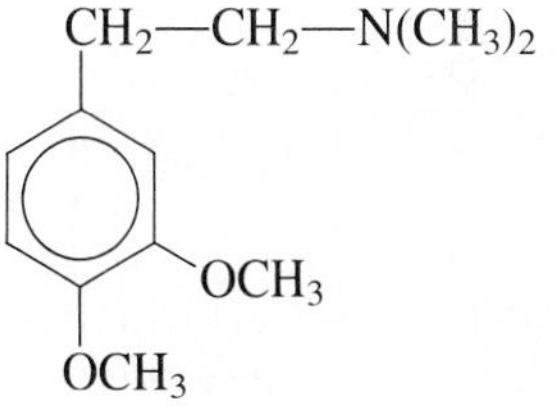

(d) 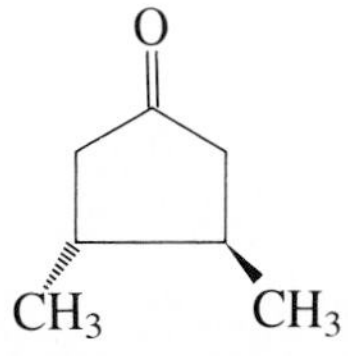   (e) Testosterone   (f)

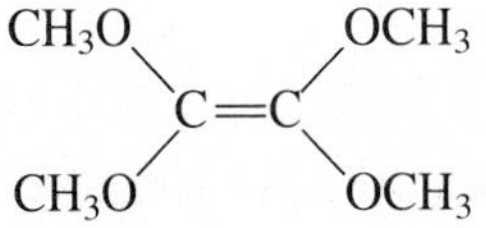

(g) Penicillin-V   (h)   (i)

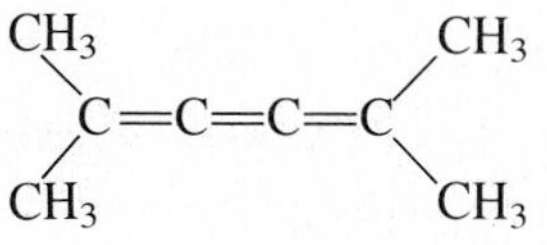

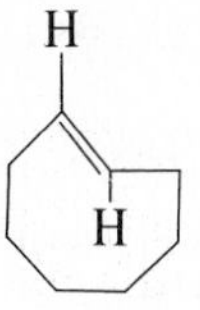

(j)

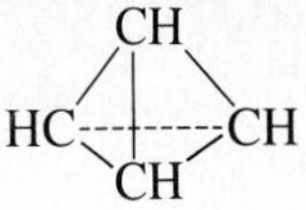

(k)

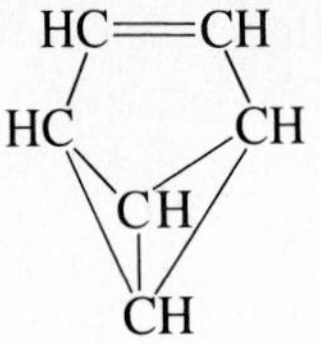

(l)

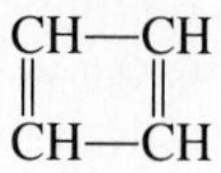

(m) 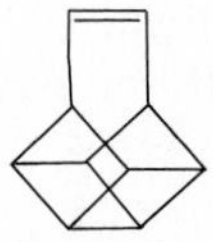 "Basketene"

(n) Prostaglandin E_2

(o) 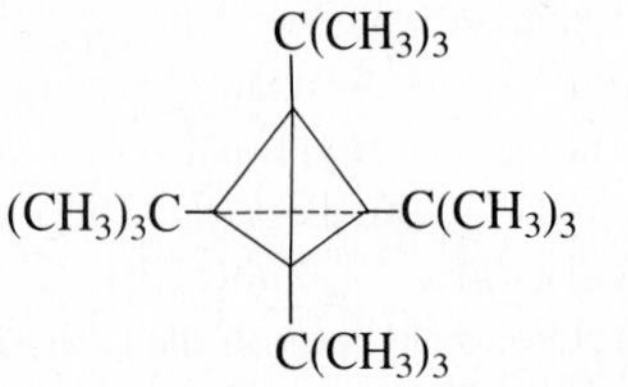 $(CH_3)_3C$—$C(CH_3)_3$ with $C(CH_3)_3$ above and $C(CH_3)_3$ below

(p) Lysergic acid diethylamide   (q) Morphine

(r) 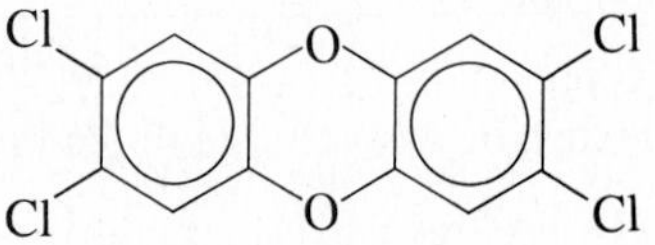 2,3,7,8-Tetrachlorodibenzodioxin

(s) Vitamin B_{12}

REFERENCES

1. Hancock, J. E. H. "An Introduction to the Literature of Organic Chemistry," *Journal of Chemical Education*, **1968,** *45*, 193–199; 260–266; 336–339.

2. March, J. *Advanced Organic Chemistry: Reactions, Mechanisms and Structure*, 2d ed. McGraw-Hill, New York, 1977, Appendix A.

3. Loewenthal, H. J. E. *Guide for the Perplexed Organic Experimentalist*. Heyden, London, 1978.

4. Luckenbach, R. "Der Beilstein," *ChemTech*, **1979,** *9*, 612–621.

5. Beilstein Institut, *How to Use Beilstein*. Springer-Verlag, New York.

index*

***Boldface** page numbers indicate pages on which the structures of compounds appear.

Multiples of Atomic Weights
To Two Decimal Places

C		H		K	
C	12.01	H_{20}	20.16	K	39.10
C_2	24.02	H_{21}	21.17	K_2	78.20
C_3	36.03	H_{22}	22.18	K_3	117.30
C_4	48.04	H_{23}	23.18		
C_5	60.05	H_{24}	24.19	N	
C_6	72.06	H_{25}	25.20	N	14.01
C_7	84.07	H_{26}	26.21	N_2	28.02
C_8	96.08	H_{27}	27.22	N_3	42.02
C_9	108.09	H_{28}	28.22	N_4	56.03
C_{10}	120.10	H_{29}	29.23	N_5	70.04
C_{11}	132.11	H_{30}	30.24	N_6	84.05
C_{12}	144.12	H_{31}	31.25		
C_{13}	156.13	H_{32}	32.26	Na	
C_{14}	168.14	H_{33}	33.26	Na	23.00
C_{15}	180.15	H_{34}	34.27	Na_2	45.99
C_{16}	192.16	H_{35}	35.28	Na_3	68.99
C_{17}	204.17	H_{36}	36.29		
C_{18}	216.18	H_{37}	37.30	O	
C_{19}	228.19	H_{38}	38.30	O	16.00
C_{20}	240.20	H_{39}	39.31	O_2	32.00
C_{21}	252.21	H_{40}	40.32	O_3	48.00
C_{22}	264.22			O_4	64.00
C_{23}	276.23	Ag		O_5	80.00
C_{24}	288.24	Ag	107.87	O_6	96.00
C_{25}	300.25	Ag_2	215.74	O_7	112.00
C_{26}	312.26			O_8	128.00
C_{27}	324.27	Br		O_9	144.00
C_{28}	336.28	Br	79.90	O_{10}	160.00
C_{29}	348.29	Br_2	159.80		
C_{30}	360.30	Br_3	239.70	P	
		Br_4	319.60	P	30.97
H		Br_5	399.50	P_2	61.94
H	1.01			P_3	92.91
H_2	2.02	Cl		P_4	123.88
H_3	3.02	Cl	35.45		
H_4	4.03	Cl_2	70.90	S	
H_5	5.04	Cl_3	106.35	S	32.06
H_6	6.05	Cl_4	141.80	S_2	64.12
H_7	7.06	Cl_5	177.25	S_3	96.18
H_8	8.06			S_4	128.24
H_9	9.07	Cu			
H_{10}	10.08	Cu	63.55	H_2O	
H_{11}	11.09	Cu_2	127.10	0.5 H_2O	9.01
H_{12}	12.10			H_2O	18.02
H_{13}	13.10	I		1.5 H_2O	27.02
H_{14}	14.11	I	126.90	2 H_2O	36.03
H_{15}	15.12	I_2	253.80	3 H_2O	54.05
H_{16}	16.13	I_3	380.71	4 H_2O	72.06
H_{17}	17.14	I_4	507.62	5 H_2O	90.08
H_{18}	18.14	I_5	634.52	6 H_2O	108.10
H_{19}	19.15				